Operator Theory: Advances and Applications
Vol. 129

Editor:
I. Gohberg

Systems, Approximation, Singular Integral Operators, and Related Topics

International Workshop on Operator Theory and Applications, IWOTA 2000

Alexander A. Borichev
Nikolai K. Nikolski
Editors

Springer Basel AG

Editors:

Alexander A. Borichev
Nikolai K. Nikolski
Département de Mathématiques
Université de Bordeaux I
351, cours de la Libération
33405 Talence
France

2000 Mathematics Subject Classification 30, 34, 35, 41, 42, 43, 46, 47, 93

A CIP catalogue record for this book is available from the
Library of Congress, Washington D.C., USA

Systems, approximation, singular integral operators, and related topics /
International Workshop on Operator Theory and Applications, IWOTA 2000.
Alexander A. Borichev ; Nikolai K. Nikolski, ed.. - Basel ; Boston ; Berlin :
Birkhäuser, 2001
(Operator theory ; Vol. 129)
ISBN 978-3-0348-9534-7 ISBN 978-3-0348-8362-7 (eBook)
DOI 10.1007/978-3-0348-8362-7

Printed on acid-free paper produced from chlorine-free pulp. TCF ∞
Cover design: Heinz Hiltbrunner, Basel

ISBN 978-3-0348-9534-7

9 8 7 6 5 4 3 2 1 http://www.birkhauser-science.ch

PREFACE

This book is devoted to some topical problems and various applications of Operator Theory and to its interplay with many other fields of analysis as modern approximation theory, theory of dynamic systems, harmonic analysis and complex analysis. It consists of 20 carefully selected surveys and research-expository papers. Their scope gives a representative status report on the field drawing a picture of a rapidly developing domain of analysis. An abundance of references completes the picture.

All papers included in the volume originate from lectures delivered at the 11th edition of the International Workshop on Operator Theory and its Applications (IWOTA-2000, June 13–16, Bordeaux). Some information about the conference, including the complete list of participants, can be found on forthcoming pages.

The editors are indebted to A.Sudakov for helping them in polishing and assembling original TeX files.

A. Borichev and N. Nikolski

Talence, May 2001

International Workshop on Operator Theory and Its Applications

(June 13–June 16, 2000, Université Bordeaux 1)

The **International Workshop on Operator Theory and its Applications** (IWOTA) is a satellite meeting of the international symposium on the Mathematical Theory of Networks and Systems (MNTS). In 2000, the MNTS is held in Perpignan, France, June 19-23. IWOTA 2000 was the *eleventh* workshop of this kind.

The previous IWOTA meetings were held in Santa Monica, California, USA (1981), Rehovot, Israel (1983), Amsterdam, The Netherlands (1985), Mesa, Arizona, USA (1987), Rotterdam, The Netherlands (1989), Sapporo, Japan (1991), Regensburg, Germany (1995), Bloomington, Indiana, USA (1996), Groningen, The Netherlands (1998).

The international Steering Committee of IWOTA consists of I. Gohberg (President, Tel Aviv), T. Ando (Sapporo), H. Bart (Rotterdam), H. Bercovici (Bloomington), R. Dijksma (Groningen), H. Dym (Rehovot), C. Foias (College Station), J. W. Helton (Vice-President, La Jolla), M. A. Kaashoek (Vice-President, Amsterdam), H. Langer (Vienna), R. Mennicken (Regensburg), L. Rodman (Williamsburg), J. G. Stampfli (Bloomington).

The purpose of IWOTA is to bring together mathematicians and engineers interested in operator theory and its applications to related fields, from differential equations, harmonic and complex analysis, to mathematical systems and control theory and signal processing. In fact, every two years IWOTA gathers leading experts from over the world for an intensive exchange of information and opinions, and for tracing the future developments in the field. The large participation of researchers from Eastern European and CIS countries is also an important feature of IWOTA meetings.

The program of IWOTA-2000 consisted of 17 plenary lectures of 50 minutes including 5 minutes for questions and remarks, and 90 section lectures of 45 and 30 minutes also including 5 minutes for questions and remarks. There were more than 170 registered participants from 29 countries of all continents (except Antarctica): Argentina – 1, Armenia – 1, Austria – 1, Azerbaijan – 1, Canada – 1, Finland – 4, France – 37, Germany – 5, Hungary – 1, Iran – 1, Israel – 14, Japan – 1, Maroc – 2, Mexico – 1, The Netherlands – 6, New Zealand – 2, Norway – 2, Poland – 3, Portugal – 2, Romania – 6, Russia – 10, Spain – 3, Sweden – 2, Turkey – 1, Ukraine – 20, United Kingdom – 4, USA – 32, Uzbekistan – 1, Venezuela – 4. The list of participants is given below.

Some financial support was provided for some participants, especially from the Eastern countries. This has been made possible through the generous sponsoring by international and French institutions listed below. We are very grateful for their help.

Special thanks. We thank the City of Bordeaux for the welcome reception on Tuesday, June 13. Warmest thanks are given to an informal team of our closest collaborators working hard for this large meeting: Geneviève Castaignede, Emmanuel Fricain, Stanislav Kupin, Cathy Metivier, Ludmila Nikolskaia, and Pascale Vitse, as well as to many others whose timely advice was so important, namely to Eric Amar, Hakim Bachir, Robert Deville, Iban Harlouchet, Andreas Hartmann, Chantal Menini, Macha Nikolskaia, David Sherman, Elisabeth Strouse, and Fouad Zarouf.

Acknowledgements. As the majority of its homologues, the IWOTA conferences have no permanent funding. Their existence and function rely on financial aid from international and national organizations supporting the development of fundamental sciences. We gratefully acknowledge the following for their financial support: INTAS (International Association for the Promotion of Cooperation with Scientists from the Independent States of the former Soviet Union), European Union; Ministère de l'Education Nationale, de la Recherche et de la Technologie, France; INRIA (Institut National de Recherche en Informatique et en Automatique, Sophia Antipolis), France; CNRS (Centre National de Recherche Scientifique), France; Ministère des Affaires Etrangères, France; Région Aquitaine; Université Bordeaux I; Université Lyon I; Université Lille I; Laboratoire de Mathématiques Pures de Bordeaux; Institut de Mathématiques de Bordeaux; Ville de Bordeaux.

Organizing Committee: Prof. Nikolai Nikolski (Chairman, Université Bordeaux-I), Prof. Laurent Baratchart (INRIA, Sophia Antipolis), Dr. Alexander Borichev (Université Bordeaux I), Dr. Gilles Cassier (Université Lyon I), Prof. Jean Esterle (Université Bordeaux I), Prof. V.-H.Vasilescu (Université Lille I)

Contents

List of Participants of IWOTA–2000

Abakumov, Evgeny, Equipe d'Analyse et de Mathematiques Appliquees, Universite de Marne-la-Vallee, Cite Descartes — 5, Bd Descartes, Champs-sup-Marne, 77454 Marne-la-Vallee CEDEX 2, France.

Adamyan, Vadim, Department of Theoretical Physics, Odessa State University, ul. Dvoryanskaya 2, 270044 Odessa, Ukraine.

Alpay, Daniel, Department of Mathematics, Ben-Gurion University of the Negev, P.O.B. 653, Beer-Sheva 84105, Israel.

Akkar, Mohamed, UFR de Mathématiques, Université Bordeaux 1, 351, cours de la Libération, 33405 Talence, France.

Amar, Eric, UFR de Mathématiques, Université Bordeaux 1, 351, cours de la Libération, 33405 Talence, France.

Ammari, Kais, Institut Elie Cartan, Departement de Mathematiques, Universite Nancy I, 54506 Vandoeuvre-les-Nancy, France.

Arazy, Jonathan, Department of Mathematics, University of Haifa, Mount Carmel, Haifa 31905, Israel.

Arlinskii, Yury, East Ukrainian State University, Kvartal Molodyozhny 20-A, Lugansk 348034, Ukraine.

Arov, Damir Z., prosp. Ak. Glushko 15, kv. 18, Odessa, 270104 Ukraine.

Avdonin, Sergei, Mathematics Department, University of Tennessee, Knoxville, TN 37996-1300, USA.

Azizov, Tomas, Dept. of Math., Voronezh State University, Universitetskaya pl. 1, Voronezh 394693, Russia.

Bachir, Mohamed, UFR de Mathématiques, Université Bordeaux 1, 351, cours de la Libération, 33405 Talence, France.

Badea, Catalin, UFR de Mathematiques, Université des Sciences, et Technologies de Lille, 59655 Villeneuve d'Ascq Cedex, France.

Bakonyi, Mihaly, Department of Mathematics, Georgia State University, Atlanta, GA 30303, USA.

Balazard, Michel, UFR de Mathématiques, Université Bordeaux 1, 351, cours de la Libération, 33405 Talence, France.

Ball, Joseph A., Department of Mathematics, Virginia Tech, Blacksburg VA 24061-0123, USA.

Baratchart, Laurent, "Projet Miaou", INRIA, 2004, route de Lucioles BP93, 06902 Sophia Antipolis Cedex, France.

Belishev, Mikhail I., St-Petersburg Dept. of the Steklov, Mathematical Institute (POMI), Fontanka 27, St. Petersburg 191011, Russia.

Belitskii, Genrich, Department of Mathematics, Ben-Gurion University of the Negev, P.O.B. 653, Beer-Sheva 84105, Israel.

Berezansky, Yurij, Institute of Mathematics, Ukrainian Acad. Sci., 3 Tereshchenkivs'ka Str., 01601 Kyiv, Ukraine.

Birman, Mikhail, Department of Physics, St. Petersburg State University, Ul'yanovskaya 1, Petrodvorets, St. Petersburg 198904, Russia.

Bodenstorfer, Bernhard, Institute of Analysis and, Technical Mathematics E114-1, Vienna University of Technology, Wiedner Hauptstrasse 8-10, A-1040 Vienna, Austria.

Borichev, Alexander, UFR de Mathématiques, Université Bordeaux 1, 351, cours de la Libération, 33405 Talence, France.

Böttcher, Albrecht, Faculty of Mathematics, Technical University, D-09107 Chemnitz, Germany.

De Branges, Louis, Department of Mathematics, Purdue University, Lafayette, IN 47907–1395, USA.

Brudnyi, Alexander, Department of Mathematics, Ben Gurion University of the Negev, P.O.Box 653, 84105, Israel.

Bruna, Joaquim, Dept. Matematiques, Universitat Autonoma de Barcelona, 08193 Bellaterra, Espagne.

Bruzual, Ramon, Apartado 47686, Los Chaguaramos, Caracas 1041-A, Venezuela.

Cassier, Gilles, Institut Desargues, Universite Lyon 1, 69622 Villeurbane Cedex, France.

Castro, Glaysar, Universidad Central de Venezuela, Facultad de Ciencias, Escuela de Fisica y Matematicas, A.P. 20513 Caracas 1020A, Venezuela.

Benhida, Chafiq, UFR de Mathematiques, Bat M2, Universite Lille 1, 59655 Villeneuve d'Ascq Cedex, France.

Cheremnikh, E., 8 Gnatjuka Str., apt. 12A, Lviv 79007, Ukraine.

Coburn, Lewis, 190 Fairlawn Drive, Amherst, NY 14226, USA.

Curtain, R.F., University of Groningen, Department of Mathematics, P.O. Box 800, 9700 AV Groningen, The Netherlands.

Davidson, Kenneth, Pure Math. Dept., University of Waterloo, Waterloo Ont. N2L-3G1, Canada.

Derfel, Gregory, Department of Mathematics and Computer Science, Ben-Gurion University, P.O.Box 653, 84105 Beer Sheva, Israel.

Derkach, Volodymyr, Department of Mathematics, Donetsk State University, Universitetskaya str. 24, 340055 Donetsk, Ukraine.

Deville, Robert, UFR de Mathématiques, Université Bordeaux 1, 351, cours de la Libération, 33405 Talence, France.

Dominguez, Marisela, Apartado 47159, Los Chaguaramos, Caracas 1041-A, Venezuela.

Douglas, Ronald G., Texas A & M University, Ninth Floor Rudder Tower, Texas A & M University, College Station, TX 77843, USA.

Dritschel, Michael, Department of Mathematics, The University of Newcastle, Newcastle upon Tyne, NE1 7RU, England.

Dyakonov, Konstantin, Departament de Matematica, Aplicada i Analisi, Universitat de Barcelona, Gran Via 585, E-08071 Barcelona, Spain.

Dym, Harry, Department of Mathematics, The Weizmann Institute of Science, Rehovot 76100, Israel.

El-Fallah, Omar, Universite Med V, Faculte des Sciences de Rabat, Departement de Mathematiques, BP 1014 Agdal Rabat, Maroc.

El Harti, Rachid, Faculte des Sciences et Techniques, Universite Hassan I, BP 577, Settat, Maroc.

Esterle, Jean, UFR de Mathématiques, Université Bordeaux 1, 351, cours de la Libération, 33405 Talence, France.

Farforovskaya, Yu., State University of Telecommunication, 191186 Moika 61, St. Petersburg, Russia.

Fedorov, Sergei, Dept. of Mathematics, University of Auckland, Private Bag 92019, Auckland, New Zealand.

Ferguson, Sarah, Department of Mathematics, Wayne State University, Detroit, MI 48202, USA.

Foias, Ciprian, Department of Mathematics, Indiana University, Rawles Hall, Bloomington, Indiana, 47405-5701, USA.

Fricain, Emmanuel, UFR de Mathématiques, Université Bordeaux 1, 351, cours de la Libération, 33405 Talence, France.

Gasimov, Zaid M., Neftyanikov ave., 67, bl. 2a, apt. 13a, 370004 Baku, Azerbaijan.

Gaspar, Dumitru, University of West Timişoara, Bul. V. Pârvan No. 4, 1900 Timişoara, Romania.

Gaspar, Păstorel, University of West Timişoara, Bul. V. Pârvan No. 4, 1900 Timişoara, Romania.

Gavruta, Pasc, "Politehnica", Str. C. Brediceanu, 21, 1900 Timisoara, Romania.

Gohberg, Israel, School of Mathematical Sciences, Tel Aviv University, 69978 Tel Aviv, Israel.

Goldberg, Seymour, University of Maryland, 1612, Peacock Lane Silver Spring, Md. 20904, USA.

Gomilko, Alexander, ul. O. Balzac 77 apt. 446, Kiev 252232, Ukraine.

Gorbachuk, Myroslav L., Institute of Mathematics, Tereshchenkivska Str. 3, Kiev 01601 MSP, Ukraine,

Gorbachuk, Valentyna I., Institute of Mathematics, Tereshchenkivska Str. 3, Kiev 01601 MSP, Ukraine,

Gubreev, Gennady, South-Ukrainian Pedagogical, University, Staroportofrankovskaya 26, Odessa 270020, Ukraine.

Harlouchet, Iban, UFR de Mathématiques, Université Bordeaux 1, 351, cours de la Libération, 33405 Talence, France.

Hartmann, Andreas, UFR de Mathématiques, Université Bordeaux 1, 351, cours de la Libération, 33405 Talence, France.

Hasanov, Mahir, Dept. Math., Istanbul Tecknical University, 80626 Maslak, Istanbul, Turkey.

Hassan, Youssfi, CMI-LATP Universite de Provence, Marseille, France.

Hassi, Seppo, Department of Mathematics and Statistics, P.O. Box 54, 00014 University of Helsinki, Finland.

Helton, W., Department of Mathematics, University of California at San Diego, La Jolla, USA.

Hryniv, Rostyslav, Institute for Applied Problems, of Mechanics and Mathematics, 3b Naukova str., 79601 Lviv, Ukraine.

Ivanov, Sergei, Russian Center of Laser Physics, St. Petersburg University, Ul'yanovskaya ul. 1, Petrodvorets, St. Petersburg 198904, Russia.

Jaming, Philippe, MAPMO — Departement Mathematiques, Universite d'Orleans, BP 6759, 45067 Orleans Cedex 2, France.

Janas, Jan, Instytut Matematyczny PAN, 31-027 Krakow, ul. Sw. Tomasza 30, Poland.

Kaashoek, Marinus A., Divisie Wiskunde en Informatica, Faculteit der Exacte Wetenschappen, Vrije Universiteit, De Boelelaan 1081a, 1081 HV Amsterdam, Pays Bas.

Kalton, Nigel, Dept. of Mathemtics, University of Missouri, Columbia, Mo. 65211, USA.

Kalyuzhniy, Dmitriy, Odessa State Academy of, Civil Engineering and Architecture, Didrihson str. 4, Odessa 270029, Ukraine.

Kapustin, Vladimir, Steklov Mathematical Institute at St. Petersburg, Fontanka 27, St. Petersburg 191011, Russia.

Karlovich, Alexei, Departamento de Matematica, Instituto Superior Tecnico, Av. Rovisco Pais, 1096 Lisboa, Portugal.

Karlovich, Yuri, CINVESTAV del I.P.N., Departamento de Matemáticas, Apartado Postal 14-740, 07000, México D.F.

Kellay, Karim, CMI-LATP, Universite de Provence, Marseille, France.

Kerchy, Laszlo, Bolyai Institute, University of Szeged, Aradi vertanuk tere 1, H-6720 Szeged, Hungary.

Kheifets, Alexander, Research Institute College of Judea and Samaria, Ariel 44837, Israel.

Kisil, Vladimir, Department of Pure Mathematics, University of Leeds, LS2 9JT Leeds, United Kingdom.

Kissin, Edward, School of Communications Technology, and Mathematical Sciences, University of North London, 166-220 Holloway Road, London N7 8DB, UK.

Klaus, Martin, Virginia Tech, Blacksburg, VA 24061, USA.

Konstantinov, Alexei, Department of Mathematics, University of Kiev, 64 Volodymyrska Str., 252033 Kiev, Ukraine.

Koshmanenko, V., Institute of Mathematics, Tereshchenkivs'ka 3, 01601 Kyiv, Ukraine.

Kosiek, Marek, Institute of Mathematics, Jagiellonian University, Reymonta 4, PL-30-059 Krakow, Poland.

Kupin, Stanislav, UFR de Mathématiques, Université Bordeaux 1, 351, cours de la Libération, 33405 Talence, France.

Kurasov, Pavel, Dept. of Mathematics, Stockholm University, 10691 Stockholm, Sweden.

Kuzhel, S., Institute of Mathematics, ul. Tereshchenkivska 3, 252601 Kiev MSP, Ukraine.

Lakaev, Saidakhmat, Department of Math., Samarkand State University, University Boulevard, 15, 703004 Samarkand, Uzbekistan.

Langer, Heinz, Institute of Analysis and Technical Mathematics, Wiedner Hauptstrasse 8-10, A-1040 Vienna, Austria.

Langer, Matthias, Institut fuer Analysis und ,Technische Mathematik, Technische Universitaet Wien, Wiedner Hauptstrasse 8-10, A-1040 Vienna, Austria.

Larson, R., Dept. of Math., College Station, Texas 77843-3368, USA.

Lasser, Rupert, Zentrum Mathematik, Technical University Munich, 80290 Munich, Germany.

Latushkin, Yuri, Department of Mathematics, University of Missouri, Columbia, MO 65311, USA.

Leblond, Juliette, INRIA, B.P. 93, 06902 Sophia-Antipolis Cedex, France.

Le Merdy, Christian, Departement de Mathematiques, Universite de Franche-Comte, 25030 Besancon Cedex, France.

Lerer, Leonid, Department of Mathematics, Technion, 32000 Haifa, Israel.

Li, Wing Suet, School of Mathematics, Georgia Tech, Atlanta, GA 30332, USA.

Litvinchuk, G. S., Departamento de Matematica, Universidade da Madeira, Largo do Colegio, 9000 Funchal, Portugal.

Lomonosov, Victor, Dept. of Math., Kent State University, Kent, OH 44242, USA.

Luger, Annemarie, Institut fuer Analysis und Technische Mathematik, Technische Universitaet Wien, Wiedner Hauptstrasse 8-10, A-1040 Vienna, Austria.

Lyubarskii, Yurii, Department of Mathematical Sciences, NTNU, 7491 Trondheim, Norway.

McCarthy, John, Dept. of Mathematics, Campus Box 1146, Washington University, St. Louis, MO 63130, USA.

Maestripieri, Alejandra, Universidad Nacional de Gral., Sarmiento, Rio de Janeiro, 670 6 B (1405), Buenos Aires, Argentina.

Malamud, Mark, Mathematics Dept., Donetsk State University, Universitetskaya Str. 24, Donetsk 340055, Ukraine.

Malinen, Jarmo, Department of Mathematics, Helsinki University of Technology, P.O.Box 1100, 02015 HUT, Finland.

Mancera, Carmen H., Escuela Superior de Ingenieros, Camino de los Descubrimientos s/n, 41092-Sevilla, Espagne.

Marcantognini, Stefania, Dept. of Mathematics, Apartado 21827, Caracas 1020A, Venezuela.

Markus, Alexander, Dept. of Mathematics, Ben-Gurion University of the Negev, Beer-Sheva 84105, Israel.

Matheron, Etienne, UFR de Mathématiques, Université Bordeaux 1, 351, cours de la Libération, 33405 Talence, France.

Matsaev, Vladimir, Dept. of Mathematics, Tel Aviv University, Ramat Aviv, 69978 Tel Aviv, Israel.

Megretski, Alexander, Department of Electrical Engineering and Computer Science, Massachusetts Institute of Technology, 77 Massachusetts Avenue, Cambridge, MA 02139-4307, USA.

Menini, Chantal, UFR de Mathématiques, Université Bordeaux 1, 351, cours de la Libération, 33405 Talence, France.

Mennicken, Reinhard, Naturwissenschaftliche Fakultaet I , Mathematik, Universitaet Regensburg, Universitaetsstr. 31, D-93053 Regensburg, Germany.

Meyer, Yves, UMR CNRS 7534, Universite Paris 9, place du M. de Lattre de Tassigny, 75775 Paris Cedex 16, France.

Minkin, Arkadi, Intel Corp., str. Tabenkin 41/13, 32801 Haifa, Israel.

Mityagin, Boris, 231, West 18th Ave, Columbus, Ohio 43210, USA.

Muminov, Mukhiddin I., Department of Math., Samarkand State University, University Boulevard 15, 703004 Samarkand, Uzbekistan.

Naboko, Sergey, Department of Math. Physics, Institute of Physics, St. Petersburg University, ul. Ulianovskaya 1, St. Petergoff, St. Petersburg 198904, Russia.

Nakazi, Takahiko, Department of Mathematics, Hokkaido Univesity, Sapporo 060, Japan.

Navarrete, Sergio B., Escuela Superior de Ingenieros, Camino de los Descubrimientos s/n, 41092-Sevilla, Espagne.

Nevanlinna, Olavi, Helsinki University of Technology, P. Box 1100, Finland.

Nikolskaia, Ludmila, UFR de Mathématiques, Université Bordeaux 1, 351, cours de la Libération, 33405 Talence, France.

Nikolski, Nikolai, UFR de Mathématiques, Université Bordeaux 1, 351, cours de la Libération, 33405 Talence, France.

Nudelman, Adolf A., Odessa State Academy of Civil Engineering and Architecture, ul. Balkovskaya 141/187, Odessa 65005, Ukraine.

Octavio, Alfredo, Depto. de Matematicas, Apartado 21827, IVIC, Caracas 1020A, Venezuela.

Olshevsky, Vadim, Department of Mathematics and Statistics, Georgia State University, Atlanta, GA 30303, USA.

Paneah, Boris, Dept. of Math., Technion, Haifa 32000, Israel.

Partington, Jonathan R., School of Mathematics, University of Leeds, Leeds LS2 9JT, U.K.

Paul, Pedro J., Escuela Superior de Ingenieros, Camino de los Descubrimientos s/n, 41092-Sevilla, Espagne.

Pavlov, Boris, P.B. 92019 Dept. Math., Univ. of Auckland, New Zealand.

Pedersen, Thomas V., UFR de Mathematiques, Université Bordeaux 1, 351, cours de la Liberation, 33405 Talence, France.

Peller, Vladimir, Kansas State University, USA.

Peterburgsky, Irina, Suffolk University, 41 Temple Street, Boston MA 02114, USA.

Petkov, Vesselin, UFR de Mathematiques, Université Bordeaux 1, 351, cours de la Liberation, 33405 Talence, France.

Pik, D.R., Vrije Universiteit Amsterdam, De Boelelaan 1081 HV, The Netherlands.

Pisier, Gilles, Equipe d'Analyse, Universite Paris VI, 4, Place Jussieu, Case 186, 75252 Paris Cedex 05, France.

Pivovarchik, Vyacheslav, ul. Preobrajenskaya 59/61 a.17, Odessa 65045, Ukraine.

Poltoratski, Alexei, Dept. of Math., Texas A&M Univ., College Station, TX 77843, USA.

Popa, Cristian C., Dept. of Teoretical Mathematics, The Weizmann Institute of Science, Rehovot 76100, Israel.

Popovici, Dan, Department of Mathematics, University of West Timisoara, B-dul V. Parvan nr. 4, 1900 Timisoara, Romania.

Pott, Sandra, Department of Mathematics, University of York, Heslington, YO 10 5DD York, UK.

Putinar, Mihai, Mathematics Department, UC Santa Barbara, CA 93106, USA.

Rodman, Leiba, Department of Mathematics, College of William and Mary, Williamsburg, VA 23187-8795, USA.

Sadosky, Cora, Dept. of Mathematics, Howard University, Washington, DC, USA.

Saff, E. B., Department of Mathematics, University of South Florida, Tampa, Florida 33620, USA.

Sakhnovich, Alexander, Belova 30 korp. 1, kv. 4, 250032 Chernigov, Ukraine.

Van Schagen, Frederik, Vrije Universiteit Amsterdam, Wiskunde en Informatica, de Boelelaan 1081A, 1081 HV Amsterdam, The Netherlands.

Schlichting, Guenter, TUM Zentrum Mathematik SCS, D-80290 München, Deutschland.

Seip, Kristian, Department of Mathematical Sciences, Norwegian University of Science and Technology, N-7491 Trondheim, Norway.

Shimorin, Serguei, Department of Mathematics, Lund University, Box 118, 221 00 Lund, Sweden.

Shkalikov, Andrei, Department of Mechanics and Mathematics, Moscow State University, Moscow 117234, Russia.

De Snoo, Henk, Department of Mathematics, University of Groningen, P.O. Box 800, 9700 AV Groningen, The Netherlands.

Solomyak, Michael, Dept. Theor. Mathematics, Weizmann Institute, Rehovot 76100, Israel.

Spitkovsky, Ilya, College of William and Mary, 117 Windsor Way, Williamsburg, VA 23188, USA.

Staffans, Olof J., Department of Mathematics, FIN-20500 Åbo, Finland.

Strouse, Elisabeth, UFR de Mathématiques, Université Bordeaux 1, 351, cours de la Libération, 33405 Talence, France.

Suciu, Nicolae, University of West Timişoara, Bul. V. Pârvan No. 4, 1900 Timişoara, Romania.

Sukhocheva, Lioudmila, Research Institute of Mathematics, Voronezh State University, Universitetskaya pl. 1, Voronezh 394693, Russia.

Thomas, Pascal J., Laboratoire Emile Picard, UMR CNRS 5580, 118 route de Narbonne, 31062 Toulouse Cedex 4, France.

Tikhonov, Alexey, Kirov st. 28, apt. 61, 95011 Simferopol, Crimea, Ukraine.

Timotin, Dan, Institut Girard Desargues, UMR 5028 Mathematiques, Universite Claude Bernard Lyon 1, 69622 Villeurbanne Cedex, France.

Tkachenko, Vadim, Department of Mathematics, and Computer Science, Ben-Gurion University of the Negev, P.O.B. 653, Beer-Sheva 84105, Israel.

Tomilov, Yuri, Department of Dynamical Systems, Institute of Mathematics, Tereshchenkivska str. 3 , 252601 Kiev-4, Ukraine.

Treil, Sergei, Mathematics Department, 151 Thayer Street/Box 1917, Providence, RI 02912, USA.

Tretter, Christiane, Dept. of Mathematics, University of Regensburg, D-93040 Regensburg, Germany.

Trunk, Carsten, MA 6-3 TU Berlin, D-10623 Berlin, Germany.

Tsekanovskii, Eduard, Dept. of Mathematics, P.O.Box 2044, Niagara University, NY 14109, USA.

Vasilescu, Florian, Laboratoire d'Analyse, UMR 8524, Departement de Mathematiques, Universite Lille 1, 59655 Villeneuve d'Ascq Cedex, France.

Vasyunin, Vasily, St.-Petersburg Departement of the Steklov Mathematical Institute, 191011 Fontanka 27, POMI, St.-Petersburg, Russia.

Verbitsky, Igor, Dept. of Math., University of Missouri-Columbia, Columbia, MO 65211, USA.

Verduyn Lunel, S. M., Mathematisch Instituut, Universiteit Leiden, P.O. Box 9512, 2300 RA Leiden, The Netherlands.

Vinnikov, Victor, Dept. of Mathematics, Ben-Gurion University of the Negev, Beer Sheva 84105, Israel.

Vitse, Pascale, UFR de Mathématiques, Université Bordeaux 1, 351, cours de la Libération, 33405 Talence, France.

Volberg, Alexander, Dept. of Mathematics, Michigan State University, East Lansing, MI 48824–1027, USA.

Winkler, Henrik, Institut f. Math. Stochastik, TU Dresden, D-01062 Dresden, Germany.

Woracek, Harald, Technical University Vienna, Wiedner Hauptstrasse 8-10 / 114.1, Austria.

Xia, Jingbo, Department of Mathematics, State University of New York at Buffalo, Buffalo, NY 14214, USA.

Xu, Quanhua, Departement de Mathematiques, Universite de Franche-Comte, 25030 Besancon Cedex, France.

Yafaev, Dmitri, IRMAR, Universite Rennes 1, Campus Beaulieu, 35042 Rennes, France.

Yger, Alain, Université Bordeaux 1, 351, cours de la Libération, 33405 Talence, France.

Yuditskii, Peter, Dept. of Mathematics, Michigan State University, East Lansing, MI 48824, USA.

Zarrabi, Mohamed, UFR de Mathématiques, Université Bordeaux 1, 351, cours de la Libération, 33405 Talence, France.

Zemanek, Jaroslav, Institute of Mathematics, Polish Academy of Sciences, P.O.Box 137, 00-950 Warsaw, Poland.

Zwart, Hans, Po.Box 217, 7500 AE Enschede, The Netherlands.

The Schur algorithm for generalized Schur functions I: coisometric realizations

Daniel Alpay, Tomas Azizov, Aad Dijksma, and Heinz Langer

Abstract. The Schur algorithm as developed by C. Chamfy [C] and J. Dufresnoy [D] (see also [BDGPS]) is related to a sequence of characteristic functions of closely outerconnected coisometric colligations in Pontryagin spaces.

1. Introduction

In this paper, $\mathbf{S}_0$ is the class of all Schur functions: these are the analytic functions s on the open unit disk $\mathbb{D}$ with the property $|s(z)| \leq 1$, $z \in \mathbb{D}$. The classical Schur algorithm as proposed by I. Schur in [S1], [S2] proceeds as follows. Given a function $s \in \mathbf{S}_0$, define a new function s_1 by the relation

$$s_1(z) := \frac{1}{z} \frac{s(z) - s(0)}{1 - s(0)^* s(z)}$$

(here and in the sequel * denotes the complex conjugate of a complex number as well as the adjoint of a matrix or operator). The function s_1 belongs again to the class $\mathbf{S}_0$, and functions $s_2, s_3, \ldots$ can be defined in the same way, in general

$$s_{j+1}(z) := \frac{1}{z} \frac{s_j(z) - s_j(0)}{1 - s_j(0)^* s_j(z)}, \quad j = 1, 2, \ldots, \tag{1.1}$$

and this procedure possibly stops at some index j_0, namely if $|s_{j_0}(0)| = 1$.

By $\mathbf{A}^0$ we denote the set of all functions which are analytic in a neighborhood of the origin in the complex plane and by $\mathbf{S}^0$ the class of generalized Schur functions in $\mathbf{A}^0$ with finitely many negative squares. For the definition we refer to Section 2. The classical Schur algorithm was generalized to functions of the class $\mathbf{S}^0$ in the papers [C], [D], and, in a very complete form, in [BDGPS, Chapter 3], see also [DGK]. In this general case, if $s \in \mathbf{S}^0$ possibly not for all indices $j = 1, 2, \ldots$, a function $s_j \in \mathbf{S}^0$ can be defined by means of (1.1), namely, there can be a different formula or there may be some gaps in the sequence of the natural numbers, say from $j_1 + 1, \ldots, j_1 + k$, such that the corresponding functions $s_{j_1+1}, \ldots, s_{j_1+k}$ are not defined in $\mathbf{S}^0$, see Section 3. The Schur sequence for $s \in \mathbf{S}^0$ will be denoted by

This research was supported in part by grants from the Netherlands Organization for Scientific Research NWO 047-008-008 and NWO 61-453, and RFBR 99-01-00391

$s_0 = s, s_{j_1}, s_{j_2}, \ldots$. The sequence of indices, a nondecreasing sequence of nonnegative integers, will be denoted by $J(s)$ and is called the ***index set*** for s. Formally the generalized Schur algorithm can also be applied to functions of the class $\mathbf{A}^0$.

The functions in $\mathbf{A}^0$ can be characterized by operator representations, see [KL], [ADRS, Chapter 2]: If $s \in \mathbf{A}^0$, there exist a Krein space $(\mathcal{K}, \langle\cdot,\cdot\rangle)$, a contraction T on $\mathcal{K}$, elements $u, v \in \mathcal{K}$, and a complex number γ such that the function

$$s_V(z) = \gamma + z\left\langle (1 - zT)^{-1}u, v\right\rangle,$$

associated with the operator matrix

$$V := \begin{pmatrix} T & u \\ \langle\cdot, v\rangle & \gamma \end{pmatrix} : \begin{pmatrix} \mathcal{K} \\ \mathbb{C} \end{pmatrix} \to \begin{pmatrix} \mathcal{K} \\ \mathbb{C} \end{pmatrix} \tag{1.2}$$

coincides with $s(z)$ in a neighborhood of $z = 0$. The operator V is called an ***operator colligation*** on $\mathcal{K} \oplus \mathbb{C}$ with entries T, u, v, and γ, and the function s_V is called its ***characteristic function***. Here the space $\mathcal{K} \oplus \mathbb{C}$ is equipped with the inner product

$$\left\langle \begin{pmatrix} x \\ \xi \end{pmatrix}, \begin{pmatrix} y \\ \eta \end{pmatrix} \right\rangle := \langle x, y\rangle + \eta^*\xi, \quad x, y \in \mathcal{K}, \quad \xi, \eta \in \mathbb{C}.$$

In this paper $\langle\cdot,\cdot\rangle$ is the notation for any inner product; it should be clear from the context on which space it is defined. The identity

$$s(z) = s_V(z) = \gamma + z\left\langle (1 - zT)^{-1}u, v\right\rangle, \text{ near } z = 0 \tag{1.3}$$

(in the sequel we will omit writing "near $z = 0$") is called the ***operator representation*** or ***realization*** of s induced by V. In the representation V can always be chosen such that either (i) V is coisometric (that is, the operator V^* is isometric) in the space $\mathcal{K} \oplus \mathbb{C}$ and ***closely outerconnected***, which means that

$$\overline{\text{span}}\,\{T^{*n}v : n = 0, 1, \ldots\} = \mathcal{K}. \tag{1.4}$$

or (ii) V is unitary in the space $\mathcal{K} \oplus \mathbb{C}$ and ***closely connected***, that is,

$$\overline{\text{span}}\,\{T^m u, T^{*n}v : m, n = 0, 1, \ldots\} = \mathcal{K}. \tag{1.5}$$

The conditions (1.4) and (1.5) are called ***minimality conditions***. If (i) holds, then (1.3) is called a ***minimal coisometric realization***, if (ii) holds it is called a ***minimal unitary realization***. If V satisfies (i) or (ii), then $\text{sq}_\pm(s_V) = \text{ind}_\pm \mathcal{K}$ (in the sense that the numbers on the left hand side and on the right hand side are either both infinite or both finite and equal), where $\text{ind}_\pm \mathcal{K}$ is the dimension of $\mathcal{K}_\pm$ in any fundamental decomposition $\mathcal{K} = \mathcal{K}_+ \oplus \mathcal{K}_-$ of the Krein space $\mathcal{K}$ and $\text{sq}_\pm(s)$ is the number of positive/negative squares of s, see Section 2. Moreover, if the equality (1.3) holds and V satisfies (i) or (ii), then V is uniquely determined up to a weak isomorphism. Here two colligations V as above and

$$V_1 := \begin{pmatrix} T_1 & u_1 \\ \langle\cdot, v_1\rangle & \gamma_1 \end{pmatrix} : \begin{pmatrix} \mathcal{K}_1 \\ \mathbb{C} \end{pmatrix} \to \begin{pmatrix} \mathcal{K}_1 \\ \mathbb{C} \end{pmatrix}$$

are called *weakly isomorphic* if there is an isometric mapping (called a weak isomorphism) $W : \mathcal{K}_1 \to \mathcal{K}$ with dense domain $\operatorname{dom} W \subset \mathcal{K}_1$ and dense range such that

$$\begin{pmatrix} T & u \\ \langle \cdot , v \rangle & \gamma \end{pmatrix} \begin{pmatrix} W & 0 \\ 0 & 1 \end{pmatrix} = \begin{pmatrix} W & 0 \\ 0 & 1 \end{pmatrix} \begin{pmatrix} T_1 & u_1 \\ \langle \cdot , v_1 \rangle & \gamma_1 \end{pmatrix} \text{ on } \begin{pmatrix} \operatorname{dom} W \\ \mathbb{C} \end{pmatrix}.$$

If V and V_1 are minimal coisometric (or minimal unitary) and weakly isomorphic, then $\operatorname{ind}_\pm \mathcal{K} = \operatorname{ind}_\pm \mathcal{K}_1$. For the proofs of these statements in the unitary case see [AI, Chapter 5] and [CDLS]; in the coisometric case they can be proved similarly.

If $s \in \mathbf{S}^0$ and $s(z) = s_V(z)$, where V is a minimal coisometric or minimal unitary colligation of the form (1.2), the space $\mathcal{K}$, called the ***state space*** of V, is a Pontryagin space. Its (by definition finite) negative index coincides with the number of negative squares of s. In this case V is uniquely determined up to unitary equivalence, that is, in the discussion above W is a unitary operator from all of $\mathcal{K}_1$ onto all of $\mathcal{K}$. In this paper we study for a function $s \in \mathbf{S}^0$ the coisometric colligations associated with the functions $(s_j)_{j \in J(s)}$. We express these colligations in terms of the colligation associated with the original function s.

In a sequel to this paper we will (1) consider the unitary realizations, (2) relate these realizations to the theory of reproducing kernel Pontryagin spaces (compare [AD] and [AG]), (3) reconsider the realizations from a more geometric point of view, by which we mean that the emphasis in the construction of the colligations lies on the main operator T and the chain of eigenfunctions and associated functions of T or some related operators at the eigenvalue zero, (4) study the sequence of Schur determinants $(\delta_j)_{j \in J(s)}$ associated with $s \in \mathbf{S}^0$ and (5) discuss the inverse problem: is it possible, and if so how, to recover the function s from the Schur parameters (also called reflection coefficients) $s_j(0)$, $j \in J(s)$?

2. Generalized Schur functions

Let κ be a nonnegative integer. A complex valued function $K(z, \zeta)$ defined on some open set $\Omega \subset \mathbb{C}$ is said to have κ positive (negative, respectively) squares if

$$K(z,\zeta)^* = K(\zeta, z), \qquad z, \zeta \in \Omega, \tag{2.1}$$

and for all choices of $n \in \mathbb{N}$, and $z_1, z_2, \dots, z_n \in \Omega$ the hermitian matrix

$$(K(z_i, z_j))_{i,j=1}^n \tag{2.2}$$

has at most κ and for at least one such choice exactly κ positive (negative, respectively) eigenvalues. If (2.1) holds and for every $m \in \mathbb{N}$ there is a matrix of the form (2.2) which has m or more positive (negative, respectively) eigenvalues, then $K(z, \zeta)$ is said to have infinitely many positive (negative, respectively) squares.

By $\mathbf{S}_\kappa$ we denote the set of all complex functions s which are meromorphic on $\mathbb{D}$ and have one of the following equivalent properties (see, for example, [KL], [ADRS]):

1. s has κ poles in $\mathbb{D}$ (counted according to their multipicities) and

$$\limsup_{r\uparrow 1}|s(r\,\mathrm{e}^{i\,t})| \leq 1, \quad \text{for almost all } t \in [0, 2\pi].$$

2. The kernel

$$S_s(z,\zeta)) := \frac{1 - s(z)s(\zeta)^*}{1 - z\zeta^*}, \quad z, \zeta \in \mathcal{D}_s, \tag{2.3}$$

has κ negative squares; here $\mathcal{D}_s$ denotes the domain of holomorphy of s.

3. The 2×2–matrix kernel

$$D_s(z,\zeta) := \begin{pmatrix} S_s(z,\zeta) & \dfrac{s(z) - s(\zeta^*)}{z - \zeta^*} \\ \dfrac{s^*(z) - s^*(\zeta^*)}{z - \zeta^*} & S_{s^*}(z,\zeta) \end{pmatrix},$$

has κ negative squares; here s^* denotes the function $s^*(z) := s(z^*)^*$.

4. The function s admits the representation

$$s(z) = \prod_{j=1}^{\kappa} \frac{1 - zz_j^*}{z - z_j}\, s_0(z)$$

with points $z_j \in \mathbb{D}$ and a function $s_0 \in \mathbf{S}_0$, $s_0(z_j) \neq 0$, $j = 1, 2, \ldots, \kappa$.

The functions of the class $\mathbf{S}_\kappa$, $\kappa = 1, 2, \ldots$, are called *generalized Schur functions with κ negative squares*, and we set $\mathbf{S} := \cup_{\kappa=0}^{\infty}\mathbf{S}_\kappa$. In the following we are particularly interested in those functions of $\mathbf{S}$ which belong to $\mathbf{A}^0$, that is, which are analytic at $z = 0$. This set is denoted by $\mathbf{S}^0$: $\mathbf{S}^0 = \mathbf{S} \cap \mathbf{A}^0$. Evidently, $\mathbf{S}_0 \subset \mathbf{S}^0$, and we set $\mathbf{S}_\kappa^0 := \mathbf{S}_\kappa \cap \mathbf{A}^0$, $\kappa = 1, 2, \ldots$. In the sequel for $s \in \mathbf{A}^0$, we set $\mathrm{sq}_\pm(s) = \kappa$ ($= \infty$, respectively) if the kernel $S_s(z,\zeta)$ in (2.3) has κ (infinitely many, respectively) positive/negative squares.

3. The generalized Schur algorithm

In this section we describe the generalized Schur algorithm for the class $\mathbf{S}^0$ as defined in [BDGPS, Chapter 3]. Let $s \in \mathbf{S}^0$ and let

$$s_0(z) := s(z) = \sigma_{00} + \sigma_{01}z + \sigma_{02}z^2 + \cdots + \sigma_{0n}z^n + \cdots$$

be its Taylor expansion at $z = 0$. With s_0 a sequence $(s_j)_{j\in J}$ of functions in $\mathbf{S}^0$, where j runs through a finite or infinite increasing sequence $J = J(s)$, the index set for s, of nonnegative integers with $0 \in J$, is defined as follows.

(1) If $s_j(z) \equiv \sigma_{j0}$ and $|\sigma_{j0}| = 1$, then the sequence stops at s_j.

(2) If

$$s_j(z) = \sigma_{j0} + \sigma_{j1}z + \sigma_{j2}z^2 + \cdots + \sigma_{jn}z^n + \cdots$$

with $|\sigma_{j0}| < 1$, define s_{j+1} by

$$s_{j+1}(z) = \frac{1}{z}\frac{s_j(z) - \sigma_{j0}}{1 - \sigma_{j0}^* s_j(z)}. \tag{3.1}$$

(3) If

$$s_j(z) = \sigma_{j0} + \sigma_{jk} z^k + \sigma_{j,k+1} z^{k+1} + \cdots ,$$

with $|\sigma_{j0}| > 1$, $k \geq 1$, and $\sigma_{jk} \neq 0$, define s_{j+k} by

$$s_{j+k}(z) = z^k \frac{1 - \sigma_{j0}^* s_j(z)}{s_j(z) - \sigma_{j0}}. \tag{3.2}$$

(4) If

$$s_j(z) = \sigma_{j0} + \sigma_{jk} z^k + \sigma_{j,k+1} z^{k+1} + \cdots ,$$

with $|\sigma_{j0}| = 1$, $k \geq 1$, and $\sigma_{jk} \neq 0$, complex numbers $c_{j0}, c_{j1}, \cdots$ are introduced by the relation

$$(s_j(z) - \sigma_{j0})(c_{j0} + c_{j1} z + c_{j2} z^2 + \cdots + c_{jn} z^n + \cdots) \equiv \sigma_{j0} z^k. \tag{3.3}$$

Consider the polynomial

$$Q_j(z) = c_{j0} + \cdots + c_{j,k-1} z^{k-1} - (c_{j,k-1}^* z^{k+1} + \cdots + c_{j0}^* z^{2k}).$$

If

$$\sigma_{j0}^* Q_j(z) s_j(z) - (Q_j(z) + z^k) = t_{j,2k+q} z^{2k+q} + t_{j,2k+q+1} z^{2k+q+1} + \cdots \tag{3.4}$$

with $t_{j,2k+q} \neq 0$ for some $q \in \{0, 1, 2, \ldots\}$, define s_{j+2k+q} by

$$s_{j+2k+q}(z) = z^q \frac{(Q_j(z) - z^k) s_j(z) - \sigma_{j0} Q_j(z)}{\sigma_{j0}^* Q_j(z) s_j(z) - (Q_j(z) + z^k)}. \tag{3.5}$$

There are two more cases possible if one restricts the considerations not to functions of the class $\mathbf{S}^0$ but to $\mathbf{A}^0$ as in [BDGPS]. These cases are:

(5) $s_j(z) \equiv \sigma_{j0}$ and $|\sigma_{j0}| > 1$. Then s_j does not belong to $\mathbf{S}^0$ since its boundary values for $|z| = 1$ are in modulus greater than one.

(6) Under the assumptions of (4) we have $\sigma_{j0}^* Q_j(z) s_j(z) - (Q_j(z) + z^k) \equiv 0$. Then with the polynomial

$$P_j(z) := c_{j0} + \cdots + c_{j,k-1} z^{k-1} \tag{3.6}$$

we obtain

$$s_j(z) = \frac{1}{\sigma_{j0}} \left(1 + \frac{1}{z^{-k} P_j(z) - z^k P_j^*(z^{-1})} \right).$$

Since for $|z| = 1$ the second term in the big brackets on the right side is purely imaginary, the modulus of the whole expression between these brackets becomes greater than one, and since $|\sigma_{j0}| = 1$, also in this case the function s_j does not belong to $\mathbf{S}^0$.

For a function $s \in \mathbf{S}_0$ with the classical Schur algorithm (see [S1], [S2]) only the steps (1) and (2) come into play, and then the index set $J(s)$ is either of the form $J(s) = \{0, 1, \ldots, n\}$ or $J(s) = \{0, 1, 2, \ldots\}$.

If we apply the generalized Schur algorithm to a function $s_0 \equiv s \in \mathbf{S}^0$, in the steps (2), (3), and (4) the new function in (3.1), (3.2) and (3.5) is obtained from the old one by a fractional linear transformation. We use the following notation. If

$$\Theta(z) = \begin{pmatrix} a(z) & b(z) \\ c(z) & d(z) \end{pmatrix} \tag{3.7}$$

is a 2×2–matrix function with $\det \Theta(z) \neq 0$, for a complex function $f(z)$ we set

$$(T_\Theta f)(z) := (a(z)f(z) + b(z))(c(z)f(z) + d(z))^{-1}$$

provided the inverse exists; sometimes for the expression on the left side we also write $T_{\Theta(z)}f(z)$. Note that T_Θ satisfies the relation

$$T_{\Theta_1\Theta_2}f = T_{\Theta_1}(T_{\Theta_2}f)$$

and that the inverse mapping is given by

$$T_{\Theta(z)}^{-1}f(z) = T_{\Theta(z)^{-1}}f(z) = (d(z)f(z) - b(z))(-c(z)f(z) + a(z))^{-1}.$$

Theorem 3.1. *If s_0 belongs to the class $\mathbf{S}^0$ then all functions s_j generated by the generalized Schur algorithm applied to s_0 belong to the class $\mathbf{S}^0$.*

Proof. Observing the characterization 2 of the functions of the classes $\mathbf{S}_\kappa$ it is sufficient to show that for each of the functions s_l on the left sides of the relations (3.1), (3.2), and (3.5) the Schur kernel $S_{s_l}(z,\zeta)$ has finitely many negative squares.

Set

$$\Psi(\sigma, z) := \frac{1}{\sqrt{1-|\sigma|^2}} \begin{pmatrix} z^{-1} & 0 \\ 0 & 1 \end{pmatrix} \begin{pmatrix} 1 & -\sigma \\ -\sigma^* & 1 \end{pmatrix}.$$

Then the formula (3.1) becomes $s_{j+1}(z) = (T_\Theta s_j)(z)$ with $\Theta(z) = \Psi(\sigma_{j0}, z)$. The formula in (3.2) can be written as

$$s_{j+k}(z) = \frac{\sigma_{j0}^*}{\sigma_{j0}} \left(\frac{1}{z^k} \frac{\dfrac{1}{s_j(z)} - \dfrac{1}{\sigma_{j0}}}{1 - \dfrac{1}{\sigma_{j0}^*} \dfrac{1}{s_j(z)}} \right)^{-1}$$

which implies that $s_{j+k}(z) = (T_\Theta s_j)(z)$ with

$$\Theta(z) = \begin{pmatrix} \sigma_{j0}^*\sigma_{j0}^{-1} & 0 \\ 0 & 1 \end{pmatrix} \begin{pmatrix} 0 & 1 \\ 1 & 0 \end{pmatrix} \Psi\left(\sigma_{j0}^{-1}, z^k\right) \begin{pmatrix} 0 & 1 \\ 1 & 0 \end{pmatrix}.$$

In (3.5) we have that $s_{j+2k+q}(z) = (T_\Theta s_j)(z)$ with

$$\Theta(z) = \begin{pmatrix} z^q(Q_j(z) - z^k) & -\sigma_{j0}z^qQ_j(z) \\ \sigma_{j0}^*Q_j(z) & -(Q_j(z) + z^k) \end{pmatrix}.$$

In all three cases the matrix function Θ has rational coefficients with poles only in $z = 0$ and is J-unitary on the unit circle, that is,

$$\Theta(z)J\Theta(z)^* = J, \quad |z| = 1, \tag{3.8}$$

where J is the matrix

$$J = \begin{pmatrix} 1 & 0 \\ 0 & -1 \end{pmatrix}.$$

Indeed, in the first two cases (3.8) can easily be verified, in the last case this relation holds because, with $P_j(z)$ as in (3.6), $Q_j(z) = P_j(z) - z^{2k} P_j^*\left(z^{-1}\right)$ and hence $z^{*k}Q_j(z) + z^k Q_j(z)^* = 0$ for $|z| = 1$.

By [AG], since Θ is rational and J–unitary on the unit circle, the linear span of the functions

$$z \mapsto \frac{J - \Theta(z)J\Theta(\zeta)^*}{1 - z\zeta^*} c, \quad \zeta \in \mathbb{D} \setminus \{0\},\ c \in \mathbb{C}^2,$$

is finite dimensional and hence the kernel

$$\frac{J - \Theta(z)J\Theta(\zeta)^*}{1 - z\zeta^*}$$

has finitely many negative and positive squares. That s_l has finitely many squares now follows from this and the formula

$$\begin{aligned} S_s(z,\zeta) = {} & \begin{pmatrix} 1 & -s(z) \end{pmatrix} \frac{J - \Theta(z)J\Theta(\zeta)^*}{1 - z\zeta^*} \begin{pmatrix} 1 & -s(\zeta) \end{pmatrix}^* \\ & + (a(z) - s(z)c(z)) S_{s_l}(z,\zeta)(a(\zeta) - s(\zeta)c(\zeta))^*, \end{aligned}$$

where $a(z)$ and $c(z)$ are the coefficients of $\Theta(z)$ as in (3.7). □

4. A model for the colligation of $s \in \mathbf{A}^0$

Given $s \in \mathbf{A}^0$ with the Taylor expansion

$$s(z) = \sigma_0 + \sum_{n=1}^{\infty} \sigma_n z^n$$

near $z = 0$. We consider the corresponding minimal coisometric colligation V from (1.2) in the space $\mathcal{K} \oplus \mathbb{C}$ with a Krein space $\mathcal{K}$:

$$V := \begin{pmatrix} T & u \\ \langle\,\cdot\,, v\rangle & \gamma \end{pmatrix}.$$

That is, the relation

$$s(z) = \gamma + z\langle (I - zT)^{-1}u, v\rangle = \sum_{n=0}^{\infty} \sigma_n z^n$$

holds, which means that $\sigma_0 = \gamma$ and $\sigma_{n+1} = \langle T^n u, v\rangle$, $n = 0, 1, 2, \ldots$.

We express the fact that the colligation V is coisometric in $\mathcal{K} \oplus \mathbb{C}$ by its entries T, u, v, γ:

$$TT^* + \langle\,\cdot\,, u\rangle u = I_{\mathcal{K}}, \tag{4.1}$$

$$\langle v, v\rangle + |\gamma|^2 = 1, \tag{4.2}$$

$$Tv + \gamma^* u = 0. \tag{4.3}$$

In case the colligation V is unitary in $\mathcal{K}\oplus\mathbb{C}$ also the following relations hold:

$$T^*T + \langle\,\cdot\,, v\rangle v = I_{\mathcal{K}}, \tag{4.4}$$

$$\langle u, u\rangle + |\gamma|^2 = 1, \tag{4.5}$$

$$T^*u + \gamma v = 0. \tag{4.6}$$

In the coisometric case, the minimality assumption (1.4) implies that the space $\mathcal{K}$ can be realized as follows. Consider the elements $\sum \alpha_n T^{*n} v \in \mathcal{K}$ for which only finitely many coefficients $\alpha_n \in \mathbb{C}$, $n = 0, 1, \dots,$ differ from zero. With such an element we associate the infinite vector $(\alpha_n)_0^\infty$. This means that we choose in $\mathcal{K}$ as a basis the sequence

$$v,\ T^*v,\ T^{*2}v, \dots, T^{*m}v, \dots.$$

Either every finite set of elements of this sequence is linearly independent, or there is a smallest nonnegative number l such that $T^{*l}v$ is a linear combination of its predecessors in the sequence. In the first case $\mathcal{K}$ is infinite dimensional and the full sequence forms a basis; in the latter case $\mathcal{K}$ is l–dimensional. In the following we consider the infinite dimensional case, leaving the changes in the other case to the reader. With respect to this basis the element v becomes $v = (\delta_{0n})_0^\infty$, where δ_{mn} for $m, n = 0, 1, \dots$ is the Kronecker symbol. In this basis, evidently, the operator T^* is the right shift, and we denote the Gram matrix which determines the inner product with respect to this basis by $\mathcal{G}$, that is, the relation

$$\left\langle \sum \alpha_n T^{*n} v, \sum \beta_n T^{*n} v \right\rangle = \left(\mathcal{G}(\alpha_n)_0^\infty, (\beta_n)_0^\infty\right)$$

holds, where the considered elements of $\mathcal{K}$ are again finite sums and $(\,\cdot\,,\,\cdot\,)$ is the inner product of the space l^2. Thus, if $\mathcal{G} = (g_{ij})_{i,j=0}^\infty$ then $g_{ij} = \left\langle T^{*j}v, T^{*i}v\right\rangle$, and the coisometricity of V in $\mathcal{K}\oplus\mathbb{C}$ implies the following formulas for $i, j = 0, 1, \dots$:

$$g_{ij} = \left\langle T^{*j}v, T^{*i}v\right\rangle = \begin{cases} 1 - |\sigma_0|^2 - |\sigma_1|^2 - \dots - |\sigma_i|^2, & i = j, \\ -\left(\sigma_0^*\sigma_{i-j} + \sigma_1^*\sigma_{i-j+1} + \cdots + \sigma_j^*\sigma_i\right), & i > j. \end{cases} \tag{4.7}$$

For example, if $i > j > 0$, then this formula follows from

$$\begin{aligned} \langle T^{*j}v, T^{*i}v\rangle &= \left\langle (TT^*)T^{*(j-1)}v, T^{*(i-1)}v\right\rangle \\ &= \left\langle T^{*(j-1)}v - \langle T^{*(j-1)}v, u\rangle u, T^{*(i-1)}v\right\rangle \\ &= \left\langle T^{*(j-1)}v, T^{*(i-1)}v\right\rangle - \sigma_j^*\sigma_i \\ &= \left\langle v, T^{*(i-j)}v\right\rangle - \left(\sigma_1^*\sigma_{i-j+1} + \dots + \sigma_j^*\sigma_i\right) \\ &= -\gamma^*\left\langle u, T^{*(i-j-1)}v\right\rangle - \left(\sigma_1^*\sigma_{i-j+1} + \dots + \sigma_j^*\sigma_i\right), \end{aligned}$$

where for the second equality we have used the relation (4.1) and for the last equality the relation (4.3). The formulas (4.7) mean that the matrix $\mathcal{G}$ can be written as

$$\mathcal{G} = I - \mathcal{A}\mathcal{A}^*, \tag{4.8}$$

where

$$\mathcal{A} := \begin{pmatrix} \sigma_0 & 0 & 0 & 0 & \cdots \\ \sigma_1 & \sigma_0 & 0 & 0 & \cdots \\ \sigma_2 & \sigma_1 & \sigma_0 & 0 & \cdots \\ \sigma_3 & \sigma_2 & \sigma_1 & \sigma_0 & \cdots \\ \vdots & \vdots & \vdots & \vdots & \ddots \end{pmatrix}.$$

By $\mathcal{G}_m$ and $\mathcal{A}_m$ we denote the $m \times m$ principal submatrices of $\mathcal{G}$ and $\mathcal{A}$. Then $\mathcal{G}_m = (I_m - \mathcal{A}_m\mathcal{A}_m^*)$, where I_m is the $m \times m$ identity matrix. The formulas (4.7) now readily imply the following two lemmas.

Lemma 4.1. *If, for $k \geq 1$, $\sigma_1 = \cdots = \sigma_{k-1} = 0$, then $\mathcal{G}_k = (1 - |\sigma_0|^2)I_k$.*

Lemma 4.2. *The following two statements are equivalent:*

(1) $|\sigma_0| = 1$ *and* $\sigma_1 = \ldots = \sigma_{k-1} = 0$.

(2) span $\{v, T^*v, \ldots, T^{*(k-1)}v\}$ *is a k-dimensional neutral subspace of $\mathcal{K}$.*

If these statements hold and $\sigma_k \neq 0$, then the space

$$\operatorname{span}\left\{v, T^*v, \ldots, T^{*(k-1)}v, \ldots, T^{*(2k-1)}v\right\}$$

is a $2k$–dimensional Pontryagin space with positive index k and negative index k.

We also need a lemma, which is essentially known, see [BGK, Corollary 2.7] and [LL, Lemma 3.1]. Nevertheless, we give a sketch of the proof not only for the sake of completeness, but also because the method of proof reappears mutatis mutandis in the proofs of the theorems below. In the sequel the anti-space of a Krein space $(\mathcal{K}, \langle\,\cdot\,,\,\cdot\,\rangle)$ is the Krein space $(\mathcal{K}', \langle\,\cdot\,,\,\cdot\,\rangle')$ where $\mathcal{K}' = \mathcal{K}$ as linear spaces and $\langle x, y\rangle' = -\langle x, y\rangle$, x, $y \in \mathcal{K}'$. Evidently, $\mathcal{K}'' = \mathcal{K}$ as Krein spaces.

Lemma 4.3. *Let f be the characteristic function of a minimal coisometric (minimal unitary, respectively) colligation V on $\mathcal{K} \oplus \mathbb{C}$ of the form (1.2):*

$$f(z) = \gamma + z\left\langle (I - zT)^{-1}u, v\right\rangle,$$

and assume $\gamma = f(0) \neq 0$. Then the function $\widehat{f}$: $\widehat{f}(z) = f(z)^{-1}$ is the characteristic function of the colligation

$$\widehat{V} = \begin{pmatrix} \widehat{A} & \widehat{u} \\ \langle\,\cdot\,, \widehat{v}\rangle' & \widehat{\gamma} \end{pmatrix}$$

on $\mathcal{K}' \oplus \mathbb{C}$, where $\mathcal{K}'$ is the anti-space of the Krein space $\mathcal{K}$, with entries

$$\widehat{T} := T - \frac{\langle\,\cdot\,, v\rangle}{\gamma}u, \quad \widehat{u} = \frac{u}{\gamma},$$

$$\widehat{v} = \frac{v}{\gamma^*}, \quad \widehat{\gamma} = \frac{1}{\gamma},$$

that is,

$$\widehat{f}(z) = \widehat{\gamma} + z\left\langle (I - z\widehat{T})^{-1}\widehat{u}, \widehat{v}\right\rangle .$$

Moreover, $\widehat{V}$ is minimal coisometric (minimal unitary, respectively).

Proof. Set $x = (1 - z\widehat{T})^{-1}\widehat{u}$. Then the following equalities can be verified, one after the other:

$$\begin{aligned}
x &= \widehat{u} + z\widehat{T}x = \frac{u}{\gamma} + zTx + z\langle x, \widehat{v}\rangle' u,\\
(1 - zT)x &= \left(\frac{1}{\gamma} + z\langle x, \widehat{v}\rangle'\right) u,\\
x &= \left(\frac{1}{\gamma} + z\langle x, \widehat{v}\rangle'\right)(1 - zT)^{-1}u,\\
\langle x, \widehat{v}\rangle' &= -\frac{1}{\gamma}\left(\frac{1}{\gamma} + z\langle x, \widehat{v}\rangle'\right)\left\langle (1 - zT)^{-1}u, v\right\rangle,\\
\langle x, \widehat{v}\rangle' &= -\frac{1}{\gamma}\frac{\langle (1 - zT)^{-1}u, v\rangle}{\gamma + z\langle (1 - zT)^{-1}u, v\rangle},
\end{aligned}$$

and hence

$$\widehat{f}(z) = \frac{1}{\gamma} + z\langle x, \widehat{v}\rangle' = \frac{1}{f(z)}.$$

The coisometricity (the unitarity, respectively) of $\widehat{V}$ follows from that of V via the formulas (4.1)–(4.3) ((4.1)–(4.6), respectively). Indeed, if they hold for T, u, v, and γ then they also hold with T, u, v, γ and the inner product $\langle\,\cdot\,,\,\cdot\,\rangle$ replaced by $\widehat{T}$, $\widehat{u}$, $\widehat{v}$, $\widehat{\gamma}$ and the inner product $\langle\,\cdot\,,\,\cdot\,\rangle'$. The verification of these equalities is straightforward and therefore omitted. The relations

$$\operatorname{span}\{v, T^*v, \cdots, T^{*n}v\} = \operatorname{span}\left\{\widehat{v}, \widehat{T}^*v, \cdots, \widehat{T}^{*n}v\right\}, \qquad n = 0, 1, \ldots,$$

readily imply the minimality statement in the case when V is coisometric. The minimality when V is unitary can be proved similarly. □

In this note, we only consider minimal coisometric colligations and minimal coisometric realizations. The minimal unitary ones will be considered elsewhere.

5. The colligation of the Schur transformation: Case (2)

In this and the following three sections we describe the coisometric colligations which correspond to the generalized Schur transformations s_{j+1}, s_{j+k}, and s_{j+2k+q} in (3.1), (3.2), and (3.5) in terms of the colligation corresponding to s_j. Instead of s_j we write simply s, and the transformed function is in this section denoted by s_1. The Schur transformation in case (2) corresponds to the following theorem. Its formulation which is more general than case (2) requires, is needed later in Section 6 when we consider case (3) where $|\sigma_0| > 1$.

Theorem 5.1. *Let $s \in \mathbf{A}^0$ and assume that in its Taylor expansion at $z = 0$,*

$$s(z) = \sigma_0 + \sigma_1 z + \cdots + \sigma_{n-1}z^{n-1} + \sigma_n z^n + \cdots,$$

the constant term $\sigma_0 = s(0)$ satisfies $|\sigma_0| < 1$. Let V, given by (1.2) with $\gamma = \sigma_0$, be a minimal coisometric colligation in the space $\mathcal{K} \oplus \mathbb{C}$, where $\mathcal{K}$ is a Krein space, such that $s(z) = s_V(z)$. Then
(i) span $\{v\}$ *is a positive subspace of $\mathcal{K}$, so the orthogonal projection P_1 in $\mathcal{K}$ onto $\mathcal{K}_1 = \{v\}^{\perp}$ exists,* $\operatorname{ind}_+ \mathcal{K}_1 = \operatorname{ind}_+ \mathcal{K} - 1$, *and* $\operatorname{ind}_- \mathcal{K}_1 = \operatorname{ind}_- \mathcal{K}$;
(ii) *the colligation*

$$V_1 := \begin{pmatrix} T_1 & u_1 \\ \langle \cdot, v_1 \rangle & \gamma_1 \end{pmatrix}$$

with entries

$$\begin{aligned} T_1 &:= P_1 T P_1, & u_1 &:= \frac{1}{\sqrt{1-|\gamma|^2}} P_1 u, \\ v_1 &:= \frac{1}{\sqrt{1-|\gamma|^2}} P_1 T^* v, & \gamma_1 &:= \frac{\langle u, v \rangle}{1-|\gamma|^2} \end{aligned}$$

is minimal coisometric in the space $\mathcal{K}_1 \oplus \mathbb{C}$;
(iii) *with $s_1(z) := s_{V_1}(z)$ it holds*

$$s_1(z) \equiv \frac{1}{z} \frac{s(z) - \sigma_0}{1 - \sigma_0^* s(z)}.$$

In particular, if $s \in \mathbf{S}^0_\kappa$ satisfies the above condition, then also $s_1 \in \mathbf{S}^0_\kappa$ and $\operatorname{sq}_+(s_1) = \operatorname{sq}_+(s) - 1$.

Proof. The relation $\langle v, v \rangle = 1 - |\gamma|^2 = 1 - |\sigma_0|^2 > 0$ (see (4.2)) implies that span $\{v\}$ is a positive subspace and hence the projection P_1 is well defined. To show that the colligation V_1 in $\mathcal{K}_1 \oplus \mathbb{C}$ is coisometric amounts to verifying that (4.1) – (4.3) imply the same formulas with T, u, v and γ replaced by T_1, u_1, v_1 and γ_1. We show this only for (4.1) and (4.3); (4.2) can be proved similarly.

Concerning (4.1), for $x \in \mathcal{K}_1$ we have

$$\begin{aligned} T_1 T_1^* x + \langle x, u_1 \rangle u_1 &= P_1 T P_1 T^* x + \frac{\langle x, u \rangle}{1-|\gamma|^2} P_1 u \\ &= P_1 T \left(T^* x - \frac{\langle T^* x, v \rangle}{\langle v, v \rangle} v \right) + \frac{\langle x, u \rangle}{1-|\gamma|^2} P_1 u \\ &= P_1 \left(x - \langle x, u \rangle u - \frac{\langle x, Tv \rangle}{\langle v, v \rangle} Tv \right) + \frac{\langle x, u \rangle}{\langle v, v \rangle} P_1 u \\ &= x - \langle x, u \rangle P_1 u - |\gamma|^2 \frac{\langle x, u \rangle}{\langle v, v \rangle} P_1 u + \frac{\langle x, u \rangle}{\langle v, v \rangle} P_1 u \\ &= x - \frac{\langle x, u \rangle}{\langle v, v \rangle} \left(\langle v, v \rangle + |\gamma|^2 - 1 \right) P_1 u = x. \end{aligned}$$

The proof of (4.3) runs as follows:

$$\begin{aligned}
T_1 v_1 + \gamma_1^* u_1 &= \\
&= \frac{1}{\sqrt{1-|\gamma|^2}} P_1 T P_1 T^* v + \frac{\langle v,u\rangle}{(1-|\gamma|^2)^{3/2}} P_1 u \\
&= \frac{1}{\sqrt{1-|\gamma|^2}} P_1 T \left(T^* v - \frac{\langle T^* v, v\rangle}{\langle v,v\rangle} v \right) + \frac{\langle v,u\rangle}{(1-|\gamma|^2)^{3/2}} P_1 u \\
&= \frac{1}{\sqrt{1-|\gamma|^2}} P_1 \left(v - \langle v,u\rangle u - \frac{\langle v, Tv\rangle}{\langle v,v\rangle} Tv \right) + \frac{\langle v,u\rangle}{(1-|\gamma|^2)^{3/2}} P_1 u \\
&= \frac{-\langle v,u\rangle}{\sqrt{1-|\gamma|^2}} P_1 u - \frac{|\gamma|^2 \langle v,u\rangle}{\langle v,v\rangle\sqrt{1-|\gamma|^2}} P_1 u + \frac{\langle v,u\rangle}{(1-|\gamma|^2)^{3/2}} P_1 u \\
&= \frac{\langle v,u\rangle}{(1-|\gamma|^2)^{3/2}} \left(-(1-|\gamma|^2) - |\gamma|^2 + 1 \right) P_1 u = 0.
\end{aligned}$$

To prove the minimality statement, we first note that there are constants α_{mj} such that for $m = 0, 1, 2, \dots$

$$(P_1 T^*)^m v = T^{*m} v + \sum_{j=0}^{m-1} \alpha_{mj} T^{*j} v.$$

This can be seen by induction and the use of (4.1)–(4.3). Thus, for $x \in \mathcal{K}_1$, we have

$$\langle T_1^{*m} v_1, x\rangle = \frac{\langle T^*(PT^*)^m v, x\rangle}{\sqrt{1-|\gamma|^2}} = \frac{\langle T^{*(m+1)} v, x\rangle}{\sqrt{1-|\gamma|^2}} + \sum_{j=0}^{m-1} \frac{\alpha_{mj}}{\sqrt{1-|\gamma|^2}} \left\langle T^{*(j+1)} v, x \right\rangle.$$

Induction arguments again show that for $x \in \mathcal{K}_1$ the equalities $\langle T_1^{*m} v_1, x\rangle = 0$, $m = 0, 1, 2, \dots$, imply $\langle T^{*m} v, x\rangle = 0$, $m = 1, 2, \dots$. Therefore the minimality of V_1 follows from that of V.

It remains to prove the formula for the characteristic function

$$s_1(z) = \gamma_1 + z \left\langle (1 - zT_1)^{-1} u_1, v_1 \right\rangle$$

of V_1. Set $x = (1 - zT_1)^{-1} u_1$. Then

$$s_1(z) = \gamma_1 + z\langle x, v_1\rangle = \frac{\langle u,v\rangle}{1-|\gamma|^2} + z \frac{\langle Tx, v\rangle}{\sqrt{1-|\gamma|^2}}. \tag{5.1}$$

Calculations as in the proof of Lemma 4.3 give

$$x = (1-zT)^{-1} u_1 - z \frac{\langle Tx, v\rangle}{\langle v,v\rangle} (1-zT)^{-1} v.$$

We apply T to both sides, take the inner product with v, and then solve for $\langle Tx, v\rangle$:

$$\langle Tx, v\rangle = \frac{\langle v,v\rangle\langle T(1-zT)^{-1} u_1, v\rangle}{\langle v,v\rangle + z\langle T(1-zT)^{-1} v, v\rangle} = \frac{\langle v,v\rangle\langle T(1-zT)^{-1} u_1, v\rangle}{\langle (1-zT)^{-1} v, v\rangle}.$$

We calculate separately the denominator and the numerator of this quotient. For the denominator we use

$$(1-zT)^{-1}v = v - \gamma^*(1-zT)^{-1}u, \tag{5.2}$$

which follows from (4.3), and we obtain

$$\begin{aligned}\langle (1-zT)^{-1}v, v\rangle &= \langle v,v\rangle - \gamma^* z\langle (1-zT)^{-1}u, v\rangle \\ &= (1-|\gamma|^2) - \gamma^*(s(z)-\gamma) \;=\; (1-\gamma^* s(z)).\end{aligned}$$

Now we consider the second factor in the numerator. With the help of (5.2) we get

$$\begin{aligned}&\langle T(1-zT)^{-1}u_1, v\rangle = \\ &= \frac{1}{z}\left\langle ((1-zT)^{-1}-1)\left(u - \frac{\langle u,v\rangle}{\langle v,v\rangle}v\right), v\right\rangle \\ &= \frac{1}{z}\left\langle (1-zT)^{-1}u - u + \gamma^* z\frac{\langle u,v\rangle}{\langle v,v\rangle}(1-zT)^{-1}u, v\right\rangle \\ &= \frac{1}{z}\left(\frac{s(z)-\gamma}{z} - \langle u,v\rangle + \gamma^*\frac{\langle u,v\rangle}{\langle v,v\rangle}(s(z)-\gamma)\right) \\ &= \frac{1}{z\langle v,v\rangle}\left(\gamma^*\langle u,v\rangle s(z) - \langle u,v\rangle + \frac{\langle v,v\rangle}{z}(s(z)-\gamma)\right).\end{aligned}$$

Inserting these formulas into the formula for $\langle Tx, v\rangle$ and observing (5.1) for $s_1(z)$ and the equality $\gamma = \sigma_0$ we obtain

$$s_1(z) = \frac{1}{z}\frac{s(z)-\sigma_0}{1-\sigma_0^* s(z)}.$$

□

Corollary 5.2. *Suppose in addition to the assumptions of Theorem* 5.1 *that for some* $l \geq 1$ *we have* $\sigma_1 = \cdots = \sigma_{l-1} = 0$. *Then*
(i) *the space* span $\{v, T^*v, \ldots, T^{*(l-1)}v\}$ *is a positive l-dimensional subspace of* $\mathcal{K}$, *so the orthogonal projection in* $\mathcal{K}$ *onto* $\mathcal{K}_l = \{v, T^*v, \ldots, T^{*(l-1)}v\}^\perp$ *exists,* $\operatorname{ind}_- \mathcal{K}_1 = \operatorname{ind}_- \mathcal{K}$, *and* $\operatorname{ind}_+ \mathcal{K}_1 = \operatorname{ind}_+ \mathcal{K} - l$;
(ii) *the colligation*

$$V_l = \begin{pmatrix} T_l & u_l \\ \langle\,\cdot\,, v_l\rangle & \gamma_l \end{pmatrix}$$

with entries

$$\begin{aligned} T_l &:= P_l T P_l, & u_l &:= \frac{1}{\sqrt{1-|\gamma|^2}}P_l u, \\ v_l &:= \frac{1}{\sqrt{1-|\gamma|^2}}P_l T^{*l} v, & \gamma_l &:= \frac{\langle u, T^{*(l-1)}v\rangle}{1-|\gamma|^2} = \left(\frac{\sigma_l}{1-|\gamma|^2}\right)\end{aligned}$$

is minimal coisometric in the space $\mathcal{K}_l \oplus \mathbb{C}$;
(iii) *with* $s_l(z) := s_{V_l}(z)$ *it holds*

$$s_l(z) \equiv \frac{1}{z^l}\frac{s(z)-\sigma_0}{1-\sigma_0^* s(z)}. \tag{5.3}$$

In particular, if $s \in \mathbf{S}_\kappa^0$ *satisfies the above conditions, then also* $s_l \in \mathbf{S}_\kappa^0$ *and* $\operatorname{sq}_+(s_1) = \operatorname{sq}_+(s) - l\,(\geq 0)$.

Proof. The proof is by induction. For $l = 1$ the corollary coincides with Theorem 5.1. We assume the corollary is true: The colligation V_l in $\mathcal{K}_l \oplus \mathbb{C}$ is coisometric and minimal and (5.3) holds. We show that if $\sigma_l = 0$ then this assumption remains true when l is replaced by $l+1$. Note that $\sigma_l = 0$ implies that $s_l(0) = 0$, hence we may apply Theorem 5.1 to $s_l(z)$: If P_{l1} denotes the projection in $\mathcal{K}_l$ onto $\{v_l\}^\perp$, then the colligation with entries

$$\begin{aligned} T_{l1} &= P_{l1}T_lP_{l1}, & u_{l1} &= P_{l1}u_l, \\ v_{l1} &= P_{l1}T_l^*v_l, & \gamma_{l1} &= \langle u_l, v_l\rangle \end{aligned}$$

is minimal and coisometric in $P_{l1}\mathcal{K}_l \oplus \mathbb{C}$ and

$$s_{l1}(z) = \gamma_{l1} + z\left\langle (1 - zT_{l1})^{-1}u_{l1}, v_{l1}\right\rangle = \frac{1}{z}\, s_l(z).$$

By Lemma 4.1 with l replaced by $l+1$,

$$T^{*l}v \in \left\{v,\, T^*v,\, \ldots, T^{*(l-1)}v\right\}^\perp,$$

hence $P_lT^{*l}v = T^{*l}v$,

$$P_l\mathcal{K} \ominus \operatorname{span}\{v_l\} = \mathcal{K} \ominus \operatorname{span}\left\{v,\, T^*v, \ldots, T^{*l}v\right\} = P_{l+1}\mathcal{K},$$

and $P_{l1}P_l = P_{l+1}$. Also the other expressions with index $l1$ coincide with the same expressions with index $l+1$. For example,

$$v_{l1} = \frac{1}{\sqrt{1-|\gamma|^2}}\, P_{l1}P_lT^*P_lT^{*l}v = \frac{1}{\sqrt{1-|\gamma|^2}}\, P_{l+1}T^{*(l+1)}v = v_{l+1}$$

and

$$\gamma_{l1} = \langle u_l, v_l\rangle = \frac{\langle u, P_lT^{*l}v\rangle}{1-|\gamma|^2} = \frac{\langle u, T^{*(l+1)}v\rangle}{1-|\gamma|^2} = \frac{\sigma_{l+1}}{1-|\gamma|^2}.$$

It follows that

$$s_{l+1}(z) = s_{l1}(z) = \frac{1}{z}\, s_l(z) = \frac{1}{z^{l+1}}\frac{s(z)-\sigma_0}{1-\sigma_0^*s(z)}.$$

□

6. The colligation of the Schur transformation: Case (3)

In this section we suppose that for the given function $s \in \mathbf{A}^0$ with Taylor expansion

$$s(z) = \sigma_0 + \sigma_1 z + \cdots + \sigma_{n-1}z^{n-1} + \sigma_n z^n + \cdots$$

at $z = 0$ we have $|\sigma_0| > 1$, that is, we are in case (3) of the generalized Schur algorithm. Let k be defined by the relation

$$k := \min\{n : n \geq 1, \sigma_n \neq 0\}. \tag{6.1}$$

Hence, if $k > 1$ then $\sigma_1 = \ldots = \sigma_{k-1} = 0$, $\sigma_k \neq 0$. Now the transformation of s is the function s_k, given by

$$s_k(z) = z^k \frac{1 - \sigma_0^* s(z)}{s(z) - \sigma_0}. \tag{6.2}$$

It can be written as

$$s_k(z) = \frac{\sigma_0^*}{\sigma_0}\left(\frac{1}{z^k}\frac{\dfrac{1}{s(z)} - \dfrac{1}{\sigma_0}}{1 - \dfrac{1}{\gamma_0^*}\dfrac{1}{s(z)}}\right)^{-1}$$

This formula suggests that in order to obtain the colligation for s_k from that of s one should take the inverse $\widehat{s}(z) := s(z)^{-1}$ and apply Corollary 5.2 to the colligation of $\widehat{s}$ (see Lemma 4.3), take the inverse again and then multiply by a number of modulus 1. This is in a nutshell the way we prove the following theorem. Again its formulation is more general than is required for case (3) in the Schur algorithm, but this is needed because we will apply the theorem in this form in Section 8 to case (4) in which $|\sigma_0| = 1$ and $q > 0$.

Theorem 6.1. *Assume that for $s \in \mathbf{A}^0$ we have $|\sigma_0| > 1$, and let k be defined by* (6.1). *Let V, given by* (1.2) *with $\gamma = \sigma_0$, be a minimal coisometric colligation in the space $\mathcal{K} \oplus \mathbb{C}$ where $\mathcal{K}$ is a Krein space such that $s(z) = s_V(z)$. Then*
(i) *the space* span $\{v, T^*v, \ldots, T^{*(k-1)}v\}$ *is a k-dimensional negative subspace of $\mathcal{K}$, so the orthogonal projection P_k in $\mathcal{K}$ onto $\mathcal{K}_k = \{v, T^*v, \ldots, T^{*(k-1)}v\}^\perp$ exists,* $\operatorname{ind}_- \mathcal{K}_k = \operatorname{ind}_- \mathcal{K} - k$, *and* $\operatorname{ind}_+ \mathcal{K}_k = \operatorname{ind}_+ \mathcal{K}$*;*
(ii) *the colligation*

$$V_k = \begin{pmatrix} T_k & u_k \\ \langle\cdot, v_k\rangle & \gamma_k \end{pmatrix}$$

with entries

$$T_k = P_kTP_k - t\frac{\langle\cdot, P_kT^{*k}v\rangle}{\sigma_k}P_ku, \quad u_k = \frac{\sqrt{|\gamma|^2 - 1}}{\sigma_k}P_ku,$$

$$v_k = \frac{\sqrt{|\gamma|^2 - 1}}{\sigma_k^*}P_kT^{*k}v, \quad \gamma_k = \frac{1 - |\gamma|^2}{\sigma_k}$$

is minimal coisometric in the space $\mathcal{K}_k \oplus \mathbb{C}$;
(iii) *with* $s_k(z) := s_{V_k}(z)$ *it holds*

$$s_k(z) \equiv z^k \frac{1-\sigma_0^* s(z)}{s(z)-\sigma_0}.$$

If, in particular, $s \in \mathbf{S}_\kappa^0$ *and* $|\sigma_0| > 1$ *then* $k \le \kappa$, $s_k \in \mathbf{S}_{\kappa-k}^0$, *and* $\mathrm{sq}_+(s_k) = \mathrm{sq}_+(s)$.

Proof. By Lemma 4.3, $\widehat{s}(z) = \dfrac{1}{s(z)}$ is the characteristic function of the colligation with entries

$$\widehat{T} = T - \frac{\langle\,\cdot\,, v\rangle}{\gamma} u, \quad \widehat{u} = \frac{u}{\gamma},$$
$$\widehat{v} = \frac{v}{\gamma^*}, \quad \widehat{\gamma} = \frac{1}{\gamma},$$

which is minimal coisometric in the space $\mathcal{K}' \oplus \mathbb{C}$. We apply Corollary 5.2 to this colligation. First note that

$$\mathrm{span}\left\{\widehat{v},\, \widehat{T}^*\widehat{v},\, \ldots,\, \widehat{T}^{*(k-1)}\widehat{v}\right\} = \mathrm{span}\left\{v\, T^* v,\, \ldots,\, T^{*(k-1)} v\right\},$$

By Lemma 4.1 the space in the right hand side is negative in $\mathcal{K}$, and so the projection P_k in $\mathcal{K}$ mentioned in the theorem is well defined. The space in the left hand side is positive in $\mathcal{K}'$ and the projection onto its orthogonal complement in $\mathcal{K}'$ coincides with P_k. We find that the colligation described in Corollary 5.2 has the entries

$$\begin{aligned}
\widehat{T}_k &= P_k \widehat{T} P_k &&= P_k T P_k, \\
\widehat{u}_k &= \frac{1}{\sqrt{1-|\widehat{\gamma}|^2}} P_k \widehat{u} &&= \frac{|\gamma|}{\gamma\sqrt{|\gamma|^2-1}} P_k u, \\
\widehat{v}_k &= \frac{1}{\sqrt{1-|\widehat{\gamma}|^2}} P_k \widehat{T}^{*k} \widehat{v} &&= \frac{|\gamma|}{\gamma^*\sqrt{|\gamma|^2-1}} P_k T^{*k} v, \\
\widehat{\gamma}_k &= \frac{\left\langle \widehat{u}, \widehat{T}^{*(k-1)}\widehat{v}\right\rangle'}{1-|\widehat{\gamma}|^2} &&= -\frac{\gamma^*\sigma_k}{\gamma(|\gamma|^2-1)},
\end{aligned}$$

it is minimal and coisometric in the space $\mathcal{K}'_k \oplus \mathbb{C}$, and has the characteristic function

$$\widehat{s}_k(z) = \frac{1}{z^k} \frac{\dfrac{1}{s(z)} - \dfrac{1}{\gamma}}{1 - \dfrac{1}{\gamma^*}\dfrac{1}{s(z)}}.$$

The first two of these formulas are easy to verify, the last two require some extra work. We only prove the third one. With

$$\tau = \frac{|\gamma|}{\gamma^*\sqrt{|\gamma|^2-1}}$$

we have

$$\begin{aligned}
\widehat{v}_k &= \frac{1}{\sqrt{1-|\widehat{\gamma}|^2}} P_k \widehat{T}^{*k}\widehat{v} \;=\; \tau P_k \left(T^* - \frac{\langle \cdot, u\rangle}{\gamma^*} v\right)^k v \\
&= \tau P_k \left(T^* - \frac{\langle \cdot, u\rangle}{\gamma^*} v\right)^{k-1} \left(T^* v - \frac{\sigma_1^*}{\gamma^*}\right) v \\
&= \tau P_k \left(T^* - \frac{\langle \cdot, u\rangle}{\gamma^*} v\right)^{k-1} T^* v \;=\; \cdots \\
&= \tau P_k \left(T^{*k} v - \frac{\sigma_k^*}{\gamma^*} v\right) \;=\; \tau P_k T^{*k} v.
\end{aligned}$$

Now we apply Lemma 4.3 to get the colligation for $\widetilde{s}_k(z) = \dfrac{1}{\widehat{s}_k(z)}$: It is minimal coisometric in the space $\mathcal{K}_k \oplus \mathbb{C}$, and its entries are given by

$$\begin{aligned}
\widetilde{T}_k &= \widehat{T}_k - \frac{\langle \cdot, \widehat{v}_k\rangle'}{\widehat{\gamma}_k} \widehat{u}_k \;=\; P_k T P_k - \frac{\langle \cdot, P_k T^{*k} v\rangle}{\sigma_k} P_k u, \\
\widetilde{u}_k &= \frac{\widehat{u}_k}{\widehat{\gamma}_k} \;=\; -\frac{\gamma\sqrt{|\gamma|^2-1}}{|\gamma|\sigma_k} P_k u, \\
\widetilde{v}_k &= \frac{\widehat{v}_k}{\widehat{\gamma}_k^*} \;=\; \frac{\gamma^*\sqrt{|\gamma|^2-1}}{|\gamma|\sigma_k^*} P_k T^{*k} v, \\
\widetilde{\gamma}_k &= \frac{1}{\widehat{\gamma}_k} \;=\; -\frac{\gamma(|\gamma|^2-1)}{\gamma^*\sigma_k}.
\end{aligned}$$

It remains to incorporate the multiplication by γ^*/γ: the colligation with entries

$$\begin{aligned}
T_k &= \widetilde{T}_k, & u_k &= -\frac{|\gamma|}{\gamma}\widetilde{u}_k, \\
v_k &= -\frac{|\gamma|}{\gamma^*}\widetilde{v}_k, & \gamma_k &= \frac{\gamma^*}{\gamma}\widetilde{\gamma}_k
\end{aligned}$$

coincides with the one mentioned in the theorem, and it is easily checked that it is minimal coisometric in the space $\mathcal{K}_k \oplus \mathbb{C}$, and that

$$\begin{aligned}
\gamma_k + z\left\langle (1-zT_k)^{-1}u_k, v_k\right\rangle &= \frac{\gamma^*}{\gamma}\left(\widetilde{\gamma}_k + z\left\langle (1-z\widetilde{T}_k)^{-1}\widetilde{u}_k, \widetilde{v}_k\right\rangle\right) \\
&= \frac{\gamma^*}{\gamma}\widetilde{s}_k(z) \;=\; s_k(z).
\end{aligned}$$

□

Remark 6.2. For $k = 1$ the Schur transformation in case (3) becomes

$$s_1(z) = z\,\frac{1-\sigma_0^* s(z)}{s(z)-\sigma_0},$$

and this formula also makes sense if $k > 1$. If $s \in \mathbf{S}_\kappa^0$, then this function s_1 belongs to the class $\mathbf{S}_{\kappa-1}$ but it has a pole at $z = 0$ and hence no operator colligation can be associated with s_1. In order to obtain s_k as in (6.2) we just have to divide $s_1(z)$ by z^{k-1}, and this function belongs to the class $\mathbf{S}^0_{\kappa-k}$.

7. The colligation of the Schur transformation: Case (4) with $q = 0$

Let $s \in \mathbf{A}^0$ have the Taylor expansion at $z = 0$ given by

$$s(z) = \sigma_0 + \sigma_1 z + \cdots + \sigma_{n-1} z^{n-1} + \sigma_n z^n + \cdots .$$

In this section we consider the case (4), that is, we assume that $|\sigma_0| = 1$, and $k\,(\geq 1)$ is again defined by (6.1), which means that $\sigma_k \neq 0$ and, if $k > 1$, $\sigma_1 = \cdots = \sigma_{k-1} = 0$. We first also assume that in (3.4) $q = 0$, which means that

$$t_{2k} := \sigma_0^* \sum_{l=0}^{k-1} c_l \sigma_{2k-l} \neq 0, \tag{7.1}$$

where the numbers c_l are defined for s as the numbers c_{jl} in (3.3) for the function s_j.

Theorem 7.1. *Assume that for $s \in \mathbf{A}^0$ we have $|\sigma_0| = 1$, and that the condition (7.1) is satisfied. Let V, given by (1.2) with $\gamma = \sigma_0$, be a minimal coisometric colligation in the space $\mathcal{K} \oplus \mathbb{C}$ where $\mathcal{K}$ is a Krein space such that $s(z) = s_V(z)$. Then*
(i) *the space* span $\{v, T^*v, \cdots, T^{*(2k-1)}\}$ *is $2k$–dimensional and its positive and negative indices are equal to k. Hence the space $\mathcal{K}_{2k} := \{v, T^*v, \cdots, T^{*(2k-1)}\}^\perp$ is a Krein space with $\operatorname{ind}_- \mathcal{K}_{2k} = \operatorname{ind}_- \mathcal{K} - k$, $\operatorname{ind}_+ \mathcal{K}_{2k} = \operatorname{ind}_+ \mathcal{K} - k$, and the orthogonal projection P_{2k} in $\mathcal{K}$ onto $\mathcal{K}_{2k}$ is well defined;*
(ii) *the colligation*

$$V_{2k} := \begin{pmatrix} T_{2k} & u_{2k} \\ \langle \cdot, v_{2k} \rangle & \gamma_{2k} \end{pmatrix}$$

with entries

$$\begin{aligned} T_{2k} &:= P_{2k} T P_{2k} - \frac{\langle \cdot, P_{2k} T^{*(2k)} v \rangle}{\sigma_k t_{2k}} P_{2k} u, & u_{2k} &:= \frac{1}{t_{2k}} P_{2k} u, \\ v_{2k} &:= \frac{1}{t_{2k}^*} P_{2k} T^{*(2k)} v, & \gamma_{2k} &:= \gamma - \frac{\sigma_k}{t_{2k}} \end{aligned}$$

is minimal coisometric in the space $\mathcal{K}_{2k} \oplus \mathbb{C}$;
(iii) *with $s_{2k}(z) := s_{V_{2k}}(z)$ it holds*

$$s_{2k}(z) \equiv \frac{(Q(z) - z^k)s(z) - \sigma_0 Q(z)}{\sigma_0^* Q(z) s(z) - (Q(z) + z^k)}.$$

In particular, if for $s \in \mathbf{S}^0_\kappa$ the above conditions are satisfied, then $s_{2k} \in \mathbf{S}^0_{\kappa-k}$ and $\operatorname{sq}_+(s_{2k}) = \operatorname{sq}_+(s) - k$.

For the proof of the theorem we first derive some formulae. We assume that the Taylor coefficients of s satisfy the conditions mentioned in the theorem. If Δ is the $k \times k$ matrix

$$\Delta := \begin{pmatrix} \sigma_k & 0 & \cdots & 0 \\ \sigma_{k+1} & \sigma_k & \cdots & 0 \\ \vdots & \ddots & \ddots & \vdots \\ \sigma_{2k-1} & \cdots & \sigma_{k+1} & \sigma_k \end{pmatrix},$$

then by formulae (4.7) and (4.8) we have that

$$\mathcal{G}_{2k} = \begin{pmatrix} 0 & -\sigma_0 \Delta^* \\ -\sigma_0^* \Delta & -\Delta\Delta^* \end{pmatrix}, \quad \mathcal{G}_{2k}^{-1} = \begin{pmatrix} I_k & -\sigma_0 \Delta^{-1} \\ -\sigma_0^* \Delta^{-*} & 0 \end{pmatrix}.$$

Now define the complex numbers $\alpha_0, \dots, \alpha_{2k-1}, \beta_0, \dots, \beta_{2k-1}$ by

$$\mathcal{G}_{2k} \begin{pmatrix} \alpha_0 \\ \vdots \\ \alpha_{k-2} \\ \alpha_{k-1} \\ \alpha_k \\ \vdots \\ \alpha_{2k-1} \end{pmatrix} = \begin{pmatrix} 0 \\ \vdots \\ 0 \\ \sigma_k \\ \sigma_{k+1} \\ \vdots \\ \sigma_{2k} \end{pmatrix}, \quad \mathcal{G}_{2k} \begin{pmatrix} \beta_0 \\ \vdots \\ \beta_{k-1} \\ \beta_k \\ \beta_{k+1} \\ \vdots \\ \beta_{2k-1} \end{pmatrix} = -\sigma_0 \begin{pmatrix} \sigma_{2k}^* \\ \vdots \\ \sigma_{k+1}^* \\ \sigma_k^* + \tau_1 \\ \tau_2 \\ \vdots \\ \tau_k \end{pmatrix},$$

where

$$\begin{pmatrix} \tau_1 \\ \vdots \\ \tau_k \end{pmatrix} = \sigma_0^* \Delta \begin{pmatrix} \sigma_{2k}^* \\ \vdots \\ \sigma_{k+1}^* \end{pmatrix}.$$

By Lemma 4.2, part (i) of the theorem is valid; in particular the projection P_{2k} exists. We shall often use that $c_0 = \sigma_0/\sigma_k$,

$$T^{2k}T^{*j}v = \begin{cases} -\sigma_0^* T^{2k-j-1}u, & j = 0, \dots k-1, \\ -\left(\sigma_0^* T^{2k-j-1}u + \sum_{i=k}^{j} \sigma_i^* T^{2k-j-1+i}u\right), & j = k, \dots 2k-1, \end{cases} \tag{7.2}$$

and that

$$\langle (1 - zT)^{-1} T^r u, v \rangle = \frac{s(z) - (\sigma_0 + \sigma_1 z + \cdots + \sigma_r z^r)}{z^{r+1}}, \tag{7.3}$$

which can be simplified since $\sigma_1 = \dots = \sigma_{k-1} = 0$.

Lemma 7.2. *Under the assumptions of Theorem* 7.1 *the following formulae hold:*

(1) $P_{2k}Tx = Tx + \dfrac{\langle x, T^{*2k}v \rangle}{\sigma_0^* \sigma_k} T^{*(k-1)}v$, $x \in \mathcal{K}_{2k}$.

(2) $P_{2k}u = u - \sum_{j=0}^{2k-1} \alpha_j T^{*j}v$.

(3) $P_{2k}T^{*2k}v = T^{*2k}v - \sum_{j=0}^{2k-1} \beta_j T^{*j}v$.

(4) $\sum_{j=0}^{s} \alpha_j \sigma_{s+k-j} = \begin{cases} -\sigma_0 \sigma_{k+s+1}, & s = 0, \dots, k-2, \\ (\sigma_k^2 - \sigma_0 \sigma_{2k}), & s = k-1. \end{cases}$

(5) $\sum_{j=\ell+k}^{2k-1} \alpha_j \sigma_{j-\ell}^* = \begin{cases} 0, & \ell = 0, \dots, k-2, \\ -\sigma_0^* \sigma_k, & \ell = k-1. \end{cases}$

(6) $\alpha_j = \begin{cases} \sigma_k c_{j+1}, & j = 0, \dots k-2, \\ \sigma_k - \sigma_0 t_{2k}, & j = k-1, \\ -\sigma_k c_{2k-j-1}^*, & j = k, \dots, 2k-1. \end{cases}$

(7) $\sigma_k Q(z) = \sigma_0 + z(\alpha_0 + \alpha_1 z + \cdots + \alpha_{2k-1} z^{2k-1}) - \alpha_{k-1} z^k.$

(8) $P_{2k} T^* x = T^* x - \dfrac{\langle x, u \rangle}{\sigma_k^*} T^{*k} v, \; x \in \mathcal{K}_{2k}.$

(9) $\langle T^{*k} v, T^{*2k} v \rangle = -(\sigma_0^* \sigma_k + \sigma_k^* \sigma_{2k}).$

(10) $\sum_{i=k}^{2k-\ell-1} \beta_{\ell+i} \sigma_i^* = \sigma_{2k-\ell}^*, \; \ell = 0, 1, \dots, k-1.$

(11) $\beta_j = \begin{cases} \sigma_0 \sigma_k^* c_j, & j = 0, \dots k-1, \\ t_{2k}^*, & j = k, \\ -\sigma_0 \sigma_k^* c_{2k-j}^*, & j = k+1, \dots, 2k-1. \end{cases}$

(12) $\langle P_{2k} T^{*2k} v, T^{*2k} v \rangle = (|t_{2k}|^2 - |t_{2k} - \sigma_0^* \sigma_k|^2).$

(13) $\sum_{j=0}^{2k-1} \beta_j \sigma_j^* = \dfrac{\sigma_0 \sigma_k^*}{\sigma_k} + \sigma_{2k}^*.$

Proof. (1) Set $P_{2k} T x = T x - \sum_{j=0}^{2k-1} \omega_j T^{*j} v$, take inner products with $T^{*i} v$, $i = 0, \dots, 2k-1$, and use vector notation to obtain

$$\begin{pmatrix} \omega_0 \\ \vdots \\ \omega_{k-1} \\ \omega_k \\ \vdots \\ \omega_{2k-2} \\ \omega_{2k-1} \end{pmatrix} = \mathcal{G}_{2k}^{-1} \begin{pmatrix} \langle x, T^* v \rangle \\ \vdots \\ \langle x, T^{*k} v \rangle \\ \langle x, T^{*(k+1)} v \rangle \\ \vdots \\ \langle x, T^{*(2k-1)} v \rangle \\ \langle x, T^{*(2k)} v \rangle \end{pmatrix} = \mathcal{G}_{2k}^{-1} \begin{pmatrix} 0 \\ \vdots \\ 0 \\ 0 \\ \vdots \\ 0 \\ \langle x, T^{*(2k-1)} v \rangle \end{pmatrix}$$

From the formula for $\mathcal{G}_{2k}^{-1}$ and the lower triangular form of Δ it readily follows that $\omega_j = 0$ if $j \neq k-1$ and $\omega_{k-1} = -\dfrac{\langle x, T^{*2k} v \rangle}{\sigma_0^* \sigma_k}$.

The relations (2), (3) and (8) can be proved similarly. To prove (4) and (5) write

$$\vec{\alpha}_1 = \begin{pmatrix} \alpha_0 \\ \vdots \\ \alpha_{k-1} \end{pmatrix}, \quad \vec{\alpha}_2 = \begin{pmatrix} \alpha_k \\ \vdots \\ \alpha_{2k-1} \end{pmatrix}, \quad \vec{\sigma}_2 = \begin{pmatrix} \sigma_{k+1} \\ \vdots \\ \sigma_{2k} \end{pmatrix},$$

and denote by $\vec{e}_j$ the j-th unit vector in the standard basis of $\mathbb{C}^k$, $j = 1, 2, \ldots, k$. Then the defining relation for the α_j's and the block matrix form of $\mathcal{G}_{2k}$ imply

$$\Delta\vec{\alpha}_1 = (-\sigma_0\vec{\sigma}_2 + \sigma_k^2\vec{e}_1), \quad \Delta^*\vec{\alpha}_2 = -\sigma_0^*\sigma_k\vec{e}_k. \tag{7.4}$$

These formulas are the equalities in (4) and (5) in vector notation.
(6) The matrix multiplications in (7.4) can also be written as multiplications of analytic functions:

$$\begin{aligned}&\left(\sigma_k + \sigma_{k+1}z + \cdots + \sigma_{2k-1}z^{2k-1}\right)\left(\alpha_0 + \alpha_1 z + \cdots + \alpha_{k-1}z^{k-1}\right)\\ &\quad = -\sigma_0\left(\sigma_{k+1} + \sigma_{k+2}z + \cdots + \sigma_{2k}z^{k-1}\right) + \sigma_k^2 z^{k-1},\end{aligned}$$

and

$$\begin{aligned}&\left(\sigma_k^* + \sigma_{k+1}^*z + \cdots + \sigma_{2k-1}^*z^{2k-1}\right)\left(\alpha_{2k-1} + \alpha_{2k-2}z + \cdots + \alpha_k z^{k-1}\right)\\ &\quad = -\sigma_0^*\sigma_k + \mathcal{O}(z^k).\end{aligned}$$

If we multiply both sides of the first equality by z and rearrange terms we get

$$\begin{aligned}&\left(\sigma_k + \sigma_{k+1}z + \cdots + \sigma_{2k-1}z^{2k-1}\right)\left(\sigma_0 + \alpha_0 z + \cdots + \alpha_{k-2}z^{k-1}\right)\\ &\quad = -\sigma_0\left(\sigma_{k+1}z + \sigma_{k+2}z^2 + \cdots + \sigma_{2k}z^k\right) + \sigma_k^2 z^k\\ &\qquad + \sigma_0\left(\sigma_k + \sigma_{k+1}z + \cdots + \sigma_{2k-1}z^{2k-1}\right)\\ &\qquad - \alpha_{k-1}z^k\left(\sigma_k + \sigma_{k+1}z + \cdots + \sigma_{2k-1}z^{2k-1}\right)\\ &\quad = \sigma_0\sigma_k + \mathcal{O}(z^k),\end{aligned}$$

and if we take complex conjugates of both sides of the second equality and replace z^* by z we obtain the equality

$$\begin{aligned}&\left(\sigma_k + \sigma_{k+1}z + \cdots + \sigma_{2k-1}z^{2k-1}\right)\left(\alpha_{2k-1}^* + \alpha_{2k-2}^*z + \cdots + \alpha_k^* z^{k-1}\right)\\ &\quad = -\sigma_0\sigma_k^* + \mathcal{O}(z^k).\end{aligned}$$

Finally, the coefficients $c_0, c_1, \ldots, c_{k-1}$ appearing in the function $Q(z)$ are determined by

$$\left(\sigma_k + \sigma_{k+1}z + \cdots + \sigma_{2k-1}z^{2k-1}\right)\left(c_0 + c_1 z + \cdots + c_{k-1}z^{k-1}\right) = \sigma_0 + \mathcal{O}(z^k). \tag{7.5}$$

Comparing coefficients of the last three equations we get

$$\sigma_k c_0 = \sigma_0 = -\frac{\sigma_k}{\sigma_k^*}\alpha_{2k-1}^*; \quad \sigma_k c_j = \alpha_{j-1} = -\frac{\sigma_k}{\sigma_k^*}\alpha_{2k-j-1}^*, \quad j = 1, \ldots, k-1.$$

From the first equality in (7.4) it can be seen that

$$\sigma_k\alpha_{k-1} = \left(\sigma_k^2 - \sigma_0\sigma_{2k}\right) + \sum_{j=1}^{k-1}\sigma_{2k-j}\alpha_{j-1}.$$

These formulas and (7.1) easily imply the equalities in (6).
The relation (7) follows from (6) and the definition of $Q(z)$, (9) follows directly

from (4.7).

(10) We use the vector notation as in the proof of (4) and (5) and set

$$\vec{\beta}_1 = \begin{pmatrix} \beta_0 \\ \vdots \\ \beta_{k-1} \end{pmatrix}, \quad \vec{\beta}_2 = \begin{pmatrix} \beta_k \\ \vdots \\ \beta_{2k-1} \end{pmatrix},$$

and

$$\vec{\gamma}_* := \begin{pmatrix} \gamma_k^* \\ \vdots \\ \gamma_1 * \end{pmatrix} \text{ if } \vec{\gamma} = \begin{pmatrix} \gamma_1 \\ \vdots \\ \gamma_k \end{pmatrix}.$$

If J denotes the $k \times k$ signature matrix whose entries on the second main diagonal are 1 and the other entries are 0, then, evidently, $J\Delta J = \Delta^t$ and $(J\vec{\gamma})^* = \vec{\gamma}_*^t$, where t stands for transposition. The defining relations for the β_j's and the block matrix form of $\mathcal{G}_{2k}$ imply

$$\Delta\vec{\beta}_1 = \sigma_0^2\sigma_k^*\vec{e}_1, \quad \Delta^*\vec{\beta}_2 = (\vec{\sigma}_2)_*. \tag{7.6}$$

The second of these equalities written out in components yields (10).

(11) Applying J to both sides of the second equality in (7.6) and taking complex conjugates of all the entries, we obtain

$$\Delta(\vec{\beta}_2)_* = \vec{\sigma}_2.$$

The first equality in (7.4) can also be written as

$$\Delta\left(\vec{\alpha}_1 - \sigma_k\vec{e}_k\right) = -\sigma_0\vec{\sigma}_2.$$

Since Δ is invertible, the last two formulas imply

$$\vec{\alpha}_1 = \sigma_k\vec{e}_k - \sigma_0(\vec{\beta}_2)_*.$$

A similar calculation involving the other equalities in (7.4) and (7.6) readily yield

$$\vec{\alpha}_2 = -\sigma_0(\vec{\beta}_1)_*.$$

These formulas imply

$$\alpha_j = \begin{cases} -\sigma_0\beta_{2k-j-1}^*, & j = 0, \ldots k-2, k \ldots, 2k-1, \\ \sigma_k - \sigma_0\beta_k^*, & j = k-1. \end{cases}$$

Expressing the β's in terms of the α's and substituting for the α's the expressions in item (6) of the lemma, we obtain (11).

(12) From (3) and (4.7) (or (7.2)) we have

$$\begin{aligned} \langle P_{2k}T^{*2k}v, T^{*2k}v\rangle &= \langle T^{*2k}v, T^{*2k}v\rangle - \sum_{j=0}^{2k-1} \beta_j \langle T^{*j}v, T^{*k}v\rangle \\ &= -\sum_{j=k}^{2k} |\sigma_j|^2 + \sigma_0^* \sum_{j=0}^{2k-1} \beta_j\sigma_{2k-j} + \sum_{j=k}^{2k-1}\sum_{i=k}^{j} \beta_j\sigma_i^*\sigma_{2k-j+i}. \end{aligned}$$

Substituting the β's given by (11) in the first series we get

$$\sigma_0^* \sum_{j=0}^{2k-1} \beta_j \sigma_{2k-j} = \sigma_0^* \sigma_k t_{2k}^* + \sigma^* \sum_{j=0}^{k-1} c_j \sigma_{2k-j} = \sigma_0^* \sigma_k t_{2k}^* + \sigma_0 \sigma^* t_{2k}.$$

Rearranging summation and applying (10), we see that the double sum is equal to

$$\sum_{\ell=0}^{k-1} \left(\sum_{i=k}^{2k-\ell-1} \beta_{\ell+i} \sigma_i^* \right) \sigma_{2k-\ell} = \sum_{\ell=0}^{k-1} |\sigma_{2k-\ell}|^2 = \sum_{j=k+1}^{2k} |\sigma_j|^2.$$

Combining these formulas we get (12).
(13) follows directly from (10) and (11). □

Proof of Theorem 7.1. Set $x = (1 - T_{2k})^{-1} u_{2k}$. Then $x \in \mathcal{K}_{2k}$ and

$$s_{2k}(z) = \gamma_{2k} + z\langle x, v_{2k}\rangle = \gamma - \frac{\sigma_k}{t_{2k}} + \frac{z}{t_{2k}} \langle x, T^{*2k} v\rangle. \tag{7.7}$$

Then, by Lemma 7.2 (1),

$$\begin{aligned} x &= (1 - zT)^{-1} u_{2k} \\ &+ z\frac{\langle x, T^{*2k} v\rangle}{\sigma_0^* \sigma_k} (1 - zT)^{-1} T^{*(k-1)} v - z\frac{\langle x, T^{*2k} v\rangle}{\sigma_k t_{2k}} (1 - zT)^{-1} P_{2k} u. \end{aligned}$$

We take the inner product of both sides with $T^{*2k}v$ and solve for $\langle x, T^{*2k}v\rangle$:

$$\langle x, T^{*2k} v\rangle = \frac{\langle (1 - zT)^{-1} u_{2k}, T^{*2k} v\rangle}{D_1(z) + D_2(z)} = \frac{1}{t_{2k}} \frac{\langle (1 - zT)^{-1} P_{2k} u, T^{*2k} v\rangle}{D_1(z) + D_2(z)}, \tag{7.8}$$

where the two summands in the denominator are

$$\begin{aligned} D_1(z) &= 1 - \frac{z}{\sigma_0^* \sigma_k} \left\langle (1 - zT)^{-1} T^{*(k-1)} v, T^{*2k} v \right\rangle, \\ D_2(z) &= \frac{z}{\sigma_k t_{2k}} \langle (1 - zT)^{-1} P_{2k} u, T^{*2k} v\rangle. \end{aligned}$$

By (7.2), (7.3), and these two formulas, we have

$$\begin{aligned} D_1(z) &= 1 - \frac{z}{\sigma_0^* \sigma_k} \left\langle (1 - zT)^{-1} T^{2k} T^{*(k-1)} v, v \right\rangle \\ &= 1 + \frac{z}{\sigma_k} \langle (1 - zT)^{-1} T^k u, v\rangle = \frac{s(z) - \sigma_0}{\sigma_k z^k}. \end{aligned}$$

According to Lemma 7.2 (2) and (7.2),

$$\begin{aligned}
D_2(z) &= \frac{z}{\sigma_k t_{2k}} \left\langle (1-zT)^{-1}T^{2k}u, v\right\rangle - \frac{z}{\sigma_k t_{2k}} \sum_{j=0}^{2k-1} \alpha_j \left\langle (1-zT)^{-1}T^{2k}T^{*j}v, v\right\rangle \\
&= \frac{z}{\sigma_k t_{2k}} \left\langle (1-zT)^{-1}T^{2k}u, v\right\rangle + \frac{z\sigma_0^*}{\sigma_k t_{2k}} \sum_{j=0}^{2k-1} \alpha_j \left\langle (1-zT)^{-1}T^{2k-j-1}u, v\right\rangle \\
&\quad + \frac{z}{\sigma_k t_{2k}} \sum_{j=k}^{2k-1} \sum_{i=1}^{j} \alpha_j \sigma_i^* \left\langle (1-zT)^{-1}T^{2k-j-1+i}u, v\right\rangle \\
&=: D_{21}(z) + D_{22}(z) + D_{23}(z).
\end{aligned}$$

Here, by (7.3),

$$D_{21} = \frac{1}{\sigma_k t_{2k} z^{2k}} \left(s(z) - (\sigma_0 + \sigma_k z^k + \cdots + \sigma_{2k} z^{2k})\right);$$

by Lemma 7.2 (4) and (7),

$$\begin{aligned}
D_{22}(z) &= \frac{z\sigma_0^*}{\sigma_k t_{2k}} \sum_{j=0}^{2k-1} \alpha_j \frac{s(z) - (\sigma_0 + \cdots + \sigma_{2k-j-1} z^{2k-j-1})}{z^{2k-j}} \\
&= \frac{\sigma_0^*}{\sigma_k t_{2k} z^{2k}} \left((s(z) - \sigma_0) \sum_{j=0}^{2k-1} \alpha_j z^{j+1} - \sum_{j=0}^{k} \sum_{r=0}^{k-j-1} \alpha_j \sigma_{r+k} z^{r+j+k+1} \right) \\
&= \frac{\sigma_0^*}{\sigma_k t_{2k} z^{2k}} \left((s(z) - \sigma_0) \sum_{j=0}^{2k-1} \alpha_j z^{j+1} - \sum_{s=0}^{k-1} \sum_{j=0}^{s} \alpha_j \sigma_{s+k-j} z^{s+k+1} \right) \\
&= \frac{\sigma_0^*}{\sigma_k t_{2k} z^{2k}} \Bigg((s(z) - \sigma_0) \left(\sigma_k Q(z) - \sigma_0 + \alpha_{k-1} z^k\right) \\
&\qquad + \sigma_0 \sum_{s=0}^{k-2} \sigma_{k+s+1} z^{s+k+1} - (\sigma_k^2 - \sigma_0 \sigma_{2k}) z^{2k} \Bigg);
\end{aligned}$$

and by Lemma 7.2 (5),

$$\begin{aligned}
D_{23}(z) &= \frac{z}{\sigma_k t_{2k}} \sum_{j=k}^{2k-1} \sum_{i=k}^{j} \alpha_j \sigma_i^* \left\langle (1-zT)^{-1}T^{2k-j-1+i}u, v\right\rangle = \\
&= \frac{z}{\sigma_k t_{2k}} \sum_{\ell=0}^{k-1} \left(\sum_{j=\ell+k}^{2k-1} \alpha_j \sigma_{j-\ell}^* \right) \left\langle (1-zT)^{-1}T^{2k-\ell-1}u, v\right\rangle \\
&= -\frac{z}{\sigma_k t_{2k}} (-\sigma_0^* \sigma_k) \left\langle (1-zT)^{-1}T^k u, v\right\rangle \\
&= -\frac{\sigma_0^*}{z^k t_{2k}} \left(s(z) - \sigma_0 - \sigma_k z^k\right).
\end{aligned}$$

Hence

$$D_1(z)+D_2(z)=D_1(z)+D_{21}(z)+D_{22}(z)+D_{23}(z)=A(z)s(z)+B(z),$$

where by (7.1) and Lemma 7.2 (6) the coefficients $A(z)$ and $B(z)$ have the form

$$\begin{aligned} A(z) &= \frac{1}{\sigma_k z^k}+\frac{1}{\sigma_k t_{2k} z^{2k}}+\frac{\sigma_0^*}{\sigma_k t_{2k} z^{2k}}\left(\sigma_k Q(z)-\sigma_0+\alpha_{k-1}z^k\right)-\frac{\sigma_0^*}{t_{2k}z^k}\\ &= \frac{\sigma_0^*}{t_{2k}z^{2k}}Q(z)+\frac{t_{2k}+\sigma_0^*\alpha_{k-1}-\sigma_0^*\sigma_k}{\sigma_k t_{2k} z^k} \;=\; \frac{\sigma_0^*}{t_{2k}z^{2k}}Q(z) \end{aligned}$$

and

$$\begin{aligned} B(z) &= -\frac{\sigma_0}{\sigma_k z^k}-\frac{\sigma_0+\sigma_k z^k+\cdots+\sigma_{2k}z^{2k}}{\sigma_k t_{2k}z^{2k}}-\frac{\sigma_k Q(z)-\sigma_0+\alpha_{k-1}z^k}{\sigma_k t_{2k} z^{2k}}\\ &\quad+\frac{\sum_{s=0}^{k-2}\sigma_{k+s+1}z^{k+s+1}-\left(\sigma_0^*\sigma_k^2-\sigma_{2k}\right)z^{2k}}{\sigma_k t_{2k}z^{2k}}+\frac{\sigma_k+\sigma_0^*\sigma_k^2 z^k}{\sigma_k t_{2k} z^k}\\ &= -\frac{\sigma_k Q(z)+\left(\sigma_0 t_{2k}+\alpha_{k-1}\right)z^k}{\sigma_k t_{2k} z^{2k}} \;=\; -\frac{1}{t_{2k}z^{2k}}\left(Q(z)+z^k\right), \end{aligned}$$

that is,

$$D_1(z)+D_2(z)=\frac{1}{t_{2k}z^{2k}}C(z),\quad C(z)=\sigma_0^*Q(z)s(z)-\left(Q(z)+z^k\right).$$

Substituting this in (7.8) we find

$$\langle x, T^{2k}v\rangle=\frac{z^{2k}\left\langle (1-zT)^{-1}P_{2k}u, T^{*2k}v\right\rangle}{C(z)}.$$

So (7.7) becomes

$$\begin{aligned} s_{2k} &= \gamma-\frac{\sigma_k}{t_{2k}}+\frac{z^{2k+1}\left\langle (1-zT)^{-1}P_{2k}u, T^{*2k}v\right\rangle}{t_{2k}C(z)} \;=\; \gamma-\frac{\sigma_k}{t_{2k}}+\frac{z^{2k}\sigma_k D_2(z)}{C(z)}\\ &= \gamma-\frac{\sigma_k}{t_{2k}}+\frac{z^{2k}\sigma_k\left(\frac{1}{t_{2k}z^{2k}}C(z)-D_1(z)\right)}{C(z)} = \sigma_0-z^{2k}\sigma_k\frac{D_1(z)}{C(z)}\\ &= \frac{\sigma_0 C(z)-z^{2k}\sigma_k D_1(z)}{C(z)}=\frac{(Q(z)-z^k)s(z)-\sigma_0 Q(z)}{\sigma_0^*Q(z)s(z)-(Q(z)+z^k)}. \end{aligned}$$

Now we prove that the colligation V_{2k} is minimal, that is,

$$\overline{\operatorname{span}}\left\{v_{2k}, T_{2k}^*v_{2k}, T_{2k}^{*2}v_{2k},\dots\right\}=\left\{v, T^*v,\dots, T^{*(2k-1)}v\right\}^{\perp}.$$

By the definition of the projection P_{2k}, the set in the left hand side is contained in the set in the right hand side. To verify the equality, we use that for $\ell=0,1,2,\dots$,

$$T_{2k}^{*\ell}v_{2k}=\frac{1}{t_{2k}^*}T^{*(2k+\ell)}v+\sum_{j=0}^{2k+\ell-1}\tau_j T^{*j}v, \tag{7.9}$$

where the coefficients τ_j are complex numbers depending on ℓ. This formula can easily be proved by induction; for $\ell = 0$ it follows from Lemma 7.2 (3). Assume $x \in \{v, T^*v, \dots, T^{*(2k-1)}v\}^\perp$ is orthogonal to all elements $T_{2k}^{*\ell}v_{2k}$, $\ell = 0, 1, 2, \dots$. Then from (7.9) and by induction, $x \perp T^{*(2k+\ell)}v$, $\ell = 0, 1, 2, \dots$. Since also $x \perp T^{*j}v$ for $j = 0, 1, \dots, 2k-1$, the minimality of V implies $x = 0$.

Finally we prove that V_{2k} is coisometric, that is that

$$T_{2k}T_{2k}^* + \langle\,\cdot\,, u_{2k}\rangle u_{2k} = I_{\mathcal{K}_{2k}}, \tag{7.10}$$
$$\langle v_{2k}, v_{2k}\rangle + |\gamma_{2k}|^2 = 1, \tag{7.11}$$
$$T_{2k}v_{2k} + \gamma_{2k}^* u_{2k} = 0. \tag{7.12}$$

In proving these relations we use that similar formulas hold for the coisometric colligation V; see (4.1)–(4.3). To prove (7.10) we write for $x \in \mathcal{K}_{2k}$,

$$T_{2k}T_{2k}^*x = \left(P_{2k}TP_2k - \frac{\langle\cdot, P_{2k}T^{*2k}v\rangle}{\sigma_k t_{2k}}P_{2k}u\right)\left(P_{2k}T^*x - \frac{\langle x, u\rangle}{\sigma_k^* t_{2k}^*}P_{2k}T^{*2k}v\right).$$

The four summands obtained after writing out the product are

$$\begin{aligned} P_{2k}TP_{2k}T^*x &= P_{2k}\left(TT^*x - \frac{\langle x, u\rangle}{\sigma_k^*}TT^{*k}v\right) \\ &= P_{2k}\left(x - \langle x, u\rangle u - \frac{\langle x, u\rangle}{\sigma_k^*}\left(T^{*(k-1)}v - \left\langle T^{*(k-1)}v, u\right\rangle u\right)\right) \\ &= x - \langle x, u\rangle P_{2k}u + \frac{\langle x, u\rangle}{\sigma_k^*}\sigma_k^* P_{2k}u = x, \end{aligned}$$

where for the first equality we used Lemma 7.2 (8),

$$\begin{aligned} -\frac{\langle P_{2k}T^*v, T^{*2k}v\rangle}{\sigma_k t_{2k}}P_{2k}u &= -\frac{1}{\sigma_k t_{2k}}\left\langle T^*x - \frac{\langle x, u\rangle}{\sigma_k^*}T^{*k}v, T^{*2k}v\right\rangle P_{2k}u \\ &= -\frac{1}{\sigma_k t_{2k}}\left(\langle x, TT^{*2k}v\rangle - \frac{\langle x, u\rangle}{\sigma_k^*}\langle T^{*k}v, T^{*2k}v\rangle\right)P_{2k}u \\ &= -\frac{1}{\sigma_k t_{2k}}\left(\left\langle x, T^{*(2k-1)}v - \sigma_{2k}^* u\right\rangle + \frac{\langle x, u\rangle}{\sigma_k^*}(\sigma_0^*\sigma_k + \sigma_k^*\sigma_{2k})\right)P_{2k}u \\ &= -\frac{\sigma_0^*}{\sigma_k^* t_{2k}}\langle x, u\rangle P_{2k}u, \end{aligned}$$

where for the third equality we used Lemma 7.2 (9),

$$-\frac{\langle x,u\rangle}{\sigma_k^* t_{2k}^*} P_{2k}TP_{2k}T^{*2k}v = -\frac{\langle x,u\rangle}{\sigma_k^* t_{2k}^*} P_{2k}\left(TT^{*2k}v - T\sum_{j=0}^{2k-1}\beta_j T^{*j}v\right)$$

$$= -\frac{\langle x,u\rangle}{\sigma_k^* t_{2k}^*} P_{2k}\left(T^{*(2k-1)}v - \sigma_{2k}^* u + \sum_{j=0}^{2k-1}\beta_j\sigma_j^* u\right)$$

$$= -\frac{\sigma_0}{\sigma_k t_{2k}^*}\langle x,u\rangle P_{2k}u,$$

where for the first and the third equalities we used Lemma 7.2 (3) and (13), and finally with the help of Lemma 7.2 (12),

$$\frac{\langle x,u\rangle}{|\sigma_k t_{2k}|^2}\left\langle P_{2k}T^{*2k}v, T^{*2k}v\right\rangle P_{2k}u$$

$$= \frac{\langle x,u\rangle}{|\sigma_k t_{2k}|^2}\left(-|\sigma_k|^2 + \sigma_0\sigma_k^* t_{2k} + \sigma_0^*\sigma_k t_{2k}^*\right) P_{2k}u$$

$$= \left(-\frac{1}{|t_{2k}|^2} + \frac{\sigma_0}{\sigma_k t_{2k}^*} + \frac{\sigma_0^*}{\sigma_k^* t_{2k}}\right)\langle x,u\rangle P_{2k}u.$$

Combining these expressions we obtain

$$T_{2k}T_{2k}^* x = x - \frac{\langle x,u\rangle}{|t_{2k}|^2} P_{2k}u = x - \langle x, u_{2k}\rangle u_{2k},$$

and this proves (7.10). Formula (7.11) follows in a simple way from Lemma 7.2 (12) and the expressions for v_{2k} and γ_{2k} in the theorem. As to (7.12),

$$T_{2k}v_{2k} + \gamma_{2k}^* u_{2k} = \frac{1}{t_{2k}^*} P_{2k}TP_{2k}T^{*2k}v$$

$$-\frac{1}{\sigma_k|t_{2k}|^2}\left\langle P_{2k}T^{*2k}v, T^{*2k}v\right\rangle P_{2k}u + \left(\sigma_0^* - \frac{\sigma_k^*}{t_{2k}^*}\right)\frac{1}{t_{2k}} P_{2k}u.$$

On account of Lemma 7.2 (3) and (11), the first summand becomes

$$\frac{1}{t_{2k}^*} P_{2k}\left(TT^{*2k}v - \sum_{j=0}^{2k-1}\beta_j TT^{*j}v\right)$$

$$= \frac{1}{t_{2k}^*} P_{2k}\left(-\sigma_{2k}^* u + \sum_{j=0}^{2k-1}\beta_j\sigma_j^* u\right) = \frac{\sigma_0\sigma_k^*}{\sigma_k t_{2k}^*} P_{2k}u,$$

and the second summand with the help of Lemma 7.2 (12) can be written as

$$-\frac{-|\sigma_k|^2 + \sigma_0\sigma_k^* t_{2k} + \sigma_0^*\sigma_k t_{2k}^*}{\sigma_k|t_{2k}|^2} P_{2k}u.$$

These two expressions added to the third summand yield zero, and this proves (7.12). □

8. The colligation of the Schur transformation: Case (4) with $q > 0$

We continue with case (4) using the notations of the previous section. But now we assume that $q > 0$. This means that (see (3.4))

$$\sigma_0^* Q_j(z)s(z) - (Q_j(z) + z^k) = t_{2k+q} z^{2k+q} + t_{2k+q+1} z^{2k+q+1} + \cdots$$

with $t_{2k+q} \neq 0$, or, equivalently, that the number t_{2k} in (7.1) is zero. We set $\tau = -\sigma_k / t_{2k+q}\,(\neq 0)$.

Theorem 8.1. *Assume that for $s \in \mathbf{A}^0$ we have $|\sigma_0| = 1$, and that $q > 0$. Let V, given by* (1.2) *with $\gamma = \sigma_0$, be a minimal coisometric colligation in the space $\mathcal{K} \oplus \mathbb{C}$, where $\mathcal{K}$ is a Krein space, such that $s(z) = s_V(z)$. Then*
(i) *the space* span $\{v, T^* v, \dots, T^{*(2k+q-1)} v\}$ *is a $(2k+q)$–dimensional Pontryagin subspace of $\mathcal{K}$ with negative index equal to $k + q$. Hence*

$$\mathcal{K}_{2k+q} = \left\{v, T^* v, \dots, T^{*(2k+q-1)} v\right\}^{\perp}$$

is a Krein space with $\operatorname{ind}_+ \mathcal{K}_{2k+q} = \operatorname{ind}_+ \mathcal{K} - k$ *and* $\operatorname{ind}_- \mathcal{K}_{2k+q} = \operatorname{ind}_- \mathcal{K} - (k+q)$, *and the orthogonal projection P_{2k+q} in $\mathcal{K}$ onto $\mathcal{K}_{2k+q}$ is well defined;*
(ii) *the colligation*

$$V_{2k+q} := \begin{pmatrix} T_{2k+q} & u_{2k+q} \\ \langle \cdot, v_{2k+q} \rangle & \gamma_{2k+q} \end{pmatrix}$$

with entries

$$\begin{aligned} T_{2k+q} &= P_{2k+q} T P_{2k+q} + \frac{\tau}{\sigma_k^2} \left\langle \cdot, P_{2k+q} T^{*(2k+q)} v \right\rangle P_{2k+q} u, \\ u_{2k+q} &= \frac{\tau}{\sigma_k} P_{2k+q} u, \\ v_{2k+q} &= \frac{\tau^*}{\sigma_k^*} P_{2k+q} T^{*(2k+q)} v, \\ \sigma_{2k+q} &= \tau \end{aligned}$$

is minimal coisometric in the space $\mathcal{K}_{2k+q} \oplus \mathbb{C}$;
(iii) *with $s_{2k+q}(z) := s_{V_{2k+q}}(z)$ it holds*

$$s_{2k+q}(z) = z^q \frac{(Q(z) - z^k)s(z) - \sigma_0 Q(z)}{\sigma_0^* Q(z)s(z) - (Q(z) + z^k)}.$$

In particular, if $s \in \mathbf{S}_\kappa^0$ satisfies the above conditions, then $s_{2k+q} \in \mathbf{S}_{\kappa-k-q}^0$ and $\operatorname{sq}_+(s_{2k+q}) = \operatorname{sq}_+(s) - k$.

Proof. By Lemma 4.3 the minimal coisometric colligation with characteristic function

$$\hat{s}(z) = \frac{1}{s(z)}$$

is given by

$$\hat{T} = T - \frac{\langle\,\cdot\,, v\rangle}{\sigma_0}u, \quad \hat{u} = \frac{1}{\sigma_0}u,$$
$$\hat{v} = \frac{1}{\sigma_0^*}v, \quad \hat{\gamma} = \frac{1}{\sigma_0}.$$

The colligation acts in the space $\mathcal{K}' \oplus \mathbb{C}$, where $\mathcal{K}'$ is the anti-space of the Krein space $\mathcal{K}$.

We define $\hat{\sigma}_j$, $\hat{Q}(z)$, $\hat{c}_j$ for $\hat{s}(z)$ as σ_j, $Q(z)$, c_j are defined for $s(z)$. The Taylor expansion of $\hat{s}(z)$ for z near 0 becomes

$$\begin{aligned}\hat{s}(z) &= \sigma_0^* - \left[\sigma_0^{*2}\left(\sigma_k z^k + \ldots + \sigma_{2k} z^{2k} + \ldots\right)\right.\\ &\quad \left. -\sigma_0^{*3}\left(\sigma_k z^k + \ldots + \sigma_{2k} z^{2k} + \ldots\right)^2 - \ldots\right]\\ &= \sigma_0^* - \sigma_0^{*2}\left[\sigma_k z^k + \ldots + \sigma_{2k-1} z^{2k-1} + \left(\sigma_{2k} - \sigma_0^*\sigma_k^2\right) z^{2k}\right] + \mathcal{O}(z^{2k+1})\end{aligned}$$

and implies

$$\hat{\sigma}_j = \begin{cases} \sigma_0^*, & j = 0,\\ 0, & j = 1, \ldots, k-1,\\ -\sigma_0^{*2}\sigma_j, & j = k, \ldots, 2k-1,\\ -\sigma_0^{*2}(\sigma_{2k} - \sigma_0^*\sigma_k^2), & j = 2k.\end{cases}$$

Hence the analog of (7.5), namely

$$\left(\hat{\sigma}_k + \hat{\sigma}_{k+1} z + \cdots + \hat{\sigma}_{2k-1} z^{2k-1}\right)\left(\hat{c}_0 + \hat{c}_1 z + \cdots + \hat{c}_{k-1} z^{k-1}\right) = \hat{\sigma}_0 + \mathcal{O}(z^k),$$

can be written as

$$\left(\sigma_k + \sigma_{k+1} z + \cdots + \sigma_{2k-1} z^{2k-1}\right)\left(\hat{c}_0 + \hat{c}_1 z + \cdots + \hat{c}_{k-1} z^{k-1}\right) = -\sigma_0 + \mathcal{O}(z^k).$$

Comparing this formula with (7.5) we observe that $\hat{c}_j = -c_j$. Hence $\hat{Q}(z) = -Q(z)$ and a simple calculation yields

$$\begin{aligned}\frac{(\hat{Q}(z) - z^k)\hat{s}(z) - \hat{\sigma}_0\hat{Q}(z)}{\hat{\sigma}_0^*\hat{Q}(z)\hat{s}(z) - (\hat{Q}(z) + z^k)} &= \frac{\sigma_0^* Q(z)s(z) - (Q(z) + z^k)}{(Q(z) - z^k)s(z) - \sigma_0 Q(z)}\\ &= \frac{t_{2k+q} z^q + \mathcal{O}(z^{q+1})}{-\sigma_k + \mathcal{O}(z)}.\end{aligned} \tag{8.1}$$

Thus the function in the left hand side does not have a pole at $z = 0$, in other words $\hat{q} = 0$. Hence (see (7.1))

$$\hat{t}_{2k} = \hat{\sigma}_0^* \sum_{j=0}^{k-1} \hat{c}_j \hat{\sigma}_{2k-j} = -\sigma_0^*\sigma_k + t_{2k} = -\sigma_0^*\sigma_k \neq 0;$$

here $t_{2k} = 0$ because we assume that $q > 0$. Thus we may apply Theorem 7.1 to the function $\hat{s}(z)$: the projection $\hat{P}_{2k}$ in $\mathcal{K}'$ onto the subspace

$$\mathcal{K}_{2k} = \left\{\hat{v}, \hat{T}^*\hat{v}, \ldots, \hat{T}^{*(2k-1)}\hat{v}\right\}^{\perp}$$

is well defined, and the minimal coisometric colligation $\hat{V}_{2k}$ in the space $\mathcal{K}'_{2k} \oplus \mathbb{C}$ whose characteristic function is

$$\hat{s}_{2k}(z) = \frac{(\hat{Q}(z) - z^k)\hat{s}(z) - \hat{\sigma}_0 \hat{Q}(z)}{\hat{\sigma}_0^* \hat{Q}(z)\hat{s}(z) - (\hat{Q}(z) + z^k)}$$

is given by

$$\begin{aligned} \hat{T}_{2k} = &= \hat{P}_{2k}\hat{T}\hat{P}_{2k} - \frac{\left\langle \cdot, \hat{P}_{2k}\hat{T}^{*(2k)}\hat{v}\right\rangle'}{\hat{\sigma}_k \hat{t}_{2k}} \hat{P}_{2k}\hat{u}, & \hat{u}_{2k} &= \frac{1}{\hat{t}_{2k}}\hat{P}_{2k}\hat{u}, \\ \hat{v}_{2k} &= \frac{1}{\hat{t}^*_{2k}}\hat{P}_{2k}\hat{T}^{*(2k)}\hat{v}, & \hat{\gamma}_{2k} &= \hat{\gamma} - \frac{\hat{\sigma}_k}{\hat{t}_{2k}}. \end{aligned}$$

The space $\mathcal{K}'_{2k}$ coincides with the space

$$\mathcal{K}_{2k} = \left\{v, T^*v, \dots, T^{*(2k-1)}v\right\}^{\perp},$$

and hence $\mathcal{K}'_{2k} = (\mathcal{K}_{2k})'$ and $\hat{P}_{2k}$ coincides with P_{2k}, the projection in Theorem 7.1. Since $\sigma_1 = \cdots = \sigma_{k-1} = 0$, there exist elements $w_j \in \mathcal{K}'$ such that

$$\hat{T}^{*(2k)} = T^{*(2k)} + \sum_{j=0}^{2k-1} \langle \cdot, w_j \rangle' \, T^{*j}v.$$

With these formulas the entries of the colligation $\hat{V}_{2k}$ can now be simplified:

$$\begin{aligned} \hat{T}_{2k} &= P_{2k}TP_{2k} - \frac{\langle \cdot, P_{2k}T^{*(2k(}v\rangle'}{\sigma_0^*\sigma_k^2} P_{2k}u, & \hat{u}_{2k} &= -\frac{1}{\sigma_k}P_{2k}u, \\ \hat{v}_{2k} &= -\frac{1}{\sigma_k^*}P_{2k}T^{*(2k)}v, & \hat{\sigma}_{2k} &= 0. \end{aligned}$$

The function

$$\hat{s}_{2k}(z) = \frac{z^q}{s_{2k+q}(z)}$$

has in $z = 0$ a zero of order q, and so we can apply Corollary 5.2 to $\hat{s}_{2k}(z)$ (with $k = q$ and $\mathcal{K}$ replaced by $\mathcal{K}'$) to remove this zero: the minimal coisometric colligation $\hat{V}_{2k+q}$ whose characteristic function is

$$\hat{s}_{2k+q}(z) = \frac{1}{z^q}\hat{s}_{2k}(z) = \frac{1}{s_{2k+q}(z)}$$

is given by

$$\begin{aligned} \hat{T}_{2k+q} &= \hat{P}_q \hat{T}_{2k} \hat{P}_q, & \hat{u}_{2k+q} &= \hat{P}_q \hat{u}_{2k}, \\ \hat{v}_{2k+q} &= \hat{P}_q \hat{T}_{2k}^{*q} \hat{v}_{2k}, & \hat{\gamma}_{2k+q} &= \frac{1}{s_{2k+q}(0)}. \end{aligned}$$

Here $\hat{P}_q$ is the orthogonal projection in the space $\mathcal{K}'_{2k}$ onto the subspace

$$\left\{\hat{v}_{2k}, \dots, \hat{T}_{2k}^{*(q-1)}\hat{v}_{2k}\right\}^{\perp}.$$

Note that by (8.1), $s_{2k+q}(0) = \tau$. Induction by m shows that there are constants α_{mj}, $j = 0, 1, \ldots, 2k+m-1$, $m = 0, 1, \ldots$, such that

$$\hat{T}_{2k}^{*m}\hat{v}_{2k} = \frac{1}{\sigma_k^*}T^{*(2k+m)}v + \sum_{j=0}^{2k+m-1}\alpha_{mj}T^{*j}v.$$

Hence for $x \in \mathcal{K}_{2k}$,

$$x \perp \left\{\hat{v}_{2k}, \ldots, \hat{T}_{2k}^{*(q-1)}\hat{v}_{2k}\right\} \Leftrightarrow x \perp \left\{v, T^*v, \ldots, T^{*(2k+q-1)}v\right\}.$$

Thus $P_{2k+q} := \hat{P}_q P_{2k}$ is the orthogonal projection in $\mathcal{K}'$ (but then also in $\mathcal{K}$) onto the subspace

$$\mathcal{K}'_{2k+q} = \left\{v, T^*v, \ldots, T^{*(2k+q-1)}v\right\}^{\perp} = (\mathcal{K}_{2k+q})',$$

where $\mathcal{K}_{2k+q}$ is as in the theorem. The entries of the colligation $\hat{V}_{2k+q}$ can now be written as

$$\begin{aligned} \hat{T}_{2k+q} &= P_{2k+q}TP_{2k+q}, & \hat{u}_{2k+q} &= -\frac{1}{\sigma_k}P_{2k+q}u, \\ \hat{v}_{2k+q} &= -\frac{1}{\sigma_k^*}P_{2k+q}T^{*(2k+q)}v, & \hat{\gamma}_{2k+q} &= \frac{1}{\tau}. \end{aligned}$$

Obtaining the formula for $\hat{T}_{2k+q}$ we used that $q > 0$, because then $P_{2k+q}T^{*(2k)}v = 0$.

Finally, we use Lemma 4.3 to determine the minimal coisometric colligation in $(\mathcal{K}'_{2k+q})' \oplus \mathbb{C} = \mathcal{K}_{2k+q} \oplus \mathbb{C}$ whose characteristic function coincides with $s_{2k+q}(z)$. Its entries are

$$\begin{aligned} T_{2k+q} &= \hat{T}_{2k+q} - \frac{\langle\cdot, \hat{v}_{2k+q}\rangle'}{\hat{\gamma}_{2k+q}}\hat{u}_{2k+q}, & u_{2k+q} &= \frac{\hat{u}_{2k+q}}{\hat{\gamma}_{2k+q}}, \\ v_{2k+q} &= \frac{\hat{v}_{2k+q}}{\hat{\gamma}_{2k+q}^*}, & \sigma_{2k+q} &= \frac{1}{\hat{\gamma}_{2k+q}}. \end{aligned}$$

If we substitute the entries of the previous colligation into this one, we get the colligation described in the theorem, apart from the minus signs in front of u_{2k+q} and v_{2k+q}. But these signs can be deleted without the colligation loosing its desired properties: coisometricity, closely outerconnectedness, and having the right characteristic function. □

9. The sequence of Schur functions

The Schur algorithm for a function $s \in \mathbf{S}^0$ gives rise to a (finite or infinite) sequence of generalized Schur functions s_j, where j runs through an increasing sequence $J(s)$ of nonnegative integers with $0 \in J(s)$, $s_0 = s$. In the previous sections we have provided formulas for the minimal coisometric colligation in the minimal coisometric realization of s_j with $j > 0$ in terms of the minimal coisometric colligation in the minimal coisometric realization of its immediate predecessor. This can also be done globally, that is, the minimal coisometric colligation in the minimal coisometric realization of s_j with $j > 0$ can be described in terms of the minimal coisometric colligation in the minimal realization of s. As a consequence we obtain that if $j \in J(s)$ is large enough, then s_j belongs to $\mathbf{S}_0$, that is, it is a classical Schur function. This result also appears in [BDGPS, Lemma 3.4.5], but we give a proof using the minimal coisometric realizations of the s_j's.

Theorem 9.1. *Consider $s \in \mathbf{A}^0$ and assume that s has the minimal coisometric representation*

$$s(z) = \gamma + z\left\langle (I - zT)^{-1}u, v\right\rangle,$$

where the colligation

$$V = \begin{pmatrix} T & u \\ \langle\,\cdot\,, v\rangle & \gamma \end{pmatrix} : \begin{pmatrix} \mathcal{K} \\ \mathbb{C} \end{pmatrix} \to \begin{pmatrix} \mathcal{K} \\ \mathbb{C} \end{pmatrix}$$

is minimal coisometric in the space $\mathcal{K} \oplus \mathbb{C}$, and $\mathcal{K}$ is a Krein space. Let $\widetilde{s}$ be one of the functions appearing in the Schur algorithm of $s(z)$. Then there exist an integer $\ell \geq 0$ and complex numbers $\widetilde{a} \neq 0$, $\widetilde{b} \neq 0$ and $\widetilde{c}$ (which can be 0) such that
(i) *the space*

$$\mathcal{P}_\ell = \operatorname{span}\left\{v, T^*v, \dots T^{*(\ell-1)}v\right\}$$

is a nondegenerate ℓ–dimensional subspace of $\mathcal{K}$, so the projection P_ℓ in $\mathcal{K}$ onto $\mathcal{K}_\ell = \mathcal{P}_\ell^\perp$ is well defined;
(ii) *$\widetilde{s}$ admits the representation*

$$\widetilde{s}(z) = \widetilde{\gamma} + z\left\langle (1 - z\widetilde{T})^{-1}\widetilde{u}, \widetilde{v}\right\rangle,$$

in which the colligation $\widetilde{V}$ with entries

$$\begin{aligned} \widetilde{T} &= \widetilde{P}T\widetilde{P} - \widetilde{c}\left\langle\,\cdot\,, \widetilde{P}T^{*\ell}v\right\rangle \widetilde{P}u, & \widetilde{u} &= \widetilde{a}\widetilde{P}u, \\ v &= \widetilde{b}\widetilde{P}T^{*\ell}v, & \sigma_k &= \widetilde{s}(0) \end{aligned}$$

is minimal coisometric in the space $\mathcal{K}_\ell \oplus \mathbb{C}$.
Hence $\operatorname{sq}_\pm(\widetilde{s}) = \operatorname{sq}_\pm(s) - \operatorname{ind}_\pm(\mathcal{P}_\ell)$; in particular if $s \in \mathbf{S}^0_\kappa$, then $\widetilde{s} \in \mathbf{S}^0_{\widetilde{\kappa}}$ with $\widetilde{\kappa} = \kappa - \operatorname{ind}_-\mathcal{P}_\ell$.

Proof. The proof is by induction by the number of the steps of the Schur algorithm. According to Corollary 5.2 and Theorems 6.1, 7.1, and 8.1, the theorem is valid if $\widetilde{s}(z)$ is the immediate successor of $s(z)$. Now assume that $\widetilde{s}(z)$ is described as in the theorem and let $\widehat{s}(z)$ be constructed from $\widetilde{s}(z)$ as in one of the results just mentioned. We show that $\widehat{s}(z)$ has a representation which is similar to that of $\widetilde{s}(z)$. The assumption implies that for $m = 0, 1, \dots$ there are complex numbers α_{mj} such that

$$\widetilde{T}^{*m}\widetilde{v} = \widetilde{\beta}T^{*(\ell+m)}v + \sum_{j=0}^{\ell+m-1} \alpha_{mj}T^{*j}v. \tag{9.1}$$

This can shown by induction by m. We set $\widetilde{\sigma}_0 = \widetilde{s}(0)$ and consider the three cases of the Schur algorithm.
Case (2): $|\widetilde{\sigma}_0| < 1$. We apply Corollary 5.2. For some $a \neq 0$ and $b \neq 0$,

$$\begin{aligned} \widehat{T} &= P_k\widetilde{T}P_k & \widehat{u} &= aP_k\widetilde{u}, \\ \widehat{v} &= bP_k\widetilde{T}^{*k}\widetilde{v}, & \widehat{\sigma} &= \widehat{s}(0), \end{aligned}$$

where P_k in $\mathcal{K}_\ell$ is the orthogonal projection onto

$$\widehat{\mathcal{K}} = \mathcal{K}_\ell \ominus \operatorname{span}\left\{\widetilde{v}, \widetilde{T}^*\widetilde{v}, \dots, \widetilde{T}^{*(k-1)}\widetilde{v}\right\}.$$

The space span $\left\{\widetilde{v}, \widetilde{T}^*\widetilde{v}, \dots, \widetilde{T}^{*(k-1)}\widetilde{v}\right\}$ is a positive subspace of $\mathcal{K}_\ell$, and it is orthogonal to span $\{v, T^*v, \dots, T^{*\ell}v\}$ in $\mathcal{K}$. It follows that the span of these two subspaces is a Pontryagin subspace of $\mathcal{K}$. By (9.1), it coincides with

$$\operatorname{span}\left\{v, T^*v, \dots T^{*(\ell+k-1)}v\right\}.$$

Hence

$$\widehat{\mathcal{K}} = \left\{v, T^*v, \dots T^{*(\ell+k-1)}v\right\}^\perp,$$

and if $\widehat{P}$ is the orthogonal projection in $\mathcal{K}$ onto $\widehat{\mathcal{K}}$, then $\widehat{P} = P_k\widetilde{P}$. It follows that, since $\widehat{P}T^{*\ell}v = 0$,

$$\begin{aligned} \widehat{T} &= \widehat{P}T\widehat{P} - c\langle\cdot, \widehat{P}T^{*\ell}v\rangle\widehat{P}u = \widehat{P}T\widehat{P}, \\ \widehat{u} &= \widehat{a}\widehat{P}u, \quad \widehat{a} = a\widetilde{a} \neq 0, \\ \widehat{v} &= \widehat{b}\widehat{P}^{*\ell}v, \quad \widehat{b} = b\widetilde{b} \neq 0, \\ \widehat{\gamma} &= \widehat{s}(0), \end{aligned}$$

which is of the desired form.

Case (3): $|\widetilde{\sigma}_0| > 1$. We apply Theorem 6.1. The reasoning and the calculations are the same as in the previous case, except that now (we use the same notation) the space

$$\operatorname{span}\left\{\widetilde{v}, \widetilde{T}^*\widetilde{v}, \dots, \widetilde{T}^{*(k-1)}\widetilde{v}\right\}$$

is a negative subspace of $\mathcal{K}_\ell$ and

$$\widehat{T} = P_k\widetilde{T}P_k - \left\langle\cdot, P_k\widetilde{T}^{*k}\widetilde{v}\right\rangle P_k\widetilde{u},$$

which implies that

$$\widehat{T} = \widehat{P}T\widehat{P} - c\left\langle\cdot, \widehat{P}T^{*\ell}\right\rangle \widehat{P}T^{r}u = \widehat{P}T\widehat{P},$$

because $\widehat{P}T^{*\ell} = 0$.

Case (4): $|\widetilde{\sigma}_0| = 1$. We apply Theorems 7.1 and 8.1. We assume that $\widetilde{s}(z)$ satisfies the conditions of Case (4). We have that

$$\begin{aligned}\widehat{T} &= P_{2k+q}\widetilde{T}P_{2k+q} - c\langle\cdot, P_{2k+q}\widetilde{T}^{*(2k+q)}\widetilde{v}\rangle P_{2k+q}\widetilde{u}\\ \widehat{u} &= aP_{2k+q}\widetilde{u}, \quad a \neq 0,\\ \widehat{v} &= bP_{2k+q}\widetilde{T}^{*(2k+q)}\widetilde{v}, \quad b \neq 0,\\ \widehat{\gamma} &= \widehat{s}(0),\end{aligned}$$

where P_{2k+q} is the orthogonal projection onto

$$\widehat{\mathcal{K}} = \mathcal{K}_\ell \ominus \operatorname{span}\left\{\widetilde{v}, \widetilde{T}^*\widetilde{v}, \dots, \widetilde{T}^{*(2k+q-1)}\widetilde{v}\right\}$$

The space span $\left\{\widetilde{v}, \widetilde{T}^*\widetilde{v}, \dots, \widetilde{T}^{*(2k+q-1)}\widetilde{v}\right\}$ is a $(2k+q)$–dimensional Pontryagin subspace of $\widetilde{\mathcal{K}}$ with negative index k and positive index $k+q$, which by (9.1) coincides with span $\{v, T^*v, \dots T^{*(\ell+2k+q-1)}\}$. Hence $\widehat{\mathcal{K}}$ is the orthogonal complement of this span in $\mathcal{K}$, and if $\widehat{P}$ is the orthogonal projection in $\mathcal{K}$ onto $\widehat{\mathcal{K}}$, then $\widehat{P} = P_{2k+q}\widetilde{P}$. It follows that

$$\begin{aligned}\widehat{T} &= \widehat{P}T\widehat{P} - \widehat{c}\left\langle\cdot, \widehat{T}^{*(\ell+2k+q)}v\right\rangle\widehat{P}u, \quad \widehat{c} = \widetilde{a}\widetilde{b}c,\\ \widehat{u} &= \widehat{a}\widehat{P}u, \quad \widehat{a} = a\widetilde{a} \neq 0,\\ \widehat{v} &= \widehat{b}PT^{*(\ell+2k+q)}v, \quad \widehat{b} = b\widetilde{b} \neq 0,\\ \widehat{\gamma} &= \widehat{s}(0),\end{aligned}$$

which shows that in this case the representation of $\widehat{s}(z)$ is also of the desired form. □

The theorem implies that after finitely many steps the generalized Schur algorithm for $s \in \mathbf{S}^0$ only yields classical Schur functions, that is, functions which are holomorphic and bounded by 1 on $\mathbb{D}$.

Corollary 9.2. *Let $(s_j)_{j\in J(s)}$ be the sequence of Schur functions obtained by applying the generalized Schur algorithm to the function $s = s_0 \in \mathbf{S}^0$. Then for sufficiently large $j \in J(s)$, s_j is a classical Schur function.*

Proof. If the sequence $J(s)$ is finite, let j_0 be its largest element. Then case (1) of the Schur algorithm prevails and $s_{j_0} \equiv \sigma_{j_0 0}$ with $|\sigma_{j_0 0}| = 1$. Evidently, this function is a classical Schur function.
Now we assume that the index set contains infinitely many indices. It follows from the proof of Theorem 9.1 that in passing from one Schur function in the algorithm to the next one, in the minimal coisometric representation ℓ increases by k in Cases (2) and (3) and by $2k+q$ in Case (4). In these cases the negative index of $\mathcal{K}_\ell$ decreases by 0, k and $k+q$, respectively; in any case it does not increase. The minimality of the colligation for s implies that for sufficiently large ℓ the subspace

$$\mathcal{P}_\ell = \operatorname{span}\left\{v, T^*v, \dots, T^{*(\ell-1)}v\right\}$$

contains a maximal negative subspace of $\mathcal{K}$, and so its orthogonal complement $\mathcal{K}_\ell$ is a Hilbert space. Thus there exists $j_0 \in J(s)$ such that for all $j \in J(s)$ with $j \geq j_0$, the state space of the minimal coisometric colligation in the realization of s_j is a Hilbert space. The corollary now follows from the fact that if the state space $\mathcal{K}$ of a minimal coisometric colligation in $\mathcal{K} \oplus \mathbb{C}$ is a Hilbert space, then its characteristic function is a classical Schur function. □

References

[AD] D. Alpay and H. Dym, *On applications of reproducing kernel spaces to the Schur algorithm and rational J-unitary factorization*, Operator Theory: Adv. Appl., **18**, Birkhäuser Verlag, Basel, 1986, 89–159.

[ADRS] D. Alpay, A. Dijksma, J. Rovnyak, and H. de Snoo, *Schur functions, Operator Colligations, and Reproducing kernel Pontryagin spaces*, Operator Theory: Adv. Appl., **96**, Birkhäuser Verlag, Basel, 1997.

[AG] D. Alpay and I. Gohberg, *Unitary rational matrix functions*, Operator Theory: Adv. Appl., **33**, Birkhäuser Verlag, Basel, 1988, 175–222.

[AI] T. Ya. Azizov and I. S. Iokhvidov, *Foundations of the theory of linear operators in spaces with an indefinite metric*, Nauka, Moscow, 1986 [Russian]; English transl. *Linear operators in spaces with an indefinite metric*, Wiley, New York, 1989.

[B] J. Ball, *Models for noncontractions*, J. Math. Anal. Appl. **52** (1975), 255–259.

[BDGPS] M. J. Bertin, A. Decomps-Guilloux, M. Grandet-Hugot, M. Pathiaux-Delfosse, and J. P. Schreiber, *Pisot and Salem numbers*, Birkhäuser Verlag, Basel, 1992.

[BGK] H. Bart, I. Gohberg, and M. A. Kaashoek, *Minimal factorization of matrix and operator functions*, Operator Theory: Adv. Appl., **1**, Birkhäuser Verlag, Basel, 1979.

[C] C. Chamfy, *Fonctions méromorphes sur le circle unité et leurs séries de Taylor*, Ann. Inst. Fourier **8** (1958), 211–251.

[CDLS] B. Ćurgus, A. Dijksma, H. Langer, and H. S. V. de Snoo, *Characteristic functions of unitary colligations and of bounded operators in Krein spaces*, Operator Theory: Adv. Appl., **41**, Birkhäuser Verlag, Basel, 1989, 125–152.

[D] J. Dufresnoy, *Sur le problème des coefficients par certaines fonctions dans le cercle unité*, Ann. Acad. Sc. Fenn. Ser. A. I, **250**, 9 (1958), 1–7.

[DGK] P. Delsarte, Y. Genin, and Y. Kamp, *Pseudo-Carathéodory functions and hermitian Toeplitz matrices*, Philips J. Res. **41** (1) (1986), 1–54.

[DLS] A. Dijksma, H. Langer, and H. S. V. de Snoo, *Unitary colligations in Krein spaces and their role in the extension theory of isometries and symmetric linear relations in Hilbert spaces*, in: *Functional Analysis II*, Proceedings Dubrovnik 1985, Lecture Notes in Mathematics, **1242**, Springer Verlag, Berlin, 1987, 1–42.

[KL] M. G. Krein and H. Langer, *Über einige Fortsetzungsprobleme, die eng mit der Theorie hermitescher Operatoren in Raume Π_κ zusammenhaängen, Teil I: Einige Funktionenklassen und ihre Darstellungen*, Math. Nachr. **77** (1977), 187–236.

[LL] H. Langer and A. Luger, *A class of* 2×2*–matrix functions*, Glasnik Matematički **35** (55) (2000), 149–160.

[S1] I. Schur, *Über die Potenzreihen, die im Innern des Einheitskreises beschränkt sind*, J. Reine Angew. Math. **147** (1917), 205–232; English translation in: *I. Schur methods in operator theory and signal processing*, Operator Theory: Adv. Appl., **18**, Birkhäuser Verlag, Basel, 1989.

[S2] I. Schur, *Über die Potenzreihen, die im Innern des Einheitskreises beschränkt sind; Fortsetzung*, J. Reine Angew. Math. **148** (1918), 122–145.

Department of Mathematics
Ben-Gurion University of the Negev
P.O. Box 653
84105 Beer Sheva, Israel
E-mail address: `dany@ivory.bgu.ac.il`

Department of Mathematics
Voronezh State University
394693 Voronezh, Russia
E-mail address: `azizov@tom.vsu.ru`

Department of Mathematics
University of Groningen
P.O. Box
9700 AV Groningen, the Netherlands
E-mail address: `dijksma@math.rug.nl`

Department of Mathematics
Vienna University of Technology
Wiedner Hauptstrasse 8–10
A-1040 Vienna, Austria
E-mail address: `hlanger@mail.zserv.tuwien.ac.at`

Hardy spaces on a finite bordered Riemann surface, multivariable operator model theory and Fourier analysis along a unimodular curve

Joseph A. Ball and Victor Vinnikov

1. Introduction

In this paper we continue the study of (in general) indefinite Hardy spaces on a finite bordered Riemann surface as reported in [5]. To put the ideas in context, we begin with the classical case, where the bordered Riemann surface is the closed unit disk $\overline{\mathbb{D}}$. We let $\mathbb{T}$ denote the unit circle, $\mathcal{E}_*$ denote a Hilbert space, and let $L^2(\mathbb{T}, \mathcal{E}_*)$ be the standard Lebesgue space of $\mathcal{E}_*$-valued, measurable functions f defined on $\mathbb{T}$ with $\|f\|_2^2 = \frac{1}{2\pi} \int_{\mathbb{T}} \|f(z)\|^2 |dz| < \infty$. Then an arbitrary element f of $L^2(\mathbb{T}, \mathcal{E}_*)$ alternatively can be presented in terms of a Fourier representation $f(z) \sim \sum_{n=-\infty}^{\infty} f_n z^n$ with Fourier coefficients f_n ($n = \dots, -1, 0, 1, \dots$) taking values in $\mathcal{E}_*$ and with $\|f\|_2^2 = \sum_{n=-\infty}^{\infty} \|f_n\|^2$. The Hardy space $H^2(\mathbb{D}, \mathcal{E}_*)$ consists of $\mathcal{E}_*$-valued analytic functions on the unit disk $\mathbb{D}$ with $\|f(z)\|^2$ possessing a harmonic majorant, and can be identified as the subspace of functions f in $L^2(\mathbb{T}, \mathcal{E}_*)$ with Fourier representation of the form $f(z) \sim \sum_{n=0}^{\infty} f_n z^n$. The orthogonal complement $H^2(\mathcal{D}, \mathcal{E}_*)^\perp$ in $L^2(\mathbb{T}, \mathcal{E}_*)$ consists of $L^2(\mathbb{T}, \mathcal{E}_*)$-functions f with Fourier series of the form $f(z) \sim \sum_{n=-\infty}^{-1} f_n z^n$, and can be identified with the Hardy space $H_0^2(\mathbb{D}_e, \mathcal{E}_*)$ consisting of functions analytic on the complement $\mathbb{D}_e$ of the closed unit disk in the extended complex plane $\mathbb{C}_\infty$ which vanish at ∞. The operator M_z of multiplication by the coordinate function z on $L^2(\mathbb{T}, \mathcal{E}_*)$ is unitary (specifically, the bilateral shift operator of multiplicity equal to $\dim \mathcal{E}_*$), and its restriction $V_{\mathcal{E}_*} := M_z|_{H^2(\mathbb{D}, \mathcal{E}_*)} \colon H^2(\mathbb{D}, \mathcal{E}_*) \to H^2(\mathbb{D}, \mathcal{E}_*)$ is an isometry (the unilateral shift operator of multiplicity equal to $\dim \mathcal{E}_*$). According to the Beurling-Lax-Halmos theorem (see, e.g., [27, 33, 35]), any subspace $\mathcal{M}$ of $H^2(\mathbb{D}, \mathcal{E}_*)$ invariant for the shift operator M_z has the form $\mathcal{M} = S \cdot H^2(\mathbb{D}, \mathcal{E})$ for another auxiliary Hilbert space $\mathcal{E}$, where S is an *inner function*, i.e., an analytic function on $\mathbb{D}$ with values in the space $\mathcal{L}(\mathcal{E}, \mathcal{E}_*)$ of operators from $\mathcal{E}$ into $\mathcal{E}_*$ such that the associated boundary-value function $S(\zeta)$ is an isometry for almost every $\zeta \in \mathbb{T}$. Inner functions form a subclass of the class $\mathcal{S}(\mathcal{E}, \mathcal{E}_*)$ of *contractive* or *Schur functions*, i.e., the class of analytic, $\mathcal{L}(\mathcal{E}, \mathcal{E}_*)$-valued functions S on $\mathbb{D}$ with $\|S(z)\| \leq 1$ for each $z \in \mathbb{D}$. In addition to having a power series representation $S(z) = \sum_{n=0}^{\infty} S_n z^n$

on $\mathbb{D}$ with coefficients $S_n \in \mathcal{L}(\mathcal{E}, \mathcal{E}_*)$, Schur-class functions are characterized by having a representation of the form

$$S(z) = D + zC(I - zA)^{-1}B \tag{1.1}$$

where the block matrix

$$U = \begin{bmatrix} A & B \\ C & D \end{bmatrix} : \begin{bmatrix} \mathcal{H} \\ \mathcal{E} \end{bmatrix} \to \begin{bmatrix} \mathcal{H} \\ \mathcal{E}_* \end{bmatrix} \tag{1.2}$$

(where $\mathcal{H}$ is an auxiliary Hilbert space—the *state space*) is *unitary*; see [17]. This representation corresponds to S being *realized* as the *transfer function* of the conservative, discrete-time, linear system

$$\begin{aligned} x(n+1) &= Ax(n) + Bu(n) \\ y(n) &= Cx(n) + Du(n). \end{aligned} \tag{1.3}$$

Here the *transfer-function* property of $S(z) = D + zC(I - zA)^{-1}B$ corresponds to the fact that $\widehat{y}(z) = S(z)\widehat{u}(z)$ if $\widehat{u}(z) = \sum_{n=-\infty}^{\infty} u(n)z^n$, $\widehat{y}(z) = \sum_{n=-\infty}^{\infty} y(n)z^n$, and $\{y(n)\}_{n\in\mathbb{Z}}$ is the l^2 output string resulting from feeding the l^2 input string $\{u(n)\}_{n\in\mathbb{Z}}$ into the recursive system of equations (with initial condition $x(-\infty) = 0$). The *conservative* property of the system (1.3) corresponds to the energy-balance relation

$$\|x(n+1)\|^2 - \|x(n)\|^2 = \|u(n)\|^2 - \|y(n)\|^2, \tag{1.4}$$

i.e., the change in the energy of the state vector from time n to time $n+1$ is equal to the net energy put into the system from the input and output ports. The case where $S(z) = D + zC(I - zA)^{-1}B$ is actually inner (i.e., has isometric boundary values almost everywhere on the unit circle and the case which we shall focus on here for simplicity) is characterized by the condition that $A^n \to 0$ in the strong operator topology as $n \to \infty$.

An operator of the form $M_z|_{S \cdot H^2(\mathbb{D}, \mathcal{E})}$ for an inner function $S \in \mathcal{S}(\mathcal{E}, \mathcal{E}_*)$ is again a unilateral shift with multiplicity equal to $\dim \mathcal{E} \le \dim \mathcal{E}_*$, and hence is completely determined up to unitary equivalence by $\dim \mathcal{E}$. However the collection of compressed-shift operators $T(S) := P_{\mathcal{H}(S)} M_z|_{\mathcal{H}(S)}$ (where we let S sweep through the collection of all inner functions), where we have set $\mathcal{H}(S)$ equal to the *model space* associated with S

$$\mathcal{H}(S) := H^2(\mathbb{D}, \mathcal{E}_*) \ominus S \cdot H^2(\mathbb{D}, \mathcal{E})$$

is quite rich: *any contraction operator T on a Hilbert space $\mathcal{H}$ for which T^{*n} tends to zero strongly as n tends to infinity is unitarily equivalent to an operator $T(S)$ for an inner function S.* This is just the $C_{\cdot 0}$ case of the Sz.-Nagy-Foiaş model theory—see [32]. S is the so called *characteristic function* of T and is simply the transfer function (1.1) obtained from the minimal embedding of $A = T^*$ into a unitary block matrix (1.2) (i.e., into a conservative discrete-time system (1.3)).

For the case of dissipative operators rather than contractions, or more generally for nonselfadjoint rather than nonunitary operators, a spectral analysis based on the theory of characteristic functions was developed by M. S. Livšic and his collaborators in the 1940s and the 1950s [18, 16], though without the use of functional models. The corresponding relation with conservative continuous-time (rather than discrete-time) systems was discovered by M. S. Livšic in the 1960s [30]; at about the same time realization formulas such as (1.1) were appearing prominently in the state space approach to system and control theory being developed by R. E. Kalman and his collaborators, but in an affine form, i.e., without such restrictions as the block matrix (1.2) being unitary (the energy-balance relation (1.4)) or the transfer function S being Schur-class (see e.g. [28]). For a good recent account of these (and other) parallel developments, see [7, 29]. An alternative model theory for contractions is due to de Branges-Rovnyak [19]; for a modern account of both model theories see [34], and for connections with system theory [11, 12].

In the 1970s Abrahamse and Douglas (see [1, 2]) introduced a generalization of some of these ideas to the setting where the unit disk is replaced by a general finitely-connected planar domain, and the unilateral shift operator is replaced by a subnormal operator with normal extension having spectrum in the boundary of the planar domain. Specifically, denote by R a finitely connected planar domain with analytic boundary ∂R. There is a notion of Hardy space $H^2(R, \mathcal{E}_*)$ and Lebesgue space $L^2(\partial R, \mathcal{E}_*)$, where one can use a conformally invariant harmonic measure on ∂R to define the L^2-inner product (see [36]). Now the operator M_z of multiplication by the coordinate function z on $L^2(\partial R, \mathcal{E}_*)$ is a normal operator with spectrum in ∂R, and $M_z|_{H^2(R,\mathcal{E}_*)}$ is a subnormal operator on $H^2(R, \mathcal{E}_*)$ with normal extension equal to M_z on $L^2(\partial R, \mathcal{E}_*)$. To obtain a model for all pure subnormal operators with normal extension having spectrum in ∂R, one must consider additional model Hardy spaces $H^2(R, \mathbf{V}_{\chi_*})$ whose elements are holomorphic sections of a vector bundle $\mathbf{V}_{\chi_*}$ over R induced by a flat, unitary factor of automorphy χ_*. Specifically, this means that χ_* is a homomorphism from the fundamental group $\pi_1(R)$ into the group $U(\mathcal{E}_*)$ of unitary operators on $\mathcal{E}_*$, and an analytic section f of $\mathbf{V}_{\chi_*}$ over R is an analytic $\mathcal{E}_*$-valued function on the universal cover $\widetilde{R}$ of R satisfying $f(T\widetilde{p}) = \chi_*(T)f(\widetilde{p})$ for all $\widetilde{p} \in \widetilde{R}$ and all deck transformations T of $\widetilde{R}$, which we identify as usual with elements of the fundamental group $\pi_1(R)$ of homotopy equivalence classes of closed curves (with fixed base point) in R. Loosely speaking, f is a multivalued function which picks up an operator multiplier $\chi_*(T)$ when we go around a closed loop T in R. Since the factor $\chi_*(T)$ is unitary, the square of the norm $\|f(z)\|^2$ is well-defined on R and the space $H^2(R, \mathbf{V}_{\chi_*})$ of all holomorphic sections f of $\mathbf{V}_{\chi_*}$ over R with $\|f(z)\|^2$ possessing a harmonic majorant can be identified, via almost-everywhere existing nontangential boundary-value $f(\zeta)$ ($\zeta \in \partial R$), as a closed subspace of the space $L^2(\partial R, \mathbf{V}_{\chi_*})$ of norm-square integrable sections of $\mathbf{V}_{\chi_*}$ over ∂R (see [1, 10] for details). Then, as shown in [1], the operators

$$V_{\mathbf{V}_{\chi_*}} := M_z|_{H^2(R,\mathbf{V}_{\chi_*})} \colon H^2(R, \mathbf{V}_{\chi_*}) \to H^2(R, \mathbf{V}_{\chi_*}),$$

where χ_* sweeps over all unitary representations of the fundamental group $\pi_1(R)$ (or equivalently, over all flat, unitary factors of automorphy), provide a model for any pure subnormal operator having minimal normal extension with spectrum in ∂R. The Beurling-Lax-Halmos theorem in this setting (due to Sarason [37], Voichick [41], and Hasumi [26] in the case $\dim \mathcal{E}_* = 1$) becomes: *any subspace $\mathcal{M}$ of $H^2(R, \mathbf{V}_{\chi_*})$ invariant for all multiplication operators M_f by rational functions f with no poles in R, is of the form $\mathcal{M} = S \cdot H^2(R, \mathbf{V}_\chi)$ for some $\chi \in Hom(\pi_1(R), U(\mathcal{E}))$* (with $\mathcal{E}$ an auxiliary Hilbert space with $\dim \mathcal{E} \leq \dim \mathcal{E}_*$), where S is an inner bundle map between $\mathbf{V}_\chi$ and $\mathbf{V}_{\chi_*}$. Here S is a holomorphic $\mathcal{L}(\mathcal{E}, \mathcal{E}_*)$-valued function on the universal cover $\widetilde{R}$ with the automorphic property

$$S(T\widetilde{p}) = \chi_*(T)S(\widetilde{p})\chi(T)^{-1}$$

and S has multivalued boundary-value function $S(\zeta)$ ($\zeta \in \partial R$) with isometric values for almost all $\zeta \in \partial R$. In [2], Abrahamse and Douglas showed that the compressed bundle shifts $T(S) = P_{\mathcal{H}(S)} M_z|_{\mathcal{H}(S)}$, where the model space $\mathcal{H}(S)$ is now given by

$$\mathcal{H}(S) = H^2(R, \mathbf{V}_{\chi_*}) \ominus S \cdot H^2(R, \mathbf{V}_\chi)$$

and where S sweeps over all inner bundle maps, provides a model for a large class (the analogue of $C_{\cdot 0}$ for this setting) of subnormal operators T having normal dilation N with $\sigma(N) \subset \partial R$. It is known only for some special cases (see [4]) whether the condition that R is a spectral set for T guarantees that T in turn has a ∂R-normal dilation.

There has been very little followup to this work of the 1970s (see [8, 44], and, in a very different vein, the recent model theory for subnormal operators with a finite rank self commutator developed by D. Yakubovich [42, 43] following some previous work of Xia). In this report we announce some progress on completing this picture; complete details will appear in [15]. Our theory is parallel to the recent spectral theory of commuting nonselfadjoint operators and related theory of overdetermined multidimensional continuous-time systems developed by M. S. Livšic and his collaborators [31] (see also [40]). However, we are dealing with nonunitary operators and discrete-time systems having many features and peculiarities of their own.

2. A model for commuting contractions

As is now well known (see e.g. [22, 20, 10, 23, 24]), one can get much insight into the function theory on a finitely-connected planar domain R by introducing the Schottky double X of R. By definition, X as a set is given by $X = R \cup \partial R \cup R_-$ where R_- is a second copy of R. The topology on R is defined by identifying the boundary of R with the boundary of R_- in the natural way. The holomorphic structure on R_- is taken to be the conjugate of the holomorphic structure on R under the identification map of a point of R_- with the corresponding point of R. In this way X becomes a compact (or closed) Riemann surface, and the map τ associating a point of R (respectively R_-) with the corresponding point of R_- (respectively R) and fixing points of ∂R is an antiholomorphic involution of X

with fixed point set $X_{\mathbb{R}}$ equal to ∂R. In addition, we see that $X \backslash X_{\mathbb{R}}$ consists of two connected components, namely R and R_-.

One of the difficulties with function theory on multiply-connected planar domains distinguishing it from function theory on the unit disk is that there are no univalent, single-valued inner functions, i.e., (single-valued) univalent holomorphic functions $\varphi \colon R \to \mathbb{D}$ with nontangential boundary-value function $\zeta \to \varphi(\zeta)$ for $\zeta \in \partial R$ having $|\varphi(\zeta)| = 1$. However, there are many such multivalent functions. In particular, there is a pair of such functions $p \to z_1(p)$ and $p \to z_2(p)$ with meromorphic continuation to all of X with the property that they generate the field $\mathcal{M}(X)$ of all meromorphic functions on X. In a change of point of view from [1], we shall consider the pair of operators (T_1, T_2) defined by

$$T_j = T_j(S) := P_{\mathcal{H}(S)} M_{z_j}|_{\mathcal{H}(S)} \text{ for } j = 1, 2$$

(where S is an inner bundle map as above) as the fundamental object of interest. Namely, we shall consider such a pair $(T_1(S), T_2(S))$ as a model for a commuting pair of contraction operators on a Hilbert space $\mathcal{H}$. The *embedding problem* is: *give conditions on a pair of commuting contraction operators* (T_1, T_2) *on a Hilbert space* $\mathcal{H}$ *which guarantee that* (T_1, T_2) *is unitarily equivalent to a pair of the form* $(T_1(S), T_2(S))$ *for some inner bundle map* S *on the domain* R. This of course is only the $C_{\cdot 0}$-case of the general theory: a similar type model can be written down for a more general contractive bundle map S, but for the sake of simplicity we do not go into such details here. The embedding issue is one of the topics to be discussed in [15].

Once we have made this step, there is no reason to restrict R to be a planar domain. More generally, R may be any open Riemann surface; one may view R as a finitely connected planar domain as before with the additional attaching of finitely many handles—the number of such handles being equal to the genus of R. Then $R \cup \partial R$ is what is known as a *bordered Riemann surface*. Just as before we may consider the double X of R, a somewhat more complicated object topologically but manageable (see e.g. [22, 5]). Abstractly, such an X is characterized simply as a *real Riemann surface of dividing type*, i.e., a compact Riemann surface with antiholomorphic involution $p \to p^\tau$ such that the complement $X \backslash X_{\mathbb{R}}$ of the fixed point set $X_{\mathbb{R}}$ of τ consists of exactly two connected components, denoted as X_+ and X_-. Again we may choose two meromorphic inner functions z_1 and z_2 on R, i.e., meromorphic functions on X which are holomorphic on $X_+ := R$ and have values of modulus 1 on $X_{\mathbb{R}} = \partial R$. A more canonical way to set up Hardy spaces in this setting is through the use of bundles defined via a factor of automorphy with a twist via a bundle Δ of half-order differentials, as we now explain; for more complete details, see [5].

In any case, the canonical set of objects for Hardy space theory in this setting is as follows. We let χ be a (not necessarily constant) factor of automorphy $(T, \widetilde{p}) \to \chi(T, \widetilde{p})$. Here T is a member of the fundamental group $\pi_0(X, p_0)$ of homotopy equivalence classes of X with base point p_0 belonging to $X_{\mathbb{R}}$; by a standard construction, $\pi_0(X, p_0)$ can be identified with the group of deck transformations for the universal cover $\widetilde{X}$ of X. The point $\widetilde{p}$ is a generic point on the universal cover $\widetilde{X}$ of X, i.e., a simply connected topological space for which there is a surjective projection map $\pi\colon \widetilde{X} \to X$ such that every closed curve on X which is not homotopically trivial lifts to an open curve on $\widetilde{X}$ (see [25]). The value $\chi(T,\widetilde{p})$ of χ at $(T,\widetilde{p})$ is assumed to lie in the space $\mathcal{GL}(\mathcal{E}_*)$ of bounded, linear, invertible operators on a Hilbert space $\mathcal{E}_*$, and the factor of automorphy property of χ is

$$\chi(T_1T_2,\widetilde{p}) = \chi(T_1, T_2(\widetilde{p}))\chi(T_2,\widetilde{p}).$$

In addition we assume that we have a holomorphic function $\widetilde{p} \to G(\widetilde{p})$ defined on all of $\widetilde{X}$ with values in $\mathcal{GL}(\mathcal{E}_*)$ such that

$$G(\widetilde{p}^\tau)^* = G(\widetilde{p}) \text{ and } \chi(T^\tau,\widetilde{p}^\tau)^* G(T\widetilde{p})\chi(T,\widetilde{p}) = G(\widetilde{p}). \tag{2.1}$$

Here χ^τ is the factor of automorphy defined by

$$\chi^\tau(T,p) = \overline{\chi(T^\tau,\widetilde{p}^\tau)}.$$

The vector bundle of interest is $H = \mathbf{V}_\chi \otimes \Delta$; here Δ is a line bundle of half-order differentials, i.e., the transition functions $g_{\alpha,\beta}$ on the intersection $U_\alpha \cap U_\beta$ of two charts (z_α, U_α) and (z_β, U_β) has the form $\sqrt{\frac{dz_\alpha}{dz_\beta}}$, and $\mathbf{V}_\chi$ is the bundle on X whose sections lift to $\mathcal{E}_*$-valued functions $\widetilde{p} \to f(\widetilde{p})$ on $\widetilde{X}$ satisfying the automorphy property

$$f(T\widetilde{p}) = \chi(T,\widetilde{p})f(\widetilde{p}).$$

The antiholomorphic involution τ induces an action on vector bundles so that $\tau\colon \mathbf{V}_\chi \to \mathbf{V}_\chi^\tau := \mathbf{V}_{\chi^\tau}$. Given G satisfying (2.1) and χ, there is a pairing $\mathbf{V}_\chi \times \mathbf{V}_\chi \to \mathcal{O}(X)$ so that, if $f(\widetilde{p})$ and $g(\widetilde{p})$ are local sections of $\mathbf{V}_\chi$ near p_0 (and hence $\overline{g(\widetilde{p}^\tau)}$ is a local section of $\mathbf{V}_{\chi^\tau}$ near p^τ)), then

$$(f,g)(p) = g(\widetilde{p}^\tau)^* G(p) f(\widetilde{p}). \tag{2.2}$$

With more attention to how coordinate charts are arranged, one can arrange that $\Delta \cong \Delta^\tau$. Then the pairing (2.2) induces a pairing $(\cdot,\cdot)\colon H \times H^\tau \to K_X$ (where $K_X = \Delta \otimes \Delta \cong \Delta \otimes \Delta^\tau$ is the canonical line bundle of holomorphic differentials on X) with action on a local section $f(\widetilde{p})$ and $g(\widetilde{p})$ of H near p (and hence $\overline{g(\widetilde{p}^\tau)}$ equal to a local section of H^τ near p^τ) given by the same formula (2.2), but with the result $(f,g)(p)$ now a differential near p. (One can arrive at this setup from a more abstract formalism, where one is given a real Riemann surface X of genus g together with a degree $g-1$ vector bundle H with only the trivial global holomorphic section and a parahermitian pairing $H \times H^\tau \to K_X$ on H; for details, see [5, 15].) This enables us to define a well-defined inner product on the space $L^2(X_\mathbb{R}, \mathbf{V}_\chi \otimes \Delta)$ of norm-square integrable sections by the formula

$$\langle f,g\rangle = \int_{X_\mathbb{R}} (f,g)(p) = \int_{X_\mathbb{R}} g(\widetilde{p}^\tau)^* G(\widetilde{p}) f(\widetilde{p}). \tag{2.3}$$

We shall assume for our purposes here that this inner product on $L^2(X_\mathbb{R}, \mathbf{V}_\chi \otimes \Delta)$ is positive-definite, making $L^2(X_\mathbb{R}, \mathbf{V}_\chi \otimes \Delta)$ a Hilbert space, although this is not required for the general theory. Details on what is required in terms of the function $G(\widetilde{p})$ for the inner product to be positive-definite are discussed in [5, 15]. A Hardy subspace of $L^2(X_{\mathbb{R}}, \mathbf{V}_\chi \otimes \Delta)$ then is the space $H^2(X_-, \mathbf{V}_\chi \otimes \Delta)$ consisting of all holomorphic sections of $\mathbf{V}_\chi \otimes \Delta$ over X_- with nontangential boundary values on $X_{\mathbb{R}}$ sitting in $L^2(X_{\mathbb{R}}, \mathbf{V}_\chi \otimes \Delta)$. The Hardy space $H^2(X_+, \mathbf{V}_\chi \otimes \Delta)$ of holomorphic sections of $\mathbf{V}_\chi \otimes \Delta$ over X_+ is defined similarly. A necessary condition for the inner product (2.3) to be nondegenerate on $H^2(X_-, \mathbf{V}_\chi \otimes \Delta)$ is that $h^0(X, \mathbf{V}_\chi \otimes \Delta) = 0$, i.e., that $\mathbf{V}_\chi \otimes \Delta$ have no global, holomorphic sections; in case (2.3) is positive definite on $L^2(X_{\mathbf{R}}, \mathbf{V}_\chi \otimes \Delta)$, then this condition $h^0(\mathbf{V}_\chi \otimes \Delta) = 0$ is automatic, and we have the orthogonal direct sum decomposition

$$L^2(X_{\mathbb{R}}, \mathbf{V}_\chi \otimes \Delta) = H^2(X_-, \mathbf{V}_\chi \otimes \Delta) \oplus H^2(X_+, \mathbf{V}_\chi \otimes \Delta).$$

One advantage (among others) of this formalism with half-order differentials is the simplicity of this decomposition: the orthogonal complement of the Hardy space $H^2(X_+, \mathbf{V}_\chi \otimes \Delta)$ over X_+ is exactly the Hardy space $H^2(X_-, \mathbf{V}_\chi \otimes \Delta)$ over X_-— one does not have to qualify with "and having a zero at infinity" as is necessary in the classical setting of the unit disk.

The spaces $H^2(X_-, \mathbf{V}_\chi \otimes \Delta)$ encompass the Abrahamse-Douglas spaces $H^2(R, \mathbf{V}_\chi)$ and it is natural to do operator theory on these spaces. In this setting there is no global uniformizer z for $X_+ \cong R$, as R is now an open Riemann surface of higher genus. Hence we have no analogue of the bundle shift over R, the central object of attention in [1, 2]. However there is a Beurling-Lax-Halmos theorem: *a subspace of $H^2(X_-, \mathbf{V}_{\chi_*} \otimes \Delta)$ invariant for M_f for all $f \in \mathcal{M}(X)$ with all poles in X_+ necessarily has the form $S \cdot H^2(X_-, \mathbf{V}_\chi \otimes \Delta)$ for an inner bundle map $S \in \mathcal{S}(\mathbf{V}_\chi, \mathbf{V}_{\chi_*})$.* Given such a bundle map S, we can introduce the associated model space

$$\mathcal{H}(S) = H^2(X_-, \mathbf{V}_{\chi_*} \otimes \Delta) \ominus S \cdot H^2(X_-, \mathbf{V}_\chi \otimes \Delta).$$

Also, we can pick out two meromorphic inner functions z_1, z_2 on X which generate the whole field $\mathcal{M}(X)$ of meromorphic functions on X as before, and then consider the pair of operators $(T_1(S), T_2(S))$ given by

$$T_j(S) = P_{\mathcal{H}(S)} M_{z_j}|_{\mathcal{H}(S)} \text{ for } j = 1, 2$$

as a model for a pair of commuting contraction operators. This is the proper context for the ($C_{\cdot 0}$ case of the) embedding problem mentioned above. These ideas give an approach to multivariable operator theory complementary to other recent approaches (see e.g. the book [21] for a collection of articles on a number of different approaches).

It is known by Ando's theorem [6] that any pair of commuting contractions has a commuting unitary dilation. In this model, one sees the unitary dilation (U_1, U_2) explicitly, namely,

$$U_j = M_{z_j} \text{ on } L^2(X_{\mathbb{R}}, \mathbf{V}_{\chi_*} \otimes \Delta) \text{ for } j = 1, 2.$$

The advantage is that one should be able to work out the geometry of the unitary dilation space in an explicit way analogous to the geometry of the unitary dilation in the Sz.-Nagy-Foiaş for the classical case. This model should provide an alternative formalism for the analysis of any number of multivariable operator theory questions, as well as for old operator-theoretic questions on multiply-connected domains, such as the spectral set question (see [4]).

Also the geometry of the unitary dilation space should relate to a multivariable Lax-Phillips scattering picture, where the inner (or more generally contractive) bundle map S plays the role of the scattering function, just as in the classical case (see [3]). In terms of the model based on an inner bundle map S, the ambient scattering space $\mathcal{K}$ for the scattering system is the whole space $L^2(X_{\mathbb{R}}, \mathbf{V}_{\chi_*} \otimes \Delta)$, there are two unitary evolution operators $U_1 = M_{z_1}$ and $U_2 = M_{z_2}$ on $\mathcal{K}$, the *incoming space* $\mathcal{G}_* = H^2(X_-, \mathbf{V}_\chi \otimes \Delta)$ is invariant under U_1^* and U_2^*, and the *outgoing subspace* $\mathcal{G} = S \cdot H^2(X_+, \mathbf{V}_\chi \otimes \Delta)$ is invariant under U_1 and U_2. These themes will also be developed in detail in [15]. A survey of a couple of other multivariable settings where the same triptych of system theory/operator model theory/Lax-Phillips scattering theory appears is given in [9].

Generalizations of the formalism to n-tuples of commuting operators with $n \geq 3$ have been worked out for the nonselfadjoint case (see [31]); a similar program should be possible here. When this is done, one arrives at a curve $\mathbf{C}$ embedded in n-dimensional complex projective space $\mathbb{P}^n$ leading to various algebro-geometric complications. It is known that, as a consequence of the failure of von Neumann's inequality in higher dimensions (see [38, 39]), Ando's theorem fails in general for the case of three or more commuting contraction operators; we expect that the algebro-geometric formalism here should give some insight into the general structure of this situation.

3. A concrete frequency domain

We now describe a more concrete model for the real Riemann surface X with vector bundle H having parahermitian pairing $H \times H^\tau \to K_X$ discussed in the previous section. We take X to be the normalizing Riemann surface for an algebraic curve $\mathbf{C}$ embedded in complex projective space $\mathbb{P}^2$. Thus there is an irreducible, homogeneous polynomial $f = f(\mu_0, \mu_1, \mu_2)$ (where $[\mu_0, \mu_1, \mu_2]$ is a generic point of $\mathbb{P}^2$) so that $\mathbf{C}$ is the zero set of f:

$$\mathbf{C} = \{[\mu_0, \mu_1, \mu_2] \in \mathbb{P}^2 \colon f(\mu_0, \mu_1, \mu_2) = 0\}.$$

To give us more structure which will naturally lead to a vector bundle, we assume that $\mathbf{C}$ (or f) has a *determinantal representation* $\mathbf{U}(\mu) = \mu_0\gamma + \mu_1\sigma_2 - \mu_2\sigma_1$, i.e., there is a positive integer r so that the polynomial f^r can be represented as

$$f(\mu)^r = \det \mathbf{U}(\mu).$$

Here σ_1, σ_2 and γ are operators between two Hilbert spaces $\mathcal{E}$ and $\widetilde{\mathcal{E}}$ assumed to be of the same finite dimension. In this situation there is associated in a natural way a vector bundle $\mathcal{E}$ over $\mathbf{C}$, in fact, a subbundle of the trivial bundle, given by

$$\mathcal{E} = \{(\mu, e) : \mathbf{U}(\mu)e = 0\}.$$

The *maximality assumption* is that the fiber of this bundle over the point μ

$$\mathcal{E}(\mu) = \{e \in \mathcal{E} : \mathbf{U}(\mu)e = 0\}$$

has maximal possible dimension r over all smooth points μ of $\mathbf{C}$. We also assume that all singular points of $\mathbf{C}$ are at worst nodes; this means that there is a surjective map $\pi : X \to \mathbf{C}$ from the normalizing Riemann surface X to the curve $\mathbf{C}$ which is holomorphic everywhere, injective over the smooth points of $\mathbf{C}$, and locally biholomorphic over the singular points $\mathbf{C}_{sing}$ of $\mathbf{C}$, with the preimage of any singular point consisting of only finitely many points. Under these assumptions, the bundle $\mathcal{E}$ over $\mathbf{C}$ naturally pulls back to a bundle over $\pi^{-1}(\mathbf{C}\backslash\mathbf{C}_{sing})$ which extends to a holomorphic vector bundle (also denoted as $\mathcal{E}$) of rank r over all of X. We will also have need of the left bundle $\mathcal{E}_\ell$ associated with the determinantal representation $\mathbf{U}(\mu)$ defined by

$$\mathcal{E}_\ell = \{(\mu, \widetilde{e}_\ell) : \mu \in \mathbf{C} \text{ and } \widetilde{e}_\ell \in \widetilde{\mathcal{E}}_\ell \text{ with } \widetilde{e}_\ell \mathbf{U}(\mu) = 0\}.$$

Here $\widetilde{\mathcal{E}}_\ell$ is the dual space of $\widetilde{\mathcal{E}}$, elements of which one can think of concretely as row vectors if one thinks of elements of $\widetilde{\mathcal{E}}$ as column vectors. As explained in [13, 14], there is a natural pairing

$$(\mathcal{E}_\ell \otimes \Delta \otimes \mathcal{O}(1)) \otimes (\mathcal{E} \otimes \Delta \otimes \mathcal{O}(1)) \to \mathcal{O}$$

given by

$$\{u_\ell, u\} = \frac{u_\ell}{\mu_0} \frac{\xi_1\sigma_1 + \xi_2\sigma_2}{\xi_1 dz_1 + \xi_2 dz_2} \frac{u}{\mu_0}. \tag{3.1}$$

Here u_ℓ and u are local holomorphic sections of $\mathcal{E}_\ell \otimes \mathcal{O}(1) \otimes \Delta$ and $\mathcal{E} \otimes \mathcal{O}(1) \otimes \Delta$ respectively near a point $p \in X$, and z_1 and z_2 are meromorphic functions on X given by $z_1 = \lambda_1 \circ \pi$, $z_2 = \lambda_2 \circ \pi$, and ξ_1, ξ_2 are arbitrary (not both zero) parameters, and the value $\{u, u_\ell\}$ of the pairing is a holomorphic scalar function near p. Here the independence of the expression (3.1) of the parameter (ξ_1, ξ_2) is related to the fact that $u(p)$ is restricted to be in $\ker \mathbf{U}(\mu(p))$ and $u_\ell(p)$ is restricted to be in $\ker_\ell \mathbf{U}(\mu(p))$. In particular, the special cases $(\xi_1, \xi_2) = (1, 0)$ and $(\xi_1, \xi_2) = (0, 1)$ give the two formulas for the same quantity

$$\{u_\ell, u\} = \frac{u_\ell}{\mu_0} \frac{\sigma_1}{dz_1} \frac{u}{\mu_0} = \frac{u_\ell}{\mu_0} \frac{\sigma_2}{dz_2} \frac{u}{\mu_0}.$$

We now wish to give X the structure of a *real* Riemann surface by having the antiholomorphic involution induced by the conjugate Cremona transformation ($(\lambda_1, \lambda_2) \to (1/\overline{\lambda_1}, 1/\overline{\lambda_2})$ in affine coordinates) restricted to the curve $\mathbf{C}$. This is analogous to the case developed in [31] and [40], where the reflection $(\lambda_1, \lambda_2) \to (\overline{\lambda_1}, \overline{\lambda_2})$ is used and one is led to a model theory for commuting pairs of nonselfadjoint operators. The goal here (and in [15]) is to take the antiholomorphic involution on the curve to be the conjugate Cremona transformation instead, and thereby arrive at a model theory for a commuting pair of nonunitary operators.

In projective coordinates, the Cremona transformation $(\lambda_1, \lambda_2) \to (1/\lambda_1, 1/\lambda_2)$ becomes the quadratic transformation

$$\mu = [\mu_0, \mu_1, \mu_2] \to \mu' = [\mu_1\mu_2, \mu_0\mu_2, \mu_0\mu_1]. \tag{3.2}$$

and the conjugate Cremona transformation is therefore

$$\mu = [\mu_0, \mu_1, \mu_2] \to \overline{\mu'} = [\overline{\mu_1\mu_2}, \overline{\mu_0\mu_2}, \overline{\mu_0\mu_1}]. \tag{3.3}$$

A complication for the theory here (not present in the case where the antiholomorphic involution is simply $\mu \to \overline{\mu}$ as in [31, 40]) is that the Cremona transformation has *base points*, i.e., points where it is not defined, namely the points $[1,0,0]$, $[0,1,0]$ and $[0,0,1]$. If the curve has degree m and is invariant under the Cremona transformation, then these base points necessarily have multiplicities i_0, i_1, i_2 such that $i_0 + i_1 + i_2 = m$; in particular, if $m > 3$, at least one of these multiplicities must be larger than 1, forcing at least one of these points to be a singular point on $\mathbf{C}$.

It is not obvious in terms of the coefficients $\sigma_1, \sigma_2, \gamma$ of the determinantal representation $\mathbf{U}(\mu)$ when the curve $\mathbf{C}$ is *unimodular*, i.e., invariant under the conjugate Cremona transformation. One way for this to happen is that there exists auxiliary operator pencils

$$\begin{aligned} \mathbf{A}(\mu) &= \mu_0 a_0 + \mu_1 a_1 + \mu_2 a_2 \colon \mathcal{E} \to \widetilde{\mathcal{E}}, \\ \widetilde{\mathbf{A}}(\mu) &= \mu_0 \widetilde{a}_0 + \mu_1 \widetilde{a}_1 + \mu_2 \widetilde{a}_2 \colon \widetilde{\mathcal{E}} \to \mathcal{E} \end{aligned}$$

so that we have the intertwining conditions

$$\begin{aligned} \mathbf{A}(\overline{\mu'})^* \mathbf{U}(\mu) &= \mathbf{U}(\overline{\mu'})^* \mathbf{A}(\mu), \\ \widetilde{\mathbf{A}}(\overline{\mu'})^* \mathbf{U}(\overline{\mu})^* &= \mathbf{U}(\mu') \widetilde{\mathbf{A}}(\mu). \end{aligned}$$

and

$$\begin{aligned} &\mathbf{A}(\mu) \text{ is injective on the fiber } \boldsymbol{\mathcal{E}}(\mu) \text{ for a generic } \mu \in \mathbf{C}, \\ &\widetilde{\mathbf{A}}(\mu) \text{ is injective on the fiber } \widetilde{\boldsymbol{\mathcal{E}}}_\ell(\mu)^* \text{ for a generic } \mu \in \mathbf{C}. \end{aligned}$$

Here $\widetilde{\boldsymbol{\mathcal{E}}}_\ell^*$ is simply the kernel bundle associated with the matrix pencil $\mathbf{U}^*(\mu) = \mathbf{U}(\overline{\mu})^*$:

$$\widetilde{\boldsymbol{\mathcal{E}}}_\ell(\overline{\mu})^* = \{\widetilde{e} \in \widetilde{\mathcal{E}} \colon \mathbf{U}(\overline{\mu})^* \widetilde{e} = 0\}.$$

In this situation, it follows that $\mathbf{A}(\mu)$ maps $\boldsymbol{\mathcal{E}}(\mu)$ into $\widetilde{\boldsymbol{\mathcal{E}}}_\ell(\overline{\mu'})^*$, or, in terms of the variable p on the normalizing Riemann surface X, $\mathbf{A}(\mu(p))$ maps $\boldsymbol{\mathcal{E}}(p)$ into $\widetilde{\boldsymbol{\mathcal{E}}}_\ell(p^\tau)^*$. In particular, it follows that $\overline{\mu'}$ is on $\mathbf{C}$ whenever μ is on $\mathbf{C}$, i.e., $\mathbf{C}$ is unimodular. Moreover, for u and v local sections of $\boldsymbol{\mathcal{E}}$ near a point p_0 and its conjugation p_0^τ respectively, $(\mathbf{A}(\overline{\mu'})v(p^\tau))^* = v(p^\tau)^* \mathbf{A}(\overline{\mu'})^*$ is a local section of $\widetilde{\boldsymbol{\mathcal{E}}}_\ell$ near p_0, and the pairing (3.1) makes sense with $u(p)$ and $u_\ell(p) = v(p^\tau)^* \mathbf{A}(\overline{\mu'})^*$. After a certain amount of algebra (see [15]), one is led to the pairing (in affine coordinates)

$$\begin{aligned}[u,v](p) &= \{v(p^\tau)^*\mathbf{A}(\overline{\mu'})^*, u(p)\} \\ &= \frac{1}{i}v(p^\tau)^*\frac{\xi_1\Sigma_1+\xi_2\Sigma_2}{\xi_1(dz_1(p)/z_1(p))+\xi_2(dz_2(p)/z_2(p))}u(p) \qquad (3.4)\end{aligned}$$

where Σ_1 and Σ_2 are given by

$$\Sigma_1 = a_2^*\sigma_1 + \sigma_1^*a_2, \qquad \Sigma_2 = a_1^*\sigma_2 + \sigma_2^*a_1. \qquad (3.5)$$

If we take u and v to be local sections of $\mathcal{E}\otimes K\otimes\mathcal{O}(1)$ near p_0 and p_0^τ respectively, the result $[u,v](p)$ is a local section of K near p_0, i.e., a 1-form near p_0, and (up to unimportant twists which can be taken care of), we have recovered the situation of the previous section in a more concrete form: namely a real Riemann surface X (here as the normalizing Riemann surface of the curve $\mathbf{C}$) together with a vector bundle H (of degree $g-1$ and no holomorphic sections after twisting with the appropriate line bundles) having a parahermitian pairing. This means that we can do Hardy space function theory in this more concrete setting, with the bundle $\mathbf{V}_\chi$ based on the factor of automorphy χ replaced (up to appropriate twists) with a kernel bundle $\mathcal{E}$ equal to a subbundle of the trivial bundle over the curve $\mathbf{C}$.

In detail, we define an inner product on the space $L^2(X_\mathbb{R},\mathcal{E}\otimes K\otimes\mathcal{O}(1))$ of norm-square integrable sections $\mathbf{u}$ and $\mathbf{v}$ of $\mathcal{E}\otimes K\otimes\mathcal{O}(1)$ over $X_\mathbb{R}$ by

$$\begin{aligned}\langle\mathbf{u},\mathbf{v}\rangle &= \int_{X_\mathbb{R}}[\mathbf{u},\mathbf{v}](p) \\ &= \int_{X_\mathbb{R}}\mathbf{v}(p^\tau)^*\Sigma_1\mathbf{u}(p)\frac{z_1(p)}{idz_1(p)} \qquad (3.6)\\ &= \int_{X_\mathbb{R}}\mathbf{v}(p^\tau)^*\Sigma_2\mathbf{u}(p)\frac{z_2(p)}{idz_2(p)}. \qquad (3.7)\end{aligned}$$

This is an in general indefinite inner product on $L^2(X_\mathbb{R},\mathcal{E}\otimes K\otimes\mathcal{O}(1))$. Although it is not necessary for a lot of the general theory, we shall assume henceforth that *this inner product is positive definite on* $L^2(X_\mathbb{R},\mathcal{E}\otimes K\otimes\mathcal{O}(1))$. A sufficient condition for this to happen is that X, as a real Riemann surface with antiholomorphic involution equal to the lift of the conjugate Cremona transformation $\mu\to\overline{\mu'}$ on $\mathbf{C}$, is of dividing type (so $X\backslash X_\mathbb{R}$ has two connected components X_- and X_+), and the functions $p\to z_1(p)$ and $p\to z_2(p)$ corresponding to coordinate evaluation in affine coordinates ($z_j(p)=\lambda_j$ for $j=1,2$ if $\pi(p)=(\lambda_1,\lambda_2)$ in affine coordinates) are inner. Once this is done, we may define the Hardy subspaces $H^2(X_+,\mathcal{E}\otimes K\otimes\mathcal{O}(1))$ and $H^2(X_-,\mathcal{E}\otimes K\otimes\mathcal{O}(1))$ to be the subspace of elements $\mathbf{u}(p)\in L^2(X_\mathbb{R},\mathcal{E}\otimes K\otimes\mathcal{O}(1))$ which arise as the almost-everywhere existing boundary-value differential of a section of $\mathcal{E}\otimes K\otimes\mathcal{O}(1)$ which is holomorphic over X_+ (respectively, holomorphic over X_- except for at worst simple poles at the poles of z_1 and z_2). Then the space $L^2(X_\mathbb{R},\mathcal{E}\otimes K\otimes\mathcal{O}(1))$ is a Hilbert space in the inner product (3.7) (rather than a Krein space as in the more general situation) with orthogonal decomposition

$$L^2(X_\mathbb{R},\mathcal{E}\otimes K\otimes\mathcal{O}(1)) = H^2(X_-,\mathcal{E}\otimes K\otimes\mathcal{O}(1))\oplus H^2(X_+,\mathcal{E}\otimes K\otimes\mathcal{O}(1)).$$

So far, the other pencil $\widetilde{\mathbf{A}}(\mu) = \mu_0\widetilde{a}_0 + \mu_1\widetilde{a}_1 + \mu_2\widetilde{a}_2$ has not played a role; this comes in when one seeks to set up a Hilbert space inner product associated with the dual bundle $\widetilde{\boldsymbol{\mathcal{E}}}$.

In this way, we have arrived at a model for the more abstract situation discussed in the previous section. Moreover, it is possible to give explicit formulas for mapping either picture over to the other. For the more general affine setting (where the Riemann surface is not equipped with an antiholomorphic involution, the curve need not be unimodular and the bundles do not come with a parahermitian pairing), this correspondence (involving objects called "Cauchy kernels" and "matrices of normalized sections of $\boldsymbol{\mathcal{E}}$") is laid out in [14]; complete details for the more structured parahermitian setting will appear in [15].

We may therefore consider operator theory problems (e.g., models for commuting pairs of contraction operators) in this more concrete kernel bundle setting. It is natural to consider another paraunitary determinantal representations $\mathbf{U}_*(\mu) = \mu_0\gamma_* + \mu_1\sigma_{2*} - \mu_2\sigma_{1*} \colon \boldsymbol{\mathcal{E}} \to \widetilde{\mathcal{E}}_*$ with associated parahermitian intertwining pencils

$$\mathbf{A}_*(\mu) = \mu_0 a_{0*} + \mu_1 a_{1*} + \mu_2 a_{2*} \colon \mathcal{E}_* \to \widetilde{\mathcal{E}}_*,$$
$$\widetilde{\mathbf{A}}_*(\mu) = \mu_0 \widetilde{a}_{0*} + \mu_1 \widetilde{a}_{1*} + \mu_2 \widetilde{a}_{2*} \colon \widetilde{\mathcal{E}}_* \to \mathcal{E}_*$$

satisfying the intertwining conditions

$$\mathbf{A}_*(\overline{\mu'})^*\mathbf{U}_*(\mu) = \mathbf{U}_*(\overline{\mu'})^*\mathbf{A}_*(\mu)$$
$$\widetilde{\mathbf{A}}_*(\overline{\mu'})^*\mathbf{U}_*(\overline{\mu})^* = \mathbf{U}_*(\mu')\widetilde{\mathbf{A}}_*(\mu)$$

in addition to $(\mathbf{U}(\mu), \mathbf{A}(\mu), \widetilde{\mathbf{A}}(\mu))$ already discussed, and a meromorphic bundle map $\mathbf{S}\colon \boldsymbol{\mathcal{E}} \to \boldsymbol{\mathcal{E}}_*$. In [13] it is shown that any such bundle map $\mathbf{S}$ arises as multiplication by a certain function in "state-space realization" form restricted to the fibers of the bundle, namely,

$$\mathbf{S}(p) = \left[D + C\left(\xi_1[z_1(p) - A_1] + \xi_2[z_2(p) - A_2]\right)^{-1}\left(\xi_1\widetilde{B}_1\sigma_1 + \xi_2\widetilde{B}_2\sigma_2\right)\right]\Big|_{\boldsymbol{\mathcal{E}}(p)} \tag{3.8}$$

Here ξ_1 and ξ_2 are free parameters (not both zero) and A_1, A_2, $\widetilde{B}_1$, $\widetilde{B}_2$, C and D are operators forming an operator

$$\begin{bmatrix} A_1 & \widetilde{B}_1\sigma_1 \\ A_2 & \widetilde{B}_2\sigma_2 \\ C & D \end{bmatrix} \colon \begin{bmatrix} \mathcal{H} \\ \mathcal{E} \end{bmatrix} \to \begin{bmatrix} \mathcal{H} \\ \mathcal{H} \\ \mathcal{E}_* \end{bmatrix} \tag{3.9}$$

(so $\widetilde{B}_1$ and $\widetilde{B}_2$ map $\widetilde{\mathcal{E}}$ to $\mathcal{H}$). Bundle mappings of the form (3.8) can be interpreted as the *transfer function* of the overdetermined 2D system

$$\begin{aligned} x(t_1+1, t_2) &= A_1 x(t_1, t_2) + \widetilde{B}_1\sigma_1 u(t_1, t_2) \\ x(t_1, t_2+1) &= A_2 x(t_1, t_2) + \widetilde{B}_2\sigma_2 u(t_1, t_2) \\ y(t_1, t_2) &= C x(t_1, t_2) + D u(t_1, t_2). \end{aligned} \tag{3.10}$$

Here $\mathcal{H}$ has the interpretation of the *state space* for the system, while $\mathcal{E}$ and $\mathcal{E}_*$ are the *input* and *output* spaces respectively, and the "time" variable (t_1, t_2) is a 2-parameter discrete-time variable residing in $\mathbb{Z} \times \mathbb{Z}$. One way of making precise this interpretation of $\mathbf{S}(p)$ as a transfer function of the system (3.10) is given in [13]; another is via the Z-transform along the curve $\mathbf{C}$ which we present in the next section.

For our application to operator model theory, we are interested in the case where the bundle map $\mathbf{S}$ is *contractive* as an operator from $H^2(X_-, \boldsymbol{\mathcal{E}} \otimes K \otimes \mathcal{O}(1))$ to $H^2(X_-, \boldsymbol{\mathcal{E}}_* \otimes K \otimes \mathcal{O}(1))$, or even *isometric* (corresponding to $\mathbf{S}$ being an *inner* bundle map). This can be characterized in terms of special properties of the pair of paraunitary determinantal representations $(\mathbf{U}, \mathbf{A}, \widetilde{\mathbf{A}})$ and $(\mathbf{U}_*, \mathbf{A}_*, \widetilde{\mathbf{A}}_*)$ for $\mathbf{C}$ and of the system operator matrix (3.9); in particular, a key property is an *energy conservation* property analogous to but more complicated than (1.4) for the classical case. This issue is one of the main themes of [15], where a pair of paraunitary determinantal representations for the curve $\mathbf{C}$ together with a system matrix (3.9) having the aforementioned special properties is dubbed a *quasiunitary vessel.* The parallel theory for the case where the antiholomorphic involution on the curve $\mathbf{C}$ is taken to be simply conjugation $\mu \to \overline{\mu}$ (where the appropriate 2D overdetermined linear system is a "quasi-Hermitian vessel" and one is modeling a pair of nonselfadjoint operators) is worked out in detail in [31] (see also [40]).

4. The Z-transform along a curve

In the previous sections, we have recovered analogues of most of the ideas present in the classical setting of function theory on the unit disk (namely, Hardy-space orthogonal decomposition of a Lebesgue space, operator model theory, unitary dilation and scattering, Schur-class function, Beurling-Lax-Halmos theorem) in the more intricate real Riemann surface/unimodular algebraic curve setting, with the exception of the notion of Fourier coefficients or inverse Fourier transform from the frequency-domain Hardy space setting to a time-domain ℓ^2 setting. We now indicate how a useful analogue of these powerful ideas exists in our more general setting as well.

Suppose that $\mathbf{u}$ is an element of the space $L^2(X_{\mathbb{R}}, \boldsymbol{\mathcal{E}} \otimes K \otimes \mathcal{O}(1))$ as defined in the previous section. We define the associated Fourier coefficient $u^{\vee}(t_1, t_2)$ of $\mathbf{u}$ at the point $(t_1, t_2) \in \mathbb{Z} \times \mathbb{Z}$ by

$$u(t_1, t_2) := \mathbf{u}^{\vee}(t_1, t_2) = \frac{1}{2\pi i} \int_{X_{\mathbb{R}}} z_1(p)^{t_1} z_2(p)^{t_2} \mathbf{u}(p). \tag{4.1}$$

As $\mathbf{u}$ is a section of $\boldsymbol{\mathcal{E}}$ (up to twisting by $K \otimes \mathcal{O}(1)$), we know that locally $\mathbf{u}(p)$ satisfies

$$z_1(p)\sigma_2 \mathbf{u}(p) - z_2(p)\sigma_1 \mathbf{u}(p) + \gamma \mathbf{u}(p) = 0.$$

Multiplying this relation by $z_1(p)^{t_1}z_2(p)^{t_2}$ and integrating over $X_{\mathbb{R}}$ then gives that $u(t_1,t_2) = \mathbf{u}^\vee(t_1,t_2)$ satisfies the difference equation

$$\sigma_2 u(t_1+1,t_2) - \sigma_1 u(t_1,t_2+1) + \gamma u(t_1,t_2) = 0. \tag{4.2}$$

Given any such $\mathcal{E}$-valued function u on $\mathbb{Z}\times\mathbb{Z}$ satisfying (4.2), let us suppose that $u(\cdot,t_2)$ is norm-square summable for each fixed $t_2 \in \mathbb{Z}$ and consider the usual Z-transform of u in the first variable:

$$\widehat{u}^1(\zeta_1,t_2) = \sum_{t_1=-\infty}^{\infty} \zeta_1^{-t_1} u(t_1,t_2). \tag{4.3}$$

Assuming that $u(\cdot,t_2) \in \ell^2$ for each $t_2 \in \mathbb{Z}$, then $\widehat{u}^1(\cdot,t_2) \in L^2(\mathbb{T},\mathcal{E})$ for each $t_2 \in \mathbb{Z}$ and we recover u from $\widehat{u}^1$ via the inversion formula

$$u(t_1,t_2) = \frac{1}{2\pi i}\int_{\mathbb{T}} \zeta_1^{t_1}\widehat{u}^1(\zeta_1,t_2)\frac{d\zeta_1}{\zeta_1}. \tag{4.4}$$

In view of (4.2) we see that $\widehat{u}^1$ satisfies

$$\zeta_1\sigma_2\widehat{u}^1(\zeta_1,t_2) - \sigma_1\widehat{u}^1(\zeta_1,t_2+1) + \gamma\widehat{u}^1(\zeta_1,t_2) = 0. \tag{4.5}$$

The difference equation (4.5) (in the variable t_2 with parameter ζ) can be solved via the method of trial solutions analogous to the more familiar theory for ordinary differential equations; one supposes a solution to be of the form $\lambda_0(\zeta_1)^{t_2}\widehat{u}_0(\zeta_1)$ to arrive at

$$[\zeta_1\lambda(\zeta_1)^{t_2}\sigma_2 - \lambda(\zeta_1)\cdot\lambda(\zeta_1)^{t_2}\sigma_1 + \lambda(\zeta_1)^{t_2}\gamma]\widehat{u}(\zeta_1) = 0.$$

Cancelling off the common factor of $\lambda(\zeta_1)^{t_2}$ then leaves us with the eigenvector-eigenvalue problem for $(u_0(\zeta_1),\lambda_0(\zeta_1))$:

$$[\sigma_2\zeta_1 + \gamma]\widehat{u}_0(\zeta_1) = \lambda(\zeta_1)\sigma_1\widehat{u}(\zeta_1). \tag{4.6}$$

We suppose that each eigenvalue $\lambda_1(\zeta_1),\dots,\lambda_{m-i_2}(\zeta_1)$ is semisimple with multiplicity r and that

$$\widehat{u}_{i,1}(\zeta_1),\dots,\widehat{u}_{i,r}(\zeta_1)$$

forms a basis for the eigenspace associated with eigenvalue $\lambda_i(\zeta_1)$ for $i = 1,\dots,m-i_2$. Thus

$$[\zeta_1\sigma_2 - \lambda_i(\zeta_1)\sigma_1 + \gamma]\widehat{u}_{i,j}(\zeta_1) = 0,$$

or, $(\zeta_1,\lambda_i(\zeta_1)) \in \mathbf{C}_0$ and $\widehat{u}_{i,j}(\zeta_1) \in \boldsymbol{\mathcal{E}}(\zeta_1,\lambda_i(\zeta_1))$ for $i = 1,\dots,m-i_2$ and $j = 1,\dots,r$. The solution $\widehat{u}^1(\zeta_1,t_2)$ of (4.5), by the general theory of linear difference equations, must be a linear combination of such special solutions, so we may write

$$\widehat{u}^1(\zeta_1,t_2) = \sum_{i=1}^{m-i_2} \widehat{u}_i(\zeta_1)\lambda_i(\zeta_1)^{t_2} \tag{4.7}$$

where $\widehat{u}_i(\zeta_1)$ is a vector in the eigenspace associated with eigenvalue $\lambda_i(\zeta_1)$. From this representation, we see in particular that

$$\widehat{u}^1(\zeta_1,0) = \sum_{i=1}^{m-i_2} \widehat{u}_i(\zeta_1).$$

In the notation of Proposition 2.5 from [13] it follows that

$$\widehat{u}_i(\zeta_1) = P(1,0,(\zeta_1,\lambda_i(\zeta_1)))\widehat{u}^1(\zeta_1,0) \tag{4.8}$$

where $P(1,0,(\zeta_1,\lambda_i(\zeta_1)))$ is the projection of $\mathcal{E}$ onto $\boldsymbol{\mathcal{E}}(\zeta_1,\lambda_i(\zeta_1))$ along the span of the spaces $\boldsymbol{\mathcal{E}}(\lambda^\ell)$, where λ^ℓ is any other point of $\mathbf{C}$ other than $(\zeta_1,\lambda_i(\zeta_1))$ which intersects the line $\lambda_1 = \zeta_1$.

Note that $z_1^{-1}(\mathbb{T}) = X_{\mathbb{R}}$ and, for a fixed point $\zeta_1 \in \mathbb{T}$, we have

$$z_1^{-1}(\zeta_1) = \pi^{-1}\{(\zeta_1,\lambda_i(\zeta_1)) \colon i = 1,\dots,m-i_0\}.$$

We can then use the change of variable $\zeta_1 \to z_1^{-1}(\zeta_1)$ to rewrite the inversion formula (4.4) as

$$\begin{aligned} u(t_1,t_2) &= \frac{1}{2\pi i}\int_{|\zeta_1|=1} \zeta_1^{t_1}\widehat{u}^1(\zeta_1,t_2)\frac{d\zeta_1}{\zeta_1} \\ &= \frac{1}{2\pi i}\int_{|\zeta_1|=1} \sum_{i=1}^{m-i_2} \widehat{u}_i(\zeta_1)\zeta_1^{t_1}\lambda_i(\zeta_1)^{t_2}\frac{d\zeta_1}{\zeta_1} \\ &= \frac{1}{2\pi}\int_{X_{\mathbb{R}}} z_1(p)^{t_1} z_2(p)^{t_2}\widehat{u}^{(1,0)}(p)\frac{dz_1}{iz_1} \end{aligned} \tag{4.9}$$

where we have set

$$\widehat{u}^{(1,0)}(p) = P(1,0,p)\sum_{t_1=-\infty}^{\infty} z_1(p)^{-t_1}u(t_1,0).$$

We shall also need the analogous expression

$$\widehat{u}^{(0,1)}(p) = P(0,1,p)\sum_{t_1=-\infty}^{\infty} z_2(p)^{-t_2}u(0,t_2).$$

By interchanging the roles of the first and second variables in the analysis above, we also recover $u(t_1,t_2)$ as

$$u(t_1,t_2) = \frac{1}{2\pi}\int_{X_{\mathbb{R}}} z_1(p)^{t_1} z_2(p)^{t_2}\widehat{u}^{(0,1)}(p)\frac{dz_2(p)}{iz_2(p)}.$$

Thus the differential $\widehat{u}^{(1,0)}(p)\frac{dz_1}{iz_1} - \widehat{u}^{(0,1)}(p)\frac{dz_2(p)}{iz_2(p)}$ restricted to $X_{\mathbb{R}}$ annihilates all monomials $z_1(p)^{t_1}z_2(p)^{t_2}$ for $t_1,t_2 \in \mathbb{Z}$, and hence must be zero. We define the Z-transform (along the curve C with paraunitary determinantal representation $(\mathbf{U}(\mu),\mathbf{A}(\mu),\widetilde{\mathbf{A}}(\mu)))$ as the common value

$$\widehat{u}(p) = \widehat{u}^{(1,0)}(p)\frac{dz_1(p)}{iz_1(p)} = \widehat{u}^{(0,1)}(p)\frac{dz_2(p)}{iz_2(p)}. \tag{4.10}$$

To give the precise setting needed for our purposes, let us now introduce some spaces. We assume that we are given two linear pencils $\mathbf{U}(\mu) = \mu_0\gamma + \mu_1\sigma_2 - \lambda_2\sigma_1$ and $\mathbf{A}(\mu) = a_0\mu_0 + \mu_1 a_1 + \mu_2 a_2$ (where the coefficients $\gamma,\sigma_1,\sigma_2,a_0,a_1,a_2$ are bounded linear operators between two finite dimensional Hilbert spaces $\mathcal{E}$ and $\widetilde{\mathcal{E}}$) such that $(\mathbf{U}(\mu),\mathbf{A}(\mu))$ is a paraunitary determinantal representation for a curve

C—the other matrix pencil $\widetilde{\mathbf{A}}(\mu)$ in the definition of paraunitary determinantal representation is not needed for present purposes). We then define operators Σ_1 and Σ_2 by

$$\Sigma_1 = a_2^*\sigma_1 + \sigma_1^* a_2, \qquad \Sigma_2 = a_1^*\sigma_2 + \sigma_2^* a_1.$$

We shall assume that Σ_1 and Σ_2 are positive semidefinite (but not necessarily strictly positive definite). We define the space $\ell^2_{\Sigma_1}(\mathbb{Z},\mathcal{E})$ to consist of all sequences $\{x(t)\}_{t\in\mathbb{Z}}$ such that

$$\sum_{t=-\infty}^{\infty} \langle \Sigma_1 x(t), x(t)\rangle < \infty.$$

The subspaces $\ell^2_{\Sigma_1}(\mathbb{Z}_+,\mathcal{E})$ and $\ell^2_{\Sigma_1}(\mathbb{Z}_-,\mathcal{E})$ are defined as the subspace of $\ell^2_{\Sigma_1}(\mathbb{Z},\mathcal{E})$ consisting of all elements u of $\ell^2_{\Sigma_1}(\mathbb{Z},\mathcal{E})$ such $u(t) \in \ker \Sigma_1$ for $t \in \mathbb{Z}_+$ ($t \in \mathbb{Z}_-$, respectively). The spaces $\ell^2_{\Sigma_2}(\mathbb{Z},\mathcal{E})$, $\ell^2_{\Sigma_2}(\mathbb{Z}_+,\mathcal{E})$ and $\ell^2_{\Sigma_2}(\mathbb{Z}_-,\mathcal{E})$ are defined analogously, with Σ_2 in place of Σ_1. Define linear spaces $\ell^2_{\mathbf{U}}$, $\ell^2_{\mathbf{U},+}$ and $\ell^2_{\mathbf{U},-}$ by

$$\begin{aligned}
\ell^2_{\mathbf{U}} &= \{u\colon \mathbb{Z}\times\mathbb{Z} \to \mathcal{E} \text{ such that } \sigma_2 u(t_1+1,t_2) - \sigma_1 u(t_1,t_2+1) + \gamma u(t_1,t_2) = 0,\\
&\qquad u(\cdot,0) \in \ell^2_{\Sigma_1}(\mathbb{Z},\mathcal{E}) \text{ and } u(0,\cdot) \in \ell^2_{\Sigma_2}(\mathbb{Z},\mathcal{E})\}\\
\ell^2_{\mathbf{U},+} &= \{u \in \ell^2_{\mathbf{U}}\colon u(\cdot,0) \in \ell^2_{\Sigma_1}(\mathbb{Z}_+,\mathcal{E})\}\\
\ell^2_{\mathbf{U},-} &= \{u \in \ell^2_{\mathbf{U}}\colon u(\cdot,0) \in \ell^2_{\Sigma_1}(\mathbb{Z}_-,\mathcal{E})\}
\end{aligned} \tag{4.11}$$

There are still quite a few details to be filled in, but the following theorem assures us that the special role played by the first variable in these definitions is only apparent, and provides the Plancherel theorem for our Z-transform $\wedge$ along the curve **C**; complete details will appear in [15].

Theorem 4.1. *Let the Z-transform $u \to \widehat{u}$ be defined as in* (4.10), *and let the spaces $\ell^2_{\mathbf{U}}$, $\ell^2_{\mathbf{U},+}$ and $\ell^2_{\mathbf{U},-}$ be defined as in* (4.11). *Then:*

1. *The spaces $\ell^2_{\mathbf{U},+}$ and $\ell^2_{\mathbf{U},-}$ can be defined equivalently as*

$$\begin{aligned}
\ell^2_{\mathbf{U},+} &= \{u \in \ell^2_{\mathbf{U}}\colon u(0,\cdot) \in \ell^2_{\Sigma_2}(\mathbb{Z}_+,\mathcal{E})\}\\
\ell^2_{\mathbf{U},-} &= \{u \in \ell^2_{\mathbf{U}}\colon u(0,\cdot) \in \ell^2_{\Sigma_2}(\mathbb{Z}_-,\mathcal{E})\}.
\end{aligned}$$

Furthermore, $\ell^2_{\mathbf{U}}$ is a Hilbert space in a well-defined norm given by

$$\|u\|^2 = \sum_{t_1=-\infty}^{\infty} \langle \Sigma_1 u(t_1,t_2), u(t_1,t_2)\rangle = \sum_{t_2=-\infty}^{\infty} \langle \Sigma_2 u(t_1,t_2), u(t_1,t_2)\rangle$$

with $\ell^2_{\mathbf{U},+}$ and $\ell^2_{\mathbf{U},-}$ equal to Hilbert subspaces.

2. *The Z-transform $u \to \widehat{u}$ is a unitary transformation of $\ell^2_{\mathbf{U}}$ onto $L^2(X_{\mathbb{R}}, \boldsymbol{\mathcal{E}} \otimes K \otimes \mathcal{O}(1))$ (with inner product equal to the negative of* (3.7)*) which takes $\ell^2_{\mathbf{U},+}$ onto $H^2(X_-, \boldsymbol{\mathcal{E}} \otimes K \otimes \mathcal{O}(1))$ and $\ell^2_{\mathbf{U},-}$ onto $H^2(X_+, \boldsymbol{\mathcal{E}} \otimes K \otimes \mathcal{O}(1))$ respectively, with inverse given by*

$$\mathbf{u}(p) \to u(t_1,t_2) = \mathbf{u}^{\vee}(t_1,t_2) = \frac{1}{2\pi}\int_{X_{\mathbb{R}}} z_1(p)^{t_1} z_2(p)^{t_2} \mathbf{u}(p). \tag{4.12}$$

3. *Each equivalence class* $[\varphi] \in \ell^2_{\Sigma_1}(\mathbb{Z}, \mathcal{E})$ *has a unique representative* φ' *so that* $\varphi'(t_1) = u(t_1, 0)$ *for a unique* $u \in \ell^2_{\mathbf{U}}$. *If* $[\varphi] \in \ell^2_{\Sigma_1}(\mathbb{Z}_+, \mathcal{E})$, *then* $u(t_1, t_2) = 0$ *for both* $t_1 < 0$ *and* $t_2 < 0$ *(but not necessarily for* $t_1 < 0$ *and* $t_2 = 0$ *or for* $t_1 = 0$ *and* $t_2 < 0$*). If* $[\varphi] \in \ell^2_{\Sigma_1}(\mathbb{Z}_-, \mathcal{E})$, *then* $\varphi'(t_1) = 0$ *for* $t_1 \geq 0$, *and* $u(t_1, t_2) = 0$ *for* $t_1 \geq 0$ *and* $t_2 \geq 0$. *Parallel results hold for* $\ell^2_{\Sigma_2}(\mathbb{Z}, \mathcal{E})$ *with the roles of the variables* t_1 *and* t_2 *interchanged.*

This Z-transform enables us to give a transfer-function interpretation to the formula (3.8). Consider the overdetermined linear system (3.10). Under the assumption that A_1 and A_2 commute, it can be shown that the system equations (3.10) are consistent as long as the input signal $u(t_1, t_2)$ satisfies the compatibility difference equation (4.2), and then the resulting output $y(t_1, t_2)$ satisfies the corresponding difference equation on the output side

$$\sigma_{2*} y(t_1 + 1, t_2) - \sigma_{1*} y(t_1, t_2 + 1) + \gamma_* y(t_1, t_2) = 0. \tag{4.13}$$

If the system operators A_1, A_2, $\widetilde{B}_1$, $\widetilde{B}_2$, C, D form part of a quasiunitary vessel with the paraunitary determinantal representations $(\mathbf{U}, \mathbf{A}, \widetilde{\mathbf{A}})$ and $(\mathbf{U}_*, \mathbf{A}_*, \widetilde{\mathbf{A}}_*)$ for the curve $\mathbf{C}$, then in fact the recurrence relations (3.10) can be used to define a well-defined input-output operator $T_\Sigma \colon \ell^2_{\mathbf{U}} \to \ell^2_{\mathbf{U}_*}$. We close this brief survey with the following result which makes precise the sense in which $\mathbf{S}(p)$ in (3.8) is the *transfer function* of the system (3.10); the proof follows in a straightforward way from applying the Z-transform to each side of the system equations (3.10) and solving for $y^\wedge(p)$ in terms of $u^\wedge(p)$, just as in the classical case.

Theorem 4.2. *Let* T_Σ *be the input-output operator for the system* (3.10) *associated with a quasiunitary vessel as above, and let* $\mathbf{S}(p)$ *be the bundle map as in* (3.8). *Then*

$$(T_\Sigma[u])^\wedge(p) = \mathbf{S}(p) \cdot u^\wedge(p).$$

References

[1] M. B. Abrahamse, and R. G. Douglas, *A class of subnormal operators related to multiply connected domains*, Advances in Math., **19** (1976), 106–148.

[2] M. B. Abrahamse and R. G. Douglas, *Operators on multiply connected domains*, Proc. of the Royal Irish Acad., **74A** (1974), 135–141.

[3] V. M. Adamjan and D. Z. Arov, *On unitary couplings of semiunitary operators*, Dokl. Akad. Nauk Arm. SSR, **XLIII** (1966) [in Russian]; English transl., Amer. Math. Soc. Transl., **95** (1970), 75–129.

[4] J. Agler, *Rational dilation on an annulus*, Annals of Mathematics, **121** (1985), 537–563.

[5] D. Alpay and V. Vinnikov, *Indefinite Hardy spaces on finite bordered Riemann surfaces*, J. Functional Analysis, **172** (2000), 221–248.

[6] T. Ando, *On a pair of commutative contractions*, Acta Sci. Math., **24** (1963), 88–90.

[7] A. C. Antoulas (ed.), *Mathematical System Theory: The Influence of R. E. Kalman*, Springer-Verlag, Berlin–New York, 1991.

[8] J. A. Ball, *Operators of class C_{00} over multiply-connected domains*, Mich. Math. J., **25** (1978), 183–196.

[9] J. A. Ball, *Linear systems, operator model theory and scattering: multivariable generalizations*, Operator Theory and its Applications (Winnipeg, MB, 1998), 151-178, Fields Inst. Commun., **25**, Amer. Math. Soc., Providence, RI, 2000.

[10] J. A. Ball and K. Clancey, *Reproducing kernels for Hardy spaces on multiply connected domains*, Integral Equations and Operator Theory, **25** (1996), 35–57.

[11] J. A. Ball and N. Cohen, *De Branges–Rovnyak operator models and systems theory: a survey*, Topics in Matrix and Operator Theory (H. Bart, I. Gohberg, and M. A. Kaashoek, eds.), OT 50, Birkhäuser Verlag, Basel, 1991, pp. 93–136.

[12] J. A. Ball and T. T. Trent, *The abstract interpolation problem and commutant lifting: a coordinate-free approach*, Operator Theory and Interpolation: International Workshop on Operator Theory and Applications, IWOTA96 (Ed. H. Bercovici, C. Foias), OT 115, Birkhäuser-Verlag, Basel–Boston, 2000, pp. 51–83.

[13] J. A. Ball and V. Vinnikov, *Zero-pole interpolation for meromorphic matrix functions on an algebraic curve and transfer functions of 2D systems*, Acta Applicandae Mathematicae, **45** (1996) 239–316.

[14] J. A. Ball and V. Vinnikov, *Zero-pole interpolation for matrix meromorphic functions on a compact Riemann surface and a matrix Fay trisecant identity*, American J. Math., **121** (1999), 841–888.

[15] J. A. Ball and V. Vinnikov, *Multidimensional discrete-time systems, algebraic curves and commuting nonunitary operators*, in preparation.

[16] M. S. Brodskii, *Triangular and Jordan representations of linear operators*, Transl. Math. Monographs, vol. 32, Amer. Math. Soc., Providence, 1971.

[17] M. S. Brodskiĭ, *Unitary operator colligations and their characteristic functions*, Russian Math. Surveys, **33** (1978), 159–191.

[18] M. S. Brodskii and M. S. Livšic, *Spectral analysis of nonselfadjoint operators and intermediate systems*, Amer. Math. Soc. Transl. Ser. 2, **13** (1960), 265–346.

[19] L. de Branges and J. Rovnyak, *Canonical models in quantum scattering theory*, Perturbation Theory and its Applications in Quantum Mechanics (C. Wilcox, ed.), Wiley, New York, 1966, pp. 295–392.

[20] K. F. Clancey, *Toeplitz operators on multiply connected domains and theta functions*, Contributions to Operator Theory and its Applications (Mesa, AZ, 1987) (Ed. I. Gohberg, J. W. Helton and L. Rodman), 311–355, OT35 Birkhäuser-Verlag, Basel–Boston, 1988.

[21] R. E. Curto, R. G. Douglas, J. D. Pincus and N. Salinas (Editors), *Multivariable Operator Theory: Papers from the Joint Summer Research Conference held at the University of Washington, Seattle, Washington, July 10–18, 1993*, Contemporary Mathematics, **185**, Amer. Math. Soc., Providence, RI, 1995.

[22] J. D. Fay, *Theta Functions on Riemann Surfaces*, Lecture Notes in Mathematics Vol. 352, Springer-Verlag, New York, 1973.

[23] S. I. Fedorov, *On harmonic analysis in a multiply connected domain and character-automorphic Hardy spaces*, St. Petersburg J., **9** (1998), 339–378.

[24] S. I. Fedorov, *On a projection from one co-invariant subspace onto another in character-automorphic Hardy space on a multiply connected domain*, Math. Nachr., **217** (2000), 53–74.

[25] H. M. Farkas and I. Kra, *Riemann Surfaces*, Second Edition, Springer-Verlag, Berlin-New York, 1991.

[26] M. Hasumi, *Invariant subspace theorems for finite Riemann surfaces*, Canad. J. Math., **18** (1966), 240–255.

[27] K. Hoffman, *Banach spaces of analytic functions*, Prentice-Hall, Englewood Cliffs, NJ, 1962.

[28] R. E. Kalman, P. Falb and M. Arbib, *Topics in Mathematical System Theory*, McGraw-Hill, New York, 1969.

[29] V. E. Katsnelson, *Right and left joint system representation of a rational matrix function in general position (system representation theory for dummies)*, Operator Theory, System Theory and Related Topics (The Moshe Livšic Anniversary Volume), Operator Theory: Adv. Appl., Birkhäuser Verlag, Basel, in press.

[30] M. S. Livšic, *Operators, oscillations, waves (open systems)*, Transl. Math. Monographs, vol. 34, Amer. Math. Soc., Providence, 1973.

[31] M. S. Livšic, N. Kravitsky, A. S. Markus and V. Vinnikov, *Theory of Commuting Nonselfadjoint Operators*, Kluwer, 1995.

[32] B. Sz.-Nagy and C. Foiaş, *Harmonic Analysis of Operators on Hilbert Space*, American Elsevier, New York, N. Y., 1970.

[33] N. Nikolski, *Treatise on the shift operator*, Springer Verlag, Berlin–Heidelberg, 1986.

[34] N. K. Nikolskii and V. I. Vasyunin, *Elements of spectral theory in terms of the free funcion model. Part I: Basic constructions*, Holomorphic Spaces and Their Operators (S. Axler, J. McCarthy, and D. Sarason, eds.), Math. Sci. Res. Inst. Publ., vol. 33, Cambridge University Press, Cambridge, 1988, pp. 351–379.

[35] M. Rosenblum and J. Rovnyak, *Hardy Classes and Operator Theory*, corrected reprint of 1985 original, Dover, Mineola, NY, 1997.

[36] W. Rudin, *Analytic functions of class H_p*, Trans. Amer. Math. Soc., **78** (1955), 46–66.

[37] D. Sarason, *The H^p spaces of an annulus*, Memoirs of the American Mathematical Society, **1**, No. 56, 1965.

[38] N. Th. Varapoulas, *On an inequality of von Neumann and an application of the metric theory of tensor products to operator theory*, J. Funct. Anal., **16** (1974), 83–100.

[39] N. Th. Varapoulas, *On a commuting family of contractions on a Hilbert space*, Rev. Roumaine Math. Pures Appl., **21** (1976), 1283–1285.

[40] V. Vinnikov, *Commuting operators and function theory on a Riemann surface*, Holomorphic Spaces (Ed. S. Axler, J. E. McCarthy, and D. Sarason), MSRI Publications, Cambridge University Press, 1998.

[41] M. Voichick, *Ideals and invariant subspaces of analytic functions*, Trans. Amer. Math. Soc., **111** (1964), 493–512.

[42] D. Yakubovich, *Subnormal operators of finite type. I. Xia's model and real algebraic curves in $\mathbb{C}^2$*, Rev. Mat. Iberoamericana, **14** (1998), 95–115.

[43] D. Yakubovich, *Subnormal operators of finite type. II. Structure theorems*, Rev. Mat. Iberoamericana, **14** (1998), 623–681.

[44] A. Zucchi, *Operators of class C_0 with spectra in multiply connected regions*, Memoirs of the American Mathematical Society, Vol. 127 No. 607 (third of 4 numbers), 1997.

2000 Mathematics Subject Classification: Primary: 47A48; Secondary: 47A20, 30F10, 14F05, 93C35

Department of Mathematics, Virginia Tech, Blacksburg, Virginia 24061
E-mail address: `ball@math.vt.edu`

Department of Mathematics, Ben Gurion University of the Negev, POB 653, 84105 Beer-Sheva, Israel
E-mail address: `vinnikov@math.bgu.ac.il`

On Hankel Operators Associated with Markov Functions

L. Baratchart, V. A. Prokhorov, and E. B. Saff

Abstract. In this paper some questions related to Hankel operators associated with Markov functions are considered. Let G be a bounded multiply connected domain with a boundary Γ consisting of closed analytic Jordan curves. We assume that G is symmetric with respect to the real axis. Let μ be a positive Borel measure with the support $\operatorname{supp}\mu = E \subset \mathbf{R}, E \subset G$. We investigate a connection between the Hankel operator A_f constructed from the Markov function

$$f(z) = \frac{1}{2\pi i}\int_E \frac{d\mu(x)}{z-x}$$

and the embedding operator $J : E_2(G) \to L_2(\mu, E)$, where $E_2(G)$ is the Smirnov class of functions analytic on G. Moreover, in the case when G is the open unit disk we state results characterizing the rate of decrease of the sequence of singular numbers of the Hankel operator A_f constructed from the Markov function with the measure μ satisfying the Szegő condition: $\operatorname{supp}\mu = [a,b] \subset (-1,1)$ and

$$\int_a^b \frac{\log(d\mu/dx)}{\sqrt{(x-a)(b-x)}}\,dx > -\infty.$$

1. Hankel and Embedding Operators

1.1. Notation and Statement of Theorem 1.1

Let $G \subseteq \mathbf{C}$ be a bounded domain symmetric with respect to the real axis whose boundary Γ consists of finitely many closed analytic Jordan curves. We assume that Γ is positively oriented with respect to G. Let $L_p(\Gamma), 1 \leq p < \infty$, be the Lebesgue space of functions φ on Γ such that

$$\|\varphi\|_p = \left(\int_\Gamma |\varphi(\xi)|^p |d\xi|\right)^{1/p} < \infty.$$

The research of these authors was supported, in part, by NSF-INRIA collaborative research grant INT-9732631 as well as (for E. B. Saff) by the NSF research grant DMS-9801677.

Denote by $L_\infty(\Gamma)$ the space of essentially bounded functions φ,

$$\|\varphi\|_\infty = \operatorname*{ess\,sup}_{\Gamma} |\varphi(\xi)| < \infty.$$

Let $E_p(G), 1 \le p \le \infty$, be the Smirnov class of analytic functions on G. Here and in what follows we consider $E_p(G)$ as a subspace of the space $L_p(\Gamma)$. The condition

$$\int_\Gamma \frac{\varphi(\xi)\, d\xi}{\xi - z} = 0 \qquad \text{for all } \ z \in \overline{\mathbf{C}} \setminus \overline{G}$$

is necessary and sufficient for a function $\varphi \in L_1(\Gamma)$ to be the boundary value of a function in the Smirnov class $E_1(G)$ (see [16] and [19] for more details about the classes $E_p(G)$).

We represent $L_2(\Gamma)$ as the direct sum $L_2(\Gamma) = E_2(G) \oplus E_2^\perp(G)$, where $E_2^\perp(G)$ is the orthogonal complement of $E_2(G)$ in $L_2(\Gamma)$. Denote by $\mathbf{P}_-$ the orthogonal projection from $L_2(\Gamma)$ onto $E_2^\perp(G)$. Let f be continuous on Γ. The Hankel operator A_f with symbol f is the map from $E_2(G)$ to $E_2^\perp(G)$ which takes $\varphi \in E_2(G)$ to $A_f\varphi = \mathbf{P}_-(\varphi f)$. Note that A_f is a compact operator.

Let μ be a positive Borel measure with the support $\operatorname{supp}\mu = E \subset \mathbf{R}, E \subset G$. We assume that the support of μ contains infinitely many points. Denote by $L_2(\mu, E)$ the Hilbert space with the inner product

$$(\varphi, \psi)_{2,\mu} = \int_E (\varphi\overline{\psi})(x) d\mu(x), \qquad \varphi, \psi \in L_2(\mu, E),$$

and the norm $\|\varphi\|_{2,\mu}$.

For the Markov function

$$f(z) = \frac{1}{2\pi i} \int_E \frac{d\mu(x)}{z - x} \tag{1}$$

there is a connection between the singular numbers $s_n(A_f)$ of the Hankel operator A_f and the singular numbers of the *embedding operator* $J : E_2(G) \to L_2(\mu, E)$,

$$J\varphi = \varphi|_E, \qquad \varphi \in E_2(G).$$

The compact operator J is given by restricting an element $\varphi \in E_2(G)$ to E.

Let X and Y be the Hilbert spaces, and let A be a compact linear operator from X to Y. For any nonnegative integer n denote by $s_n(A)$ the *n-th singular number* of the operator A:

$$s_n(A) = \inf_K \|A - K\|,$$

where the infimum is taken over the collection of all linear operators $K : X \to Y$ of rank at most n, and $\|\cdot\|$ is the norm of the corresponding linear operator. It can be seen that the sequence $\{s_n(A)\}$, $n = 0, 1, 2, \ldots$, coincides with the sequence of eigenvalues (counting multiplicity) of the operator $(A^*A)^{1/2}$, where $A^* : Y \to X$ is the adjoint of A, and

$$s_n(A) = \inf_{X_{-n}} \|A|_{X_{-n}}\|,$$

where X_{-n} runs over all possible subspaces of codimension n of X (see [12] for more details about singular numbers).

We now formulate the main result of this section establishing a connection between the operators $A_f^* A_f$ and $(J^*J)^2$.

Theorem 1.1. *Let G be a domain symmetric with respect to real axis and let f be the Markov function of* (1), *where μ is a positive Borel measure with the support* $\operatorname{supp}\mu = E \subset \mathbf{R}$, *$E \subset G$, containing infinitely many points. Then*

$$A_f^* A_f = (J^*J)^2. \tag{2}$$

Consequently,

$$s_n(J)^2 = s_n(A_f), \qquad n = 0, 1, 2, \dots. \tag{3}$$

Remark 1.2. *In the case when G is the open unit disk $\{z : |z| < 1\}$ and $E \subset (-1, 1)$ it is not hard to prove (see* [4]*) that for $|z| < 1$*

$$(A_f^* A_f)(\varphi)(z) = (J^*J)^2(\varphi)(z) = \frac{1}{4\pi^2}\int_E\int_E \frac{\varphi(x)}{(1-yx)(1-yz)}d\mu(x)d\mu(y).$$

1.2. Some Properties of A_f

In this subsection we investigate some properties of Hankel operators. We first state a result giving necessary and sufficient conditions for a function a belonging to the space $L_2(\Gamma)$ to be an element of the subspace $E_2^{\perp}(G)$ (see [17]).

Suppose that $a \in L_2(\Gamma)$. Then $a \in E_2^{\perp}(G)$ if and only if there exists a function $b \in E_2(G)$ such that

$$\overline{a}(\xi)\,|d\xi| = b(\xi)\,d\xi \quad \text{a.e. on} \quad \Gamma. \tag{4}$$

Let $\varphi \in E_2(G)$. Since $A_f\varphi = \mathbf{P}_-(\varphi f)$,

$$A_f\varphi = \varphi f - \psi = u, \tag{5}$$

where $u \in E_2^{\perp}(G)$ and $\psi \in E_2(G)$. On the basis of the fact that $A_f^* u = \mathbf{P}_+(u\overline{f})$, where $\mathbf{P}_+$ is the orthogonal projection of $L_2(\Gamma)$ onto $E_2(G)$, we obtain that

$$A_f^* u = u\overline{f} - v = \omega, \tag{6}$$

where $\omega \in E_2(G)$ and $v \in E_2^{\perp}(G)$. Since $u, v \in E_2^{\perp}(G)$, we can assert (see (4)) that there exist $\alpha, \beta \in E_2(G)$ such that

$$\overline{u}(\xi)|d\xi| = \alpha(\xi)d\xi \quad \text{and} \quad \overline{v}(\xi)|d\xi| = \beta(\xi)d\xi$$

a.e. on Γ. Therefore, by (5) and (6), we get

$$(\varphi f - \psi)(\xi)d\xi = \overline{\alpha(\xi)}\ |d\xi|, \text{ and } (\alpha f - \beta)(\xi)d\xi = \overline{\omega(\xi)}\ |d\xi|$$

a.e. on Γ, and so for the Markov function (1) we have

$$\left(\varphi(\xi)\frac{1}{2\pi i}\int_E \frac{d\mu(x)}{\xi - x} - \psi(\xi)\right)\ d\xi = \overline{\alpha(\xi)}\ |d\xi| \quad \text{a.e. on} \quad \Gamma, \tag{7}$$

$$\left(\alpha(\xi)\frac{1}{2\pi i}\int_E \frac{d\mu(x)}{\xi - x} - \beta(\xi)\right) d\xi = \overline{\omega(\xi)}\ |d\xi| \quad \text{a.e. on} \quad \Gamma. \tag{8}$$

Remark 1.3. It follows immediately from (4) and (5) that α and ψ are uniquely determined by equation (7). Namely, if for $\tilde{\psi}, \tilde{\alpha} \in E_2(G)$ we have

$$\left(\varphi(\xi)\frac{1}{2\pi i}\int_E \frac{d\mu(x)}{\xi - x} - \tilde{\psi}(\xi)\right) d\xi = \overline{\tilde{\alpha}(\xi)}|d\xi|, \quad \text{a.e. on} \quad \Gamma,$$

then $\tilde{\psi} = \psi$ and $\tilde{\alpha} = \alpha$. Analogously, β and ω are uniquely determined by (8).

1.3. The Embedding Operator J

Let $J : E_2(G) \to L_2(\mu, E)$ be the embedding operator $J\varphi = \varphi|_E, \varphi \in E_2(G)$. By the Cauchy formula, for any function $\varphi \in E_2(G)$ we have

$$(J\varphi)(x) = \frac{1}{2\pi i}\int_\Gamma \frac{\varphi(\xi)d\xi}{\xi - x}, \quad x \in E.$$

The adjoint operator $J^* : L_2(\mu, E) \to E_2(G)$ of J can be represented in the form

$$(J^*g)(\xi) = \mathbf{P}_+(\Phi), \qquad g \in L_2(\mu, E), \tag{9}$$

where

$$\Phi(\xi) = \left(\frac{1}{2\pi i}\int_E \frac{g(x)d\mu(x)}{x - \bar{\xi}}\right)\frac{|d\xi|}{d\xi}.$$

Let $\varphi \in E_2(G)$. Equality (9) implies that

$$(J^*J)\varphi(\xi) = (\Phi - \Psi)(\xi) = \alpha_1(\xi), \quad \text{a.e. on} \quad \Gamma,$$

where $\alpha_1 \in E_2(G)$ and $\Psi \in E_2^{\perp}(G)$. By (4), there exists a function $\psi_1 \in E_2(G)$ such that $\overline{\Psi}(\xi)|d\xi| = \psi_1(\xi)d\xi$ a.e. on Γ. Therefore, we get

$$\left(\frac{1}{2\pi i}\int_E \frac{\overline{\varphi(x)}d\mu(x)}{\xi - x} - \psi_1(\xi)\right) d\xi = \overline{\alpha_1(\xi)}|d\xi| \quad \text{a.e. on} \quad \Gamma. \tag{10}$$

Analogously, using the formula $(J^*J)\alpha_1(\xi) = \omega_1(\xi)$, where $\omega_1 \in E_2(G)$, we obtain that there exists $\beta_1 \in E_2(G)$ such that

$$\left(\frac{1}{2\pi i}\int_E \frac{\overline{\alpha_1(x)}d\mu(x)}{\xi - x} - \beta_1(\xi)\right) d\xi = \overline{\omega_1(\xi)}|d\xi| \quad \text{a.e. on} \quad \Gamma. \tag{11}$$

1.4. Proof of Theorem 1.1

We shall prove that $(J^*J)^2\varphi = (A_f^*A)\varphi$ for every $\varphi \in E_2(G)$. Let us consider equation (10). Replacing ξ by $\bar{\xi}$ and then taking the conjugate, we get

$$\left(-\frac{1}{2\pi i}\int_E \frac{\varphi(x)d\mu(x)}{\xi - x} - \psi_1^*(\xi)\right)(-d\xi) = \overline{\alpha_1^*(\xi)}|d\xi| \quad \text{a.e. on} \quad \Gamma,$$

where $\psi^*(\xi) = \overline{\psi_1(\bar{\xi})} \in E_2(G)$ and $\alpha_1^*(\xi) = \overline{\alpha_1(\bar{\xi})} \in E_2(G)$. Therefore,

$$\left(\frac{1}{2\pi i}\int_E \frac{\varphi(x)d\mu(x)}{\xi - x} + \psi_1^*(\xi)\right) d\xi = \overline{\alpha_1^*(\xi)}|d\xi| \quad \text{a.e. on} \quad \Gamma. \tag{12}$$

We now note that for $\varphi \in E_2(G)$ the function

$$\int_E \frac{\varphi(x)d\mu(x)}{\xi - x} - \varphi(\xi)\int_E \frac{d\mu(x)}{\xi - x}$$

belongs to $E_2(G)$. Then, from (7) and (12) we can conclude with the help of Remark 1.3 that $\alpha_1^* = \alpha$. Using (11) and the fact $\overline{\alpha_1(x)} = \alpha_1^*(x)$ for $x \in E$, we obtain

$$\left(\frac{1}{2\pi i}\int_E \frac{\alpha(x)d\mu(x)}{\xi - x} - \beta_1(\xi)\right) d\xi = \overline{\omega_1(\xi)}|d\xi| \quad \text{a.e. on} \quad \Gamma.$$

From this, on account of (8) and Remark 1.3, we get $\omega_1 = \omega$. Consequently, for any $\varphi \in E_2(G)$ we have $(J^*J)^2\varphi = (A_f^* A_f)\varphi$, which implies (2) and (3).

□

1.5. Connection between the Best Meromorphic Approximation Problem, the Theory of Hankel Operators, and n-widths

There is a connection between the singular numbers $s_n(A_f)$ of the Hankel operator A_f with symbol $f \in C(\Gamma)$ and the error

$$\Delta_{n,\infty} = \Delta_{n,\infty}(f;G) = \inf_{h \in \mathcal{M}_{n,\infty}(G)} \|f - h\|_\infty$$

in best meromorphic approximation of f in $L_\infty(\Gamma)$ in the class

$$\mathcal{M}_{n,\infty}(G) = \{h = \alpha/\beta : \alpha \in E_\infty(G),\ \beta \text{ is a polynomial, } \deg\beta \le n, \beta \not\equiv 0\}.$$

This connection was first investigated by Adamyan, Arov, and Kreĭn (see [2], [3]) and is known as the AAK method. The Adamyan-Arov-Kreĭn theorem relates to the situation when the Hankel operator is constructed from a function given on the boundary Γ of the unit disk G.

Theorem 1.4 (Adamyan, Arov, Kreĭn). *Let f be continuous on the unit circle $\Gamma = \{z : |z| = 1\}$. Then*

$$\Delta_{n,\infty}(f;G) = s_n(A_f), \quad n = 0, 1, 2, \ldots,$$

and the unique best approximant $h_n \in \mathcal{M}_{n,\infty}(G)$ satisfying

$$\Delta_{n,\infty}(f;G) = \|f - h_n\|_\infty$$

is given by $h_n = \mathbf{P}_+(Q_n f)/Q_n$, where Q_n is the first element of some $(n+1)$-st Schmidt pair of A_f, $\|Q_n\|_2 = 1$.

As shown in [17] we have the following connection between the quantities $\Delta_n(f;G)$ and the singular numbers $s_n(A_f)$ in the case when G is a N-connected domain.

Let G be a N-connected domain with a boundary Γ consisting of closed analytic Jordan curves and let f be continuous on Γ. Then, for all integers $n \ge N - 1$,

$$\Delta_{n+N-1}(f;G) \le s_n(A_f) \le \Delta_n(f;G). \tag{13}$$

Let G be a bounded domain whose boundary Γ consists of N closed analytic Jordan curves. In this case the Smirnov class $E_p(G), 1 \le p \le \infty$, coincides with the Hardy space $H_p(G)$ of analytic functions on G (see [11], [13], [20] for more details about the Hardy spaces). Let μ be a positive Borel measure with support $\operatorname{supp}\mu = E \subset G$. There is a link between the singular numbers $s_n(J)$ of the embedding operator J and n-widths of the unit ball of the Hardy space $H_2(G)$ in $L_2(\mu, E)$ as we now describe.

Let X be a Banach space and $\mathbf{A}$ be a convex, compact, centrally symmetric subset of X. The *Kolmogorov n-width* of $\mathbf{A}$ in X is given by

$$d_n(\mathbf{A}, X) := \inf_{X_n} \sup_{\varphi \in \mathbf{A}} \inf_{g \in X_n} \|\varphi - g\|,$$

where the infimum is taken over all n dimensional subspaces X_n of X. The *Gel'fand n-width* of $\mathbf{A}$ in X is defined as follows:

$$d^n(\mathbf{A}, X) := \inf_{X_{-n}} \sup_{x \in X_{-n} \cap \mathbf{A}} \|x\|,$$

where X_{-n} is an arbitrary subspace of X of codimension n. The *linear n-width* of $\mathbf{A}$ in X is given by

$$\delta_n(\mathbf{A}, X) := \inf_{K} \sup_{\varphi \in \mathbf{A}} \|\varphi - K\varphi\|,$$

where $K : X \to X$ varies over all linear operators of rank n (see [15] for more details about n-widths).

Let A_2 be the restriction to E of the closed unit ball of $H_2(G)$. We have (see, for example, [10], [15])

$$s_n(J) = d_n(A_2, L_2(\mu, E)) = d^n(A_2, L_2(\mu, E)) \le \delta_n(A_2, L_2(\mu, E)).$$

It is proved by S.D. Fisher [10] that there is a constant $C > 0$ depending on G and E but not n or μ such that

$$\frac{1}{C}\gamma_n \le d_n(A_2, L_2(\mu, E)) \le \delta_n(A_2, L_2(\mu, E)) \le C\gamma_n,$$

where

$$\gamma_n = \inf_{\{z_1, \ldots, z_n\} \subset G} \sup\{\|\varphi\|_{2,\mu} : \varphi \in A_2 \text{ and } \varphi(z_k) = 0, k = 1, \ldots, n\}.$$

Moreover, if E is small enough (see [10, Theorem 4]), then

$$\begin{aligned} s_n(J) &= d_n(A_2, L_2(\mu, E)) = d^n(A_2, L_2(\mu, E)) \\ &= \delta_n(A_2, L_2(\mu, E)) = \gamma_n. \end{aligned}$$

In the case when G is the open unit disk, S.D. Fisher and C.A. Micchelli (see [8], [9], and [7]) proved that always

$$\begin{aligned} s_n(J) &= d_n(A_2, L_2(\mu, E)) = d^n(A_2, L_2(\mu, E)) \\ &= \delta_n(A_2, L_2(\mu, E)) = \gamma_n. \end{aligned}$$

2. Best Meromorphic Approximation of Markov Functions on the Unit Circle

2.1. Some Formulas

Here and in what follows we consider the case when G is the open unit disk with the center at 0. Let μ be a positive Borel measure whose support $\operatorname{supp}\mu = E \subset (-1,1)$ contains infinitely many points and let f be the Markov function associated with μ. In this case the singular values $s_n(A_f)$ of A_f are strictly decreasing (i.e. the eigenspaces are one dimensional). This follows from Theorem 1.1 and results of S. D. Fisher and C.A. Micchelli (see [8], [9], and [7]), where it is proved that the eigenvalues for $J^*J : H_2(G) \to H_2(G)$,

$$(J^*J)g(z) = \frac{1}{2\pi}\int_E \frac{g(x)}{1-xz}d\mu(x), \quad g \in H_2(G), \quad |z| < 1,$$

are simple, the corresponding eigenspaces are one dimensional and the $(n+1)$-st eigenfunction has exactly n zeros in G. Thus (cf. [5]) an eigenfunction $Q_n = Q_{n,\infty}$, $\|Q_n\|_2 = 1$, for A_f corresponding to the singular value $s_n(A_f)$ satisfies the equation

$$s_n(A_f)Q_n(z) = \frac{1}{2\pi}\int_E \frac{Q_n(x)}{1-zx}d\mu(x), \quad |z| < 1. \tag{14}$$

We remark that Q_n is unique up to multiplication by a unimodular scalar. As shown in [5], equation (14) implies that the polynomial $w_n^*(z) = \prod_{k=1}^n (z - x_{k,n})$ constructed from the zeros of Q_n is an orthogonal polynomial with a varying weight function (see Subsection 2.2). It is possible to show further that all n zeros $x_{1,n}, \ldots, x_{n,n}$ of Q_n are simple, lie on the smallest closed interval containing the support of μ and that Q_n can be analytically continued from G to $\overline{\mathbf{C}} \setminus E^{-1}$, where $E^{-1} = \{1/x : x \in E\}$ (see [4] for more details about properties of Q_n).

2.2. Orthogonality

We can represent Q_n in the form $Q_n = B_n\varphi_n$, where φ_n is an outer function and B_n is a Blaschke product constructed from the zeros Q_n:

$$B_n(z) = \prod_{k=1}^n \frac{z - x_{k,n}}{1 - x_{k,n}z} = \frac{w_n^*(z)}{w_n(z)},$$

$$w_n^*(z) = \prod_{k=1}^n (z - x_{k,n}), \quad w_n(z) = \prod_{k=1}^n (1 - x_{k,n}z).$$

By (14),

$$\int_E \frac{Q_n(x)}{1 - x_{k,n}x}d\mu(x) = 0 \quad \text{for} \quad k = 1, \ldots, n.$$

Thus the following orthogonality relations are valid:

$$\int_E T(x)\frac{Q_n(x)}{w_n(x)}d\mu(x) = 0$$

and

$$\int_E T(x)w_n^*(x)\frac{\varphi_n(x)}{w_n^2(x)}d\mu(x)=0 \tag{15}$$

for any polynomial T, $\deg T \le n-1$.

3. Potential Theory

To describe the asymptotic behavior of the singular values and corresponding singular functions for a Hankel operator whose symbol is a Markov function, we need to recall some fundamental quantities from potential theory.

Let $g(z,\xi)$ be the Green function for the unit disk G with pole at $\xi \in G$:

$$g(z,\xi)=\log\left|\frac{1-\bar{\xi}z}{z-\xi}\right|.$$

Let $E\subset G$ be a compact set. Denote by $\mathcal{P}(E)$ the set of all positive unit Borel measures σ with support $\operatorname{supp}\sigma\subset E$. The Green potential of a measure $\sigma\in\mathcal{P}(E)$ is denoted by

$$V_g^\sigma(z)=\int_E g(z,\xi)d\sigma(\xi).$$

We consider the energy of a measure $\sigma\in\mathcal{P}(E)$ with respect to the Green potential

$$I_g^\sigma=\int_E V_g^\sigma(z)d\sigma(z)=\int_E\int_E g(z,\xi)d\sigma(\xi)d\sigma(z).$$

It is a well known fact (see, for example, [18]) that if logarithmic capacity $\operatorname{cap}(E)>0$, then there exists a unique measure $\omega\in\mathcal{P}(E)$ such that

$$I_g^\omega=\inf\{I_g^\sigma:\sigma\in\mathcal{P}(E)\}.$$

Moreover,

$$V_g^\omega(z)=I_g^\omega \quad \text{q.e. on} \quad E,$$

where q.e. (quasi-everywhere) means neglecting sets of zero logarithmic capacity. The measure ω is called the *Green equilibrium measure*. The *condenser capacity* is $C(E,\Gamma):=1/I_g^\omega$. Let $r=e^{-1/C(E,\Gamma)}=e^{-I_g^\omega}$

Let us consider the case when $E=[a,b]\subset(-1,1)$. Then

$$I_g^\omega=\frac{\pi K}{2K'} \quad \text{or} \quad r=\exp\left(-\frac{\pi K}{2K'}\right),$$

where

$$K=\int_0^1\frac{dx}{\sqrt{(1-x^2)(1-\tau^2x^2)}},\quad \tau^2=\frac{(1-a^2)(1-b^2)}{(1-ab)^2},$$

and K' is the corresponding elliptic integral for $\tau'=\sqrt{1-\tau^2}$ (see [1], [18]). Moreover, the Green equilibrium measure is

$$d\omega(x)=\frac{kdx}{\sqrt{(x-a)(b-x)(1-ax)(1-bx)}},\quad x\in[a,b],$$

where

$$k = \frac{(1-ab)}{2K'}.$$

4. Statements of Convergence Theorems

4.1. Szegő Function $D_\psi(z)$

We assume that the support of μ is $E = [a, b] \subset (0, 1)$ and the measure μ satisfies the Szegő condition

$$\int_a^b \frac{\log(d\mu/dx)}{\sqrt{(x-a)(b-x)}} dx > -\infty.$$

Let $E^{-1} = [1/b, 1/a]$. Let $\Phi : \overline{\mathbf{C}} \setminus (E \cup E^{-1}) \to \{r < |z| < 1/r\}, \Phi(1) = 1$, be the conformal mapping of the region $\overline{\mathbf{C}} \setminus (E \cup E^{-1})$ onto the annulus with the interior radius r and the exterior radius $1/r$. Then $|\Phi| = r$ on E and $r = e^{-1/C(E,\Gamma)}$, where $C(E, \Gamma)$ is the capacity of the condenser (E, Γ).

Let $d\omega$ be the Green equilibrium measure for $E = [a, b]$. We can write the Riesz decomposition of μ

$$d\mu = \psi d\omega + d\mu_s,$$

where $\psi = d\mu/d\omega$ denotes the Radon-Nikodym derivative of μ with respect to ω, and μ_s is a singular measure. We also need the *geometric mean* $\mathcal{G}_\omega(\psi)$ of ψ with respect to ω:

$$\mathcal{G}_\omega(\psi) = \exp\left(\int_E \log \psi \, d\omega\right)$$

and the Szegő function D_ψ for a doubly-connected region $\overline{\mathbf{C}} \setminus (E \cup E^{-1})$ (see [14]). The Szegő function $\mathcal{D}_\psi(z)$ for $\overline{\mathbf{C}} \setminus (E \cup E^{-1})$ has the following properties:

1): $\mathcal{D}_\psi(z)$ is analytic and non-vanishing in $\overline{\mathbf{C}} \setminus (E \cup E^{-1})$;
2): the increment $\Delta_\Gamma \arg \mathcal{D}_\psi$ of the argument of $\mathcal{D}_\psi$ along Γ is equal to 0;
3): $|\mathcal{D}_\psi(z)|^2 = \psi(z)$ a.e. on E; more precisely, the nontangential limits of $|\mathcal{D}_\psi(\xi)|^2$ as ξ approaches $z \in E$ equal $\psi(z)$ a.e.;
4): $|\mathcal{D}_\psi(z)|^2 = \mathcal{G}_\omega(\psi)$ on Γ.

We have (cf. [6])

$$\begin{aligned}\mathcal{D}_\psi(z) &= \sqrt{\mathcal{G}_\omega(\psi)} \exp\Bigg(\sqrt{(z-a)(z-b)(1-az)(1-bz)} \\ &\quad \times \frac{1}{2\pi} \int_a^b \frac{\log(\psi(x)/\mathcal{G}_\omega(\psi))(1-2xz+x^2)dx}{\sqrt{(x-a)(b-x)(1-ax)(1-bx)(z-x)(1-xz)}} \Bigg).\end{aligned}$$

4.2. Main Results

In this subsection we formulate theorems characterizing the limit distribution of poles and degree of convergence of best meromorphic approximants $h_n = h_{n,p}$, $1 \leq p \leq \infty$, to the Markov function of (1) in the space $L_p(\Gamma)$. The methods used are based on an investigation of Szegő type asymptotics for the orthogonal polynomials (cf. (15)) associated with the poles of h_n.

Let $1 \leq p \leq \infty$ and let $\mathcal{M}_{n,p}(G)$ be the following class of meromorphic functions on the unit disk G:

$$\mathcal{M}_{n,p}(G) = \{h = \alpha/\beta : \alpha \in H_p(G),\ \beta \text{ is a polynomial, } \deg \beta \leq n, \beta \not\equiv 0\}.$$

The deviation of $f \in L_p(\Gamma)$ from the class $\mathcal{M}_{n,p}(G)$ will be denoted by $\Delta_{n,p}$:

$$\Delta_{n,p} = \Delta_{n,p}(f;G) = \inf_{h \in \mathcal{M}_{n,p}(G)} \|f-h\|_p. \tag{16}$$

Let $1/p + 1/q = 1$. There exists a best approximant $h_n = h_{n,p}$ in $\mathcal{M}_{n,p}(G)$ to the function f in (1) in the space $L_p(\Gamma)$:

$$\Delta_{n,p} = \|f - h_n\|_p$$

such that all n poles of h_n in G are simple and belong to E. The function h_n can be represented in the form $h_n = P_n/Q_n$, where $P_n = P_{n,p} \in H_p(G)$ and $Q_n = Q_{n,p} \in H_{2q}(G)$, $\|Q_n\|_{2q} = 1$. We have $Q_n = B_n\varphi_n$, where B_n is a Blaschke product of degree n constructed by the zeros of Q_n, and $\varphi_n \in H_{2q}(G)$, $\|\varphi_n\|_{2q} = 1$, φ_n is positive on $(-1,1)$. It is possible to show that φ_n^q and $B_n\varphi_n^{2-q}(f-h_n)$ can be extended analytically to $\overline{\mathbf{C}} \setminus E^{-1}$ and $\overline{\mathbf{C}} \setminus E$, respectively (see [4]).

Let $\varphi^* \in H_{2q}(G), \varphi^*(0) > 0$, satisfy

$$\mathcal{G}_\omega(|\varphi^*|^2) = \sup_{\varphi \in H_{2q}(G), \|\varphi\|_{2q}=1} \mathcal{G}_\omega(|\varphi|^2).$$

It is easy to prove that $\varphi^*(z) \equiv 1$ for $p = 1$. In the case when $1 < p \leq \infty$ we have

$$\varphi^*(z) = \frac{k_1}{\sqrt[2q]{(1-az)(1-bz)}}, \quad z \in G,$$

where $k_1 = \sqrt[2q]{k/2}$.

We now state a theorem characterizing the degree of convergence of $\Delta_{n,p}$ to zero as $n \to \infty$ and the limit behavior of B_n and φ_n (see [4]). In the following theorems we assume that f is a Markov function generated by a measure μ satisfying the Szegő condition in Subsection 4.1.

Theorem 4.1. *We have for* $1 \leq p \leq \infty$,

(i):

$$\frac{\Delta_{n,p}}{2r^{2n}\mathcal{G}_\omega(\psi)\mathcal{G}_\omega((\varphi^*)^2)} \to 1 \quad as \quad n \to \infty. \tag{17}$$

In particular, for $p = \infty$, (17) *describes the asymptotic behavior of the singular values* $s_n(A_f) = \Delta_{n,\infty}$.

(ii): *For $1 < p \leq \infty$, the sequence $B_n^2 \varphi_n^2 d\mu / \Delta_{n,p}$ converges weak* to $d\omega$:*

$$\frac{B_n^2 \varphi_n^2}{\Delta_{n,p}} d\mu \xrightarrow{*} d\omega \quad as \quad n \to \infty.$$

Moreover,

$$\frac{(B_n \mathcal{D}_\psi \mathcal{D}_{(\varphi^*)^2})(z)}{\Phi^n(z) \sqrt{\mathcal{G}_\omega(\psi) \mathcal{G}_\omega(\varphi^*)}} \to 1$$

uniformly on compact subsets of $\overline{\mathbb{C}} \setminus (E \cup E^{-1})$ as $n \to \infty$, and

$$\varphi_n(z) \to \varphi^*(z)$$

and

$$\varphi_n^q(z) \to (\varphi^*)^q(z)$$

uniformly on compact subsets of G and $\overline{\mathbb{C}} \setminus E^{-1}$, respectively, as $n \to \infty$.

The next theorem describes the convergence of h_n to f (see [4]).

Theorem 4.2. *Let $1 < p \leq \infty$. We have for $z \in G \setminus E$,*

$$\frac{(f - h_n)(z)}{\frac{2(\mathcal{D}_\psi^2 \mathcal{D}_{\varphi^{*2}}^2)(z) r^{2n}}{i\Phi^{2n}(z)} \sqrt{\frac{(1-az)(1-bz)}{(z-a)(z-b)}} \left(\frac{(1-az)(1-bz)}{k/2} \right)^{(1-q)/q}} \to 1,$$

and for $z \in \overline{\mathbb{C}} \setminus (E \cup E^{-1})$

$$\frac{(\varphi_n^{2-q}(f - h_n))(z)}{\frac{2(\mathcal{D}_\psi^2 \mathcal{D}_{\varphi^{*2}}^2)(z) r^{2n}}{\Phi^{2n}(z)}} \to \frac{\sqrt{k/2}}{i\sqrt{(z-a)(z-b)}},$$

where both limits as $n \to \infty$ are locally uniform.

References

[1] N. I. Achiezer, *Elements of the Theory of Elliptic Functions*, 2nd rev. ed., "Nauka", Moscow, 1970; English transl., Amer. Math. Soc., Providence, RI, 1990.

[2] V. M. Adamyan, D. Z. Arov, and M. G. Kreĭn, *Infinite Hankel matrices and generalized Carathéodory-Fejér and Riesz problem*, Functional Anal. Appl., **2** (1968), 1–18.

[3] V. M. Adamyan, D. Z. Arov, and M. G. Kreĭn, *Analytic properties of Schmidt pairs, Hankel operators, and the generalized Schur-Takagi problem*, Mat. Sb., **86 (128)** (1971), 34–75; English transl. in Math. USSR Sb., **15** (1971).

[4] L. Baratchart, V. A. Prokhorov, and E. B. Saff, *Best meromorphic approximation of Markov functions on the unit circle* (to appear).

[5] L. Baratchart, J. Leblond, F. Mandréa, and E.B. Saff, *How can meromorphic approximation help to solve some 2D inverse problems for the Laplacian?*, Inverse Problems, **15** (1999), 79–90.

[6] L. Baratchart, H. Stahl and F. Wielonsky, *Asymptotic uniqueness of best rational approximants of given degree to Markov functions in $\mathcal{L}^2$ of the circle*, Constr. Approx. (to appear).

[7] S. D. Fisher, *Function Theory on Planar Domains*, Wiley, New York, 1983.

[8] S. D. Fisher and C.A. Micchelli, *Optimal sampling of holomorphic functions*, Amer. J. Math., **103** (1984), 593-609.

[9] S. D. Fisher and C. A. Micchelli, *Optimal sampling of holomorphic functions II*, Math. Ann., **273** (1985), 131–147.

[10] S. D. Fisher, *Widths and optimal sampling in spaces of analytic functions*, Constr. Approx. **12** (1996), 463–480.

[11] J. B. Garnett, *Bounded Analytic Functions*, Academic Press, New York, 1981.

[12] I. Ts. Gokhberg [Israel Gohberg] and M. G. Kreĭn, *Introduction to the Theory of Linear Nonselfadjoint Operators in Hilbert Space*, "Nauka", Moscow, 1965; English transl., Amer. Math. Soc., Providence, RI.

[13] P. Koosis, *Introduction to H^p-spaces*, Cambridge Univ. Press, Cambridge, 1980.

[14] A. L. Levin, E. B. Saff, *Szegő type asymptotics for minimal Blaschke products*, Progress in Approximation Theory (A. A. Gonchar and E. B. Saff, eds.), Springer-Verlag, (1992), 105–126.

[15] A. Pinkus, *N-widths in Approximation Theory*, Springer–Verlag, New York, 1985.

[16] I. I. Privalov, *Boundary Properties of Analytic Functions*, 2nd ed., GITTL, Moscow, 1950; German transl., VEB Deutscher Verlag Wiss., Berlin, 1956.

[17] V. A. Prokhorov, *On a theorem of Adamyan, Arov, and Kreĭn*, Mat. Sb., **184** (1993), 89–104; English transl. in Russian Acad. Sci. Sb. Math.. **78** (1994).

[18] E. B. Saff and V. Totik, *Logarithmic Potentials with External Fields*, Springer-Verlag, Heidelberg, 1997.

[19] G. Ts. Tumarkin and S. Ya. Khavinson, *On the definition of analytic functions of class E_p in multiply connected domains*, Uspekhi Mat. Nauk, **13** (1958), no. 1 (79), 201–206 (Russian).

[20] M. Voichick and L. Zalcman, *Inner and outer functions on Riemann surfaces*, Proc. Amer. Math. Soc., **16** (1965), 1200–1204.

Key words and phrases: Meromorphic approximation, Best approximation, Hankel operator, Embedding operator, Singular numbers, n-width, Orthogonal polynomials

AMS Classification: 41A20, 30E10, 47B35, 47A10

INRIA,
2004 Route des Lucioles B.P.93,
06902 Sophia Antipolis Cedex, France
E-mail address: baratcha@sophia.inria.fr

Department of Mathematics and Statistics,
University of South Alabama,
Mobile, Alabama 36688-0002, U.S.A.
E-mail address: prokhorov@mathstat.usouthal.edu

Institute for Constructive Mathematics,
Department of Mathematics,
University of South Florida,
Tampa, Florida 33620, U.S.A.
E-mail address: esaff@math.usf.edu

Threshold Effects near the Lower Edge of the Spectrum for Periodic Differential Operators of Mathematical Physics

Michael Birman and Tatyana Suslina

Abstract. In $L_2(\mathbb{R}^d)$, we consider vector periodic DO $\mathcal{A}$ admitting a factorization $\mathcal{A} = \mathcal{X}^*\mathcal{X}$, where $\mathcal{X}$ is a homogeneous DO of first order. Many operators of mathematical physics have this form. The effects that depend only on a rough behavior of the spectral decomposition of $\mathcal{A}$ in a small neighborhood of zero are called threshold effects at $\lambda = 0$. An example of a threshold effect is the behavior of a DO in the small period limit. Another example is related to the negative discrete spectrum of the operator $\mathcal{A} - \alpha V$, $\alpha > 0$, where $V(\mathbf{x}) \geq 0$ and $V(\mathbf{x}) \to 0$ as $|\mathbf{x}| \to \infty$. The "effective characteristics", namely, the homogenized medium, the effective mass, the effective Hamiltonian, etc. arise in these problems. We propose a general approach to these problems based on the spectral perturbation theory for operator-valued functions admitting analytic factorization. A great deal of considerations is done in abstract terms.

0. Introduction

0.1.

Periodic differential operators (DO) acting on $\mathbb{R}^d$, $d \geq 1$, can be partially diagonalized with the help of the Gelfand transformation (see Section 3). Then, the initial DO is represented as the direct integral of a family of DO's, each of the latter acts on the cell of periods Ω glued in a torus. This family depends on the parameter $\mathbf{k} \in \mathbb{R}^d$ (called the quasimomentum). We study *semibounded selfadjoint* DO's.

Further analysis (the Floquet-Bloch decomposition) is possible if an operator acting in $L_2(\Omega)$ has compact resolvent that depends on $\mathbf{k}$ continuously. This condition is valid for many DO's of mathematical physics. Then, the spectrum of the initial DO has a *band structure*. It is convenient to assume that the lower edge of the spectrum coincides with the point $\lambda = 0$. The solution of some questions may happen to require the knowledge of the approximate spectral resolution for the initial DO only near the lower edge of the spectrum. In such cases we speak about the *threshold effects* at the point $\lambda = 0$. Usually, it is not an easy task to detect whether or not a given effect is threshold [1]. For instance, we mention the question about the discrete spectrum to the left of the point $\lambda = 0$ for a periodic DO perturbed by a negative potential vanishing at infinity. If the potential does not decay too rapidly, then the threshold effect dominates; on the contrary, if a perturbation is rapidly decaying, then the high-energy part of the initial DO that corresponds to large values of λ plays the main role. (See Subsection 9.3 below, where we give references.) Another important example of the threshold effect at the point $\lambda = 0$ is the behavior of a periodic DO in $L_2(\mathbb{R}^d)$ in the small period limit. Below, we pay significant attention to this problem.

0.2.

Our main goal is to give a description, concise and convenient for applications, of the spectral characteristics of periodic DO's near the threshold $\lambda = 0$. By using partial diagonalization, we reduce the problem to some questions of the analytic perturbation theory for the discrete spectrum. The difficulties are related to the fact that usually the unperturbed eigenvalue is *multiple* and the parameter $\mathbf{k}$ is *multidimensional.* Such cases cannot be treated by means of the classical perturbation theory, and one should look for a roundabout way. (If at least one of the two above reasons does not occur, the problem simplifies significantly.) We take $t = |\mathbf{k}|$ as the perturbation parameter; at the same time, we must make our constructions and estimates uniform with respect to $\boldsymbol{\theta} = t^{-1}\mathbf{k}$.

We start with an abstract operator theory method (see Sections 1, 2), distinguishing the case where the family in question admits an *analytic factorization.* This additional structure allows us to advance unexpectedly far in abstract terms. Often, the operators in applications admit the required factorization from the very beginning, in other cases we can introduce the factorization forcedly into the problem; see Sections 6-8. If the operator does not admit the required factorization, it is much more difficult to study the threshold effects, and it is useful to understand this clearly. See Subsection 9.4 for more details.

Keeping in mind the applications to (vector) DO's of second order, in the abstract method we restrict ourselves to the quadratic dependence on the parameter. The crucial place is to distinguish and to study the notion of the *spectral germ of an operator family at the point* $t = 0$ (see Section 1). In general, we believe that an effect should be classified as threshold if, in the first order of approximation, it can be described in terms of the corresponding spectral germ. In the abstract form, the threshold effects are considered in Section 2. For example, in essence, estimate (2.21) gives an abstract solution to the question of the behavior of periodic DO's in the small period limit. This approach is applied to the DO's in Section 5, where the specific properties of DO's are used only slightly.

[1] Threshold effects may be related to the edges of internal spectral gaps, but we will speak only about the threshold $\lambda = 0$.

0.3.

The threshold effects bring about the so-called *effective characteristics*. We believe that the mechanism of their appearance is the following. Since a threshold effect is determined only by the spectral germ, the initial periodic DO can be replaced (in the description of this effect) by any other periodic DO with the same germ. Among these "equivalent" DO's, there may be simple operators, sometimes, DO with constant coefficients. That is why the notions of effective mass, effective Hamiltonian, etc., appear in the quantum mechanics problems and the notions of effective (homogenized) medium, homogenized DO appear in the problems concerning periodic structures with vanishing period. Advantages of description via the effective characteristics are obvious. Usually, the very idea of the existence of effective characteristics has a physical origin. At the same time, when using the effective characteristics, one should remember that, in fact, they only contain information about the spectral germ, moreover, in a disguised form. For example, comparison of (5.10) with (5.12), (5.13) makes this quite clear.

0.4.

In Section 3, we distinguish a comparatively wide class of elliptic periodic DO's of second order acting in $L_2(\mathbb{R}^d; \mathbb{C}^n)$. This class includes a number of operators of mathematical physics, though it does not satisfy all the needs of applications (see Subsection 0.7 below and also Subsection 9.4). Under the Gelfand transformation, each operator $\mathcal{A}$ of this class generates an operator family $A(t, \boldsymbol{\theta})$, $t\boldsymbol{\theta} = \mathbf{k}$, in $L_2(\Omega; \mathbb{C}^n)$, which admits a factorization analytic in $t = |\mathbf{k}|$. Now, the spectral germ $S(\boldsymbol{\theta})$ for $A(t, \boldsymbol{\theta})$ at $t = 0$ depends on the parameter $\boldsymbol{\theta}$, which is essential for what follows.

On the basis of general results of Sections 1, 2, in Section 4, we introduce the *effective characteristics* for each $\mathcal{A}$. They are defined directly via the corresponding germ $S(\boldsymbol{\theta})$. In Section 5, on the basis of Corollary 2.5, we show that the same characteristics arise in the homogenization procedure for DO in the small period limit.

0.5.

The study of periodic problems with rapidly varying medium parameters (with small period) is a broad field of theoretical and applied science. There are a variety of methods specific for this field, and a lot of significant results. Limit procedures have been studied for the boundary problems in bounded domains; methods have been developed for constructing full asymptotic (with respect to the small period) expansions; the homogenization procedures have been analyzed for nonselfadjoint operators, for nonstationary operators, for nonlinear problems, etc. A lot of surveys and monographs are devoted to these topics. It was especially useful for us to acquaint with the remarkable books [2], [3], [11]. We are far from the idea to revise all this large area by our methods. At the same time, we would like to pay the reader's attention to the following.

0.6.

In the homogenization theory, finding the effective characteristics is an initial problem, which is usually solved by direct methods. Even in the cases where the Floquet-Bloch decomposition and the analytic perturbation theory (usually, for a simple eigenvalue) are employed for this purpose, these methods are viewed as purely technical and are studied in no relation to the general notion of a threshold effect. The essence of our approach is distinguishing the notion of the spectral germ at $\lambda = 0$. The germ directly determines the effective characteristics, which inevitably appear in any specific threshold effect. As has already been mentioned, homogenization in the small period limit is one of these effects; another one corresponds to the problem about the negative discrete spectrum. In Sections 6-8, we illustrate our unified approach by the traditional operators of mathematical physics. Probably, some concrete facts described there are new.

0.7.

This paper does not include the material concerning the periodic Maxwell operator and periodic operators in domains of cylinder or layer type, etc. Though the general approach of Sections 1, 2 is applicable to such problems, the corresponding operators do not belong to the class introduced in Section 3. Probably, the possibilities of our approach are not exhausted in some other aspects.

We postpone further (rather heterogeneous) comments and explanations till Section 9. Necessary references are also given there.

0.8. Notation

Let $\mathfrak{H}$ and $\mathfrak{G}$ be two separable Hilbert spaces. The symbols $(\cdot,\cdot)_{\mathfrak{H}}$ and $\|\cdot\|_{\mathfrak{H}}$ stand for the inner product and the norm in $\mathfrak{H}$; the symbol $\|\cdot\|_{\mathfrak{H}\to\mathfrak{G}}$ stands for the norm of a bounded operator from $\mathfrak{H}$ to $\mathfrak{G}$. Sometimes we omit the indices, if this does not lead to confusion. Let $I = I_{\mathfrak{H}}$ denote the unit operator in $\mathfrak{H}$. If $\mathfrak{N}$ is a subspace in $\mathfrak{H}$, then $\mathfrak{N}^{\perp} := \mathfrak{H} \ominus \mathfrak{N}$. If P is the orthoprojector of $\mathfrak{H}$ onto $\mathfrak{N}$, then $P^{\perp}$ stands for the orthoprojector onto $\mathfrak{N}^{\perp}$. For a selfadjoint operator T in $\mathfrak{H}$, $\sigma(T)$ denotes its spectrum. The symbol $\langle\cdot,\cdot\rangle$ denotes the standard inner product in $\mathbb{C}^n$, $|\cdot|$ is the norm of a vector in $\mathbb{C}^n$; the unit $(n\times n)$-matrix is denoted by $\mathbf{1} = \mathbf{1}_n$. Next, $\mathbf{x} = (x^1,\dots,x^d) \in \mathbb{R}^d$, $iD_j = \partial_j = \partial/\partial x^j$, $j = 1,\dots,d$, $\nabla = \operatorname{grad} = (\partial_1,\dots,\partial_d)$, $\mathbf{D} = -i\nabla = (D_1,\dots,D_d)$, $\nabla^* = -\operatorname{div}$. The classes L_p of $\mathbb{C}^n$-valued functions in a domain $\Omega \subseteq \mathbb{R}^d$ are denoted by $L_p(\Omega;\mathbb{C}^n)$, $1 \le p \le \infty$. The Sobolev classes of order s with integrability index p of $\mathbb{C}^n$-valued functions in a domain $\Omega \subseteq \mathbb{R}^d$ are denoted by $W_p^s(\Omega;\mathbb{C}^n)$. For $p = 2$ we abbreviate this to $H^s(\Omega;\mathbb{C}^n) = W_2^s(\Omega;\mathbb{C}^n)$. For $n = 1$ we write simply $W_p^s(\Omega)$, $H^s(\Omega)$.

0.9. Acknowledgements

We are grateful to V. Ivrii and P. Kuchment for fruitful scientific contacts that stimulated our interest to the topic of this paper.

1. Operator families admitting factorization

Sections 1, 2 contain the necessary material on the analytic perturbation theory of selfadjoint operator families, which lies in the foundation of further considerations. Our goal is to specify the case where the positive operator family admits a factorization (as compared with the general perturbation theory). We study operator families near the lower edge of the spectrum.

1.1. Operator families admitting factorization

Let $\mathfrak{H}$, $\mathfrak{H}_*$ be complex separable Hilbert spaces. Suppose that the operator $X_0 : \mathfrak{H} \to \mathfrak{H}_*$ is densely defined and closed, and that the operator $X_1 : \mathfrak{H} \to \mathfrak{H}_*$ is bounded. Then the operator

$$X(t) := X_0 + tX_1, \quad t \in \mathbb{R}, \tag{1.1}$$

is closed on the domain $\operatorname{Dom} X(t) := \operatorname{Dom} X_0$. We have $X(t)^* = X_0^* + tX_1^*$ on the domain $\operatorname{Dom} X(t)^* := \operatorname{Dom} X_0^*$. The selfadjoint positive operator family

$$A(t) := X(t)^* X(t) \tag{1.2}$$

in $\mathfrak{H}$ is our main object. The operator (1.2) is generated by the closed quadratic form $\|X(t)u\|^2_{\mathfrak{H}_*}$, $u \in \operatorname{Dom} X_0$. For the operator $A_0 := A(0) = X_0^* X_0$, we put

$$\mathfrak{N} := \operatorname{Ker} A_0 = \operatorname{Ker} X_0. \tag{1.3}$$

In what follows, we suppose that the following condition is true.

Condition 1.1. *The point $\lambda = 0$ is an isolated point of the spectrum of the operator A_0 and*

$$(0 <)\, n := \dim \mathfrak{N} < \infty. \tag{1.4}$$

We denote the spectral projector of the operator $A(t)$ for a (closed) interval $[0, \sigma]$ by $F(t, \sigma)$, and put $\mathfrak{F}(t, \sigma) := F(t, \sigma)\mathfrak{H}$. We fix a number $\delta > 0$, $8\delta < d^0$, where d^0 is the distance from the point $\lambda = 0$ to the other part of the spectrum of A_0. We often write $F(t) := F(t, \delta)$ and $\mathfrak{F}(t) := \mathfrak{F}(t, \delta)$. The following statement can be easily checked with the help of the spectral theorem.

Proposition 1.2. *We have*

$$\dim \mathfrak{F}(t, \delta) = \dim \mathfrak{F}(t, 3\delta) = n, \quad |t| \le t^0 = t^0(\delta) = \delta^{1/2} \|X_1\|^{-1}. \tag{1.5}$$

We also mention that *the projector $F(t)$ is a real-analytic function of t for $|t| \le t^0$.*

Simultaneously with (1.2), it is convenient to consider the selfadjoint operator family

$$A_*(t) := X(t) X(t)^* \tag{1.6}$$

in $\mathfrak{H}_*$. We put $A_{*0} := X_0 X_0^*$,

$$\mathfrak{N}_* := \operatorname{Ker} A_{*0} = \operatorname{Ker} X_0^*, \quad n_* := \dim \mathfrak{N}_*.$$

The operators $A(t)$ and $A_*(t)$ have the same nonzero spectrum. In general, the numbers n and n_* are distinct. It may happen that $n_* = \infty$. We suppose [2] that $n_* > 0$. Next, *suppose that*

$$n \leq n_*. \tag{1.7}$$

We can always satisfy (1.7), *changing the roles of* $A(t)$ *and* $A_*(t)$.

1.2. The operator R

Introduce the notation $\mathcal{D} := \operatorname{Dom} X_0 \cap \mathfrak{N}^\perp$. Since the point $\lambda = 0$ is isolated in the spectrum of A_0, the form $(X_0\varphi, X_0\zeta)_{\mathfrak{H}_*}$, $\varphi, \zeta \in \mathcal{D}$, defines an inner product in $\mathcal{D}$, converting $\mathcal{D}$ into the Hilbert space. Let $z \in \mathfrak{H}_*$. Let $\widehat{\varphi} \in \mathcal{D}$ satisfy the equation $X_0^*(X_0\widehat{\varphi} - z) = 0$ understood in the weak sense. In other words, we look for $\widehat{\varphi} \in \mathcal{D}$ satisfying the identity

$$(X_0\widehat{\varphi}, X_0\zeta)_{\mathfrak{H}_*} = (z, X_0\zeta)_{\mathfrak{H}_*}, \quad \forall\, \zeta \in \mathcal{D}. \tag{1.8}$$

Since the right-hand side of (1.8) is an antilinear continuous functional of $\zeta \in \mathcal{D}$, there exists the unique element $\widehat{\varphi} \in \mathcal{D}$ satisfying (1.8). Moreover, $\|X_0\widehat{\varphi}\|_{\mathfrak{H}_*} \leq \|z\|_{\mathfrak{H}_*}$. Obviously, $X_0\widehat{\varphi} - z \in \mathfrak{N}_*$.

Now, let

$$\omega \in \mathfrak{N}, \quad z := -X_1\omega. \tag{1.9}$$

Consider $\widehat{\varphi} = \widehat{\varphi}(\omega)$ satisfying (1.8), and put

$$\omega_* := X_0\widehat{\varphi} + X_1\omega \in \mathfrak{N}_*. \tag{1.10}$$

In accordance with (1.8)-(1.10), the linear operator from $\mathfrak{N}$ to $\mathfrak{N}_*$, which takes ω into ω_*, is denoted by R:

$$R : \mathfrak{N} \to \mathfrak{N}_*, \quad R\omega = \omega_*. \tag{1.11}$$

Let P be the orthoprojector of $\mathfrak{H}$ onto $\mathfrak{N}$, and let P_* be the orthoprojector of $\mathfrak{H}_*$ onto $\mathfrak{N}_*$. By (1.10), $X_1\omega = \omega_* - X_0\widehat{\varphi}$, where $\omega_* \in \mathfrak{N}_*$ and $X_0\widehat{\varphi} \in \operatorname{Ran} X_0 \subset \mathfrak{N}_*^\perp$. Hence, $\omega_* = P_*X_1\omega$. This yields the "explicit" representation

$$R = P_*X_1|_{\mathfrak{N}}, \tag{1.12}$$

which is equivalent to

$$P_*X_1P = R \oplus \mathbb{O}. \tag{1.13}$$

Here we extend R onto $\mathfrak{N}^\perp$ by zero.

The continuous operator $R_* : \mathfrak{N}_* \to \mathfrak{N}$ is defined similarly. We have

$$R_* = PX_1^*|_{\mathfrak{N}_*}, \quad PX_1^*P_* = R_* \oplus \mathbb{O}. \tag{1.14}$$

By (1.12)-(1.14), it is clear that

$$R_* = R^*. \tag{1.15}$$

---

[2] The condition $n_* > 0$ holds automatically, if $A(t)F(t)$ has at least one positive eigenvalue for small $t > 0$.

1.3. The spectral germ S of a family $A(t)$ at $t=0$

Definition 1.3. *The selfadjoint operator*

$$S := R^*R : \mathfrak{N} \to \mathfrak{N} \tag{1.16}$$

is called the spectral germ of the operator family (1.2) *at the point* $t=0$.

From (1.13)-(1.15) it follows that

$$S = PX_1^*P_*X_1|_{\mathfrak{N}}, \quad S \oplus \mathbb{O} = PX_1^*P_*X_1P. \tag{1.17}$$

The operator $S_* := RR^* : \mathfrak{N}_* \to \mathfrak{N}_*$ is the spectral germ of the family (1.6). The part of the operator S acting in $(\operatorname{Ker} S)^\perp$ is unitarily equivalent to the part of S_* acting in $(\operatorname{Ker} S_*)^\perp$. Below (see Section 2), we will clarify the role of the operator S. In particular, we will show that S *is independent of the concrete choice of factorization* (1.2) *for* $A(t)$.

Definition 1.4. *The spectral germ S of the operator family* (1.2) *at $t=0$ is called nondegenerate if* $\operatorname{Ker} S = \{0\}$.

Obviously, the nondegeneracy of S is equivalent to the condition $\operatorname{Ker} R = \{0\}$, or, equivalently, to the condition $\operatorname{rank} R = n$.

Let $\widetilde{A}(t) = \widetilde{X}(t)^*\widetilde{X}(t)$ be one more operator family in $\mathfrak{H}$, subject to the same conditions (see Subsection 1.1) as $A(t)$. Let $\widetilde{\mathfrak{N}} := \operatorname{Ker} \widetilde{A}(0)$ and let $\widetilde{S} : \widetilde{\mathfrak{N}} \to \widetilde{\mathfrak{N}}$ be the spectral germ of the family $\widetilde{A}(t)$ at $t=0$.

Definition 1.5. *Operator families $A(t)$ and $\widetilde{A}(t)$ are called spectral-equivalent at the point $t=0$ if $\mathfrak{N} = \widetilde{\mathfrak{N}}$ and $S = \widetilde{S}$.*

Obviously, the defined relation is equivalence relation on the set of operator families of the form (1.2).

1.4. Estimates for the operator S

Upper estimates. Let $\mathfrak{N}_* \subset \mathfrak{G}_* \subset \mathfrak{H}_*$, where $\mathfrak{G}_*$ is a subspace in $\mathfrak{H}_*$, and let Π_* be the orthoprojector onto $\mathfrak{G}_*$. Obviously, (1.17) implies that

$$S \oplus \mathbb{O} \le PX_1^*\Pi_*X_1P, \tag{1.18}$$

or, equivalently,

$$(S\zeta,\zeta)_{\mathfrak{H}} \le \|\Pi_*X_1\zeta\|^2_{\mathfrak{H}_*}, \quad \zeta \in \mathfrak{N}. \tag{1.19}$$

In particular, if $\Pi_* = I_{\mathfrak{H}_*}$, then inequalities (1.18), (1.19) take the form

$$S \oplus \mathbb{O} \le PX_1^*X_1P,$$

$$(S\zeta,\zeta)_{\mathfrak{H}} \le \|X_1\zeta\|^2_{\mathfrak{H}_*}, \quad \zeta \in \mathfrak{N}. \tag{1.20}$$

Finally, (1.20) implies the rougher estimate

$$S \le \|X_1\|^2 I_{\mathfrak{N}}.$$

Estimate (1.19) gives an opportunity for refinement of estimates for the operator S on the basis of the Ritz method.

Lower estimates for S. Let $\mathfrak{G}'_*$ be a subspace in $\mathfrak{N}_*$ and let Π'_* be the orthoprojector onto $\mathfrak{G}'_*$. Then

$$S \oplus \mathbb{O} \geq P X_1^* \Pi'_* X_1 P,$$

$$(S\zeta, \zeta)_{\mathfrak{H}} \geq \|\Pi'_* X_1 \zeta\|^2_{\mathfrak{H}_*}, \quad \zeta \in \mathfrak{N}. \tag{1.21}$$

The upper and lower estimates for S_* can be obtained in a similar way.

1.5. The case of a family $X(t) = \widehat{X}(t)M$

Let $\widehat{\mathfrak{H}}$ be one more Hilbert space, and let $\widehat{X}(t) = \widehat{X}_0 + t\widehat{X}_1 : \widehat{\mathfrak{H}} \to \mathfrak{H}_*$ be a family of the form (1.1) and satisfying the assumptions of Subsection 1.1. Suppose that $M : \mathfrak{H} \to \widehat{\mathfrak{H}}$ is an isomorphism. Let $X(t) = \widehat{X}(t)M = X_0 + tX_1 : \mathfrak{H} \to \mathfrak{H}_*$, where $X_0 = \widehat{X}_0 M$, $X_1 = \widehat{X}_1 M$. All the objects related to $\widehat{X}(t)$ are marked by the upper index "$\widehat{\ }$". We note that $\widehat{\mathfrak{N}} = M\mathfrak{N}$, $\widehat{\mathfrak{N}}_* = \mathfrak{N}_*$, $\widehat{n} = n$, $\widehat{n}_* = n_*$. Besides, $R = P_* X_1|_{\mathfrak{N}} = \widehat{P}_* \widehat{X}_1 M|_{\mathfrak{N}}$, i. e.,

$$R = \widehat{R} M|_{\mathfrak{N}}. \tag{1.22}$$

For $\zeta \in \mathfrak{N}$ and $\widehat{\zeta} = M\zeta$, we have

$$(S\zeta, \zeta)_{\mathfrak{H}} = \|R\zeta\|^2_{\mathfrak{H}_*} = \|\widehat{R}\widehat{\zeta}\|^2_{\mathfrak{H}_*} = (\widehat{S}\widehat{\zeta}, \widehat{\zeta})_{\widehat{\mathfrak{H}}} = (M^*\widehat{S}M\zeta, \zeta)_{\mathfrak{H}}. \tag{1.23}$$

Hence,

$$S = PM^*\widehat{S}M|_{\mathfrak{N}}. \tag{1.24}$$

From (1.23) it follows that

$$\frac{(S\zeta, \zeta)_{\mathfrak{H}}}{\|\zeta\|^2_{\mathfrak{H}}} = \frac{(\widehat{S}\widehat{\zeta}, \widehat{\zeta})_{\widehat{\mathfrak{H}}}}{\|\widehat{\zeta}\|^2_{\widehat{\mathfrak{H}}}} \frac{\|M\zeta\|^2_{\widehat{\mathfrak{H}}}}{\|\zeta\|^2_{\mathfrak{H}}} \leq \|M\|^2 \frac{(\widehat{S}\widehat{\zeta}, \widehat{\zeta})_{\widehat{\mathfrak{H}}}}{\|\widehat{\zeta}\|^2_{\widehat{\mathfrak{H}}}}, \tag{1.25}$$

and, similarly, $(\widehat{S}\widehat{\zeta}, \widehat{\zeta})_{\widehat{\mathfrak{H}}} \|\widehat{\zeta}\|^{-2}_{\widehat{\mathfrak{H}}} \leq \|M^{-1}\|^2 (S\zeta, \zeta)_{\mathfrak{H}} \|\zeta\|^{-2}_{\mathfrak{H}}$. Hence, the eigenvalues γ_l of the operator S and the eigenvalues $\widehat{\gamma}_l$ of the operator $\widehat{S}$ satisfy the relations

$$\|M^{-1}\|^{-2}\widehat{\gamma}_l \leq \gamma_l \leq \|M\|^2 \widehat{\gamma}_l, \quad l = 1, \ldots, n.$$

The role of the numbers γ_l in the perturbation theory for the family $A(t)$ will be clarified below in Subsection 2.1.

Now, let $n = 1$. Then (1.25) implies the following relation for $\gamma = \gamma_1$, $\widehat{\gamma} = \widehat{\gamma}_1$:

$$\gamma = \widehat{\gamma} \|M\zeta\|^2_{\widehat{\mathfrak{H}}} \|\zeta\|^{-2}_{\mathfrak{H}}. \tag{1.26}$$

2. Perturbation theory

2.1. Eigenvalue problem for the operator S

According to the general analytic perturbation theory, for $|t| \leq t^0$, there exist real-analytic functions $\lambda_l(t)$ (the branches of eigenvalues) and real-analytic $\mathfrak{H}$-valued functions $\varphi_l(t)$ (the branches of eigenvectors) such that

$$A(t)\varphi_l(t) = \lambda_l(t)\varphi_l(t), \quad l = 1, \ldots, n, \quad |t| \leq t^0 = t^0(\delta), \tag{2.1}$$

and $\varphi_l(t)$, $l = 1, \dots, n$, form an orthonormal basis in $\mathfrak{F}(t)$. Moreover, for sufficiently small t_*, we have the convergent power series expansions

$$\lambda_l(t) = \gamma_l t^2 + \cdots, \quad \gamma_l \geq 0, \quad l = 1, \dots, n, \quad |t| \leq t_* \leq t^0, \tag{2.2}$$

$$\varphi_l(t) = \omega_l + t\widehat{\varphi}_l + t^2 \varphi_l^{(2)} + \cdots, \quad l = 1, \dots, n, \quad |t| \leq t_* \leq t^0, \tag{2.3}$$

and $\omega_l := \varphi_l(0)$, $l = 1, \dots, n$, form an orthonormal basis in $\mathfrak{N}$.

Relations (2.1) are equivalent to

$$(X(t)\varphi_l(t), X(t)\zeta)_{\mathfrak{H}_*} = \lambda_l(t)(\varphi_l(t), \zeta)_{\mathfrak{H}}, \quad \forall\, \zeta \in \operatorname{Dom} X_0. \tag{2.4}$$

The values ω_l, $\widehat{\varphi}_l$, $\varphi_l^{(2)}, \dots$, and $\gamma_l, \dots$, can be found from (2.4) and from the relation $\|\varphi_l(t)\|^2_{\mathfrak{H}} = 1$, by comparing the terms with the same power of t. Comparing the terms of first order, we obtain (see (1.8), (1.9))

$$\widehat{\varphi}_l - \widehat{\varphi}(\omega_l) \in \mathfrak{N}, \quad l = 1, \dots, n. \tag{2.5}$$

Comparing the terms with t^2, we see that

$$(X_0\widehat{\varphi}_l + X_1\omega_l, X_1\zeta)_{\mathfrak{H}_*} + (X_0\varphi_l^{(2)} + X_1\widehat{\varphi}_l, X_0\zeta)_{\mathfrak{H}_*} = \gamma_l(\omega_l, \zeta)_{\mathfrak{H}}, \quad \forall\, \zeta \in \operatorname{Dom} X_0. \tag{2.6}$$

In accordance with (1.10), (1.11), (2.5), $X_0\widehat{\varphi}_l + X_1\omega_l = R\omega_l$. For $\zeta \in \mathfrak{N}$, relation (2.6) takes the form

$$(R\omega_l, X_1\zeta)_{\mathfrak{H}_*} = \gamma_l(\omega_l, \zeta)_{\mathfrak{H}}, \quad \forall\, \zeta \in \mathfrak{N}. \tag{2.7}$$

By (1.12), (1.17),

$$(R\omega_l, X_1\zeta)_{\mathfrak{H}_*} = (PX_1^* R\omega_l, \zeta)_{\mathfrak{H}} = (S\omega_l, \zeta)_{\mathfrak{H}}.$$

Then (2.7) yields the following proposition.

Proposition 2.1. *The numbers γ_l and the elements ω_l defined by* (2.1)-(2.3) *are eigenvalues and eigenvectors of the operator S:*

$$S\omega_l = \gamma_l\omega_l, \quad l = 1, \dots, n. \tag{2.8}$$

This clarifies the meaning of eigenvalue problem for the operator S.

Series (2.2), (2.3) are determined by the family $A(t)$ directly and independently of factorization (1.2). Therefore, relations (2.8) can be viewed as another definition of the operator S. Thus, S does not depend on the choice of factorization. At the same time, relation (1.16) can be interpreted as "inheritance" of the chosen factorization for $A(t)$ by the operator S. The nondegeneracy of S is equivalent to the inequalities

$$\gamma_l \geq c_* > 0, \quad l = 1, \dots, n, \tag{2.9}$$

for the numbers γ_l defined by (2.2).

If all the eigenvalues γ_l are simple, then (2.8) determines the initial elements ω_l in (2.3) (up to a phase factor). If there are multiple eigenvalues among γ_l, then, in general, the knowledge of S is insufficient for this purpose.

2.2. Approximation of $F(t)$ and $A(t)F(t)$

Both operator-valued functions $F(t)$ and $A(t)F(t)$ are real-analytic for $|t| \leq t^0(\delta)$. We have the following power series expansions:

$$F(t) = P + tF_1 + \cdots, \quad |t| \leq t_*, \tag{2.10}$$

$$A(t)F(t) = t^2 SP + \cdots, \quad |t| \leq t_*. \tag{2.11}$$

Explain that, in accordance with (2.1) and Proposition 1.2,

$$A(t)F(t) = \sum_{l=1}^{n} \lambda_l(t)(\cdot, \varphi_l(t))_{\mathfrak{H}}\, \varphi_l(t), \quad |t| \leq t^0(\delta). \tag{2.12}$$

Combining (2.12) and (2.2), (2.3), we obtain (2.11). Relation (2.11) is significant, since S admits representations (1.16), (1.17), which do not require the knowledge of eigenvectors $\omega_1, \ldots, \omega_n$.

However, representations (2.10), (2.11) are not completely sufficient for our purposes. We put

$$F(t) - P =: t\Phi(t), \quad |t| \leq t^0(\delta), \tag{2.13}$$

$$A(t)F(t) - t^2 SP =: t^3 \Psi(t), \quad |t| \leq t^0(\delta). \tag{2.14}$$

Proposition 2.2. *We have*

$$\|\Phi(t)\| \leq C_1, \quad \|\Psi(t)\| \leq C_2, \quad |t| \leq t^0(\delta) = \delta^{1/2}\|X_1\|^{-1}, \tag{2.15}$$

where $C_j = C_j(\delta, \|X_1\|)$, *and* $C_j(\delta, \cdot)$ *are monotone increasing functions,* $j = 1, 2$.

Recall (cf. Proposition 1.2), that we choose the number δ such that $8\delta < d^0$, and next we fix δ. As compared with (2.10), (2.11), estimates (2.15) are valid on a wider interval, and the constants in estimates are well controled, which is essential for us.

Unfortunately, estimates (2.15) do not follow from the ready results of the spectral perturbation theory. However, for the lack of space, we do not give a rather technical proof here.

2.3. Approximation of the resolvent

Now, we establish the asymptotic estimate, which will directly lead to solution of the homogenization problem for a periodic DO with small period (see Section 5). We rely on estimates (2.15) supplemented by the following condition.

Condition 2.3. *There exists a constant* c_* *such that*

$$\lambda_l(t) \geq c_* t^2, \quad l = 1, \ldots, n, \quad |t| \leq t^0(\delta). \tag{2.16}$$

Obviously, (2.16) implies (2.9) with the same constant.

Recall that in (2.13), (2.14), $F(t) = F(t, \delta)$ is the spectral projector of $A(t)$ introduced in Subsection 1.1.

Theorem 2.4. *We put*

$$(A(t)+\varepsilon^2 I)^{-1}F(t) - P(t^2SP+\varepsilon^2 I)^{-1} =: \Xi_\varepsilon(t), \quad 0<\varepsilon\le 1. \tag{2.17}$$

Under Condition 2.3, *we have*

$$\varepsilon\|\Xi_\varepsilon(t)\| \le C_3 =: 4C_1 c_*^{-1/2} + C_2 c_*^{-3/2}, \quad |t|\le t^0(\delta), \tag{2.18}$$

where C_1, C_2 *are constants from* (2.15).

Proof. Temporarily, we will write A instead of $A(t)$, F instead of $F(t)$, etc. We start with the following representation of the operator (2.17), which can be checked by direct calculation:

$$\Xi_\varepsilon = J_1 + J_2, \quad J_1 = (A+\varepsilon^2 I)^{-1}FP^\perp,$$

$$J_2 = (A+\varepsilon^2 I)^{-1}FP - P(t^2SP+\varepsilon^2 I)^{-1}P = \\ (A+\varepsilon^2 I)^{-1}(F-P)P + (F-P)(t^2SP+\varepsilon^2 I)^{-1}P + J_3,$$

$$J_3 = (A+\varepsilon^2 I)^{-1}P - F(t^2SP+\varepsilon^2 I)^{-1}P = \\ (A+\varepsilon^2 I)^{-1}\left((t^2SP+\varepsilon^2 I)P - (A+\varepsilon^2 I)F\right)(t^2SP+\varepsilon^2 I)^{-1}P = \\ (A+\varepsilon^2 I)^{-1}(t^2SP - AF)(t^2SP+\varepsilon^2 I)^{-1}P + \\ \varepsilon^2(A+\varepsilon^2 I)^{-1}(P-F)(t^2SP+\varepsilon^2 I)^{-1}P.$$

Thus, taking account of (2.13), (2.14), we have

$$\Xi_\varepsilon = J_1 + J_{21} + J_{22} + J_{31} + J_{32}, \tag{2.19}$$

$$J_{21} = t(A+\varepsilon^2 I)^{-1}\Phi P, \quad J_{22} = t\Phi(t^2SP+\varepsilon^2 I)^{-1}P,$$

$$J_{31} = -t^3(A+\varepsilon^2 I)^{-1}\Psi(t^2SP+\varepsilon^2 I)^{-1}P, \quad J_{32} = -\varepsilon^2 t(A+\varepsilon^2 I)^{-1}\Phi(t^2SP+\varepsilon^2 I)^{-1}P.$$

The operators J_{21}, J_{22}, J_{31}, J_{32} are estimated in the same way. One should only take into account that, by (2.9), (2.16),

$$\|(A+\varepsilon^2 I)^{-1}\| \le (c_* t^2+\varepsilon^2)^{-1}, \quad \|P(t^2SP+\varepsilon^2 I)^{-1}\| \le (c_* t^2+\varepsilon^2)^{-1}.$$

Combining this with (2.15), we see that

$$\varepsilon\|J_{21}+J_{22}+J_{31}+J_{32}\| \le 3C_1 c_*^{-1/2} + C_2 c_*^{-3/2}. \tag{2.20}$$

For estimating J_1, in addition, we note that $FP^\perp = (P+t\Phi)P^\perp = t\Phi P^\perp$. Then $\varepsilon\|J_1\| \le C_1 c_*^{-1/2}$. In combination with (2.19), (2.20), this yields (2.18). □

Note that, by Proposition 1.2, $\|(A+\varepsilon^2 I)^{-1}F^\perp\| \le (3\delta)^{-1}$ for $|t|\le t^0$. Therefore, Theorem 2.4 implies the following corollary.

Corollary 2.5. *Under the assumptions of Theorem* 2.4, *we have*

$$\varepsilon^2\|(A(t)+\varepsilon^2 I)^{-1} - P(t^2SP+\varepsilon^2 I)^{-1}\| \le C_4\varepsilon, \\ C_4 = C_3 + (3\delta)^{-1}, \quad 0<\varepsilon\le 1, \quad |t|\le t^0(\delta). \tag{2.21}$$

3. Periodic differential operators of second order. Main notions

3.1. Differential operators in $L_2(\mathbb{R}^d;\mathbb{C}^n)$ with periodic coefficients

Let $m, n \in \mathbb{N}$, and let $m \geq n$. Consider the linear homogeneous differential operator $b(\mathbf{D}) : L_2(\mathbb{R}^d;\mathbb{C}^n) \to L_2(\mathbb{R}^d;\mathbb{C}^m)$ with constant coefficients. Let $\widetilde{\mathbf{e}}_1,\dots,\widetilde{\mathbf{e}}_n$ be the standard orthonormal basis in $\mathbb{C}^n$. The standard orthonormal basis in $\mathbb{C}^m$ is denoted by $\mathbf{e}_1,\dots,\mathbf{e}_m$. By $\mathbf{e}'_1,\dots,\mathbf{e}'_d$ we denote the standard orthonormal basis in $\mathbb{R}^d$. Then the *symbol* of the operator $b(\mathbf{D})$ is a $(m \times n)$-matrix $b(\boldsymbol{\xi})$, which is a linear homogeneous function of $\boldsymbol{\xi} \in \mathbb{R}^d$. Suppose that

$$\operatorname{rank} b(\boldsymbol{\xi}) = n, \quad 0 \neq \boldsymbol{\xi} \in \mathbb{R}^d. \tag{3.1}$$

Relation (3.1) yields

$$\alpha_0 \mathbf{1}_n \leq b(\boldsymbol{\theta})^* b(\boldsymbol{\theta}) \leq \alpha_1 \mathbf{1}_n, \quad |\boldsymbol{\theta}| = 1, \quad 0 < \alpha_0 \leq \alpha_1 < \infty. \tag{3.2}$$

Note that

$$b(\boldsymbol{\xi}) = \sum_{j=1}^{d} \xi^j b_j, \quad \boldsymbol{\xi} = \sum_{j=1}^{d} \xi^j \mathbf{e}'_j, \tag{3.3}$$

where b_j are constant $(m \times n)$-matrices of rank n.

Let Γ be a lattice in $\mathbb{R}^d$. Let vectors $\mathbf{a}_1,\dots,\mathbf{a}_d \in \mathbb{R}^d$ constitute a basis of Γ, and let

$$\Omega := \{\mathbf{x} \in \mathbb{R}^d : \ \mathbf{x} = t_1\mathbf{a}_1 + \dots + t_d\mathbf{a}_d, \quad 0 < t_j < 1, \quad j = 1,\dots,d\}$$

be the elementary cell of the lattice Γ. Suppose that invertible $(n \times n)$-matrix $f(\mathbf{x})$ and $(m \times m)$-matrix $h(\mathbf{x})$ are Γ-periodic functions of $\mathbf{x} \in \mathbb{R}^d$, and

$$f,\ f^{-1} \in L_\infty; \quad h,\ h^{-1} \in L_\infty. \tag{3.4}$$

Consider the periodic differential operator

$$\mathcal{X} := hb(\mathbf{D})f : L_2(\mathbb{R}^d;\mathbb{C}^n) \to L_2(\mathbb{R}^d;\mathbb{C}^m),$$

$$\operatorname{Dom} \mathcal{X} = \{\mathbf{u} \in L_2(\mathbb{R}^d;\mathbb{C}^n) : \ f\mathbf{u} \in H^1(\mathbb{R}^d;\mathbb{C}^n)\}. \tag{3.5}$$

The operator $\mathcal{X}$ is closed on the domain (3.5). The selfadjoint operator $\mathcal{A} := \mathcal{X}^*\mathcal{X}$ in $L_2(\mathbb{R}^d;\mathbb{C}^n)$ is generated by the quadratic form

$$a[\mathbf{u},\mathbf{u}] := \|\mathcal{X}\mathbf{u}\|^2_{L_2(\mathbb{R}^d;\mathbb{C}^m)}, \quad \mathbf{u} \in \operatorname{Dom} \mathcal{X}. \tag{3.6}$$

Formally, $\mathcal{A} = f(\mathbf{x})^* b(\mathbf{D})^* g(\mathbf{x}) b(\mathbf{D}) f(\mathbf{x})$, where $g(\mathbf{x}) := h(\mathbf{x})^* h(\mathbf{x})$. Using the Fourier transformation and conditions (3.1), (3.4), it is easily seen that

$$c_0 \int_{\mathbb{R}^d} |\nabla(f\mathbf{u})|^2\, d\mathbf{x} \leq a[\mathbf{u},\mathbf{u}] \leq c_1 \int_{\mathbb{R}^d} |\nabla(f\mathbf{u})|^2\, d\mathbf{x}, \quad \mathbf{u} \in \operatorname{Dom} \mathcal{X}, \quad 0 < c_0 \leq c_1 < \infty. \tag{3.7}$$

The constants c_0, c_1 are estimated via the L_∞-norms of matrix-valued functions h, h^{-1} from (3.4) and via the constants α_0, α_1 from (3.2).

3.2. Differential operators in $L_2(\Omega;\mathbb{C}^n)$

Consider the operator family

$$\mathcal{X}(\mathbf{k}) := hb(\mathbf{D}+\mathbf{k})f : L_2(\Omega;\mathbb{C}^n) \to L_2(\Omega;\mathbb{C}^m), \tag{3.8}$$

$$\mathfrak{d} := \operatorname{Dom}\mathcal{X}(\mathbf{k}) = \{\mathbf{u} \in L_2(\Omega;\mathbb{C}^n) : \ f\mathbf{u} \in \widetilde{H}^1(\Omega;\mathbb{C}^n)\}, \tag{3.9}$$

depending on the *quasimomentum* $\mathbf{k} \in \mathbb{R}^d$. Here $\widetilde{H}^1(\Omega;\mathbb{C}^n)$ *is formed by the functions of class* $L_2(\Omega;\mathbb{C}^n)$, *whose periodic extension belongs to* $H^1_{\mathrm{loc}}(\mathbb{R}^d;\mathbb{C}^n)$. The operators (3.8) are closed on the domain (3.9). The selfadjoint operator $\mathcal{A}(\mathbf{k}) := \mathcal{X}(\mathbf{k})^*\mathcal{X}(\mathbf{k})$ in $L_2(\Omega;\mathbb{C}^n)$ is generated by the quadratic form

$$a(\mathbf{k})[\mathbf{u},\mathbf{u}] := \|\mathcal{X}(\mathbf{k})\mathbf{u}\|^2_{L_2(\Omega;\mathbb{C}^m)}, \quad \mathbf{u} \in \mathfrak{d}. \tag{3.10}$$

Formally, $\mathcal{A}(\mathbf{k}) = f(\mathbf{x})^* b(\mathbf{D}+\mathbf{k})^* g(\mathbf{x}) b(\mathbf{D}+\mathbf{k}) f(\mathbf{x})$. Using the Fourier series expansion for $f\mathbf{u}$ and conditions (3.1), (3.4), we see that

$$c_0 \int_\Omega |(\mathbf{D}+\mathbf{k})(f\mathbf{u})|^2\, d\mathbf{x} \le a(\mathbf{k})[\mathbf{u},\mathbf{u}] \le c_1 \int_\Omega |(\mathbf{D}+\mathbf{k})(f\mathbf{u})|^2\, d\mathbf{x}, \quad \mathbf{u} \in \mathfrak{d}. \tag{3.11}$$

Here the constants c_0, c_1 are the same as in (3.7). Let

$$\mathfrak{N} := \operatorname{Ker}\mathcal{A}(0) = \operatorname{Ker}\mathcal{X}(0). \tag{3.12}$$

Relation (3.11) with $\mathbf{k} = 0$ implies that

$$\mathfrak{N} = \{\mathbf{u} \in L_2(\Omega;\mathbb{C}^n) : \ f\mathbf{u} = \mathbf{c} \in \mathbb{C}^n\}, \quad \dim\mathfrak{N} = n. \tag{3.13}$$

Since the embedding of $\widetilde{H}^1(\Omega;\mathbb{C}^n)$ into $L_2(\Omega;\mathbb{C}^n)$ is compact, we can deduce from (3.4), (3.11) that the *resolvent of* $\mathcal{A}(\mathbf{k})$ *is compact.* Equivalently, the spectrum of $\mathcal{A}(\mathbf{k})$ is discrete. Let

$$E_1(\mathbf{k}) \le E_2(\mathbf{k}) \le \cdots \le E_s(\mathbf{k}) \le \cdots, \quad \mathbf{k} \in \mathbb{R}^d, \tag{3.14}$$

be the consecutive eigenvalues of $\mathcal{A}(\mathbf{k})$ counted with multiplicities. Let $\widetilde{\Gamma}$ be the lattice dual with respect to Γ. *The band functions* (3.14) are $\widetilde{\Gamma}$-periodic in $\mathbf{k}$. The eigenvalues can be enumerated so that these functions become *continuous.* Relations (3.12), (3.13) show that

$$\min_{\mathbf{k}\in\mathbb{R}^d} E_j(\mathbf{k}) = E_j(0) = 0, \quad j = 1,\dots,n. \tag{3.15}$$

Let $\mathbf{b}_1,\dots,\mathbf{b}_d$ be the basis of the lattice $\widetilde{\Gamma}$ such that $\mathbf{b}_j\mathbf{a}_l = 2\pi\delta_{jl}$, $j,l = 1,\dots,d$, and let

$$\widetilde{\Omega} := \{\mathbf{k} \in \mathbb{R}^d : \ \mathbf{k} = \tau_1\mathbf{b}_1 + \cdots + \tau_d\mathbf{b}_d, \quad -1 < 2\tau_j < 1, \quad j = 1,\dots,d\}.$$

Inequalities (3.11) allow us to establish a number of useful estimates for the functions (3.14). Let

$$\mathcal{K}(r) = \{|\mathbf{k}| \le r\}, \quad r \le r_0 := \operatorname{dist}\ \{0, \partial\widetilde{\Omega}\}. \tag{3.16}$$

Then

$$E_l(\mathbf{k}) \ge c_*|\mathbf{k}|^2, \quad c_* > 0, \quad l = 1,\dots,n, \quad \mathbf{k} \in \mathcal{K}(r_0), \tag{3.17}$$

$$E_{n+1}(\mathbf{k}) \ge d_* > 0, \quad \mathbf{k} \in \widetilde{\Omega}, \tag{3.18}$$

$$E_1(\mathbf{k}) \geq d(r) > 0, \quad \mathbf{k} \in \widetilde{\Omega} \setminus \mathcal{K}(r), \quad 0 < r \leq r_0. \tag{3.19}$$

First, these estimates are proved (with the help of the Fourier series expansion) in the case of the form

$$\int_{\Omega} |(\mathbf{D} + \mathbf{k})\mathbf{v}|^2 \, d\mathbf{x}, \quad \mathbf{v} \in \widetilde{H}^1(\Omega; \mathbb{C}^n).$$

Next, combining (3.4), (3.11) and the variational principle, we carry the estimates over to the case of the form (3.10) of general type.

3.3. The direct integral decomposition for the operator $\mathcal{A}$

The operators $\mathcal{A}(\mathbf{k})$ allow us to partially diagonalize the operator $\mathcal{A}$. Consider the mapping (*the Gelfand transformation*)

$$\mathcal{U} : L_2(\mathbb{R}^d; \mathbb{C}^n) \to \int_{\widetilde{\Omega}} \oplus L_2(\Omega) \, d\mathbf{k} =: \mathcal{H}, \tag{3.20}$$

that is defined on the functions of the Schwartz class by the formula

$$(\mathcal{U}\mathbf{v})(\mathbf{k}, \mathbf{x}) = (\operatorname{mes} \widetilde{\Omega})^{-1/2} e^{-i\langle \mathbf{k}, \mathbf{x} \rangle} \sum_{\mathbf{l} \in \Gamma} e^{-i\langle \mathbf{k}, \mathbf{l} \rangle} \mathbf{v}(\mathbf{x} + \mathbf{l}),$$

and is extended by continuity to a unitary mapping from $L_2(\mathbb{R}^d; \mathbb{C}^n)$ onto $\mathcal{H}$. It turns out that, for $\mathbf{u} \in \operatorname{Dom} \mathcal{X}$,

$$(\mathcal{U}\mathbf{u})(\mathbf{k}, \cdot) \in \mathfrak{d}, \quad \text{for a.e. } \mathbf{k} \in \widetilde{\Omega}, \tag{3.21}$$

$$a[\mathbf{u}, \mathbf{u}] = \int_{\widetilde{\Omega}} a(\mathbf{k})[(\mathcal{U}\mathbf{u})(\mathbf{k}, \cdot), (\mathcal{U}\mathbf{u})(\mathbf{k}, \cdot)] \, d\mathbf{k}. \tag{3.22}$$

Conversely, if for $\mathbf{u} \in L_2(\mathbb{R}^d; \mathbb{C}^n)$ we have (3.21) and the integral (3.22) is finite, then $\mathbf{u} \in \operatorname{Dom} \mathcal{X}$. Consequently, in the direct integral $\mathcal{H}$ (see (3.20)) the operator $\mathcal{A}$ acts as multiplication by the operator-valued function $\mathcal{A}(\mathbf{k})$, $\mathbf{k} \in \widetilde{\Omega}$. Briefly, all this can be expressed by the formula

$$\mathcal{U}\mathcal{A}\mathcal{U}^{-1} = \int_{\widetilde{\Omega}} \oplus \mathcal{A}(\mathbf{k}) \, d\mathbf{k}. \tag{3.23}$$

Thus, the mapping $\mathcal{U}$ transforms $\mathcal{A}$ into the operator of multiplication by the operator-valued function $\mathcal{A}(\mathbf{k})$.

The spectrum $\sigma(\mathcal{A})$ of the operator $\mathcal{A}$ is the union of segments (bands), that are the ranges of band functions (3.14). Relation (3.15) implies that the lower edge of the spectrum of $\mathcal{A}$ is the point $\lambda = 0$: $\inf \sigma(\mathcal{A}) = 0$. The behaviour of the band functions near the lower edge of the spectrum is of great interest. In what follows, we pay the main attention to this question.

3.4. Application of the method of Section 1 to $\mathcal{A}(\mathbf{k})$

For $d > 1$, the operator family $\mathcal{A}(\mathbf{k})$ depends on the *multi-dimensional* parameter $\mathbf{k}$. In this case, the analytic perturbation theory gives satisfactory results only for *simple* eigenvalues. The role of the unperturbed operator A_0 is played by $\mathcal{A}(0)$. Now, by (3.12), (3.13), we have $\dim \mathfrak{N} = n \geq 1$. Thus, when $n > 1$, the eigenvalue $\lambda = 0$ for $\mathcal{A}(0)$ is multiple. To avoid this difficulty, for $\mathbf{k} \in \mathbb{R}^d$ we put $\mathbf{k} = t\boldsymbol{\theta}$, $t = |\mathbf{k}|$, $|\boldsymbol{\theta}| = 1$, and view t as the perturbation parameter. At the same time, all the constructions will depend on the additional parameter $\boldsymbol{\theta}$, which will be often marked in the notation. Thus, we apply the method of Section 1 with

$$\mathfrak{H} = L_2(\Omega; \mathbb{C}^n), \quad \mathfrak{H}_* = L_2(\Omega; \mathbb{C}^m),$$

$$X(t) = X(t, \boldsymbol{\theta}) = \mathcal{X}(t\boldsymbol{\theta}), \quad A(t) = A(t, \boldsymbol{\theta}) = \mathcal{A}(t\boldsymbol{\theta}), \tag{3.24}$$

$$X_0 = \mathcal{X}(0) = hb(\mathbf{D})f, \quad \operatorname{Dom} X_0 = \mathfrak{d}, \tag{3.25}$$

$$X_1 = X_1(\boldsymbol{\theta}) = hb(\boldsymbol{\theta})f.$$

The notation (3.12) is coordinated with (1.3), the number n in (1.4) is the same as in (3.13). It is clear that $d^0 = E_{n+1}(0) \geq d_*$, where d_* is the constant from (3.18). We choose the number δ such that $0 < 8\delta < d_*$ and

$$\widetilde{t} = \widetilde{t}(\delta) := \delta^{1/2} \left(\max_{\boldsymbol{\theta} \in \mathbb{S}^{d-1}} \|X_1(\boldsymbol{\theta})\| \right)^{-1} \leq r_0, \tag{3.26}$$

where r_0 is defined by (3.16). Next, we fix δ. Note (see (1.5)) that $\widetilde{t} \leq t^0(\delta)$.

The kernel $\mathfrak{N}_*$ of X_0^* has the form

$$\mathfrak{N}_* = \{\mathbf{q} \in L_2(\Omega; \mathbb{C}^m) : \ h^*\mathbf{q} \in \widetilde{H}^1(\Omega; \mathbb{C}^m), \ b(\mathbf{D})^* h(\mathbf{x})^* \mathbf{q} = 0\}. \tag{3.27}$$

Combining this with the condition $m \geq n$, we see that for the operator (3.25) the following *alternative* realizes: either $n_* = \infty$ (if $m > n$) or $n_* = n$ (if $m = n$). Condition (1.7) holds automatically.

The branches of eigenvalues $\lambda_l(t, \boldsymbol{\theta})$ and eigenvectors $\varphi_l(t, \boldsymbol{\theta})$ (see (2.1)) are analytic in t and depend on $\boldsymbol{\theta}$. In general, for $t \in [0, \widetilde{t}]$, the first band functions $E_s(t\boldsymbol{\theta})$, $s = 1, \ldots, n$, coincide with $\lambda_l(t, \boldsymbol{\theta})$, $l = 1, \ldots, n$, only partially, since the analytic branches λ_l are not necessarily enumerated in the nondecreasing order. At the same time, (3.17) yields

$$\lambda_l(t, \boldsymbol{\theta}) \geq c_* t^2, \quad l = 1, \ldots, n, \quad c_* > 0, \quad t \in [0, \widetilde{t}]. \tag{3.28}$$

Here c_* does not depend on $\boldsymbol{\theta}$. Relation (3.28) implies that

$$\gamma_l(\boldsymbol{\theta}) \geq c_* > 0, \quad l = 1, \ldots, n,$$

i. e., (2.9) is true. Thus, *the spectral germ* $S(\boldsymbol{\theta})$ *of the family* (3.24) $A(t, \boldsymbol{\theta})$ *is nondegenerate for all* $\boldsymbol{\theta}$.

In what follows, we employ the results of Sections 1, 2 for $|\mathbf{k}| = t \in [0, \widetilde{t}]$.

3.5. Absolute continuity of the spectrum of the operator $\mathcal{A}$ near $\lambda = 0$

The following proposition follows from relations (3.16)-(3.19), (3.28) and from the analyticity of eigenvalues $\lambda_l(t, \boldsymbol{\theta})$ and eigenvectors $\varphi_l(t, \boldsymbol{\theta})$, $l = 1, \dots, n$, in t.

Proposition 3.1. *Let $\sigma = c_* \widetilde{t}^2$, where c_* is a constant from* (3.17) *and $\widetilde{t}$ is defined by* (3.26). *Let $\mathcal{E}_\sigma$ be the spectral projector of the operator $\mathcal{A}$ corresponding to the interval $[0, \sigma]$, and let $\mathcal{A}_\sigma$ be the part of $\mathcal{A}$ acting in the subspace $\mathcal{E}_\sigma L_2(\mathbb{R}^d; \mathbb{C}^n)$. Then the spectrum of the operator $\mathcal{A}_\sigma$ is absolutely continuous.*

For the lack of space, we omit a routine but bulky proof.

4. Periodic differential operators of second order. Effective characteristics

We are going to construct the operators $R = R(\boldsymbol{\theta})$ and $S = S(\boldsymbol{\theta})$ for the operator family $A(t, \boldsymbol{\theta})$ (see (3.24)). It is convenient to consider the case where $f = \mathbf{1}$ first, and then to proceed to the general case on the basis of Subsection 1.5. *In the case where $f = \mathbf{1}$, we agree to mark all the corresponding objects by the upper index "$\widehat{\ }$".* Relation (3.13) with $f = \mathbf{1}$ yields

$$\widehat{\mathfrak{N}} = \{\mathbf{u} \in L_2(\Omega; \mathbb{C}^n) : \ \mathbf{u} = \mathbf{c} \in \mathbb{C}^n\}, \quad \widehat{n} = n. \tag{4.1}$$

Note that (see (3.27)) $\widehat{\mathfrak{N}}_* = \mathfrak{N}_*$.

4.1. The operator $\widehat{R}(\boldsymbol{\theta})$

Let $\mathfrak{M} = \{\mathbf{q} \in L_2(\Omega; \mathbb{C}^m) : \ \mathbf{q} = \mathbf{C} \in \mathbb{C}^m\}$ be the subspace of constant vector-valued functions. In accordance with Subsection 1.2, for each $\mathbf{C} \in \mathfrak{M}$ there exists a unique element $\mathbf{v} \in \widetilde{H}^1(\Omega; \mathbb{C}^n)$ such that $\int_\Omega \mathbf{v}\, d\mathbf{x} = 0$ and (cf. (1.8) with $z = -h\mathbf{C}$)

$$(gb(\mathbf{D})\mathbf{v}, b(\mathbf{D})\mathbf{w})_{L_2(\Omega;\mathbb{C}^m)} = -(g\mathbf{C}, b(\mathbf{D})\mathbf{w})_{L_2(\Omega;\mathbb{C}^m)}, \quad \forall\, \mathbf{w} \in \widetilde{H}^1(\Omega; \mathbb{C}^n). \tag{4.2}$$

We put $B\mathbf{C} := h(b(\mathbf{D})\mathbf{v}+\mathbf{C}) \in \mathfrak{N}_* \subset L_2(\Omega; \mathbb{C}^m)$. Obviously, the operator $B : \mathfrak{M} \to L_2(\Omega; \mathbb{C}^m)$ does not depend on the parameter $\boldsymbol{\theta}$. Note that $\int_\Omega h^{-1} B\mathbf{C}\, d\mathbf{x} = |\Omega|\mathbf{C}$. This implies that $\operatorname{Ker} B = \{0\}$ or, equivalently, $\operatorname{rank} B = m$. Now, let $\mathbf{c} \in \widehat{\mathfrak{N}}$ and $\mathbf{C} = b(\boldsymbol{\theta})\mathbf{c}$. Then (see (1.8)-(1.11)) $\mathbf{v}$ plays the role of $\widehat{\varphi}(\mathbf{c})$ and $\widehat{R}(\boldsymbol{\theta})\mathbf{c} = B(b(\boldsymbol{\theta})\mathbf{c})$. Thus,

$$\widehat{R}(\boldsymbol{\theta}) : \widehat{\mathfrak{N}} \to \mathfrak{N}_*, \quad \widehat{R}(\boldsymbol{\theta}) = Bb(\boldsymbol{\theta}), \quad \boldsymbol{\theta} \in \mathbb{S}^{d-1}. \tag{4.3}$$

By (3.1), from (4.3) and from the relation $\operatorname{Ker} B = \{0\}$ it follows that $\operatorname{Ker} \widehat{R}(\boldsymbol{\theta}) = \{0\}$ or, equivalently,

$$\operatorname{rank} \widehat{R}(\boldsymbol{\theta}) = n. \tag{4.4}$$

4.2. The operator $\widehat{S}(\boldsymbol{\theta})$. Effective matrix

Below, it is convenient to indicate the dependence of objects on the matrix g (when $b(\boldsymbol{\xi})$ is fixed) in the notation explicitly: $\widehat{A}(t,\boldsymbol{\theta}) = \widehat{A}(t,\boldsymbol{\theta};g)$, $\widehat{R}(\boldsymbol{\theta}) = \widehat{R}(\boldsymbol{\theta};g)$, $\widehat{S}(\boldsymbol{\theta}) = \widehat{S}(\boldsymbol{\theta};g)$. By (4.3), it is clear that the operator $\widehat{S}(\boldsymbol{\theta};g) = \widehat{R}(\boldsymbol{\theta};g)^*\widehat{R}(\boldsymbol{\theta};g) : \mathfrak{N} \to \mathfrak{N}$ satisfies the relation $\widehat{S}(\boldsymbol{\theta};g) = b(\boldsymbol{\theta})^* B^* B b(\boldsymbol{\theta})$. Relation (4.4) shows again (see Subsection 3.4) that the *operator $\widehat{S}(\boldsymbol{\theta};g)$ is nondegenerate. Let g^0 be a constant (independent of $\boldsymbol{\theta}$) positive matrix corresponding to the operator $B^*B : \mathfrak{M} \to \mathfrak{M}$ in the standard basis $\mathbf{e}_1,\dots,\mathbf{e}_m$.* Thus, $\widehat{S}(\boldsymbol{\theta};g) = b(\boldsymbol{\theta})^* g^0 b(\boldsymbol{\theta})$. In the case of the operator family $\widehat{A}(t,\boldsymbol{\theta};g^0)$ that corresponds to a constant matrix $g = g^0 = (h^0)^*h^0$, relation (4.2) yields $\mathbf{v} = 0$. Then $\widehat{R}(\boldsymbol{\theta};g^0) = h^0 b(\boldsymbol{\theta})$, $\widehat{S}(\boldsymbol{\theta};g^0) = b(\boldsymbol{\theta})^* g^0 b(\boldsymbol{\theta})$. Hence,

$$\widehat{S}(\boldsymbol{\theta};g) = \widehat{S}(\boldsymbol{\theta};g^0) = b(\boldsymbol{\theta})^* g^0 b(\boldsymbol{\theta}), \quad \boldsymbol{\theta} \in \mathbb{S}^{d-1}. \tag{4.5}$$

In particular, (4.5) shows that the families $\widehat{A}(t,\boldsymbol{\theta};g)$ and $\widehat{A}(t,\boldsymbol{\theta};g^0)$ *are spectral-equivalent at $t = 0$*. We put

$$\widehat{S}(\mathbf{k};g^0) := t^2 \widehat{S}(\boldsymbol{\theta};g^0) = b(\mathbf{k})^* g^0 b(\mathbf{k}), \quad \mathbf{k} \in \mathbb{R}^d.$$

Then $\widehat{S}(\mathbf{k};g^0)$ is the symbol of the DO

$$\widehat{\mathcal{A}}^0 := b(\mathbf{D})^* g^0 b(\mathbf{D}) \tag{4.6}$$

with constant coefficients.

Definition 4.1. *A constant positive $(m\times m)$-matrix g_{eff} is called the effective matrix for an operator family $\widehat{A}(t,\boldsymbol{\theta};g)$ if $\widehat{S}(\boldsymbol{\theta};g) = \widehat{S}(\boldsymbol{\theta};g_{\text{eff}})$.*

If g_{eff} is the effective matrix, then

$$\widehat{S}(\boldsymbol{\theta};g) = \widehat{S}(\boldsymbol{\theta};g_{\text{eff}}) = b(\boldsymbol{\theta})^* g_{\text{eff}} b(\boldsymbol{\theta}), \quad \boldsymbol{\theta} \in \mathbb{S}^{d-1}. \tag{4.7}$$

We have proved the following proposition.

Proposition 4.2. *For an operator family $\widehat{A}(t,\boldsymbol{\theta};g)$ there exists at least one effective matrix. Namely, the matrix g^0 corresponding to the operator B^*B is effective.*

Let $\mathbf{v}_j$ be solution of (4.2) with $\mathbf{C} = \mathbf{e}_j$, $j = 1,\dots,m$. We put $\mathbf{u}_j := B\mathbf{e}_j = h(b(\mathbf{D})\mathbf{v}_j + \mathbf{e}_j)$. Then $|\Omega|g^0$ is the Gram matrix, i. e., $g^0 = |\Omega|^{-1}\{(\mathbf{u}_j,\mathbf{u}_l)_{L_2(\Omega;\mathbb{C}^m)}\}$, $j,l = 1,\dots,m$.

If matrices $g(\mathbf{x})$ and b_j, $j = 1,\dots,d$, (see (3.3)) have real-valued entries, then the solutions $\mathbf{v}_j$ are purely imaginary, and the vector-functions $\mathbf{u}_j$ are real. In this case, the matrix g^0 also has real-valued entries.

4.3. Uniqueness condition of the effective matrix

Relations (4.5), (4.7) show that the effective matrix g_{eff} is defined uniquely (and coincides with g^0) if and only if $\cup_{\boldsymbol{\theta}\in\mathbb{S}^{d-1}} \operatorname{Ran} b(\boldsymbol{\theta}) = \mathbb{C}^m$. By (3.3), the latter condition is equivalent to coincidence of the linear span of $\operatorname{Ran} b_j$, $j = 1,\dots,d$, with $\mathbb{C}^m$:

$$\mathcal{L}\{\operatorname{Ran} b_j,\ j = 1,\dots,d\} = \mathbb{C}^m. \tag{4.8}$$

Consider the block $(m \times nd)$-matrix $\mathbf{B} := \{b_1, \dots, b_d\}$. Obviously, relation (4.8) is equivalent to the condition

$$\operatorname{rank} \mathbf{B} = m. \tag{4.9}$$

By (3.1), condition (4.9) is valid automatically if $m = n$. Thus, we have proved the following proposition.

Proposition 4.3. *The effective matrix g_{eff} for an operator family $\widehat{A}(t, \boldsymbol{\theta}; g)$ is unique if and only if relation* (4.9) *is true. Under condition* (4.9), $g_{\text{eff}} = g^0$. *The latter statement is a fortiori valid if $m = n$.*

Note that condition (4.9) is a fortiori violated if $m > nd$.

4.4. Estimates for the effective matrix

From the general estimate (1.20) it follows that

$$(\widehat{S}(\boldsymbol{\theta})\mathbf{c}, \mathbf{c})_{L_2(\Omega;\mathbb{C}^n)} \le \|hb(\boldsymbol{\theta})\mathbf{c}\|^2_{L_2(\Omega;\mathbb{C}^m)}, \quad \mathbf{c} \in \widehat{\mathfrak{N}}. \tag{4.10}$$

By (4.5), the left-hand side of (4.10) takes the form $|\Omega|\langle g^0 b(\boldsymbol{\theta})\mathbf{c}, b(\boldsymbol{\theta})\mathbf{c}\rangle$. The right-hand side of (4.10) coincides with $|\Omega|\langle \overline{g}\, b(\boldsymbol{\theta})\mathbf{c}, b(\boldsymbol{\theta})\mathbf{c}\rangle$, where $\overline{g} := |\Omega|^{-1}\int_\Omega g(\mathbf{x})\, d\mathbf{x}$. Thus,

$$\langle g^0 b(\boldsymbol{\theta})\mathbf{c}, b(\boldsymbol{\theta})\mathbf{c}\rangle \le \langle \overline{g}\, b(\boldsymbol{\theta})\mathbf{c}, b(\boldsymbol{\theta})\mathbf{c}\rangle, \quad \mathbf{c} \in \mathbb{C}^n. \tag{4.11}$$

Under condition (4.9), we deduce from (4.11) that

$$g^0 \le \overline{g} := |\Omega|^{-1} \int_\Omega g(\mathbf{x})\, d\mathbf{x}. \tag{4.12}$$

Let $\mathfrak{G}'_*$ be the subspace in $\mathfrak{N}_*$ formed by the functions of the form $(h^*)^{-1}\mathbf{C}$, $\mathbf{C} \in \mathbb{C}^m$, and let Π'_* be the orthoprojector of $L_2(\Omega; \mathbb{C}^m)$ onto $\mathfrak{G}'_*$. By the general estimate (1.21),

$$(\widehat{S}(\boldsymbol{\theta})\mathbf{c}, \mathbf{c})_{L_2(\Omega;\mathbb{C}^n)} \ge \|\Pi'_* hb(\boldsymbol{\theta})\mathbf{c}\|^2_{L_2(\Omega;\mathbb{C}^m)}, \quad \mathbf{c} \in \widehat{\mathfrak{N}}. \tag{4.13}$$

Mention that, if $m = n$, then $\mathfrak{G}'_* = \mathfrak{N}_*$, $\Pi'_* = P_*$, and inequality (4.13) turns into the equality. Introduce the notation $\underline{g} := \left(|\Omega|^{-1}\int_\Omega g^{-1}\, d\mathbf{x}\right)^{-1}$. Then

$$\Pi'_* \mathbf{w} = |\Omega|^{-1}(h^*)^{-1}\underline{g} \int_\Omega h^{-1}\mathbf{w}\, d\mathbf{x}, \quad \mathbf{w} \in L_2(\Omega; \mathbb{C}^m),$$

and $\Pi'_* hb(\boldsymbol{\theta})\mathbf{c} = (h^*)^{-1}\underline{g}b(\boldsymbol{\theta})\mathbf{c}$. Then the right-hand side of (4.13) takes the form $|\Omega|\langle \underline{g}b(\boldsymbol{\theta})\mathbf{c}, b(\boldsymbol{\theta})\mathbf{c}\rangle$. Thus, (4.13) yields

$$\langle g^0 b(\boldsymbol{\theta})\mathbf{c}, b(\boldsymbol{\theta})\mathbf{c}\rangle \ge \langle \underline{g}b(\boldsymbol{\theta})\mathbf{c}, b(\boldsymbol{\theta})\mathbf{c}\rangle, \quad \mathbf{c} \in \mathbb{C}^n. \tag{4.14}$$

Under condition (4.9), estimate (4.14) implies the inequality

$$g^0 \ge \underline{g} := \left(|\Omega|^{-1} \int_\Omega g^{-1}\, d\mathbf{x}\right)^{-1}, \tag{4.15}$$

which turns into the equality when $m = n$.

We have proved the following proposition.

Proposition 4.4. a) *If* $m > n$, *then, under condition* (4.9), *the effective matrix* $g_{\mathrm{eff}} = g^0$ *for an operator family* $\widehat{A}(t,\boldsymbol{\theta};g)$ *satisfies estimates* (4.12), (4.15). b) *If* $m = n$, *then the effective matrix coincides with* $\underline{g}$:

$$g_{\mathrm{eff}} = g^0 = \underline{g} \quad \text{for} \quad m = n. \tag{4.16}$$

4.5. The general case ($f \neq 1$)

Now we proceed to the general case of the operator family $A(t,\boldsymbol{\theta})$ defined by (3.8), (3.24). We rely on the results of Subsection 1.5. Now, the role of an isomorphism $M : L_2(\Omega;\mathbb{C}^n) \to L_2(\Omega;\mathbb{C}^n)$ is played by the operator of multiplication by the matrix-valued function f. We will indicate the dependence of objects on the matrix g (when $b(\boldsymbol{\xi})$ and f are fixed) in the notation: $A(t,\boldsymbol{\theta}) = A(t,\boldsymbol{\theta};g)$, $R(\boldsymbol{\theta}) = R(\boldsymbol{\theta};g)$, $S(\boldsymbol{\theta}) = S(\boldsymbol{\theta};g)$. By (1.22), $R(\boldsymbol{\theta};g) = \widehat{R}(\boldsymbol{\theta};g)f|_{\mathfrak{N}}$, where $\mathfrak{N}$ is defined by (3.13). By (1.24),

$$S(\boldsymbol{\theta};g) = Pf^*\widehat{S}(\boldsymbol{\theta};g)f|_{\mathfrak{N}}. \tag{4.17}$$

Obviously, the kernel $\mathfrak{N}$ (see (3.13)) does not depend on g. Definition 4.1 is carried over to the general case of the family $A(t,\boldsymbol{\theta};g)$.

Definition 4.5. *A constant positive* $(m\times m)$*-matrix* g_{eff} *is called the effective matrix for an operator family* $A(t,\boldsymbol{\theta};g)$ *if* $S(\boldsymbol{\theta};g) = S(\boldsymbol{\theta};g_{\mathrm{eff}})$.

Relation (4.17) directly implies the following proposition.

Proposition 4.6. *A matrix* g_{eff} *is effective for* $A(t,\boldsymbol{\theta};g)$ *if and only if it is effective for* $\widehat{A}(t,\boldsymbol{\theta};g)$.

Thus, g_{eff} *does not depend on* f. In particular, the matrix g^0 defined in Subsection 4.2 is effective also for the family $A(t,\boldsymbol{\theta};g)$. It is clear that *Propositions* 4.2-4.4 *are automatically carried over to the general case of a family* $A(t,\boldsymbol{\theta};g)$.

4.6.

Distinguish the case where $n = 1$. Then $\widehat{S}(\boldsymbol{\theta};g)$ (see (4.5)) is the operator of multiplication by the number $\widehat{\gamma}(\boldsymbol{\theta}) = b(\boldsymbol{\theta})^*g^0b(\boldsymbol{\theta})$. By (1.26) and (4.17), $S(\boldsymbol{\theta};g)$ is the operator of multiplication by the number

$$\gamma(\boldsymbol{\theta}) = \widehat{\gamma}(\boldsymbol{\theta})|\Omega|\,\|f^{-1}\|^{-2}_{L_2(\Omega)}. \tag{4.18}$$

4.7.

It is natural to associate the notion of the *effective matrix* g_{eff} with the initial DO $\mathcal{A}$ defined in Subsection 3.1. The differential operator

$$\mathcal{A}_{\mathrm{eff}} := f(\mathbf{x})^*b(\mathbf{D})^*g_{\mathrm{eff}}b(\mathbf{D})f(\mathbf{x}) \tag{4.19}$$

in $L_2(\mathbb{R}^d;\mathbb{C}^n)$ is called the *effective DO* for $\mathcal{A}$. If in (4.19) $g_{\mathrm{eff}} = g^0$, then $\mathcal{A}_{\mathrm{eff}}$ is denoted by $\mathcal{A}^0$:

$$\mathcal{A}^0 := f(\mathbf{x})^*b(\mathbf{D})^*g^0b(\mathbf{D})f(\mathbf{x}). \tag{4.20}$$

In accordance with (4.6),

$$\mathcal{A}^0 = f(\mathbf{x})^* \widehat{\mathcal{A}}^0 f(\mathbf{x}).$$

If $f = \mathbf{1}$ and condition (4.9) is valid, *the matrix* g^0 is called the *matrix of effective medium*, and the operator $\mathcal{A}^0 = \widehat{\mathcal{A}}^0$ is called the *operator of effective medium.*

5. Homogenization of periodic differential operators with small period

5.1. The homogenization of differential operators

Usually, the notion of effective medium is associated with the homogenization procedure for DO's, i. e., with the analysis of behavior of periodic DO's in the small period limit. In the Hilbert space $\mathfrak{H} = L_2(\mathbb{R}^d; \mathbb{C}^n)$, we consider operators of the form

$$\widehat{\mathcal{A}}_\varepsilon = b(\mathbf{D})^* g(\varepsilon^{-1}\mathbf{x}) b(\mathbf{D}), \quad \varepsilon > 0, \tag{5.1}$$

or of more general form

$$\mathcal{A}_\varepsilon = f(\varepsilon^{-1}\mathbf{x})^* b(\mathbf{D})^* g(\varepsilon^{-1}\mathbf{x}) b(\mathbf{D}) f(\varepsilon^{-1}\mathbf{x}), \quad \varepsilon > 0. \tag{5.2}$$

Conditions on b, f, g are the same as in Subsection 3.1.

The initial question of the homogenization theory (as applied to the operator (5.1)) is the study of the resolvent convergence (in some sense):

$$(\widehat{\mathcal{A}}_\varepsilon + I)^{-1} \underset{\varepsilon \to 0}{\longrightarrow} (\widehat{\mathcal{A}}^0 + I)^{-1}, \tag{5.3}$$

where $\widehat{\mathcal{A}}^0$ is defined by (4.6). In the homogenization theory, the constant matrix g^0 (the matrix of effective medium) arises exactly in passing to the limit (5.3).

Our approach to obtaining results of this type has a number of special features. In order to construct g^0, we do not involve introducing the parameter ε and passing to the limit. The matrix g^0 is constructed beforehand, on the basis of different considerations (see Subsection 4.2). We distinguish the main term of the resolvent $(\mathcal{A}_\varepsilon + I)^{-1}$ for small ε, using the results of the operator theory (see Theorem 2.4 and Corollary 2.5). We consider the general operator (5.2). The main term of the resolvent is expressed in terms of the germ $S(\boldsymbol{\theta})$ for the operator family $A(t, \boldsymbol{\theta})$ generated by the operator

$$\mathcal{A} = f(\mathbf{x})^* b(\mathbf{D})^* g(\mathbf{x}) b(\mathbf{D}) f(\mathbf{x}). \tag{5.4}$$

This directly implies the limit relation (5.3) for the operator (5.1). In the case where the matrix f is nonconstant, the resolvent of the operator (5.2) does not have a limit. It is convenient to postpone further comments till Subsection 5.4; see also Subsection 9.1.

5.2.

Let T_ε be the unitary rescaling operator in $\widetilde{\mathfrak{H}} = L_2(\mathbb{R}^d; \mathbb{C}^n)$ defined by $T_\varepsilon : \mathbf{u}(\cdot) \mapsto \varepsilon^{d/2}\mathbf{u}(\varepsilon\cdot)$. Let $\mathcal{U}$ be the Gelfand transformation (see Subsection 3.3) and let $\mathcal{A}(\mathbf{k})$ be the family of operators acting in $\mathfrak{H} = L_2(\Omega; \mathbb{C}^n)$ and related to $\mathcal{A}$ by (3.23). We put

$$\mathcal{B}^{(\varepsilon)}(\mathbf{k}) = \varepsilon^2(\mathcal{A}(\mathbf{k}) + \varepsilon^2 I)^{-1}. \tag{5.5}$$

The operator of multiplication by the operator-valued function (5.5), which acts in the direct integral $\mathcal{H}$ (see (3.20)), is denoted by $\mathcal{B}^{(\varepsilon)}$:

$$\mathcal{B}^{(\varepsilon)} = \int_{\widetilde{\Omega}} \oplus \mathcal{B}^{(\varepsilon)}(\mathbf{k})\, d\mathbf{k}.$$

Then

$$(\mathcal{A}_\varepsilon + I)^{-1} = T_\varepsilon^* \mathcal{U}^* \mathcal{B}^{(\varepsilon)} \mathcal{U} T_\varepsilon, \tag{5.6}$$

and the problem reduces to the study of operators (5.5) with $\mathbf{k} \in \widetilde{\Omega}$.

Let $\widetilde{t}$ be defined by (3.26). By (3.19),

$$\|\mathcal{B}^{(\varepsilon)}(\mathbf{k})\|_{\mathfrak{H}\to\mathfrak{H}} \leq \varepsilon^2(d(r))^{-1}, \quad \mathbf{k} \in \widetilde{\Omega} \setminus \mathcal{K}(r), \quad 0 < r \leq \widetilde{t}. \tag{5.7}$$

Now, let $\mathbf{k} \in \mathcal{K}(\widetilde{t})$. We use the notation of Subsection 3.4. In particular, the family $A(t, \boldsymbol{\theta})$ corresponds to the operator $\mathcal{A}(\mathbf{k}) = \mathcal{A}(t\boldsymbol{\theta})$, where $t = |\mathbf{k}|$ and $\boldsymbol{\theta} \in \mathbb{S}^{d-1}$ plays the role of parameter.

Introduce the operator

$$\mathcal{L}^{(\varepsilon)}(\mathbf{k}) = \varepsilon^2(t^2 S(\boldsymbol{\theta})P + \varepsilon^2 I)^{-1}P, \quad t\boldsymbol{\theta} = \mathbf{k} \in \mathcal{K}(\widetilde{t}), \tag{5.8}$$

in $\mathfrak{H}$, where P is the orthoprojector onto $\mathfrak{N}$ (see (3.13)) and $S(\boldsymbol{\theta})$ is the spectral germ of the family $A(t, \boldsymbol{\theta})$ at $t = 0$. The operator (5.8) acts nontrivially only on $\mathfrak{N}$. We employ Corollary 2.5, replacing $t^0(\delta)$ by the number $\widetilde{t} = \widetilde{t}(\delta) \leq t^0(\delta)$. By (3.28), Condition 2.3 is valid. Now, (2.21) yields the estimate

$$\|\mathcal{B}^{(\varepsilon)}(\mathbf{k}) - \mathcal{L}^{(\varepsilon)}(\mathbf{k})\|_{\mathfrak{H}\to\mathfrak{H}} \leq C_4\varepsilon, \quad t\boldsymbol{\theta} = \mathbf{k} \in \mathcal{K}(\widetilde{t}), \quad 0 < \varepsilon \leq 1, \tag{5.9}$$

where the constant C_4 does not depend on $\boldsymbol{\theta}$. Let us explain the latter statement. In accordance with (2.18), (2.21), $C_4 = 4C_1 c_*^{-1/2} + C_2 c_*^{-3/2} + (3\delta)^{-1}$. The number δ (see (3.26)) is fixed; the constant c_* from (3.17) does not depend on $\boldsymbol{\theta}$. It remains to take $C_j = C_j(\delta, \max_{\boldsymbol{\theta}} \|X_1(\boldsymbol{\theta})\|)$, $j = 1, 2$ (see Proposition 2.2).

5.3.

Let χ_r be the characteristic function of the ball $\mathcal{K}(r)$, and let $\mathcal{L}_r^{(\varepsilon)}$ be the operator in $\mathcal{H}$ of multiplication by the operator-valued function $\chi_r(\cdot)\mathcal{L}^{(\varepsilon)}(\cdot)$:

$$\mathcal{L}_r^{(\varepsilon)} = \int_{\mathcal{K}(r)} \oplus \mathcal{L}^{(\varepsilon)}(\mathbf{k})\, d\mathbf{k}, \quad 0 < r \leq \widetilde{t}.$$

We put

$$\Lambda_r^{(\varepsilon)} = T_\varepsilon^* \mathcal{U}^* \mathcal{L}_r^{(\varepsilon)} \mathcal{U} T_\varepsilon, \quad 0 < r \leq \widetilde{t}.$$

Now (5.6), (5.7), (5.9) directly imply that

$$\|(\mathcal{A}_\varepsilon + I)^{-1} - \Lambda_r^{(\varepsilon)}\|_{\mathfrak{H}\to\mathfrak{H}} \le (C_4 + d(r)^{-1})\varepsilon, \quad 0 < r \le \widetilde{t}, \quad 0 < \varepsilon \le 1. \tag{5.10}$$

Thus, $\Lambda_r^{(\varepsilon)}$ *is the leading term of the resolvent* $(\mathcal{A}_\varepsilon + I)^{-1}$ *as* $\varepsilon \to 0$, and the estimate of error is well controled. Essentially, the operator $\Lambda_r^{(\varepsilon)}$ is determined by the operator (5.8). The latter is defined via the spectral germ $S(\boldsymbol{\theta})$ only. The number r in (5.10) can be taken arbitrarily small. At the same time, the constant in estimate (5.10) grows, since the constant $d(r)$ from (3.19) tends to zero as $r \to 0$. However, we recall that the estimate (5.7) was roughened: in (5.10) $d(r)^{-1}$ can be replaced by $\varepsilon d(r)^{-1}$. Then, along with (5.10), we obtain

$$\limsup_{\varepsilon\to 0} \varepsilon^{-1}\|(\mathcal{A}_\varepsilon + I)^{-1} - \Lambda_r^{(\varepsilon)}\|_{\mathfrak{H}\to\mathfrak{H}} \le C_4, \quad 0 < r \le \widetilde{t},$$

where the constant C_4 does not depend on r.

5.4.

Now, let us replace the matrix $g(\mathbf{x})$ in (5.4) by the effective matrix g^0. We apply the estimate of the form (5.10) to the corresponding operator $\mathcal{A}^0$ (see (4.20)):

$$\|(\mathcal{A}_\varepsilon^0 + I)^{-1} - \Lambda_r^{(\varepsilon)}\|_{\mathfrak{H}\to\mathfrak{H}} \le (C_4^0 + d^0(r)^{-1})\varepsilon, \quad 0 < r \le \widetilde{t}^0, \quad 0 < \varepsilon \le 1. \tag{5.11}$$

Here C_4^0, $d^0(r)$, $\widetilde{t}^0$ are the analogs of the constants C_4, $d(r)$, $\widetilde{t} = \widetilde{t}(\delta)$ corresponding to the operator $\mathcal{A}^0$. The operators $\mathcal{A}$ and $\mathcal{A}^0$ have the same spectral germ $S(\boldsymbol{\theta})$ for all $\boldsymbol{\theta}$. Hence, the operator $\Lambda_r^{(\varepsilon)}$ in (5.10) and (5.11) is the same. Combining (5.10) and (5.11), we see that

$$\|(\mathcal{A}_\varepsilon + I)^{-1} - (\mathcal{A}_\varepsilon^0 + I)^{-1}\|_{\mathfrak{H}\to\mathfrak{H}} \le C_5\varepsilon, \quad 0 < \varepsilon \le 1, \tag{5.12}$$

$$C_5 = C_4 + C_4^0 + d(t_+)^{-1} + d^0(t_+)^{-1}, \quad t_+ = \min\{\widetilde{t}, \widetilde{t}^0\}.$$

Estimate (5.12) is the required result of the homogenization theory. In particular, if the role of $\mathcal{A}$ is played by the operator

$$\widehat{\mathcal{A}} = b(\mathbf{D})^* g(\mathbf{x}) b(\mathbf{D}),$$

then the operator $\widehat{\mathcal{A}}^0$ defined by (4.6) has constant coefficients. Therefore, $\widehat{\mathcal{A}}_\varepsilon^0 = \widehat{\mathcal{A}}^0$, and (5.12) takes the form

$$\|(\widehat{\mathcal{A}}_\varepsilon + I)^{-1} - (\widehat{\mathcal{A}}^0 + I)^{-1}\|_{\mathfrak{H}\to\mathfrak{H}} \le C_5\varepsilon, \quad 0 < \varepsilon \le 1. \tag{5.13}$$

Estimate (5.13) shows the exact meaning of relation (5.3).

If condition (4.9) of the uniqueness of effective matrix does not hold, the operator $\mathcal{A}_\varepsilon^0$ in (5.12) can be replaced by the operator $(\mathcal{A}_{\mathrm{eff}})_\varepsilon$ corresponding to (4.19).

6. The periodic acoustic operator and Schrödinger operator

6.1. The acoustic operator

The simplest example of a periodic operator $\mathcal{A}$ considered in Sections 3-5 is the operator

$$\widehat{\mathcal{A}} = \mathbf{D}^* g(\mathbf{x})\mathbf{D} = \nabla^* g(\mathbf{x})\nabla \tag{6.1}$$

in $L_2(\mathbb{R}^d)$, $d \geq 1$, describing a periodic acoustic medium. Here $g(\mathbf{x})$ is a Γ-periodic $(d \times d)$-matrix with real-valued entries and such that

$$g(\mathbf{x}) > 0, \quad g + g^{-1} \in L_\infty. \tag{6.2}$$

Let h be a Γ-periodic $(d \times d)$-matrix satisfying (3.4) and such that $g(\mathbf{x}) = h(\mathbf{x})^* h(\mathbf{x})$. (For instance, one can put $h(\mathbf{x}) = g(\mathbf{x})^{1/2}$.) The operator (6.1) coincides with the operator $\mathcal{A}$ defined in Subsection 3.1 with $n = 1$, $m = d$, $f = 1$, $b(\boldsymbol{\xi}) = \boldsymbol{\xi}$. Obviously, condition (3.1) (with $n = 1$) is valid. Now, the kernels (4.1) and (3.27) are given by the relations

$$\widehat{\mathfrak{N}} = \{u \in L_2(\Omega) : \ u = \text{const}\}, \quad \dim \widehat{\mathfrak{N}} = n = 1, \quad d \geq 1;$$

$$\widehat{\mathfrak{N}}_* = \{\mathbf{q} \in L_2(\Omega; \mathbb{C}^d) : \ h^* \mathbf{q} \in \widetilde{H}^1(\Omega; \mathbb{C}^d), \ \ \operatorname{div} h^* \mathbf{q} = 0\}, \quad n_* = \infty, \quad \text{for } \ d \geq 2; \tag{6.3}$$

$$\widehat{\mathfrak{N}}_* = \{q \in L_2(\Omega) : \ h^* q = \text{const}\}, \quad n_* = 1, \quad \text{for } \ d = 1. \tag{6.4}$$

In accordance with Subsections 1.2, 1.3, we construct the operators $\widehat{R}(\boldsymbol{\theta}) : \widehat{\mathfrak{N}} \to \widehat{\mathfrak{N}}_*$ and $\widehat{S}(\boldsymbol{\theta}) : \widehat{\mathfrak{N}} \to \widehat{\mathfrak{N}}$. Let $v = v(\boldsymbol{\theta}) \in \widetilde{H}^1(\Omega)$ be solution of the equation $\mathbf{D}^* g(\mathbf{D}v + \boldsymbol{\theta}) = 0$ (understood in the weak sense). We put $\mathbf{u}(\boldsymbol{\theta}) = h(\mathbf{D}v(\boldsymbol{\theta}) + \boldsymbol{\theta}) \in \widehat{\mathfrak{N}}_*$. Then $\widehat{R}(\boldsymbol{\theta})c = \mathbf{u}(\boldsymbol{\theta})c$, $c \in \widehat{\mathfrak{N}}$, and $\widehat{S}(\boldsymbol{\theta})$ is the operator of multiplication by the number

$$\widehat{\gamma}(\boldsymbol{\theta}) = |\Omega|^{-1} \|\mathbf{u}(\boldsymbol{\theta})\|^2_{L_2(\Omega;\mathbb{C}^d)}, \quad \boldsymbol{\theta} \in \mathbb{S}^{d-1}. \tag{6.5}$$

Now, we construct the matrix g^0 for the operator (6.1). In accordance with Subsection 4.2, let $v_j \in \widetilde{H}^1(\Omega)$ be the solution of the equation $\mathbf{D}^* g(\mathbf{D}v_j + \mathbf{e}_j) = 0$, such that $\int_\Omega v_j \, d\mathbf{x} = 0$, $j = 1, \dots, d$. We put $\mathbf{u}_j = h(\mathbf{D}v_j + \mathbf{e}_j)$. Then

$$g^0 = |\Omega|^{-1} \{(\mathbf{u}_j, \mathbf{u}_l)_{L_2(\Omega;\mathbb{C}^d)}\}, \quad j, l = 1, \dots, d. \tag{6.6}$$

It is easily seen that now the $(d \times d)$-matrix $\mathbf{B}$ (see Subsection 4.3) is the unit matrix. Therefore, condition (4.9) is valid. By Proposition 4.3, this implies that the effective matrix is unique and is equal to g^0. By (4.5),

$$\widehat{\gamma}(\boldsymbol{\theta}) = \boldsymbol{\theta}^* g^0 \boldsymbol{\theta} = \langle g^0 \boldsymbol{\theta}, \boldsymbol{\theta} \rangle. \tag{6.7}$$

We put

$$\widehat{\gamma}(\mathbf{k}) := t^2 \widehat{\gamma}(\boldsymbol{\theta}) = \langle g^0 \mathbf{k}, \mathbf{k} \rangle, \quad \mathbf{k} \in \mathbb{R}^d. \tag{6.8}$$

By Proposition 4.4(a), we have

$$\langle \underline{g} \mathbf{k}, \mathbf{k} \rangle \leq \widehat{\gamma}(\mathbf{k}) \leq \langle \overline{g} \mathbf{k}, \mathbf{k} \rangle.$$

In accordance with Subsection 4.7, g^0 is the *matrix of effective medium* and the operator $\widehat{\mathcal{A}}^0 = \widehat{\gamma}(\mathbf{D}) = \mathbf{D}^* g^0 \mathbf{D}$ is the *operator of effective medium.*

If $d = 1$, then $m = n = 1$ and, by Proposition 4.4(b), for (1×1)-matrix g^0 we have

$$g^0 = \underline{g} \quad \text{for} \quad d = 1. \tag{6.9}$$

In this case, by (6.7) and (6.9), the number $\widehat{\gamma}(\boldsymbol{\theta})$ does not depend on $\boldsymbol{\theta}$ (if $d = 1$, $\boldsymbol{\theta}$ has only two values) and coincides with $\underline{g}$:

$$\widehat{\gamma}(\boldsymbol{\theta}) = \widehat{\gamma} = \underline{g} \quad \text{for} \quad d = 1. \tag{6.10}$$

6.2. The Schrödinger operator

Now, we consider the periodic Schrödinger operator with a variable metric

$$H = \mathbf{D}^* \tilde{g}(\mathbf{x}) \mathbf{D} + p(\mathbf{x}) = \nabla^* \tilde{g}(\mathbf{x}) \nabla + p(\mathbf{x}), \quad \mathbf{x} \in \mathbb{R}^d, \tag{6.11}$$

where a $(d \times d)$-matrix $\tilde{g}(\mathbf{x})$ and a function $p(\mathbf{x})$ are real and Γ-periodic in $\mathbb{R}^d$. Suppose that

$$\tilde{g}(\mathbf{x}) > 0, \quad \tilde{g} + \tilde{g}^{-1} \in L_\infty, \tag{6.12}$$

$$p \in L_s(\Omega), \quad 2s > d \quad \text{for} \quad d \geq 2; \quad s = 1 \quad \text{for} \quad d = 1. \tag{6.13}$$

We introduce the quadratic form

$$\mathfrak{h}[u, u] = \int\limits_{\mathbb{R}^d} (\langle \tilde{g} \mathbf{D} u, \mathbf{D} u \rangle + p|u|^2)\, d\mathbf{x}, \quad u \in H^1(\mathbb{R}^d), \tag{6.14}$$

in $L_2(\mathbb{R}^d)$. Under conditions (6.12), (6.13), the form (6.14) is closed and lower semibounded. In $L_2(\mathbb{R}^d)$, the form (6.14) generates a selfadjoint operator H formally given by the expression (6.11). By adding an appropriate constant to $p(\mathbf{x})$, we can assume that the lower edge of the spectrum of H is the point $\lambda = 0$: $\inf \sigma(H) = 0$. Then (see, for instance, [8, Section 4]) the *equation* $\nabla^* \tilde{g} \nabla \omega + p\omega = 0$ *has a positive* Γ*-periodic solution* ω *of the class* $H^1_{\mathrm{loc}}(\mathbb{R}^d) \cap \mathrm{Lip}\,\tau$ (with some $\tau > 0$). Here the equation is understood in the weak sense, i. e., as the identity

$$\int\limits_{\Omega} (\langle \tilde{g} \mathbf{D} \omega, \mathbf{D} \zeta \rangle + p\omega \overline{\zeta})\, d\mathbf{x} = 0, \quad \forall\, \zeta \in \widetilde{H}^1(\Omega).$$

Besides, the function ω is a *multiplicator* in $H^1(\mathbb{R}^d)$ and in $\widetilde{H}^1(\Omega)$. By substitution $u = \omega v$, the form (6.14) turns into

$$\mathfrak{h}[u, u] = \int\limits_{\mathbb{R}^d} \omega^2 \langle \tilde{g} \mathbf{D} v, \mathbf{D} v \rangle\, d\mathbf{x}, \quad u = \omega v, \quad v \in H^1(\mathbb{R}^d). \tag{6.15}$$

This means that the operator H admits the representation

$$H = \omega^{-1} \mathbf{D}^* \omega^2 \tilde{g} \mathbf{D} \omega^{-1} = \omega^{-1} \mathbf{D}^* g \mathbf{D} \omega^{-1}, \quad \text{where} \quad g := \omega^2 \tilde{g}, \tag{6.16}$$

understood in the sense of forms. We can take (6.16) as the *definition* of the operator H, and it is sufficient to assume that ω is a measurable Γ-periodic function such that

$$\omega(\mathbf{x}) > 0, \quad \omega + \omega^{-1} \in L_\infty.$$

The precise definition of H is given via the quadratic form (6.15), but now ω is not a multiplicator in $H^1(\mathbb{R}^d)$ anymore. We see that H coincides with the operator $\mathcal{A}$ defined in Subsection 3.1 with $n=1$, $m=d$, $f=\omega^{-1}$, $g=\omega^2\tilde{g}=h^*h$, $b(\boldsymbol{\xi})=\boldsymbol{\xi}$.

6.3.

In accordance with Subsection 4.5, the results for the operator (6.16) can be recalculated from the results for the operator (6.1). The kernel (3.13) takes the form

$$\mathfrak{N}=\{u\in L_2(\Omega):\ u=c\omega,\ c\in\mathbb{C}\},\quad \dim\mathfrak{N}=n=1,$$

while $\mathfrak{N}_*=\widehat{\mathfrak{N}}_*$ (see (6.3), (6.4)). For the operator $R(\boldsymbol{\theta}):\mathfrak{N}\to\mathfrak{N}_*$ we have

$$R(\boldsymbol{\theta})c\omega=\mathbf{u}(\boldsymbol{\theta})c,\quad c\omega\in\mathfrak{N},$$

where the function $\mathbf{u}(\boldsymbol{\theta})$ is defined in Subsection 6.1. According to Subsection 4.6, $S(\boldsymbol{\theta}):\mathfrak{N}\to\mathfrak{N}$ is the operator of multiplication by the number

$$\gamma(\boldsymbol{\theta})=\widehat{\gamma}(\boldsymbol{\theta})|\Omega|\|\omega\|^{-2}_{L_2(\Omega)},\tag{6.17}$$

where $\widehat{\gamma}(\boldsymbol{\theta})$ is defined by (6.5).

By Proposition 4.6, the matrix g^0 defined by (6.6) is the effective matrix for the operator (6.16). In accordance with Subsection 4.7, the operator

$$H^0=\omega^{-1}\mathbf{D}^*g^0\mathbf{D}\omega^{-1}=\omega^{-1}\widehat{\gamma}(\mathbf{D})\omega^{-1}\tag{6.18}$$

is the effective DO for H.

There is no loss of generality in assuming that the function ω in (6.16) satisfies the normalization condition $\|\omega\|^2_{L_2(\Omega)}=|\Omega|$. Then (6.17) takes the form $\gamma(\boldsymbol{\theta})=\widehat{\gamma}(\boldsymbol{\theta})$. In this case, according to the quantum mechanics terminology, the matrix $(2g^0)^{-1}$ corresponds to the *tensor of effective masses*, and the operator (6.18) is called the *effective Hamiltonian for the threshold* $\lambda=0$.

Let us discuss the case where $d=1$. Then $m=n=1$ and, by (6.10), $\widehat{\gamma}=\underline{g}=|\Omega|\,\|\tilde{g}^{-1/2}\omega^{-1}\|^{-2}_{L_2(\Omega)}$. Combining this with (6.17), we see that the number $\gamma(\boldsymbol{\theta})$ does not depend on $\boldsymbol{\theta}$ and is given by

$$\gamma=|\Omega|^2\|\omega\|^{-2}_{L_2(\Omega)}\|\tilde{g}^{-1/2}\omega^{-1}\|^{-2}_{L_2(\Omega)}.$$

In particular,

$$\gamma=|\Omega|^2\|\omega\|^{-2}_{L_2(\Omega)}\|\omega^{-1}\|^{-2}_{L_2(\Omega)}\quad\text{for}\ d=1,\ \tilde{g}=1.\tag{6.19}$$

6.4.

Both examples in this section correspond to the case where $n=1$. Then the corresponding operator family in $L_2(\Omega)$ is analytic not only in $t=|\mathbf{k}|$, but also in the quasimomentum $\mathbf{k}\in\mathbb{R}^d$. However, we did not employ the latter property.

7. The two-dimensional periodic Pauli operator

7.1. The Pauli operator

Let the *magnetic potential* $\mathbf{A} = \{A_1, A_2\}$ be a Γ-periodic $\mathbb{R}^2$-valued function on $\mathbb{R}^2$ such that

$$\mathbf{A} \in L_r(\Omega), \quad r > 2. \tag{7.1}$$

As usual,

$$\sigma_1 = \begin{pmatrix} 0 & 1 \\ 1 & 0 \end{pmatrix}, \quad \sigma_2 = \begin{pmatrix} 0 & -i \\ i & 0 \end{pmatrix}, \quad \sigma_3 = \begin{pmatrix} 1 & 0 \\ 0 & -1 \end{pmatrix}$$

are the Pauli matrices. Consider the operator

$$\mathcal{D} := (D_1 - A_1)\sigma_1 + (D_2 - A_2)\sigma_2, \quad \operatorname{Dom} \mathcal{D} = H^1(\mathbb{R}^2; \mathbb{C}^2), \tag{7.2}$$

in $L_2(\mathbb{R}^2; \mathbb{C}^2)$. By definition, the Pauli operator $\mathcal{P}$ is the square of $\mathcal{D}$:

$$\mathcal{P} := \mathcal{D}^2 = \begin{pmatrix} P_- & 0 \\ 0 & P_+ \end{pmatrix}. \tag{7.3}$$

The precise definition of the operator $\mathcal{P}$ in $L_2(\mathbb{R}^2; \mathbb{C}^2)$ is given via the closed form $\|\mathcal{D}\mathbf{u}\|^2_{L_2(\mathbb{R}^2;\mathbb{C}^2)}$, $\mathbf{u} \in H^1(\mathbb{R}^2; \mathbb{C}^2)$. If the potential $\mathbf{A}$ is sufficiently smooth, the blocks $P_\pm$ satisfy the relations $P_\pm = (\mathbf{D} - \mathbf{A})^2 \pm (\partial_1 A_2 - \partial_2 A_1)$.

We rely on the known factorization (see [7, 8]) for the operators (7.2), (7.3). By the gauge transformation, the potential $\mathbf{A}$ can be subject to the conditions

$$\operatorname{div} \mathbf{A} = 0, \quad \int_\Omega \mathbf{A}\, d\mathbf{x} = 0,$$

without violating (7.1) (see [7]). Then there exists a real-valued *periodic* function φ such that

$$\nabla\varphi = \{A_2, -A_1\}, \quad \int_\Omega \varphi\, d\mathbf{x} = 0. \tag{7.4}$$

Relations (7.1), (7.4) imply that

$$\varphi \in \widetilde{W}^1_r(\Omega) \subset \operatorname{Lip} \tau, \quad \tau = 1 - 2r^{-1}. \tag{7.5}$$

Here $\widetilde{W}^1_r(\Omega)$ is the subspace of $W^1_r(\Omega)$ formed by the functions whose Γ-periodic extension belongs to $W^1_{r,\mathrm{loc}}(\Omega)$. Introduce the notation $\omega_\pm := \exp(\pm\varphi)$,

$$f := \begin{pmatrix} \omega_+ & 0 \\ 0 & \omega_- \end{pmatrix}, \tag{7.6}$$

$$Q_\pm := D_1 \pm iD_2, \quad \mathcal{Q} := \begin{pmatrix} 0 & Q_- \\ Q_+ & 0 \end{pmatrix}.$$

By (7.5), $\omega_\pm \in \widetilde{W}^1_r(\Omega)$. The operators (7.2), (7.3) can be written as

$$\mathcal{D} = f\mathcal{Q}f, \tag{7.7}$$

$$\mathcal{P} = f\mathcal{Q}f^2\mathcal{Q}f. \tag{7.8}$$

Then the blocks $P_\pm$ of the operator (7.3) admit the representations

$$P_+ = Y^*Y, \quad P_- = YY^*, \quad \text{where } Y := \omega_+ Q_- \omega_-. \tag{7.9}$$

It is convenient to take (7.7)-(7.9) as ***definitions*** of the operators $\mathcal{D}$, $\mathcal{P}$, $P_\pm$, assuming that $\omega_\pm$ are arbitrary Γ-periodic functions such that

$$\omega_\pm(\mathbf{x}) > 0, \quad \omega_+\omega_- = 1, \quad \omega_+, \omega_- \in L_\infty.$$

Precisely, $\mathcal{D}$ is defined by the relation (7.7) on the domain

$$\operatorname{Dom}\mathcal{D} = \{\mathbf{u} \in L_2(\mathbb{R}^2;\mathbb{C}^2) : \ f\mathbf{u} \in H^1(\mathbb{R}^2;\mathbb{C}^2)\}. \tag{7.10}$$

The precise definition of the operator (7.8) is given via the quadratic form $\|f\mathcal{Q}f\mathbf{u}\|^2_{L_2(\mathbb{R}^2;\mathbb{C}^2)}$, $\mathbf{u} \in \operatorname{Dom}\mathcal{D}$. The blocks $P_\pm$ are defined via the quadratic forms $\|\omega_\pm Q_\mp \omega_\mp u\|^2_{L_2(\mathbb{R}^2)}$, $\omega_\mp u \in H^1(\mathbb{R}^2)$. It is clear that $\operatorname{Ker} Y = \operatorname{Ker} Y^* = \{0\}$. Therefore, (7.9) implies that the operator P_+ is unitarily equivalent to P_-. Obviously, the study of the operator $\mathcal{P}$ reduces to the study of the blocks $P_\pm$.

7.2.

The operators $P_\pm$ coincide with the operator $\mathcal{A}$ defined in Subsection 3.1, if $m = n = 1$, and if the roles of f, h, $b(\boldsymbol{\xi})$ are played by $f_\pm = \omega_\mp$, $h_\pm = \omega_\pm$, $b_\pm(\boldsymbol{\xi}) = \xi^1 \mp i\xi^2$, correspondingly. Now, the kernels (3.13) and (3.27) are given by the relations

$$\mathfrak{N}_\pm = \{u \in L_2(\Omega) : \ u = c\omega_\pm\}, \quad \dim \mathfrak{N}_\pm = n = 1,$$

$$(\mathfrak{N}_*)_\pm = \mathfrak{N}_\mp, \quad n_* = \dim (\mathfrak{N}_*)_\pm = 1.$$

By Proposition 4.4(b), the effective (1×1)-matrix $g^0_\pm$ is unique and is equal to $\underline{g}_\pm$:

$$g^0_\pm = |\Omega| \, \|\omega_\mp\|^{-2}_{L_2(\Omega)}. \tag{7.11}$$

According to Subsection 4.6, for the numbers $\widehat{\gamma}_\pm(\boldsymbol{\theta})$ we have

$$\widehat{\gamma}_\pm(\boldsymbol{\theta}) = (\theta^1 \pm i\theta^2) g^0_\pm (\theta^1 \mp i\theta^2) = g^0_\pm. \tag{7.12}$$

Therefore, the numbers $\widehat{\gamma}_\pm(\boldsymbol{\theta}) = \widehat{\gamma}_\pm$ do not depend on $\boldsymbol{\theta}$. From (4.18), (7.11) and (7.12), it is clear that the numbers $\gamma_\pm(\boldsymbol{\theta}) = \gamma_\pm$ also do not depend on $\boldsymbol{\theta}$ and coincide with each other:

$$\gamma := \gamma_+ = \gamma_- = |\Omega|^2 \|\omega_-\|^{-2}_{L_2(\Omega)} \|\omega_+\|^{-2}_{L_2(\Omega)}. \tag{7.13}$$

Hence, $S_\pm : \mathfrak{N}_\pm \to \mathfrak{N}_\pm$ are operators of multiplication by the number γ. Mention that the structure of (7.13) copies the structure of (6.19). In this sense, the two-dimensional Pauli operator inherits the properties of the one-dimensional Schrödinger operator.

By (4.20) and (7.11), the effective operators $P^0_\pm$ are given by the relations

$$P^0_\pm = |\Omega| \, \|\omega_\mp\|^{-2}_{L_2(\Omega)} \omega_\mp(-\Delta)\omega_\mp. \tag{7.14}$$

7.3. The Pauli operator with a variable metric

Now, we consider a generalization of the operator (7.3). Let $\tilde{g}(\mathbf{x})$ be a Γ-periodic (2×2)-matrix with real-valued entries and satisfying condition (6.12). Let $\tilde{h}(\mathbf{x})$ be a periodic (2×2)-matrix such that $\tilde{h}$, $\tilde{h}^{-1}\in L_\infty$ and $\tilde{g}(\mathbf{x})=\tilde{h}(\mathbf{x})^*\tilde{h}(\mathbf{x})$. We put

$$\mathcal{P}_{\tilde{g}} := \mathcal{D}\tilde{g}\mathcal{D}, \tag{7.15}$$

where $\mathcal{D}$ is defined by (7.7), (7.10). The precise definition of the operator $\mathcal{P}_{\tilde{g}}$ is given via the quadratic form $\|\tilde{h}\mathcal{D}\mathbf{u}\|^2$, $\mathbf{u}\in \operatorname{Dom}\mathcal{D}$. By (7.7) and (7.15), the operator $\mathcal{P}_{\tilde{g}}$ admits the representation

$$\mathcal{P}_{\tilde{g}} = f\mathcal{Q}f\tilde{g}f\mathcal{Q}f. \tag{7.16}$$

We see that the operator (7.16) coincides with the operator $\mathcal{A}$ from Subsection 3.1 if $m=n=2$, f is given by (7.6), $h=\tilde{h}f$, $b(\mathbf{D})=\mathcal{Q}$. Then $g=f\tilde{g}f$,

$$b(\boldsymbol{\xi}) = \begin{pmatrix} 0 & \xi^1 - i\xi^2 \\ \xi^1 + i\xi^2 & 0 \end{pmatrix}.$$

The kernels (3.13) and (3.27) are given by the relations

$$\mathfrak{N} = \{\mathbf{u}\in L_2(\Omega;\mathbb{C}^2):\ f\mathbf{u}=\mathbf{c}\in\mathbb{C}^2\},\quad \dim\mathfrak{N}=n=2,$$

$$\mathfrak{N}_* = \{\mathbf{v}\in L_2(\Omega;\mathbb{C}^2):\ f\tilde{h}^*\mathbf{v}=\mathbf{c}\in\mathbb{C}^2\},\quad \dim\mathfrak{N}_*=n_*=2.$$

By Proposition 4.4(b), the effective matrix g^0 is unique and is equal to $\underline{g}$:

$$g^0 = |\Omega|\left(\int_\Omega f^{-1}\tilde{g}^{-1}f^{-1}\,d\mathbf{x}\right)^{-1}. \tag{7.17}$$

By (4.5) and (4.17), the operator $S(\boldsymbol{\theta}):\mathfrak{N}\to\mathfrak{N}$ is given by the relation

$$S(\boldsymbol{\theta}) = Pfb(\boldsymbol{\theta})g^0b(\boldsymbol{\theta})f, \tag{7.18}$$

where P is the orthoprojector onto $\mathfrak{N}$. According to the variational principle, the eigenvalues $\gamma_1(\boldsymbol{\theta})$, $\gamma_2(\boldsymbol{\theta})$ of the operator (7.18) are equal to extrema of the following quotient of quadratic forms:

$$\frac{(S(\boldsymbol{\theta})f^{-1}\mathbf{c},f^{-1}\mathbf{c})_{L_2(\Omega;\mathbb{C}^2)}}{\|f^{-1}\mathbf{c}\|^2_{L_2(\Omega;\mathbb{C}^2)}} = \frac{|\Omega|\langle g^0b(\boldsymbol{\theta})\mathbf{c},b(\boldsymbol{\theta})\mathbf{c}\rangle}{\langle\mathcal{F}\mathbf{c},\mathbf{c}\rangle},\quad \mathbf{c}\in\mathbb{C}^2, \tag{7.19}$$

where $\mathcal{F}:=\int_\Omega f^{-2}\,d\mathbf{x}$. We substitute $\mathbf{C}=b(\boldsymbol{\theta})\mathbf{c}$ and note that $b(\boldsymbol{\theta})\mathcal{F}b(\boldsymbol{\theta})=\mathcal{F}_*$, where $\mathcal{F}_*:=\int_\Omega f^2\,d\mathbf{x}$. Then (7.19) turns into

$$\frac{|\Omega|\langle g^0\mathbf{C},\mathbf{C}\rangle}{\langle\mathcal{F}_*\mathbf{C},\mathbf{C}\rangle},\quad \mathbf{C}\in\mathbb{C}^2.$$

Thus, *the numbers $\gamma_1(\boldsymbol{\theta})$, $\gamma_2(\boldsymbol{\theta})$ do not depend on $\boldsymbol{\theta}$ and coincide with the eigenvalues γ_1, γ_2 of the matrix*

$$|\Omega|f_*g^0f_*,\quad\text{where}\quad f_*:=\mathcal{F}_*^{-1/2}=\operatorname{diag}\{\|\omega_+\|^{-1}_{L_2(\Omega)},\|\omega_-\|^{-1}_{L_2(\Omega)}\}. \tag{7.20}$$

According to (4.20), the effective operator $\mathcal{P}^0_{\tilde g}$ is of the form

$$\mathcal{P}^0_{\tilde g} = f\mathcal{Q}g^0\mathcal{Q}f. \tag{7.21}$$

In the case where $\tilde g = \mathbf{1}$, the operator (7.15) coincides with (7.3). Then (7.17) yields

$$g^0 = |\Omega| \operatorname{diag}\{\|\omega_-\|^{-2}_{L_2(\Omega)}, \|\omega_+\|^{-2}_{L_2(\Omega)}\},$$

and the matrix (7.20) turns into $\gamma\mathbf{1}$, where γ is defined by (7.13). Thus, in this case, $\gamma_1 = \gamma_2 = \gamma$, which agrees with the results of Subsections 7.1, 7.2. The effective operator (7.21) with $\tilde g = \mathbf{1}$ takes the form

$$\mathcal{P}^0 = \begin{pmatrix} P^0_- & 0 \\ 0 & P^0_+ \end{pmatrix},$$

where $P^0_\pm$ are defined by (7.14).

8. The periodic operator of elasticity theory

8.1. Anisotropic elastic medium

The operator U in $L_2(\mathbb{R}^d; \mathbb{C}^d)$ describing the periodic elastic (in general, anisotropic) medium can be written as the operator $\mathcal{A}$ from Subsection 3.1. For this, we represent the strain tensor as a column, taking into account the tensor's symmetry explicitly. Then $n = d$, $2m = d(d+1)$, $f(\mathbf{x}) = \mathbf{1}$, $g(\mathbf{x})$ is a Γ-periodic $(m \times m)$-matrix with real-valued entries and satisfying condition (6.2). Explain that the matrix $g(\mathbf{x})$ corresponds to an "economical" form of the Hooke tensor (we take into account its symmetries). Let $h(\mathbf{x})$ be a periodic $(m \times m)$-matrix such that $g(\mathbf{x}) = h(\mathbf{x})^* h(\mathbf{x})$ and h, $h^{-1} \in L_\infty$. For simplicity, we consider the cases where $d = 2$ or $d = 3$ only. For $d = 2$ we have $n = 2$, $m = 3$, and the (3×2)-matrix $b(\boldsymbol{\xi})$ is given by

$$b(\boldsymbol{\xi}) = \begin{pmatrix} \xi^1 & 0 \\ \xi^2 & \xi^1 \\ 0 & \xi^2 \end{pmatrix}, \quad d = 2. \tag{8.1}$$

For $d = 3$ we have $n = 3$, $m = 6$, and the (6×3)-matrix $b(\boldsymbol{\xi})$ is of the form

$$b(\boldsymbol{\xi}) = \begin{pmatrix} \xi^1 & 0 & 0 \\ \xi^2 & \xi^1 & 0 \\ \xi^3 & 0 & \xi^1 \\ 0 & \xi^2 & 0 \\ 0 & \xi^3 & \xi^2 \\ 0 & 0 & \xi^3 \end{pmatrix}, \quad d = 3. \tag{8.2}$$

Obviously, condition (3.1) is valid. We write down the matrices b_j (see (3.3)):

$$b_1 = \begin{pmatrix} 1 & 0 \\ 0 & 1 \\ 0 & 0 \end{pmatrix}, \quad b_2 = \begin{pmatrix} 0 & 0 \\ 1 & 0 \\ 0 & 1 \end{pmatrix}, \quad d = 2;$$

$$b_1 = \begin{pmatrix} 1 & 0 & 0 \\ 0 & 1 & 0 \\ 0 & 0 & 1 \\ 0 & 0 & 0 \\ 0 & 0 & 0 \\ 0 & 0 & 0 \end{pmatrix}, \quad b_2 = \begin{pmatrix} 0 & 0 & 0 \\ 1 & 0 & 0 \\ 0 & 0 & 0 \\ 0 & 1 & 0 \\ 0 & 0 & 1 \\ 0 & 0 & 0 \end{pmatrix}, \quad b_3 = \begin{pmatrix} 0 & 0 & 0 \\ 0 & 0 & 0 \\ 1 & 0 & 0 \\ 0 & 0 & 0 \\ 0 & 1 & 0 \\ 0 & 0 & 1 \end{pmatrix}, \quad d = 3.$$

It is clear that the block matrix $\mathbf{B}$ (see Subsection 4.3) contains all the standard basis vectors as its columns. Therefore, condition (4.9) is valid.

The kernel $\mathfrak{N}$ (see (4.1)) is given by the relation

$$\mathfrak{N} = \{\mathbf{v} \in L_2(\Omega; \mathbb{C}^d) : \mathbf{v} = \mathbf{c} \in \mathbb{C}^d\}, \quad \dim \mathfrak{N} = d, \quad d = 2, 3,$$

and the kernel $\mathfrak{N}_*$ is given by (3.27); $n_* = \dim \mathfrak{N}_* = \infty$.

By Proposition 4.3, the effective $(m \times m)$-matrix g^0 is unique. We construct g^0 in accordance with Subsection 4.2. Let $\mathbf{v}_j \in \widetilde{H}^1(\Omega; \mathbb{C}^d)$ be solution of the equation $b(\mathbf{D})^* g(b(\mathbf{D})\mathbf{v}_j + \mathbf{e}_j) = 0$, $j = 1, \dots, m$. We put $\mathbf{u}_j = h(b(\mathbf{D})\mathbf{v}_j + \mathbf{e}_j)$. Then

$$g^0 = |\Omega|^{-1}\{(\mathbf{u}_j, \mathbf{u}_l)_{L_2(\Omega;\mathbb{C}^m)}\}, \quad j, l = 1, \dots, m.$$

The matrix g^0 satisfies estimates (4.12), (4.15).

By (4.5), the operator $S(\boldsymbol{\theta}) : \mathfrak{N} \to \mathfrak{N}$ satisfies the relation

$$S(\boldsymbol{\theta}) = b(\boldsymbol{\theta})^* g^0 b(\boldsymbol{\theta}).$$

In accordance with Subsection 4.7, g^0 is the matrix of effective medium and

$$U^0 = b(\mathbf{D})^* g^0 b(\mathbf{D})$$

is the operator of effective medium.

8.2. Isotropic elastic medium

In the case of the *isotropic* elastic medium, the matrix $g(\mathbf{x})$ is expressed via only two characteristics, namely, the shear modulus $\mu(\mathbf{x})$ and the modulus of compression $K(\mathbf{x})$. Assume that μ and K are Γ-periodic measurable functions such that

$$0 < \mu_0 \le \mu(\mathbf{x}) \le \mu_1 < \infty, \quad 0 < K_0 \le K(\mathbf{x}) \le K_1 < \infty. \tag{8.3}$$

The matrix $g(\mathbf{x})$ has the form

$$g(\mathbf{x}) = g_{\mu,K}(\mathbf{x}) := \begin{pmatrix} K+\mu & 0 & K-\mu \\ 0 & \mu & 0 \\ K-\mu & 0 & K+\mu \end{pmatrix}, \quad d = 2; \tag{8.4}$$

$$g(\mathbf{x}) = g_{\mu,K}(\mathbf{x}) := \frac{1}{3}\begin{pmatrix} 3K+4\mu & 0 & 0 & 3K-2\mu & 0 & 3K-2\mu \\ 0 & 3\mu & 0 & 0 & 0 & 0 \\ 0 & 0 & 3\mu & 0 & 0 & 0 \\ 3K-2\mu & 0 & 0 & 3K+4\mu & 0 & 3K-2\mu \\ 0 & 0 & 0 & 0 & 3\mu & 0 \\ 3K-2\mu & 0 & 0 & 3K-2\mu & 0 & 3K+4\mu \end{pmatrix}, \quad d = 3. \tag{8.5}$$

Under conditions (8.3), the matrix $g_{\mu,K}$ is positively definite. The inverse matrix has the form

$$g_{\mu,K}^{-1} = \frac{1}{4}\begin{pmatrix} K^{-1}+\mu^{-1} & 0 & K^{-1}-\mu^{-1} \\ 0 & 4\mu^{-1} & 0 \\ K^{-1}-\mu^{-1} & 0 & K^{-1}+\mu^{-1} \end{pmatrix}, \quad d=2;$$

and for $d = 3$

$$g_{\mu,K}^{-1} = \tfrac{1}{18}\times$$

$$\begin{pmatrix} 2K^{-1}+6\mu^{-1} & 0 & 0 & 2K^{-1}-3\mu^{-1} & 0 & 2K^{-1}-3\mu^{-1} \\ 0 & \mu^{-1} & 0 & 0 & 0 & 0 \\ 0 & 0 & \mu^{-1} & 0 & 0 & 0 \\ 2K^{-1}-3\mu^{-1} & 0 & 0 & 2K^{-1}+6\mu^{-1} & 0 & 2K^{-1}-3\mu^{-1} \\ 0 & 0 & 0 & 0 & \mu^{-1} & 0 \\ 2K^{-1}-3\mu^{-1} & 0 & 0 & 2K^{-1}-3\mu^{-1} & 0 & 2K^{-1}+6\mu^{-1} \end{pmatrix}.$$

We put

$$\overline{\mu} := |\Omega|^{-1}\int_\Omega \mu(\mathbf{x})\,d\mathbf{x}, \quad \overline{K} := |\Omega|^{-1}\int_\Omega K(\mathbf{x})\,d\mathbf{x},$$

$$\underline{\mu} := \left(|\Omega|^{-1}\int_\Omega \mu(\mathbf{x})^{-1}\,d\mathbf{x}\right)^{-1}, \quad \underline{K} := \left(|\Omega|^{-1}\int_\Omega K(\mathbf{x})^{-1}\,d\mathbf{x}\right)^{-1}.$$

Obviously,

$$\overline{g} = \overline{g_{\mu,K}} = g_{\overline{\mu},\overline{K}}, \quad \underline{g} = \underline{g_{\mu,K}} = g_{\underline{\mu},\underline{K}},$$

In general, the effective matrix $g^0 = g^0_{\mu,K}$ for an isotropic medium is not of "isotropic" form (8.4), (8.5). However, by general estimates (4.12), (4.15), $g^0_{\mu,K}$ is estimated via the "isotropic" matrices $g_{\underline{\mu},\underline{K}}$, $g_{\overline{\mu},\overline{K}}$ with averaged characteristics:

$$g_{\underline{\mu},\underline{K}} \le g^0_{\mu,K} \le g_{\overline{\mu},\overline{K}}.$$

8.3. The Hill body, $d = 2$

Consider the case of the isotropic medium *with a constant shear modulus* $\mu = \text{const}$. This case is known in the elasticity theory as the "Hill body". In this case, the operator U can be written as $U = \operatorname{rot}\mu\operatorname{rot} - \nabla\nu\operatorname{div}$, where $\nu = K+\mu$ for $d = 2$, $\nu = K + 4\mu/3$ for $d = 3$. Consequently, the operator U admits the more "economical" factorization. The two-dimensional case is especially simple. For $d = 2$, instead of (8.1), (8.4), we can use the factorization for U with $m = n = 2$, $f = \mathbf{1}$,

$$g(\mathbf{x}) = \begin{pmatrix} K(\mathbf{x})+\mu & 0 \\ 0 & \mu \end{pmatrix}, \quad b(\boldsymbol{\xi}) = \begin{pmatrix} \xi^1 & \xi^2 \\ \xi^2 & -\xi^1 \end{pmatrix}. \tag{8.6}$$

Let $h(\mathbf{x}) = g(\mathbf{x})^{1/2}$. Now, the kernels (4.1) and (3.27) are given by the relations

$$\mathfrak{N} = \{\mathbf{v} \in L_2(\Omega;\mathbb{C}^2) : \mathbf{v} = \mathbf{c} \in \mathbb{C}^2\}, \quad \dim\mathfrak{N} = n = 2,$$

$$\mathfrak{N}_* = \{\mathbf{u} \in L_2(\Omega;\mathbb{C}^2) : \mathbf{u} = h^{-1}\mathbf{c} \in \mathbb{C}^2\}, \quad \dim\mathfrak{N}_* = n_* = 2.$$

By Proposition 4.4(b), the effective matrix g^0 is unique and is equal to $\underline{g}$:

$$g^0 = \underline{g} = \operatorname{diag}\{\underline{K+\mu}, \mu\}, \quad \underline{K+\mu} := \left(|\Omega|^{-1} \int_\Omega (K+\mu)^{-1}\, d\mathbf{x} \right)^{-1}.$$

Putting $K_* := \underline{(K+\mu)} - \mu$, we obtain

$$g^0 = \begin{pmatrix} K_* + \mu & 0 \\ 0 & \mu \end{pmatrix}. \tag{8.7}$$

By (4.5), the operator $S(\boldsymbol{\theta}) : \mathfrak{N} \to \mathfrak{N}$ is represented in the form $S(\boldsymbol{\theta}) = b(\boldsymbol{\theta}) g^0 b(\boldsymbol{\theta})$. In accordance with (4.6), the operator of effective medium is given by the relation

$$U^0 = b(\mathbf{D}) g^0 b(\mathbf{D}).$$

Formula (8.7) shows (cf. (8.6)) that *in the case where* $\mu =$ const *and* $d = 2$, *the effective medium is isotropic and is defined by the modules* K_* *and* μ.

8.4. The Hill body, $d = 3$

Now, let $d = 3$ and $\mu =$ const. Then U can be written as the operator $\mathcal{A}$ from Subsection 3.1 with $n = 3$, $m = 4$, $f = \mathbf{1}$,

$$g(\mathbf{x}) = \begin{pmatrix} K(\mathbf{x}) + 4\mu/3 & 0 & 0 & 0 \\ 0 & \mu & 0 & 0 \\ 0 & 0 & \mu & 0 \\ 0 & 0 & 0 & \mu \end{pmatrix}, \quad b(\boldsymbol{\xi}) = \begin{pmatrix} \xi^1 & \xi^2 & \xi^3 \\ 0 & \xi^3 & -\xi^2 \\ \xi^3 & 0 & -\xi^1 \\ \xi^2 & -\xi^1 & 0 \end{pmatrix}. \tag{8.8}$$

Let $h(\mathbf{x}) = g(\mathbf{x})^{1/2}$. For the kernel $\mathfrak{N}$ we have

$$\mathfrak{N} = \{\mathbf{v} \in L_2(\Omega; \mathbb{C}^3) : \ \mathbf{v} = \mathbf{c} \in \mathbb{C}^3\}, \ \dim \mathfrak{N} = n = 3,$$

and the kernel $\mathfrak{N}_*$ is given by (3.27), $n_* = \dim \mathfrak{N}_* = \infty$. According to Subsection 4.2, we construct the matrix g^0. Let $\mathbf{v}_j \in \widetilde{H}^1(\Omega; \mathbb{C}^3)$ be solution of the equation $b(\mathbf{D})^* g(\mathbf{x})(b(\mathbf{D})\mathbf{v}_j + \mathbf{e}_j) = 0$, $j = 1, \dots, 4$. We put $\mathbf{u}_j = h(b(\mathbf{D})\mathbf{v}_j + \mathbf{e}_j)$. Then

$$g^0 = |\Omega|^{-1}\{(\mathbf{u}_j, \mathbf{u}_l)_{L_2(\Omega;\mathbb{C}^4)}\}, \quad j, l = 1, \dots, 4.$$

Consider $\mathbf{v}_1$. We introduce the notation $\Phi = (K + 4\mu/3)(D_1 v_1^1 + D_2 v_1^2 + D_3 v_1^3 + 1)$, $\Psi^1 = \mu(D_2 v_1^3 - D_3 v_1^2)$, $\Psi^2 = \mu(D_3 v_1^1 - D_1 v_1^3)$, $\Psi^3 = \mu(D_1 v_1^2 - D_2 v_1^1)$, $\Psi = (\Psi^1, \Psi^2, \Psi^3)$. Then the equation for $\mathbf{v}_1$ takes the form $\nabla \Phi = \operatorname{rot} \Psi$. This implies that the periodic function Φ satisfies the Laplace equation $\Delta \Phi = 0$. Hence, $\Phi = c =$ const. Integrating the relation $D_1 v_1^1 + D_2 v_1^2 + D_3 v_1^3 + 1 = (K + 4\mu/3)^{-1} c$ over Ω, we obtain

$$c = \left(|\Omega|^{-1} \int_\Omega (K + 4\mu/3)^{-1}\, d\mathbf{x} \right)^{-1} =: \underline{(K + 4\mu/3)}.$$

Then

$$g_{11}^0 = |\Omega|^{-1}\|\mathbf{u}_1\|^2_{L_2(\Omega;\mathbb{C}^4)} = |\Omega|^{-1}(g(b(\mathbf{D})\mathbf{v}_1 + \mathbf{e}_1), \mathbf{e}_1)_{L_2(\Omega;\mathbb{C}^4)} = |\Omega|^{-1}\int_\Omega \Phi\, d\mathbf{x} = c = \underline{(K + 4\mu/3)}.$$

Next,

$$g_{12}^0 = |\Omega|^{-1}(\mathbf{u}_1, \mathbf{u}_2)_{L_2(\Omega;\mathbb{C}^4)} = |\Omega|^{-1}(g(b(\mathbf{D})\mathbf{v}_1 + \mathbf{e}_1), \mathbf{e}_2)_{L_2(\Omega;\mathbb{C}^4)} = |\Omega|^{-1}\int_\Omega \mu(D_3 v_1^2 - D_2 v_1^3)\, d\mathbf{x} = 0.$$

Similarly, all the off-diagonal elements of the matrix g^0 are equal to zero. Next,

$$g_{22}^0 = |\Omega|^{-1}\|\mathbf{u}_2\|^2_{L_2(\Omega;\mathbb{C}^4)} = |\Omega|^{-1}(g(b(\mathbf{D})\mathbf{v}_2 + \mathbf{e}_2), \mathbf{e}_2)_{L_2(\Omega;\mathbb{C}^4)} = |\Omega|^{-1}\int_\Omega \mu(D_3 v_2^2 - D_2 v_2^3 + 1)\, d\mathbf{x} = \mu.$$

Similarly, $g_{33}^0 = g_{44}^0 = \mu$. We put $K_* := \underline{(K + 4\mu/3)} - 4\mu/3$. We have proved that

$$g^0 = \operatorname{diag}\{K_* + 4\mu/3, \mu, \mu, \mu\}. \tag{8.9}$$

It is easily seen that now condition (4.9) is valid. Hence, the matrix of effective medium is unique. According to (4.6), the operator of effective medium is given by the relation

$$U^0 = b(\mathbf{D})^* g^0 b(\mathbf{D}).$$

Formula (8.9) shows that, as in the two-dimensional case, *for the Hill body the effective medium is isotropic and is characterized by the modules* K_* *and* μ.

Similar results (including the results for the Hill body) are valid for the elasticity theory in arbitrary dimension $d \geq 2$.

9. Comments

9.1. Applications considered in Sections 6-8 are rather different

The acoustic operator (see Subsection 6.1) directly falls under considerations of Sections 3-5, since it is factorized from the very beginning. The same is "nearly true" for the operator of elasticity theory studied in Section 8. We write the strain tensor as a column, taking the tensor's symmetry into account explicitly. The description of the operator becomes more concise and makes it possible to apply the method of Sections 3-5 directly. We need to *reshape* the Schrödinger operator (see Subsection 6.2) to a factorized form especially. The same is related to the two-dimensional Pauli operator considered in Section 7. After factorizing, there are no obstructions to applying the method.

For the acoustic and elastic operators, we have $f = \mathbf{1}$. The passage from the matrix g to the effective matrix g^0 allows us to replace the periodic DO by a DO with constant coefficients. For the Schrödinger and Pauli operators, a nontrivial factor f occurs in the factorization. As a result, though g is replaced by g^0, the effective DO keeps the variable coefficients of type f. In the homogenization problem, this corresponds to the difference between estimates (5.13) and (5.12).

9.2. Alternative factorizations

It turns out that for the Hill body ($\mu = \text{const}$) with $d = 2$ (see Subsection 8.3), along with the "general" factorization (8.1), (8.4), the more economical factorization (8.6) with $m = n = 2$ is possible. This directly leads to (8.7), on the basis of Proposition 4.4(b). For the Hill body with $d = 3$ (see Subsection 8.4), we also employ the more economical factorization (8.8) instead of (8.2), (8.5), but in this case $4 = m \neq n = 3$.

Another example of an alternative factorization is connected with the two-dimensional Pauli operator considered in Subsection 7.1. Along with the factorization (7.9), the blocks $P_\pm$ can be written in a different factorized form, namely, $P_\pm = \omega_\mp \mathbf{D}^* \omega_\pm (\mathbf{1} + \sigma_2) \omega_\pm \mathbf{D} \omega_\mp$. However, this representation only formally agrees with the form of DO in Subsection 3.1, since the matrix $\mathbf{1} + \sigma_2$ is singular.

9.3. The negative discrete spectrum of the operator $H - \alpha V$

Let H be a periodic operator (6.11), and let H^0 be the effective Hamiltonian (6.18), calculated under the normalization condition $\|\omega\|^2_{L_2(\Omega)} = |\Omega|$. Let $V(\mathbf{x}) \geq 0$, $V \in L_\infty$ and

$$V(\mathbf{x}) = O(|\mathbf{x}|^{-2}(\log|\mathbf{x}|)^{-1/q}), \quad |\mathbf{x}| \to \infty, \quad 2q > d \geq 3. \tag{9.1}$$

The number of negative eigenvalues of $H - \alpha V$, $\alpha > 0$, is denoted by $N(\alpha) = N(\alpha; H, V)$. It turns out that $N(\alpha) = O(\alpha^q)$, $\alpha \to \infty$. We put $N_0(\alpha) = N(\alpha; H^0, V) = N(\alpha; \mathbf{D}^* g^0 \mathbf{D}, \omega^2 V)$. In [5] it was shown[3] that

$$\lim_{\alpha\to\infty} \left\{ \begin{matrix} \sup \\ \inf \end{matrix} \right\} \alpha^{-q} N(\alpha) = \lim_{\alpha\to\infty} \left\{ \begin{matrix} \sup \\ \inf \end{matrix} \right\} \alpha^{-q} N_0(\alpha). \tag{9.2}$$

Relation (9.2) shows that, under condition (9.1), the main term of $N(\alpha)$, $\alpha \gg 1$, is determined by the *threshold effect* near $\lambda = 0$. To the contrary, if $V \in L_{d/2}(\mathbb{R}^d)$, $d \geq 3$, the threshold $\lambda = 0$ does not influence the main term of asymptotics of $N(\alpha)$ as $\alpha \to \infty$.

The questions mentioned above were studied in [4], [5], [6], [10]. In these papers the Floquet-Bloch decomposition was also used, and the role of the threshold effects was studied. This is the only relationship with the material presented here.

[3]In [5], in many aspects more complete and more general results were obtained. In particular, the discrete spectrum in the internal gaps was studied.

9.4. The limits of applicability of the method

The examples given in Sections 6-8 do not exhaust the domain of applicability of the method proposed. In particular, on the basis of general results of Sections 1, 2, we can study the threshold effects for the Maxwell operator. For this, first we should *extend* the system, so as to eliminate the differential constraints (solenoidal conditions), and then we should *take the square of the operator*. The DO of second order arising in this way is factorized, but it does not belong to the class of DO's described in Section 3. This does not hinder the study of this operator nearly in the same way. The periodic problems in domains of cylinder or layer type, etc., bring a lot of examples of application of our method. For such problems, the momentum dimension is less than the coordinate dimension, which leads to new properties of the effective characteristics. This material will be presented elsewhere.

Now, we mention the cases where our method is not applicable.

1) Consider *the matrix Schrödinger operator* $H_n = -\Delta + p(\mathbf{x})$ in $L_2(\mathbb{R}^d; \mathbb{C}^n)$, where p is a periodic Hermitian $(n \times n)$−matrix-valued function. Now we do not have a factorized representation similar to (6.16). Assume that the point $\lambda = 0$ is the lower edge of the spectrum for H_n, and let $H_n(\mathbf{k})$ be the corresponding operator family in $L_2(\Omega; \mathbb{C}^n)$. Now, the point $\mathbf{k} = 0$ does not play a distinguished role; it may happen that $\operatorname{Ker} H_n(0) = \{0\}$. As a result, we loose the possibility to study the threshold effects efficiently.

2) *The periodic Pauli operator for* $d = 3$ is defined by the formula $\mathcal{P}_3 = \mathcal{D}_3^2$, where $\mathcal{D}_3 = (D_1 - A_1)\sigma_1 + (D_2 - A_2)\sigma_2 + (D_3 - A_3)\sigma_3$. This operator does not admit a factorization similar to (7.8). Moreover, we cannot describe the kernel $\operatorname{Ker} \mathcal{D}_3(0)$ for $\mathcal{D}_3(\mathbf{k})$ in a reasonable way.

3) *The magnetic periodic Schrödinger operator* is defined by the formula $\mathcal{M} = (\mathbf{D} - \mathbf{A}(\mathbf{x}))^*(\mathbf{D} - \mathbf{A}(\mathbf{x}))$, $d \geq 2$, where $\operatorname{div} \mathbf{A} = 0$, $\int_\Omega \mathbf{A}\, d\mathbf{x} = 0$. Now, the lower edge of the spectrum is a positive number. After shifting, we loose factorization. It is impossible to study the lower edge of the spectrum by our method even for $d = 2$.

9.5.

Two-sided estimates (4.12), (4.15) for g^0 in specific cases (the acoustic operator, the elastic operator, the Maxwell operator) are well known and are called the Voight-Reuss bracketing (for instance, see [11]). In Subsection 4.4, we clarify the general nature of such estimates: they follow from the abstract estimates for the germ S (see Subsection 1.4). The latter estimates give an opportunity for refining the "bracketing". Apparently, it was not recognized earlier that (4.16) is a consequence of the relation $m = n$.

The effective mass (see (7.13)) for the Pauli operators $P_\pm$ was recently calculated in [8] in a slightly different way. In that paper, the reasons of isotropy (the independence of γ on $\boldsymbol{\theta}$) remained unclear. The Pauli operator (7.15) with a metric is introduced here for the first time. The authors do not know if this model admits a reasonable physical interpretation. From the mathematical point of view, the model is rather natural. In [8], a metric was introduced in the Pauli operator in a different way, which, apparently, was less apt.

Concerning factorization for the Schrödinger operator, see [12] and [8, Section 4]. Relation (6.17) was not distinguished before.

The other examples in Sections 6-8 illustrate our approach by the known material. Concerning the classical methods of homogenization in the problems of acoustics and elasticity theory, see [3], [11] and references therein. About the Hill body, see [11, Chapter 12].

9.6.

The specific methods of solving the problems of the homogenization theory are well developed. The idea of applicability of the analytic perturbation theory methods to the homogenization theory (on the basis of the Floquet-Bloch decomposition) stands by itself. However, this idea is not new at all. Some material of this kind can be found in [3, Chapter 4] and in [11, Chapter 2]. Apparently, this way was not employed consistently and intensively enough. We think that there was a lack of the exact definition of a threshold effect. Also, the operators admitting an appropriate factorization were not distinguished. By these reasons, the present paper does not directly intersect with papers of other authors that use the Floquet-Bloch decomposition and the perturbation theory. Among such papers, we mention [13], where the acoustic operator with $g \in C^\infty$ for $d \geq 3$ was considered. By means of the perturbation theory, the full asymptotic expansion in powers of ε for solution of the equation $\mathbf{D}^* g(\varepsilon^{-1}\mathbf{x})\mathbf{D}u = q \in C_0^\infty(\mathbb{R}^d)$ was obtained. In [9], for the acoustic operator, formula (6.8) (in our notation) was proved, where the left-hand side was defined in terms of the perturbation theory and in the right-hand side g^0 was the effective matrix arising in the classical homogenization theory. At last, mention [1], where the Dirichlet problem for the operator $\mathbf{D}^* g(\varepsilon^{-1}\mathbf{x})\mathbf{D}$ in a bounded domain with smooth boundary was considered. Let $\nu_j(\varepsilon)$ be the eigenvalues of this problem. The interesting question of the behavior of $\varepsilon^2\nu_j(\varepsilon)$ as $\varepsilon \to 0$ was studied. In all these papers $n = 1$, which simplifies considerations significantly.

References

[1] G. Allaire and C. Conca, *Bloch wave homogenization and spectral asymptotic analysis*, J. Math. Pures Appl., **77** (1998), 153–208.

[2] N. S. Bakhvalov and G. P. Panasenko, *Homogenization: averaging processes in periodic media. Mathematical problems in the mechanics of composite materials,* (Russian), Nauka, Moscow, 1984, 352 pp.; English transl.: Mathematics and its Applications (Soviet Series), **36**, Kluwer Academic Publishers Group, Dordrecht, 1989, 366 pp.

[3] A. Bensoussan, J. L. Lions and G. Papanicolaou, *Asymptotic analysis for periodic structures,* North Holland Publishing Company, Amsterdam, 1978, 700 pp.

[4] M. Sh. Birman, *The discrete spectrum of the periodic Schrödinger operator perturbed by a decreasing potential,* (Russian) Algebra i Analiz, **8** (1996), no. 1, 3-20; English transl.: St. Petersburg Math. J., **8** (1997), no. 1, 1-14.

[5] M. Sh. Birman, *Discrete spectrum in the gaps of the perturbed periodic Schrödinger operator. II. Non-regular perturbations,* (Russian) Algebra i Analiz, **9** (1997), no. 6, 62-89; English transl.: St. Petersburg Math. J., **9** (1998), no. 6, 1073-1095.

[6] M. Sh. Birman, A. Laptev and T. A. Suslina, *Discrete spectrum of the two-dimensional periodic elliptic second order operator perturbed by a decaying potential. I. Semibounded gap,* (Russian), Algebra i Analiz, **12** (2000), no. 4, 36-78; English transl.: St. Petersburg Math. J., **12** (2001), no. 4.

[7] M. Sh. Birman and T. A. Suslina, *Absolute continuity of the two-dimensional periodic magnetic Hamiltonian with discontinuous vector-valued potential,* (Russian) Algebra i Analiz, **10** (1998), no. 4, 1-36; English transl.: St. Petersburg Math. J., **10** (1999), no. 4, 579-601.

[8] M. Sh. Birman and T. A. Suslina, *Two-dimensional periodic Pauli operator. The effective masses at the lower edge of the spectrum,* Oper. Theory Adv. Appl., **108** (1999), Birkhäuser, Basel, 13-31.

[9] C. Conca and M. Vanninathan, *Homogenization of periodic structures via Bloch decomposition,* SIAM J. Appl. Math., **57** (1997), no. 6, 1639–1659.

[10] V. Ivrii, *Accurate spectral asymptotics for periodic operators,* Journées "Equations aux derivées partielles", Saint-Jean-de-Monts (1999), 1-11.

[11] V. V. Jikov, S. M. Kozlov and O. A. Oleinik, *Homogenization of differential operators and integral functionals,* (Russian), Fizmatgiz, Moscow, 1993, 464 pp.; English transl.: Springer-Verlag, Berlin, 1994, 570 pp.

[12] W. Kirsh and B. Simon, *Comparison theorems for the gap of Schrödinger operators,* J. Funct. Anal., **75** (1987), no. 2, 396–410.

[13] E. V. Sevost'yanova, *Asymptotic expansion of the solution of a second-order elliptic equation with periodic rapidly oscillating coefficients,* (Russian) Mat. Sbornik, **115** (1981), 204-222; English transl.: Math. USSR-Sb., **43** (1982), no. 2, 181-198.

Key words and phrases: Periodic differential operators, analytic perturbation theory, threshold effect, homogenization, effective medium, effective Hamiltonian

Department of Physics,
St.Petersburg State University,
1, Ul'yanovskaya,
Petrodvorets,
St.Petersburg,
198904
RUSSIA
E-mail address: tanya@petrov.stoic.spb.su

Cauchy's Singular Integral Operator and Its Beautiful Spectrum

A. Böttcher and Yu. I. Karlovich

Abstract. The Cauchy singular integral operator, S, is one of the main actors in the theory of Fourier convolutions, Toeplitz operators, Riemann-Hilbert problems, Wiener-Hopf and singular integral equations. While the boundedness of S has been studied for many decades, final results on the spectrum of S were obtained only in recent times. During the last few years, it was discovered that there is a surprising and undreamt-of metamorphosis of the (local) spectra of S from circular arcs via horns and logarithmic double-spirals to so-called logarithmic leaves with a halo. This article is a survey of this fascinating development. We hope it is both a feast for the eyes and an illustration of Nick Trefethen's motto: The spectrum gives an operator a personality!

Throughout what follows Γ is a simple arc in the plane, that is, a plane set which is homeomorphic to the line segment $[0, 1]$. We suppose that Γ is rectifiable and oriented. The Cauchy singular integral operator S_Γ is formally defined by

$$(S_\Gamma f)(t) = \frac{1}{\pi i} \int_\Gamma \frac{f(\tau)}{\tau - t} d\tau \quad (t \in \Gamma).$$

Here the integral is understood in the following sense:

$$\int_\Gamma \frac{f(\tau)}{\tau - t} d\tau := \lim_{\varepsilon \to 0} \int_{\Gamma \setminus \Gamma(t,\varepsilon)} \frac{f(\tau)}{\tau - t} d\tau,$$

where the "portion" $\Gamma(t, \varepsilon)$ is $\Gamma(t, \varepsilon) := \{\tau \in \Gamma : |\tau - t| < \varepsilon\}$.

1. Boundedness

We equip Γ with length measure $|d\tau|$. A weight is a function

$$w : \Gamma \to [0, \infty]$$

such that $w^{-1}(\{0, \infty\})$ has measure zero. We consider S_Γ as an operator on the weighted Lebesgue space $L^p(\Gamma, w)$ with the norm

$$\|f\| := \left(\int_\Gamma |f(\tau)|^p w(\tau)^p \, |d\tau| \right)^{1/p}.$$

The following theorem is one of the great achievements of mathematical analysis of the 20th century.

The operator S_Γ is bounded on $L^p(\Gamma, w)$
$\Longleftrightarrow$ $1 < p < \infty$,
Γ *is a Carleson arc,*
w *is a Muckenhoupt weight.*

This theorem grew out of the work of

Hardy, Littlewood, M. Riesz, Mikhlin, Babenko,
Khvedelidze, Helson, Szegö, Widom, Forelli, Danilyuk,
Shelepov, Paatashvili, Khuskivadze, ... ,

the deciding steps of the proof were done by

Calderón, David,
Hunt, Muckenhoupt, Wheeden,

and more recent developments (including new proofs) are connected with the names

Coifman, McIntosh, Meyer, Jones, Journé, Semmes, Murai,
Dynkin, Osilenker, Mattila, Melnikov, Verdera, ...

(see [5, 9, 10, 11, 12, 13, 14, 15, 18, 23, 24, 25, 26, 27, 31]).

The arc Γ is said to be a *Carleson arc* (= *Ahlfors-David arc*) if there is a constant $C_\Gamma < \infty$ such that

$$|\Gamma(t, \varepsilon)| \le C_\Gamma \varepsilon \text{ for all } \varepsilon > 0 \text{ and } t \in \Gamma, \tag{1}$$

where $|\Gamma(t,\varepsilon)|$ is the (length) measure of $\Gamma(t,\varepsilon)$. A weight w on Γ is referred to as a *Muckenhoupt weight* if there exists a constant $C_w < \infty$ such that

$$\left(\int_{\Gamma(t,\varepsilon)} w(\tau)^p |d\tau|\right)^{1/p} \left(\int_{\Gamma(t,\varepsilon)} w(\tau)^{-q} |d\tau|\right)^{1/q} \le C_w \varepsilon \text{ for all } \varepsilon > 0 \text{ and } t \in \Gamma, \tag{2}$$

where here and in what follows $1/p + 1/q = 1$. We remark that, by Hölder's inequality,

$$|\Gamma(t,\varepsilon)| \le \left(\int_{\Gamma(t,\varepsilon)} w(\tau)^p |d\tau|\right)^{1/p} \left(\int_{\Gamma(t,\varepsilon)} w(\tau)^{-q} |d\tau|\right)^{1/q},$$

so that in fact (1) is implied by (2). Thus, on writing A_p for the set of all pairs (Γ, w) satisfying (2), we can restate the above theorem in the form

$$S_\Gamma \in \mathcal{B}(L^p(\Gamma, w)) \Longleftrightarrow 1 < p < \infty \quad \text{and} \quad (\Gamma, w) \in A_p.$$

2. Carleson arcs

Here are a few examples of Carleson arcs.

The arc Γ is called a C^1 arc if, after suitable rotation, it can be given by

$$\Gamma = \{\tau \in \mathbf{C} : \tau = x + if(x), a \le x \le b\}$$

with $f \in C^1[a,b]$. If the arc Γ is a finite union of C^1 arcs, then Γ is said to be a piecewise C^1 arc. We remark that the only possible "singularities" of piecewise C^1 arcs are corners and cusps (Figure 1).

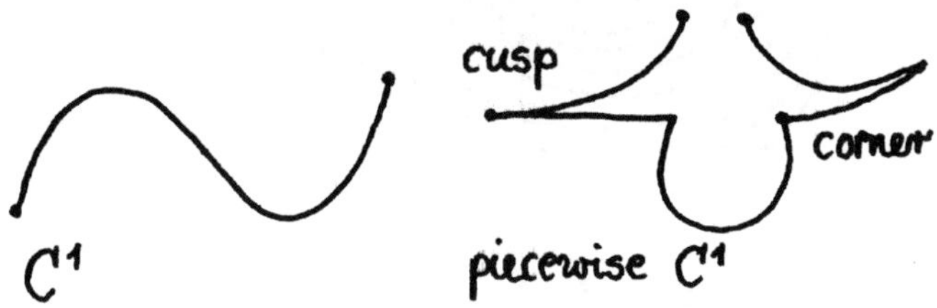

Fig. 1

A sufficiently interesting class of Carleson arcs is constituted by arcs of the form

$$\Gamma = \{0,1\} \cup \{\tau \in \mathbf{C} : \tau = re^{i\theta(r)}, 0 < r < 1\}, \tag{3}$$

where

$$\theta(r) = h(\log|\log r|)|\log r|$$

and $h \in C^1(\mathbf{R})$ is a real-valued function for which h and h' are bounded on $\mathbf{R}$ (see [5]). If $h(x) = 0$ for all x, then (3) is simply the line segment $[0,1]$. The choice $h(x) = \delta$ ($\ne 0$) for all $x \in \mathbf{R}$ yields a logarithmic spiral (see Figure 2).

Fig. 2

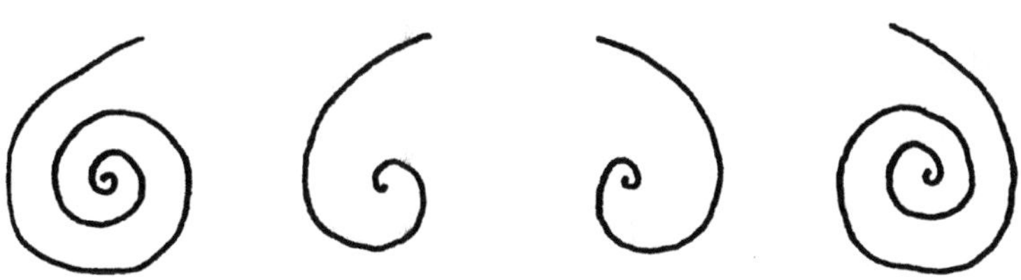

$\delta < 0$	$\delta < 0$	$\delta > 0$	$\delta > 0$
$\lvert\delta\rvert$ small	$\lvert\delta\rvert$ large	large	small

Letting $h(x) = \delta + \gamma \sin(\lambda x)$, we get Carleson arcs as in Figures 3 and 4. Moreover, a "logarithmic spiral composed of logarithmic spirals" like in Figure 5 can be shown to be a Carleson arc. On replacing each logarithmic spiral in Figure 5 by an arc similar to the whole arc of Figure 5, we arrive at a Carleson arc as in Figure 6.

Fig. 3

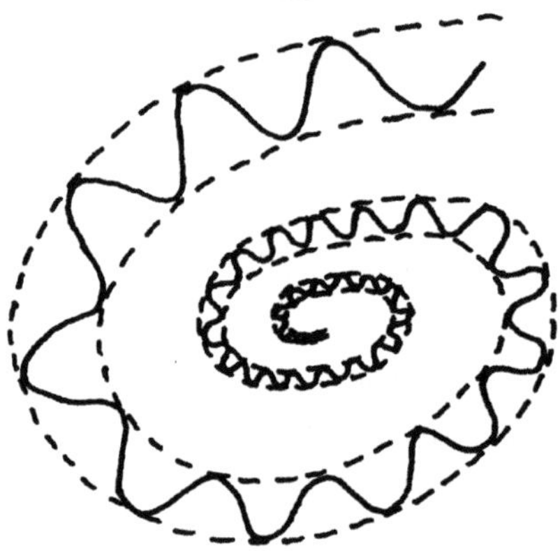

Fig. 4

3. Muckenhoupt weights

Let Γ be a Carleson arc. A weight w of the form

$$w(\tau) = \prod_{j=1}^{n} |\tau - t_j|^{\mu_j} \quad (\tau \in \Gamma)$$

is referred to as a power weight; here $t_1, \ldots, t_n$ are distinct points on Γ and $\mu_1, \ldots, \mu_n$ are real numbers. One can show (see, e.g., [5, 22]) that

$$(\Gamma, w) \in A_p \iff -\frac{1}{p} < \mu_j < \frac{1}{q} \quad \text{for all} \quad j.$$

Obviously, the only "singularities" of power weights are zeros or poles (Figure 7).

Fig. 5

Fig. 6

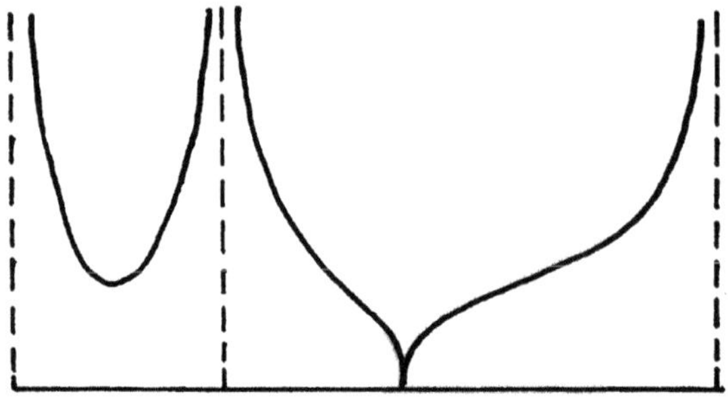

Fig. 7

A wealth of nontrivial Muckenhoupt weights can be constructed as follows. Suppose Γ is given by (3) and put

$$w\left(re^{i\theta(r)}\right) = e^{v(r)} \quad (0 < r < 1), \tag{4}$$

where

$$v(r) = g(\log|\log r|)\log r$$

and $g \in C^2(\mathbf{R})$ is a real-valued function such that g, g', g'' are bounded on $\mathbf{R}$. Clearly, the "singularity" of w is at $t = 0$, and the behaviour of w near $t = 0$ is completely determined by the behaviour of $g(x)$ as $x \to +\infty$. It can be shown (see [5]) that $(\Gamma, w) \in A_p$ if and only if

$$-\frac{1}{p} < \liminf_{x\to+\infty}\left(g(x) + g'(x)\right) \le \limsup_{x\to+\infty}\left(g(x) + g'(x)\right) < \frac{1}{q}.$$

In particular, for $g(x) = \mu + \varepsilon\sin(\eta x)$ we obtain that $(\Gamma, w) \in A_p$ exactly if

$$-\frac{1}{p} < \mu - |\varepsilon|\sqrt{\eta^2+1} \le \mu + |\varepsilon|\sqrt{\eta^2+1} < \frac{1}{q}. \tag{5}$$

If $g(x) = \mu$ for all $x \in \mathbf{R}$, we get the power weight

$$w\left(re^{i\theta(r)}\right) = e^{\mu\log r} = r^\mu,$$

and (5) says that $(\Gamma, w) \in A_p$ if and only if $-1/p < \mu < 1/q$. The choice $g(x) = \varepsilon\sin x$ yields an oscillating weight as the one in Figure 8, and (5) tells us that $(\Gamma, w) \in A_p$ if and only if $|\varepsilon|\sqrt{2} < \min(1/p, 1/q)$.

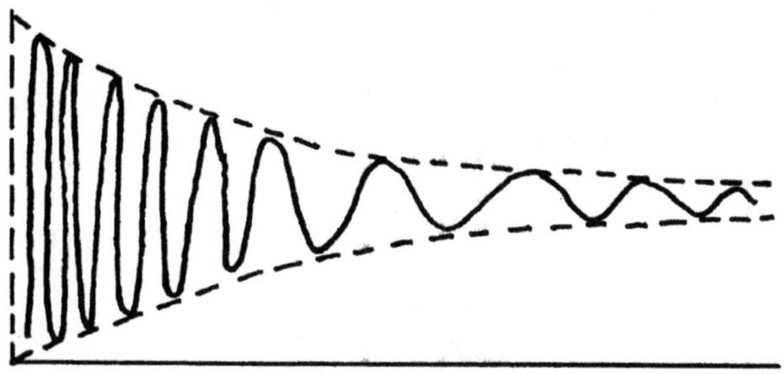

Fig. 8

Finally, we should never forget that an oscillating weight as in Figure 8 may "live" on oscillating arcs like those of Figures 2 to 6.

4. Spectrum and local spectra

The *spectrum* of S_Γ is the set

$$\operatorname{sp} S_\Gamma := \{\lambda \in \mathbf{C} : S_\Gamma - \lambda I \text{ is not invertible in } \mathcal{B}(L^p(\Gamma, w))\},$$

and the *essential spectrum* of S_Γ is defined as the set

$$\operatorname{sp}_{\text{ess}} S_\Gamma := \{\lambda \in \mathbf{C} : S_\Gamma - \lambda I \text{ is not invertible modulo compact operators in } \mathcal{B}(L^p(\Gamma, w))\}.$$

Clearly, $\operatorname{sp}_{\text{ess}} S_\Gamma \subset \operatorname{sp} S_\Gamma$.

We always assume that the arc Γ contains its two endpoints, so that Γ is a compact set. At each point $t \in \Gamma$, one can associate a so-called local spectrum $\operatorname{sp}_t S_\Gamma$ with the operator S_Γ. Given $\varphi \in C(\Gamma)$, we denote by M_φ the operator of multiplication by φ on $L^p(\Gamma, w)$. An operator $A \in \mathcal{B}(L^p(\Gamma, w))$ is said to be locally invertible at $t \in \Gamma$ if there exist operators B and D in $\mathcal{B}(L^p(\Gamma, w))$ and a function $\varphi \in C(\Gamma)$ such that $\varphi(t) = 1$ and both $M_\varphi AB - M_\varphi$ and $DAM_\varphi - M_\varphi$ are compact. The *local spectrum* of S_Γ at $t \in \Gamma$ is defined by

$$\operatorname{sp}_t S_\Gamma := \{\lambda \in \mathbf{C} : S_\Gamma - \lambda I \text{ is not locally invertible at } t\}.$$

The local principle of Simonenko [33] (also see [5, 16, 22]) implies that

$$\boxed{\operatorname{sp}_{\text{ess}} S_\Gamma = \bigcup_{t \in \Gamma} \operatorname{sp}_t S_\Gamma.}$$

Thus, in order to find the essential spectrum of S_Γ, it suffices to identify the local spectra of S_Γ.

Let a and b be the endpoints of the arc Γ. It is well known that

$$\mathrm{sp}_t\, S_\Gamma = \{-1, 1\} \text{ for } t \in \Gamma \setminus \{a, b\}.$$

Hence, the actual problem is the determination of the local spectra $\mathrm{sp}_a\, S_\Gamma$ and $\mathrm{sp}_b\, S_\Gamma$. We will see that the last two spectra are always compact and connected sets which contain the points -1 and 1. Figure 9 shows an example.

Fig. 9

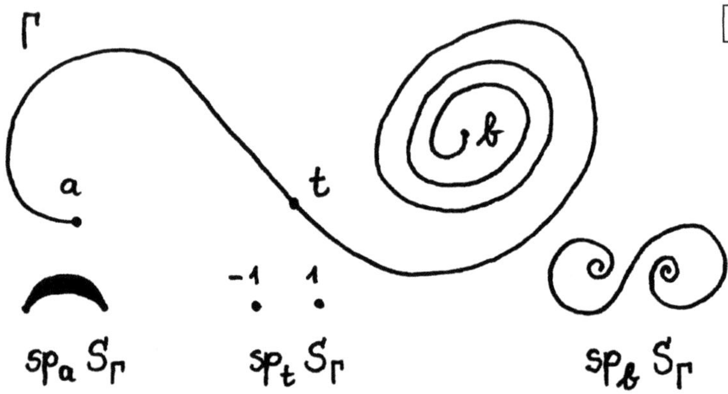

It follows that

$$\mathrm{sp}_{\mathrm{ess}}\, S_\Gamma = \mathrm{sp}_a\, S_\Gamma \cup \mathrm{sp}_b\, S_\Gamma.$$

To obtain $\mathrm{sp}\, S_\Gamma$, choose any continuous arc C_a between -1 and 1 in $\mathrm{sp}_a\, S_\Gamma$ and any continuous arc C_b between 1 and -1 in $\mathrm{sp}_b\, S_\Gamma$. Give C_a the orientation from -1 to 1, and orient C_b from 1 to -1. Then $C_a \cup C_b$ is a closed continuous and oriented curve in the plane. For $\lambda \notin C_a \cup C_b$, let $\mathrm{wind}_{a,b}(\lambda)$ denote the winding number of $C_a \cup C_b$ about λ. One can show (see, e.g., [5, 22]) that

$$\mathrm{sp}\, S_\Gamma = \mathrm{sp}_{\mathrm{ess}}\, S_\Gamma \cup \{\lambda \in \mathbf{C} \setminus \mathrm{sp}_{\mathrm{ess}}\, S_\Gamma : \mathrm{wind}_{a,b}(\lambda) \neq 0\}.$$

Figure 10 illustrates the procedure, and in Figure 11 we see the several types of spectra.

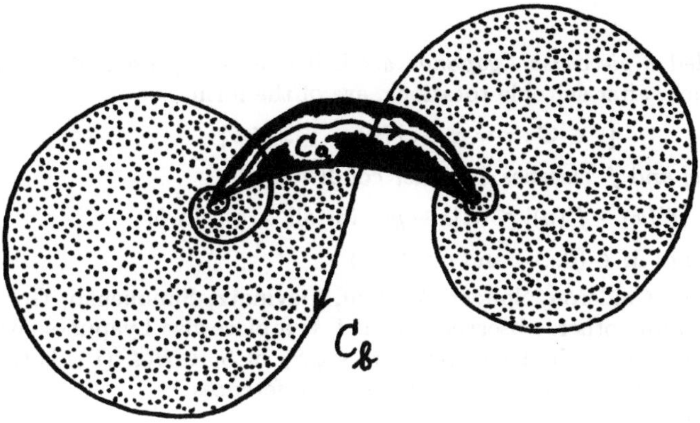

Fig. 10

Fig. 11

In summary, the identification of $\operatorname{sp} S_\Gamma$ can be completely reduced to the problem of describing the local spectrum of S_Γ at the endpoints of the arc Γ.

The beginning

The arc Γ is called a *piecewise Lyapunov arc* if it can be represented as a finite union of arcs which, after suitable rotations, are of the form

$$\{\tau \in \mathbf{C} : \tau = x + if(x), a \le x \le b\}$$

where $f \in C^1[a,b]$ and f' satisfies a Hölder condition,

$$|f'(x) - f'(y)| \le M|x-y|^{\varepsilon} \text{ for all } x, y \in [a,b]$$

with some constants $M \in (0,\infty)$ and $\varepsilon \in (0,1)$.

In the sixties, Calderón, Spitzer, Widom, Devinatz, Simonenko, Gohberg, Krupnik, and certainly others observed that if Γ is a piecewise Lyapunov arc and t is an endpoint of Γ, then the local spectrum $\mathrm{sp}_t\, S_\Gamma$ of $S_\Gamma \in \mathcal{B}(L^2(\Gamma))$ is the line segment $[-1,1]$ (Figure 12); see [5, 22] for precise references.

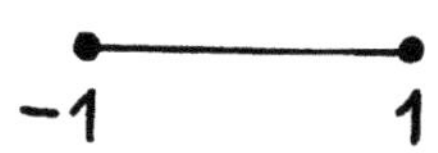

Fig. 12

Metamorphosis 1

In the late sixties and early seventies, Widom, Gohberg, and Krupnik studied S_Γ on $L^p(\Gamma, w)$ in the case where Γ is piecewise Lyapunov and w is a power weight (see [20, 22, 36]). Let t be an endpoint of Γ and suppose

$$w(\tau) = |\tau - t|^{\mu} \quad (-1/p < \mu < 1/q).$$

They showed that $\mathrm{sp}_t\, S_\Gamma$ is a certain circular arc between -1 and 1 whose shape is completely determined by the value of $1/p + \mu$ and the orientation of Γ (Figure 13).

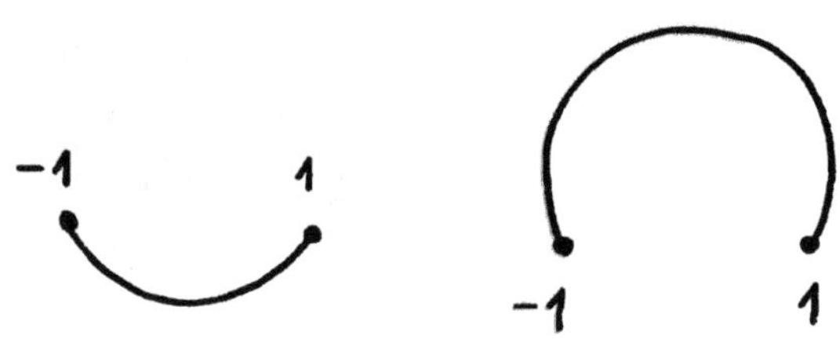

Fig. 13

Metamorphosis 2

After the discovery of the circular arcs, the development paused a long time. In 1990, Spitkovsky [35] surprised the community with the identification of $\mathrm{sp}_t\, S_\Gamma$ for the space $L^p(\Gamma, w)$ in the case where t is an endpoint of a piecewise Lyapunov arc

Γ and w is an arbitrary Muckenhoupt weight. He proved that in this case the local spectrum is always a so-called *horn*, that is, the closed set between two circular arcs joining -1 and 1 (see Figure 14).

Fig. 14

Metamorphosis 3

In 1993, inspired by the Spitkovsky result and by certain 1988 results of Aizenshtat, Litvinchuk, and one of the authors [1] on conformal welding and singular integrals on curves that look like logarithmic spirals, we turned our attention to the local spectrum $\mathrm{sp}_t\, S_\Gamma$ of $S_\Gamma \in \mathcal{B}(L^p(\Gamma, w))$ at the endpoint of an arbitrary Carleson arc Γ.

We first considered so-called spiralic Carleson arcs. The definition is as follows. We can write

$$\tau - t = |\tau - t| e^{i \arg(\tau - t)} \text{ for } \tau \in \Gamma \setminus \{t\},$$

and the argument can be chosen so that it is continuous on $\Gamma \setminus \{t\}$. Seifullayev [30] showed that for an arbitrary Carleson arc the estimate

$$\arg(\tau - t) = O\Big(\big|\log|\tau - t|\big|\Big) \quad (\tau \to t)$$

holds. A *spiralic Carleson arc* is a Carleson arc satisfying the stronger condition

$$\arg(\tau - t) = \delta \big|\log|\tau - t|\big| + O(1) \quad (\tau \to t)$$

with some constant $\delta \in \mathbf{R}$. In 1994, we were able to prove that if t is an endpoint of a spiralic Carleson arc Γ, then $\mathrm{sp}_t\, S_\Gamma$ of $S_\Gamma \in \mathcal{B}(L^p(\Gamma, w))$ is a logarithmic double spiral between -1 and 1 if w is a power weight and a so-called *spiralic horn* (= closed set bounded by two logarithmic double spirals between -1 and 1) if w is an arbitrary Muckenhoupt weight [3]. See Figure 15.

Fig. 15

Metamorphosis 4

In 1995, the case of an arbitrary Carleson arc with a power weight was disposed of in our paper [4]. The local spectra are so-called *logarithmic leaves with a medium separating point.* Examples are shown in Figure 16.

Fig. 16

Metamorphosis 5

A few months later, also in 1995, we finally succeeded in settling the case of arbitrary Carleson arcs with arbitrary Muckenhoupt weights [6]. It turned out that in this most general case the local spectra are what we called *logarithmic leaves with a halo.* Figure 17 shows two examples.

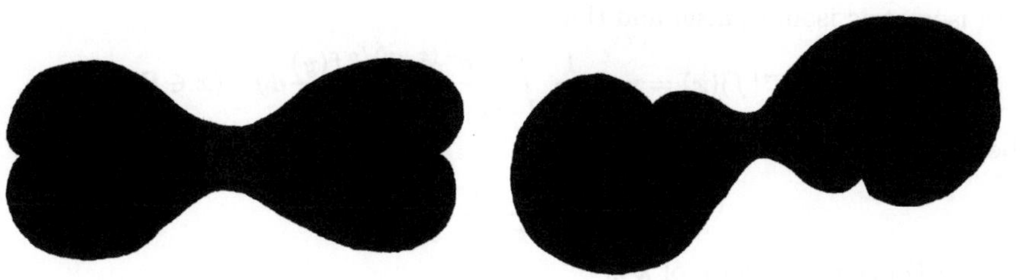

Fig. 17

In what follows we will say more on the origin and the precise shapes of the local spectra sketched in the figures.

5. Fourier techniques

To get a feeling for the matter, we consider the cases where Γ is the real line $\mathbf{R}$ or the half-line $\mathbf{R}_+ := [0,\infty)$ (although neither $\mathbf{R}$ nor $\mathbf{R}_+$ are rectifiable and thus arcs in the above sense).

The Cauchy singular integral operator on $L^p(\mathbf{R})$ is given by the formula

$$(S_{\mathbf{R}}f)(x) = \text{v.p.}\,\frac{1}{\pi i}\int_{-\infty}^{\infty}\frac{f(y)}{y-x}dy \quad (x\in\mathbf{R}).$$

Hence, $S_{\mathbf{R}}$ is the convolution by the kernel $k(t) = -1/(\pi i t)$, and therefore, at least formally, the spectrum of $S_{\mathbf{R}}$ is the essential range of the Fourier transform Fk of k. Since

$$(Fk)(\xi) := \int_{-\infty}^{\infty} k(t)e^{it\xi}dt = -\operatorname{sign}\xi \quad (\xi\in\mathbf{R}),$$

it results that the spectrum of $S_{\mathbf{R}}$ is the doubleton $\{-1,1\}$. As $\mathbf{R}$ looks locally like Γ at the points $t\in\Gamma\setminus\{a,b\}$, it becomes plausible that $\operatorname{sp}_t S_\Gamma = \{-1,1\}$ whenever $t\in\Gamma\setminus\{a,b\}$.

To understand $\operatorname{sp}_t S_\Gamma$ at an endpoint of the arc Γ, we turn to the Cauchy singular integral operator on $L^p(\mathbf{R}_+)$,

$$(S_{\mathbf{R}_+}f)(x) = \text{v.p.}\,\frac{1}{\pi i}\int_{0}^{\infty}\frac{f(y)}{y-x}dy \quad (x\in\mathbf{R}_+).$$

This is no longer a pure convolution. Some sort of machinery for tackling operators like $S_{\mathbf{R}_+}$ was created by Eskin, Dynin, and Duduchava (see [17, 19, 22]). The basic idea is that the map

$$\Phi : L^p(\mathbf{R}_+)\to L^p(\mathbf{R}),\quad (\Phi f)(x) = e^{-x/p}f(e^{-x})$$

is an isometric isomorphism and that

$$(\Phi S_{\mathbf{R}_+}\Phi^{-1}f)(x) = \text{v.p.}\,\frac{1}{\pi i}\int_{-\infty}^{\infty}\frac{e^{-(x-y)/p}f(y)}{1-e^{-(x-y)}}dy \quad (x\in\mathbf{R}).$$

The last operator is the convolution by

$$k(t) = \frac{1}{\pi i}\frac{e^{-t/p}}{1-e^{-t}},$$

and the Fourier transform of k is

$$(Fk)(\xi) = \coth\pi(\xi + i/p) \quad (\xi\in\mathbf{R}).$$

Thus, again at least formally, we arrive at the conclusion that

$$\operatorname{sp} S_{\mathbf{R}_+} = \left\{\coth\pi(\xi+i/p) : \xi\in\overline{\mathbf{R}}\right\}, \tag{6}$$

where $\overline{\mathbf{R}} := \mathbf{R}\cup\{-\infty,+\infty\}$. We have

$$\coth\pi(\xi+i/p) = \frac{e^{2\pi(\xi+i/p)}+1}{e^{2\pi(\xi+i/p)}-1} = M_{-1,1}\left(e^{2\pi(\xi+i/p)}\right),$$

where $M_{-1,1}$ is the Möbius transform

$$M_{-1,1}(\zeta) := \frac{\zeta+1}{\zeta-1}.$$

As ξ moves on $\mathbf{R}$ from $-\infty$ to $+\infty$, the point $\xi + i/p$ traces out a horizontal straight line in the complex plane, the map $z\mapsto e^{2\pi z}$ maps this horizontal line into a ray starting at the origin, and the Möbius transform $\zeta\mapsto M_{-1,1}(\zeta)$ maps this ray into a circular arc between $M_{-1,1}(0) = -1$ and $M_{-1,1}(\infty) = 1$ (see Figure 18). Consequently, set (6) is a circular arc! Despite the (technical and substantial) problems that arise when transferring the above argument to the case of the endpoint of a Carleson arc with a Muckenhoupt weight, this argument reveals at least the emergence of circular arcs in the spectral theory of singular integral operators and, as will be seen now, it also anticipates the way in which the sets in Figures 15 to 17 can be described very lucidly.

6. Anatomy of leaves

We use the name *leaf* as a collective name for the sets which appear as local spectra $\operatorname{sp}_t S_\Gamma$ of $S_\Gamma\in\mathcal{B}(L^p(\Gamma,w))$ at an endpoint t of Γ. The shape of a leaf is always determined by a so-called indicator set and two subsequent transformations of this set.

The *indicator set* is a certain subset of $\mathbf{C}$. It depends on the value of $p\in(1,\infty)$, on the behaviour of the arc Γ near its endpoint t (and also on whether t is the starting or terminating point of the oriented arc Γ), and on the behaviour of the weight w in a neighbourhood of t. The exponential map

$$z\mapsto e^{2\pi z}$$

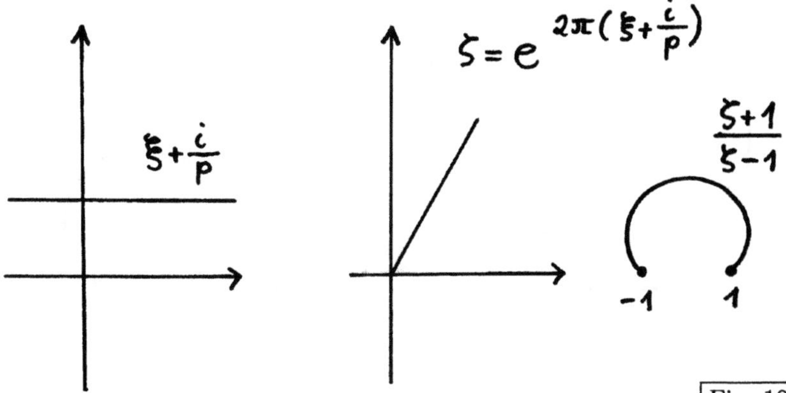

Fig. 18

transforms the indicator set into a set we call the *intermediate set*. Finally, the Möbius transform

$$\zeta \mapsto \frac{\zeta+1}{\zeta-1}$$

sends the intermediate set to a set which turns out to be just the local spectrum $\mathrm{sp}_t\, S_\Gamma$ and which is referred to as a leaf.

indicator set		intermediate set		leaf
z	$\mapsto$	$e^{2\pi z}$	$\mapsto$	$\dfrac{e^{2\pi z}+1}{e^{2\pi z}-1}$

Metamorphosis 1. Suppose Γ is nice (piecewise Lyapunov) and w is nice (say, $w(\tau)=|\tau-t|^{\mu}$). In that case the situation is as in Figure 19: the indicator set is the horizontal straight line $\operatorname{Im} z = 1/p+\mu$, the intermediate set is a ray between the origin and infinity, and the leaf is a circular arc between -1 and 1. See Figure 19.

Metamorphosis 2. Let Γ be nice and let w be an arbitrary Muckenhoupt weight. One can associate two so-called indices of powerlikeness μ and ν with w, and the indicator set becomes the closed stripe

$$1/p+\mu \le \operatorname{Im} z \le 1/p+\nu.$$

We may think of the indicator set as a union of horizontal straight lines. Figure 20 then shows that the intermediate set is a union of rays, that is, a sector, and that the leaf is a union of circular arcs, or, in other words, a horn. Figure 20 illustrates things.

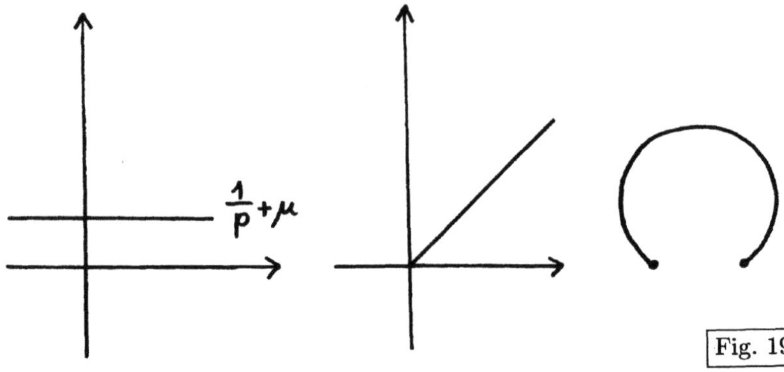

Fig. 19

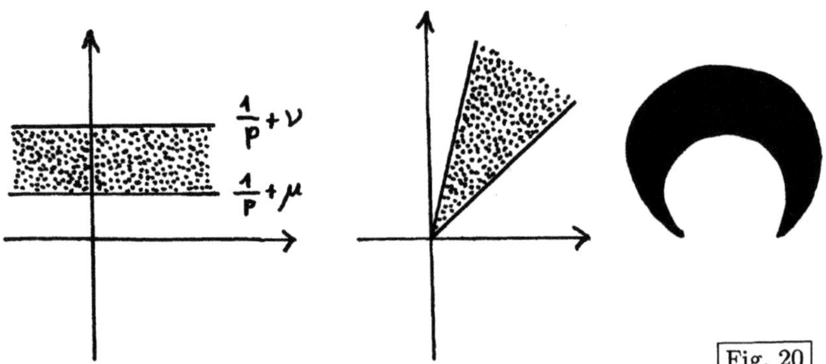

Fig. 20

Metamorphosis 3. Let Γ be a spiralic curve and let first $w(\tau) = |\tau - t|^{\mu}$. Now the indicator set can be shown to be a straight line of the form

$$y = \delta x + \mu + 1/p \quad (z = x + iy).$$

The exponential map transforms a straight line with nonzero slope into a logarithmic spiral wriggling out of the origin and scrolling up at the point at infinity. The Möbius transform $M_{-1,1}$ eventually maps this logarithmic spiral to a logarithmic double spiral between -1 and 1. See Figure 21.

If Γ is spiralic and w is an arbitrary Muckenhoupt weight, then the indicator set is a stripe with a slope, and the further evolution is as in Figure 22, which can be obtained as a superposition of pictures like those in Figure 21.

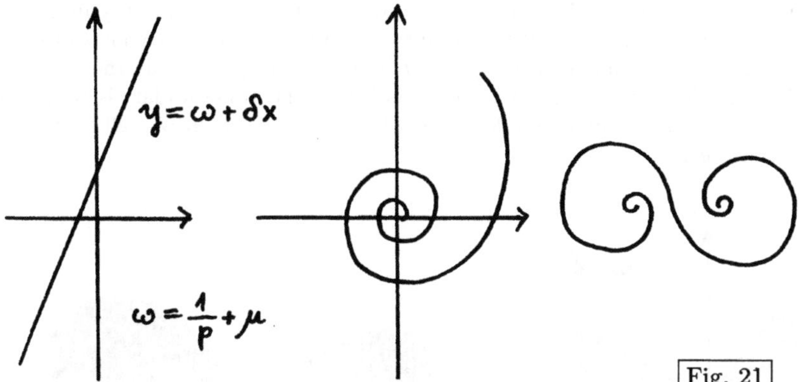

Fig. 21

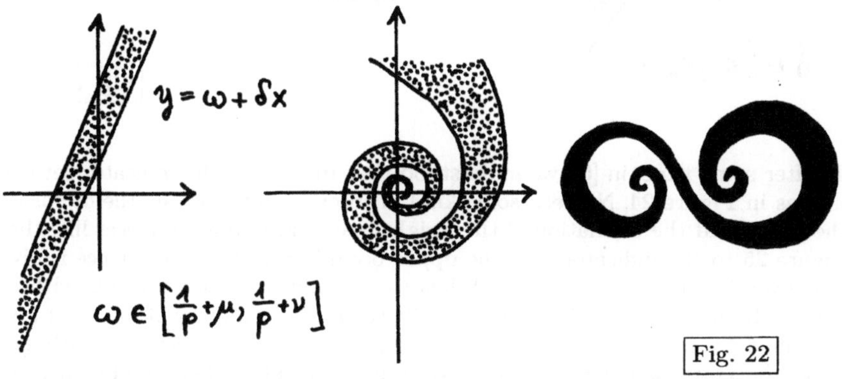

Fig. 22

Metamorphosis 4. Let Γ be an arbitrary Carleson arc, but suppose w is nice (power weight). The indicator set can be shown to be the closed set between two straight lines meeting the imaginary axis at the same point,

$$\{z = x + iy : y = \omega + \delta x, \delta_- \leq \delta \leq \delta_+\}.$$

The two extreme slopes δ_- and δ_+ can be interpreted as the spirality indices of Γ at t. After applying the exponential map and the Möbius transform to this indicator set, we get a leaf as in Figure 23 (the intermediate set is omitted in this figure).

Metamorphosis 5. Finally, suppose Γ is an arbitrary Carleson arc and w is an arbitrary Muckenhoupt weight. Figures 20 and 23 make us suspect that the situation should be as in Figure 24. There are indeed cases in which the indicator set is as in Figure 24: it is completely characterised by two slopes (the spirality indices δ_- and δ_+) and the ordinates of the four points at which the boundary lines meet the imaginary axis (the numbers $1/p + \mu_\pm, 1/p + \nu_\pm$, where $\mu_\pm, \nu_\pm$ are the indices of powerlikeness). The resulting leaf is shown is Figure 24. Its boundary is comprised of four pieces of logarithmic double spirals, by virtue of which we call such a leaf a *logarithmic leaf.*

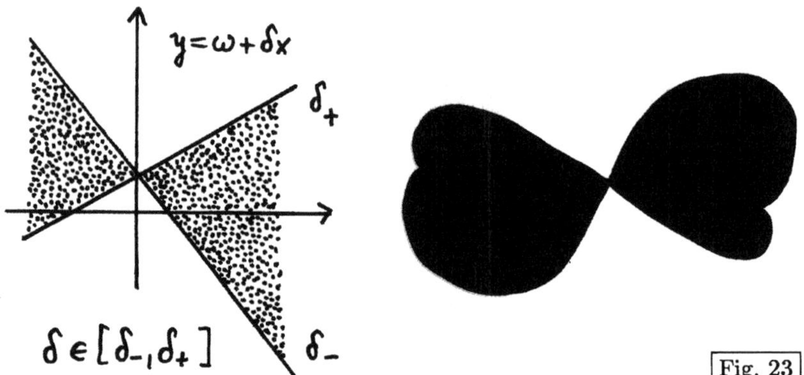

Fig. 23

Only after some time, in [6], we understood that in general the indicator set need not be as in Figure 24. Namely, some kind of interference between the oscillation of the curve and the oscillation of the weight may add two more pieces like those in Figure 25 to the indicator set. The upper boundary of the upper piece is given by a convex function $y = \beta(x)$ which has two asymptotes that coincide with the two upper boundary lines of Figure 24, while the lower boundary of the lower piece is described by a concave function $y = \alpha(x)$ the two asymptotes of which exist and are the two lower boundary lines of Figure 24. These two additional pieces add something to the logarithmic leaf shown in Figure 24 we call a *halo.* We also remark that we always have

$$-1/p < \alpha(0) \leq \beta(0) < 1/q. \tag{7}$$

In particular, the line segment cut out of the imaginary axis by the indicator set is of length less than 1.

The above remarks characterise the shapes of leaves completely. Viz, we can prove the following: given any convex function $y = \beta(x)$ and any concave function $y = \alpha(x)$ which have asymptotes as in Figure 25 and which satisfy (7), there exist a Carleson arc and a Muckenhoupt weight such that the graphs of these two functions are just the upper and lower boundaries of the corresponding indicator set.

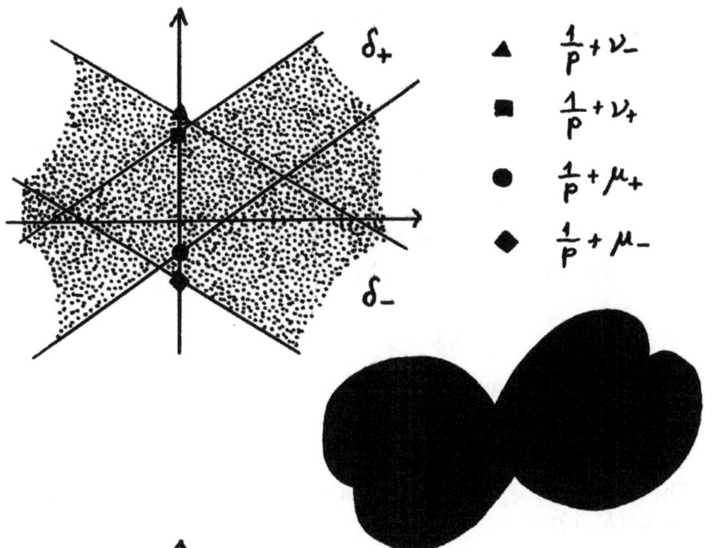
δ_+
$\frac{1}{p}+\nu_-$
$\frac{1}{p}+\nu_+$
$\frac{1}{p}+\mu_+$
$\frac{1}{p}+\mu_-$
δ_-

Fig. 24

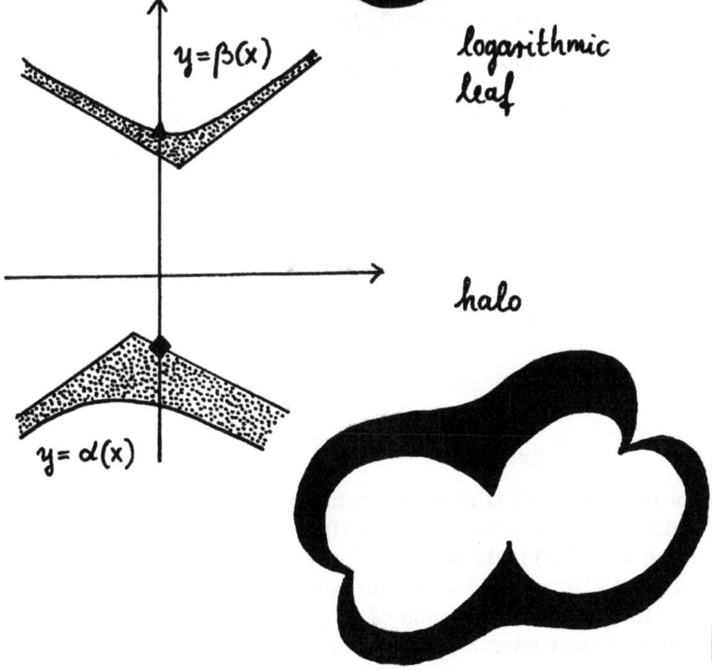
$y=\beta(x)$
logarithmic leaf
halo
$y=\alpha(x)$

Fig. 25

The problem we are left with is to find the indicator set for given $L^p(\Gamma, w)$ and a given endpoint t of Γ. This problem leads us into the mathematics behind all the beautiful pictures and first requires an interlude on Toeplitz operators.

7. Toeplitz operators

Let K be a Jordan curve in the plane, that is, a curve which is homeomorphic to the unit circle $\mathbf{T}$. Suppose K is rectifiable and a Carleson curve, and let w be a Muckenhoupt weight on K (which means that (1) and (2) are satisfied with K in place of Γ). Thus, $(K, w) \in A_p$. We orient K counter-clockwise.

The Cauchy singular integral operator S_K is bounded on $L^p(K, w)$ and one can show that

$$S_K^2 = I$$

(whence $\operatorname{sp} S_K = \{-1, 1\}$). Consequently, the operator

$$P_K := \frac{1}{2}(I + S_K)$$

is a projection. The image of P_K on $L^p(K, w)$ is denoted by $H^p(K, w)$ and called the *Hardy space* of $L^p(K, w)$:

$$H^p(K, w) := P_K L^p(K, w).$$

The *Toeplitz operator* $T(a)$ generated by a function $a \in L^\infty(K)$ is the operator

$$T(a) : H^p(K, w) \to H^p(K, w), \quad f \mapsto P_K(af).$$

We remark that if K is the complex unit circle, then the matrix representation of $T(a) \in \mathcal{B}(L^2(K))$ with respect to the standard orthonormal basis $\{e^{in\theta}/\sqrt{2\pi}\}_{n=0}^\infty$ is the Toeplitz matrix

$$\begin{pmatrix} a_0 & a_{-1} & a_{-2} & \dots \\ a_1 & a_0 & a_{-1} & \dots \\ a_2 & a_1 & a_0 & \dots \\ \dots & \dots & \dots & \dots \end{pmatrix}$$

constituted by the Fourier coefficients of a:

$$a_n = \frac{1}{2\pi} \int_0^{2\pi} a(e^{i\theta}) e^{-in\theta} \, d\theta \quad (n \in \mathbf{Z}).$$

Now suppose $a \in L^\infty(K)$ is piecewise continuous, $a \in PC(K)$, which means that the one-sided limits $a(t-0)$ and $a(t+0)$ exist at every point $t \in K$. The essential spectrum $\operatorname{sp}_{\text{ess}} T(a)$ is the set of all $\lambda \in \mathbf{C}$ for which $T(a) - \lambda I = T(a - \lambda)$ is not invertible on $H^p(K, w)$ modulo compact operators, and one can show that $\operatorname{sp}_{\text{ess}} T(a)$ results from the essential range of a by filling in certain sets X_t between $a(t-0)$ and $a(t+0)$ whenever a has a jump at $t \in K$. The set X_t is called the *local spectrum* of $T(a)$ at the point t and is denoted by $\mathrm{sp}_t\, T(a)$. On defining $\mathrm{sp}_t\, T(a) = \{a(t)\}$ if a is continuous at t, we can write

$$\mathrm{sp}_{\mathrm{ess}}\, T(a) = \bigcup_{t\in K} \mathrm{sp}_t\, T(a).$$

What is $\mathrm{sp}_t\, T(a)$?

Let $\varphi : \mathbf{T} \to K$ be any parametrisation of K such that $\varphi(z)$ traces out K counter-clockwise as z rounds $\mathbf{T}$ counter-clockwise. Suppose $\varphi(e^{i\theta_0}) = t$ and fix any $\varepsilon \in (0, 2\pi)$. The curve

$$\Gamma_t := \{\varphi(e^{i\theta}) : \theta_0 \le \theta \le \theta_0 + \varepsilon\} \tag{8}$$

is a simple arc with the starting point t. One can show (see, e.g., [5]) that

$$\mathrm{sp}_t\, T(a) = \left\{ \frac{a(t-0) + a(t+0)}{2} + \lambda \frac{a(t+0) - a(t-0)}{2} : \lambda \in \mathrm{sp}_t\, S_{\Gamma_t} \right\},$$

where $\mathrm{sp}_t\, S_{\Gamma_t}$ is the local spectrum of the operator $S_{\Gamma_t} \in \mathcal{B}(L^p(\Gamma_t, w|\Gamma_t))$ at the point t. In other terms, $\mathrm{sp}_t\, T(a)$ is a homothetic copy of $\mathrm{sp}_t\, S_{\Gamma_t}$: $\mathrm{sp}_t\, T(a) = h(\mathrm{sp}_t\, S_{\Gamma_t})$ with $h(z) = \alpha + \beta z$, which is determined by the requirements $h(-1) = a(t-0)$ and $h(1) = a(t+0)$. Figure 26 illustrates things.

Moral:

> *The problem of finding the local spectra of Toeplitz operators with piecewise continuous symbols*
>
> *is equivalent to*
>
> *the problem of describing the local spectra of the Cauchy singular integral operator.*

8. The mathematics behind the scene

In fact, we solved the problem of identifying the local spectra of the Cauchy singular integral operator by solving the problem of determining the local spectra of Toeplitz operators.

One of the questions arising in this connection is whether every simple Carleson arc Γ can be extended to a Carleson Jordan curve K (Figure 27). Clearly, this is not quite obvious for arcs as in Figures 3 to 6. The answer is nevertheless yes. This was shown in our paper [2] with Bishop, Spitkovsky by having recourse to some subtle results of geometric function theory.

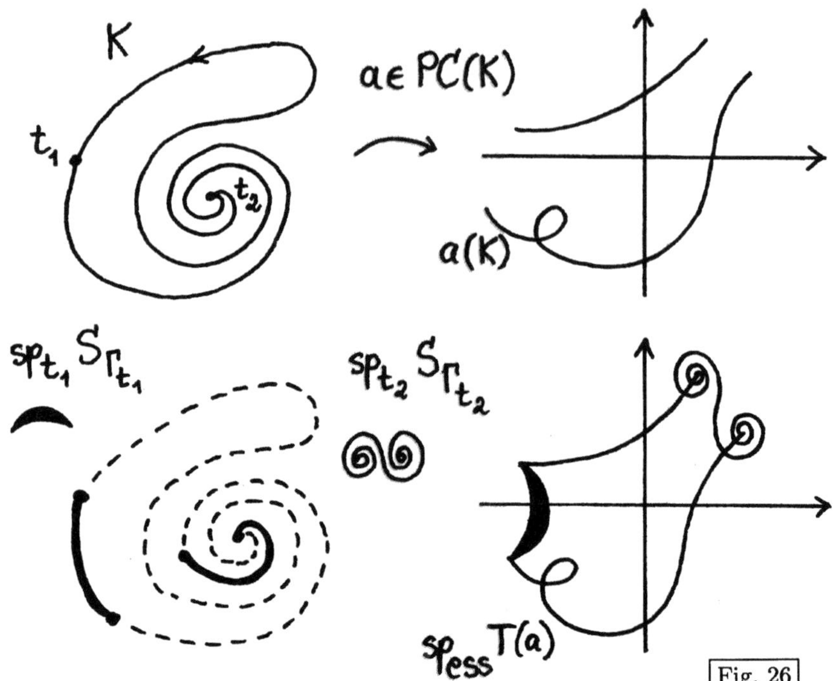

Fig. 26

Let now K be a Carleson Jordan curve and let w be a Muckenhoupt weight on K: $(K, w) \in A_p$. A 1960 result by Simonenko [32] states that if $a \in L^\infty(K)$, then $T(a)$ is invertible modulo compact operators on $H^p(K, w)$ if and only if a admits a so-called *Wiener-Hopf factorisation*

$$a(\tau) = a_-(\tau)\tau^\kappa a_+(\tau) \quad (\tau \in K). \tag{9}$$

Here κ is an integer, the factors $a_\pm$ and their inverses belong to certain classes of analytic functions, and, most importantly, $|a_+^{-1}|w$ must be a Muckenhoupt weight on K:

$$(K, |a_+^{-1}|w) \in A_p. \tag{10}$$

Now suppose $a \in PC(K)$ and $t \in K$. Localisation techniques reduce the problem of finding $\mathrm{sp}_t\, T(a)$ to the problem of describing the essential spectrum of $T(b_t)$ for certain canonical piecewise continuous functions b_t. A good choice of the b_t's is complex powers: $b_t(\tau) = \tau^{-\gamma}$ ($\tau \in K$), where $\gamma \in \mathbf{C}$; by appropriately fixing γ and the argument, we can achieve that b_t has a prescribed jump at t and is

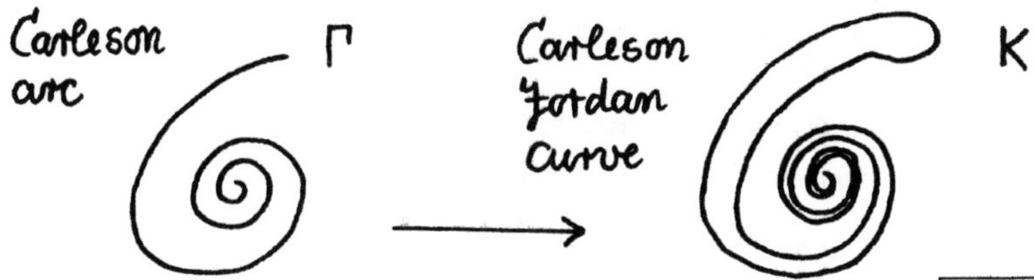

Fig. 27

continuous on $K \setminus \{t\}$. A factorisation of type (9) is given by

$$\tau^{-\gamma} = \left(1 - \frac{t}{\tau}\right)^{\kappa+\gamma} \tau^{\kappa}(\tau - t)^{-\gamma} \quad (\tau \in K)$$

(with suitably chosen $\kappa \in \mathbf{Z}$), and condition (10) amounts to the requirement

$$\Big(K, |(\cdot - \tau)^{\gamma}| w(\cdot)\Big) \in A_p. \tag{11}$$

Finally, consideration of $T(a-\lambda)$ for all $\lambda \in \mathbf{C}$ leads to the investigation of inclusion (11) for all $\gamma \in \mathbf{C}$. We call the set

$$N_t(K, w, p) := \Big\{\gamma \in \mathbf{C} : \Big(K, |(\cdot - \tau)^{\gamma}| w(\cdot)\Big) \in A_p\Big\} \tag{12}$$

the *modified indicator set* of K, w, p at t. One can prove that the modified indicator set is always an open set which contains the origin and which is bounded by the graph of a convex function $\widetilde{\alpha}_t$ from below and the graph of a concave function $\widetilde{\beta}_t$ from above (see Figure 28).

What we are looking for is the two functions α and β of Figure 25. These functions are given by

$$\alpha(x) = -\frac{1}{p} - \widetilde{\alpha}_t(x), \quad \beta(x) = \frac{1}{q} - \widetilde{\beta}_t(x). \tag{13}$$

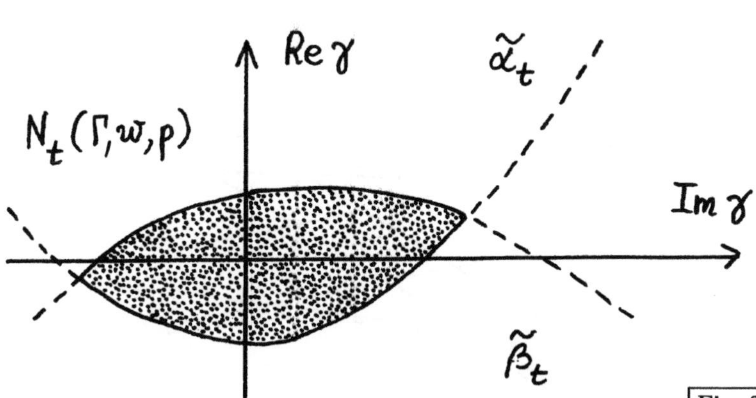

Fig. 28

Thus, it remains to find the modified indicator set (12). To do this, we associate a certain submultiplicative function ϱ with K, w, p, t, that is, a function $\varrho : (0,\infty) \to (0,\infty)$ satisfying

$$\varrho(xy) \le \varrho(x)\varrho(y) \ \text{ for all } \ x, y \in (0,\infty)$$

(we omit the construction of ϱ here). The numbers

$$\alpha := \lim_{x\to 0} \frac{\log \varrho(x)}{\log x}, \qquad \beta := \lim_{x\to\infty} \frac{\log \varrho(x)}{\log x}$$

are called the Boyd indices of ϱ. A central result of our entire theory is the following alternative characterisation of Muckenhoupt weights:

$$(K, w) \in A_p \Longleftrightarrow -\frac{1}{p} < \alpha \le \beta < \frac{1}{q}.$$

Applying this result to (11), that is, to the weight

$$\left|(\tau - t)^{y+ix}\right| w(\tau) \quad (\tau \in K),$$

we get the two functions $y = \widetilde{\alpha}_t(x)$ and $y = \widetilde{\beta}_t(x)$ of Figure 28 and thus the two desired functions (13). Details of the material of this section can be found in our book [5].

9. Work with Rabinovich

In 1996, V.S. Rabinovich surprised us with the observation that the indicator set can always be represented as a union of straight lines (see Figures 19 to 25 and also Figure 29). Let us denote the indicator set by Y, and for given real numbers ω and δ, let $\Pi(\omega,\delta)$ stand for the straight line $\{z = x + iy : y = \omega + \delta x\}$. The problem can now be formulated as follows. Given Γ, w, p and an endpoint t of Γ, find a set $\mathcal{M} \subset \mathbf{R}^2$ such that

$$Y = \bigcup_{(\omega,\delta)\in\mathcal{M}} \Pi(\omega,\delta). \tag{14}$$

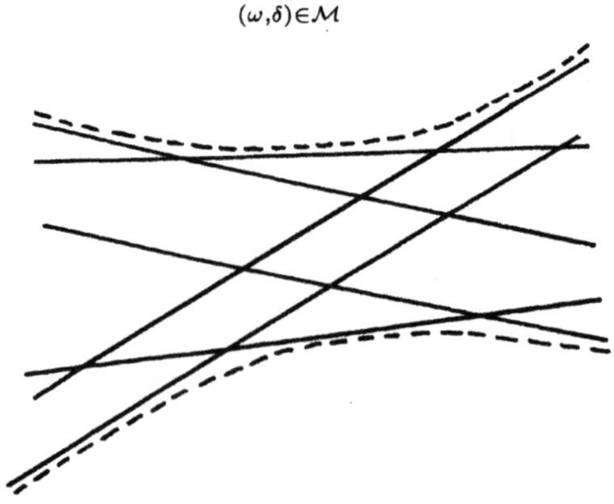

Fig. 29

When pursuing this idea, Rabinovich and we developed two more approaches to the spectral theory of the Cauchy singular integral operator. The first of them is based on Fourier(-Mellin) techniques [7], the second makes use of the theory of so-called limit operators [8], and both eventually reduce things to pseudodifferential operators on the real line with slowly oscillating symbols. For example, one can associate a family

$$\{A_{(\omega,\delta)}\}_{(\omega,\delta)\in\mathcal{M}}$$

of limit operators with a given operator A so that A is locally invertible if and only if all its limit operators are locally invertible:

$$\mathrm{sp}_t\, A = \bigcup_{(\omega,\delta)\in\mathcal{M}} \mathrm{sp}_t\, A_{(\omega,\delta)}. \tag{15}$$

It turns out that, under certain assumptions, the limit operators $A_{(\omega,\delta)}$ are sufficiently simple, which means that their indicator sets are straight lines. Clearly, this reveals that (14) is nothing but (15) in disguise.

In contrast to the approach originally worked out in [3, 4, 6], the two approaches we elaborated with Rabinovich are not applicable to all pairs (Γ, w) in A_p but only to pairs (Γ, w) in some proper subset A_p^0 of A_p. However, A_p^0 is sufficiently large: one can show that all possible leaves are attained at pairs (Γ, w) in A_p^0.

To illustrate the kind of results we obtained with Rabinovich, suppose Γ and w are of the form (3) and (4), that is,

$$\Gamma = \{0,1\} \cup \{\tau \in \mathbf{C} : \tau = re^{i\theta(r)}, 0 < r < 1\}, \tag{16}$$

$$w\left(re^{i\theta(r)}\right) = e^{v(r)} \quad (0 < r < 1), \tag{17}$$

where θ and v are real-valued functions in $C^\infty(0,1]$ which are subject to some (natural) additional restrictions concerning their oscillations as $r \to 0$. Consider the map

$$(0,1) \to \mathbf{R}^2, \quad r \mapsto \left(\frac{1}{p} + rv'(r), r\theta'(r)\right) \tag{18}$$

and denote by $\mathcal{P}$ the set of the partial limits of this map as $r \to 0$. It turns out that (14) holds with $\mathcal{M}$ equal to the convex hull of $\mathcal{P}$:

$$\boxed{Y = \bigcup_{(\omega,\delta)\in\operatorname{conv}\mathcal{P}} \Pi(\omega,\delta). \qquad (19)}$$

Moreover, the various indices we have encountered (in Figure 24, for example) can almost immediately be read-off from $\operatorname{conv}\mathcal{P}$. This is illustrated by Figure 30. Let us examine two cases.

Fig. 30

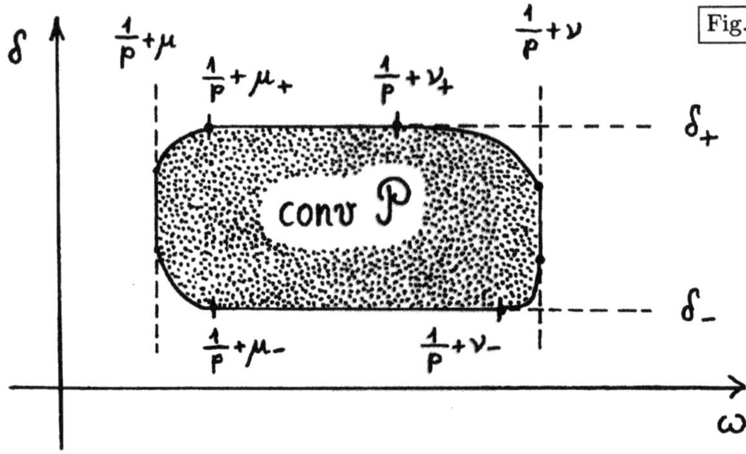

Metamorphosis 2. Suppose $\theta(r)$ is constant, so that Γ is a line segment, but make no further restrictions on the $v(r)$ in (17). In this case (18) is the map

$$(0,1) \to \mathbf{R}^2, \quad r \mapsto (1/p + rv'(r), 0),$$

and hence $\mathcal{P}$ is a horizontal line segment:

$$\mathcal{P} = [1/p + \mu, 1/p + \nu] \quad (= \operatorname{conv} \mathcal{P})$$

for certain $\mu \le \nu$. From (19) we obtain

$$Y = \bigcup_{1/p+\mu \le \omega \le 1/p+\nu} \Pi(\omega, 0) = \{x + iy \in \mathbf{C} : 1/p + \mu \le y \le 1/p + \nu\},$$

which is in accordance with Figure 20. The set $\operatorname{conv} \mathcal{P}$, the indicator set Y, and the leaf $\mathrm{sp}_t\, S_\Gamma$ are shown in Figure 31.

Fig. 31

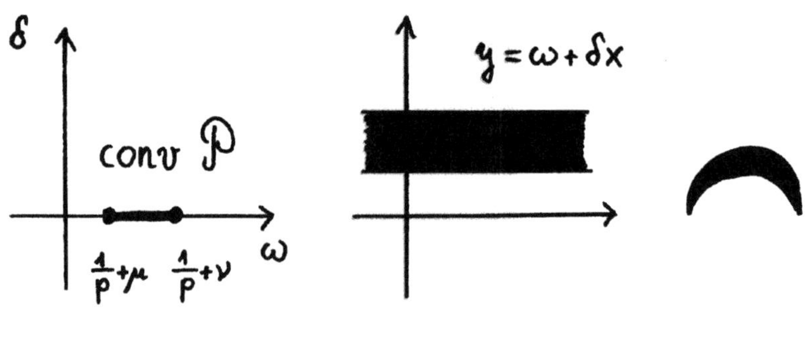

Metamorphosis 5. Suppose both $\theta(r)$ and $v(r)$ are arbitrary. If the set $\operatorname{conv}\mathcal{P}$ is a rectangle, then the indicator set is as in Figure 24, and the corresponding leaf is a logarithmic leaf (see Figure 32). However, in general $\operatorname{conv}\mathcal{P}$ is not a rectangle, and the deviation of $\operatorname{conv}\mathcal{P}$ from a rectangle causes a halo (see Figures 25 and 33).

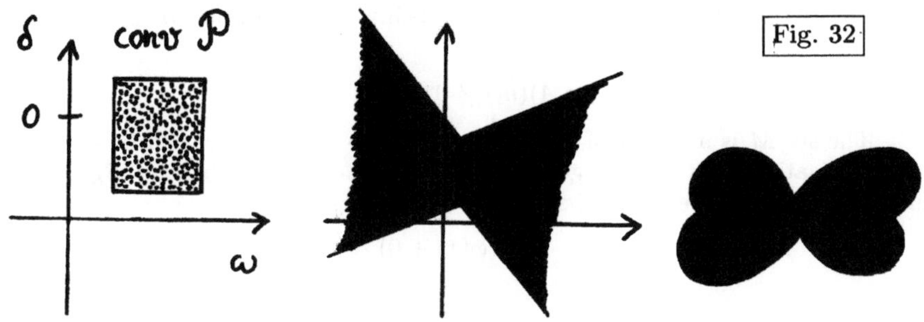

Fig. 32

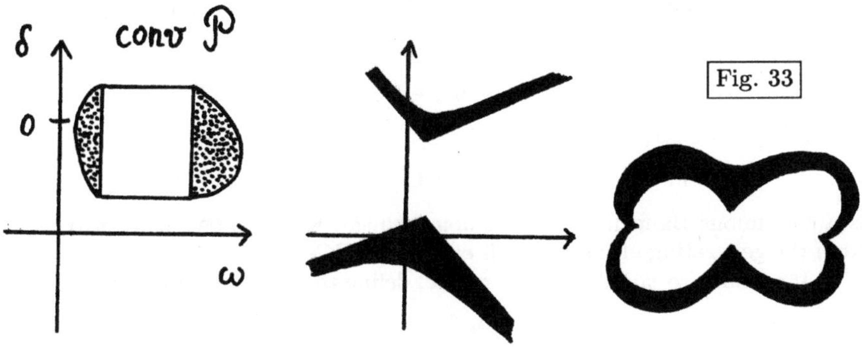

Fig. 33

10. Algebras of singular integral operators

Let K be a counter-clockwise oriented Jordan curve, let w be a weight on K, and suppose $(K,w) \in A_p$. As above, we let $\mathcal{B}(L^p(K,w))$ stand for the Banach algebra of all bounded linear operators on $L^p(K,w)$. The *algebra* $\operatorname{alg}(S_K, PC(K))$ *of singular integral operators with piecewise continuous coefficients* is the smallest closed subalgebra of $\mathcal{B}(L^p(K,w))$ which contains S_K and all multiplication operators by functions in $PC(K)$. We write aI for the operator of multiplication by a, and in case aI follows another operator, say B, we abbreviate aIB to aB. For example, such operators as

$$(aI + bS_K cS_K)(dI + eS_k) + fS_K gS_K hS_K$$

belong to $\operatorname{alg}(S_K, PC(K))$.

For operators in $\operatorname{alg}(S_K, PC(K))$, invertibility modulo compact operators can be checked by means of a so-called *symbol calculus.* The symbol calculus is based on a set M and a map

$$\operatorname{Sym} : \operatorname{alg}(S_K, PC(K)) \to C^{2\times 2}(M)$$

of $\operatorname{alg}(S_K, PC(K))$ into the algebra of all 2×2 matrix functions on M. The result says that an operator $A \in \operatorname{alg}(S_K, PC(K))$ is invertible modulo compact operators if and only if

$$\det(\operatorname{Sym} A)(m) \neq 0 \text{ for all } m \in M.$$

The set M is a subset of $\mathbf{T} \times \mathbf{C}$. Let $\varphi : \mathbf{T} \to K$ be an orientation-respecting parametrisation of K. At the generating elements of the algebra $\operatorname{alg}(S_K, PC(K))$ the values of Sym are as follows:

$$(\operatorname{Sym} aI)(\tau, z) = \begin{pmatrix} a(\varphi(\tau)+0) & 0 \\ 0 & a(\varphi(\tau)-0) \end{pmatrix},$$

$$(\operatorname{Sym} S_K)(\tau, z) = \begin{pmatrix} z & (1-z^2)/2 \\ 2 & -z \end{pmatrix} \quad \text{for } z \notin \{-1, 1\}$$

$$(\operatorname{Sym} S_K)(\tau, z) = \begin{pmatrix} z & 0 \\ 0 & -z \end{pmatrix} \quad \text{for } z \in \{-1, 1\}$$

For each $(\tau, z) \in M$, the map

$$\operatorname{Sym} : \operatorname{alg}(S_K, PC(K)) \to \mathbf{C}^{2\times 2}, \quad A \mapsto (\operatorname{Sym} A)(\tau, z)$$

is a continuous Banach algebra homomorphism, which allows us to extend Sym from the generating elements to all of $\operatorname{alg}(S_K, PC(K))$.

It remains to specify M. For $\tau \in \mathbf{T}$, define the arc $\Gamma_{\varphi(\tau)}$ by (8):

$$\Gamma_{\varphi(\tau)} := \{\varphi(e^{i\theta}) : \theta_0 \leq \theta \leq \theta_0 + \varepsilon\},$$

where $\tau = e^{i\theta_0}$. One can show (see, e.g., [5]) that

$$\boxed{M = \bigcup_{\tau \in \mathbf{T}} \left(\{\tau\} \times \operatorname{sp}_{\varphi(\tau)} S_{\Gamma_{\varphi(\tau)}}\right).}$$

Thus, M is something that might be called a *leaf bundle.*

In the case of nice curves K and nice weights w, this identification of M is due to Gohberg and Krupnik [21, 22]. For instance, if K is piecewise Lyapunov, we know that

$$\operatorname{sp}_{\varphi(\tau)} S_{\Gamma_{\varphi(\tau)}} = \text{line segment } [-1, 1] \text{ for } L^2(K),$$

$$\operatorname{sp}_{\varphi(\tau)} S_{\Gamma_{\varphi(\tau)}} = \text{circular arc } \mathcal{A}_p \text{ between } -1 \text{ and } 1 \text{ for } L^p(K).$$

Thus, in these two cases M may be identified with the cylinder $\mathbf{T} \times [-1,1]$ and the barrel $\mathbf{T} \times \mathcal{A}_p$, respectively. If w is a power weight of the form

$$w(t) = \prod_{j=1}^{n} |t - t_j|^{\mu_j} \quad (t \in K),$$

then the n arcs $\{\varphi^{-1}(t_j)\} \times \mathcal{A}_p$ of the barrel $\mathbf{T} \times \mathcal{A}_p$ must be replaced certain other arcs (whose shape is determined by $1/p + \mu_j$). Figure 34 shows the sets M in these cases. For more complicated curves and weights the set M is as in Figure 35.

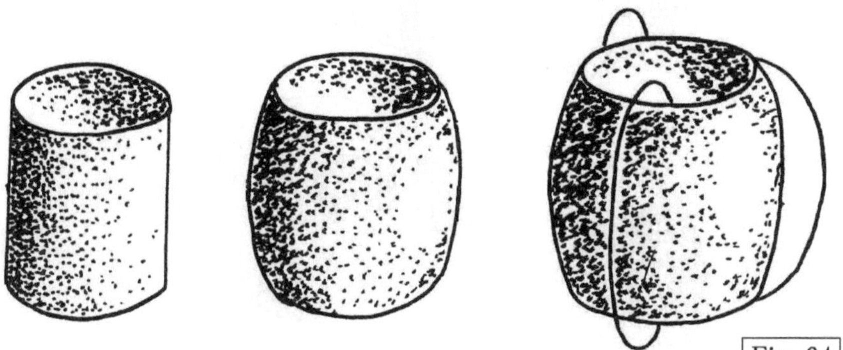

Fig. 34

11. Composed curves

Let finally K be a composed Carleson curve, that is, suppose K is a connected set which may be represented as a union of finitely many simple Carleson arcs each pair of which have at most endpoints in common (see Figure 36 for an example). Let further w be a Muckenhoupt weight on K, $(K,w) \in A_p$.

The spectrum of S_K on $L^p(K,w)$ can be found by an index (winding number) argument from the essential spectrum $\mathrm{sp}_{\mathrm{ess}}\, S_K$, which in turn is the union of the local spectra,

$$\mathrm{sp}_{\mathrm{ess}}\, S_K = \bigcup_{t \in K} \mathrm{sp}_t\, S_K.$$

Fix a point $t \in K$. At this point, there are n_+ outgoing arcs of K and n_- incoming arcs of K. The difference $\varepsilon := n_+ - n_-$ is called the *valency* of t. Pick one of the $n_+ + n_-$ arcs having t as an endpoint and orient this arc so that t becomes its starting point. We denote the arc obtained in this way by Γ_t. Let Y_t be the indicator set associated with Γ_t.

Fig. 35

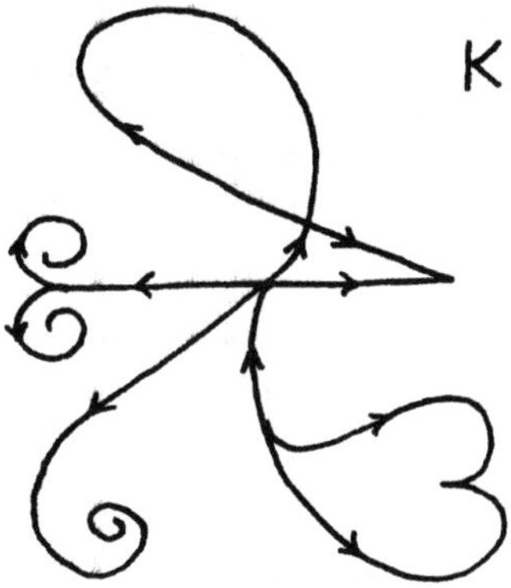

Fig. 36

It turns out (see, e.g., [5]) that $\mathrm{sp}_t\, S_K = \{-1, 1\}$ if $\varepsilon = 0$ and, for $\varepsilon \neq 0$,

$$\mathrm{sp}_t\, S_K = \left\{ z \in \mathbf{C} : \left(\frac{z+1}{z-1}\right)^{\varepsilon} = e^{2\pi\gamma} \text{ with some } \gamma \in Y_t \right\} \cup \{-1, 1\}.$$

Two typical local spectra at a point of the valency 3 are shown in Figure 37.

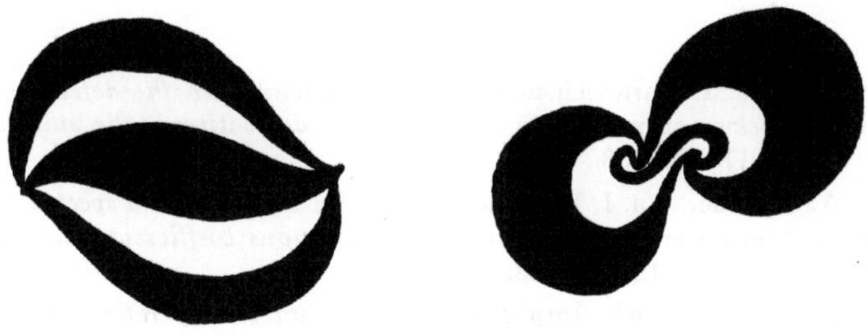

Fig. 37

We remark that the spectral theory of the singular integral operator on composed curves is based on three pillars. First, we have to know the local spectra of the Cauchy singular integral operator at the endpoints of simple arcs; this was the main topic of this lecture. Secondly, when employing localisation techniques at a point with $N := n_+ + n_-$ adjacent arcs, one is led to a Banach algebra generated by N idempotents. In the case where K is a Jordan curve, we always have $N = 2$ and can therefore have recourse to well known results on Banach algebras generated by two idempotents (Douglas [16] and Roch and Silbermann [29]). In the general case things are more complicated. An "N projections theorem" that fits perfectly for our purposes was established in 1996 in our joint paper [28] with Gohberg, Krupnik, Roch, Silbermann, Spitkovsky (many projections required many authors). Thirdly, when tackling composed curves one arrives at the problem of extending a Carleson star to a Carleson flower and of simultaneously continuing a Muckenhoupt weight on the star to a Muckenhoupt weight on the flower (see Figure 38). This problem was solved in 1996 by Chris Bishop, Spitkovsky, and ourselves in [2] by using methods and results of geometric function theory.

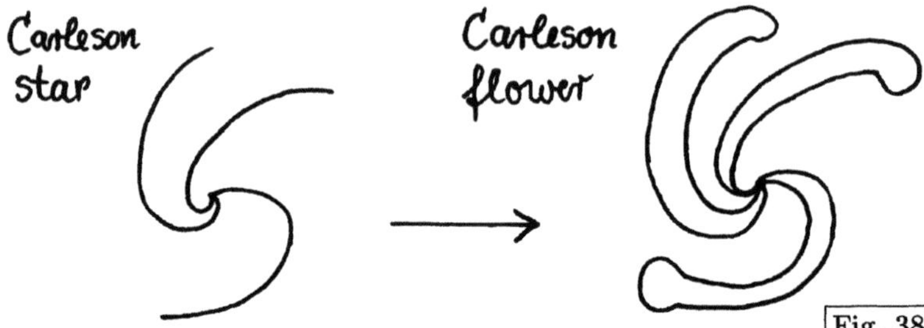

Fig. 38

Finally, once all the results mentioned in the preceding paragraph are available, one can construct a symbol calculus for operators in the Banach algebra alg $(S_K, PC(K))$ associated with a composed curve K. Details are in [5].

References

[1] A. V. Aizhenstat, Yu. I. Karlovich and G. S. Litvinchuk, *On the defect numbers of the Khveselava-Vekua operator with discontinuous derivative in the shift,* Soviet Math. Dokl., **43** (1991), 633–638.

[2] C. J. Bishop, A. Böttcher, Yu. I. Karlovich and I. Spitkovsky, *Local spectra and index of singular integral operators with piecewise continuous coefficients on composed curves,* Math. Nachr., **206** (1999), 5–83.

[3] A. Böttcher and Yu. I. Karlovich, *Toeplitz and singular integral operators on Carleson curves with logarithmic whirl points,* Integral Equations and Operator Theory, **22** (1995), 127–161.

[4] A. Böttcher and Yu. I. Karlovich, *Toeplitz and singular integral operators on general Carleson Jordan curves,* Operator Theory: Advances and Applications, **90** (1996), 119–152.

[5] A. Böttcher and Yu. I. Karlovich, *Carleson Curves, Muckenhoupt Weights, and Toeplitz Operators,* Progress in Mathematics, **154**, Birkhäuser Verlag, Basel 1997.

[6] A. Böttcher and Yu. I. Karlovich, *Toeplitz operators with PC symbols on general Carleson Jordan curves with arbitrary Muckenhoupt weights,* Trans. A. M. S., **351** (1999), 3143–3196.

[7] A. Böttcher, Yu. I. Karlovich and V. S. Rabinovich, *Mellin pseudodifferential operators with slowly varying symbols and singular integrals on Carleson curves with Muckenhoupt weights,* Manuscripta Math., **95** (1998), 363–376.

[8] A. Böttcher, Yu. I. Karlovich and V. S. Rabinovich, *The method of limit operators for one-dimensional singular integrals with slowly oscillating data,* J. Operator Theory, **43** (2000), 171–198.

[9] A. P. Calderón, *Cauchy integrals on Lipschitz curves and related operators,* Proc. Nat. Acad. Sci. USA, **74** (1977), 1324–1327.

[10] R. R. Coifman, P. W. Jones and S. Semmes, *Two elementary proofs of the L^2 boundedness of Cauchy integrals on Lipschitz curves,* Journ. A. M. S., **2** (1989), 553–564.

[11] R. R. Coifman, A. McIntosh and Y. Meyer, *L'intégrale de Cauchy définit un opérateur borné sur L^2 pour les courbes Lipschitziennes,* Ann. Math., **116** (1982), 361–387.

[12] G. David, *Opérateurs intégraux singuliers sur certaines courbes du plan complexe,* Ann. Sci. École Norm. Super., **17** (1984), 157–189.

[13] G. David and J.-L. Journé, *A boundedness criterion for general Calderón-Zygmund operators,* Ann. Math., **120** (1984), 371–397.

[14] G. David, J.-L. Journé and S. Semmes, *Opérateurs de Calderón-Zygmund, fonctions para-accrétives et interpolation,* Rev. Matem. Iberoamericana, **4** (1985), 1–56.

[15] G. David and S. Semmes, *Analysis of and on Uniformly Rectifiable Sets,* Math. Surveys and Monographs, **38**, A. M. S., Providence, RI, 1993.

[16] R. G. Douglas, *Local Toeplitz operators,* Proc. L. M. S., **36** (1978), 234–276.

[17] R. V. Duduchava, *Integral Equations with Fixed Singularities,* B.G. Teubner Verlagsgesellschaft, Leipzig 1979.

[18] E. M. Dynkin, *Methods of the theory of singular integrals* (*Hilbert transform and Calderón-Zygmund theory*), In: Itogi Nauki Tekh., Sovr. Probl. Matem., Fund. Napravl., **15**, Moscow 1987, 197–292 [Russian].

[19] G. I. Eskin, *Boundary Value Problems for Elliptic Pseudodifferential Operators,* Nauka, Moscow 1971 [Russian].

[20] I. Gohberg and N. Krupnik, *On the spectrum of singular integral operators on the space L^p,* Studia Math., **31** (1968), 347–362 [Russian].

[21] I. Gohberg and N. Krupnik, *Singular integral operators with piecewise continuous coefficients and their symbols,* Math. USSR Izv., **5** (1971), 955–979.

[22] I. Gohberg and N. Krupnik, *One-Dimensional Linear Singular Integral Equations,* Vols. I and II, Birkhäuser Verlag, Basel, Boston, Berlin 1992.

[23] R. Hunt, B. Muckenhoupt and R. Wheeden, *Weighted norm inequalities for the conjugate function and Hilbert transform,* Trans. A. M. S., **176** (1973), 227–251.

[24] J.-L. Journé, *Calderón-Zygmund Operators, Pseudo-Differential Operators and the Cauchy Integral of Calderón,* Lect. Notes Math., **1384**, Springer-Verlag, Berlin 1989.

[25] P. Mattila, M. S. Melnikov and J. Verdera, *The Cauchy integral, analytic capacity, and uniform rectifiability,* Ann. Math., **144** (1996), 127–136.

[26] M. S. Melnikov and J. Verdera, *A geometric proof of the L^2 boundedness of the Cauchy integral on Lipschitz graphs,* Internat. Math. Research Notes, **7** (1995), 325–331.

[27] V. A. Paatashvili and G. A. Khuskivadze, *On the boundedness of the Cauchy singular integral on Lebesgue spaces in the case of non-smooth contours,* Trudy Tbil. Matem. Inst. Akad Nauk GSSR, **69** (1982), 93–107 [Russian].

[28] S. Roch et al., *Banach algebras generated by N idempotents and applications,* Operator Theory: Advances and Applications, **90** (1996) 19-54.

[29] S. Roch and B. Silbermann, *Algebras generated by idempotent and the symbol calculus for singular integral operators,* Integral Equations Operator Theory, **11** (1988), 385–419.

[30] R. K. Seifullayev, *The Riemann boundary value problem on non-smooth open curves,* Matem. Sb., **112** (1980), 147–161 [Russian].

[31] S. Semmes, *Square function estimates and the $T(b)$ theorem,* Proc. A. M. S., **110** (1990), 721–726.

[32] I. B. Simonenko, *The Riemann boundary value problem with measurable coefficients,* Dokl. Akad. Nauk SSSR, **135** (1969), 279–281.

[33] I. B. Simonenko, *A new general method of studying linear operator equations of the type of singular integral equations,* Part I: Izv. Akad. Nauk SSSR, Ser. Mat., **29** (1965), 567–586; Part II: ibidem, 757–782 [Russian].

[34] I. B. Simonenko, *Some general questions of the theory of the Riemann boundary value problem,* Math. USSR Izv., **2** (1968), 1091–1099.

[35] I. M. Spitkovsky, *Singular integral operators with PC symbols on the spaces with general weights,* J. Funct. Anal., **105** (1992), 129–143.

[36] H. Widom, *Singular integral equations on L^p,* Trans. A. M. S., **97** (1960), 131–160.

Fakultät für Mathematik,
Technische Universität Chemnitz,
09107 Chemnitz, GERMANY
E-mail address: `aboettch@mathematik.tu-chemnitz.de`

Departamento de Matemáticas,
CINVESTAV del I.P.N.,
Apartado Postal 14-740,
07000 México, D.F., MÉXICO
E-mail address: `karlovic@math.cinvestav.mx`

Approximation problems in some holomorphic spaces, with applications

I. Chalendar, J. Leblond, and J. R. Partington

Abstract. We review a selection of extremal problems to do with constrained approximation in certain Banach spaces of holomorphic functions, including the classical Hardy spaces and Paley–Wiener spaces. In many cases the solutions are best interpreted in terms of linear operators. Applications of the problems under discussion to systems identification, signal processing, inverse problems for partial differential equations, and operator theory are presented.

1. Introduction

In recent years much research effort has been focused on certain approximation and recovery problems in classes of analytic functions. These include the following classes of problem, which we review in this paper because of their applications in other fields of analysis. The following preliminary descriptions are necessarily brief and fuller details follow later.

1. Approximation: given a function defined on some set, find a holomorphic function which approximates it on this set. Further constraints may be imposed upon the approximating function.
2. Extension: given a function defined on some set, extend its definition to a larger set so that the extended function is close to a holomorphic function.
3. Recovery: given the values of a holomorphic function on a subset of its domain of definition, recover the function.
4. Correction: given an unbounded holomorphic function, approximate it by a bounded function.

The spaces of holomorphic functions which principally concern us are the Hardy spaces (either on a disc or on a multiply-connected domain) and the Paley–Wiener spaces. These spaces arise naturally in a variety of applications, of which we discuss the identification of linear systems, signal processing, inverse problems for partial differential equations, and factorization results for linear operators. The solutions to these problems are operator-theoretic in nature, and we make use of Hankel and Toeplitz operators and their variants.

The layout of this paper is as follows. Section 2 treats various extremal problems, mostly in the context of Hilbert function spaces. Section 3 analyses the Toeplitz-like operators which appear in the solution of many such problems, with an emphasis on spectral theory. The applications to systems identification are described in Section 4. We continue by describing some variants of the extremal problems, where complex functions are constrained by their real or imaginary parts, in Section 5. More applications follow: to signal processing in Section 6, and to inverse problems for partial differential equations in Section 7. Finally, in Section 8, we discuss further approximation results in Hardy spaces, and their applications, which include abstract spectral theory of linear operators.

2. Bounded extremal problems

2.1. Bounded extremal problems in Hardy classes

The systematic approach to extremal problems in Hardy spaces H^p of the disc goes back to [48] and [68] in around 1950, and has been extensively developed since then, see [36, chap. 8], [43, chap. IV], [59] and references therein. For integers p, q, with $1 \leq p \leq \infty$ and $1/p+1/q = 1$, given a function $f \in L^p(\mathbb{T})$, these extremal problems concern both sides of the following duality relation (a consequence of the Hahn–Banach theorem):

$$\min_{g \in H^p} \|f - g\|_{L^p(\mathbb{T})} = \sup_{\substack{\phi \in H_0^q \\ \|\phi\|_{L^q(\mathbb{T})} \leq 1}} \frac{1}{2\pi} \int_{\mathbb{T}} f(z)\, \phi(z)\, dz \,. \tag{1}$$

The above left-hand side minimization issue is usually called the *dual* extremal problem associated to f and it is known that:

- it always admits a solution,
- for $1 \leq p < \infty$, the solution is unique (computational aspects are clear for $p = 2$ but not in general, see [19, 36]),
- for $p = \infty$ (Nehari extension), the solution is unique and the error has constant modulus if $f \in H^\infty + C(\mathbb{T})$. In this case, there is an associated Hankel operator Γ_f which is compact and singular value decomposition provides a way to get the solution (although this is effective only for rational f and some care has to be taken with continuity properties of best approximation operators [80]). This is still true for the problem of best uniform meromorphic approximation, known as the Adamjan–Arov–Krein (AAK, for short) problem, see [1] and Section 4.1.

The Hankel operator mentioned above, with *symbol* $f \in L^\infty(\mathbb{T})$, can be defined by the equation

$$\Gamma_f u = P_{\overline{H}_0^2}(fu), \qquad u \in H^2, \tag{2}$$

where $P_{\overline{H}_0^2} : L^2(\mathbb{T}) \to \overline{H}_0^2$ denotes the orthogonal projection (see, for example, [75, 92]). Let us write $\mathbb{T} = I \cup J$, the union of two disjoint arcs both of positive

Lebesgue measure, and assume that we are given a function $f \in L^p(I)$ that we want to approximate by traces on I of H^p functions. For such an issue to be well-posed, we must add some constraint concerning the behaviour of the approximants on the complementary part J (this follows from density properties of $H^p_{|I}$ in $L^p(I)$, see [13, prop. 1], [11, prop. 3]). For $\Psi \in L^p(J)$, and M a positive real number, we define the subset $\mathcal{B}^p_{M,\Psi}$ of H^p by:

$$\mathcal{B}^p_{M,\Psi} := \{g \in H^p : \|\Psi - g\|_{L^p(J)} \leq M\}.$$

With these notations, the Bounded Extremal Problem we consider can be stated as follows.

Problem 2.1. *For f, Ψ, M as above, find $g_p \in \mathcal{B}^p_{M,\Psi}$ such that*

$$\|f - g_p\|_{L^p(I)} = \min_{g \in \mathcal{B}^p_{M,\Psi}} \|f - g\|_{L^p(I)} := \beta_p .$$

Instances of such a problem for $p = 2$ have been treated in [64] with $f = 0$ and in [5] for $\Psi = 0$. We next review some properties of Problem 2.1 which have been established in [11, 13] and which generalize the ones related to dual extremal problems on $\mathbb{T}$ and recalled above. Note also that band-limited versions of Nevanlinna–Pick or Carathéodory–Fejér interpolation issues can also be rephrased as Problem 2.1: see the unpublished report [14].
Although a duality relation is no longer available on I for the class $\mathcal{B}^p_{M,\Psi}$ of approximants, it can be replaced by uniform convexity properties for $1 < p < \infty$, by combining differentiation with more refined results on Douglas algebras for $p = 1$, and by rephrasing the problem as a weighted one on the whole of $\mathbb{T}$ when $p = \infty$. Put $\mathcal{C}^p_{M,\Psi} := \{g_{|I} : \ g \in \mathcal{B}^p_{M,\Psi}\}$. Here, and later, the notation $p \vee q$ denotes a function equal to p on I and to q on J.

Theorem 2.2. *(i) For $1 \leq p < \infty$, there exists a unique solution g_p to Problem 2.1; moreover, $\|\Psi - g\|_{L^p(J)} = M$ whenever $f \notin \mathcal{C}^p_{M,\Psi}$.*
(ii) For $p = \infty$, if $f \vee \Psi \in H^\infty + C(\mathbb{T})$, then there exists a solution g_∞ to Problem 2.1; if moreover $f \notin \mathcal{C}^\infty_{M,\Psi}$, then g_∞ is unique and $f - g_\infty$ and $\Psi - g_\infty$ have constant modulus β_∞ and M a.e. on I and J respectively.

Related constructive aspects and computational issues are described in Section 3.1 in the Hilbert case $p = 2$ which is our main concern in the remainder of this work except, however, for Section 4 where approximation issues in the uniform norm are again considered.

2.2. Bounded extremal problems in Hilbert spaces

Suppose that we are given a Hilbert space H that can be decomposed into two distinct orthogonal direct sums of subspaces as:

$$H = H_+ \oplus H_- = K_+ \oplus K_- . \tag{3}$$

Given a function in K_+, we want to approximate it as well as possible by the projection on K_+ of an element of H_+ whose projection on K_- has controlled norm. Two decompositions that we have in mind here (and we discuss others later) are the Hardy space decomposition discussed in Section 3.1:

$$L^2(\mathbb{T}) = H^2 \oplus \overline{H_0^2} = L^2(I) \oplus L^2(J), \tag{4}$$

with $J = \mathbb{T} \setminus I$, and the Paley–Wiener space decomposition discussed in Section 6:

$$L^2(\mathbb{R}) = PW(\Omega) \oplus PW(\mathbb{R} \setminus \Omega) = L^2(J) \oplus L^2(\mathbb{R} \setminus J),$$

In most applications we think of H_+ as a class of holomorphic functions, and P_{K_+} as a restriction mapping. We shall discuss only the case when decomposition (3) holds with $K_+ \cap H_- = \{0\}$, when it follows easily that $P_{K_+}H_+$ is dense in K_+.

Problem 2.3. *Let $M > 0$ be a constant, and let $k_2 \in K_-$ be such that*

$$S_M := \{h \in H_+ : \|P_{K_-}h - k_2\|_H \leq M\} \neq \emptyset.$$

For $k_1 \in K_+$, with $k_1 \notin P_{K_+}S_M$ find $h_0 \in S_M$ such that

$$\|P_{K_+}h_0 - k_1\|_H = \min\{\|P_{K_+}h - k_1\|_H : h \in S_M\} =: \beta, \quad \text{say}.$$

Such a problem has been handled in [66] and provides a suitable framework in order to unify the formulation of various approximation issues that involve restrictions of analytic functions to subsets of the boundary of their domain.

Theorem 2.4. [66] *Problem 2.3 has a unique solution h_0, the constraint is saturated (i.e., $\|P_{K_-}h_0 - k_2\|_H = M$), and the solution is*

$$h_0 = (I + \lambda T)^{-1} P_{H_+}(k_1 + (1 + \lambda)k_2),$$

where $T : H_+ \to H_+$ is the self-adjoint operator $P_{H_+}P_{K_-}$, and $\lambda > -1$ is a constant such that $\|P_{K_-}h_0 - k_2\|_H = M$.

Companion to the above is an Extension Problem (or Bounded Completion Problem), which can be seen as one of extending the definition of a partly-specified function in order that the extended function be as close to a holomorphic function as possible.

Problem 2.5. *Given $k_2 \in K_-$ and $M > 0$, let $R_M = \{k \in K_- : \|k - k_2\|_H \leq M\}$. Given $k_1 \in K_+$, find $k_0 \in R_M$ such that*

$$\operatorname{dist}(k_1 + k_0, H_+) = \min_{k \in R_M} \operatorname{dist}(k_1 + k, H_+).$$

This problem is analysed in [66] in the general Hilbertian framework, having been treated for H^2 in [14, 18]. The analogous H^∞ problem has also been solved [15]. In any of these situations the appropriate solution can be constructively deduced from that of the related Bounded Extremal Problem 2.3.

We mention also an extrapolation result from [66], which generalizes a theorem of Patil (see [78, 79]), on the recovery of functions in H^2 from their values on a subset of the unit circle. The approximants produced by the next result coincide with those constructed by Patil, which are discussed in fuller detail in Section 3.2.

Theorem 2.6. [66] *Given a decomposition* (3) *with* $K_+ \cap H_- = \{0\}$ *and* $K_- \cap H_+ = \{0\}$, *and given* $k_1 \in K_+$, *let*

$$f_\lambda = (I + \lambda T)^{-1} P_{H_+} k_1 .$$

Then $k_1 \in P_{K_+} H_+$ *if and only if* (f_λ) *remains bounded as* $\lambda \to -1$, *in which case* (f_λ) *tends in norm to the unique* $h \in H_+$ *such that* $P_{K_+} h = k_1$.

Constructive aspects of Problem 2.3 have been approached in a number of particular instances of the decomposition (3). For Decomposition (4), $P_{K_+} g = \chi_I g$ for all $g \in H$, and $\|f - g\|_{L^2(I)} = \|\chi_I (f - g)\|_{L^2(\mathbb{T})}$; the problem coincides with Problem 2.1 for $p = 2$ and is discussed further in Section 3.1. Hardy spaces for circular domains also come up in this set-up, see Section 3.3, as well as the Paley–Wiener classes already referred to, see Section 6.

Decomposition (3) may also be considered in more general Banach spaces. For example, one can write

$$L^p(\mathbb{T}) = H^p \oplus \overline{H}_0^p = L^p(I) \oplus L^p(J), \qquad 1 < p < \infty .$$

(see [53, chap. 9]) and thus return to Problem 2.1. Another potentially interesting decomposition involves the notion of a Decomposing Banach Algebra [33, 56], of which the Wiener algebra of all absolutely convergent Fourier series is the most familiar example. However, little general analysis has been done in the non-Hilbertian case.

3. Toeplitz operators and spectra

3.1. Hardy classes in the disc

Without loss of generality we take $I = (e^{-ia}, e^{ia})$ for some a with $0 < a < \pi$. We work with the particular form (4) of decomposition (3):

$$L^2(\mathbb{T}) = H^2 \oplus \overline{H}_0^2 = L^2(I) \oplus L^2(J),$$

and we present some analysis given in [9]. In this case, T is the Toeplitz operator $T : H^2 \to H^2$ with symbol χ_J. Following the constructive diagonalization procedure for self-adjoint Toeplitz operators given in [84], we find that T is unitarily equivalent to multiplication M_x by the independent variable on $L^2([0,1], d\rho)$, where $d\rho(x) = \pi^{-1}(\sin a)dx$. Indeed, $M_x = VTV^{-1}$, where $V : H^2 \to L^2([0,1], d\rho)$ is the unitary transformation whose effect on Cauchy kernels $k_\alpha(z) = 1/(1 - \overline{\alpha} z)$ is given by

$$(V k_\alpha)(x) = [\overline{\psi_x(\alpha)}(1 - \overline{\alpha} e^{ia})^{1/2}(1 - \overline{\alpha} e^{-ia})^{1/2}]^{-1},$$

and where, for $0 < x < 1$, ψ_x is the unique outer function such that $\psi_x(0) > 0$ and such that $|\psi_x|^2$ is equal to x on I and to $1 - x$ on J.

Calculations using the transformation V enable one to compare the quantities $e = e(\lambda) := \beta_2$ and $M = M(\lambda)$, in the following version of the Bounded Extremal Problem 2.1 with $p = 2$, which also coincides with Problem 2.3 for the decomposition (4).

Problem 3.1. *Given functions $f \in L^2(I)$, $\Psi \in L^2(J)$, and a positive constant M, find $g_2 \in \mathcal{B}^2_{M,\Psi}$ such that*

$$e := \|f - g_2\|_{L^2(I)} = \min_{g \in \mathcal{B}^2_{M,\Psi}} \|f - g\|_{L^2(I)} .$$

Here λ gives an implicit parametrization of the optimal solutions g_2; the error estimates satisfy $e(\lambda) \to 0$ and, if $f \notin H^2_{|I}$, $M(\lambda) \to \infty$ as $\lambda \to -1$. Note that the optimal $g_2 = g_\lambda$ is given by

$$g_\lambda = (1 + \lambda T)^{-1} P_{H^2}(f \vee (1 + \lambda)\Psi), \tag{5}$$

for the unique $\lambda > -1$ such that $\|\Psi - g_\lambda\|_{L^2(J)} = M$ when $f \notin \mathcal{C}^2_{M,\Psi}$, and for $\lambda = -1$ if $f \in \mathcal{C}^2_{M,\Psi}$. A weighted version of Problem 3.1 has been considered in [65] while a smoother one, namely in Hardy–Sobolev spaces, has been treated in [10].
The following result from [9] is a sample of what can be proven by translating the expressions for the approximation errors into integrals and then estimating them. Earlier results in this direction can be found in [5, 11].

Theorem 3.2 ([9]). *Suppose that f satisfies*

$$(1 - e^{-i\theta}e^{ia})^{-1/2}(1 - e^{-i\theta}e^{-ia})^{-1/2} f(e^{i\theta}) \in L^1(I),$$

and

$$(1 - e^{-i\theta}e^{ia})^{1/2}(1 - e^{-i\theta}e^{-ia})^{1/2} f(e^{i\theta}) \in \mathcal{W}^{1,1}(I)$$

(the Sobolev class consisting of functions with derivatives in $L^1(I)$). Then

$$M(\lambda)^2 = o((\lambda + 1)^{-1} \log^{-2}(\lambda + 1)),$$

as $\lambda \to -1$, while

$$e(\lambda)^2 = o(|\log^{-1}(1 + \lambda)|).$$

3.2. Recovery problems

Faster convergence rates are available if f is already the restriction of an H^2 function (to be denoted again by f). In that case, at least if M is large enough, Problem 3.1 amounts to reconstructing $f \in H^2$ from the knowledge of $f_{|I}$. We touched upon these ideas in Section 2.2, and now we explore them further.

Recovery issues for analytic functions from traces on the boundary were originally approached by Carleman, Goluzin and Krylov (see [3, 79] and the bibliography therein). It appears that recovery procedures of the Carleman type are strongly related to the above approximation scheme to $f \in L^2(I)$ and that the boundedness on J of such approximants provides criteria to decide whether f is the trace of some H^2 function or not, a so-called Loewner type question that was taken up in [3, 6, 64, 67, 79, 83], for instance.

If $f \in H^2_{|I}$ and $\Psi = 0$, then the solution g_λ given by (5) coincides with the Patil's sequence of approximants (f_α) defined below by (7) and built for recovery purposes rather than approximation, [11, 78, 79] . Put first

$$w_\rho(z) = \exp\left\{\frac{1}{2\pi}\int_I \frac{e^{it}+z}{e^{it}-z}\log\rho(e^{it})\,dt\right\} : \quad z \in \mathbb{D}, \tag{6}$$

for the outer H^∞ function with modulus $\rho \in L^\infty(\mathbb{T})$, $\rho \geq 0$, and $\log\rho \in L^1(\mathbb{T})$ with positive value at the origin. Defining then $\varphi = w_{\varrho \vee 1}$ for the outer "quenching" function with constant modulus $\varrho > 1$ on I and 1 on J, we get that

$$\begin{aligned} g_\lambda(z) &= f_\alpha(z) := \frac{1}{2i\pi}\int_I \left(\frac{\varphi(\xi)}{\varphi(z)}\right)^\alpha f(\xi)\frac{d\xi}{\xi - z} \quad \text{for} \quad z \in \mathbb{D}, \\ &= \frac{1}{\varphi(z)^\alpha}(P_{H^2}\,\chi_I\,\varphi^\alpha f)(z), \quad \text{where} \quad \alpha = -\frac{\log(\lambda+1)}{2\log\varrho}. \end{aligned} \tag{7}$$

For $f \in H^2_{|I}$ and the collection (f_α) of H^2 functions defined by (7), it is already known from [3, Theorems 1.1, 1.11] that, as $\alpha \to \infty$:
(i) f_α converges to f in H^2; and hence
(ii) $f_\alpha(z)$ tends to $f(z)$ uniformly over compact subsets of $\mathbb{D}$.

Point *(i)* is Patil's result from [79] for $p = 2$ while point *(ii)* follows from the Goluzin–Krylov local convergence result, see also [78, 79]. However, these two points are improved by the analysis performed in [9] which is to the effect that, if $f \in H^2_{|I}$:
(i)′ we have the following estimate of the H^2 convergence rate:
$\|f - f_\alpha\|_{L^2(\mathbb{T})} = O(\exp(-\alpha\log\varrho)) \to 0$ as $\alpha \to \infty$;
(ii)′ if f admits an absolutely continuous derivative on $\mathbb{T}$, then $f_\alpha(z)$ tends to $f(z)$ almost everywhere on $\mathbb{T}$.

Point *(i)′* follows from the estimate $\|f - g_\lambda\|_{L^2(\mathbb{T})} = O((1+\lambda)^{1/2}) \to 0$ as $\lambda \to -1$, even without further assumptions on f. Point *(ii)′* is obtained by further improving estimates of this H^2 convergence rate under the additional smoothness hypothesis. This enables one to conclude that there is pointwise convergence of g_λ to f almost everywhere on $\mathbb{T}$; we refer to [9] for further details.
A further extension of point *(i)* comes from the fact that, for all $f \in L^2(I)$, $e(\lambda) = \|f - g_\lambda\|_{L^2(I)} \to 0$ as $\lambda \to -1$, see Section 3.1. Hence, even if f does not already belong to $H^2_{|I}$,
(i)″ f_α converges to f in $L^2(I)$ as α goes to infinity.

Those considerations lead to the following characterization of traces on I of H^2 functions, see [11]. Analogous results hold in general H^p spaces.

Construct for each $f \in L^2(I)$ a family (g_ρ) of H^2 functions such that $g_{\rho|_I}$ converges to f in $L^2(I)$ when ρ tends to some value ρ_0 and g_ρ converges in H^2 as $\rho \to \rho_0$ whenever $f \in H^2_{|I}$; then $f \in H^2_{|I}$ if and only if $\|g_\rho\|_{L^2(J)}$ remains bounded.

If one takes for instance $g_\rho = g_\lambda = f_\alpha$ defined by (7), this is illustrated by the fact that, as $\lambda \to -1$, either $f \notin H^2_{|I}$ and $M(\lambda) \to \infty$, or $f \in H^2_{|I}$ and $M(\lambda) \to \|f\|_{L^2(J)}$ (point (i)), while $\|f - g_\lambda\|_{L^2(I)} \to 0$ in both cases (point $(i)''$).
The above principle can also be viewed as a consequence of [6, cor. 1] or [83, I, Theorem 1] which assert that a given function $f \in L^2(I)$ is the trace of some H^2 function of $L^2(\mathbb{T})$ norm less than $M_\mathbb{T}$ if and only if, for all $\phi \in L^2(I)$,

$$|(f\, , \phi)_{L^2(I)}| \leq M_\mathbb{T} \, \|P_{H^2} \, \chi_I \, \phi\|_{L^2(\mathbb{T})} \, .$$

See also [66, Theorem 4.2] for an abstract version of this characterization result in the Hilbertian framework (3). Such issues are also examined in [6] for rather general classes of reproducing kernel Hilbert spaces of analytic functions, including Hardy–Sobolev spaces for example.

3.3. Hardy classes of circular domains and strips

Let G be a bounded, finitely-connected domain in $\mathbb{C}$, whose boundary consists of the union of a finite number of disjoint Jordan loops. We know [45, Theorem 2, p. 237] that such a domain is conformally equivalent to a *circular domain*, that is, a domain consisting of the open unit disc from which a finite number of pairwise disjoint closed discs have been removed, and from now on we assume that G is itself a circular domain:

$$G = \mathbb{D} \setminus \bigcup_{j=1}^{N} (a_j + r_j \overline{\mathbb{D}}), \tag{8}$$

with the obvious inequalities satisfied by the a_j and r_j for $j = 1, \ldots , N$. We write $D_j = a_j + r_j\mathbb{D}$ for $1 \leq j \leq n$. Let Γ denote the boundary of G, so that

$$\Gamma = \Gamma_0 \cup \Gamma_1 \cup \ldots \cup \Gamma_N,$$

where $\Gamma_0 = \partial\,\mathbb{D}$, and, for $j = 1, \ldots , N$, we have $\Gamma_j = a_j + r_j \partial\,\mathbb{D}$. We normalize the Lebesgue measure on Γ so that each circle Γ_j is given unit measure. For $1 \leq j \leq N$, we write $\Omega_j = \mathbb{C} \setminus \{a_j + r_j\overline{\mathbb{D}}\}$, so that $G = \mathbb{D} \cap \bigcap_{j=1}^{N} \Omega_j$.

The Hardy spaces $H^p(G)$ on a circular domain G were defined by Rudin [85] in terms of analytic functions f such that $|f(z)|^p$ has a *harmonic majorant* on G, that is, a real harmonic function $u(z)$ such that $|f(z)|^p \leq u(z)$ on G. Rudin then showed that $H^p(G)$ has a natural decomposition into "elementary" Hardy spaces. That is, for $1 \leq p < \infty$ the Hardy space $H^p(G)$ can equivalently be defined by

$$H^p(G) = H^p(\mathbb{D}) \oplus \bigoplus_{j=1}^{N} H^p_0(\Omega_j), \tag{9}$$

where for $j \geq 1$ we define the Hardy space $H^p_0(\Omega_j)$ of functions analytic in Ω_j, with a zero limit at infinity in the obvious way. It is known (cf. [28]) that the sum

in (9) is a topological direct sum, that is, that the projections onto the summands are bounded operators.
It is also possible to define the Hardy spaces $H^p(\partial G)$ for $1 \le p < \infty$ as the closure in $L^p(\partial G)$ of the set R_G of rational functions whose poles lie in the complement of $\overline{G}$. This approach, similar to one in [8], was taken in [28]. The spaces $H^p(G)$ and $H^p(\partial G)$ are then isomorphic in a natural way, and so we identify the two spaces. In the case $p = \infty$, we can regard $H^\infty(G)$, the space of bounded analytic functions on G, as a direct sum of spaces $H^\infty(\mathbb{D}) \oplus \bigoplus_{j=1}^N H_0^\infty(\Omega_j)$ (cf. Theorem 7.1 of [32]), and hence we may regard it as a closed subspace $H^\infty(\partial G)$ of $L^\infty(\partial G)$. We give the subspaces $H^p(\partial G)$ the norm inherited from the spaces $L^p(\partial G)$.
The most completely analysed example of a circular domain is the annulus $G = \mathbb{D} \setminus s\mathbb{D}$ for some fixed s with $0 < s < 1$. Here the Lebesgue measure on ∂G is normalized so that the circles $\mathbb{T}$ and $s\mathbb{T}$ each have unit measure. It can be shown easily that the sequence of functions $(e_n)_{n\in\mathbb{Z}}$, defined by $e_n(z) = z^n/\sqrt{1+s^{2n}}$, is an orthonormal basis of $H^2(\partial G)$.

Let us now work with the set-up (3), where $H = L^2(\partial G)$, $H_+ = H^2(\partial G)$, $H_- = H \ominus H_+$, $K_+ = L^2(\mathbb{T})$ and $K_- = L^2(s\mathbb{T})$. The solution to the standard Bounded Extremal Problem 2.3 can now be expressed as follows.

Theorem 3.3. [90] *Let $f_0 \in L^2(\mathbb{T}) \setminus P_{L^2(\mathbb{T})}H^2(\partial G)$, $f_1 \in L^2(s\mathbb{T})$ and $M > 0$. Then there exists a function $g_0 \in H^2(\partial G)$ such that $\|g_0 - f_1\|_{L^2(s\mathbb{T})} \le M$ and*

$$\|f_0 - g_0\|_{L^2(\mathbb{T})} = \inf\{\|f_0 - g\|_{L^2(\mathbb{T})} : g \in H^2(\partial G), \|g - f_1\|_{L^2(s\mathbb{T})} \le M\}.$$

Indeed, if $f_0(z) \sim \sum_{n\in\mathbb{Z}} a_n z^n$ for $z \in \mathbb{T}$ and $f_1(z) \sim \sum_{n\in\mathbb{Z}} b_n z^n$ for $z \in s\mathbb{T}$, then

$$g(z) = \sum_{n\in\mathbb{Z}} \frac{a_n + \alpha s^{2n} b_n}{1 + \alpha s^{2n}} z^n,$$

where $\alpha > 0$ is the unique constant such that

$$\sum_{n\in\mathbb{Z}} \frac{|a_n - b_n|^2 s^{2n}}{(1 + \alpha s^{2n})^2} = M^2.$$

The key to the proof of the above theorem is that the Toeplitz operator $P_{H_+}P_{K_-}$ on H_+ is diagonalizable in a very simple way: the functions z^n are eigenvectors for $n \in \mathbb{Z}$, and the corresponding eigenvalues are $s^{2n}/(1+s^{2n})$.
The analysis of such problems has applications in the theory of inverse Dirichlet–Neumann problems, such as those involved in the detection of faults in materials from electrical or thermal measurements. We refer to [12] for more details, see also Section 7.
A similar analysis was performed in [90] for the Hardy space on the strip

$$X = \{z \in \mathbb{C} : 0 < \operatorname{Re} z < 1\},$$

that is, taking $H = L^2(\partial X)$, $H_+ = H^2(\partial X)$, the closure of the subspace consisting of rational functions with poles $\mathbb{C} \setminus \overline{X}$, $K_+ = L^2(i\mathbb{R})$ and $K_- = L^2(1 + i\mathbb{R})$.

An analogous extremal problem can be formulated and solved as a corollary to the results of [66] recalled in Section 2.2. In this case the Toeplitz operator $T = P_{H_+}P_{K_-}$ on H_+ can be written as $T = \mathcal{L}^{-1}M_\phi\mathcal{L}$, where $\mathcal{L}$ is the bilateral Laplace transform, and M_ϕ is multiplication by the function $\phi(t) = e^{-t}/(e^t + e^{-t})$ on $L^2(\mathbb{R})$. The spectrum of T is now the whole interval $[0,1]$, rather than a discrete set, and very explicit solutions to questions such as the corresponding Bounded Extremal Problem can be derived.

4. Applications to systems identification

Approximation in Hardy spaces (particularly H^2 and H^∞) has been of great use recently in the identification of linear time-invariant systems. We recall that if T is a bounded operator on ℓ^2 that commutes with the shift then, by means of the unitary z-transformation

$$Z : \ell^2 \to H^2, \qquad Z((a_n)) = \sum_{n=0}^{\infty} a_n z^n,$$

T can be regarded as a multiplication operator on the Hardy space H^2. More precisely, if $Tu = y$, then $Y(z) = G(z)U(z)$ for some $G \in H^\infty$ with $\|G\|_\infty = \|T\|$, where Y denotes Zy and U denotes Zu.
Similarly, continuous time-invariant systems T (operators commuting with all right shifts) on $L^2(0,\infty)$ correspond to multiplication operators on $H^2(\mathbb{C}_+)$, the Hardy space of the right-hand half plane, by means of the Laplace transform,

$$\mathcal{L} : L^2(0,\infty) \to H^2(\mathbb{C}_+), \qquad \mathcal{L}f(s) = \int_0^\infty f(t)e^{-st}\,dt.$$

Again, if $Tu = y$, then $Y(s) = G(s)U(s)$, for some $G \in H^\infty(\mathbb{C}_+)$, with $\|G\|_\infty = \|T\|$, where Y denotes $\mathcal{L}y$ and U denotes $\mathcal{L}u$. In the language of systems theory, u is called the ***input***, y the ***output***, and G the ***transfer function*** of the system. We refer to [41, 52] for introductions to this subject. See also [37, 50] for more on the links between operator theory and engineering and [44, 91] for the intriguing connexions between shift-invariant operators and transfer functions.
The problem of identification is to obtain an approximation or *model* of an unknown system by means of a finite number of measurements. These can be input–output measurements (which we do not discuss here), or transfer function measurements, on the circle or imaginary axis as appropriate, which can be obtained by applying periodic inputs and measuring the corresponding outputs. The norms most commonly used for measuring the error of an approximating system are the H^∞ and H^2 norms. Worst-case identification assumes that the measured values are perturbed by disturbances (for example, nonlinearities or noise) which are small in some sense; it is to be contrasted with stochastic identification, which assumes some statistical model for the disturbances. We refer to the books [30, 78, 87] for more about worst-case identification, as well as to the survey articles [69, 71, 72].

Some early papers in the subject of worst-case frequency-domain identification are are [46, 49, 76, 77].

4.1. Solutions to uniform Bounded Extremal Problems

We return to considerations about uniform meromorphic approximation issues of AAK type with constraint in modulus that generalize Problem 2.1 for $p = \infty$ and have been considered in [15]. Classes of uniform extremal problems are also considered in [51].
Define $R_N \subset L^\infty(\mathbb{T})$ to be the set of rational functions with no poles on $\mathbb{T}$ and at most N poles in $\mathbb{D}$. For Ψ, $M \in L^\infty(J)$ and $M > 0$ a.e. on J, we define the subset $\mathcal{B}_{M,\Psi}^{N,\infty}$ of $H^\infty + R_N$ by:

$$\mathcal{B}_{M,\Psi}^{N,\infty} := \{g \in H^\infty + R_N : \ |\Psi - g| \leq M \text{ a.e. on } J\}.$$

Problem 4.1. *For $f \in L^\infty(I)$, Ψ, M as above, find $g_N \in \mathcal{B}_{M,\Psi}^{N,\infty}$ such that*

$$\|f - g_N\|_{L^\infty(I)} = \min_{g \in \mathcal{B}_{M,\Psi}^{N,\infty}} \|f - g\|_{L^\infty(I)} := \beta_N(f \vee \psi, M).$$

Note that the particular case where $N = 0$ and M is constant coincides with Problem 2.1 for $p = \infty$. We now present a brief account of theoretical and computational results about Problem 4.1. Put $\mathcal{C}_{M,\Psi}^{N,\infty} := \{g_{|I} : \ g \in \mathcal{B}_{M,\Psi}^{N,\infty}\}$ and let $w_{\beta_N \vee M} \in H^\infty$ be the outer function defined by (6).

Theorem 4.2. [15] *If $f \vee \Psi \in H^\infty + C(\mathbb{T})$, then*
(i) there exists a solution g_N to Problem 4.1.
If moreover $f \notin \mathcal{C}_{M,\Psi}^{N,\infty}$, then:
(ii) g_N is unique and $|f - g_N| = \beta_N$ while $|\Psi - g_N| = M$ a.e. on I and J respectively;
(iii) we have $g_N = v_N\, w_{\beta_N \vee M}$ for the solution v_N to the AAK problem:

$$\|(f \vee \Psi) w_{\beta_N \vee M}^{-1} - v_N\|_{L^\infty(\mathbb{T})} = \min_{v \in H^\infty + R_N} \|(f \vee \Psi) w_{\beta_N \vee M}^{-1} - v\|_{L^\infty(\mathbb{T})}, \qquad (10)$$

whose value is equal to 1.

Again if $N = 0$ and M is constant, (i) and (ii) above correspond to Theorem 2.2, (ii). Note that the assumptions $\Psi \in (H^\infty + C(\mathbb{T}))_{|J}$ and $M > 0$ are sufficient to ensure that $\mathcal{B}_{M,\Psi}^{\infty} \neq \emptyset$, see [13, Theorem 1]; hence $\mathcal{B}_{M,\Psi}^{N,\infty} \neq \emptyset$. This need not be true in general. Observe also that $\beta_N > 0$, for $f \notin \mathcal{C}_{M,\Psi}^{N,\infty}$; hence $w_{\beta_N \vee M}$ is invertible in H^∞. Point (ii) is a direct consequence of (iii) together with properties of AAK problems on $\mathbb{T}$ recalled in Section 2.1, that ensure existence and uniqueness of v_N in (10). Furthermore, under the above assumption, the Hankel operator Γ with symbol $(f \vee \psi) w_{\beta_N \vee M}^{-1}$ (defined in (2)) is compact and the value of Problem (10) is equal to its $(N+1)$th singular value σ_{N+1}. Recall that, in this case, σ_{N+1} coincides with the positive square root of the $(N+1)$st largest eigenvalue of the selfadjoint H^2 operator $\Gamma^* \Gamma$; call ν_{N+1} the associated eigenvector.
Theorem 4.2 (iii) efficiently reduces Problem 4.1 to AAK approximation and provides a way to solve it whenever f, Ψ, M satisfy the hypotheses.

1. Compute the unique positive value β_N such that $\sigma_{N+1} = 1$; this can be done by using a dichotomy procedure on the singular value computation.
2. Obtain the solution v_N to Problem (10):
$$v_N = (f \vee \psi) w_{\beta_N \vee M}^{-1} - \frac{\Gamma \nu_{N+1}}{\nu_{N+1}}.$$
3. Build g_N accordingly to *(iii)*.

This is almost a constructive procedure, except for the fact that AAK problems can be effectively solved only when the function to be approximated is rational, which is not the case of $(f \vee \psi) w_{\beta_N \vee M}^{-1}$ in Problem (10). It raises a rather bad aspect of AAK problems which is the lack of uniform continuity of the associated best approximation operator [73]. In other spaces of functions, however, continuity does hold under some sufficient conditions [80], and this leads to a generic computational algorithm which is described in more detail in [15]. By performing a preliminary rational approximation step in the ***separable*** Hölder–Zygmund class, followed by the necessary AAK approximation step, and using Theorem 4.2, it is possible to solve Problem 4.1 in the generic sense that the algorithm is valid for an open dense subset of the possible data.

4.2. Robust identification from band-limited frequency data

One well-established algorithm for frequency-domain identification of linear systems, which is an inverse problem for a convolution operator, uses a two-stage process [46, 49, 76, 77]. The measured data consists of a finite set of values, corresponding to corrupted measurements of an unknown transfer function $h \in A(\mathbb{D})$ at a finite set of points on the unit circle $\mathbb{T}$. In the first step a continuous function $f \in C(\mathbb{T})$ is constructed, which interpolates the given data in an appropriate sense (for numerical purposes it is usual to work with a trigonometric polynomial). This is then followed by a Nehari approximation scheme, approximating f optimally by an H^∞ function g. The validity of this approach follows from the easily-proven inequality
$$\|g - h\|_{L^\infty(\mathbb{T})} \leq \|g - f\|_{L^\infty(\mathbb{T})} + \|f - h\|_{L^\infty(\mathbb{T})} \leq 2\|f - h\|_{L^\infty(\mathbb{T})}.$$
In practical applications, h is the transfer function of a system that behaves linearly ***only*** in some frequency range, and approximate values of h may not be available at all parts of the unit circle. This problem is seen most clearly in the continuous-time case, when measurements are to be taken on the imaginary axis, and cannot be taken at arbitrarily high frequencies; by a conformal mapping, we can transform the mathematical analysis to the disc.

In [16, 17] the techniques of band-limited approximation were applied to some data provided by the French National Space Agency (CNES). The algorithm is based on the following construction:

1. Given approximate values of an unknown function $g \in A(\mathbb{D})$ on a subinterval I of the circle, construct a smooth function $f \in C(I)$ interpolating the data.

2. Given $M > 0$ solve the Bounded Extremal Problem 4.1 with $N = 0$, namely, find $g_0 \in H^\infty$ such that $\|g_0\|_{L^\infty(J)} \leq M$ and

$$\|g_0 - f\|_{L^\infty(I)} = \min\{\|g - f\|_{L^\infty(I)} : g \in H^\infty, \|g\|_{L^\infty(J)} \leq M\}.$$

As this stands, the algorithm produces a model which is unsuitable for practical use (for instance, it is not normally in $A(\mathbb{D})$), but various rational approximation steps can be used to circumvent this problem. Fuller technical details may be found in [16, 17].

5. Approximation with constraints on real or imaginary part

5.1. Passive systems

We now give an account of another extremal problem with applications in Systems Theory, which was solved by McCarthy and Jacobson in [70] and also by Fisher and Micchelli in the unpublished report [38]. Let $\mathcal{P}$ denote the set of H^2 functions with non-negative real part, i.e.,

$$\mathcal{P} = \{g \in H^2 : \operatorname{Re} g(z) \geq 0 \quad \text{for all } z \in \mathbb{D}\}.$$

The problem under consideration is as follows.

Problem 5.1. *Given $f \in H^2$, find $g_0 \in \mathcal{P}$ such that*

$$\|f - g_0\|_{L^2(\mathbb{T})} = \inf\{\|f - g\|_{L^2(\mathbb{T})} : g \in \mathcal{P}\}.$$

This infimum is attained, and g_0 is unique, because $\mathcal{P}$ is a closed convex subset of a Hilbert space. Let us write f and g_0 in terms of their real and imaginary parts thus: $f = p + iq$, $g_0 = u + iv$. Without loss of generality we may take $q(0) = v(0) = 0$. Let μ denote normalized Lebesgue measure on $\mathbb{T}$.

Theorem 5.2. [70] *Under the above conditions, the optimal approximant $g_0 = u+iv$ satisfies $u = max(p + c, 0)$, where $c = (1/2)\int_{\mathbb{T}}(u - p)\, d\mu$ is the unique point at which the decreasing function*

$$x \mapsto x\mu\{z : p(z) > -x\} - \int_{p \leq -x} p\, d\mu - 2x$$

on $\mathbb{R}^+$ changes sign.

Given u we can compute v as the Hilbert transform of u. The authors of [70] give applications to the study of passive linear systems, which are those that can be modelled by transfer functions in $\mathcal{P}$. The physical importance of such systems is explained in [34, 47]. Applications of the result above can be found to the model matching problem and systems identification.

5.2. Approximation with constrained imaginary part

Consider now the following constrained approximation problem, where, as before, $\mathbb{T}$ is written as the disjoint union of two sets I and J of positive measure.

Problem 5.3. *Given $f \in L^2(I)$, $\Psi \in L^2(J)$ and $M \geq 0$, find $g_0 \in H^2$ such that $\|\operatorname{Im} g_0 - \Psi\|_{L^2(J)} \leq M$ and*

$$\|g_0 - f\|_{L^2(I)} = \min\{\|g - f\|_{L^2(I)} : g \in H^2, \|\operatorname{Im} g - \Psi\|_{L^2(J)} \leq M\}.$$

Boundary parameter identification for the 2D Laplace equation may be efficiently formulated as Problem 5.3 since the constraint can be used to impose the condition that the imaginary part of g_0 should be close to some known function Ψ on J that may be available through boundary measurements, as in [12, 26], see Section 7. The above problem was analysed in [55], and its solution can be expressed implicitly in terms of the solution to the more standard Bounded Extremal Problem 3.1. Indeed, constraints on J in both problems are linked by $\|\operatorname{Im} g - \Psi\|_{L^2(J)} = \|g - (\operatorname{Re} g - i\Psi)\|_{L^2(J)}$; hence g_0 is related to the solution to Problem 3.1 associated to the implicit gauge function $\operatorname{Re} g - i\Psi$ on J.
A direct solution can also be given in terms of the real-linear operator $\tau_\lambda : H^2 \to H^2$ defined (for $\lambda > 0$) by

$$\tau_\lambda g = P_{H^2}(g\chi_I) + i\lambda P_{H^2}(\operatorname{Im} g\chi_J)\,.$$

It turns out that, for every $\lambda > 0$, τ_λ is a monotone operator (see, for example, [24] for the properties of such operators) and hence is invertible.

Theorem 5.4. [55] *Let $M > 0$ and suppose that $f \in L^2(I)$ is not already the trace of an H^2 function g with $\|\operatorname{Im} g - \Psi\|_{L^2(J)} \leq M$. Then the unique solution g_0 to Problem 5.3 is given by*

$$\tau_\lambda g_0 = P_{H^2}(f \vee i\lambda\Psi),$$

for the unique $\lambda > 0$ such that $\|\operatorname{Im} g_0 - \Psi\|_{L^2(J)} = M$. Moreover, if f is the trace on I of an H^2 function g with $\|\operatorname{Im} g - \Psi\|_{L^2(J)} \leq M$, then $g_0 = f$ and $\tau_0 g_0 = P(f \vee 0)$.

Note that τ_λ actually involves both Toeplitz and Hankel operators, for we also have:

$$\tau_\lambda g = g + \left(\frac{\lambda}{2} - 1\right) P_{H^2}(g\chi_J) - \frac{\lambda}{2} P_{H^2}(\overline{g}\chi_J)\,,$$

where $\overline{g}$ stands for the function $z \to \overline{g(\overline{z})}$.

6. Applications to Signal Processing

Let $\Omega = [-b, b]$ and $J = [-T, T]$ be subintervals of the real line. An important decomposition in the context of signal processing is the following:

$$L^2(\mathbb{R}) = PW(\Omega) \oplus PW(\mathbb{R} \setminus \Omega) = L^2(J) \oplus L^2(\mathbb{R} \setminus J), \tag{11}$$

where $PW(\Omega)$ is the Paley–Wiener space of band-limited signals, namely the subspace consisting of those functions f in $L^2(\mathbb{R})$ whose Fourier transform is supported on Ω, that is,

$$f(t) = \frac{1}{2\pi} \int_\Omega \hat{f}(\omega) e^{i\omega t}\, dt,$$

where $\hat{f} \in L^2(\Omega)$. See [78], for example.

More generally, let G be any locally compact Abelian group with dual group Γ, and let $\mathcal{F} : L^2(G) \to L^2(\Gamma)$ be the Fourier–Plancherel transform, which can be defined as a unitary operator [86]. We may consider the decomposition of the form (3), where $H = L^2(G)$, $H_+ = \mathcal{F}^{-1}L^2(A)$, $H_- = \mathcal{F}^{-1}L^2(\Gamma \setminus A)$, $K_+ = L^2(B)$, and $K_- = L^2(G \setminus B)$, for fixed Borel sets $A \subseteq \Gamma$ and $B \subseteq G$. The spectral theory of the corresponding Toeplitz operators is simpler in this case.

Theorem 6.1. [90] *Suppose that A and B have finite Haar measure; then the Toeplitz operator $S := P_{H_+}P_{K_+}$ defined on H_+ is Hilbert–Schmidt, and hence $T := P_{H_+}P_{K_-}$ has an orthonormal basis of eigenvectors.*

In the case of the Paley–Wiener decomposition (11), the eigenvectors of T are known as Prolate Spheroidal Wave Functions, and are described in [74, 89]. They form an orthonormal basis (ψ_n) of $PW(\Omega)$, and $(\psi_n \chi_J/\sqrt{\lambda_n})$ forms an orthonormal basis of $L^2(J)$, where (λ_n) is the sequence of eigenvalues of S.

This enables us to provide an explicit solution to various problems in signal analysis, linked with approximating time-limited signals by the restriction of band-limited signals (see [88] for an interesting discussion of such problems). The following result, due originally to Slepian and Pollak [89] and derived by Smith [90], along with additional results on constrained approximation, as a corollary of the results of [66], is typical.

Theorem 6.2. *Suppose that $f \in L^2(J) \setminus P_{L^2(J)}PW(\Omega)$ and $M > 0$. Then there exists $g_0 \in PW(\Omega)$ such that $\|g_0\|_{L^2(\mathbb{R}\setminus J)} \leq M$ and*

$$\|g_0 - f\|_{L^2(J)} = \inf\{\|g - f\|_{L^2(J)} : g \in PW(\Omega), \|g\|_{L^2(\mathbb{R}\setminus J)} \leq M\}.$$

Moreover, writing $f = \sum_{n=0}^\infty a_n \psi_n \chi_J$ for some coefficients (a_n), then we have

$$g_0 = \sum_{n=0}^\infty \frac{\lambda_n a_n}{\mu(1-\lambda_n) + \lambda_n} \psi_n,$$

where $\mu > 0$ is the unique constant such that

$$\sum_{n=0}^\infty \frac{(1-\lambda_n)\lambda_n^2 |a_n|^2}{(\mu(1-\lambda_n) + \lambda_n)^2} = M^2.$$

A good general reference for Toeplitz and Hankel operators on Paley–Wiener spaces is [82].

7. Applications to 2D inverse problems for Laplace operator

For various partial differential equations, inverse problems that consist in recovering missing data on some part J of the boundary ∂D of a domain D from data on the complementary part $I = \partial D \setminus J$ arise in many applications of non-destructive control (tomography and medical imaging, fault detection and control, underground prospecting, etc.), see [54]. Particular two-dimensional instances of such issues, namely the ones for $D \subset \mathbb{C}$ that involve the Laplace operator and harmonic functions (see e.g. [4, 7]), can be efficiently rephrased in terms of constrained extremal problems in spaces of analytic or meromorphic functions as in [12, 26].

In order to illustrate this approach, we consider here the following inverse problem of parameter identification in $\mathbb{D}$, see [26]. Note that, by making a conformal transformation, it can be handled in general simply connected Hölder-smooth domains D. Such an issue arises for example in corrosion detection by electrical impedance tomography. In the simple linear case, the parameter characterizing the damage caused by corrosion is a Robin exchange coefficient, the direct problem to be solved being a Laplace equation.

Problem 7.1. *Given a prescribed (current) flux $\phi \in L^2(I)$, $\phi \not\equiv 0$ together with measurements (of the voltage potential) $u_m \in W^{1,2}(I)$, find a function φ on J such that the solution u of*

$$\left\{ \begin{array}{rcll} \Delta u & = & 0 & \text{in} \quad \mathbb{D}, \\ \dfrac{\partial u}{\partial n} & = & \phi & \text{on} \quad I, \\ \dfrac{\partial u}{\partial n} + \varphi u & = & 0 & \text{on} \quad J, \end{array} \right. \tag{12}$$

also satisfies $u_{|_I} = u_m$.

We stick here to the classical notation where n denotes the outer unit normal vector to $\mathbb{T}$. Smoothness and identifiability (uniqueness) results from [25, 26] ensure that if $\phi \in L^2(I)$ and φ_1, φ_2 belong to some class of continuous functions on J, then the associated solutions u_1, u_2 of (12) belong to the Sobolev space $W^{1,2}(\mathbb{T})$ of functions that admit an $L^2(\mathbb{T})$ derivative; moreover, it holds that if $u_{1|_I} = u_{2|_I}$, then $\varphi_1 = \varphi_2$ on J.

Solving the Inverse Problem 7.1 with the use of tools from Section 3 first requires us to go from harmonic to analytic functions. From the knowledge of the flux $\phi \in L^2(I)$ and of the voltage potential $u_m \in W^{1,2}(I)$ in system 7.1, we can build the trace on I of a function g analytic in $\mathbb{D}$. Indeed,

$$f = u_m + i \int \phi \, d\theta \qquad \text{on} \quad I, \tag{13}$$

is the trace on I of a (unique) function g analytic in $\mathbb{D}$: $f = g_{|I}$. This holds because any harmonic conjugate function v to u in $\mathbb{D}$ (such that $f = u + iv$ is an analytic function in $\mathbb{D}$) satisfies the Cauchy–Riemann equations:

$$\frac{\partial v}{\partial \theta} = \frac{\partial u}{\partial n} \quad \text{on} \quad \mathbb{T}\,.$$

That g actually belongs to the Hardy–Sobolev $H^2 \cap \mathcal{W}^{1,2}(\mathbb{T})$ follows from smoothness preserving properties of the Hilbert transform (harmonic conjugation operator), for instance the M. Riesz theorem, see [2, 21, 43]. Our aim is then to recover g on J from the knowledge of its trace f on I. This would solve Problem 7.1, since

$$u = \operatorname{Re} g \quad \text{in} \quad \mathbb{D},$$

and then

$$\varphi = -\frac{1}{\operatorname{Re} g}\frac{\partial \operatorname{Im} g}{\partial \theta} \quad \text{on} \quad J\,, \tag{14}$$

is the expected solution. Note that the above equality should be properly understood (non-tangential limits on the right hand side). However, we want the recovery procedure to be convergent in order to carry some stability and robustness properties. Our concern here is that, in practice, one may not know exactly u_m nor f on I: values of u_m might only be available through experimental devices that necessarily induce perturbations while the knowledge of f also requires that of a primitive of ϕ, which may be computed numerically. Hence, f does not coincide in general with the trace on I of an H^2 function. This is the reason why the following approach is taken, with a resolution algorithm which solves a Bounded Extremal Problem rather than an extrapolation problem [26].

1. From the data u_m and ϕ, build f on I through (13).
2. Solve Problem 3.1 for f, $\Psi = 0$, and a suitable value of the constraint M (it can be computed by cutting I into two disjoint parts I_1, I_2 and iteratively solving a bounded extremal problem on I_1, in order to minimize a criterion on I_2).This gives a function g_2.
3. Finally, applying (14) to g_2 gives the desired coefficient φ.

In fact, two additional steps around Step 2 above are needed in order for the above algorithm to converge, which is in accordance with a global stability result concerning Problem 7.1, [26]. Indeed, the procedure has to be applied at a higher order: Step 2 must be applied to the tangential derivative of f on I and be followed by the computation of an appropriate primitive g_2 of the solution to Problem 3.1. Note that this is a way to solve Bounded Extremal Problems in Hardy–Sobolev spaces [10], and that continuity properties can be deduced from boundary estimates in these classes, see [20].

Various other 2D inverse problems can be handled by analytic, meromorphic, or rational approximation techniques, such as crack detection from boundary data as in [12], where it shown how approximants provide information about the singularities of the solution. Also, Hardy spaces of circular domains, see Section 3.3, may be of use to model the presence of inclusions or defaults in some material. Concerning for instance the recovery of a damaged and unknown part of the boundary, extremal problems with constraint on the imaginary part, see Section 5.2, may furnish a suitable framework, since this imaginary part (flux) may be data of the problem on the whole boundary.

8. Interpolation between $H^2(\partial G)$ and $H^\infty(\partial G)$ with applications

8.1. Interpolation spaces

In this section we observe that the $H^p(\partial G)$ spaces ($2 < p < \infty$), defined in Section 3.3, are interpolation spaces between $H^2(\partial G)$ and $H^\infty(\partial G)$ in a sense to be made precise. This result is a version of the celebrated theorem of Jones [57] that established the same result for the disc. Other important interpolation results in this area are due to Pisier [81] and Bourgain [23]. Proving versions of the Jones result valid for $H^p(G)$ (and there seems to be no way to deduce them directly), can be achieved by making constructions similar to one found in the work of Kisliakov and Xu [62, 63].

Theorem 8.1. *Suppose that G is the annulus $\{z \in \mathbb{C} : s < |z| < 1\}$, for fixed $s > 0$. Let $f \in H^2(\partial G)$. Suppose that $f = f_0 + f_1$ where $f_0 \in L^2(\partial G)$ and $f_1 \in L^\infty(\partial G)$. Then, for any $\mu > 0$, there exist $g_0 \in H^2(\partial G)$ and $g_1 \in H^\infty(\partial G)$ satisfying $f = g_0 + g_1$ with*

$$\|g_0\|_{H^2(\partial G)} \leq (2 + 4\mu)\|f_0\|_{L^2(\partial G)} \quad \text{and} \quad \|g_1\|_{H^\infty(\partial G)} \leq (1 + 1/\mu)\|f_1\|_{L^\infty(\partial G)}.$$

We may also generalize Theorem 8.1 to arbitrary circular domains although the precise constants involved are no longer so simple.

Theorem 8.2. *Suppose that G is a circular domain defined by (8). For each $\mu > 0$, there exists a positive constant $C(G, \mu)$ with the following properties. Let $f \in H^2(\partial G)$, and suppose that $f = f_0 + f_1$ where $f_0 \in L^2(\partial G)$ and $f_1 \in L^\infty(\partial G)$. Then there exists $g_0 \in H^2(\partial G)$ and $g_1 \in H^\infty(\partial G)$ satisfying $f = g_0 + g_1$ with*

$$\|g_0\|_{H^2(\partial G)} \leq (2 + C(G, \mu))\|f_0\|_{L^2(\partial G)}$$

and

$$\|g_1\|_{H^\infty(\partial G)} \leq (1 + 1/\mu)\|f_1\|_{L^\infty(\partial G)}.$$

We can reformulate the above results in the language of interpolation spaces. Let A_0, A_1 be a couple of interpolation (Banach) spaces. For $x \in A_0 + A_1$ and $t > 0$, let

$$K_t(x, A_0, A_1) = \inf\{\|x_0\|_{A_0} + t\|x_1\|_{A_1} : x = x_0 + x_1, x_0 \in A_0, x_1 \in A_1\}.$$

The real interpolation space $(A_0, A_1)_{\theta,p}$ is defined to be the space of all $x \in A_0 + A_1$ such that $\|x\|_{\theta,p} < \infty$ where

$$\|x\|_{\theta,p} = \left(\int_0^\infty (t^{-\theta} K_t(x, A_0, A_1))^p \frac{dt}{t}\right)^{1/p}.$$

This notion is motivated by the well-known result $L^p = (L^2, L^\infty)_{\theta,p}$ with $2 < p < \infty$ and $\frac{1}{p} = \frac{1-\theta}{2}$, [22, Section 5.2]. Peter Jones [57] gave the following interpolation results for the unit circle:

$$H^p = (H^2, H^\infty)_{\theta,p}, \qquad \text{where} \qquad 2 < p < \infty \qquad \text{and} \qquad \frac{1}{p} = \frac{1-\theta}{2},$$

which we now extend to $H^p(\partial G)$.

Proposition 8.3. *Let $2 < p < \infty$ and $f \in H^p(\partial G)$. Then there exists a positive constant C_G such that*

$$\begin{aligned} C_G K_t(f, H^2(\partial G), H^\infty(\partial G)) &\leq K_t(f, L^2(\partial G), L^\infty(\partial G)) \\ &\leq K_t(f, H^2(\partial G), H^\infty(\partial G)), \end{aligned}$$

and hence

$$H^p(\partial G) = (H^2(\partial G), H^\infty(\partial G))_{\theta,p}, \qquad \text{where} \qquad \frac{1}{p} = \frac{1-\theta}{2}.$$

8.2. Simultaneous H^2 and H^∞ approximation

The first application of $H^2(G)$–$H^\infty(G)$ interpolation is in providing a very explicit version of the Nehari approximation theorem in order to obtain an analytic approximant to a given function which is to be nearly optimal in both L^∞ and L^2 norms. In the case of the disc, such approximants have significant implications in Systems and Control Theory, and were first provided by Kaftal, Larson and Weiss [58] in an operator-algebraic context. Further extensions were given in [39, 40, 42] by Frazho and his co-workers. The results have been expressed in terms of central interpolants and the more abstract context of the Commutant Lifting theorem. The following is a key result along these lines.

Theorem 8.4. [58] *Given $f \in L^\infty(\mathbb{T})$ and a number $\delta > 1$, one can find a function h in H^∞ such that simultaneously*

$$\|f - h\|_{L^\infty(\mathbb{T})} \leq \delta \operatorname{dist}_\infty(f, H^\infty)$$

and

$$\|f - h\|_{L^2(\mathbb{T})} \leq \frac{\delta}{\sqrt{\delta^2 - 1}} \operatorname{dist}_2(f, H^2),$$

where $\operatorname{dist}_\infty(f, H^\infty)$ is the distance of f to H^∞, and similarly $\operatorname{dist}_2(f, H^2)$ is the distance of f to H^2.

The extensions we discuss here are for Hardy spaces on circular domains.

Theorem 8.5. *Suppose that G is a circular domain defined by (8). For each $\mu > 0$, there exists a positive constant $C(G, \mu)$ with the following properties. Let $f \in L^\infty(\partial G)$. Then there exists $g \in H^\infty(\partial G)$ such that*

$$\|f - g\|_{L^\infty(\partial G)} \leq (2 + 1/\mu) \operatorname{dist}_\infty(f, H^\infty(\partial G))$$

and

$$\|f-g\|_{L^2(\partial G)} \leq (3+C(G,\mu))\,\mathrm{dist}_2(f,H^2(\partial G)).$$

Remark that in the particular case where G is an annulus, we can take $C(G,\mu) = 4\mu$.

8.3. "Correction" results

The second application of Theorem 8.2 is in providing so-called "correction" theorems, analogous to those of of Kisliakov [60, 61], whereby an H^p function for $2 < p < \infty$ is approximated by an H^∞ function with small H^2 error.

Theorem 8.6. *Suppose that G is a circular domain defined by (8). Then there exists a constant K_G with the following properties. Let $p > 2$ and $f \in H^p(\partial G)$. For any $\varepsilon > 0$ there exist $g_0 \in H^2(\partial G)$ and $g_1 \in H^\infty(\partial G)$ such that $f = g_0 + g_1$,*

$$\|g_0\|_{H^2(\partial G)} \leq \varepsilon, \qquad \text{and} \qquad \|g_1\|_{H^\infty(\partial G)} \leq 3\left(\frac{2+K_G}{\varepsilon}\right)^{2/(p-2)}.$$

It is also possible to give a direct proof of Theorem 8.6 which does not rely on Theorem 8.2. This is of particular interest in the case of the unit disc and we obtain thereby an improvement in the constants involved (cf. Proposition 2.2 of [27] which applies only for $p > 4$).

Proposition 8.7. [29, Prop. 3.1] *Let $p > 2$ and $f \in H^p$. For any $\varepsilon > 0$ there exist $g_0 \in H^2$ and $g_1 \in H^\infty$ such that $f = g_0 + g_1$,*

$$\|g_0\|_{H^2} \leq \varepsilon, \qquad \text{and} \qquad \|g_1\|_{H^\infty} \leq \left(\frac{3}{\varepsilon}\right)^{2/(p-2)}.$$

The following corollary of Theorem 8.6 can also be proved directly and in a stronger form (with $(1/\varepsilon)^\delta$ replaced by $\log(1/\varepsilon)$ in (15)) using a more general result of Kisliakov [60, 61]: these rely on weak-type $(1-1)$ estimates for operators given by Calderón–Zygmund kernels.

Corollary 8.8. [28, Cor. 2.1] *Let $\varepsilon > 0$ be sufficiently small, let $\delta > 0$, and let $f \in L^\infty(\partial G)$. Then there exist $g^+ \in H^\infty(G)$ and $g^- \in H^\infty(\mathbb{C}\setminus\overline{G})$ such that*

$$\|f-(g^+ + g^-)\|_{L^2(\partial G)} < \varepsilon\|f\|_{L^\infty(\partial G)}$$

and

$$\|g^+\|_{H^\infty(\partial G)} + \|g^-\|_{H^\infty(\mathbb{C}\setminus\overline{G})} < C(1/\varepsilon)^\delta\|f\|_{L^\infty(\partial G)}, \tag{15}$$

where C is a constant that depends only on G and δ.

8.4. Factorization results

Let $\mathcal{H}$ be a separable, infinite dimensional, complex Hilbert space, and let $\mathcal{L}(\mathcal{H})$ denote the algebra of all bounded linear operators on $\mathcal{H}$. Suppose that $T \in \mathcal{L}(\mathcal{H})$ is such that its spectrum is included in G^- (the closure of G). Denote by R_G the set of rational functions whose poles lie in the complement of G^-. Suppose also that

$$\|f(T)\| \leq \sup\{|f(\lambda)| : \lambda \in G\}, \qquad f \in R_G.$$

By virtue of [35], we may assume without loss of generality with respect to properties we consider that T admits a G-rational normal boundary dilation, that is, there exists a Hilbert space $\mathcal{K} \supset \mathcal{H}$, a normal operator $N \in \mathcal{L}(\mathcal{K})$ such that $\sigma(N) \subset \partial G$ and for any $r \in R_G$ and any $x \in \mathcal{H}$, $r(T)x = P_{\mathcal{H}} r(N)x$. In the particular case where N is absolutely continuous (i.e., its spectral measure is absolutely continuous with respect to Lebesgue measure), we say that $T \in \mathcal{D}N_a(G)$. Let β_0 denote the position function (i.e., $\beta_0(z) = z$) in any of the spaces we consider. Any absolutely continuous normal operator $N \in \mathcal{L}(\mathcal{K})$ is (unitarily equivalent to) a part of the operator M_{β_0} of multiplication by β_0 on some space $L^2(\Gamma, \mathcal{D})$ of $\mathcal{D}$-valued functions on the Borel set Γ ($\mathcal{D}$ being a Hilbert space).
For $x, y \in \mathcal{K}$, one defines the function $x \overset{T}{\cdot} y$ by $x \overset{T}{\cdot} y(\xi) = (x(\xi), y(\xi))_{\mathcal{D}}$. Thus for $T \in \mathcal{D}N_a(G)$ we can define a sesquilinear map $(x, y) \in \mathcal{H} \times \mathcal{H} \to x \overset{T}{\cdot} y \in L^1(\Gamma)$ which does not depend on any choice of the normal boundary dilation [31, p. 347]. We denote by $\mathcal{P}_*(\Gamma)$ the predual of $H^\infty(\Gamma)$: $\mathcal{P}_*(\Gamma) = L^1(\Gamma)/^{\perp}H^\infty(\Gamma)$ and, for $f \in L^1(\Gamma)$, by $[f]$ the image of f in $\mathcal{P}_*(\Gamma)$ via the quotient map. Using Corollary 8.8 one can prove the following result which generalizes [27, Theorem 3.2].

Theorem 8.9. [28, Thm 3.1] *Let $T \in \mathcal{D}N_a(G)$ and let $(x_n)_{n \geq 1}$ be a sequence of elements of $\mathcal{H}$. The following assertions are equivalent:*

$$(i) \quad \lim_{n\to\infty} \|x_n \overset{T}{\cdot} w\|_{L^1(\Gamma)} = 0, \qquad w \in \mathcal{H}$$

$$(ii) \quad \lim_{n\to\infty} (\|[x_n \overset{T}{\cdot} w]\|_{\mathcal{P}_*(\Gamma)} + \|[w \overset{T}{\cdot} x_n]\|_{\mathcal{P}_*(\Gamma)}) = 0, \quad w \in \mathcal{H}.$$

Denote by Φ_T the $H^\infty(\Gamma)$ functional calculus for $T \in \mathcal{D}N_a(G)$ defined by $\Phi_T(h) = h(T)$, where

$$h(T)x = P_{\mathcal{H}} h(N)x, \qquad h \in H^\infty(\Gamma), \qquad x \in \mathcal{H}.$$

Though N is not necessarily unique, this functional calculus is uniquely defined as shown by Theorem 2.3 in [31].

Definition 8.10. *For a circular domain G, the class $\mathbb{A}^G$ is the subclass of $\mathcal{D}N_a(G)$ consisting of those $T \in \mathcal{L}(\mathcal{H})$ for which Φ_T is isometric. The class $\mathbb{A}^G_{\aleph_0}$ consists of those $T \in \mathbb{A}^G$ such that, for any infinite array $(f_{i,j})_{i,j\geq 1}$ of functions in $L^1(\Gamma)$, there exist some sequences $(x_i)_{i\geq 1}$ and $(y_j)_{j \geq 1}$ in $\mathcal{H}$ such that:*

$$[f_{i,j}] = [x_i \overset{T}{\cdot} y_j], \qquad i, j \geq 1.$$

Our final result generalizes Corollary 4.4 in [27] and is an immediate consequence of Theorem 8.9.

Corollary 8.11. [28, Cor. 3.1] *Let $T \in \mathcal{L}(\mathcal{H})$ be in the class $\mathbb{A}^{G}_{\aleph_0}$. Then, for any infinite array of functions in $L^1(\Gamma)$, there exist bounded sequences $(x_i)_{i\geq 1}$ and $(y_j)_{j\geq 1}$ in $\mathcal{H}$ such that:*

$$f_{i,j} = x_i \overset{T}{\cdot} y_j, \qquad i,j \geq 1.$$

References

[1] V. M. Adamjan, D. Z. Arov, and M. G. Krein, *Analytic properties of Schmidt pairs for a Hankel operator and the generalized Schur–Takagi problem*, Math. USSR Sbornik **15** (1971), 31–73.

[2] L. V. Ahlfors, *Complex analysis*, McGraw-Hill, 1953.

[3] L. Aizenberg, *Carleman's formulas in complex analysis*, Kluwer Academic Publishers, 1993.

[4] G. Alessandrini, *Examples of instability in inverse boundary-value problems*, Inverse Problems **13** (1997), 887–897.

[5] D. Alpay, L. Baratchart, and J. Leblond, *Some extremal problems linked with identification from partial frequency data*, 10th conference on analysis and optimization of systems, Sophia–Antipolis, 1992 (R. F. Curtain, A. Bensoussan, and J. L. Lions, eds.), L.N.C.I.S., vol. 185, Springer-Verlag, 1993, pp. 563–573.

[6] D. Alpay and J. Leblond, *Traces of Hardy functions and reproducing kernel Hilbert spaces*, Arch. Math. **64** (1995), 490–499.

[7] S. Andrieux and A. Ben Abda, *Identification of planar cracks by complete overdetermined data: inversion formulae*, Inverse Problems **12** (1996), 553–563.

[8] J. A. Ball and K. F. Clancey, *Reproducing kernels for Hardy spaces on multiply connected domains*, Integr. Equat. Oper. Th. **25** (1996), 35–57.

[9] L. Baratchart, J. Grimm, J., Leblond, and J. R. Partington, *Asymptotic estimates for interpolation and constrained approximation in H^2 by diagonalization of Toeplitz operators*, in preparation.

[10] L. Baratchart and J. Leblond, *Identification harmonique et trace des classes de Hardy sur un arc de cercle*, Actes du Colloque en l'honneur du 60e anniversaire du professeur Jean Céa, Sophia–Antipolis, 1992, Cépaduès–Editions, 1993, pp. 17–29.

[11] L. Baratchart and J. Leblond, *Hardy approximation to L^p functions on subsets of the circle with $1 \leq p < \infty$*, Constructive Approximation **14** (1998), 41–56.

[12] L. Baratchart, J. Leblond, F. Mandréa, and E. B. Saff, *How can meromorphic approximation help solve some 2D inverse problems for the Laplacian?*, Inverse problems **15** (1999), 79–90.

[13] L. Baratchart, J. Leblond, and J. R. Partington, *Hardy approximation to L^∞ functions on subsets of the circle*, Constructive Approximation **12** (1996), 423–436.

[14] L. Baratchart, J. Leblond, and J. R. Partington, *Constrained H^p interpolation on subsets of the circle*, INRIA Research Report no. 3763, (1999).

[15] L. Baratchart, J. Leblond, and J. R. Partington, *Problems of Adamjan–Arov–Krein type on subsets of the circle and minimal norm extensions*, Constructive Approximation **16** (2000), 333–357.

[16] L. Baratchart, J. Leblond, J. R. Partington, and N. Torkhani, *Robust identification from partial frequency data*, Proc. of the 33rd CDC, IEEE, 1994, pp. 3900–3905.

[17] L. Baratchart, J. Leblond, J. R. Partington, and N. Torkhani, *Robust identification from band-limited data*, IEEE Trans. Automat. Control **42** (1997), 1318–1325.

[18] L. Baratchart, J. Leblond, and N. Torkhani, *Best bounded H^2 extension of partial frequency data*, Proc. 3rd European Control Conference, Rome (A. Isidori, S. Bittanti, E. Mosca, A. De Luca, M. D. Di Benedetto, and G. Oriolo, eds.), vol. 2, 1995, pp. 1330–1335.

[19] L. Baratchart and F. Seyfert, *An L^p analog to AAK theory for $p \geq 2$*, J. Functional Anal. (2001), to appear.

[20] L. Baratchart and M. Zerner, *On the recovery of functions from pointwise boundary values in a Hardy-Sobolev class of the disk*, J. Comput. Appl. Math. **46** (1993), 255–269.

[21] H. G. W. Begehr, *Complex analytic methods for partial differential equations*, World Scientific, 1994.

[22] J. Bergh and J. Löfström, *Interpolation spaces*, Springer Verlag, Berlin, 1976, Grundlehren der mathematischen Wissenschafte vol. 223.

[23] J. Bourgain, *Some consequences of Pisier's approach to interpolation*, Israel J. Math. **77** (1992), 165–185.

[24] H. Brézis, *Analyse fonctionnelle*, Dunod, 1999.

[25] S. Chaabane and M. Jaoua, *Identification of Robin coefficients by the means of boundary measurements*, Inverse Problems **15** (1999), 1425–1438.

[26] S. Chabaane, M. Jaoua, and J. Leblond, *Parameter identification for Laplace equation and approximation in Hardy classes*, in preparation, short versions in Proc. CARI and CDC, 2000.

[27] I. Chalendar and J. Esterle, *L^1-factorization for C_{00}-contractions with isometric functional calculus*, J. Funct. Anal. **154** (1998), 174–194.

[28] I. Chalendar and J. R. Partington, *Approximation problems and representations of Hardy spaces in circular domains*, Studia Math. **136** (1999), 255–269.

[29] I. Chalendar and J. R. Partington, *Interpolation between Hardy spaces on circular domains with applications to approximation*, Arch. Math. (2001), to appear.

[30] J. Chen and G. Gu, *Control-oriented system identification: an H_∞ approach*, John Wiley, 2000.

[31] B. Chevreau and W. S. Li, *On certain representations of $H^\infty(G)$ and the reflexivity of associated operator algebras*, J. Funct. Anal. **128** (1995), no. 2, 341–373.

[32] B. Chevreau, C. Pearcy, and A. L. Shields, *Finitely connected domrins G, representations of $H^\infty(G)$, and invariant subspaces*, J. Operator Theory **6** (1981), 375–405.

[33] K. Clancey and I. Gohberg, *Factorization of matrix functions and singular integral operators*, OT, Birkhäuser Verlag, Basel, 1981.

[34] C. A. Desoer and M. Vidyasagar, *Feedback systems: input–output properties*, Academic Press, 1975.

[35] R. G. Douglas and V. Paulsen, *Completely bounded maps and hypo-Dirichlet algebras*, Acta Sci. Math. (Szeged) **50** (1986), 143–157.

[36] P. L. Duren, *Theory of H^p spaces*, Academic Press, 1970.

[37] A. Feintuch, *Robust control theory in Hilbert space*, Springer-Verlag, 1998.

[38] S. D. Fisher and C. A. Micchelli, *Minimal norm interpolation in H^2 with nonnegative real part*, IBM Research Center Technical Report RC 12351, 1986.

[39] C. Foias, A. Frazho, and A. Tannenbaum, *On combined H^∞-H^2 suboptimal interpolants*, Linear Algebra Appl. **203/204** (1994), 443–469.

[40] C. Foias and A. E. Frazho, *Commutant lifting and simultaneous H^∞ and L^2 suboptimization*, SIAM J. Math. Anal. **23** (1992), 984–994.

[41] B. A. Francis, *A course in H_∞ control theory*, Springer-Verlag, 1987.

[42] A. E. Frazho and S. M. Kerat, *Applications of the Schur algorithm to mixed H^2 and H^∞ Nehari problems*, J. Math. Anal. Appl. **178** (1993), 488–508.

[43] J. B. Garnett, *Bounded analytic functions*, Academic Press, 1981.

[44] T. T. Georgiou and M. C. Smith, *Graphs, causality and stabilizability: Linear, shift-invariant systems on $L_2[0,\infty)$*, Math. Control Signals Systems **6** (1993), 195–223.

[45] G. M. Goluzin, *Geometric theory of functions of a complex variable*, A.M.S., Providence, 1969, Translated from the Russian.

[46] G. Gu and P. P. Khargonekar, *Linear and nonlinear algorithms for identification in H_∞ with error bounds*, IEEE Trans. Automat. Control **37** (1992), 953–963.

[47] E. A. Guillemin, *Synthesis of passive networks*, John Wiley, 1957.

[48] S. Ja. Havinson, *On some extremal problems of the theory of analytic functions*, A.M.S. Translations **32** (1951), 139–154.

[49] A. J. Helmicki, C. A. Jacobson, and C. N. Nett, *Control oriented system identification : a worst-case / deterministic approach in H_∞*, IEEE Trans. Automat. Control **36** (1991), no. 10, 1163–1176.

[50] J. W. Helton, *Operator theory, analytic functions, matrices and electrical engineering*, American Mathematical Society, 1987.

[51] J. W. Helton and O. Merino, *Conditions for optimality over H^∞*, SIAM J. Control Optim. **31** (1993), 1379–1415.

[52] J. W. Helton and O. Merino, *Classical control using H^∞ methods: theory, optimization, and design*, SIAM, 1998.

[53] K. Hoffman, *Banach spaces of analytic functions*, Dover, 1988.

[54] V. Isakov, *Uniqueness and stability in multi-dimensional inverse problems*, Inverse Problems **9** (1993), 579–621.

[55] B. Jacob, J. Leblond, J.-P. Marmorat and J.R. Partington, *A constrained approximation problem arising in parameter identification*, preprint, 2001.

[56] B. Jacob and J. R. Partington, *On the boundedness and continuity of the spectral factorization mapping*, SIAM J. Control Optim. (2001), to appear.

[57] P. Jones, *L^∞ estimates for the $\overline{\partial}$-problem on a half plane*, Acta Math. **150** (1983), 137–152.

[58] V. Kaftal, D. Larson, and G. Weiss, *Quasitriangular subalgebras of semi-finite von Neumann algebras are closed*, J. Funct. Anal. **107** (1992), 387–401.

[59] S. Ya. Khavinson, *Two papers on extremal problems in Complex Analysis*, American Mathematical Society Translations (2) **129**, 1986.

[60] S. V. Kisliakov, *Quantitative aspect of correction theorems*, Zap. Nauchn. Sem. LOMI **92** (1979), 182–191.

[61] S. V. Kisliakov, *A sharp correction theorem*, Studia Math. **113** (1995), no. 2, 177–196.

[62] S. V. Kisliakov, *Interpolation of H^p spaces: some recent developments*, Function spaces, interpolation spaces, and related topics (Haifa, 1995) (Y. Brudnyi, M. Cwikel, and Y. Sagher, eds.), Israel Math. Conf. Proc., vol. 13, Bar-Ilan Univ, Ramat Gan, 1999, pp. 102–140.

[63] S. V. Kisliakov and Q. Xu, *Sur l'interpolation des espaces H^p*, C. R. Acad. Sci. Paris, Série I **313** (1991), 249–254.

[64] M. G. Krein and P. Y. Nudel'man, *Approximation of $L^2(\omega_1, \omega_2)$ functions by minimum-energy transfer functions of linear systems*, Problemy Peredachi Informatsii **11** (1975), no. 2, 37–60, English translation.

[65] J. Leblond and M. Olivi, *Weighted H^2 approximation of transfer functions*, Math. Control Signals Systems **11** (1998), 28–39.

[66] J. Leblond and J. R. Partington, *Constrained approximation and interpolation in Hilbert function spaces*, J. Math. Anal. Appl. **234** (1999), 500–513.

[67] K. Loewner, *Über monotone Matrixfunktionen*, Math. Zeitschrift **38** (1934), 177–216.

[68] A. J. Macintyre and W. W. Rogosinski, *Extremum problems in the theory of analytic functions*, Acta Math. **82** (1950), 275–325.

[69] P. M. Mäkilä, J. R. Partington, and T.K. Gustafsson, *Worst-case control-relevant identification*, Automatica **31** (1995), 1799–1819.

[70] J. E. McCarthy and C. A. Jacobson, *A note on H_2 optimal control problems with passivity constraints*, IEEE Trans. Automat. Control **45** (2000), no. 2, 286–290.

[71] M. Milanese and A. Vicino, *Information based complexity and nonparametric worst-case system identification*, J. Complexity **9** (1993), 427–446.

[72] B. Ninness and G. C. Goodwin, *Estimation of model quality*, Automatica **31** (1995), 1771–1797.

[73] M. Papadimitrakis, *Continuity of the operator of best uniform approximation by bounded analytic functions*, Bull. London Math. Soc. **25** (1993), 44–48.

[74] A. Papoulis, *Signal analysis*, McGraw–Hill, 1977.

[75] J. R. Partington, *An introduction to Hankel operators*, Cambridge University Press, 1988.

[76] J. R. Partington, *Robust identification and interpolation in H_∞*, Int. J. Control **54** (1991), 1281–1290.

[77] J. R. Partington, *Robust identification in H_∞*, J. Math. Anal. Appl. **166** (1992), 428–441.

[78] J. R. Partington, *Interpolation, identification and sampling*, Oxford University Press (Clarendon Press), 1997.

[79] D. J. Patil, *Representation of H^p functions*, Bull. Amer. Math. Soc. **78** (1972), 617–620.

[80] V. V. Peller, *Hankel operators and continuity properties of the operators of best approximation*, Leningrad Math. J. **2** (1991), no. 1, 139–160.

[81] G. Pisier, *Interpolation between H^p spaces and non-commutative generalizations I*, Pacific J. Math. **155** (1992), 341–368.

[82] R. Rochberg, *Toeplitz and Hankel operators on the Paley–Wiener space*, Integr. Equat. Oper. Th. **10** (1987), 187–235.

[83] M. Rosenblum and J. Rovnyak, *Restrictions of analytic functions I, II, III*, Proc. American Mathematical Society **48–51–52** (1975).

[84] M. Rosenblum and J. Rovnyak, *Hardy classes and operator theory*, Oxford University Press, 1985.

[85] W. Rudin, *Analytic functions of class* H_p, Trans. Amer. Math. Soc. **78** (1955), 46–66.

[86] W. Rudin, *Fourier analysis on groups*, John Wiley, 1962.

[87] R. Sánchez-Peña and M. Sznaier, *Robust systems theory and applications*, John Wiley, 1998.

[88] D. Slepian, *On bandwidth*, Proc. IEEE **64** (1976), 292–300.

[89] D. Slepian and H. O. Pollak, *Prolate spheroidal wave functions, Fourier analysis and uncertainty – I*, Bell Syst. Tech. J. **40** (1961), no. 1, 43–63.

[90] M. Smith, *The spectral theory of Toeplitz operators applied to approximation problems in Hilbert function spaces*, Preprint, University of Leeds, 2000.

[91] G. Weiss, *Representation of shift-invariant operators on* L^2 *by* H^∞ *transfer functions: an elementary proof, a generalization to* L^p*, and a counterexample for* L^∞, Math. Control Signals Systems **4** (1991), 193–203.

[92] N. J. Young, *An introduction to Hilbert space*, Cambridge University Press, 1988.

Institut Girard Desargues, UFR de Mathématiques, Université Claude Bernard Lyon 1, 69622 Villeurbanne Cedex, France
E-mail address: chalenda@desargues.univ-lyon1.fr

INRIA, BP 93, 2004 route des Lucioles, 06902 Sophia–Antipolis Cedex, France
E-mail address: leblond@sophia.inria.fr

School of Mathematics, University of Leeds, Leeds LS2 9JT, U.K.
E-mail address: J.R.Partington@leeds.ac.uk

The Bargmann isometry and Gabor-Daubechies wavelet localization operators

L. A. Coburn

Abstract. I consider the relationship between *Berezin–Toeplitz* operators T_φ and the *Gabor–Daubechies* wavelet localization operators L_φ^w. For Gaussian "window" w_0, it is easy to check that $\beta L_\varphi^{w_0}\beta^{-1} = T_\varphi$ with β the Bargmann isometry. For more general w, and some interesting classes of "symbols" φ, I discuss some new equivalences of the form $\beta L_\varphi^w \beta^{-1} = T_{(I+D)\varphi}$, where $D = D(w)$ is a constant-coefficient linear differential operator with constant term 0. One consequence is that there is a φ of modulus one with $L_\varphi^{w_1} = 0$ for w_1 a normalized Hermite function with polynomial part of degree one. A detailed discussion of the Bargmann isometry is provided. Proof sketches of new results are provided with details to appear elsewhere.

1. Introduction

For f in the Segal–Bargmann Hilbert space $H^2(\mathbb{C}^n, d\mu)$ of Gaussian square-integrable entire functions on complex n-space, [BC1] began a systematic study of densely defined operators of the form:

$$(T_g f)(z) = \int_{\mathbb{C}^n} K(z,w)g(w)f(w)d\mu(w)$$

$$(H_g f)(z) = \int_{\mathbb{C}^n} K(z,w)\{g(z) - g(w)\}f(w)d\mu(w)$$

where $z \cdot w = z_1\bar{w}_1 + \cdots + z_n\bar{w}_n$, $|z|^2 = z \cdot z$ and $K(z,w) = e^{z\cdot w/2}$ is the Bergman kernel function for $H^2(d\mu)$,

$$d\mu(z) = (2\pi)^{-n}\exp\{-|z|^2/2\}dv(z)$$

(dv is Lebesgue volume measure), and $K(\cdot, z)g(\cdot)$ is assumed to be in $L^2(\mathbb{C}^n, d\mu)$ for all z in $\mathbb{C}^n$. Note that the *Hankel* operator H_g maps $H^2(\mathbb{C}^n, d\mu)$ into $L^2(\mathbb{C}^n, d\mu)$ while the *Berezin–Toeplitz* operator T_g maps $H^2(\mathbb{C}^n, d\mu)$ into itself. It is known that $T_g = 0$ implies $g = 0$. Moreover, $T_{\bar{z}_1} = 2\partial_1$ ($\partial_j = \frac{\partial}{\partial z_j}$, $\bar{\partial}_j = \frac{\partial}{\partial \bar{z}_j}$). For f in

Research partially supported by a grant of the NSF.

$L^2(d\mu)$, the orthogonal projection of f onto the closed subspace $H^2(d\mu)$ is given by

$$(Pf)(z) = \int_{\mathbb{C}^n} e^{z\cdot \overline{w}/2} f(w) d\mu(w).$$

Thus, for M_g the operator of "multiplication by g" on $L^2(d\mu)$, we see that $H_g = (I-P)M_gP$ while $T_g = PM_gP$. The study of such *Hankel* and *Toeplitz* operators on Bergman and Hardy spaces of square-integrable holomorphic functions on bounded domains in $\mathbb{C}^n$ has a considerable history.

Let $\mathbb{H}_n$ be the Heisenberg group. $\mathbb{H}_n$ can be regarded spatially as $\mathbb{C}^n \times \mathbb{R}$, with $\mathbb{R}$ the real numbers and with multiplication given by

$$(a,s)(b,t) = (a+b, s+t+\mathrm{Im}(b\cdot \overline{a})/2).$$

The Heisenberg group is a nilpotent Lie group which plays a key role in quantum mechanics and in the representation theory of nilpotent Lie groups. It is standard [F, p. 19] that Lebesgue measure on $\mathbb{C}^n \times \mathbb{R}$ is biinvariant Haar measure on $\mathbb{H}_n$.

The irreducible strongly continuous unitary representations of $\mathbb{H}_n$ on Hilbert space are well-known [F, p. 37]. Here, I focus on the Segal–Bargmann representation on $H^2(\mathbb{C}^n, d\mu)$, with $d\mu$ the Gaussian measure described above. This representation is given by

$$\rho(a,t) = e^{it} W_a$$

with

$$(W_a f)(z) = k_a(z) f(z-a),$$

where $k_a(z) = \exp\{z\cdot \overline{a}/2 - |a|^2/4\}$ is the normalized Bergman reproducing kernel function for "evaluation at a". The connection between representation theory of $\mathbb{H}_n$ and Berezin–Toeplitz operators is given by [BC1] as

$$\rho(a,t) = e^{it} W_a = e^{it} e^{|a|^2/4} T_{\chi_a(\cdot)},$$

where

$$\chi_a(z) = \exp\{i\mathrm{Im}(z\cdot \overline{a})\}.$$

There is an interesting connection between Berezin–Toeplitz operators on $H^2(\mathbb{C}^n, d\mu)$ and *Gabor–Daubechies* wavelet localization operators on $L^2(\mathbb{R}^n, dv)$. Precisely, the *Gabor–Daubechies* localization operator, L^w_φ with "symbol" (or "weight function") φ, is defined on $L^2(\mathbb{R}^n, dv)$ by

$$\begin{aligned}\langle L^w_\varphi f, g\rangle &= (2\pi)^{-n} \int_{\mathbb{C}^n} \varphi(a) \langle f, \rho^s(a,0)w\rangle \langle \rho^s(a,0)w, g\rangle dv(a) \\ &= (2\pi)^{-n} \int_{\mathbb{C}^n} \varphi(a) \langle \beta f, W_a \beta w\rangle \langle W_a\beta w, \beta g\rangle dv(a)\end{aligned}$$

where β is the Bargmann isometry [B], [F, pp. 40, 47],

$$\beta : L^2(\mathbb{R}^n, dv) \to H^2(\mathbb{C}^n, d\mu),$$

which is surjective, with $\beta^{-1}\rho(a,t)\beta \equiv \rho^s(a,t)$, the Schrödinger representation of $\mathbb{H}_n$. Here, we use the "undotted product" notation

$$wz = w_1z_1 + w_2z_2 + \cdots + w_nz_n, \quad w^2 = ww$$

for w, z in $\mathbb{C}^n$ or $\mathbb{R}^n$ and, for p, q, x in $\mathbb{R}^n$, with $f(x)$ in $L^2(\mathbb{R}^n, dv)$ and $a = p + iq$, we have

$$(\beta f)(z) = \pi^{-\frac{n}{4}} \int_{\mathbb{R}^n} f(x) e^{zx - \frac{1}{2}x^2 - \frac{1}{4}z^2} dv(x),$$

$$[\rho^s(p+iq,t)f](x) = e^{it - iqx + i\frac{1}{2}pq} f(x-p).$$

Typically, the "window" w is chosen to be a unit vector in L^2. For $w_0 = \pi^{-\frac{n}{4}} e^{-\frac{1}{2}|x|^2}$, we have $\beta w_0 = 1$ and it is not hard to check that $\beta L_\varphi^{w_0}\beta^{-1} = T_\varphi$. Thus, the Bargmann isometry allows the amalgamation of the substantial work already done in analyzing T_φ [Be, Bo, BC1, BC2, C1, C2, CX, G, XZ] and $L_\varphi^{w_0}$ [D1, D2, D3, DFN, DW, FN].

There has been significant recent progress [C2] toward determining the largest class of functions φ, ψ for which there is a "twisted multiplication" $\varphi \diamond \psi$ so that $T_\varphi T_\psi = T_{\varphi\diamond\psi}$ where T_φ is the Berezin–Toeplitz operator associated to φ. This work has some interesting applications to the analysis of the Gabor–Daubechies operators described above. In particular, for φ, ψ in $B_a(\mathbb{C}^n)$, the algebra of Fourier–Stieltjes transforms of compactly supported, complex, bounded regular Borel measures on $\mathbb{C}^n$, $T_\varphi T_\psi$ is a Berezin–Toeplitz operator $T_{\varphi\diamond\psi}$ and I was able to obtain the "Moyal-type" formula

$$\varphi \diamond \psi = \sum_k \frac{(-2)^{|k|}}{k!} (\partial^k \varphi)(\bar{\partial}^k \psi), \tag{1}$$

where k ranges over all n-tuples of non-negative integers, $\partial^k = \partial_1{}^{k_1}\partial_2{}^{k_2}\cdots\partial_n{}^{k_n}$, $k! = k_1!k_2!\cdots k_n!$, $|k| = k_1 + \cdots + k_n$, and the series on the right converges uniformly and absolutely to a function in $B_a(\mathbb{C}^n)$. In [C2], I also checked that for φ, ψ arbitrary polynomials in z and $\bar{z}$, $T_\varphi T_\psi$ is a (generally unbounded) Berezin–Toeplitz operator $T_{\varphi\diamond\psi}$, with $\varphi \diamond \psi$ given again by (1).

For windows which are more general than w_0, the Bargmann isometry should also be useful in the analysis of L_φ^w. For $z^k = z_1{}^{k_1}z_2{}^{k_2}\cdots z_n{}^{k_n}$ and $\{e_k = (2^{|k|}k!)^{-1/2}z^k\}$ the standard orthonormal basis for $H^2(\mathbb{C}^n, d\mu)$, we have $\beta^{-1}e_k = w_k$ where w_k is the k-th normalized Hermite function [F, p. 51], [S, pp. 105, 106]. Hence, the windows $w = w_k$, where

$$w_k = \pi^{-\frac{n}{4}}(2^{|k|}k!)^{-\frac{1}{2}} H_k(x) e^{-\frac{1}{2}x^2},$$

with

$$H_k(x) = (-1)^{|k|} e^{x^2} D^k e^{-x^2}$$

for $D^k = (\frac{\partial}{\partial x_1})^{k_1} \cdots (\frac{\partial}{\partial x_n})^{k_n}$, provide a natural prototype for our analysis. I have obtained some interesting results for

$$w_1 = w_{(1,0,0,\ldots,0)} = 2^{1/2}\pi^{-n/4} x_1 \exp\{-|x|^2/2\}.$$

In this case,

$$\beta L_\varphi^{w_1}\beta^{-1} = (1/2)H_{\bar{z}_1}^* M_\varphi H_{\bar{z}_1},$$

where M_φ is "multiplication by φ", P is the orthogonal projection operator from $L^2(d\mu)$ onto $H^2(d\mu)$ and $H_{\bar{z}_1} = (I-P)M_{\bar{z}_1}P$ is a Hankel operator from $H^2(d\mu)$ into $L^2(d\mu)$. It is an easy consequence of the "boson commutation relations" that $(1/2)H_{\bar{z}_1}^* H_{\bar{z}_1} = P$.

For φ an arbitrary polynomial in z and $\bar{z}$, $L_\varphi^{w_1}$ is densely defined and, using results of [C2], I have checked that

$$\beta L_\varphi^{w_1}\beta^{-1} = T_{\varphi+2\bar{\partial}_1\partial_1\varphi}.$$

Since $I + 2\bar{\partial}_1\partial_1$ is an invertible linear transformation on the linear space of all polynomials in z and $\bar{z}$, it follows that $L_\varphi^{w_1} = 0$ if and only if $\varphi = 0$ and

$$L_\varphi^{w_1}L_\psi^{w_1} = L_{\varphi*\psi}^{w_1},$$

where $\varphi, \psi, \varphi * \psi$ are polynomials in $z, \bar{z}$ with

$$\varphi * \psi = \sum_{j=0}^{\infty}(-2\bar{\partial}_1\partial_1)^j[(\varphi + 2\bar{\partial}_1\partial_1\varphi) \diamond (\psi + 2\bar{\partial}_1\partial_1\psi)],$$

and with the associative product $\varphi \diamond \psi$ defined by (1).

A contrast to the case of polynomial φ is provided by the algebra of bounded, uniformly continuous functions $B_a(\mathbb{C}^n)$, described above. For φ in $B_a(\mathbb{C}^n)$, we can also check that

$$\beta L_\varphi^{w_1}\beta^{-1} = T_{\varphi+2\bar{\partial}_1\partial_1\varphi}.$$

Here, we have

$$\varphi(z) = \int_{\mathbb{C}^n} \chi_a(z)d\sigma(a)$$

and so

$$\varphi + 2\bar{\partial}_1\partial_1\varphi = \int_{\mathbb{C}^n} \chi_a(\cdot)(1 - |a_1|^2/2)d\sigma(a).$$

It follows that $\varphi+2\bar{\partial}_1\partial_1\varphi = 0$ whenever support(σ) is contained in $\{a : |a_1|^2 = 2\}$. The choice $\sigma(a) = \delta_{(\sqrt{2},0,0,\dots,0)}(a)$ shows that $L_\varphi^{w_1} = 0$ when $\varphi = \exp\{(z_1 - \bar{z}_1)/\sqrt{2}\}$. *This result also provides a clear contrast to the situation for $L_\varphi^{w_0}$, where $L_\varphi^{w_0} = 0$ implies $\varphi = 0$ for arbitrary φ.*

For the window $w_2 = w_{(2,0,0,\dots,0)} = 2^{-1/2}\pi^{-n/4}(2x_1{}^2 - 1)\exp\{-|x|^2/2\}$, we have $\beta w_2 = 8^{-1/2}z_1{}^2$ and there is a similar, but more complicated, analysis. Here, we have

$$\beta L_\varphi^{w_2}\beta^{-1} = (1/8)A^*M_\varphi A,$$

where

$$A = -2(I-P)M_{\bar{z}_1}PM_{\bar{z}_1}P + (I-P)M_{\bar{z}_1^2}P.$$

Again using results from [C2] and direct (but brutal) calculation, we find for φ an arbitrary polynomial in z and $\bar{z}$, that

$$\beta L_\varphi^{w_2}\beta^{-1} = T_{\varphi+4\bar{\partial}_1\partial_1\varphi+2(\bar{\partial}_1\partial_1)^2\varphi}. \tag{2}$$

There is a corresponding formula for the composition $L^{w_2}_{\varphi} L^{w_2}_{\psi}$. We also have (2) for φ in $B_a(\mathbb{C}^n)$.

In the next section of this paper, I provide a full discussion of the Bargmann isometry and the relation it provides between the Schrödinger and Segal–Bargmann representations of the Heisenberg group. In the final section, I provide an outline of the proofs of the results described above. Detailed proofs and more general results will appear elsewhere.

I thank Samuel D. Schack and Jingbo Xia for useful discussions.

2. The Bargmann isometry and representation theory for the Heisenberg group

Because of notational differences between this treatment, [F], and the original [B] as well as [D3, p. 34] and [FN], it seems useful to expand on the discussion of the role of the Bargmann isometry. Recall that $w_0 = \pi^{-\frac{n}{4}} e^{-\frac{1}{2}x^2}$. After [F], we *define*

$$V(f, w_0)(p, q) = \langle f, \rho^s(-(p + iq), 0)w_0\rangle$$

for f in $L^2(\mathbb{R}^n, dv)$ and p, q in $\mathbb{R}^n$.

Lemma. *For $z = p + iq$, we have*

$$V(f, w_0)(-p, -q) = e^{-\frac{1}{4}|z|^2}(\beta f)(z).$$

It follows that

$$\int_{\mathbb{C}^n} |(\beta f)(z)|^2 d\mu(z) = \|f\|^2.$$

Proof. For $z = p + iq$, we have

$$\begin{aligned} V(f, w_0)(-p, -q) &= \langle f, \rho^s(p + iq, 0)w_0\rangle \\ &= \pi^{-\frac{n}{4}} \int_{\mathbb{R}^n} f(x) e^{iqx - \frac{1}{2}ipq - \frac{1}{2}(x-p)^2} dv(x) \\ &= \pi^{-\frac{n}{4}} e^{-\frac{1}{4}|z|^2} \int_{\mathbb{R}^n} f(x) e^{-\frac{1}{2}x^2 + x(p+iq) - \frac{1}{4}(p+iq)^2} dv(x) \\ &= e^{-\frac{1}{4}|z|^2}(\beta f)(z). \end{aligned}$$

A modification of [F, p. 31] now shows that

$$\begin{aligned} \|f\|^2 &= (2\pi)^{-n} \int_{\mathbb{R}^{2n}} |V(f, w_0)(-p, -q)|^2 dpdq \\ &= \int_{\mathbb{C}^n} |(\beta f)(z)|^2 d\mu(z). \end{aligned}$$

and this completes the proof. □

Proposition 2.1. *For $a = c + id$ with c, d in $\mathbb{R}^n$, we have*

$$\beta[\rho^s(c+id,0)f](z) = W_a(\beta f)(z).$$

Proof. By the Lemma, for $z = p + iq$ we have

$$\begin{aligned}\beta[\rho^s(c+id,0)f](z) &= e^{\frac{1}{4}|z|^2} V(\rho^s(c+id,0)f, w_0)(-p,-q)\\ &= e^{\frac{1}{4}|z|^2} \langle \rho^s(-(p+iq),0)\rho^s(c+id,0)f, w_0\rangle\\ &= e^{\frac{1}{4}|z|^2} e^{\frac{1}{2}i(qc-pd)} e^{-\frac{1}{4}\{(c-p)^2+(d-q)^2\}}(\beta f)(z-a)\\ &= k_a(z)(\beta f)(z-a)\\ &= W_a(\beta f)(z).\square\end{aligned}$$

Corollary 2.2. *We have*

$$\beta[\rho^s(c+id,t)f](z) = \rho(a,t)(\beta f)(z).$$

Proof. Immediate. □

Corollary 2.3. *We have*

$$\beta[\rho^s(c+id,0)w_0](z) = k_a(z)$$

for $a = c + id$.

Proof. It is easy to check that $\beta w_0 = 1$. □

To complete this discussion, we check that, for $n = 1$, $\beta^{-1}e_k = w_k$ is the k-th normalized Hermite function. The corresponding result for general n follows immediately. This result originates in [B]. We recall that the normalized Hermite functions form an orthonormal basis for $L^2(\mathbb{R}, dx)$ and are given by [S]

$$w_k = \pi^{-\frac{1}{4}}(2^k k!)^{-\frac{1}{2}} H_k(x) e^{-\frac{1}{2}x^2},$$

where

$$H_k(x) = (-1)^k e^{x^2} D^k e^{-x^2}$$

with $D = \frac{d}{dx}$.

Proposition 2.4. *For $n = 1$, we have*

$$\beta w_k = e_k = (2^k k!)^{-\frac{1}{2}} z^k.$$

Proof. For $f(x)$ of the form $p(x)e^{-\frac{1}{2}x^2}$ with $p(x)$ a polynomial, integration by parts gives

$$\begin{aligned}\beta(Df)(z) &= \pi^{-\frac{1}{4}} \int_{\mathbb{R}} f'(x) e^{zx-\frac{1}{2}x^2-\frac{1}{4}z^2} dx\\ &= \pi^{-\frac{1}{4}} \int_{\mathbb{R}} f(x)(x-z) e^{zx-\frac{1}{2}x^2-\frac{1}{4}z^2} dx\\ &= \beta(M_x f)(z) - z(\beta f)(z),\end{aligned}$$

for M_x the operator of "multiplication by x". Hence,

$$\beta[(M_x - D)f](z) = z(\beta f)(z) \tag{3}$$

and iteration of (3) gives

$$\beta[(M_x - D)^k f](z) = z^k(\beta f)(z).$$

In particular, for $f(x) = \pi^{-\frac{1}{4}}e^{-\frac{1}{2}x^2}$, we have $(\beta f)(z) = 1$ and

$$\pi^{-\frac{1}{4}}\beta[(M_x - D)^k e^{-\frac{1}{2}x^2}](z) = z^k.$$

Next, we check that

$$(M_x - D)^k e^{-\frac{1}{2}x^2} = H_k(x)e^{-\frac{1}{2}x^2}.$$

This is an easy induction on k, using the known recursion relation

$$H_{k+1}(x) = 2xH_k(x) - H_k{}'(x).$$

It follows that

$$\beta w_k = (2^k k!)^{-\frac{1}{2}} z^k = e_k.$$

and the proof is complete. □

3. Computations for Hermite windows

Here, I present an outline of the proofs of the new results described above. Detailed proofs of these and some more general results will appear elsewhere. For $p(z)$ an arbitrary polynomial in $H^2(\mathbb{C}^n, d\mu)$, and β the Bargmann isometry discussed above, we have

Theorem 3.1. *For φ such that $e^{(\cdot)\cdot a/2}\varphi(\cdot)$ is in $L^2(\mathbb{C}^n,\ d\mu)$ for all a in $\mathbb{C}^n$, we have*

$$\begin{aligned}\beta L_\varphi^{w_1}\beta^{-1}p(z) &= \frac{1}{2}H_{\bar{z}_1}^* M_\varphi H_{\bar{z}_1}p(z)\\ &= \frac{1}{2}\{T_{\bar{z}_1}T_\varphi T_{z_1} + T_{z_1}T_\varphi T_{\bar{z}_1} - T_\varphi T_{z_1}T_{\bar{z}_1} - T_{z_1}T_{\bar{z}_1}T_\varphi\}p(z).\end{aligned}$$

There are several interesting consequences of Theorem 3.1.

Theorem 3.2. *For φ an arbitrary polynomial in z and $\bar{z}$, we have*

$$\beta L_\varphi^{w_1}\beta^{-1}p(z) = T_{\varphi+2\bar{\partial}_1\partial_1\varphi}p(z)$$

for all $p(z)$ polynomial in z.

Corollary 3.3. *For φ an arbitrary polynomial in z and $\bar{z}$, $L_\varphi^{w_1} = 0$ if and only if $\varphi = 0$.*

Corollary 3.4. *For φ, ψ polynomials in $z, \bar{z}$, we have $L_\varphi^{w_1} L_\psi^{w_1} = L_{\varphi * \psi}^{w_1}$ on the dense domain $\{\beta^{-1}p : p$ polynomial in $H^2(\mathbb{C}^n, d\mu)\}$ where*

$$\varphi * \psi = \sum_{j=0}^{\infty} (-2\bar{\partial}_1 \partial_1)^j [(\varphi + 2\bar{\partial}_1\partial_1\varphi) \diamond (\psi + 2\bar{\partial}_1\partial_1\psi)],$$

and with the associative product $\varphi \diamond \psi$ defined by (1).

For some bounded φ, we have analogous results *to some extent.*

Theorem 3.5. *For φ in $B_a(\mathbb{C}^n)$, we have*

$$\beta L_\varphi^{w_1} \beta^{-1} = T_{\varphi + 2\bar{\partial}_1\partial_1\varphi}.$$

Proof. Using Theorem 3.1 and the identity $W_a = e^{|a|^2/4} T_{\chi_a(\cdot)}$ from [BC1], we see that

$$\begin{aligned}
&\frac{1}{2} H_{\bar{z}_1}^* M_{\chi_a(\cdot)} H_{\bar{z}_1} p(z) \\
&\quad = \frac{1}{2} e^{-|a|^2/4} \{T_{\bar{z}_1} W_a T_{z_1} + T_{z_1} W_a T_{\bar{z}_1} - W_a T_{z_1} T_{\bar{z}_1} - T_{z_1} T_{\bar{z}_1} W_a\} p(z) \\
&\quad = e^{-|a|^2/4} \{\partial_1 [k_a(z)(z_1 - a_1) p(z-a)] + z_1 k_a(z) (\partial_1 p)(z-a) \\
&\qquad\qquad - k_a(z)(z_1 - a_1)(\partial_1 p)(z-a) - z_1 \partial_1 [k_a(z) p(z-a)]\} \\
&\quad = (1 - |a_1|^2/2) e^{-|a|^2/4} W_a p(z),
\end{aligned}$$

so that

$$\begin{aligned}
\frac{1}{2} H_{\bar{z}_1}^* M_{\chi_a(\cdot)} H_{\bar{z}_1} &= (1 - |a_1|^2/2) e^{-|a|^2/4} W_a \\
&= (1 - |a_1|^2/2) T_{\chi_a(\cdot)}.
\end{aligned}$$

Now, integrating both sides of this operator equation with respect to $d\sigma(a)$ for σ any complex-valued, compactly supported, bounded regular Borel measure on $\mathbb{C}^n$ gives

$$\frac{1}{2} H_{\bar{z}_1}^* M_\varphi H_{\bar{z}_1} = T_{\varphi + 2\bar{\partial}_1\partial_1\varphi}$$

for

$$\varphi(z) = \int_{\mathbb{C}^n} \chi_a(z) d\sigma(a),$$

and φ, $\bar{\partial}_1\partial_1\varphi$ are both in $B_a(\mathbb{C}^n)$. As usual [BC1], operator integrals are understood in the weak sense and judicious use of Fubini's Theorem is required. Note that

$$\varphi(z) + 2\bar{\partial}_1\partial_1\varphi(z) = \int_{\mathbb{C}^n} (1 - |a_1|^2/2) \chi_a(z) d\sigma(a)$$

by differentiation "under the integral". The desired result follows immediately from a further application of Theorem 3.1. □

Corollary 3.6. *For φ in $B_a(\mathbb{C}^n)$ with $\varphi(z) = \int_{\mathbb{C}^n} \chi_a(z)d\sigma(a)$ and support(σ) contained in $\{a : |a_1|^2 = 2\}$, we have $\varphi + 2\bar{\partial}_1\partial_1\varphi = 0$ so that $L_\varphi^{w_1} = 0$.*

Next, we turn our attention to the window

$$w_2 = w_{(2,0,0,\dots,0)} = 2^{-1/2}\pi^{-n/4}(2x_1{}^2 - 1)\exp\{-|x|^2/2\}.$$

Here, we have $\beta w_2 = 8^{-1/2}z_1{}^2$ and it is instructive to compare the analysis with the similar, but easier, case $w = w_1$. For $p(z)$ an arbitrary polynomial in $H^2(\mathbb{C}^n, d\mu)$, and β the Bargmann isometry discussed above, we can check

Theorem 3.7. *Let $A = -2(I-P)M_{\bar{z}_1}PM_{\bar{z}_1}P + (I-P)M_{\bar{z}_1^2}P$. For φ such that $e^{(\cdot)\cdot a/2}\varphi(\cdot)$ is in $L^2(\mathbb{C}^n, d\mu)$ for all a in $\mathbb{C}^n$, we have*

$$\begin{aligned}\beta L_\varphi^{w_2}\beta^{-1}p(z) &= \frac{1}{8}A^*M_\varphi Ap(z)\\ &= \frac{1}{8}\{4T_{z_1}T_{\bar{z}_1}T_\varphi T_{z_1}T_{\bar{z}_1} - 2T_{z_1}T_{\bar{z}_1}T_{\bar{z}_1}T_\varphi T_{z_1} - 2T_{\bar{z}_1}T_\varphi T_{z_1}T_{z_1}T_{\bar{z}_1}\\ &\quad - 2T_{z_1}T_{z_1}T_{\bar{z}_1}T_\varphi T_{\bar{z}_1} - 2T_{z_1}T_\varphi T_{z_1}T_{\bar{z}_1}T_{\bar{z}_1} + T_{\bar{z}_1}T_{\bar{z}_1}T_\varphi T_{z_1}T_{z_1}\\ &\quad + T_\varphi T_{z_1}T_{z_1}T_{\bar{z}_1}T_{\bar{z}_1} + T_{z_1}T_{z_1}T_\varphi T_{\bar{z}_1}T_{\bar{z}_1} + T_{z_1}T_{z_1}T_{\bar{z}_1}T_{\bar{z}_1}T_\varphi\}p(z).\end{aligned}$$

We can now obtain analogues of our earlier results for $w = w_1$.

Theorem 3.8. *For φ an arbitrary polynomial in z and $\bar{z}$, we have*

$$\beta L_\varphi^{w_2}\beta^{-1}p(z) = T_{\varphi + 4\bar{\partial}_1\partial_1\varphi + 2(\bar{\partial}_1\partial_1)^2\varphi}p(z),$$

for all $p(z)$ polynomial in z.

Corollary 3.9. *For φ an arbitrary polynomial in z and $\bar{z}$, $L_\varphi^{w_2} = 0$ if and only if $\varphi = 0$.*

It is elementary to obtain a formula for $L_\varphi^{w_2}L_\psi^{w_2}$ analogous to that obtained for $L_\varphi^{w_1}L_\psi^{w_1}$ in Corollary 3.4 to Theorem 3.2. I now provide, for the record, the analogue of Theorem 3.5.

Theorem 3.10. *For φ in $B_a(\mathbb{C}^n)$, we have*

$$\beta L_\varphi^{w_2}\beta^{-1} = T_{\varphi + 4\bar{\partial}_1\partial_1\varphi + 2(\bar{\partial}_1\partial_1)^2\varphi}.$$

Corollary 3.11. *For φ in $B_a(\mathbb{C}^n)$ with $\varphi(z) = \int_{\mathbb{C}^n} \chi_a(z)d\sigma(a)$ and support(σ) contained in $\{a : |a_1|^2 = 4 \pm 2\sqrt{2}\}$, we have $\varphi + 4\bar{\partial}_1\partial_1\varphi + 2(\bar{\partial}_1\partial_1)^2\varphi = 0$ so that $L_\varphi^{w_2} = 0$.*

References

[B] V. Bargmann, *On a Hilbert space of analytic functions and an associated integral transform*, Communications on Pure and Applied Mathematics, **14** (1961), 187–214.

[Be] F. A. Berezin, *Covariant and contravariant symbols of operators*, Math. USSR Izv., **6** (1972), 1117–1151.

[BC1] C. A. Berger and L. A. Coburn, *Toeplitz operators on the Segal–Bargmann space*, Transactions AMS, **301** (1987), 813–829.

[BC2] C. A. Berger and L. A. Coburn, *Heat Flow and Berezin–Toeplitz Estimates*, American Journal of Mathematics, **116** (1994), 563–590.

[Bo] D. Borthwick, *Microlocal techniques for semiclassical problems in geometric quantization*, Perspectives on Quantization (editors: L. A. Coburn and M. A. Rieffel), Contemporary Math., **214**, AMS, Providence, R.I., 1998, 23–37.

[C1] L. A. Coburn, *The measure algebra of the Heisenberg group*, Journal of Functional Analysis, **161** (1999), 509–525.

[C2] L. A. Coburn, *On the Berezin–Toeplitz calculus*, to appear in Proceedings of the AMS.

[CX] L. A. Coburn and J. Xia, *Toeplitz algebras and Rieffel deformations*, Communications in Mathematical Physics, **168** (1995), 23–38.

[D1] I. Daubechies, *Time frequency localization operators: A geometric phase space approach*, IEEE Transactions on Information Theory, **34** (1988), 605–612.

[D2] I. Daubechies, *The wavelet transform, time-frequency localization and signal analysis*, IEEE Transactions on Information Theory, **36** (1990), 961–1005.

[D3] I. Daubechies, *Ten Lectures on Wavelets*, CBMS-NSF Regional Conference Series 6, SIAM, Philadelphia, 1992.

[DFN] F. De Mari, H. G. Feichtinger and K. Nowak, *Uniform eigenvalue estimates for time-frequency localization operators*, preprint.

[DW] J. Du and M. W. Wong, *Gaussian functions and Daubechies operators*, Integral Equations and Operator Theory, **38** (2000), 1–8.

[FN] H. G. Feichtinger and K. Nowak, *A Szego type theorem for Gabor–Toeplitz localization operators*, preprint.

[F] G. B. Folland, *Harmonic analysis in phase space*, Annals of Mathematics Studies, Princeton Univ. Press, Princeton, 1989.

[G] V. Guillemin, *Toeplitz operators in n-dimensions*, Integral Equations and Operator Theory, **7** (1984), 145–205.

[S] G. Szegö, *Orthogonal polynomials*, AMS Colloquium Publications, **23**, AMS, Providence, RI., 1975.

[XZ] J. Xia and D. Zheng, *Standard deviation and Schatten class Hankel operators on the Segal–Bargmann space*, preprint.

2000 Mathematics Subject Classification: 47B35

Department of Mathematics,
SUNY at Buffalo,
Buffalo, New York 14260, USA
E-mail address: lcoburn@acsu.buffalo.edu

Stabilization of irrational transfer functions by controllers with internal loop

Ruth F. Curtain, George Weiss, and Martin Weiss

Abstract. Transfer functions are called well-posed if they are bounded and analytic on some right half-plane. Two concepts of L^2-stabilization are analyzed for well-posed transfer functions: the usual one, as in Figure 3, and a more general one called stabilization with internal loop, see Figure 4. In both cases we obtain a complete parametrization of all stabilizing controllers in terms of a doubly coprime factorization of the original transfer function. Moreover, the connection between the two stabilization concepts is clarified. We analyze two special subclasses of stabilizing controllers with internal loop, called canonical and dual canonical controllers. We show that, in a certain sense, every stabilizing controller with internal loop is equivalent to a canonical controller, and also to a dual canonical controller.

1. An intriguing control problem

We consider the following problem: For an unstable plant with transfer function

$$\mathbf{P}(s) = 3\frac{s+1}{s-1},$$

design a controller that performs the following two tasks:

(1) it stabilizes $\mathbf{P}$,

(2) even if an external (unpredictable) disturbance d is added to the input u of the plant, the controller ***predicts exactly*** the output y of the plant one second ahead. In other words, at some auxiliary output of our controller we can at any moment of time t read off the value $y(t+1)$.

This might seem to be impossible at first sight since, as we said, d is not predictable. The problem is illustrated in Figure 1.

Let us connect our plant to a system consisting of a delay line and a summation point, as shown in Figure 2. This interconnection has two inputs and two outputs: one input is the disturbance d and one output is the output y of the plant. The other input is denoted by ζ_i and the other output is denoted by ζ_o. Realizing this interconnection changes hardly anything: the plant still works in

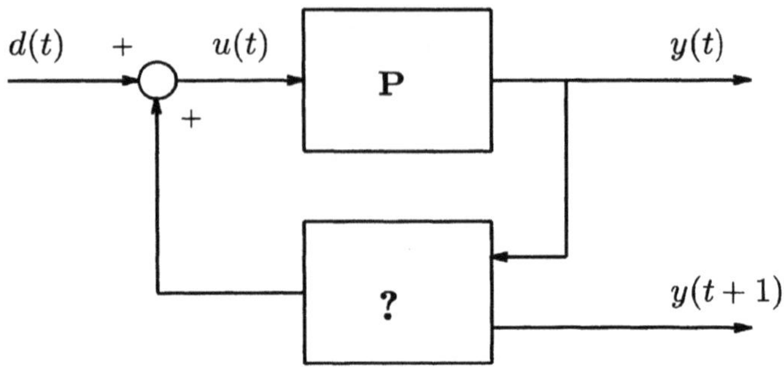

FIGURE 1. A seemingly impossible problem

open loop, with the input $u = d + \zeta_i$. The transfer function from ζ_i to ζ_o is $\mathbf{H}(s) = 1 - e^{-s} + \mathbf{P}(s)$.

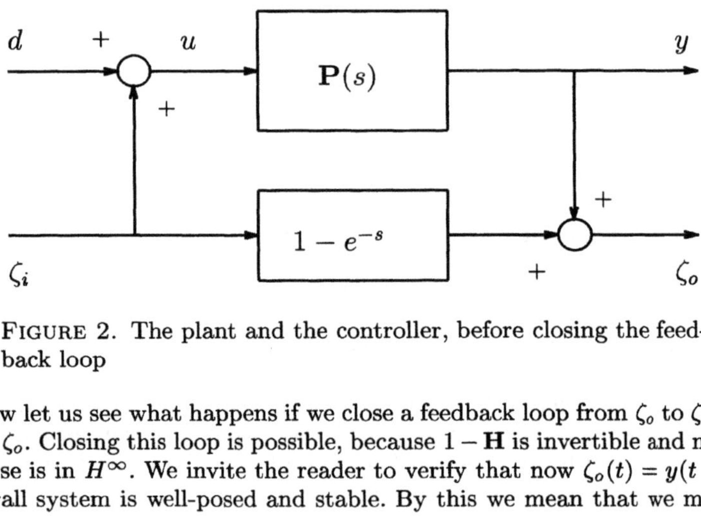

FIGURE 2. The plant and the controller, before closing the feedback loop

Now let us see what happens if we close a feedback loop from ζ_o to ζ_i, i.e., we set $\zeta_i = \zeta_o$. Closing this loop is possible, because $1 - \mathbf{H}$ is invertible and moreover, its inverse is in H^∞. We invite the reader to verify that now $\zeta_o(t) = y(t+1)$ and the overall system is well-posed and stable. By this we mean that we may inject additional input signals at any point, and the transfer function from any such input signal to any other signal will be in H^∞. Thus, we have solved our original problem.

We emphasize the importance of proceeding in stages: first connecting the device to the plant, and only afterwards closing the feedback loop. Proceeding in the wrong order, i.e., trying to connect ζ_i to ζ_o first, will not work: it is an ill-posed feedback. In closed loop, the transfer function from y to ζ_o is e^s, but realizing this transfer function in open loop (i.e., without being connected to the plant) is clearly impossible. This is an example of a so-called ***controller with internal loop***, introduced in Weiss and Curtain [17]. We study the structure of such controllers in the next sections, and we return to this example at the end of Section 4.

2. Stabilization with internal loop

In this paper, we investigate the concept of a stabilizing controller, as well as the problem of parametrizing all stabilizing controllers for a well-posed (possibly irrational) transfer function. Our analysis is entirely in the frequency domain.

Throughout this paper, U and Y are Hilbert spaces, called the ***input space*** and the ***output space***, respectively. An $\mathcal{L}(U,Y)$-valued transfer function is called ***well-posed*** if it is bounded and analytic on some right half-plane (where Re $s > \alpha$). This class of functions is a natural generalization of the proper rational transfer functions and moreover, it coincides with the transfer functions of well-posed linear systems, a class for which there exists a well developed state space theory (see Salamon [7, 8], Staffans [10, 11, 12], Weiss [15, 16] and Weiss and Rebarber [18]). We do not distinguish between two transfer functions defined on two different right half-planes if one function is a restriction of the other (thus, by a transfer function we mean, in fact, an equivalence class of analytic functions). We mention that matrix-valued transfer functions with entries in the factor space of H^∞, used for example in Georgiou and Smith [6] and in Foias, Özbay and Tannenbaum [4], are neither contained in, nor do they contain the matrix-valued well-posed transfer functions.

For any Banach space Z, we denote by $H^\infty(Z)$ the Banach space of Z-valued bounded analytic functions on the usual right half-plane (where Re $s > 0$), with the sup norm. We write H^∞ instead of $H^\infty(Z)$, if the space Z is clear from the context. An $\mathcal{L}(U,Y)$-valued transfer function $\mathbf{P}$ is called ***stable*** if $\mathbf{P} \in H^\infty(\mathcal{L}(U,Y))$. This concept of stability is also called ***L^2-stability*** (or ***input-output stability***), because it is equivalent to the property that input functions $u \in L^2([0,\infty),U)$ are mapped into output functions $y \in L^2([0,\infty),Y)$, via the formula $\hat{y} = \mathbf{P}\hat{u}$, where a hat is used to indicate the Laplace transformation.

Definition 2.1. *Assume that* $\mathbf{P}$ *and* $\mathbf{C}$ *are well-posed transfer functions with values in* $\mathcal{L}(U,Y)$ *and* $\mathcal{L}(Y,U)$*, respectively. We say that* $\mathbf{C}$ *is an* ***admissible feedback transfer function*** *for* $\mathbf{P}$ *if* $I-\mathbf{CP}$ *(or equivalently* $I-\mathbf{PC}$*) has a well-posed inverse. In particular, we say that* $\mathbf{C}$ ***stabilizes*** $\mathbf{P}$ *if*

$$\begin{bmatrix} I & -\mathbf{C} \\ -\mathbf{P} & I \end{bmatrix}^{-1} \in H^\infty(\mathcal{L}(U \times Y)).$$

The intuitive interpretation of the last condition can be given by the block diagram in Figure 3, where the transfer function from $\begin{bmatrix} v_p \\ v_c \end{bmatrix}$ to $\begin{bmatrix} u_p \\ u_c \end{bmatrix}$ is

$$\mathbf{T}_{\mathbf{P},\mathbf{C}} = \begin{bmatrix} I & -\mathbf{C} \\ -\mathbf{P} & I \end{bmatrix}^{-1} = \begin{bmatrix} (I-\mathbf{CP})^{-1} & \mathbf{C}(I-\mathbf{PC})^{-1} \\ \mathbf{P}(I-\mathbf{CP})^{-1} & (I-\mathbf{PC})^{-1} \end{bmatrix}. \tag{2.1}$$

A few words about the connection with state space theory: If **P** and **C** are the transfer functions of two well-posed linear systems and their feedback connection is exponentially stable, then **C** stabilizes **P**, as is easy to see. The more surprizing fact is that the converse is true as well, provided that the two systems satisfy certain natural assumptions, such as regularity, stabilizability and detectability, see Section 4 of Weiss and Curtain [17]. Recently, these assumptions have been replaced by the weaker conditions of optimizability and detectability (without assuming regularity), see Section 6 of Weiss and Rebarber [18]. For the state space theory of coprime factorizations we refer to Staffans [10] and to our paper [3].

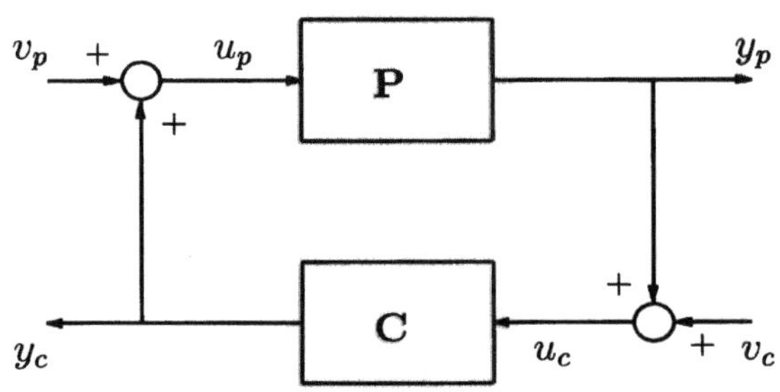

FIGURE 3. The feedback connection of **P** and **C**

While extending the theory of dynamic stabilization to regular linear systems (a subclass of the well-posed linear systems) in [17], it became apparent that a more general concept of stabilizing controller is needed. Indeed, it was shown in Example 6.5 of [17] that even the standard observer-based controller is not, in general, a stabilizing controller in the usual sense; that is, it is not a well-posed linear system and, correspondingly, its transfer function is not well-posed. However, the observer-based controller may be regarded in a natural way as a stable well-posed linear system with two inputs and two outputs. First we connect one input and one output to the plant, obtaining a new well-posed linear system. Next we connect the remaining input of the controller to its remaining output, so obtaining a stable closed-loop system. For the details of this procedure see [17].

Motivated by this fact, a new type of controller was introduced, the so-called *stabilizing controller with internal loop*, see Definition 4.10 in [17]. This was later used in [3], Townley *et al* [13] and [18]. We now formulate the frequency domain counterpart of this concept. In fact, we introduce two new concepts, which generalize the two concepts from Definition 2.1.

Definition 2.2. *Let U, Y and R be Hilbert spaces. Let* $\mathbf{P}$ *and* $\mathbf{K} = \begin{bmatrix} \mathbf{K}_{11} & \mathbf{K}_{12} \\ \mathbf{K}_{21} & \mathbf{K}_{22} \end{bmatrix}$ *be well-posed transfer functions with values in* $\mathcal{L}(U,Y)$ *and in* $\mathcal{L}(Y \times R, U \times R)$, *respectively, and denote* $\mathbf{G} = \begin{bmatrix} \mathbf{P} & 0 \\ 0 & \mathbf{K} \end{bmatrix}$. *We say that* $\mathbf{K}$ *is an* admissible feedback transfer function with internal loop *for* $\mathbf{P}$ *if* $I - F\mathbf{G}$ *has a well-posed inverse, where*

$$F = \begin{bmatrix} 0 & I & 0 \\ I & 0 & 0 \\ 0 & 0 & I \end{bmatrix}, \qquad F \in \mathcal{L}(Y \times U \times R, U \times Y \times R). \tag{2.2}$$

In particular, we say that $\mathbf{K}$ stabilizes $\mathbf{P}$ with internal loop *if*

$$(I - F\mathbf{G})^{-1} \in H^\infty(\mathcal{L}(U \times Y \times R)).$$

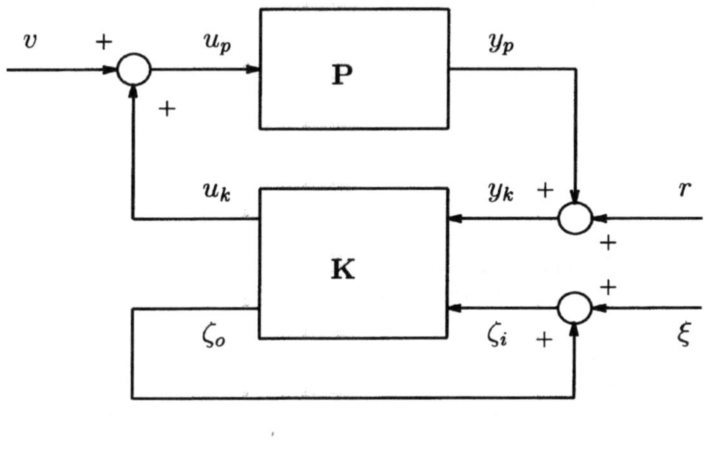

FIGURE 4. The plant $\mathbf{P}$ connected to a controller $\mathbf{K}$ with internal loop

The intuitive interpretation of Definition 2.2 is the following: $\mathbf{P}$ represents the plant and $\mathbf{K}$ is the transfer function of the controller in Figure 4 from $\begin{bmatrix} y_k \\ \zeta_i \end{bmatrix}$ to $\begin{bmatrix} u_k \\ \zeta_o \end{bmatrix}$, when all the connections are open. The connection from ζ_o to ζ_i is the so-called *internal loop*. The admissibility part of the definition means that it is possible to close all the connections in Figure 4, and obtain well-posed transfer functions from the three external inputs (v, r and ξ) to all the other signals. The stabilization part of the definition means that in fact all these closed-loop transfer functions are stable. To understand that this is indeed what the definition says, it is helpful to redraw Figure 4 by clearly indicating the role of F as a feedback operator for $\mathbf{G}$, the parallel connection of $\mathbf{P}$ and $\mathbf{C}$, as is done in Figure 5.

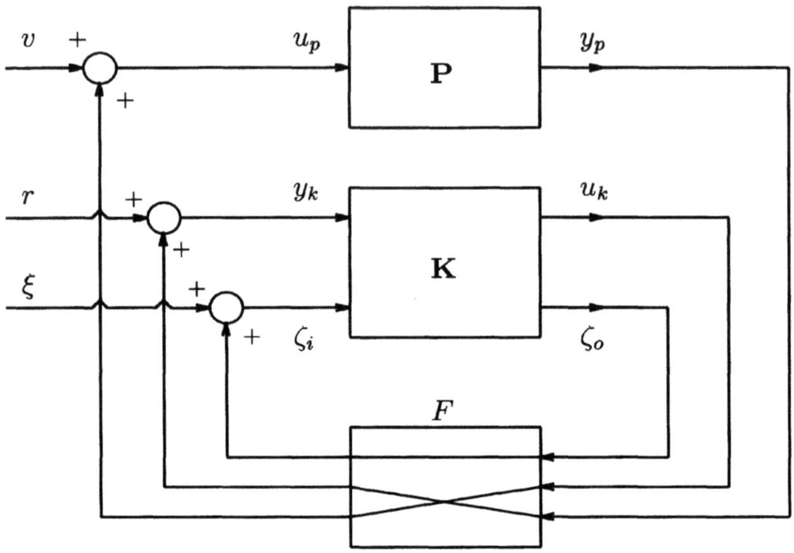

FIGURE 5. The parallel connection of $\mathbf{P}$ and $\mathbf{K}$ with the feedback operator F. This diagram is equivalent to the one in Figure 4.

If $I - \mathbf{K}_{22}$ has a well-posed inverse, then the two concepts introduced in Definition 2.2 reduce to those introduced in Definition 2.1. Indeed, in this case the internal loop can be closed first (i.e., before connecting the controller to the plant), to obtain a conventional controller with the following transfer function from y_k to u_k:

$$\mathbf{C} = \mathbf{K}_{11} + \mathbf{K}_{12}(I - \mathbf{K}_{22})^{-1}\mathbf{K}_{21}. \tag{2.3}$$

In many cases, the controller $\mathbf{C}$ defined by (2.3) will not be well-posed in the sense of our definition: for example, it could be improper or even anticausal. However, the internal loop construction is not merely an artifice for allowing improper or anticausal controllers. It includes controllers for which the expresssion (2.3) *is not defined at all* (this can happen if $I - \mathbf{K}_{22}$ is nowhere invertible). A finite-dimensional single input-single output (SISO) and physically meaningful example of a stabilizing controller with internal loop for which $\mathbf{C}$ does not exist was given in Section 1 of [17]. Another, infinite-dimensional SISO example follows in this section.

The connection of the stabilization concept in Definition 2.2 to the state space theory of stabilizing controllers with internal loop (see Definition 4.10 in [17]) is similar to the one we explained for the stabilization concept in Definition 2.1: If $\mathbf{P}$ is the transfer function of a well-posed plant and $\mathbf{K}$ is the transfer function of a stabilizing controller with internal loop for the same plant, then $\mathbf{K}$ stabilizes $\mathbf{P}$ with internal loop. The converse holds whenever the two systems satisfy certain natural assumptions, see Proposition 4.11 in [17] (where the product FG is denoted by $\mathbf{L}$). A stronger version of this result from [17] is Theorem 6.4 in [18], where the assumptions are optimizability and estimatability (but not regularity).

For the remainder of this paper, if $\mathbf{P}$ and $\mathbf{K}$ are as in Definition 2.2, we shall call $\mathbf{K}$ a *stabilizing controller with internal loop for* $\mathbf{P}$. This clearly contradicts the terminology of [17], since $\mathbf{K}$ is only the transfer function of a controller, and not the controller itself. However, since our discussion here is only in terms of transfer functions, this will (hopefully) not lead to confusions.

Example 2.3. *We take $U = Y = R = \mathbb{C}$ and*

$$\mathbf{P}(s) = \frac{2}{1+e^{-2s}}, \qquad \mathbf{K} = \begin{bmatrix} 0 & 1 \\ -1 & 1 \end{bmatrix}.$$

It is easy to see that the transfer function (2.3) *of the controller is undefined since $I - \mathbf{K}_{22} = 0$. However, it is not difficult to check that $\mathbf{K}$ stabilizes $\mathbf{P}$ with internal loop (this verification can be simplified considerably by using Theorem 4.2). The plant is unstable: $\mathbf{P}$ has infinitely many poles on the imaginary axis. All these facts can be understood intuitively from the following physical interpretation.*

Consider a circuit composed of a unit resistor connected in series with a lossless transmission line of unit length, with unit distributed capacitance and unit distributed inductance. One end of the transmission line is earthed, as shown in Figure 6. The input of this system is the current through the resistor denoted by $u(t)$, whereas the observation is chosen to be the input voltage of the circuit $y(t)$.

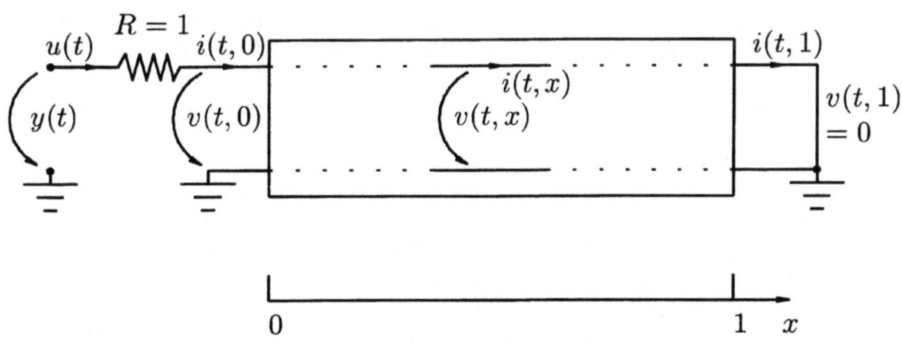

FIGURE 6. Transmission line with resistor

The local current $i(t,x)$ through and the local voltage $v(t,x)$ across the transmission line are related by the equations

$$\frac{\partial i}{\partial t} = -\frac{\partial v}{\partial x}, \qquad \frac{\partial v}{\partial t} = -\frac{\partial i}{\partial x}, \tag{2.4}$$

with the boundary conditions

$$i(t,0) = u(t),$$

$$v(t,1) = 0 \quad \textit{(because of the earthing)},$$

$$y(t) = u(t) + v(t,0) \quad \textit{(u(t) is added because of the resistor)}.$$

It is readily verified that the transfer function from u to y, the impedance of the circuit, is $\mathbf{P}(s) = \frac{2}{1+e^{-2s}}$. The infinite number of poles on the imaginary axis can be explained by the fact that disconnecting the input of the circuit, i.e., $u(t) = 0$, creates infinitely many undamped oscillatory modes on the line, due to the lossless nature of the line. The only dissipative element, the resistor, cannot dissipate the energy of these oscillations if no current traverses it. A simple interpretation of the proposed controller is the earthing of the input of the circuit. This causes $y(t) = 0$, something that is clearly not achievable by a conventional controller. It is clear that the system does become stable under the new boundary condition, since by earthing the input, the resistor will now be able to dissipate the energy from the circuit. All these intuitive conclusions will be justified rigorously in Section 2.

Another motivation for introducing controllers with internal loop is to obtain a clean and elegant Youla parametrization. In the Youla parametrization of conventional controllers, if the plant is not strictly proper, it is difficult to see a priori how to choose the parameter in such a way that the resulting controller will be well-posed. Even if we choose to ignore well-posedness, as some researchers do, we still have to ensure that the denominator in the Youla parametrization is invertible. Thus, the parametrization is not clean; there is always an extra condition on the parameter. This in turn makes it awkward to use this parametrization to solve, for example, the H^∞-control problem for well-posed linear systems. By contrast, we do obtain a clean parametrization for all stabilizing canonical or dual canonical controllers. A stabilizing controller with internal loop $\mathbf{K}$ is called *canonical* if $\mathbf{K}_{11} = 0$, $\mathbf{K}_{12} = I$ and $\mathbf{K}_{21}, \mathbf{K}_{22} \in H^\infty$. It is called *dual canonical* if $\mathbf{K}_{11} = 0$, $\mathbf{K}_{21} = I$ and $\mathbf{K}_{12}, \mathbf{K}_{22} \in H^\infty$. We shall prove that, in a certain sense, it is sufficient to consider canonical or dual canonical controllers (Theorem 5.2).

The organization of the paper is as follows. In Section 3, we recall the usual Youla parameterization of all stabilizing controllers in terms of a doubly coprime factorization (cf. Francis [5], Vidyasagar [14] and Baras [1]). However, since we consider the input and output spaces to be Hilbert spaces, certain technical difficulties arise and we actually need to redo the proofs for this situation. We also give two examples to illustrate the invertibility problems with this parametrization.

In Section 4 we introduce canonical and dual canonical controllers. We show that a plant $\mathbf{P}$ is stabilizable with internal loop by a canonical (dual canonical) controller if and only if $\mathbf{P}$ has a right-coprime (left-coprime) factorization. We give a complete parameterization of all (dual) canonical stabilizing controllers with internal loop. Moreover, the relationship between (conventional) stabilization and stabilization with internal loop by a (dual) canonical controller is clarified.

In Section 5 we introduce the concept of equivalence for stabilizing controllers with internal loop. We show that if the plant $\mathbf{P}$ has a doubly coprime factorization, then any stabilizing controller with internal loop for $\mathbf{P}$ is equivalent to a canonical one as well as to a dual canonical one.

3. Coprime factorization

This section contains more or less well-known material on doubly coprime factorizations, only the statements are slightly different, and the proofs are a little more involved, because we consider operator-valued functions. Thus, we must be careful to distinguish between right and left invertibility, to assume the existence of coprime factorizations, etc. The reader who is not interested in such technicalities may skip to the examples at the end of the section.

Definition 3.1. *Let* $\mathbf{P}$ *be a well-posed transfer function. A* right-coprime factorization of $\mathbf{P}$ over H^∞ *is an ordered set of four operator-valued functions* M, N, $\tilde{R}$, $\tilde{S}$ *in* H^∞ *such that* M *has a well-posed inverse,* $\mathbf{P} = NM^{-1}$, *and* $\tilde{S}M - \tilde{R}N = I$.

A left-coprime factorization of $\mathbf{P}$ over H^∞ *is an ordered set of four operator-valued functions* $\tilde{M}$, $\tilde{N}$, R, S *in* H^∞ *such that* $\tilde{M}$ *has a well-posed inverse,* $\mathbf{P} = \tilde{M}^{-1}\tilde{N}$, *and* $\tilde{M}S - \tilde{N}R = I$.

A doubly coprime factorization of $\mathbf{P}$ *over* H^∞ *is an ordered set of eight operator-valued functions* $M, N, R, S, \tilde{M}, \tilde{N}, \tilde{R}, \tilde{S}$ *in* H^∞ *such that, on some right half-plane,* M *and* $\tilde{M}$ *have well-posed inverses,* $\mathbf{P}$ *has the factorizations*

$$\mathbf{P} = NM^{-1} = \tilde{M}^{-1}\tilde{N}, \tag{3.5}$$

and

$$\begin{bmatrix} M & R \\ N & S \end{bmatrix}^{-1} = \begin{bmatrix} \tilde{S} & -\tilde{R} \\ -\tilde{N} & \tilde{M} \end{bmatrix}. \tag{3.6}$$

If a left- and a right-coprime factorization of $\mathbf{P}$ are given, as in (3.5), then it is easy to construct $R, S, \tilde{R}, \tilde{S} \in H^\infty$ such that we have a doubly coprime factorization, as in (3.6), see Theorem 60 in Vidyasagar [14] or Lemma 4.3 in Staffans [10]. For a matrix-valued function, (3.6) is often written in the form

$$\begin{bmatrix} \tilde{S} & -\tilde{R} \\ -\tilde{N} & \tilde{M} \end{bmatrix}\begin{bmatrix} M & R \\ N & S \end{bmatrix} = \begin{bmatrix} I & 0 \\ 0 & I \end{bmatrix}. \tag{3.7}$$

Indeed, for matrix-valued functions, (3.7) is equivalent to (3.6), but in general it is weaker than (3.6). Most arguments about doubly coprime factorizations rely on (3.7), but occasionally we need the factors from (3.7) in reversed order, see for example the proof of Theorem 3.4.

Notice that Definition 3.1 extends easily to transfer functions that are not necessarily well-posed, but are matrix-valued with components in the quotient field of H^∞, denoted by H^∞/H^∞. Smith [9] showed that a transfer function in this class is stabilizable if and only if it has left- and right-coprime factorizations. For operator-valued transfer functions, no similar result seems to be known. If a matrix-valued well-posed transfer function has a left- or a right-coprime factorization, or if it is stabilizable, then obviously its entries are in H^∞/H^∞.

Explicit formulas for coprime factorizations in terms of state space realizations are known for a wide class of infinite-dimensional systems called regular linear systems, see our paper [3]. More general but (unavoidably) somewhat less explicit formulas for well-posed linear systems were given by Staffans [10]. A modest, but useful formula in terms of transfer functions is the following.

Proposition 3.2. *Let* $\mathbf{P}$ *and* $\mathbf{C}$ *be well-posed transfer functions with values in* $\mathcal{L}(U,Y)$ *and* $\mathcal{L}(Y,U)$ *respectively. If*

$$\mathbf{C} \in H^\infty \quad \textit{and} \quad \mathbf{P}(I-\mathbf{CP})^{-1} \in H^\infty, \tag{3.8}$$

then denoting

$$\mathbf{L} = \begin{bmatrix} \mathbf{C} \\ I \end{bmatrix} \mathbf{P}(I-\mathbf{CP})^{-1}[I \ \ -\mathbf{C}],$$

the following is a doubly coprime factorization of $\mathbf{P}$*:*

$$\begin{bmatrix} M & R \\ N & S \end{bmatrix} = I + \mathbf{L}, \qquad \begin{bmatrix} \tilde{S} & -\tilde{R} \\ -\tilde{N} & \tilde{M} \end{bmatrix} = I - \mathbf{L}. \tag{3.9}$$

Proof. It is easy to see that $\mathbf{L}^2 = 0$, so that $(I+\mathbf{L})(I-\mathbf{L}) = (I-\mathbf{L})(I+\mathbf{L}) = I$, which is (3.6). To check (3.5), we note that $M = (I-\mathbf{CP})^{-1}$ and $N = \mathbf{P}(I-\mathbf{CP})^{-1}$, whence $\mathbf{P} = NM^{-1}$. The identity $\mathbf{P} = \tilde{M}^{-1}\tilde{N}$ is verified similarly. □

Lemma 3.3. *Let* $\mathbf{P}$ *and* $\mathbf{C}$ *be well-posed transfer functions with values in* $\mathcal{L}(U,Y)$ *and* $\mathcal{L}(Y,U)$ *respectively and suppose that* $\mathbf{P}$ *has a left coprime factorization* $\mathbf{P} = \tilde{M}^{-1}\tilde{N}$ *and* $\mathbf{C}$ *has a right-coprime factorization* $\mathbf{C} = WV^{-1}$*. Then* $\mathbf{C}$ *stabilizes* $\mathbf{P}$ *if and only if* $\Delta = \tilde{M}V - \tilde{N}W$ *is invertible over* H^∞*.*

There is an obvious dual statement in which right is replaced by left.

Proof. Let $R, S, \tilde{R}_c, \tilde{S}_c$ be such that

$$\tilde{M}S - \tilde{N}R = I \qquad \text{and} \qquad \tilde{S}_c W - \tilde{R}_c V = I. \tag{3.10}$$

Notice that $\Delta = \tilde{M}V - \tilde{N}W = \tilde{M}(I-\mathbf{PC})V$.

Suppose first that $\mathbf{C}$ stabilizes $\mathbf{P}$. Then clearly Δ^{-1} exists and

$$\Delta^{-1} = V^{-1}(I-\mathbf{PC})^{-1}\tilde{M}^{-1}.$$

It is readily verified that, with the notation from (2.1),

$$\begin{aligned} \mathbf{T}_{\mathbf{P},\mathbf{C}} &= \begin{bmatrix} I+\mathbf{C}(I-\mathbf{PC})^{-1}\mathbf{P} & \mathbf{C}(I-\mathbf{PC})^{-1} \\ (I-\mathbf{PC})^{-1}\mathbf{P} & (I-\mathbf{PC})^{-1} \end{bmatrix} \\ &= \begin{bmatrix} I+W\Delta^{-1}\tilde{N} & W\Delta^{-1}\tilde{M} \\ V\Delta^{-1}\tilde{N} & V\Delta^{-1}\tilde{M} \end{bmatrix} \end{aligned} \tag{3.11}$$

and since $\mathbf{T}_{\mathbf{P},\mathbf{C}} \in H^\infty$, all components are in H^∞.

Now post-multiply the second identity in (3.10) by $\Delta^{-1}\tilde{M}$ to obtain

$$\tilde{S}_c W \Delta^{-1}\tilde{M} - \tilde{R}_c V \Delta^{-1}\tilde{M} = \Delta^{-1}\tilde{M},$$

which shows that $\Delta^{-1}\tilde{M} \in H^\infty$. Similarly, post-multiplying the second identity in (3.10) by $\Delta^{-1}\tilde{N}$ shows that $\Delta^{-1}\tilde{N} \in H^\infty$. Next, pre-multiplying the first identity in (3.10) by Δ^{-1} gives

$$\Delta^{-1}\tilde{M}S - \Delta^{-1}\tilde{N}R = \Delta^{-1}$$

and hence $\Delta^{-1} \in H^\infty$.

Conversely, if $\Delta^{-1} \in H^\infty$, then we see from $\Delta = \tilde{M}(I - \mathbf{PC})V$ that $(I - \mathbf{PC})^{-1}$ exists, and (3.11) shows that $\mathbf{C}$ stabilizes $\mathbf{P}$. □

The following theorem is the well known Youla-Bongiorno parameterization.

Theorem 3.4. *Suppose that the well-posed transfer function* $\mathbf{P}$ *has a doubly coprime factorization over* H^∞, *as in* (3.5) and (3.6).

(1) *Assume that* $\mathbf{C}$ *is a well-posed transfer function that stabilizes* $\mathbf{P}$. *Then* $\mathbf{C}$ *has a right-coprime factorization over* H^∞ *if and only if* $\mathbf{C}$ *has a left-coprime factorization over* H^∞. *In this case, there exists a unique* $Q \in H^\infty$ *such that* $S + NQ$ *and* $\tilde{S} + Q\tilde{N}$ *have well-posed inverses and*

$$\mathbf{C} = (R + MQ)(S + NQ)^{-1} = (\tilde{S} + Q\tilde{N})^{-1}(\tilde{R} + Q\tilde{M}) \tag{3.12}$$

are right- and left-coprime factorizations of $\mathbf{C}$.

(2) *Conversely, if* $Q \in H^\infty$ *is such that one of* $S + NQ$ *or* $\tilde{S} + Q\tilde{N}$ *has a well-posed inverse, then the other also has a well-posed inverse and the equality in* (3.12) *holds. In this case, the transfer function* $\mathbf{C}$ *defined in* (3.12) *stabilizes* $\mathbf{P}$.

Proof. We prove part (1). Suppose that $\mathbf{C}$ stabilizes $\mathbf{P}$ and $\mathbf{C} = WV^{-1}$ is a right-coprime factorization. We show that $\mathbf{C}$ can be represented as in the first factorization from (3.12). By Lemma 3.3, $\Delta = \tilde{M}V - \tilde{N}W$ is invertible over H^∞, so

$$Q = M^{-1}(W\Delta^{-1} - R) \tag{3.13}$$

is well defined and, moreover,

$$R + MQ = W\Delta^{-1}. \tag{3.14}$$

We calculate

$$\begin{aligned}
S + NQ &= S + NM^{-1}(W\Delta^{-1} - R) && \text{(from (3.13))}\\
&= S + \tilde{M}^{-1}\tilde{N}(W\Delta^{-1} - R) && \text{(since } \tilde{M}^{-1}\tilde{N} = NM^{-1}\text{)}\\
&= \tilde{M}^{-1}(\tilde{M}S + \tilde{N}W\Delta^{-1} - \tilde{N}R) &&\\
&= \tilde{M}^{-1}(I + \tilde{N}W\Delta^{-1}) && \text{(from (3.6))}\\
&= \tilde{M}^{-1}(\tilde{M}V - \tilde{N}W + \tilde{N}W)\Delta^{-1} && \text{(by the definition of } \Delta\text{).}
\end{aligned}$$

Thus

$$S + NQ = V\Delta^{-1} \tag{3.15}$$

and $S+NQ$ has a well-posed inverse, since V has one and $\Delta \in H^\infty$.

Furthermore, from (3.14) and (3.15) we obtain

$$\begin{aligned}\mathbf{C} = WV^{-1} &= W\Delta^{-1}\Delta V^{-1}\\ &= (R+MQ)(S+NQ)^{-1}.\end{aligned}$$

We have to prove that $Q \in H^\infty$. This follows from

$$\begin{aligned} Q &= (\tilde{S}M - \tilde{R}N)Q && \text{(from (3.6))}\\ &= \tilde{S}(W\Delta^{-1} - R) - \tilde{R}(V\Delta^{-1} - S) && \text{(from (3.14) and (3.15))}\\ &= (\tilde{S}W - \tilde{R}V)\Delta^{-1} && \text{(from (3.6))}.\end{aligned}$$

Next, we show that $\tilde{S} + Q\tilde{N}$ has a well-posed inverse and the two factorizations in (3.12) are equal. From (3.7), pre-multiplying by $\begin{bmatrix} I & -Q \\ 0 & I \end{bmatrix}$ and post-multiplying by $\begin{bmatrix} I & Q \\ 0 & I \end{bmatrix}$, we obtain

$$\begin{bmatrix} \tilde{S}+Q\tilde{N} & -\tilde{R}-Q\tilde{M} \\ -\tilde{N} & \tilde{M} \end{bmatrix} \begin{bmatrix} M & R+MQ \\ N & S+NQ \end{bmatrix} = \begin{bmatrix} I & 0 \\ 0 & I \end{bmatrix}. \tag{3.16}$$

By reversing the order of the factors in (3.7) and using a similar device we can obtain a version of (3.16) with the factors in the reverse order. This proves that $X = \begin{bmatrix} \tilde{S}+Q\tilde{N} & -\tilde{R}-Q\tilde{M} \\ -\tilde{N} & \tilde{M} \end{bmatrix}$ has a well-posed inverse. (Note that this does not follow from (3.16) alone, since we work on infinite-dimensional spaces.)

Using (3.7), (3.11), the definition of Δ and the expression $Q = (\tilde{S}W - \tilde{R}V)\Delta^{-1}$, it is readily verified that

$$X\mathbf{T}_{\mathbf{P},\mathbf{C}} = \begin{bmatrix} \tilde{S}+Q\tilde{N} & 0 \\ 0 & \tilde{M} \end{bmatrix}. \tag{3.17}$$

Since X and $\mathbf{T}_{\mathbf{P},\mathbf{C}}$ have well-posed inverses, $\tilde{S}+Q\tilde{N}$ must have one. The (1,2) block of (3.16) shows that the two factorizations in (3.12) are equal. Now we have to prove the coprimeness of the factorizations in (3.12). The following identity which follows from (3.6) shows that $R+MQ$ and $S+NQ$ are right-coprime:

$$\tilde{M}(S+NQ) - \tilde{N}(R+MQ) = I. \tag{3.18}$$

The fact that the other factorization is left-coprime is proved similarly.

Finally, to prove the uniqueness of Q, first notice that Q from (3.13) is independent of the choice of the factors in the right-coprime factorization $\mathbf{C} = WV^{-1}$. Assume that there is a $Q_1 \in H^\infty$ such that $\mathbf{C} = (R+MQ_1)(S+NQ_1)^{-1}$. Denote $W_1 = R+MQ_1$ and $V_1 = S+NQ_1$ to obtain $\Delta_1 = \tilde{M}V_1 - \tilde{N}W_1 = I$. Substituting this into (3.13) shows that $Q = Q_1$. Thus we have proved part (1), assuming the existence of a right-coprime factorization for $\mathbf{C}$. A similar argument proves (1) if we start from the assumption that $\mathbf{C}$ has a left-coprime factorization.

To prove part (2), we assume that $Q \in H^\infty$ is such that $S + NQ$ has a well-posed inverse, and we define $\mathbf{C} = (R + MQ)(S + NQ)^{-1}$. The identity (3.18) obtained from (3.6) shows that this factorization of $\mathbf{C}$ is right-coprime. We apply Lemma 3.3 with $V = S + NQ$, $W = R + MQ$ and $\Delta = I$ to conclude that $\mathbf{C}$ stabilizes $\mathbf{P}$. The remaining statements in part (2) follow from part (1). If we start from a $Q \in H^\infty$ such that $\tilde{S} + Q\tilde{N}$ has a well-posed inverse, the proof of part (2) is similar. □

Let us apply this theorem to the transfer function $\mathbf{P}$ from Example 2.3.

Example 3.5. *A doubly coprime factorization of* $\mathbf{P}(s) = \frac{2}{1+e^{-2s}}$ *is given by* $N(s) = \tilde{N}(s) = 2$, $M(s) = \tilde{M}(s) = 1 + e^{-2s}$, $S(s) = \tilde{S}(s) = 1$, $R(s) = \tilde{R}(s) = \frac{1}{2}e^{-2s}$. *The Youla parametrization* (3.12) *yields the controller*

$$\mathbf{C}(s) = \left(\frac{1}{2}e^{-2s} + (1 + e^{-2s})Q(s)\right)(1 + 2Q(s))^{-1}$$

for any $Q \in H^\infty$, *and introducing* $\tilde{Q} = 1 + 2Q$ *we obtain the more transparent form*

$$\mathbf{C}(s) = \frac{1}{2}(1 + e^{-2s}) - \frac{1}{2}\tilde{Q}^{-1}. \tag{3.19}$$

The above formula yields well-posed controllers provided that $\tilde{Q}$ *has a well-posed inverse. On the other hand, the controller suggested in Example 2.3 from physical considerations is not accounted for by* (3.19) *(see also Example 4.7).*

In the above example, it was easy to see which parameters $\tilde{Q}$ should be chosen to obtain a well-posed controller. However, in general it is not a priori clear which $Q \in H^\infty$ will guarantee that $S + NQ$ has a well-posed inverse. We illustrate this point by another example.

Example 3.6. *This is the frequency domain version of Example 6.5 in* [17]. *We take* $U = Y = \ell^2$, *the space of square summable sequences* $x = (x_1, x_2, \dots)$. *The transfer function of the plant is* $\mathbf{P}(s) = \mathrm{diag}\,(P_n(s))$, $n = 1, 2, 3, \dots$, *where*

$$P_n(s) = \frac{(0.64n - 1)s - n}{s(s + n)}.$$

This has poles of multiplicity one at $s = -n$, $n = 1, 2, 3, \dots$, *a pole of infinite multiplicity at* $s = 0$ *and* $\mathbf{P}$ *is well-posed. An observer-based controller for this plant has the transfer function* $\mathbf{C}(s) = \mathrm{diag}\,(C_n(s))$, *where*

$$C_n(s) = \frac{(1 - 0.64n)s + n}{s^2 + s(2 - 0.28n) + 2n},$$

as computed in [17]. *At first glance this seems fine;* $C_n(s)$ *is a strictly proper function for every* n. *However, the real parts of the poles of* $C_n(s)$ *converge to*

$+\infty$, as is easy to see. Consequently, $\mathbf{C}(s)$ cannot be uniformly bounded in norm on any right half-plane and it is therefore not well-posed.

One can use this observer-based controller to obtain a doubly coprime factorization $M = \tilde{M} = diag\,(M_n)$, $N = \tilde{N} = diag\,(N_n)$, $R = \tilde{R} = diag\,(R_n)$, $S = \tilde{S} = diag\,(S_n)$, with $N_n(s) = \tilde{N}_n(s) = \frac{(0.64n-1)s-n}{s^2+s(1+0.36n)+2n}$, $M_n = \tilde{M}_n = 1 + N_n$, $S_n = \tilde{S}_n = 1 - N_n$, $R_n = \tilde{R}_n = -N_n$. According to the Youla parametrization (3.12) a set of stabilizing controllers for $\mathbf{P}$ is given by $\mathbf{C}(s) = diag\,(C_n(s))$, where

$$C_n = (R_n + M_n Q)(S_n + N_n Q)^{-1}, \tag{3.20}$$

and $Q \in H^\infty$ is such that $(S+NQ)^{-1}$ is well-posed. As we have seen in the case of the observer-based controller ($Q = 0$), this is not always the case. With some care, we can choose the parameter $Q \in H^\infty$ such that $\mathbf{C}$ is well-posed. For example, $Q(s) = 1$ yields $\mathbf{C}(s) = I_{\ell^2}$. However, it is not a priori clear which $Q \in H^\infty$ will lead to a well-posed $\mathbf{C}$.

4. Canonical and dual canonical controllers

In [17], a procedure was developed to design stabilizing controllers with internal loop (in the state-space framework, assuming regularity, stabilizability and detectability). The transfer functions of the controllers obtained there were of the form

$$\mathbf{K} = \begin{bmatrix} 0 & I \\ \mathbf{K}_{21} & \mathbf{K}_{22} \end{bmatrix}, \quad \text{with } \mathbf{K}_{21}, \mathbf{K}_{22} \in H^\infty. \tag{4.21}$$

Since this class plays a special role, we call the controllers of the form (4.21) *canonical controllers*. Analogously, controllers of the form

$$\mathbf{K} = \begin{bmatrix} 0 & \mathbf{K}_{12} \\ I & \mathbf{K}_{22} \end{bmatrix}, \quad \text{with } \mathbf{K}_{12}, \mathbf{K}_{22} \in H^\infty, \tag{4.22}$$

will be called *dual canonical controllers* (they can be obtained by a dual design procedure, as explained in Section 5 of [17]).

We analyze the properties of (dual) canonical controllers in some detail. First we recall Proposition 4.8 from [17].

Proposition 4.1. *Using the notation of Definition 2.2, suppose that $\mathbf{K}_{11}$ is an admissible feedback transfer function for $\mathbf{P}$. Then the following two conditions are equivalent:*

(1) *$I - \mathcal{F}(\mathbf{K}, \mathbf{P})$ has a well-posed inverse, where*

$$\mathcal{F}(\mathbf{K}, \mathbf{P}) = \mathbf{K}_{22} + \mathbf{K}_{21}\mathbf{P}(I - \mathbf{K}_{11}\mathbf{P})^{-1}\mathbf{K}_{12}.$$

(2) *$\mathbf{K}$ is an admissible feedback transfer function with internal loop for $\mathbf{P}$.*

In fact, Proposition 4.8 from [17] only states that (1) implies (2), but the converse is also true, with practically the same proof.

Theorem 4.2. *The canonical controller* $\mathbf{K}$ *stabilizes* $\mathbf{P}$ *with internal loop iff*

$$\Delta = I - \mathbf{K}_{22} - \mathbf{K}_{21}\mathbf{P}$$

is invertible on some right half-plane and Δ^{-1}, $\mathbf{P}\Delta^{-1} \in H^\infty$.

If $\mathbf{P}$ *has a right-coprime factorization* $\mathbf{P} = NM^{-1}$, *then* $\mathbf{K}$ *stabilizes* $\mathbf{P}$ *with internal loop iff* $D = M - \mathbf{K}_{22}M - \mathbf{K}_{21}N$ *is invertible over* H^∞.

Note that Δ and Δ^{-1} might be defined only on some right half-plane of the form $\operatorname{Re} s > \alpha$ with $\alpha > 0$, but this is not a problem if Δ^{-1} has an analytic extension to a function in H^∞ (see our convention at the beginning of Section 2).

Proof. First we prove the first statement. According to Proposition 4.1, $I - F\mathbf{G}$ has a well-posed inverse iff $I - \mathcal{F}(\mathbf{K}, \mathbf{P}) = \Delta$ has a well-posed inverse. Moreover,

$$(I - F\mathbf{G})^{-1} = \begin{bmatrix} I + \Delta^{-1}\mathbf{K}_{21}\mathbf{P} & \Delta^{-1}\mathbf{K}_{21} & \Delta^{-1} \\ \mathbf{P}(I + \Delta^{-1}\mathbf{K}_{21}\mathbf{P}) & I + \mathbf{P}\Delta^{-1}\mathbf{K}_{21} & \mathbf{P}\Delta^{-1} \\ \Delta^{-1}\mathbf{K}_{21}\mathbf{P} & \Delta^{-1}\mathbf{K}_{21} & \Delta^{-1} \end{bmatrix}$$

and so $(I - F\mathbf{G})^{-1} \in H^\infty$ iff the expressions

$$\Delta^{-1}, \quad \mathbf{P}\Delta^{-1}, \quad \Delta^{-1}\mathbf{K}_{21}\mathbf{P}, \quad \mathbf{P}(I + \Delta^{-1}\mathbf{K}_{21}\mathbf{P})$$

are all in H^∞. We show that if the first two of these expressions are in H^∞, then so are the other two. We have

$$\begin{aligned} I + \Delta^{-1}\mathbf{K}_{21}\mathbf{P} &= \Delta^{-1}(\Delta + \mathbf{K}_{21}\mathbf{P}) \\ &= \Delta^{-1}(I - \mathbf{K}_{22}) \in H^\infty \end{aligned}$$

and

$$\mathbf{P}(I + \Delta^{-1}\mathbf{K}_{21}\mathbf{P}) = \mathbf{P}\Delta^{-1}(I - \mathbf{K}_{22}) \in H^\infty.$$

Thus, $\mathbf{K}$ stabilizes $\mathbf{P}$ iff $\Delta^{-1} \in H^\infty$ and $\mathbf{P}\Delta^{-1} \in H^\infty$.

Let us prove the second assertion in the theorem. If $\mathbf{P} = NM^{-1}$ and $D^{-1} \in H^\infty$, then from the formulae

$$\Delta^{-1} = MD^{-1}, \quad \mathbf{P}\Delta^{-1} = ND^{-1},$$

we see that $\Delta^{-1}, \mathbf{P}\Delta^{-1} \in H^\infty$. Conversely, assuming that M, N, $\tilde{R}$ and $\tilde{S}$ are as in Definition 3.1 and $\Delta^{-1}, \mathbf{P}\Delta^{-1} \in H^\infty$, then

$$\tilde{S}\Delta^{-1} - \tilde{R}\mathbf{P}\Delta^{-1} = (\tilde{S}M - \tilde{R}N)(M - \mathbf{K}_{22}M - \mathbf{K}_{21}N)^{-1} = D^{-1}.$$

Since $\tilde{S}$ and $\tilde{R}$ are in H^∞, we see that $D^{-1} \in H^\infty$. □

We have the following interesting connection with right-coprime factorizations.

Corollary 4.3. $\mathbf{P}$ *has a right-coprime factorization if and only if* $\mathbf{P}$ *is stabilizable with internal loop by a canonical controller.*

Proof. If $\mathbf{K}$ stabilizes $\mathbf{P}$, then by Theorem 4.2, Δ^{-1} and $\mathbf{P}\Delta^{-1}$ are in H^∞. So $M = \Delta^{-1}$ and $N = \mathbf{P}\Delta^{-1}$ is a right-coprime factorization, since

$$(I - \mathbf{K}_{22})M - \mathbf{K}_{21}N = I.$$

Conversely, if $\mathbf{P} = NM^{-1}$ is a right-coprime factorization such that $\tilde{S}M - \tilde{R}N = I$, then $\mathbf{K} = \begin{bmatrix} 0 & I \\ \tilde{R} & I - \tilde{S} \end{bmatrix}$ stabilizes $\mathbf{P}$ with internal loop. Indeed, from $\Delta = I - (I - \tilde{S}) - \tilde{R}NM^{-1} = (\tilde{S}M - \tilde{R}N)M^{-1} = M^{-1}$, we see that $\Delta^{-1} = M$ and $\mathbf{P}\Delta^{-1} = N$, and we can apply Theorem 4.2. □

Proposition 4.4. *Suppose that* $\mathbf{P}$ *has a doubly coprime factorization as in* (3.5) *and* (3.6). *Then all canonical controllers that stabilize* $\mathbf{P}$ *with internal loop are parameterized by*

$$\mathbf{K} = \begin{bmatrix} 0 & I \\ \tilde{E}(\tilde{R} + Q\tilde{M}) & I - \tilde{E}(\tilde{S} + Q\tilde{N}) \end{bmatrix},$$

where $Q, \tilde{E} \in H^\infty$ *and* $\tilde{E}$ *is invertible over* H^∞.

Proof. For $\mathbf{K}$ as above we show that it stabilizes $\mathbf{P}$ with internal loop. We compute, using (3.6),

$$\begin{aligned} D &= M - \mathbf{K}_{22}M - \mathbf{K}_{21}N \\ &= \tilde{E}(\tilde{S} + Q\tilde{N})M - \tilde{E}(\tilde{R} + Q\tilde{M})N \\ &= \tilde{E}(\tilde{S}M - \tilde{R}N + Q(\tilde{N}M - \tilde{M}N)) = \tilde{E}. \end{aligned}$$

Since $\tilde{E}$ is invertible over H^∞, Theorem 4.2 shows that $\mathbf{K}$ stabilizes $\mathbf{P}$.

Conversely, suppose that $\mathbf{K} = \begin{bmatrix} 0 & I \\ \mathbf{K}_{21} & \mathbf{K}_{22} \end{bmatrix}$ stabilizes $\mathbf{P}$ with internal loop. Theorem 4.2 shows that $D = M - \mathbf{K}_{22}M - \mathbf{K}_{21}N$ is invertible over H^∞. Defining

$$Q = (D^{-1}\mathbf{K}_{21} - \tilde{R})\tilde{M}^{-1}$$

we evaluate

$$\begin{aligned} \tilde{S} + Q\tilde{N} &= \tilde{S} + D^{-1}\mathbf{K}_{21}\tilde{M}^{-1}\tilde{N} - \tilde{R}\tilde{M}^{-1}\tilde{N} \\ &= (\tilde{S}M - \tilde{R}N)M^{-1} + D^{-1}\mathbf{K}_{21}NM^{-1} \\ &= D^{-1}(\mathbf{K}_{21}N + D)M^{-1} \\ &= D^{-1}(M - \mathbf{K}_{22}M)M^{-1} \\ &= D^{-1}(I - \mathbf{K}_{22}) \end{aligned}$$

and

$$\tilde{R} + Q\tilde{M} = \tilde{R} + (D^{-1}\mathbf{K}_{21} - \tilde{R}) = D^{-1}\mathbf{K}_{21}.$$

Thus, denoting $\tilde{E} = D$, we have that $\mathbf{K}_{21} = \tilde{E}(\tilde{R}+Q\tilde{M})$ and $\mathbf{K}_{22} = I - \tilde{E}(\tilde{S}+Q\tilde{N})$ as claimed. It remains to prove that $Q \in H^\infty$. From (3.6) and our formulae for

$\tilde{R} + Q\tilde{M}$ and $\tilde{S} + Q\tilde{N}$, we have

$$\begin{aligned} Q &= Q(\tilde{M}S - \tilde{N}R) \\ &= (D^{-1}\mathbf{K}_{21} - \tilde{R})S - [D^{-1}(I - \mathbf{K}_{22}) - \tilde{S}]R \\ &= D^{-1}\mathbf{K}_{21}S - D^{-1}(I - \mathbf{K}_{22})R, \end{aligned}$$

and it is clear that both terms are in H^∞. □

We call the function Q appearing in Proposition 4.4, the *Youla parameter* of $\mathbf{K}$. We mention that by using (one half of) Proposition 4.4, it is possible to give an alternative, shorter proof of Theorem 5.4 (the main result) in [17].

The following corollary contains the dual statements of Theorem 4.2, Corollary 4.3 and Proposition 4.4.

Corollary 4.5. *Let* $\mathbf{P}$ *be a well-posed transfer function.*

a. *The dual canonical controller* $\mathbf{K}$ *stabilizes* $\mathbf{P}$ *with internal loop iff* $\Delta = I - \mathbf{K}_{22} - \mathbf{P}\mathbf{K}_{12}$ *is invertible on some right half-plane and* $\Delta^{-1}, \Delta^{-1}\mathbf{P} \in H^\infty$.
b. *If* $\mathbf{P}$ *has a left-coprime factorization* $\mathbf{P} = \tilde{M}^{-1}\tilde{N}$, *then* $\mathbf{K}$ *stabilizes* $\mathbf{P}$ *with internal loop iff* $D = \tilde{M} - \tilde{M}\mathbf{K}_{22} - \tilde{N}\mathbf{K}_{12}$ *is invertible over* H^∞.
c. $\mathbf{P}$ *has a left-coprime factorization if and only if* $\mathbf{P}$ *is stabilizable with internal loop by a dual canonical controller.*
d. *If* $\mathbf{P}$ *has a doubly coprime factorization as in* (3.5) *and* (3.6), *then all dual canonical controllers which stabilize* $\mathbf{P}$ *with internal loop are given by*

$$\mathbf{K} = \begin{bmatrix} 0 & (R + MQ)E \\ I & I - (S + NQ)E \end{bmatrix},$$

where $Q, E \in H^\infty$ *and* E *is invertible over* H^∞.

Again, we call Q appearing in Corollary 4.5 the *Youla parameter* of $\mathbf{K}$. The formulae for the parameterization of all stabilizing (dual) canonical controllers are reminiscent of the Youla parameterization (3.12). Consequently, we expect a strong relationship between stabilization with internal loop and the usual concept of stabilization. This is clarified in the following proposition.

Proposition 4.6. *Let* $\mathbf{P}$ *and* $\mathbf{C}$ *be well-posed transfer functions with values in* $\mathcal{L}(U, Y)$ *and* $\mathcal{L}(Y, U)$, *respectively. Assume that* $\mathbf{C}$ *has a left-coprime factorization* $\mathbf{C} = (I - \mathbf{K}_{22})^{-1}\mathbf{K}_{21}$. *Then* $\mathbf{P}$ *is stabilized by* $\mathbf{C}$ *iff* $\mathbf{P}$ *is stabilized with internal loop by the canonical controller* $\mathbf{K} = \begin{bmatrix} 0 & I \\ \mathbf{K}_{21} & \mathbf{K}_{22} \end{bmatrix}$.

Proof. Suppose that $\mathbf{C} = (I - \mathbf{K}_{22})^{-1}\mathbf{K}_{21}$ stabilizes $\mathbf{P}$. Define

$$\begin{aligned} \Delta &= I - \mathbf{K}_{22} - \mathbf{K}_{21}\mathbf{P} \\ &= (I - \mathbf{K}_{22})(I - \mathbf{C}\mathbf{P}). \end{aligned}$$

Then, according to Definition 2.1 and formula (2.1),

$$\Delta^{-1}\mathbf{K}_{21} = (I - \mathbf{C}\mathbf{P})^{-1}\mathbf{C} \in H^\infty,$$

$$\Delta^{-1}(I - \mathbf{K}_{22}) = (I - \mathbf{CP})^{-1} \in H^\infty.$$

Since $\mathbf{K}_{21}$ and $I - \mathbf{K}_{22}$ are left-coprime, there exist R and S in H^∞ such that $(I - \mathbf{K}_{22})S - \mathbf{K}_{21}R = I$, whence

$$\Delta^{-1}(I - \mathbf{K}_{22})S - \Delta^{-1}\mathbf{K}_{21}R = \Delta^{-1}.$$

Since both terms on the left-hand side are in H^∞, we see that $\Delta^{-1} \in H^\infty$.

Similarly, $\mathbf{P}\Delta^{-1} \in H^\infty$ follows from coprimeness and the facts that

$$\mathbf{P}\Delta^{-1}(I - \mathbf{K}_{22}) = \mathbf{P}(I - \mathbf{CP})^{-1} \in H^\infty,$$

$$\mathbf{P}\Delta^{-1}\mathbf{K}_{21} = \mathbf{P}(I - \mathbf{CP})^{-1}\mathbf{C} \in H^\infty.$$

According to Theorem 4.2, $\mathbf{K}$ stabilizes $\mathbf{P}$ with internal loop.

Conversely, suppose that $\mathbf{K}$ stabilizes $\mathbf{P}$ with internal loop. Then, from

$$\begin{aligned}
(I - \mathbf{CP})^{-1} &= \Delta^{-1}(I - \mathbf{K}_{22}) \\
\mathbf{P}(I - \mathbf{CP})^{-1} &= \mathbf{P}\Delta^{-1}(I - \mathbf{K}_{22}) \\
(I - \mathbf{CP})^{-1}\mathbf{C} &= \Delta^{-1}\mathbf{K}_{21} \\
(I - \mathbf{PC})^{-1} &= I + \mathbf{P}(I - \mathbf{CP})^{-1}\mathbf{C} = I + \mathbf{P}\Delta^{-1}\mathbf{K}_{21}
\end{aligned}$$

and from Theorem 4.2 we see that all entries of $\mathbf{T}_{\mathbf{P},\mathbf{C}}$ from (2.1) are in H^∞. □

Notice that if a canonical $\mathbf{K}$ stabilizes $\mathbf{P}$ with internal loop, then $\mathbf{K}_{21}$ and $I - \mathbf{K}_{22}$ are left-coprime, since $(I - \mathbf{K}_{22})\Delta^{-1} - \mathbf{K}_{21}\mathbf{P}\Delta^{-1} = I$. Naturally, Proposition 4.6 has a dual statement for right-coprime factorizations of $\mathbf{C}$.

An interesting feature of the last result is that it remains true without the requirement that $\mathbf{C}$ be well-posed, if we modify Definitions 2.1 and 3.1 accordingly, replacing everywhere well-posedness by analyticity.

We complete this section by returning to two earlier examples.

Example 4.7. *Using the results in Example 3.5 and Proposition 4.4, we obtain that all canonical controllers that stabilize* $\mathbf{P}(s) = \frac{2}{1+e^{-2s}}$ *are*

$$\mathbf{K}(s) = \begin{bmatrix} 0 & 1 \\ \frac{1}{2}\tilde{E}(s)\left[-1 + (1 + e^{-2s})\tilde{Q}(s)\right] & 1 - \tilde{E}(s)\tilde{Q}(s) \end{bmatrix},$$

where $\tilde{E}, \tilde{E}^{-1}, \tilde{Q} \in H^\infty$. *(We have used, as in Example 3.5,* $\tilde{Q} = 1 + 2Q$.*) If* $\tilde{Q}^{-1}$ *is well-posed, then we obtain the controllers* (3.19) *from Example 3.5, and if* $\tilde{E}(s) = 2$ *and* $\tilde{Q} = 0$, *then we obtain the controller suggested in Example 2.3.*

Example 4.8. *The plant in the "intriguing control problem" of Section 1 and its controller are determined by the following transfer functions:*

$$\mathbf{P}(s) = 3\frac{s+1}{s-1}, \qquad \mathbf{K}(s) = \begin{bmatrix} 0 & 1 \\ 1 & 1 - e^{-s} \end{bmatrix}.$$

Note that $\mathbf{K}$ is canonical (and also dual canonical). To check that $\mathbf{K}$ stabilizes $\mathbf{P}$ with internal loop, we use Theorem 4.2. We compute $\Delta = e^{-s} - 3\frac{s+1}{s-1}$, so that

$$\Delta^{-1} = -\frac{1}{3} \cdot \frac{s-1}{(1-\frac{e^{-s}}{3})s + (1+\frac{e^{-s}}{3})}.$$

First we show that Δ^{-1} has no poles in the closed right half-plane. Indeed, if such a pole p were to exist, it would have to satisfy

$$p = -\frac{1+\frac{e^{-p}}{3}}{1-\frac{e^{-p}}{3}}.$$

But for $Re\, p \geq 0$, the right-hand side above is in the open left half-plane, which is a contradiction. Thus, Δ^{-1} is analytic on an open set containing the closed right half-plane. To prove that $\Delta^{-1} \in H^\infty$, it only remains to show that it does not behave badly at infinity. It is easy to see that $|\Delta^{-1}(s)| < 1$ for $|s|$ sufficiently large (we consider only s with $Re\, s > 0$). The fact that $\mathbf{P}\Delta^{-1} \in H^\infty$ follows similarly. Computing the transfer function from d to y and from d to ζ_o, we see that the first is equal to the second delayed by 1 (i.e., multiplied by e^{-s}).

5. Reduction to a (dual) canonical controller

In this section, we show that the canonical controllers are not as special as they seem; any stabilizing controller with internal loop is, in a certain sense, equivalent to a canonical controller. To do this, let us analyze the connection in Figure 4 in more detail. We use the auxiliary inputs v and r from Figure 4, so that $u_p = u_k + v$ and $y_k = y_p + r$, while assuming that $\xi = 0$. If $\mathbf{K}$ stabilizes $\mathbf{P}$ with internal loop, then we can define the transfer function $\mathbf{T}_{\mathbf{P},\mathbf{K}} \in H^\infty$ by

$$\begin{bmatrix} \hat{u}_p \\ \hat{y}_k \end{bmatrix} = \mathbf{T}_{\mathbf{P},\mathbf{K}} \begin{bmatrix} \hat{v} \\ \hat{r} \end{bmatrix}.$$

We call $\mathbf{T}_{\mathbf{P},\mathbf{K}}$ the *compensation operator* of the pair $(\mathbf{P},\mathbf{K})$. This is analogous to the transfer function $\mathbf{T}_{\mathbf{P},\mathbf{C}}$ introduced in (2.1). In fact, $\mathbf{T}_{\mathbf{P},\mathbf{K}}$ is the left upper 2×2 corner of the matrix $(I - F\mathbf{G})^{-1}$ from Definition 2.2.

Definition 5.1. *Let $\mathbf{P}$ be well-posed and suppose that $\mathbf{K}_1$ and $\mathbf{K}_2$ are stabilizing controllers with internal loop for $\mathbf{P}$. We call $\mathbf{K}_1$ and $\mathbf{K}_2$* equivalent *if the compensation operators of $(\mathbf{P},\mathbf{K}_1)$ and $(\mathbf{P},\mathbf{K}_2)$ are equal.*

The main result of this section is the following.

Theorem 5.2. *Suppose that $\mathbf{P}$ has a doubly coprime factorization. If $\mathbf{K}$ is a stabilizing controller with internal loop for $\mathbf{P}$, then there exists a canonical controller $\tilde{\mathbf{K}}$ such that $\tilde{\mathbf{K}}$ stabilizes $\mathbf{P}$ with internal loop and $\tilde{\mathbf{K}}$ is equivalent to $\mathbf{K}$.*

Proof. We use the notation from (3.5) and (3.6). By Proposition 4.4, for every $Q \in H^\infty$, the following $\tilde{\mathbf{K}}$ stabilizes $\mathbf{P}$ with internal loop:

$$\tilde{\mathbf{K}} = \begin{bmatrix} 0 & I \\ \tilde{R} + Q\tilde{M} & I - (\tilde{S} + Q\tilde{N}) \end{bmatrix}. \tag{5.23}$$

We shall find a Youla parameter $Q \in H^\infty$ such that the compensation operator of $(\mathbf{P}, \tilde{\mathbf{K}})$ coincides with that of $(\mathbf{P}, \mathbf{K})$. First, we explain how this Q will be constructed. We consider the extended transfer function

$$\tilde{\mathbf{P}} = \begin{bmatrix} -M^{-1}R & M^{-1} \\ \tilde{M}^{-1} & \mathbf{P} \end{bmatrix}.$$

We denote the inputs of $\tilde{\mathbf{P}}$ by $\begin{bmatrix} \tilde{w}_p \\ \tilde{u}_p \end{bmatrix}$ and its outputs by $\begin{bmatrix} \tilde{v}_p \\ \tilde{y}_p \end{bmatrix}$, as in Figure 7.

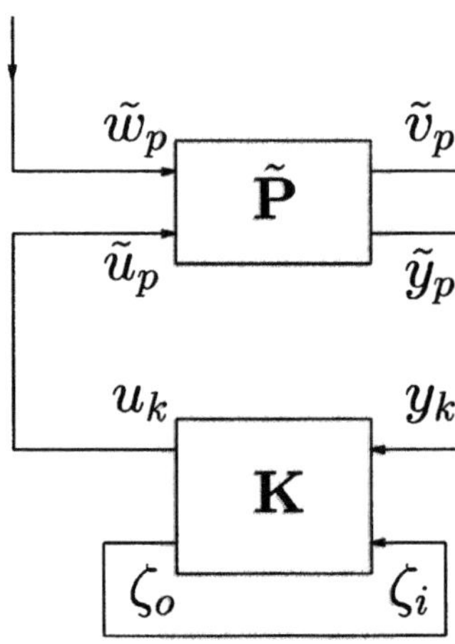

FIGURE 7. The extended system $\tilde{\mathbf{P}}$ connected to the controller $\mathbf{K}$ with internal loop. The transfer function from $\tilde{w}_p$ to $\tilde{v}_p$ is the desired Youla parameter.

We shall prove that if $\mathbf{K}$ stabilizes $\mathbf{P}$, then the interconnection of $\tilde{\mathbf{P}}$ and $\mathbf{K}$ shown in Figure 7 is stable. Then we take Q to be the H^∞ transfer function from $\tilde{w}_p$ to $\tilde{v}_p$ in the interconnection in Figure 7. There are two assertions that have to be proven in order to complete the proof of the theorem:

Assertion 1: If $\mathbf{K}$ stabilizes $\mathbf{P}$, then the system in Figure 7 is stable.

Assertion 2: If Q is chosen to be the transfer function from $\tilde{w}_p$ to $\tilde{v}_p$ in the interconnection in Figure 7, and $\tilde{\mathbf{K}}$ is given by (5.23), then $\mathbf{T}_{\mathbf{P},\mathbf{K}} = \mathbf{T}_{\tilde{\mathbf{P}},\tilde{\mathbf{K}}}$.

Proof of Assertion 1: Suppose that $\mathbf{K}$ stabilizes $\mathbf{P}$ with internal loop. According to Definition 2.2, this means that the parallel connection of $\mathbf{P}$ and $\mathbf{K}$ as independent channels, denoted by $\mathbf{G}$ in the definition, is such that $(I - F\mathbf{G})^{-1} \in H^\infty$. Since F is invertible, this is equivalent to the fact that $\mathbf{G}$ is stabilized by the static output feedback

$$\begin{bmatrix} u_p \\ y_k \\ \zeta_i \end{bmatrix} = \begin{bmatrix} 0 & I & 0 \\ I & 0 & 0 \\ 0 & 0 & I \end{bmatrix} \begin{bmatrix} y_p \\ u_k \\ \zeta_o \end{bmatrix} = F \begin{bmatrix} y_p \\ u_k \\ \zeta_o \end{bmatrix}.$$

We have to prove that the parallel connection of $\tilde{\mathbf{P}}$ and $\mathbf{K}$,

$$\tilde{\mathbf{G}} = \begin{bmatrix} \tilde{\mathbf{P}} & 0 \\ 0 & \mathbf{K} \end{bmatrix} = \begin{bmatrix} -M^{-1}R & M^{-1} & 0 & 0 \\ \tilde{M}^{-1} & \mathbf{P} & 0 & 0 \\ 0 & 0 & \mathbf{K}_{11} & \mathbf{K}_{12} \\ 0 & 0 & \mathbf{K}_{21} & \mathbf{K}_{22} \end{bmatrix},$$

is stabilized by the static output feedback

$$\begin{bmatrix} \tilde{w}_p \\ \tilde{u}_p \\ y_k \\ \zeta_i \end{bmatrix} = \begin{bmatrix} 0 & 0 & 0 & 0 \\ 0 & 0 & I & 0 \\ 0 & I & 0 & 0 \\ 0 & 0 & 0 & I \end{bmatrix} \begin{bmatrix} \tilde{v}_p \\ \tilde{y}_p \\ u_k \\ \zeta_o \end{bmatrix} = \tilde{F} \begin{bmatrix} \tilde{v}_p \\ \tilde{y}_p \\ u_k \\ \zeta_o \end{bmatrix}.$$

Equivalently, we have to prove that the four transfer functions from (2.1), $(I - \tilde{\mathbf{G}}\tilde{F})^{-1}$, $(I - \tilde{\mathbf{G}}\tilde{F})^{-1}\tilde{\mathbf{G}}$, $\tilde{F}(I - \tilde{\mathbf{G}}\tilde{F})^{-1}$ and $(I - \tilde{F}\tilde{\mathbf{G}})^{-1}$, are in H^∞. Because $\tilde{F}$ is a constant transfer function, it is enough to prove that the first two are in H^∞, and the other two will follow immediately (for the last one, use the identity $(I - \tilde{F}\tilde{\mathbf{G}})^{-1} = I + \tilde{F}(I - \tilde{\mathbf{G}}\tilde{F})^{-1}\tilde{\mathbf{G}}$). Notice that

$$\tilde{\mathbf{G}} = \left[\begin{array}{c|c} -M^{-1}R & \begin{bmatrix} M^{-1} & 0 & 0 \end{bmatrix} \\ \hline \begin{bmatrix} \tilde{M}^{-1} \\ 0 \\ 0 \end{bmatrix} & \mathbf{G} \end{array}\right]$$

and

$$\tilde{F} = \left[\begin{array}{c|c} 0 & \begin{bmatrix} 0 & 0 & 0 \end{bmatrix} \\ \hline \begin{bmatrix} 0 \\ 0 \\ 0 \end{bmatrix} & F \end{array}\right].$$

It is easy to see that

$$(I - \tilde{\mathbf{G}}\tilde{F})^{-1} = \left[\begin{array}{c|c} I & \begin{bmatrix} 0 & M^{-1} & 0 \end{bmatrix}(I - \mathbf{G}F)^{-1} \\ \hline \begin{bmatrix} 0 \\ 0 \\ 0 \end{bmatrix} & (I - \mathbf{G}F)^{-1} \end{array}\right].$$

Since F stabilizes $\mathbf{G}$, $(I-\mathbf{G}F)^{-1}$ is in H^∞. Therefore, $(I-\tilde{\mathbf{G}}\tilde{F})^{-1}$ is in H^∞, if we show that $\begin{bmatrix} 0 & M^{-1} & 0 \end{bmatrix}(I-\mathbf{G}F)^{-1}$ is in H^∞. Let us denote by T_{jk} the constituent blocks of $(I-\mathbf{G}F)^{-1}$, i.e.,

$$(I-\mathbf{G}F)\begin{bmatrix} T_{11} & T_{12} & T_{13} \\ T_{21} & T_{22} & T_{23} \\ T_{13} & T_{23} & T_{33} \end{bmatrix} = \begin{bmatrix} I & 0 & 0 \\ 0 & I & 0 \\ 0 & 0 & I \end{bmatrix}.$$

From the upper row of this equality, we deduce that

$$T_{11} - I = \mathbf{P}T_{21}, \quad T_{12} = \mathbf{P}T_{22}, \quad T_{13} = \mathbf{P}T_{23}.$$

Recall from (3.6) that $\tilde{S}M - \tilde{R}N = I$ and compute

$$\begin{aligned} \tilde{S}T_{21} - \tilde{R}(T_{11} - I) &= \tilde{S}T_{21} - \tilde{R}NM^{-1}T_{21} \\ &= (\tilde{S}M - \tilde{R}N)M^{-1}T_{21} \\ &= M^{-1}T_{21}. \end{aligned} \tag{5.24}$$

Similarly,

$$\tilde{S}T_{22} - \tilde{R}T_{12} = \tilde{S}T_{22} - \tilde{R}NM^{-1}T_{22} = M^{-1}T_{22} \tag{5.25}$$

and

$$\tilde{S}T_{23} - \tilde{R}T_{13} = \tilde{S}T_{23} - \tilde{R}NM^{-1}T_{23} = M^{-1}T_{23}. \tag{5.26}$$

This shows that

$$\begin{bmatrix} 0 & M^{-1} & 0 \end{bmatrix}(I-\mathbf{G}F)^{-1} = M^{-1}\begin{bmatrix} T_{21} & T_{22} & T_{23} \end{bmatrix} \in H^\infty.$$

So we have shown that $(I-\tilde{\mathbf{G}}\tilde{F})^{-1}$ is in H^∞.

Let us now show that $(I-\tilde{\mathbf{G}}\tilde{F})^{-1}\tilde{\mathbf{G}}$ is in H^∞. We can compute, using the notation introduced before, that

$$(I-\tilde{\mathbf{G}}\tilde{F})^{-1}\tilde{\mathbf{G}}$$
$$= \left[\begin{array}{c|c} -M^{-1}R + M^{-1}T_{21}\tilde{M}^{-1} & \begin{bmatrix} M^{-1} & 0 & 0 \end{bmatrix} + M^{-1}\begin{bmatrix} T_{21} & T_{22} & T_{23} \end{bmatrix}\mathbf{G} \\ \hline (I-\mathbf{G}F)^{-1}\begin{bmatrix} \tilde{M}^{-1} \\ 0 \\ 0 \end{bmatrix} & (I-\mathbf{G}F)^{-1}\mathbf{G} \end{array}\right] \tag{5.27}$$

Since F stabilizes $\mathbf{G}$, we have that $(I-\mathbf{G}F)^{-1}\mathbf{G}$ is in H^∞. We have to check that the other three blocks in (5.27) are in H^∞. For this, we consider the identity

$$\begin{bmatrix} T_{11} & T_{12} & T_{13} \\ T_{21} & T_{22} & T_{23} \\ T_{13} & T_{23} & T_{33} \end{bmatrix}(I-\mathbf{G}F) = \begin{bmatrix} I & 0 & 0 \\ 0 & I & 0 \\ 0 & 0 & I \end{bmatrix}.$$

From the second row and the second column of the previous identity, we deduce

$$T_{12} = T_{11}\mathbf{P}, \quad T_{22} - I = T_{21}\mathbf{P}, \quad T_{32} = T_{31}\mathbf{P}, \tag{5.28}$$

$$T_{22}\mathbf{K}_{11} + T_{23}\mathbf{K}_{21} = T_{21}, \quad T_{22}\mathbf{K}_{12} + T_{23}(\mathbf{K}_{22} - I) = 0. \tag{5.29}$$

We can now proceed to show that the remaining three blocks in (5.27) are in H^∞.

Block (2,1): We have to prove that

$$(I - \mathbf{G}F)^{-1}\begin{bmatrix} \tilde{M}^{-1} \\ 0 \\ 0 \end{bmatrix} = \begin{bmatrix} T_{11}\tilde{M}^{-1} \\ T_{21}\tilde{M}^{-1} \\ T_{31}\tilde{M}^{-1} \end{bmatrix} \in H^\infty.$$

Recall from (3.6) that $\tilde{M}S - \tilde{N}R = I$ and use the identities (5.28) to compute

$$\begin{aligned} T_{11}S - T_{12}R &= T_{11}S - T_{11}\mathbf{P}R \\ &= T_{11}\tilde{M}^{-1}(\tilde{M}S - \tilde{N}R) = T_{11}\tilde{M}^{-1}, \end{aligned}$$

and similarly

$$\begin{aligned} T_{21}S + (I - T_{22})R &= T_{21}S - T_{21}\mathbf{P}R \\ &= T_{21}\tilde{M}^{-1}(\tilde{M}S - \tilde{N}R) = T_{21}\tilde{M}^{-1}, \end{aligned}$$

$$T_{31}S - T_{32}R = T_{31}S - T_{31}\mathbf{P}R = T_{31}\tilde{M}^{-1}.$$

These identities show that block (2,1) in (5.27) is in H^∞.

Block (1,1): We have, using identity (5.24) and (3.6), that

$$\begin{aligned} -M^{-1}R + M^{-1}T_{21}\tilde{M}^{-1} &= -M^{-1}R + (\tilde{S}T_{21} - \tilde{R}T_{11} + \tilde{R})\tilde{M}^{-1} \\ &= \tilde{S}T_{21}\tilde{M}^{-1} - \tilde{R}T_{11}\tilde{M}^{-1}. \end{aligned}$$

As we previously proved, $T_{21}\tilde{M}^{-1}$ and $T_{11}\tilde{M}^{-1}$ are in H^∞, and so block (1,1) in (5.27) is in H^∞.

Block (1,2): A little computation shows that block (2,1) is equal to

$$\begin{bmatrix} M^{-1} + M^{-1}T_{21}\mathbf{P} & M^{-1}(T_{22}\mathbf{K}_{11} + T_{23}\mathbf{K}_{21}) & M^{-1}(T_{22}\mathbf{K}_{12} + T_{23}\mathbf{K}_{22}) \end{bmatrix}.$$

Using (5.28), we have

$$M^{-1} + M^{-1}T_{21}\mathbf{P} = M^{-1} + M^{-1}(T_{22} - I) = M^{-1}T_{22},$$

which is in H^∞ by (5.25). Using (5.29), we obtain

$$M^{-1}(T_{22}\mathbf{K}_{11} + T_{23}\mathbf{K}_{21}) = M^{-1}T_{21},$$

which is in H^∞ by (5.24), and

$$M^{-1}(T_{22}\mathbf{K}_{12} + T_{23}\mathbf{K}_{22}) = M^{-1}T_{23},$$

which is in H^∞ by (5.26). This concludes the proof of Assertion 1.

Proof of Assertion 2: We have chosen Q to be the transfer function from $\tilde{w}_p$ to $\tilde{v}_p$ in Figure 7. By Assertion 1, this is in H^∞. We must show that, if we choose $\tilde{\mathbf{K}}$ from (5.23) with this Q, then $\mathbf{T}_{\mathbf{P},\mathbf{K}} = \mathbf{T}_{\mathbf{P},\tilde{\mathbf{K}}}$.

$\tilde{\mathbf{K}}$ from (5.23) can be realized via the system $\mathbf{K}_G$ with outputs $\tilde{u}_k$, $\tilde{w}_k$, $\tilde{\zeta}_o$, inputs $\tilde{y}_k$, $\tilde{v}_k$, $\tilde{\zeta}_i$ and connection $\tilde{v}_k = Q\tilde{w}_k$, as in Figure 8, with

$$\mathbf{K}_G = \begin{bmatrix} 0 & 0 & I \\ \tilde{M} & 0 & -\tilde{N} \\ \tilde{R} & I & I-\tilde{S} \end{bmatrix}.$$

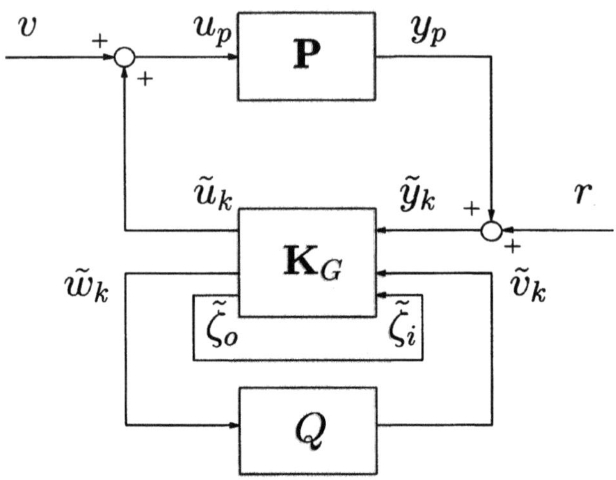

FIGURE 8. The system in Figure 4, with $\tilde{\mathbf{K}}$ in place of $\mathbf{K}$, after the realization of $\tilde{\mathbf{K}}$ as the feedback interconnection of $\mathbf{K}_G$ with Q (and with $\xi = 0$)

To see this, consider

$$\begin{bmatrix} \tilde{u}_k \\ \tilde{w}_k \\ \tilde{\zeta}_o \end{bmatrix} = \begin{bmatrix} 0 & 0 & I \\ \tilde{M} & 0 & -\tilde{N} \\ \tilde{R} & I & I-\tilde{S} \end{bmatrix} \begin{bmatrix} \tilde{y}_k \\ \tilde{v}_k \\ \tilde{\zeta}_i \end{bmatrix}, \qquad \tilde{v}_k = Q\tilde{w}_k.$$

These yield $\tilde{u}_k = \tilde{\zeta}_i$ and $\tilde{\zeta}_o = (\tilde{R} + Q\tilde{M})\tilde{y}_k + (I - \tilde{S} - Q\tilde{N})\tilde{\zeta}_i$, i.e.,

$$\begin{bmatrix} \tilde{u}_k \\ \tilde{\zeta}_o \end{bmatrix} = \tilde{\mathbf{K}} \begin{bmatrix} \tilde{y}_k \\ \tilde{\zeta}_i \end{bmatrix}.$$

The realization of $\tilde{\mathbf{K}}$ as in Figure 8 is more convenient to show that $(\mathbf{P}, \tilde{\mathbf{K}})$ has the same compensation operator as $(\mathbf{P}, \mathbf{K})$, i.e.,

$$\begin{bmatrix} u_p \\ \tilde{y}_k \end{bmatrix} = \mathbf{T}_{\mathbf{P},\mathbf{K}} \begin{bmatrix} v \\ r \end{bmatrix}.$$

Let us expand Q in Figure 8 into its components to obtain Figure 9.

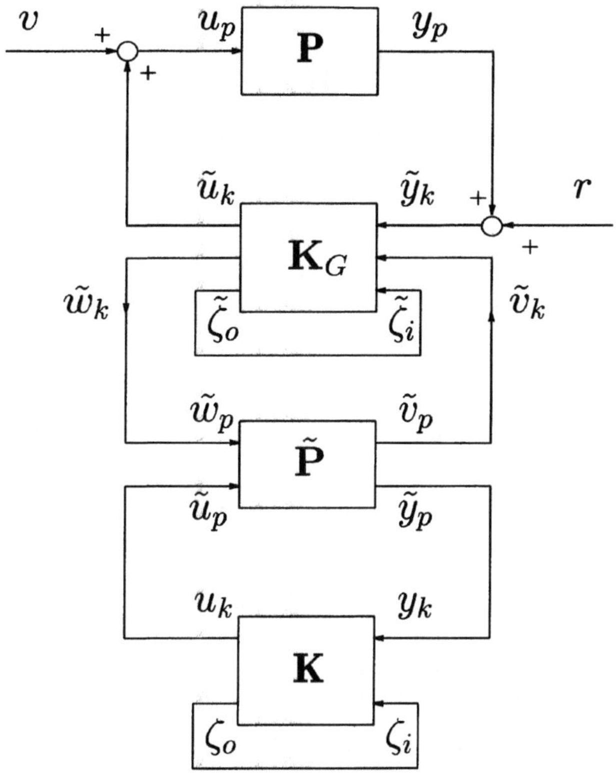

FIGURE 9. Expanding Q in Figure 8, using that Q is obtained in Figure 7 as the transfer function from $\tilde{w}_p$ to $\tilde{v}_p$

The idea is now to prove that $y_k = \tilde{y}_k$ and $u_k = \tilde{u}_k$. If this is so, then $y_k = y_p + r$ and $u_p = u_k + v$, and this implies that the transfer function from $\begin{bmatrix} v \\ r \end{bmatrix}$ to $\begin{bmatrix} u_p \\ y_k \end{bmatrix}$ is $\mathbf{T}_{\mathbf{P},\mathbf{K}}$. Since $y_k = \tilde{y}_k$, this is enough to prove our claim. So it remains only to prove that $y_k = \tilde{y}_k$ and $u_k = \tilde{u}_k$ in Figure 9.

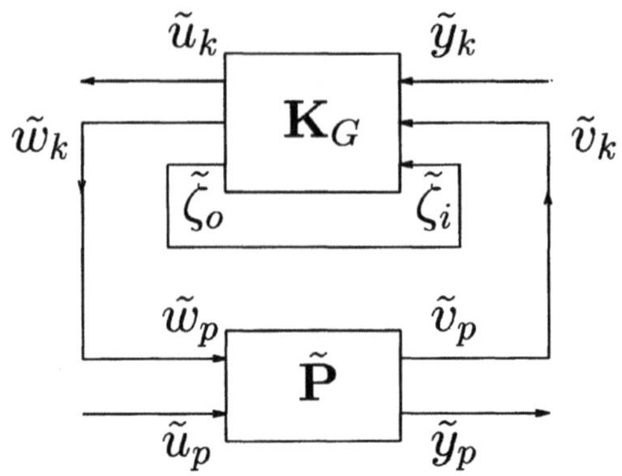

FIGURE 10. Connection between $\mathbf{K}_G$ and $\tilde{\mathbf{P}}$ in Figure 9

First we show that the connection between $\mathbf{K}_G$ and $\tilde{\mathbf{P}}$ in Figure 9 (represented separately in Figure 10) is well-posed, i.e., the output feedback $\bar{F}$ given by

$$\begin{bmatrix} \tilde{y}_k \\ \tilde{v}_k \\ \tilde{\zeta}_i \\ \tilde{w}_p \\ \tilde{u}_p \end{bmatrix} = \begin{bmatrix} 0 & 0 & 0 & 0 & 0 \\ 0 & 0 & 0 & I & 0 \\ 0 & 0 & I & 0 & 0 \\ 0 & I & 0 & 0 & 0 \\ 0 & 0 & 0 & 0 & 0 \end{bmatrix} \begin{bmatrix} \tilde{u}_k \\ \tilde{w}_k \\ \tilde{\zeta}_o \\ \tilde{v}_p \\ \tilde{y}_p \end{bmatrix} = \bar{F} \begin{bmatrix} \tilde{u}_k \\ \tilde{w}_k \\ \tilde{\zeta}_o \\ \tilde{v}_p \\ \tilde{y}_p \end{bmatrix}$$

is admissible for the parallel connection of $\mathbf{K}_G$ and $\tilde{\mathbf{P}}$,

$$\bar{\mathbf{G}} = \begin{bmatrix} \mathbf{K}_G & 0 \\ 0 & \tilde{\mathbf{P}} \end{bmatrix}.$$

Straightforward computations using the identity (3.6) yield

$$(I - \bar{\mathbf{G}}\bar{F})^{-1} = \begin{bmatrix} I & -R & M & M & 0 \\ 0 & I + \tilde{N}R & -\tilde{N}M & -\tilde{N}M & 0 \\ 0 & -R & M & M & 0 \\ 0 & -\tilde{S}R & -I + \tilde{S}M & \tilde{S}M & 0 \\ 0 & S & -N & -N & I \end{bmatrix}.$$

So we have well-posedness and, in addition, $(I - \bar{\mathbf{G}}\bar{F})^{-1} \in H^\infty$. Moreover,

$$(I - \bar{\mathbf{G}}\bar{F})^{-1}\bar{\mathbf{G}} = \begin{bmatrix} 0 & * & * & * & I \\ ** & * & * & * & * \\ ** & * & * & * & * \\ ** & * & * & * & * \\ I & * & * & * & 0 \end{bmatrix},$$

in which a star denotes an irrelevant entry. This shows that

$$\begin{bmatrix} \tilde{y}_k \\ \tilde{u}_p \end{bmatrix} = \begin{bmatrix} 0 & I \\ I & 0 \end{bmatrix} \begin{bmatrix} \tilde{u}_k \\ \tilde{y}_p \end{bmatrix}$$

and hence $y_k = \tilde{y}_k$ and $u_k = \tilde{u}_k$, as claimed. □

Example 5.3. *It is interesting to calculate the canonical equivalent to a dual canonical controller. Suppose that* $\mathbf{K} = \begin{bmatrix} 0 & \mathbf{K}_{12} \\ I & \mathbf{K}_{22} \end{bmatrix}$ *stabilizes the plant* $\mathbf{P}$ *with internal loop. If* $\mathbf{P}$ *admits a doubly coprime factorization, then the procedure in the proof of Theorem 5.2 produces the canonical controller*

$$\tilde{\mathbf{K}} = \begin{bmatrix} 0 & I \\ M^{-1}\mathbf{K}_{12}\tilde{\Delta}^{-1} & I - M^{-1}(I + \mathbf{K}_{12}\tilde{\Delta}^{-1}\mathbf{P}) \end{bmatrix},$$

where $\tilde{\Delta} = I - \mathbf{K}_{22} - \mathbf{P}\mathbf{K}_{12}$, *and* M *comes from the coprime factorization* (3.5).

It is clear that the following dual to Theorem 5.2 holds.

Corollary 5.4. *Suppose that* $\mathbf{P}$ *has a doubly coprime factorization. If* $\mathbf{K}$ *is a stabilizing controller with internal loop for* $\mathbf{P}$, *then there exists a dual canonical controller* $\tilde{\mathbf{K}}$ *such that* $\tilde{\mathbf{K}}$ *stabilizes* $\mathbf{P}$ *with internal loop and* $\tilde{\mathbf{K}}$ *is equivalent to* $\mathbf{K}$.

Theorem 5.5. *Suppose that* $\mathbf{P}$ *has a doubly coprime factorization as in* (3.5), (3.6), *and* $\mathbf{K}$ *is a stabilizing controller with internal loop for* $\mathbf{P}$. *Then there exists a unique* $Q \in H^\infty$ *such that* $\mathbf{K}$ *is equivalent to the canonical controller*

$$\mathbf{K}_l = \begin{bmatrix} 0 & I \\ \tilde{R} + Q\tilde{M} & I - (\tilde{S} + Q\tilde{N}) \end{bmatrix},$$

and is also equivalent to the dual canonical controller

$$\mathbf{K}_r = \begin{bmatrix} 0 & R + MQ \\ I & I - (S + NQ) \end{bmatrix}.$$

The proof is the same as for Theorem 5.2.

6. Conclusions

In this paper, we have examined two concepts of stabilization for the class of well-posed transfer functions. The motivation for choosing this class was that its members have realizations as well-posed linear systems (see Salamon [8], Staffans [12]). We have proposed a new and more general type of controller, controllers with an internal loop. The need for this extension is demonstrated by several examples of such controllers, which cannot be reduced to conventional controllers. We have introduced canonical and dual canonical controllers, which are controllers with internal loop of a special (simple) structure. We have found that these are closely related to (doubly) coprime factorizations of the plant transfer function. For the case that $\mathbf{P}$ has a doubly coprime factorization, we have given a complete parameterization of all stabilizing controllers with internal loop which are (dual) canonical. We have proven that any stabilizing controller with internal loop is equivalent to a canonical one and also to a dual canonical one. Finally, we remark that, although our motivation was to develop a new theory for the well-posed class of irrational transfer functions, the concept of stabilization with internal loop is new even for rational transfer functions.

References

[1] J. S. Baras, *Frequency domain design of linear distributed systems*, Proc. of the 19th IEEE Decision and Control Conference, pp. 728–732, 1980.

[2] F. M. Callier and C. A. Desoer, *An algebra of transfer functions for distributed linear time-invariant systems*, IEEE Trans. Circuits and Systems, **25** (1978), 651–663; (Corrections: **26** (1979), p. 320).

[3] R. F. Curtain, G. Weiss, and M. Weiss, *Coprime factorization for regular linear systems*, Automatica, **32** (1996), 1519–1531.

[4] C. Foias, H. Özbay and A. Tannenbaum, *Robust Control of Infinite Dimensional Systems*, Springer-Verlag, LNCIS 209, 1996.

[5] B. A. Francis, *A Course in H^∞ Control Theory*, Springer-Verlag, LNCIS vol. 88, Berlin, 1987.

[6] T. T. Georgiou and M. C. Smith, *Graphs, causality and stabilizability: linear, shift-invariant systems on $L_2[0,\infty)$*, Math. for Control, Signals and Systems, **6** (1993), 195–223.

[7] D. Salamon, *Infinite-dimensional systems with unbounded control and observation: A functional analytic approach*, Trans of the Amer. Math. Soc., **300** (1987), 383–431.

[8] D. Salamon, *Realization theory in Hilbert space*, Math. Systems Theory, **21** (1989), 147–164.

[9] M. C. Smith, *On stabilization and the existence of coprime factorizations*, IEEE Trans. on Automatic Control, **34** (1989), 1005–1007.

[10] O. J. Staffans, *Coprime factorizations and well-posed linear systems*, SIAM J. Control and Optim., **36** (1998), 1268–1292.

[11] O. J. Staffans, *Quadratic optimal control of well-posed linear systems*, SIAM J. Control and Optim., **37** (1998), 131–164.

[12] O. J. Staffans, *Admissible factorizations of Hankel operators induce well-posed linear systems*, Systems and Control Letters, **37** (1999), 301–307.

[13] S. Townley, G. Weiss and Y. Yamamoto, *Discretizing continuous-time controllers for infinite-dimensional linear systems*, Proc. of the MTNS Symposium, Padova, Italy, July 1998, pp. 547–550.

[14] M. Vidyasagar, *Control Systems Synthesis: A Factorization Approach*, M.I.T. Press, 1985.

[15] G. Weiss, *Transfer functions of regular linear systems. Part I: characterizations of regularity*, Trans. Amer. Math. Soc., **342** (1994), 827–854.

[16] G. Weiss, *Regular linear systems with feedback*, Mathematics for Control, Signals and Systems, **7** (1994), 23–57.

[17] G. Weiss and R. F. Curtain, *Dynamic stabilization of regular linear systems*, IEEE Trans. Aut. Control, **42** (1997), 1–18.

[18] G. Weiss and R. Rebarber, *Optimizability and estimatability for infinite-dimensional linear systems*, SIAM. J. Control and Optim., to appear in 2000.

Dept. of Mathematics,
University of Groningen,
9700 AV Groningen,
The Netherlands

Dept. of Electr. & Electronic Eng.,
Imperial College of Sci. & Technol.,
Exhibition Rd., London SW7 2BT,
United Kingdom

Physics and Electronics Laboratory,
TNO, Oude Waalsdorperweg 63,
2509 JG The Hague,
The Netherlands

Free Semigroup Algebras
A Survey

Kenneth R. Davidson

Abstract. This is a survey of recent results on free semigroup algebras, which are the WOT-closed algebras generated by n isometries with pairwise orthogonal ranges.

Free semigroup algebras are WOT-closed algebras generated by n isometries with pairwise orthogonal ranges. They were introduced in [27] as an interesting class of operator algebras in their own right. The prototype algebra, obtained from the left regular representation of the free semigroup on n letters, was introduced by Popescu [45] in connection with multi-variable non-commutative dilation theory. This algebra has a great deal of analytic structure associated to the unit ball in $\mathbb{C}^n$ which justifies its name as the *non-commutative analytic Toeplitz algebra*. The general free semigroup algebras contain interesting computable information about the unitary invariants for the n-tuple of generators. This has allowed the classification of large classes of representations of the Cuntz algebra. Such classifications are important in various applications of C^*-algebras. In particular, the work of Bratteli and Jorgensen [15] uses such representations to generate wavelets, and unitary invariants for special classes of representations are central to their work. In this article, we will survey results about free semigroup algebras themselves, with passing reference to various applications.

1. Connections to C^*-algebras

Let $S_1, \dots, S_n$ denote an n-tuple of isometries with pairwise orthogonal ranges. The orthogonality relations are given algebraically by $S_i^* S_j = \delta_{ij} I$ for $1 \le i, j \le n$, or equivalently by

$$S_i^* S_i = I \quad \text{for} \quad 1 \le i \le n \qquad \text{and} \qquad \sum_{i=1}^n S_i S_i^* \le I.$$

The C^*-algebra generated by such an n-tuple was introduced by Cuntz [21]. There are only two possibilities. When $\sum_{i=1}^n S_i S_i^* = I$, the C^*-algebra is $*$-isomorphic to the Cuntz algebra $\mathcal{O}_n$, which is simple and purely infinite. When $\sum_{i=1}^n S_i S_i^* < I$,

The author was partially supported by an NSERC grant.

there is again only one $*$-isomorphism class of C^*-algebras, known as the Cuntz–Toeplitz algebra $\mathcal{E}_n$. This algebra contains an ideal $\mathcal{K}$ generated by $I - \sum_{i=1}^n S_i S_i^*$ which is isomorphic to the compact operators $\mathfrak{K}$. This is the only proper ideal, and $\mathcal{E}_n/\mathcal{K}$ is isomorphic to $\mathcal{O}_n$.

Perhaps because there are only two possibilities for this C^*-algebra, it is difficult to use the C^*-algebraic structure to determine spatial invariants (up to unitary equivalence). A famous theorem of Glimm [32] states that, at least considering the parameterization of (cyclic) representations via the state space and the GNS construction, it is essentially impossible to determine complete unitary invariants. More precisely, Glimm's theorem states that for any non-type I C^*-algebra, there is no countable family of Borel functions on the state space which distinguish the corresponding representations up to unitary equivalence. As the only simple type I algebras are the finite matrix algebras $\mathfrak{M}_n$ and the compact operators $\mathfrak{K}$, it follows that this classification is impossible for most C^*-algebras.

Nevertheless, there are good reasons for wanting to do this. We mention two. Bob Powers [50] introduced the study of WOT-continuous $*$-endomorphisms of $\mathcal{B}(\mathcal{H})$. Such a map is determined by what it does to $\mathfrak{K}$, which has only one irreducible representation up to unitary equivalence, the identity map. So the restriction of this endomorphism π to $\mathfrak{K}$ is just a multiple, say n, of the identity map. This extended integer n is known as the *Powers index*. Note then that there are n isometries S_i for $1 \le i \le n$ so that $\pi(A) = \sum_{i=1}^n S_i A S_i^*$. These isometries are not unique, but they are unique up to an action of the $n \times n$ unitary group $\mathfrak{U}_n$, which amounts to the choice of a basis for the range of $\pi(E_{11})$, where E_{11} is a matrix unit of $\mathfrak{K}$. From this one finds that the C^*-algebra generated by these $S_1, \dots, S_n$ is unique, and the choice of generators is determined up to a so-called *gauge automorphism*. See [37, 18, 14] for further information.

The second application is work of Bratteli, Jorgensen and others [14, 15, 17, 34, 13] using representations of the Cuntz algebra to generate wavelets. They introduced [14] a class of representations known as *finitely correlated representations* in connection with the endomorphisms above. We shall see can that they be completely classified by exploiting a connection with dilation theory in our work. Their isometries act on $L^2(\mathbb{T})$, and have the form $S_i f(z) = m_i(z) f(z^n)$ where m_i are functions of modulus 1 satisfying the orthogonality relations

$$\sum_{k=1}^n m_i(z\omega^k)\overline{m_j(z\omega^k)} = \delta_{ij} n$$

where $\omega = e^{2\pi i/n}$ [16]. These are used to generate wavelet bases of $L^2(\mathbb{R})$ with $n-1$ mother wavelets under integer translation and n-fold dilation. The interested reader is referred to the papers cited in the references.

We make an elementary observation about the representation theory of $\mathcal{E}_n$. Suppose that π is a $*$-representation of $\mathcal{E}_n$. The restriction to $\mathcal{K}$ must be equivalent to a multiple α of the identity representation because $\mathcal{K}$ is isomorphic to the compact operators. Standard C^*-algebra theory (c.f.[23]) shows that the restriction of π to $\overline{\pi(\mathcal{K})\mathcal{H}}$ is also unitarily equivalent to $\mathrm{id}^{(\alpha)}$, and the restriction to the complement factors through $\mathcal{E}_n/\mathcal{K} \simeq \mathcal{O}_n$. Hence $\pi \simeq \mathrm{id}^{(\alpha)} \oplus \sigma q$ where q is the quotient map onto $\mathcal{O}_n$ and σ is a representation of $\mathcal{O}_n$. Below we shall see that this decomposition is equivalent to the Wold decomposition.

2. Definitions and Some Examples

The *free semigroup algebra* generated by an n-tuple of isometries with orthogonal ranges $S = (S_1, \dots, S_n)$ is the unital WOT-closed algebra $\mathfrak{S} = \overline{\mathrm{Alg}\{S_1, \dots, S_n\}}^{\mathrm{WOT}}$.

The fundamental example is the left regular representation of the free semigroup $\mathbb{F}_n^+$ consisting of all words in n non-commuting letters $\{1, 2, \dots, n\}$. The empty word $\varnothing$ is the identity element for $\mathbb{F}_n^+$. The Hilbert space $\mathcal{K}_n = \ell^2(\mathbb{F}_n^+)$ has the orthonormal basis $\{\xi_w : w \in \mathbb{F}_n^+\}$. Isometries L_v are defined for every word $v \in \mathbb{F}_n^+$ by $L_v \xi_w = \xi_{vw}$. In particular, the generators determine the n-tuple of isometries $L_1, \dots, L_n$. Notice that the L_i have orthogonal ranges spanned by all words beginning with the letter i for $1 \le i \le n$, and that the vector $\xi_\varnothing$ spans the complement of the sum $\sum_{i=1}^n L_i \mathcal{K}_n$. The free semigroup algebra generated by $L = (L_1, \dots, L_n)$ is denoted by $\mathfrak{L}_n$ and is called the *non-commutative analytic Toeplitz algebra*.

The semigroup $\mathbb{F}_n^+$ is graded by the length function $|w|$ of a word w, since $|vw| = |v|\,|w|$. The space $\mathcal{K}_n$ is known as Fock space, and may be decomposed into a direct sum of subspaces $\mathcal{K}_n = \sum_{k \ge 0}^{\oplus} \mathcal{H}_k$ where $\mathcal{H}_k = \mathrm{span}\{\xi_w : |w| = k\}$. Then $\dim \mathcal{H}_k = n^k$. Thus $\mathcal{H}_0 = \mathbb{C}$, $\mathcal{H}_1 \simeq \ell_n^2$ and $\mathcal{H}_k \simeq \mathcal{H}_1^{\otimes k}$, the tensor product of k copies of n-dimensional space. The isometries L_i are known as creation operators in this context, and are written by mathematical physicists as $L_i \zeta = \xi_i \otimes \zeta$ in this presentation of $\mathcal{K}_n$.

Since $P_0 := I - \sum_{i=1}^n L_i L_i^* = \xi_0 \xi_0^*$ is non-zero, it follows that $\mathrm{C}^*(L) = \mathcal{E}_n$. Note also that $\mathbb{F}_n^+$ has a right regular representation given by $R_i \xi_w = \xi_{wi}$ for $1 \le i \le n$. It is routine to verify that $R_v \xi_w = \xi_{w\tilde{v}}$ where $\tilde{v}$ is the word v in reverse order. Clearly the free semigroup algebra $\mathfrak{R}_n$ generated by $R = (R_1, \dots, R_n)$ commutes with $\mathfrak{L}_n$. Moreover this algebra is unitarily equivalent to $\mathfrak{L}_n$ via the unitary $W \xi_w = \xi_{\tilde{w}}$.

When $n = 1$, this representation just yields the unilateral shift. In this case, $\mathfrak{L}_1$ is just the classical analytic Toeplitz algebra on H^2 isomorphic to H^∞ of the unit disk. Also $\mathfrak{R}_1 = \mathfrak{L}_1 = \mathfrak{L}_1'$. When $n \ge 2$, there is much analogous analytic structure associated to the unit ball in $\mathbb{C}^n$ which will be examined in detail later. At this point, we content ourselves with a simple observation. Each element $A \in \mathfrak{L}_n$ is determined by $A\xi_\varnothing = \sum_{v \in \mathbb{F}_n^+} a_v \xi_v$ because

$$A\xi_w = AR_{\tilde{w}}\xi_\varnothing = R_{\tilde{w}} A \xi_\varnothing = \sum_{v \in \mathbb{F}_n^+} a_v \xi_{vw}.$$

At least when this is a finite sum, it follows that $A = \sum_{v \in \mathbb{F}_n^+} a_v L_v$. In general, we think of this infinite sum as a Fourier series of analytic type.

In fact, if we define maps $\Phi_k(A) = \sum_{|v|=k} a_w L_w$, then Cesaro means may be defined by $\Sigma_k(A) = \sum_{j=0}^{k-1} (1 - \frac{j}{k}) \Phi_j(A)$. These are completely positive unital maps of $\mathfrak{L}_n$ into itself which converge to the identity map in the pointwise strong operator topology [27, Lemma 1.1].

Now suppose that $S = (S_1, \dots, S_n)$ is an arbitrary set of isometries with orthogonal ranges. We shall write S_v for the monomial $v(S) = S_{i_1} \dots S_{i_k}$ for any word $v = i_1 \dots i_k \in \mathbb{F}_n^+$. A *wandering subspace* is a subspace $\mathcal{W}$ such that the subspaces $\{S_v \mathcal{W} : v \in \mathbb{F}_n^+\}$ are pairwise orthogonal. The span of these subspaces is the $\mathfrak{S}$–invariant subspace $\mathfrak{S}[\mathcal{W}]$ generated by $\mathcal{W}$. Clearly the restriction of the isometries S_i to $\mathfrak{S}[\mathcal{W}]$ is unitarily equivalent to a direct sum of $\dim \mathcal{W}$ copies of the left regular representation. We shall see that such subspaces are omnipresent in all known examples.

Popescu [42] established the analogue of the Wold decomposition in this context, and the proof is essentially the same as for a single isometry. Let $\mathcal{W} = (I - \sum_{i=1}^n S_i S_i^*)\mathcal{H}$. It is readily verified that $\mathcal{W}$ is a wandering space and that $\mathfrak{S}[\mathcal{W}]$ is a *reducing* subspace. The restriction to $\mathfrak{S}[\mathcal{W}]$ is a multiple of the left regular representation, as noted above, and the restriction to the complement yields a representation of the Cuntz algebra because the sum of the ranges there is the whole space. The C^*-algebraic view of this was mentioned in the previous section.

A representation which is a multiple of the left regular representation is called *pure*, and a representation which generates the Cuntz algebra is said to be of *Cuntz type*. There are many representations of the latter type. We provide a few to keep in mind.

Example 2.1. *Let $u = i_1 \dots i_k$ be a word in $\mathbb{F}_n^+$ and let $\lambda \in \mathbb{T}$ be a scalar of modulus one. Define a Hilbert space $\mathcal{K}_u \simeq \mathbb{C}^k \oplus \mathcal{K}_n^{k(n-1)}$ with orthonormal basis $\zeta_1, \dots, \zeta_k$ for $\mathbb{C}^k$ and index the copies of $\mathcal{K}_n$ by (s,j), where $1 \le s \le k$, $1 \le j \le n$ and $j \ne i_s$, with basis $\{\xi_{s,j,w} : w \in \mathbb{F}_n^+\}$. Define a representation $\sigma_{u,\lambda}$ of $\mathbb{F}_n^+$ and isometries $S_i = \sigma_{u,\lambda}(i)$ by*

$$\begin{aligned} S_i \zeta_s &= \zeta_{s-1} && \text{if } \ i = i_s,\ s > 1 \\ S_i \zeta_1 &= \lambda \zeta_k && \text{if } \ i = i_1 \\ S_i \zeta_s &= \xi_{s,i,\varnothing} && \text{if } \ i \ne i_s \\ S_i \xi_{s,j,w} &= \xi_{s,j,iw} && \text{for all } \ i, s, j, w \end{aligned}$$

Notice that each $\xi_{s,j,\varnothing}$ is a wandering vector generating the (s,j)th copy of $\mathcal{K}_n$. The sum of the ranges of the S_i's is the whole space, so this is a Cuntz representation. The copy of $\mathbb{C}^k$ is invariant for the S_i^*'s and each basis vector ζ_s is a cyclic vector for $\mathfrak{S}$. Think of this as a ring of k basis vectors being permuted around the ring by the appropriate S_{i_s} at each point. Off of each node ζ_s there are $n-1$ copies of Fock space, one for each $i \ne i_s$.

The reader can quickly verify that every wandering vector is orthogonal to $\operatorname{span}\{\zeta_1, \dots, \zeta_k\}$. *Thus the span of all wandering vectors is precisely* $\operatorname{span}\{\zeta_1, \dots, \zeta_k\}^\perp$. *We shall be able to detect this in $\mathfrak{S}$ because it turns out that the projection*

P onto span$\{\zeta_1, \dots, \zeta_k\}$ *belongs to $\mathfrak{S}$, and is the largest projection in $\mathfrak{S}$ such that $P\mathfrak{S}P$ is self-adjoint. This will allow us to completely describe the algebra $\mathfrak{S}$, and to decompose the corresponding representation of $\mathcal{O}_n$ into a direct sum of irreducible representations. In particular, notice that the restriction of $\mathfrak{S}$ to $P^\perp\mathcal{H}$ is unitarily equivalent to the direct sum of $k(n-1)$ copies of $\mathfrak{L}_n$. Since $P \in \mathfrak{S}$, we need only understand $\mathfrak{S}P$ to have a complete picture of the algebra $\mathfrak{S}$. This turns out to be $\mathfrak{W}P$ where $\mathfrak{W}$ is the von Neumann algebra generated by S. Moreover because $P\mathfrak{W}P$ is the finite dimensional C^*-algebra on $P\mathcal{H}$ generated by the restrictions of the S_i^*'s, it will be straightforward to completely analyze this example.*

Example 2.2. *Let $x = i_1 i_2 i_3 \dots$ be an infinite word in $\{1, \dots, n\}$. Define a sequence $x_m = i_1 i_2 \dots i_m$ for $m \geq 0$. Let $\mathbb{F}_n^+ x^{-1}$ denote the collection of words in the free group on n generators of the form $v = u x_m^{-1}$ for u in $\mathbb{F}_n^+$ and some $m \geq 0$. Identify words which are the same after cancellation, namely $u x_m^{-1} = (u i_{m+1}) x_{m+1}^{-1}$. Let $\mathcal{H}_x$ be the Hilbert space with orthonormal basis $\{\xi_v : v \in \mathbb{F}_n^+ x^{-1}\}$. Define a representation π_x of $\mathbb{F}_n^+$ and isometries $S_i = \pi_x(i)$ on $\mathcal{H}_x$ by $\pi_x(w)\xi_v = \xi_{wv}$ for $v \in \mathbb{F}_n^+ x^{-1}$ and $w \in \mathbb{F}_n^+$.*

This is also a Cuntz representation because it is readily apparent that each basis vector is the image of some other basis vector, whence the sum of the ranges of the S_i's is the whole space. Indeed, notice that each $\xi_{x_m^{-1}}$ is a wandering vector which generates $\ell^2(\mathbb{F}_n^+ x_m^{-1})$. Consequently $\mathcal{H}_x$ is expressed as the increasing union of a nested sequence of invariant subspaces on which $\mathfrak{S}$ acts like the left regular representation. Such Cuntz representations are in a natural sense inductive limits of (possibly multiples of) the left regular representation, and hence are said to be of inductive type. In particular, it is easy to convince yourself that the algebra $\mathfrak{S}$ is canonically isometrically isomorphic to $\mathfrak{L}_n$.

3. Dilation Theory

The Sz.-Nagy dilation theorem [56] shows that an arbitrary contraction on a Hilbert space $\mathcal{H}$ dilates to an isometry on a larger space. Frahzo [31] and Bunce [19] observed that there is a natural analogue for a contractive n-tuple $A = (A_1, \dots, A_n)$. There is no commutativity required, only the norm condition $\|A\| = \big\| \sum A_i A_i^* \big\|^{1/2} \leq 1$. Perhaps the simplest proof uses the Schaeffer construction. Namely consider $D_A = (I_n - A^*A)^{1/2}$ which acts on the direct sum $\mathcal{H}^{(n)}$ of n copies of $\mathcal{H}$ (where I_n is the identity operator on $\mathcal{H}^{(n)}$). Then $\begin{bmatrix} A \\ D_A \end{bmatrix}$ is an isometry from $\mathcal{H}^{(n)}$ into $\mathcal{H} \oplus \mathcal{H}^{(n)}$. Thus the n columns $\begin{bmatrix} A_i \\ X_i \end{bmatrix}$ of this isometry, for $1 \leq i \leq n$, are isometries from $\mathcal{H}$ into $\mathcal{H} \oplus \mathcal{H}^{(n)}$ with pairwise orthogonal ranges. Consider the Hilbert space $\mathcal{K} = \mathcal{H} \oplus \big(\mathcal{H}^{(n)} \otimes \mathcal{K}_n\big)$, and define isometries

$$S_i = \begin{bmatrix} A_i & 0 \\ X_i & I_n \otimes L_i \end{bmatrix}$$

There is a slight abuse of notation where we identify X_i with its composition with the natural inclusion of $\mathcal{H}^{(n)}$ onto $\mathcal{H}^{(n)} \otimes \mathbb{C}\xi_\varnothing$ as a subspace of $\mathcal{H}^{(n)} \otimes \mathcal{K}_n$.

Popescu [42] showed that there is a unique minimal dilation, and extends this argument to a countable row contraction. The minimal dilation is obtained from the construction above by restricting each S_i to $\operatorname{span}\{S_v\mathcal{H} : v \in \mathbb{F}_n^+\}$, which is evidently the minimal invariant subspace for the S_i's containing $\mathcal{H}$ in this dilation.

Popescu pursues the analogy with the one-variable case and finds many parallels with the Sz.-Nagy–Foiaş theory. In particular, he establishes the analogue of the commutant lifting theorem [42]: if X is a contraction commuting with $A_1, \dots, A_n$, then X dilates to a contraction Y on the larger space which commutes with the dilation $S_1, \dots, S_n$.

There is a natural analogue [45] of the von Neumann inequality. The point is the same as Sz.-Nagy's proof in one variable, namely if p is a non-commuting polynomial in n-variables, then the $1,1$ entry of $p(S_1, \dots, S_n)$ is just $p(A_1, \dots, A_n)$. Hence

$$\|p(A_1, \dots, A_n)\| \leq \|p(S_1, \dots, S_n)\|.$$

Moreover the right hand side does not depend on the particular dilation! Indeed, this norm may be computed in $\mathrm{C}^*(S)$, and so there are only two possibilities, namely $\mathcal{E}_n$ and $\mathcal{O}_n$. Since $\mathcal{O}_n$ is a quotient of $\mathcal{E}_n$, the norm is apparently at least as large in $\mathcal{E}_n$. Thus

$$\|p(A_1, \dots, A_n)\| \leq \|p(L_1, \dots, L_n)\|.$$

However on the norm closed unital algebra $\mathcal{A}_n$ generated by $L_1, \dots, L_n$, the quotient map is completely isometric. Popescu calls this algebra $\mathcal{A}_n$ the *non-commutative disk algebra.*

Evidently von Neumann's inequality allows the immediate construction of a functional calculus on $\mathcal{A}_n$ for every row contraction A. Following Arveson's approach to non-commutative dilation theory [6, 7], it follows that there is a unique unital completely positive map from $\mathcal{O}_n$ onto $\mathrm{C}^*(A_1, \dots, A_n)$ which sends S_i onto A_i for $1 \leq i \leq n$. In [49], Popescu develops an analogue of the Poisson kernel to provide an explicit formula for this map. Since $\mathcal{O}_n$ is simple, it is the smallest C^*-algebra which contains $\mathcal{A}_n$ completely isometrically. By Hamana's Theorem [33], there is a unique minimal C^*algebra containing any unital operator algebra completely isometrically, which Arveson named the *C^*-envelope.* Hence $\mathcal{O}_n$ is the C^*-envelope of $\mathcal{A}_n$. This shows in particular that the algebras $\mathcal{A}_n$ are not isomorphic for different values of n [48].

In the one variable case, it is desirable when possible to extend the disk algebra functional calculus to $H^\infty(\mathbb{D})$. This requires the contraction to dilate to an absolutely continuous isometry (meaning that the unitary part of the Wold decomposition has a spectral measure which is absolutely continuous with respect to Lebesgue measure on the circle). Popescu [45, 46] identifies the corresponding class of row contractions and extends the functional calculus to $\mathfrak{L}_n$, which Popescu calls $\mathcal{F}_n^\infty$. A special case which is especially tractable occurs when $\|A\| = r < 1$.

In this instance, Bunce [19] basically showed that there is a unique way to extend the functional calculus which sends L_i to A_i to all of $\mathfrak{L}_n$ by setting $\Phi(X) = \sum_{w\in\mathbb{F}_n^+} x_w w(A)$ for $X = \sum_{w\in\mathbb{F}_n^+} x_w L_w$ in $\mathfrak{L}_n$. The strict contraction condition may be used to show that $\| \sum_{|w|=k} x_w w(A)\| \le r^k \|X\|$, and thus $\Phi(X)$ is a norm convergent series.

We also mention [41, 43] as related papers of Popescu developing Sz.-Nagy–Foiaş models.

4. The Spatial Structure of $\mathfrak{L}_n$

In this section, we develop the spatial structure theory of the non-commutative analytic Toeplitz algebra $\mathfrak{L}_n$. The results of this section are due to Popescu [43, 47], myself and Pitts [27] and Arias and Popescu [4]. Our goal is to stress the analogue with the analytic Toeplitz algebra $\mathcal{T}(H^\infty)$ generated by the unilateral shift.

Suppose that A is an operator commuting with the right regular algebra $\mathfrak{R}_n$. Then $A\xi_\varnothing = \sum_{v\in\mathbb{F}_n^+} a_v \xi_v$. The Cesaro means defined in section 2 suggest that we consider the sequence $\sum_{|v|<k} \big(1 - \frac{|v|}{k}\big) a_v L_v$. These converge strongly to A, which shows that the commutant of $\mathfrak{R}_n$ is just $\mathfrak{L}_n$. Conversely $\mathfrak{L}_n' = \mathfrak{R}_n$ as these algebras are unitarily equivalent.

This leads to several immediate consequences, such as the fact that $\mathfrak{L}_n$ is inverse closed, i.e. if $A \in \mathfrak{L}_n$ is invertible in $\mathcal{B}(\mathcal{K}_n)$, then the inverse lies in $\mathfrak{L}_n$. Also since $\mathfrak{R}_n$ contains isometries with orthogonal ranges, it is immediate that $\|A\|_e = \|A\|$ for all $A \in \mathfrak{L}_n$. In particular this establishes the fact that the quotient from $\mathcal{E}_n$ to $\mathcal{O}_n$ is (completely) isometric on $\mathcal{A}_n$.

With a bit more work, we see that $\mathfrak{L}_n$ contains no normal elements which are not scalar. To see this, notice that $\xi_\varnothing$ is in the kernel of each L_i^*, and thus is an eigenvector for every A^* with $A \in \mathfrak{L}_n$. Indeed, $A^*\xi_\varnothing = \overline{a_\varnothing}\xi_\varnothing$. Suppose that A were normal. Then $\xi_\varnothing$ would also be an eigenvector for A and so $A\xi_\varnothing = a_\varnothing \xi_\varnothing$. From the Fourier series, it follows that $A = a_\varnothing I$.

A more careful study shows [27, Theorem 1.7] that every non-zero element of $\mathfrak{L}_n$ is injective and has non-zero connected essential spectrum $\sigma_e(A) = \sigma(A)$. In particular, $\mathfrak{L}_n$ contains no proper projections or non-zero quasinilpotents. Hence $\mathfrak{L}_n$ is a semisimple algebra.

The first deep connection to function theory is an analogue of the Beurling Theorem for $\mathfrak{L}_n$ [43] (c.f. [27]). The Beurling Theorem [10] says that every invariant subspace of the unilateral shift represented on H^2 as the Toeplitz operator T_z has the form wH^2 where w is an inner function ($w \in H^\infty(\mathbb{D})$ and $|w(e^{i\theta})| = 1$ *a.e.*). These subspaces are always cyclic (w is the cyclic vector) and they are the range of the isometry T_w. Now the only isometries in the Toeplitz algebra $\mathcal{T}(H^\infty)$ are of the form T_w for w inner, and $\mathcal{T}(H^\infty)$ is maximal abelian. So Beurling's theorem may be restated as saying that the invariant subspaces of the unilateral shift are the ranges of isometries in the commutant. This is the form which generalizes.

Theorem 4.1. *Every invariant subspace of $\mathfrak{L}_n$ is the direct sum of cyclic subspaces. The cyclic invariant subspaces of $\mathfrak{L}_n$ are precisely the ranges of isometries in $\mathfrak{R}_n$; and the choice of isometry is unique up to a scalar.*

The proof starts with an invariant subspace $\mathcal{M}$ and forms the wandering space $\mathcal{W} = \mathcal{M} \ominus \left(\sum_i L_i \mathcal{M}\right)$. The grading on Fock space makes it straightforward to see that there can be no Cuntz part, and thus $\mathcal{M}$ splits into a direct sum of cyclic subspaces obtained by choosing an orthonormal basis for $\mathcal{W}$. For each such vector ζ, one constructs an isometry R_ζ in $\mathfrak{R}_n$ by setting $R_\zeta \xi_w = L_w \zeta$ and extending by linearity. This is the desired isometry onto the cyclic subspace $\mathfrak{L}_n[\zeta]$.

This naturally leads to an inner–outer factorization as follows. Take any $A \in \mathfrak{L}_n$. The subspace $\overline{\mathrm{Ran}(A)}$ is invariant for $\mathfrak{R}_n$ with cyclic vector $A\xi_\varnothing$. Thus it is the range of an isometry $L \in \mathfrak{L}_n$. Therefore A factors as $A = LB$ where $B = L^*A$. One verifies that B commutes with each R_i, and thus lies in $\mathfrak{L}_n$. Moreover B evidently has dense range, and by analogy is called an *outer operator.*

Like $\mathcal{T}(\mathcal{H}^\infty)$, the algebra $\mathfrak{L}_n$ has a rich collection of invariant subspaces of co-dimension one. These correspond to eigenvectors for the adjoint algebra $\mathfrak{L}_n^*$. When $n = 1$, each point $\lambda \in \mathbb{D}$ determines a kernel function $k_\lambda = (1 - |\lambda|^2)^{1/2}(1 - \overline{\lambda} z)^{-1}$ which are the eigenvectors for the backward shift $T_z^* k_\lambda = \overline{\lambda} k_\lambda$. This vector yields a weak-$*$ continuous point evaluation $h(\lambda) = \langle T_h k_\lambda, k_\lambda\rangle$ for $h \in H^\infty$. Exactly the same thing occurs for $n \geq 2$ corresponding to each point λ in the unit ball $\mathbb{B}_n$ of $\mathbb{C}^n$ [27, Theorem 2.6].

Theorem 4.2. *The eigenvectors for $\mathfrak{L}_n^*$ are the vectors*

$$\nu_\lambda = (1 - \|\lambda\|^2)^{1/2} \sum_{w \in \mathbb{F}_n^+} \overline{w(\lambda)} \xi_w = (1 - \|\lambda\|^2)^{1/2} \Big(I - \sum_{i=1}^n \overline{\lambda_i} L_i\Big)^{-1} \xi_\varnothing$$

for λ in the unit ball $\mathbb{B}_n$. They satisfy $L_i^* \nu_\lambda = \overline{\lambda_i} \nu_\lambda$ for $1 \leq i \leq n$.

Moreover $\langle p(L)\nu_\lambda, \nu_\lambda\rangle = p(\lambda)$ for every polynomial $p = \sum_w a_w w$ in the semi-group algebra $\mathbb{C}\mathbb{F}_n^+$. This extends to a WOT*–continuous multiplicative linear functional on $\mathfrak{L}_n$ given by $\varphi_\lambda(A) = \langle A\nu_\lambda, \nu_\lambda\rangle$.*

The vector ν_λ is cyclic for $\mathfrak{L}_n$. The subspace $\{\nu_\lambda\}^\perp$ is $\mathfrak{L}_n$–invariant, and its wandering subspace is n-dimensional.

The general issue of the structure of weak-$*$ continuous functionals on operator algebras and the connection to reflexivity has its roots in the Scott Brown technique c.f.[11]. The fact that $\mathfrak{L}_n' = \mathfrak{R}_n$ contains isometries with orthogonal ranges makes the following result straightforward [29]. A somewhat more delicate analysis [12] shows that $\mathfrak{L}_n$ has the related property $X_{0,1}$ which we will not define.

Theorem 4.3. *Let φ be a weak-$*$ continuous linear functional on the spatial tensor product $\mathcal{B}(\mathcal{H}) \otimes \mathfrak{L}_n$ with $\|\varphi\| < 1$. Then there are unit vectors η and ζ so that $\varphi(A) = \langle A\eta, \zeta\rangle$ for all $A \in \mathcal{B}(\mathcal{H}) \otimes \mathfrak{L}_n$.*

In particular, every weak-$*$ continuous functional on $\mathfrak{L}_n$ is given by a vector functional. So one obtains the important consequence that the weak-$*$ and WOT topologies coincide on $\mathfrak{L}_n$.

We now have in hand ample information to study reflexivity. An operator algebra $\mathfrak{A}$ is *reflexive* if the algebra can be recovered from its invariant subspace lattice $\mathcal{L} = \mathrm{Lat}(\mathfrak{A})$ as the set $\mathrm{Alg}(\mathcal{L})$ of all operators leaving each subspace invariant. The lattice $\mathcal{L}$ determines a seminorm on $\mathcal{B}(\mathcal{H})$ by

$$\beta_{\mathcal{L}}(T) := \sup_{L\in\mathcal{L}} \|L^{\perp}TL\|.$$

Clearly, $\beta_{\mathcal{L}}(T) = 0$ precisely when T belongs to $\mathrm{Alg}(\mathcal{L})$. Moreover, it is elementary to show that

$$\beta_{\mathcal{L}}(T) \le \mathrm{dist}(T, \mathrm{Alg}(\mathcal{L})) \quad \text{for all} \quad T \in \mathcal{B}(\mathcal{H}).$$

The algebra is said to be *hyper-reflexive* if these norms are comparable. In this case, the hyper-reflexivity constant is the smallest number C such that

$$\mathrm{dist}(T, \mathrm{Alg}(\mathcal{L})) \le C\beta_{\mathcal{L}}(T) \quad \text{for all} \quad T \in \mathcal{B}(\mathcal{H}).$$

When an operator $\mathfrak{A}$ is the commutant of another algebra, there is another natural measure of the distance to $\mathfrak{A}$. Define the derivation δ_T by $\delta_T(X) = XT - TX$. Notice that the restriction of δ_T to $\mathfrak{A}'$ is zero if and only if T belongs to $\mathfrak{A}'' = \mathfrak{A}$. Suppose that $\mathrm{dist}(T, \mathfrak{A}) = d$ and choose $A \in \mathfrak{A}$ such that $\|T - A\| = d$. This is possible because $\mathfrak{A}$ is WOT-closed. Then

$$\|\delta_T|_{\mathfrak{A}'}\| = \|\delta_{T-A}|_{\mathfrak{A}'}\| \le 2\|T - A\| = 2d.$$

On the other hand, it is not automatic that the distance to $\mathfrak{A}$ is bounded by a constant times $\|\delta_T|_{\mathfrak{A}'}\|$.

The list of algebras known to be hyper-reflexive is rather short. It includes nest algebras which have constant 1 due to the Arveson Distance Formula [8]. Christensen [20] showed that injective von Neumann algebras have constant at most 4. Von Neumann algebras with abelian commutant have constant at most 2, as do abelian von Neumann algebras [52]. Von Neumann algebras are also commutants. Since von Neumann algebras are spanned by their projections, it is easy to relate the quantities $\beta_{\mathrm{Lat}\,\mathfrak{A}}(T)$ and $\|\delta_T|_{\mathfrak{A}'}\|$ in this situation. Lastly, the case most closely related to our study is the analytic Toeplitz algebra $\mathcal{T}(H^\infty)$, which has distance constant at most 19 [22].

The non-commutative analytic Toeplitz algebras $\mathfrak{L}_n$ are reflexive [4]. Indeed they are hyper-reflexive [27]. The constant we found was 51. However Bercovici [12] obtained a beautiful general result that yields distance constant 3 for any algebra with property $X_{0,1}$. We have:

Theorem 4.4. *The algebras $\mathfrak{L}_n$ are hyper-reflexive. Moreover, for all T in $\mathcal{B}(\mathcal{K}_n)$,*

$$\frac{1}{3}\,\mathrm{dist}(T, \mathfrak{L}_n) \le \sup_{L\in\mathrm{Lat}(\mathfrak{L}_n)} \|L^{\perp}TL\| \le \|\delta_T|_{\mathfrak{R}_n}\| \le 2\,\mathrm{dist}(T, \mathfrak{L}_n).$$

5. The Algebraic Structure of $\mathfrak{L}_n$

This section deals with [28] which develops the algebraic structure of $\mathfrak{L}_n$, culminating in a description of the automorphism group, and ties the algebra even more strongly to analytic function theory on the complex n-ball.

5.1. Ideals

First consider WOT-closed ideals. Let $\mathrm{Id}_r(\mathfrak{L}_n)$, $\mathrm{Id}_l(\mathfrak{L}_n)$ and $\mathrm{Id}(\mathfrak{L}_n)$ denote the sets of all WOT-closed right, left and two-sided ideals respectively. Suppose that $\mathfrak{J}$ belongs to $\mathrm{Id}_r(\mathfrak{L}_n)$. Observe that the subspace $\overline{\mathfrak{J}\xi_\varnothing}$ belongs to $\mathrm{Lat}\,\mathfrak{R}_n$. To see this, note that

$$\mathfrak{R}_n\overline{\mathfrak{J}\xi_\varnothing} = \overline{\mathfrak{J}\mathfrak{R}_n\xi_\varnothing} = \overline{\mathfrak{J}\mathcal{K}_n} = \overline{\mathfrak{J}\mathfrak{L}_n\xi_\varnothing} = \overline{\mathfrak{J}\xi_\varnothing}.$$

Thus $\overline{\mathfrak{J}\xi_\varnothing} = \overline{\mathfrak{J}\mathcal{K}_n}$ is the range of $\mathfrak{J}$ and is $\mathfrak{R}_n$ invariant. Similarly when $\mathfrak{J}$ belongs to $\mathrm{Id}_\ell(\mathfrak{L}_n)$, we have $\mathfrak{L}_n\mathfrak{J}\xi_\varnothing = \mathfrak{J}\xi_\varnothing$; so $\overline{\mathfrak{J}\xi_\varnothing}$ is $\mathfrak{L}_n$ invariant. Hence when $\mathfrak{J}$ is a two-sided ideal, $\overline{\mathfrak{J}\xi_\varnothing}$ belongs to $\mathrm{Lat}(\mathfrak{L}_n) \cap \mathrm{Lat}(\mathfrak{R}_n)$. We define the map $\mu(\mathfrak{J}) = \overline{\mathfrak{J}\xi_\varnothing}$.

For the moment, we consider right ideals. Left ideals are *not* handled in the same way. Suppose that $\mathcal{M}$ is an invariant subspace for $\mathfrak{R}_n$. Define $\iota(\mathcal{M}) = \{J \in \mathfrak{L}_n : J\xi_\varnothing \in \mathcal{M}\}$. Clearly this is a WOT-closed subspace of $\mathfrak{L}_n$. For any $J \in \iota(\mathcal{M})$ and $A \in \mathfrak{L}_n$,

$$JA\xi_\varnothing \in \overline{J\mathcal{K}_n} = \overline{J\mathfrak{R}_n\xi_\varnothing} = \overline{\mathfrak{R}_n J\xi_\varnothing} \subset \mathcal{M}.$$

Whence $\iota(\mathcal{M})$ is a right ideal. Likewise, if $\mathcal{M} \in \mathrm{Lat}(\mathfrak{L}_n)$, then $\iota(\mathcal{M})$ is a WOT-closed left ideal.

It is easy to see that μ respects sums: $\mu(\mathfrak{J}_1 + \mathfrak{J}_2) = \mu(\mathfrak{J}_1) \vee \mu(\mathfrak{J}_2)$. However it also respects intersections, which relies on the fact that μ turns out to be a bijection. This fact relies on the Beurling Theorem 4.1 and Theorem 4.3 on weak-$*$ continuous functionals.

Theorem 5.1. *Let* $\mu : \mathrm{Id}_r(\mathfrak{L}_n) \to \mathrm{Lat}(\mathfrak{R}_n)$ *be given by* $\mu(\mathfrak{J}) = \overline{\mathfrak{J}\xi_\varnothing}$. *Then* μ *a complete lattice isomorphism. The restriction of* μ *to the set* $\mathrm{Id}(\mathfrak{L}_n)$ *is a complete lattice isomorphism onto* $\mathrm{Lat}\,\mathfrak{L}_n \cap \mathrm{Lat}\,\mathfrak{R}_n$. *The inverse map is* ι.

It turns out that there are significant differences between right and left ideals. The tight correspondence between right ideals and invariant subspaces, which also works well for two sided ideals, has no good corresponding result for left ideals. The reason is that there are good factorization results for elements of $\mathfrak{L}_n$ with isometries on the left, but not on the right. Indeed, notice that if L is an isometry, then $\|LA\| = \|A\|$, but $\|AL\|$ may be much smaller. Kribs [35] found significant pathology in the structure of left ideals which arose because of strange factorization results. For example, the isometry L_1 cannot be properly factored as $L_1 = AB$ if $\|A\| = \|B\| = 1$. However it can be properly factored in $\mathfrak{L}_n$ if the norm condition is dropped. Results of this kind are used to show that the map μ is neither one-to-one nor surjective from $\mathrm{Id}_l(\mathfrak{L}_n)$ to $\mathrm{Lat}(\mathfrak{L}_n)$.

A weak-$*$ continuous character on $\mathfrak{L}_n$ has a kernel which is a weak-$*$ closed two-sided ideal of co-dimension one. As the weak-$*$ and WOT topologies coincide, it is WOT-closed. Its range is an invariant subspace of co-dimension one, and thus it determines an eigenvector for $\mathfrak{L}_n^*$. This leads to the conclusion that the only weak-$*$ continuous characters are the maps φ_λ for $\lambda \in \mathbb{B}_n$ which we found in the previous section.

One important ideal which plays a central role is the WOT-closed commutator ideal $\mathfrak{C}$. Let $\mathcal{K}_n^s$ denote symmetric Fock space, which consists of all vectors of the form $\zeta = \sum_{w \in \mathbb{F}_n^+} a_w \xi_w$ such that $a_w = a_v$ if $w(z) = v(z)$, where evaluation is taken at a commuting n-tuple $z = (z_1, \dots, z_n)$ of indeterminates. In other words, if $|w| = k$ and $\sigma \in S_k$ is a permutation of k symbols and if $\sigma(w)$ denotes the permutation of the k letters in w via σ, then $a_{\sigma(w)} = a_w$. In particular, the eigenvectors ν_λ are evidently symmetric, and in fact they span $\mathcal{K}_n^s$. The main result about $\mathfrak{C}$ is that

$$\mathfrak{C} = \bigcap_{\lambda \in \mathbb{B}_n} \ker \varphi_\lambda \qquad \text{and} \qquad \mu(\mathfrak{C}) = \mathcal{K}_n^{s\perp}.$$

Another set of important ideals are denoted $\mathfrak{L}_n^{0,k}$. The ideal $\mathfrak{L}_n^0 = \ker \varphi_0$ consists of those elements whose leading Fourier coefficient vanishes. The ideals $\mathfrak{L}_n^{0,k}$ are the kth powers of this ideal, and consist of elements whose Fourier coefficients vanish up to order $k-1$. It is clear that the range $\mu(\mathfrak{L}_n^{0,k}) = \operatorname{span}\{\xi_w : |w| \geq k\}$. This subspace has wandering dimension n^k. The ideal structure can be used to show that every element of $\mathfrak{L}_n^{0,k}$ factors as a sum $A = \sum_{|w|=k} L_w A_w = LB$ where L is the row isometry with n^k entries L_w for $|w| = k$, and B, the column operator with the n^k entries A_w, has norm $\|A\|$. This norm control is crucial in later developments.

5.2. Representations

In the category of unital operator algebras, we take the viewpoint that the natural representations are completely contractive and unital. Given an operator algebra $\mathfrak{A}$, for each $1 \leq k < \infty$, we let $\operatorname{Rep}_k(\mathfrak{A})$ denote the set of completely contractive representations of $\mathfrak{A}$ into $\mathcal{B}(\mathcal{H}_k)$, where $\mathcal{H}_k$ is a fixed Hilbert space of dimension k. Put the topology of pointwise convergence on this space. Even though $\mathfrak{L}_n$ has a natural weak-$*$ topology, there are good reasons to consider norm continuous representations which are not weak-$*$ continuous at this juncture. This space is a reasonable topological object because k is finite, and we can obtain a lot of information from these finite dimensional representations.

There are many irreducible representations of $\mathfrak{L}_n$ (even WOT-continuous ones) on spaces of every dimension. Indeed, take any n-tuple $A = (A_1, \dots, A_n)$ of $k \times k$ matrices such that $\|A\| = r < 1$ and so that the A_i's generate the full matrix algebra $\mathfrak{M}_k$. Then the functional calculus of Bunce and Popescu which comes out of the dilation theory yields a WOT-continuous completely contractive representation sending L_i to A_i.

Conversely, suppose that Φ is a completely contractive representation of $\mathfrak{L}_n$ on $\mathcal{B}(\mathcal{H})$ with $\dim \mathcal{H} = k < \infty$. Then

$$\Phi(L) = \begin{bmatrix}\Phi(L_1) & \dots & \Phi(L_n)\end{bmatrix} =: \begin{bmatrix}A_1 & \dots & A_n\end{bmatrix}$$

must be a contraction. This determines a natural continuous projection $\pi_{n,k}$ of $\mathrm{Rep}_k(\mathfrak{L}_n)$ into the set $\overline{\mathbb{B}}_{n,k}$ of all contractive n-tuples of $k \times k$ matrices.

Both $\mathrm{Rep}_k(\mathfrak{L}_n)$ and $\overline{\mathbb{B}}_{n,k}$ are compact Hausdorff spaces and $\pi_{n,k}$ is continuous. Moreover we observed above that the open ball $\mathbb{B}_{n,k}$ is in the range of $\pi_{n,k}$. Hence this map is surjective. The Bunce argument may be modified to show uniqueness of the functional calculus when $\|A\| = r < 1$, and it follows that the restriction of $\pi_{n,k}^{-1}$ to $\mathbb{B}_{n,k}$ is a homeomorphism. However, over the boundary points, the projection is generally not injective, and the fibre $\pi_{n,k}^{-1}(A)$ may be very large.

When $n = 1$, $\pi_{1,1}$ is the natural projection from the maximal ideal space of H^∞ onto the closed $\overline{\mathbb{D}}$. Over each point λ in the open disk $\mathbb{D}$, there is the unique functional of point evaluation at λ. However over each point on the boundary circle, there is a huge fibre.

When $n > 1$ and $k = 1$, the points $\lambda \in \mathbb{B}_n$ correspond to the point evaluations φ_λ; so the set of weak-$*$ continuous characters on $\mathfrak{L}_n$ is homeomorphic to $\mathbb{B}_n$. Thus each element $A \in \mathfrak{L}_n$ determines a function $\hat{A}(\lambda) = \varphi_\lambda(A)$ on $\mathbb{B}_n$ via the Gelfand transform. Moreover it is clear that $\hat{L_i} = z_i$ is the ith coordinate function. Moreover WOT-convergent sequences in $\mathfrak{L}_n$ are sent to sequences of functions which converge uniformly on compact subsets of the ball. Hence it follows that $\hat{A}$ is analytic, and that the map from A to $\hat{A}$ is a contractive map of $\mathfrak{L}_n$ into $H^\infty(\mathbb{B}_n)$. This map is not surjective, and the exact description of the range will be dealt with in section 8.

In the $k = 1$ case, all the boundary points are essentially the same, so the fibre $\pi_{n,1}^{-1}(1, 0, \dots, 0)$ is typical. There is a natural WOT-continuous map of $\mathfrak{L}_n$ onto $H^\infty(\mathbb{D})$ with kernel generated by $\{L_i : i \geq 2\}$. Composing this map with any multiplicative linear function of H^∞ sending z to 1 yields an element of the fibre $\pi_{n,1}^{-1}(1, 0, \dots, 0)$. In particular, this fibre is huge! It is an open question whether there are any other multiplicative linear functionals in the fibre which do not factor in this manner.

When $n \geq 2$ and $k \geq 2$, a boundary point of $\overline{\mathbb{B}}_{n,k}$ may have a unique preimage or it may be huge, depending on the point.

5.3. Automorphisms

It turns out that the automorphism group of $\mathfrak{L}_n$ exhibits significant analytic structure.

A natural first question is about continuity. Every automorphism Θ of $\mathfrak{L}_n$ is automatically norm and WOT-continuous. Norm continuity is easy, and follows from a *gliding hump* argument. That is, under the assumption that the map is unbounded, one constructs a single element A which encodes an appropriate sequence on which the norm blows up, allowing one to show that $\|\Theta(A)\|$ is infinite, reaching a contradiction.

The WOT-continuity is more subtle, and relies in part on the equivalence of the WOT and weak-$*$ topologies. Indeed, for a dual topology, one can apply the Krein–Smulian Theorem which says that a subspace is weak-$*$ closed if its intersection with the unit ball is weak-$*$ closed. This is used to show that $\Theta(\mathfrak{L}_n^0)$ is weak-$*$ closed. This leads to the conclusion that $\varphi = \varphi_0\Theta^{-1}$ is a weak-$*$ continuous character, and thus equals some φ_λ. Similarly, for each $\lambda \in \mathbb{B}_n$, there is a $\mu \in \mathbb{B}_n$ so that $\varphi_\lambda\Theta^{-1} = \varphi_\mu$. This determines a map from the ball into itself given by $\tau_\Theta(\lambda) = \mu$. A computation shows that

$$\tau_\Theta(\lambda) = \varphi_\lambda\Theta^{-1}(L) := (\varphi_\lambda\Theta^{-1}(L_1), \dots, \varphi_\lambda\Theta^{-1}(L_n)) = \hat{T}(\lambda)$$

where $T = \Theta^{-1}(L) = \begin{bmatrix}\Theta^{-1}(L_1) & \dots & \Theta^{-1}(L_n)\end{bmatrix}$ is a $1 \times n$ row operator. Thus this map is analytic. Another computation shows that $\tau_{\Theta_1}\tau_{\Theta_2} = \tau_{\Theta_1\Theta_2}$. In particular, $\tau_{\Theta^{-1}} = \tau_\Theta^{-1}$. So each τ_Θ is a biholomorphic homeomorphism of $\mathbb{B}_n$. Thus we have constructed a homomorphism τ from $\mathrm{Aut}(\mathfrak{L}_n)$ into $\mathrm{Aut}(\mathbb{B}_n)$, the group of conformal automorphisms of the ball.

The object now is to compute the kernel of this map, and to show that τ is surjective. The kernel is straightforward because of the weak-$*$ continuity of Θ. Clearly $\tau_\Theta = \mathrm{id}$ precisely when $\Theta(L_i) - L_i$ belongs to $\bigcap \ker(\varphi_\lambda)$, which we saw earlier is the commutator ideal $\mathfrak{C}$. However this easily implies that $\Theta(A) - A \in \mathfrak{C}$ for every polynomial in the generators. The weak-$*$ continuity allows us to extend this to all of $\mathfrak{L}_n$. Hence Θ is trivial modulo the commutator ideal. Such automorphisms are called *quasi-inner*, and the subgroup of quasi-inner automorphisms is denoted q-Inn$(\mathfrak{L}_n)$.

Now consider the question of surjectivity. The conformal maps of $\mathbb{B}_n$ which fix the origin are just the unitary maps on $\mathbb{C}^n$. A class of automorphisms known as *gauge automorphisms* of the Cuntz–Toeplitz algebra are well-known from quantum mechanics. For any unitary U in $\mathfrak{U}_n$, construct a unitary on Fock space by $\widetilde{U} = 1 \oplus \sum_{k\geq 1} \oplus U^{\otimes k}$ where 1 acts on $\mathbb{C}$, U acts on the n-dimensional Hilbert space $\mathcal{H} \simeq \mathbb{C}^n$, and $U^{\otimes k} = U \otimes \cdots \otimes U$ acts on $\mathcal{H}^{\otimes k} = \mathcal{H} \otimes \cdots \otimes \mathcal{H}$. It is an easy calculation to show that the left creation operator L_ζ which tensors on the left by a vector $\zeta \in \mathcal{H}$ is conjugated by $\widetilde{U}$ as $\widetilde{U}L_\zeta\widetilde{U}^* = L_{U\zeta}$. Thus $\mathrm{Ad}\,\widetilde{U}$ satisfies $\mathrm{Ad}\,\widetilde{U}(L_i) = L_{\zeta_i}$ where $\zeta_1, \dots, \zeta_n$ is an orthonormal basis for $\mathcal{H}$. Thus it follows that $\mathrm{span}\{L_1, \dots, L_n\}$ is fixed by $\mathrm{Ad}\,\widetilde{U}$. Hence $\mathrm{Ad}\,\widetilde{U}$ implements an automorphism Θ_U of $\mathfrak{L}_n$.

An easy calculation shows that $\Theta_U\Theta_V = \Theta_{UV}$; so this is a homomorphism of the unitary group $\mathcal{U}_n$ into the automorphism group $\mathrm{Aut}(\mathfrak{L}_n)$. However a bit of a complication appears since $\tau_{\Theta_U} = \overline{U}$, the coordinate-wise conjugate of U, rather than U itself.

Now it is easy to see that any transitive subgroup of $\mathrm{Aut}(\mathbb{B}_n)$ containing $\mathfrak{U}_n$ is the whole group [53]. So it suffices to construct automorphisms with $\tau_\Theta(0) = \lambda$ for every $\lambda \in \mathbb{B}_n$. One way to accomplish this by a unitarily implemented automorphism is to use the fact that $\{\nu_\lambda\}^\perp$ has an n-dimensional wandering space $\mathcal{W}_\lambda$. An orthonormal basis $\zeta_1, \dots, \zeta_n$ for this subspace yields a set of isometries L_{ζ_i} which generate $\mathfrak{L}_n$. The automorphism sending L_i to L_{ζ_i} does the job, although the argument is a bit long. We describe another approach used by Voiculescu [57] to construct unitarily implemented automorphisms of the Cuntz–Toeplitz algebra.

Consider the Lie group $U(1,n)$ consisting of those $(n+1)\times(n+1)$ matrices X such that $X^*JX = J$, where $J = \left[\begin{smallmatrix} -1 & 0 \\ 0 & I_n \end{smallmatrix}\right]$. These matrices have the form $X = \left[\begin{smallmatrix} x_0 & \eta_1^* \\ \eta_2 & X_1 \end{smallmatrix}\right]$ where the coefficients satisfy the relations:

1. $\|\eta_1\|^2 = \|\eta_2\|^2 = |x_0|^2 - 1$
2. $X_1\eta_1 = \overline{x_0}\eta_2$ and $X_1^*\eta_2 = x_0\eta_1$
3. $X_1^*X_1 = I_n + \eta_1\eta_1^*$ and $X_1X_1^* = I_n + \eta_2\eta_2^*$.

The group $\mathrm{Aut}(\mathbb{B}_n)$ is a quotient of $U(1,n)$, which is obtained by representing the conformal maps as fractional linear transformations. Indeed the map $X \to \theta_X$ given by

$$\theta_X(\lambda) = \frac{X_1\lambda + \eta_2}{x_0 + \langle \lambda, \eta_1\rangle} \quad \text{for} \quad \lambda \in \mathbb{B}^n.$$

is the desired homomorphism, and the kernel is the subgroup $\mathbb{T}$ of scalars of modulus one.

Voiculescu observed that there is a related automorphism of $\mathcal{E}_n$ determined by its action on generators:

$$\Theta_X(L_\zeta) = (x_0 I - L_{\eta_2})^{-1}(L_{X_1\zeta} - \langle \zeta, \eta_1\rangle I).$$

Moreover he constructs a unitary operator U_X by

$$U_X(A\xi_\varnothing) = \Theta_X(A)(x_0I - L_{\eta_2})^{-1}\xi_\varnothing \quad \text{for all} \quad A \in \mathcal{A}_n$$

so that $\Theta_X(A) = U_X A U_X^*$ for all A in $\mathcal{A}_n$. Unitarily implemented automorphisms are WOT-continuous, and thus this maps $\mathfrak{L}_n$ into itself. The beauty of this approach is that once one finds this lovely formula, it is routine to verify that the map is a homomorphism, and hence each Θ_X is invertible and so is an automorphism.

Again the computation of τ has a twist. It turns out that $\tau(\Theta_X) = \theta_{\overline{X}}$ where $\overline{X}$ is the pointwise conjugate of X. This is evidently enough to show that τ is surjective. As a bonus, we have constructed a section from $\mathrm{Aut}(\mathbb{B}_n)$ into the subgroup $\mathrm{Aut}_u(\mathfrak{L}_n)$ of all unitarily implemented automorphisms of $\mathfrak{L}_n$. In fact, this is the full subgroup. To see this, note that if Θ is just isometric and $\tau_\Theta = \mathrm{id}$, then $1 = \Theta(L_i) = L_i + C_i$ where $C_i \in \mathfrak{C}$. However

$$1 \geq \|(L_i + C_i)\xi_\varnothing\|^2 = \|\xi_i + C_i\xi_\varnothing\|^2 = 1 + \|C_i\xi_\varnothing\|^2.$$

It follows that $C_i\xi_\varnothing = 0$ and so $C_i = 0$. Hence $\Theta = \mathrm{Id}$.

The complete result is

Theorem 5.2. *There is a natural short exact sequence*

$$0 \longrightarrow \text{q-Inn}(\mathfrak{L}_n) \longrightarrow \mathrm{Aut}(\mathfrak{L}_n) \xrightarrow{\ \tau\ } \mathrm{Aut}(\mathbb{B}_n) \longrightarrow 0.$$

The map τ takes Θ to $\tau_\Theta(\lambda) = \varphi_\lambda\Theta^{-1}(L)$ for $\lambda \in \mathbb{B}_n$. Moreover, τ has a continuous section which carries $\mathrm{Aut}(\mathbb{B}_n)$ onto the subgroup $\mathrm{Aut}_u(\mathfrak{L}_n)$ of unitarily implemented automorphisms. So $\mathrm{Aut}(\mathfrak{L}_n)$ is a semidirect product.

6. Classification of Special Free Semigroup Algebras

We now turn to the analysis of more general free semigroup algebras. In this section, we will be concerned with the classification of certain special classes of algebras. The atomic representations are classified in [27] and the finitely correlated representations are classified in [25]. In the next section, we shall see how much of this apparently special structure actually extends to all free semigroup algebras.

6.1. Atomic Representations

Say that an n-tuple of isometries $S = (S_1, \dots, S_n)$ with orthogonal ranges is *atomic* if there is an orthonormal basis $\{\xi_k\}$ for $\mathcal{H}$ which is permuted up to scalars by each S_i. This requires that there be endomorphisms $\pi_i : \mathbb{N} \to \mathbb{N}$ for $1 \leq i \leq n$ and scalars $\lambda_{i,k} \in \mathbb{T}$ such that $S_i\xi_k = \lambda_{i,k}\xi_{\pi_i(k)}$. Equivalently, this says that there is an atomic masa containing all the range projections $P_w = w(S)w(S)^*$ for $w \in \mathbb{F}_n^+$. We call the corresponding representation of $\mathcal{E}_n$ and the free semigroup algebra $\mathfrak{S}$ atomic as well.

There is a connection between these representations and the permutation representations of $\mathcal{O}_n$ recently introduced and studied by Bratteli and Jorgensen in [15]. Permutation representations are the subclass of atomic representations for which all scalars $\lambda_{i,k} = 1$ and $\sum_{i=1}^n S_iS_i^* = I$. Bratteli and Jorgensen were interested in decomposing permutation representations into direct sums of irreducible representations. However, the condition that $\lambda_{i,k} = 1$ forced them to make certain restrictive assumptions. In general, to obtain a decomposition into irreducible representations, arbitrary scalars are needed as we shall see.

The basic idea of this analysis is easy. Start with any basis vector ξ, and see where it is sent by the action of the semigroup $\mathbb{F}_n^+$. Either it is a wandering vector, or there is some word which maps ξ back to a scalar multiple of itself. When it returns, one shows that one obtains a reducing subspace on which the representation looks like Example 2.1. When ξ is wandering, one checks to see if it is in the range of one of the S_i's. One begins pulling back until one reaches a basis vector orthogonal to the range of all of the S_i's, one enters a recurring loop as above, or the chain continues indefinitely. In the first case, this produces a wandering vector which generates a reducing subspace on which the algebra is unitarily equivalent to $\mathfrak{L}_n$. And the latter case produces a summand of inductive type unitarily equivalent to Example 2.2.

This shows that every atomic representation decomposes into a direct sum of the basic building blocks which we have already identified. The discussion now turns to deciding when these representations are irreducible, what the free semigroup looks like, and how this decomposition may be accomplished.

Consider Example 2.1 first. An easy observation is that when $u = v^p$ for $p \geq 2$ is a power of a smaller word v, the ring structure has p-fold symmetry. This allows the representation $\sigma_{u,\lambda}$ to be split into p summands σ_{v,μ_j} where μ_j are the p pth roots of λ. So even if $\lambda = 1$, other roots of unity are needed in this decomposition.

On the other hand, one needs to establish that $\sigma_{u,\lambda}$ is irreducible when u is *primitive*, meaning that it is not a power of a smaller word. It is here that the free semigroup algebra $\mathfrak{S}_{u,\lambda}$ that it generates is useful. Consider the polynomial $(\overline{\lambda}u(S))^k$. This is an isometry in $\mathfrak{S}_{u,\lambda}$ which maps the vector ζ_k onto itself. Consider what it does to the other vectors in this basis. If it is another basis vector in the ring, it cannot be mapped back to itself, for that would imply symmetry. So eventually it is mapped out of the ring onto a wandering vector, and then it is mapped off into the far reaches of one of the wandering spaces. The same happens to each basis vector which is not in the ring. Hence $(\overline{\lambda}u(S))^k$ converges in the WOT topology to the rank one projection $P = \zeta_k\zeta_k^*$ onto $\mathbb{C}\zeta_k$. So P belongs to $\mathfrak{S}_{u,\lambda}$.

Suppose that this representation splits into a sum $\mathcal{M} \oplus \mathcal{M}^\perp$ of reducing subspaces. Either $\mathcal{M}$ or $\mathcal{M}^\perp$ is not orthogonal to ζ_k, say $\mathcal{M}$. Since $\mathcal{M}$ is invariant for P, it must contain ζ_k, which is a cyclic vector. Hence $\mathcal{M} = \mathcal{H}$, and therefore this representation is irreducible.

Consider what the free semigroup algebra looks like in this case. Since it contains P, it also contains $v_1(S)Pv_2(S)$. If v_1 and v_2 are chosen properly, we can build the set of matrix units $\zeta_i\zeta_j^*$ for $1 \le i,j \le k$. Let $Q = \sum_{i=1}^k \zeta_i\zeta_i^*$. Then $Q\mathfrak{S}_{u,\lambda}Q$ is isomorphic to $\mathfrak{M}_k$, the algebra of $k \times k$ matrices. Because each ζ_i is cyclic, every non-zero vector in the range of Q is cyclic. So $\mathfrak{S}_{u,\lambda}Q = \mathcal{B}(\mathcal{H})Q$. Finally, $Q^\perp\mathcal{H}$ is the direct sum of $k(n-1)$ copies of Fock space on which $\mathfrak{S}_{u,\lambda}$ looks like $k(n-1)$ copies of $\mathfrak{L}_n$. Hence $\mathfrak{S}_{u,\lambda}$ has the form

$$\mathfrak{B}_{n,k} = \begin{bmatrix} \mathfrak{M}_k & 0 \\ Q^\perp\mathcal{B}(\mathcal{H})Q & \mathfrak{L}_n^{(k(n-1))} \end{bmatrix}$$

with respect to the decomposition of $\mathcal{H} = Q\mathcal{H} \oplus Q^\perp\mathcal{H}$. Notice that the algebra (rather than the specific representation) depends only on n and $|u| = k$.

Now consider how to identify a summand of type (u, λ) in an arbitrary atomic representation. Again study the action of the polynomial $(\overline{\lambda}u(S))^k$ on a basis vector. If it lives in the kth position of a ring corresponding to the word primitive u, then it is mapped back to a multiple of itself. In a representation $\sigma_{v,\mu}$ where v is not a cyclic permutation of u, every standard basis vector is mapped off to infinity by these polynomials. Likewise in any pure or inductive type representation, it is equally evident that each basis vector is sent off to infinity. However in a representation $\sigma_{u,\mu}$ for $\mu \ne \lambda$, ζ_k is sent to $(\overline{\lambda}\mu)^k\zeta_k$. On any other basis vector ζ, the sequence $(\overline{\lambda}u(S))^k\zeta$ heads off to infinity as before. One deals with this added complication by setting $p_m(x) = \frac{1}{m!}\sum_{j=1}^{m!} x^{m!+j}$, and considering $p_m(\overline{\lambda}u(S))$. This sequence converges in the strong operator topology to the projection $P_{u,\lambda}$ onto the span of the vectors corresponding to ζ_k in each summand of type $\sigma_{u,\lambda}$.

Thus the multiplicity of $\sigma_{u,\lambda}$ is computed as the rank of $P_{u,\lambda}$. In particular we now see that $\sigma_{u,\lambda}$ and $\sigma_{v,\mu}$ are unitarily equivalent only if v is a cyclical permutation of u and $\mu = \lambda$.

Summing $P_{u,\lambda}$ over all pairs (u, λ) yields a projection Q in the free semigroup algebra $\mathfrak{S}$ corresponding to the projection onto all of the rings. Observe that $Q\mathfrak{S}Q$ is a (type I) von Neumann algebra. Every standard basis vector in $Q^\perp\mathcal{H}$ is a wandering vector. Indeed, $Q^\perp\mathfrak{S}Q^\perp$ is the direct sum of copies of $\mathfrak{L}_n$ and of inductive algebras, which were observed in Example 2.2 to be (completely) isometrically isomorphic to $\mathfrak{L}_n$. Thus it follows that $Q^\perp\mathfrak{S}Q^\perp$ is completely isometrically isomorphic to $\mathfrak{L}_n$. As $Q^\perp\mathcal{H}$ is invariant for $\mathfrak{S}$, the algebra has a lower triangular form. The 2, 1 entry is specified by the structure of the pieces coming from each ring algebra. In fact, $\mathfrak{S}Q = \mathfrak{W}Q$ where $\mathfrak{W}$ is the von Neumann algebra generated by $\mathfrak{S}$, as we see in the next section.

Lastly consider the representations of inductive type. It is clear that if one can delete a finite number of terms from the beginning of two words in order to make them equal, then the two representations are unitarily equivalent. In this case, the two sequences are called *shift tail equivalent.* Otherwise they are inequivalent. They are irreducible except when the word is equivalent to a periodic word. The representation π_x where $x = uuuuu\dots$ is equivalent to a direct integral of the representation $\sigma_{u,\lambda}$ over the unit circle with respect to Lebesgue measure.

To complete the picture, one may compute the multiplicity of the pure part as $\operatorname{rank}(I - \sum_{i=1}^n S_iS_i^*)$ and of each inductive representation π_x. Recall from Example 2.2 the sequence x_m. Then note that $x_m(S)s_m(S)^*$ is a projection onto the range of $x_m(S)$ spanned by a subset of the standard basis. As m increases, one obtains precisely those basis vectors which can be pulled back indefinitely along this sequence. When it is not periodic, there is one vector in each summand of π_x and none in any other. Whence the multiplicity is the rank of $\textsc{sot}\text{–}\lim x_m(S)s_m(S)^*$. In the periodic case, one obtains instead a subspace on which $u(S)$ is a bilateral shift, and one computes its multiplicity.

6.2. Finitely Correlated Representations

Our approach here ties dilation theory strongly to the classification of an important class of representations of $\mathcal{E}_n$. Bratteli and Jorgensen [14] introduced this class to study endomorphism of $\mathcal{B}(\mathcal{H})$ and then showed them to be of central importance in generating wavelets [16, 17]. Their class of representations is obtained by the GNS construction from a state ψ which has the property that the cyclic vector ξ_ψ generates a *finite dimensional* subspace $\mathcal{V} = \operatorname{span}\{w(S)^*\xi_\psi : w \in \mathbb{F}_n^+\}$ for the adjoints S_i^* of the generating isometries. Our slightly weaker definition does not insist that the representation have a cyclic vector, only that $\mathcal{V}$ be cyclic for $\mathfrak{S}$. That is, a free semigroup $\mathfrak{S}$ is *finitely correlated* if there is a finite dimensional subspace $\mathcal{V}$ which is invariant for $\mathfrak{S}^*$ and is cyclic for $\mathfrak{S}$. In particular this class contains all of the atomic ring representations mentioned above.

Our viewpoint about where such representations arise comes from dilation theory. Let $A_i = (S_i^*|_\mathcal{V})^*$ be the compressions of each S_i to $\mathcal{V}$. A simple calculation shows that

$$\sum_{i=1}^n A_iA_i^* = P_\mathcal{V} \sum_{i=1}^n S_iS_i^*|_\mathcal{V} \le I_\mathcal{V}$$

Thus $A = (A_1, \dots, A_n)$ is a row contraction; and it is a row isometry if (and only if) S is a Cuntz representation. Notice that S is an isometric dilation of the row contraction A. Moreover since $\mathcal{V}$ is cyclic, this is a minimal dilation. By the uniqueness of the minimal isometric dilation, we see that S is completely determined by A. Our goal is to start with A and find complete unitary invariants for the set of isometries S. In particular, we will be able to decide if A and B determine unitarily equivalent representations.

For convenience we will consider only the Cuntz case $AA^* = I$. For the general case, we mention only that the multiplicity of the pure part may be computed directly from A as $\operatorname{rank}(I - AA^*)$. The additional complication in decomposing the representation can be found in [25].

It is easy to see that $\mathcal{V}^\perp$ is of pure type, so that the restriction of $\mathfrak{S}$ to this subspace is a multiple $\mathfrak{L}_n^{(\alpha)}$ of $\mathfrak{L}_n$. However $\mathcal{V}^\perp$ is not normally a maximal subspace of this type. It does place $\mathfrak{S}$ inside the algebra $\mathcal{B}(\mathcal{H})P_\mathcal{V} + P_\mathcal{V}^\perp \mathfrak{S} P_\mathcal{V}^\perp$, which is unitarily equivalent to $\mathfrak{B}_{n,d}$ where $d = \dim \mathcal{V}$. In particular, we see immediately that the weak-$*$ and WOT topologies coincide.

We search for a (finite rank) projection in $\mathfrak{S}$. We focus on the algebra $\mathfrak{A} = \operatorname{Alg}\{A_1, \dots, A_n\}$ of matrices generated by A. The key technical result is that for every non-zero vector $x \in \mathcal{H}$, the subspace $\mathfrak{S}^*[x]$ always intersects $\mathcal{V}$. In the special case in which $\mathfrak{A} = \mathcal{B}(\mathcal{V})$, one can now show that $P_\mathcal{V}$ belongs to $\mathfrak{S}$. From this we deduce, as in the atomic ring case, that the representation is irreducible. The free semigroup algebra in this case is exactly the algebra $\mathfrak{B}_{n,d}$ above.

More generally, one finds that whenever $\mathcal{M}$ is a minimal invariant subspace for $\mathfrak{A}^*$, that $\mathfrak{S}[\mathcal{M}]$ is a reducing subspace. The restriction of $\mathfrak{S}$ to this subspace is irreducible, and it contains the projection $P_\mathcal{M}$. The restriction of $\mathfrak{A}^*$ to $\mathcal{M}$ is an algebra of matrices with no proper invariant subspace. So by Burnside's Theorem, $\mathfrak{A}^*|_\mathcal{M} = \mathcal{B}(\mathcal{M})$. We deduce that $\mathfrak{S}_{\mathfrak{S}[\mathcal{M}]} \simeq \mathfrak{B}_{n,d}$ where $d = \dim \mathcal{M}$.

We then form a maximal family $\{\mathcal{M}_i\}$ of pairwise orthogonal minimal $\mathfrak{A}^*$–invariant subspaces. This yields an orthogonal decomposition of the free semigroup algebra into a direct sum of irreducible algebras on $\mathfrak{S}[\mathcal{M}_i]$. This procedure does not depend on how this maximal family is obtained. In particular the span of the $\mathcal{M}_i$'s is the subspace $\widetilde{\mathcal{V}}$ spanned by all minimal $\mathfrak{A}^*$–invariant subspaces.

At this stage, you might suspect that because of the structure $\mathfrak{A}^*|_{\mathcal{M}_i} = \mathcal{B}(\mathcal{M}_i)$ that $\mathfrak{A}^*|_{\widetilde{\mathcal{V}}}$ is a C^*-algebra. Remarkably this is true, and once this is established, one can use the Wedderburn Theorem to decompose it into a sum of full matrix algebras. This yields a corresponding orthogonal decomposition of $\mathfrak{S}$.

Thus the decomposition problem now turns on deciding when two such irreducible representations are unitarily equivalent. This suggests studying intertwining maps between two such representations. The tool that makes this possible is the completely positive map Φ on $\mathcal{B}(\mathcal{V})$ given by

$$\Phi(X) = \sum_{i=1}^{n} A_i X A_i^*.$$

Suppose that there is an invertible map T between two minimal $\mathfrak{A}^*$–invariant subspaces $\mathcal{V}_1$ and $\mathcal{V}_2$ which commutes with $\mathfrak{A}$. Write $B_i := P_{\mathcal{V}_1} A_i|_{\mathcal{V}_1}$ and $C_i := P_{\mathcal{V}_2} A_i|_{\mathcal{V}_2}$. Since $TB_iT^{-1} = C_i$ and

$$\sum_{i=1}^{n} B_i B_i^* = I_{\mathcal{V}_1} \qquad \text{and} \qquad \sum_{i=1}^{n} C_i C_i^* = I_{\mathcal{V}_2},$$

we compute that

$$I_{\mathcal{V}_2} = \sum_{i=1}^{n} (TB_iT^{-1})(TB_iT^{-1})^* = T\Phi(T^{-1}T^{*-1})T^*.$$

Therefore

$$\Phi(T^{-1}T^{*-1}) = T^{-1}T^{*-1}.$$

This leads to the discovery of two interesting facts about completely positive maps on $\mathfrak{M}_k$. The first is that a non-scalar operator X such that $\Phi(X) = X$ implies that $\mathfrak{A}^* = \mathrm{Alg}\{A_1^*, \dots, A_n^*\}$ has two pairwise orthogonal minimal invariant subspaces. This had gone unnoticed because no-one had ever started with a completely positive map Φ on $\mathfrak{M}_k$, found a Stinespring decomposition which leads to the form as above and then looked at the algebra generated by the A_i^*'s. It is not at all obvious that the uniqueness in the Stinespring condition would make this an interesting object; but it is!

The second consequence is even more surprising. When $\mathcal{V}$ is the orthogonal direct sum of minimal $\mathfrak{A}^*$–invariant subspaces, then $\mathfrak{A}$ is a C^*-algebra and the fixed point set of Φ coincides with the commutant of $\mathfrak{A}$. The following example puts this result in perspective.

Example 6.1. *Let*

$$A_1 = \begin{bmatrix} 1/\sqrt{2} & 0 & 0 \\ 1/2\sqrt{2} & 1/2 & 1/2\sqrt{2} \\ 0 & 0 & 1/\sqrt{2} \end{bmatrix} \quad \textit{and } A_2 = \begin{bmatrix} 1/\sqrt{2} & 0 & 0 \\ -1/2\sqrt{2} & 1/2 & -1/2\sqrt{2} \\ 0 & 0 & 1/\sqrt{2} \end{bmatrix}.$$

Note that $A_1A_1^* + A_2A_2^* = I$. A calculation shows that the fixed point set of Φ is the set of matrices $X = [x_{ij}]$ such that $x_{12} = x_{21} = x_{23} = x_{32} = 0$ and $x_{11} + x_{13} + x_{31} + x_{33} = 2x_{22}$. In particular, this is not an algebra. The algebra $\mathfrak{A}^*$ has two minimal invariant subspaces, $\mathbb{C}e_1$ and $\mathbb{C}e_3$. Note that the compression of $\mathfrak{A}$ to $\mathrm{span}\{e_1, e_3\}$ *consists of scalar matrices, and the fixed point set of the restricted completely positive map is the full 2×2 matrix algebra.*

Putting all of this information together yields the complete picture:

Theorem 6.2. *Let $A_1, \dots, A_n$ be operators on a finite dimensional space $\mathcal{V}$ such that $\sum_{i=1}^{n} A_iA_i^* = I$, and let $S_1, \dots, S_n$ be their joint isometric dilation. Let $\widetilde{\mathcal{V}}$ be the subspace of $\mathcal{V}$ spanned by all minimal $\mathfrak{A}^*$–invariant subspaces. Then the compression $\widetilde{\mathfrak{A}}$ of $\mathfrak{A}$ to $\widetilde{\mathcal{V}}$ is a C^*-algebra. Let $\widetilde{\mathfrak{A}}$ be decomposed as $\sum_{g\in G}^{\oplus} \mathfrak{M}_{d_g} \otimes \mathbb{C}^{m_g}$*

with respect to a decomposition $\widetilde{\mathcal{V}} = \sum_{g\in G}^{\oplus} \mathcal{V}_g^{(m_g)}$, *where* $\mathcal{V}_g$ *has dimension* d_g *and multiplicity* m_g. *Then the dilation acts on the space* $\mathcal{H} = \sum_{g\in G}{}^{\oplus} \mathcal{H}_g^{(m_g)}$ *where* $\mathcal{H}_g = \mathcal{V}_g \oplus \mathcal{K}_n^{(d_g(n-1))}$. *The algebra* $\mathfrak{S}$ *decomposes as* $\mathfrak{S} \simeq \sum_{g\in G}^{\oplus} \mathcal{B}_{n,d_g}^{(m_g)}$.

In particular, the algebra $\mathfrak{S}$ contains the projection $P_{\widetilde{\mathcal{V}}}$, and $P_{\widetilde{\mathcal{V}}}\mathfrak{S}P_{\widetilde{\mathcal{V}}}$ is the C^*-algebra $\widetilde{\mathfrak{A}}$. This is what makes the decomposition possible. This description looks a bit complicated, but it yields some useful and easily implemented consequences.

Corollary 6.3. *Let* $\mathfrak{S}$ *be the algebra determined by the joint isometric dilation of a contractive* n*-tuple* A *on a finite dimensional space* $\mathcal{V}$ *such that* $AA^* = I$. *Then* $\mathfrak{S}$ *is irreducible if and only if the fixed point set of* $\Phi(X) = \sum_{i=1}^n A_i X A_i^*$ *is trivial, i.e.* $\{X : \Phi(X) = X\} = \mathbb{C}I$.

And the general question of unitarily equivalent representations of this type reduces to a (possibly difficult) finite dimensional problem.

Theorem 6.4. *Let* $A = (A_1, \dots, A_n)$ *and* $B = (B_1, \dots, B_n)$ *be contractive* n*-tuples on finite dimensional spaces* $\mathcal{V}_A$ *and* $\mathcal{V}_B$ *respectively such that* $AA^* = I_{\mathcal{V}_A}$ *and* $BB^* = I_{\mathcal{V}_B}$. *Let* $S = (S_1, \dots, S_n)$ *and* $T = (T_1, \dots, T_n)$ *be their joint minimal isometric dilations on Hilbert spaces* $\mathcal{H}_A$ *and* $\mathcal{H}_B$*; and let* σ_A *and* σ_B *be the induced representations of* $\mathcal{O}_n$. *Let* $\widetilde{\mathcal{V}}_A$ *be the subspace spanned by all minimal* $\mathfrak{A}^*$*–invariant subspaces; and similarly define* $\widetilde{\mathcal{V}}_B$. *Then* σ_A *and* σ_B *are unitarily equivalent if and only if* $A^*|_{\widetilde{\mathcal{V}}_A}$ *is unitarily equivalent to* $B^*|_{\widetilde{\mathcal{V}}_B}$.

A result of Kribs [36] that shows that one can detect the $\mathfrak{A}^*$–invariant subspaces from the completely positive map Φ without computing the A_i's explicitly. He shows that $\mathcal{M}$ is $\mathfrak{A}^*$–invariant if and only if $P_{\mathcal{M}} \le \Phi(P_{\mathcal{M}})$.

The structure of these algebras is so tight that one can explicitly describe many of their invariant subspaces. In particular, the column $\mathfrak{S}P_{\mathcal{V}}$ is a 'slice' of a type I von Neumann algebra, and thus is hyper-reflexive with distance constant 4 by Christensen's result [20]. The restriction to $\widetilde{\mathcal{V}}^\perp$ is a multiple of $\mathfrak{L}_n$, and thus has hyper-reflexivity constant 3 by Bercovici [12]. Combining these estimates yields:

Corollary 6.5. *The algebra* $\mathfrak{S}$ *determined by the joint isometric dilation of a contractive* n*-tuple on a finite dimensional space is hyper-reflexive with distance constant at most 5.*

7. Structure Theory

Our attention earlier has been on special classes which have some useful extra property. However it is now possible to extend some of the structural picture to all free semigroup algebras. The focus in on several features which have emerged as important in these special classes. The first feature is the central role played by wandering vectors, because the restriction of S to a subspace generated by a wandering vector is unitarily equivalent to the left regular representation L. Secondly there is the role of projections which occur in the algebra $\mathfrak{S}$. Since $\mathfrak{L}_n$ has no proper projections itself, projections in $\mathfrak{S}$ have made it possible to decompose the algebra into pieces.

We shall say that an algebra $\mathfrak{S}$ is *type L* if it is canonically isomorphic to $\mathfrak{L}_n$, meaning that the map sending S_i to L_i may be extended to an algebra isomorphism. In addition to multiples of $\mathfrak{L}_n$, we have also seen that atomic representations of inductive type are type L. More generally any inductive representation (meaning that $\mathcal{H}$ is the increasing union of invariant subspaces on which $\mathfrak{S}$ is pure) must be type L. In fact these algebras are completely isometrically isomorphic and weak-$*$ homeomorphic by the canonical WOT-continuous map connecting them. This turns out to be true for all type L algebras.

An important tool in this general analysis is the ideal $\mathfrak{S}_0$ of $\mathfrak{S}$ generated by $\{S_1, \dots, S_n\}$. This ideal is either co-dimension 1 if it does not contain I or it is all of $\mathfrak{S}$. If $\mathfrak{S}$ has a wandering vector ξ, then $\varphi(A) = \langle A\xi, \xi\rangle$ is a unital WOT-continuous functional with kernel $\mathfrak{S}_0$. Whence $\mathfrak{S}_0 \neq \mathfrak{S}$. Conversely, if $\mathfrak{S}_0 \neq \mathfrak{S}$, then there is a WOT-continuous functional φ with $\ker\varphi = \mathfrak{S}_0$. This functional may be represented as $\varphi(A) = \sum_{i=1}^{p} \langle A\zeta_i, \eta_i\rangle$. Hence on the p-fold ampliation $\mathfrak{S}^{(p)}$ acting on $\mathcal{H}^{(p)}$, one has $\varphi(A) = \langle A^{(p)}\zeta, \eta\rangle$ where ζ has coefficients ζ_i and η has coefficients η_i. Consequently ζ is orthogonal to $\overline{\mathfrak{S}_0^{(p)}\zeta}$. The projection of ζ onto the orthogonal complement will be a wandering vector for the subspace $\mathfrak{S}^{(p)}[\zeta]$.

On the other hand, if $\mathfrak{S}_0 = \mathfrak{S}$, then I is the WOT-limit of a net A_α in $\mathfrak{S}_0$. It follows easily that $S_i^* A$ is a net in $\mathfrak{S}$ which converges WOT to S_i^*. Thus $\mathfrak{S}$ contains each S_i^*, and so is the von Neumann algebra $\mathfrak{W}$ generated by S.

There are two important open questions related to this circle of ideas:

Question 7.1. *Can $\mathfrak{S}$ be a von Neumann algebra?*

Question 7.2. *Does every (type L) representation have a wandering vector?*

We have some partial information which pushes the ampliation argument significantly further, and adds some information to the second question.

Theorem 7.3. *Suppose that $\mathfrak{S}$ is type L. Then for p sufficiently large, the space $\mathcal{H}^{(p)}$ is spanned by wandering vectors of $\mathfrak{S}^{(p)}$.*

It follows from our argument above that whenever $\mathfrak{S}_0 \neq \mathfrak{S}$, there is a WOT-continuous map from $\mathfrak{S}$ into $\mathfrak{L}_n$ sending each generator S_i to L_i, just by sending A to $A^{(p)}|_{\mathfrak{S}^{(p)}[\zeta]}$. This map is always surjective. Indeed,

Theorem 7.4. *Suppose that $\Phi : \mathfrak{S} \to \mathfrak{L}_n$ is a WOT-continuous homomorphism such that $\Phi(S_i) = L_i$ for $1 \le i \le n$. Then Φ is surjective, and $\mathfrak{S}/\ker(\Phi)$ is completely isometrically isomorphic and weak-$*$ homeomorphic to $\mathfrak{L}_n$.*

Clearly the map Φ is unique when it exists. We need to identify its kernel. It is not difficult to convince yourself that $\ker\Phi = \cap_{k\ge1}\mathfrak{S}_0^k =: \mathfrak{J}$. The important observation is that $\mathfrak{J}$ is invariant under left multiplication by S_i^* as well as by S_i, and thus it is a WOT-closed left ideal in $\mathfrak{W}$. Consequently $\mathfrak{J}$ contains a self-adjoint projection P so that $\mathfrak{J} = \mathfrak{W}P$. This is a crucial step in the main structure theorem that we are seeking.

Theorem 7.5. *Let $\mathfrak{S}$ be a free semigroup algebra, and let $\mathfrak{W}$ be its enveloping von Neumann algebra. Then there is a largest projection P in $\mathfrak{S}$ such that $P\mathfrak{S}P$ is self-adjoint. It has the following properties:*

1. $\mathfrak{W}P = \bigcap_{k\geq 1} \mathfrak{S}_0^k$,
2. $P^\perp \mathcal{H}$ *is invariant for* $\mathfrak{S}$,
3. *if* $P \neq I$, *then* $\mathfrak{S}P^\perp$ *is completely isometrically isomorphic and weak-* homeomorphic to* $\mathfrak{L}_n$ *via the canonical* WOT*-continuous homomorphism* Φ *with* $\Phi(S_i) = L_i$ *for* $1 \leq i \leq n$, *and*
4. $\mathfrak{S} = \mathfrak{W}P + P^\perp \mathfrak{S} P^\perp$.

Note that the description of P is given only in terms of $\mathfrak{S}$, and hence P is an invariant of the algebra and is not dependent on a choice of generators. The theorem yields a canonical decomposition of $\mathfrak{S}$ into a lower triangular form where the first column is a slice of a von Neumann algebra and the $(2,2)$ entry is type L.

Since type L algebras contain no proper projections, we may conclude that any free semigroup algebra which is merely algebraically isomorphic to a subalgebra of $\mathfrak{L}_n$ is automatically type L, and thus is completely isometrically isomorphic and weak-* homeomorphic to $\mathfrak{L}_n$ via the canonical map.

As an immediate corollary, we can characterize the radical.

Corollary 7.6. *With $n \geq 2$ and notation as above, the radical of $\mathfrak{S}$ is $P^\perp \mathfrak{S} P$. Thus the following are equivalent:*

1. $\mathfrak{S}$ *is semisimple and is not self-adjoint*
2. $\mathfrak{S}$ *is type L*
3. $\mathfrak{S}$ *has no non-scalar idempotents*
4. $\mathfrak{S}$ *has no non-zero quasinilpotent elements.*

One interesting and non-trivial consequence of the structure theorem is information about the geometry of the unit ball. The Russo–Dye Theorem [54] states that in any C^*-algebra, the convex hull of all unitary elements is the whole unit ball. Also the algebra H^∞ is the convex hull of the inner functions [38]. In a free semigroup algebra which may contain no non-scalar unitaries at all (when it is type L), we instead consider the rich collection of isometries. One useful consequence of the information about wandering vectors in ampliations is that whenever V is an isometry in $\mathfrak{L}_n$ and Φ is the canonical map of a type L algebra $\mathfrak{S}$ onto $\mathfrak{L}_n$, then $\Phi^{-1}(V)$ is also an isometry. So there are many isometries in any free semigroup algebra. We obtain an analogue of the Russo–Dye Theorem in our context.

Theorem 7.7. *The convex hull of* $\operatorname{Isom}(\mathfrak{S})$ *contains the open unit ball of* $\mathfrak{S}$. *Moreover, if* $\|A\| < 1 - \frac{1}{k}$ *for* $k > 0$ *an even integer, then* A *is the average of* $6k$ *isometries.*

Examples 7.8. *We conclude this section by considering a few examples.*

Consider the finitely correlated representations studied in the previous section. The projection P is the projection onto $\widetilde{\mathcal{V}}$ and $P\mathfrak{S}P = \widetilde{\mathfrak{A}}$ is a C^*-algebra. The type L portion is in fact pure.

Now consider the atomic representations. The projection P has range equal to the direct sum of all of the rings. The type L portion consists of the span of all wandering vectors, which includes all of the inductive representations.

Here is an example which yields a large class of inductive type L representations. Let $S = (S_1, \dots, S_n)$ be any n-tuple of isometries with orthogonal ranges acting on $\mathcal{H}$. Let U be the unitary of multiplication by z on $L^2(\mathbb{T})$. Define a new n-tuple of isometries on $\mathcal{H} \otimes L^2(\mathbb{T})$ by $S \otimes U = (S_1 \otimes U, \dots, S_n \otimes U)$. Evidently

$$\sum_{i=1}^{n} (S_i \otimes U)(S_i \otimes U)^* = \sum_{i=1}^{n} S_i S_i^* \otimes I.$$

So these isometries have orthogonal ranges. In addition, if S is of Cuntz type, then so is the tensored n-tuple.

However, this new representation has a spanning set of wandering vectors of the form $\xi \otimes z^k$ for any $\xi \in \mathcal{H}$ and $k \in \mathbb{Z}$, as a simple calculation shows. Thus this representation has type L. In fact it is inductive, since the restriction to $\mathcal{H} \otimes \bar{z}^k H^2(\mathbb{T})$ is pure with wandering space $\mathcal{H} \otimes \mathbb{C}\bar{z}^k$.

Consider the representation π_{1^∞}. This is an atomic representation of inductive type, and hence is of type L. Because 1^∞ is periodic, it is also a direct integral $\pi_{1^\infty} \simeq \int_{\mathbb{T}} \sigma_{1,\lambda}\, d\lambda$. Indeed, let $\mathcal{K} = \mathbb{C} \oplus \mathcal{K}_n$. The representation $\sigma_{1,\lambda}$ is determined by generators

$$S_1^\lambda = \begin{bmatrix} \lambda & 0 \\ 0 & L_1 \end{bmatrix} \quad \text{and} \quad S_2^\lambda = \begin{bmatrix} 0 & 0 \\ \xi_\varnothing & L_2 \end{bmatrix}.$$

Thus the representation π_{1^∞} may be represented on $\mathcal{H}_T := L^2(\mathbb{T}) \otimes \mathcal{K}$ by

$$S_1 = \begin{bmatrix} U & 0 \\ 0 & I \otimes L_1 \end{bmatrix} \quad \text{and} \quad S_2 = \begin{bmatrix} 0 & 0 \\ I \otimes \xi_\varnothing & I \otimes L_2 \end{bmatrix}$$

where U is multiplication by z on $L^2(\mathbb{T})$.

Let E be a measurable subset of $\mathbb{T}$ with positive measure. Let V denote $U|_{L^2(E)}$ and $J = I_{L^2(E)}$. Now consider the representation ρ_E on $\mathcal{H}_E = L^2(E) \otimes \mathcal{K}$ by

$$T_1 = \begin{bmatrix} V & 0 \\ 0 & J \otimes L_i \end{bmatrix} \quad \text{and} \quad T_2 = \begin{bmatrix} 0 & 0 \\ J \otimes \xi_\varnothing & J \otimes L_2 \end{bmatrix}.$$

It is evident that any vector of the form $0 \oplus (f \otimes \xi_\varnothing)$ is a wandering vector. It can be shown that the restriction of a type L representation to an invariant subspace containing a wandering vector remains type L. In fact in this example, $\mathcal{H}_E$ is spanned by wandering vectors.

There are a number of open questions raised by this example about sub-inductive representations. Is $\mathcal{H}_E$ of inductive type? Is $\mathcal{H}_E^{(\infty)}$ of inductive type?

Is $\mathcal{H}_T \oplus \mathcal{H}_E$ of inductive type? Also more generally if the restriction of any type L representation to an invariant subspace still type L?

Consider the case $n = 1$. We are given an isometry S, which decomposes using the Wold decomposition and the spectral theory of unitary operators as $S \simeq U_+^{(\alpha)} \oplus U_a \oplus U_s$ where U_+ is the unilateral shift, U_a is a unitary with spectral measure absolutely continuous with respect to Lebesgue measure m, and U_s is a singular unitary. Let m_a and m_s denote scalar measures equivalent to the spectral measures of U_a and U_s respectively. If $\alpha > 0$ or if $m_a = m$, then by [58]

$$\mathfrak{S} = W(S) \simeq H^\infty(U_+^{(\alpha)} \oplus U_a) \oplus L^\infty(m_s)(U_s).$$

The von Neumann algebra it generates is

$$\mathfrak{W} = W^*(S) = \mathcal{B}(\mathcal{H})^{(\alpha)} \oplus L^\infty(U_a) \oplus L^\infty(m_s)(U_s).$$

The projection P of the Structure Theorem is just the projection onto the singular part. In this case, it is always a direct summand.

If $\alpha = 0$ and the essential support of U_a is a proper measurable subset of the circle, then $\mathfrak{S} = \mathfrak{W}$ is self-adjoint. Note that this may occur even though the spectrum of S is the whole circle.

The $n = 1$ case exhibits two phenomena which we cannot seem to replicate in the non-commutative case, and remain important open questions.

The first is the situation just noted that $\mathfrak{S}$ can be self-adjoint. The second is that there are isometries S_1 and S_2 such that $W(S_1 \oplus S_2) \simeq H^\infty$, the type L case, yet neither $W(S_i)$ are type L, and in fact are von Neumann algebras. One simply takes S_i to be multiplication by z on the upper and lower half circles respectively.

8. Interpolation and the Commutative Theory

In this section, we will examine some natural interpolation questions [29, 5] about the algebra $\mathfrak{L}_n$ including describing the image of the Gelfand map from $\mathfrak{L}_n$ into $H^\infty(\mathbb{B}_n)$. The complete picture requires making a connection to the theory of commuting row contractions. We will connect the dilation theory for this class to that of the non-commuting class, and will use it to describe the image algebra precisely. Finally we will show that we obtain a reproducing kernel Hilbert space, and obtain yet another view of this interpolation.

8.1. Interpolation

Consider the Gelfand map which takes $A \in \mathfrak{L}_n$ to the function $\hat{A}(\lambda) = \varphi_\lambda(A) = \langle A\nu_\lambda, \nu_\lambda\rangle$ on the unit ball $\mathbb{B}_n$. Fix finite subsets $\{\alpha_1, \dots, \alpha_k\}$ and $\{c_1, \dots, c_n\}$ of $\mathbb{B}_n$. Consider the Nevanlinna–Pick question of whether there is an element A in the unit ball of $\mathfrak{L}_n$ such that $\hat{A}(\alpha_j) = c_j$ for $1 \le j \le k$. If the norm condition is dropped, it is elementary to verify that there are solutions in $\mathfrak{L}_n$ for any choice of constants. Following Sarason [55], we can formulate interpolation questions for $\mathfrak{L}_n$ as operator theoretic questions. The set of all solutions is a coset of $\mathfrak{J} = \bigcap_{1\le j\le k} \varphi_{\alpha_j}$, which is a

WOT-closed ideal of $\mathfrak{L}_n$. Thus the smallest possible norm of a solution is $\operatorname{dist}(A, \mathfrak{J})$ where A is any particular solution. Moreover compactness in the WOT-topology shows that this norm is attained. The issue then is to decide if this quotient has norm at most 1.

We consider the general problem of computing the distance to a WOT-closed right ideal $\mathfrak{J}$. Recall that the range $\mathcal{M} = \overline{\mathfrak{J}\xi_\varnothing} = \overline{\mathfrak{J}\mathcal{K}_n}$ is invariant for $\mathfrak{R}_n$. Clearly for any $J \in \mathfrak{J}$,

$$\|A + J\| \geq \|P_{\mathcal{M}}^\perp (A + J)\| = \|P_{\mathcal{M}}^\perp A\|.$$

The converse is also completely correct [29].

Theorem 8.1. *Let $\mathfrak{J}$ be a* WOT*-closed right ideal in $\mathfrak{L}_n$; and let $\mathcal{M}$ denote its range. Then $\mathfrak{L}_n/\mathfrak{J}$ is completely isometric to $P_{\mathcal{M}}^\perp \mathfrak{L}_n$. That is, for every matrix $A = [A_{ij}]$ in $\mathfrak{M}_p(\mathfrak{L}_n)$, $p \geq 1$,*

$$\operatorname{dist}(A, \mathfrak{M}_p(\mathfrak{J})) = \|(P_{\mathcal{M}}^\perp \otimes I_p) A\|.$$

The basic idea is to use the Hahn–Banach Theorem and the fact that weak-$*$ continuous functionals on $\mathfrak{L}_n \otimes \mathcal{B}(\mathcal{H})$ are given by rank-one vector functionals. These vectors corresponding to functionals which annihilate $\mathfrak{J}$ are generally not supported on $\mathcal{M}^\perp$. However, using the Beurling Theorem, one is able to construct a new representation of this functional which is no longer rank one, but is supported on $\mathcal{M}^\perp$.

The desired conclusion for two-sided ideals is immediate because in this case $\mathcal{M}$ is also invariant for $\mathfrak{L}_n$. So $P_{\mathcal{M}}^\perp A = P_{\mathcal{M}}^\perp A P_{\mathcal{M}}^\perp$, and the compression to $\mathcal{M}^\perp$ is a (completely contractive, unital, WOT-continuous) homomorphism.

Corollary 8.2. *Let $\mathfrak{J}$ be a* WOT*-closed (two-sided) ideal in $\mathfrak{L}_n$; and let $\mathcal{M} = \mu(\mathfrak{J}) = \overline{\mathfrak{J}\xi_\varnothing}$ denote its range. Then $\mathfrak{L}_n/\mathfrak{J}$ is completely isometrically isomorphic and weak-$*$ homeomorphic to the compression $P_{\mathcal{M}}^\perp \mathfrak{L}_n|_{P_{\mathcal{M}}^\perp}$ of $\mathfrak{L}_n$ to $\mathcal{M}^\perp$.*

Now apply this to the Nevanlinna–Pick problem. The range of the ideal $\mathfrak{J} = \bigcap_{1 \leq j \leq k} \varphi_{\alpha_j}$ has complement $\mathcal{M}^\perp = \operatorname{span}\{\nu_{\alpha_j} : 1 \leq j \leq k\}$. Thus there is a contractive solution to $\hat{A}(\alpha_j) = c_j$ if and only of $\|P_{\mathcal{M}}^\perp A P_{\mathcal{M}}^\perp\| \leq 1$ for any solution A. This is equivalent to the condition $P_{\mathcal{M}}^\perp (I - AA^*) P_{\mathcal{M}}^\perp \geq 0$. Since we have a basis for $\mathcal{M}^\perp$, it is a standard result that this is equivalent to the positivity of the $k \times k$ matrix with entries $\langle (I - AA^*)\nu_{\alpha_i}, \nu_{\alpha_j} \rangle$. A routine calculation factoring out a few things yields the classic Pick matrix condition

$$\left[\frac{1 - c_i \bar{c}_j}{1 - \langle \alpha_i, \alpha_j \rangle} \right] \geq 0.$$

Taking a limit, one obtains a characterization of the image in $H^\infty(\mathbb{B}_n)$: an analytic function h on $\mathbb{B}_n$ is in the image of the unit ball of $\mathfrak{L}_n$ if and only if

$$\left[\frac{1 - h(\alpha_i)\overline{h(\alpha_j)}}{1 - \langle \alpha_i, \alpha_j \rangle} \right] \geq 0$$

for all finite subsets of $\mathbb{B}_n$. In this case, the ideal is $\bigcap_{\alpha\in\mathbb{B}_n}\varphi_\alpha$, which is the commutator ideal $\mathfrak{C}$. The quotient $\mathfrak{L}_n/\mathfrak{C}$ is abelian, but it is not all of $H^\infty(\mathbb{B}_n)$. Indeed one may take $A_k = \sum_{\{w\in\mathbb{F}_2 : w(z_1,z_2)=z_1^k z_2^k\}} L_w$ and explicitly compute $\|\hat{A}_k\|_\infty/\|A+\mathfrak{C}\|$. This tends to 0, and thus the injective map from $\mathfrak{L}_n/\mathfrak{C}$ into $H^\infty(\mathbb{B}_n)$ is not surjective.

One obtains Carathéodory type conditions as well. For example, suppose that coefficients a_w are specified for $|w| \le k$. Is there an element A in the unit ball of $\mathfrak{L}_n$ whose initial Fourier coefficients are a_w? This amounts to computing $\operatorname{dist}(\sum_{|w|\le k} a_w L_w, \mathfrak{L}_0^{k+1})$. The distance estimate reduces this to a single norm condition $\|P_k \sum_{|w|\le k} a_w L_w\| \le 1$ where P_k projects onto $\operatorname{span}\{\xi_w : |w| \le k\}$.

Sarason's original approach was somewhat different, in that he established his estimates using (a prototype of) the commutant lifting theorem. Arias and Popescu [5] independently established these interpolation results, and give a proof based on Popescu's commutant lifting theorem [42].

8.2. Commuting Dilations

To better understand the abelian algebra $\mathfrak{L}_n/\mathfrak{C}$, we turn to commutative dilation theory. The appropriate models live on symmetric Fock space $\mathcal{K}_n^s$. Consider all n-tuples $k = (k_1, \dots, k_n) \in \mathbb{N}_0^n$, and write z^k for $z_1^{k_1} \dots z_n^{k_n}$ where z_i are commuting variables. Let $\mathcal{P}_k := \{w \in \mathbb{F}_n^+ : w(z) = z^k\}$. Consider an orthogonal but unnormalized basis for $\mathcal{K}_n^s$ given by

$$\zeta^k = \frac{1}{|\mathcal{P}_k|} \sum_{w\in\mathcal{P}_k} \xi_w .$$

Indeed $|\mathcal{P}_k| = \binom{|k|!}{k_1!\,k_2!\cdots k_n!}$ and $\|\zeta^k\| = |\mathcal{P}_k|^{-1/2}$.

Consider the operators $M_i\zeta^k = \zeta^{k+\delta_i}$, where δ_i has a 1 in the ith entry and 0's elsewhere. If one normalizes the basis ζ^k, one sees that the M_i's are certain weighted shifts.

We observe that M_i is just the compression of the left creation operator L_i to symmetric Fock space. Indeed,

$$\begin{aligned}
\langle L_i\zeta^k, \zeta^l\rangle = \langle \zeta^k, L_i^*\zeta^l\rangle &= \frac{1}{|\mathcal{P}_k|\,|\mathcal{P}_l|}\Big\langle \sum_{w\in\mathcal{P}_k} \xi_w, L_i^* \sum_{v\in\mathcal{P}_l} \xi_v\Big\rangle \\
&= \frac{1}{|\mathcal{P}_k|\,|\mathcal{P}_l|}\Big\langle \sum_{w\in\mathcal{P}_k} \xi_w, \sum_{u\in\mathcal{P}_{l-\delta_i}} \xi_u\Big\rangle \\
&= \begin{cases} 0 & \text{if } \ l \ne k+\delta_i \\ \frac{1}{|\mathcal{P}_k|\,|\mathcal{P}_l|}|\mathcal{P}_k| = \langle \zeta^l, \zeta^l\rangle & \text{if } \ l = k+\delta_i \end{cases} \\
&= \langle \zeta^{k+\delta_i}, \zeta^l\rangle
\end{aligned}$$

Thus $M_i = P_s L_i|_{\mathcal{K}_n^s}$ where P_s is the projection onto $\mathcal{K}_n^s$. It follows immediately that

$$MM^* = \sum_{i=1}^n M_i M_i^* = P_s \sum_{i=1}^n L_i L_i^*|_{\mathcal{K}_n^s} = I - \xi_\varnothing \xi_\varnothing^* \leq I.$$

The main dilation result in this context is due to Drury [30]. It says that every strict commuting contraction dilates to a multiple of $M = [M_1, \dots, M_n]$.

Theorem 8.3. *Let $A = [A_1, \dots, A_n]$ be a commuting n-tuple of operators on a Hilbert space $\mathcal{H}$ with $\|A\| < 1$. Then there is an isometry V of $\mathcal{H}$ into $\mathcal{K}_n^s \otimes \mathcal{H}$ such that $A_i V^* = V^* M_i$ for $1 \leq i \leq n$.*

We sketch a proof based on the Frahzo–Bunce–Popescu dilation, which yields an isometry V into $\mathcal{K}_n \otimes \mathcal{H}$ such that $A_i V^* = V^*(L_i \otimes I)$ for $1 \leq i \leq n$. It suffices to show that the range of V is contained in $\mathcal{K}_n^s \otimes \mathcal{H}$. However the intertwining relation implies that the range $\mathcal{V} = V\mathcal{H}$ is invariant for each $L_i^* \otimes I$, and the commutativity of the A_i's implies that $L_u^* L_v^* \xi = L_v^* L_u^* \xi$ for all $u, v \in \mathbb{F}_n^+$ and all $\xi \in \mathcal{V}$. So if we write $\xi = \sum_{w \in \mathbb{F}_n^+} \xi_w \otimes x_w$ where $x_w \in \mathcal{H}$, then we obtain $x_{uvw} = x_{vuw}$ for all $u, v, w \in \mathbb{F}_n^+$. At bit of thought shows that this forces $x_u = x_v$ if $u(z) = v(z)$, and thus ξ is symmetric as claimed.

Now Arveson's dilation theory yields a canonical completely positive map φ from $\mathrm{C}^*(M)$ into $\mathrm{C}^*(A)$ such that $\varphi(M_i) = A_i$. In fact, $\varphi(M_u M_v^*) = A_u A_v^*$ for all $u, v \in \mathbb{F}_n^+$. This readily extends to all commuting contractions by considering $rA = (rA_1, \dots, rA_n)$ for $r < 1$ and letting r increase to 1. In particular, one obtains Drury's von Neumann inequality:

Corollary 8.4. *Let $A = [A_1, \dots, A_n]$ be a commuting n-tuple of operators on a Hilbert space $\mathcal{H}$ with $\|A\| \leq 1$. Then $\|p(A)\| \leq \|p(M)\|$ for all polynomials p in n commuting variables.*

Arveson [9] goes a step further and shows that $\mathrm{C}^*(M)$ is the C^*-envelope of M. Indeed, an easy calculation shows that $\mathrm{C}^*(M)$ is irreducible and contains the compact operators. The commutators of M_i and M_j^* are easily computed, and are compact. Thus the quotient $\mathrm{C}^*(M)/\mathfrak{K}$ is an abelian C^*-algebra and the images $Z_i = M_i + \mathfrak{K}$ satisfy $ZZ^* = \sum_{i=1}^n |Z_i|^2 = I$. Thus the maximal ideal space is contained in the unit sphere $\partial\mathbb{B}_n$. One can show that the spectrum is invariant under the action of the unitary group $\mathfrak{U}_n$. Indeed, each $U \in \mathfrak{U}_n$ determines the unitary operator $\widetilde{U}$ on Fock space, and this preserves $\mathcal{K}_n^s$. The restriction $\widetilde{U}|_{\mathcal{K}_n^s}$ implements an automorphism of $\mathrm{C}^*(M)$. Thus $\mathrm{C}^*(M)/\mathfrak{K} \simeq C(\partial\mathbb{B}_n)$.

8.3. Reproducing kernel Hilbert spaces

To make the connection to interpolation, we observe that $\mathcal{K}_n^s$ is a reproducing kernel Hilbert space. The eigenvectors ν_λ for $\lambda \in \mathbb{B}_n$ span all of $\mathcal{K}_n^s$, and

$$\nu_\lambda = (1 - \|\lambda\|^2)^{1/2} \sum_{w \in \mathbb{F}_n^+} \overline{w(\lambda)} \xi_w = (1 - \|\lambda\|^2)^{1/2} \sum_{k \in \mathbb{N}_0^n} \overline{\lambda}^k |\mathcal{P}_k| \zeta^k.$$

We renormalize for convenience and set

$$u_\lambda = (1 - \|\lambda\|^2)^{-1/2}\nu_\lambda = \sum_{k\in\mathbb{N}_0^n} \overline{\lambda}^k |\mathcal{P}_k| \zeta^k.$$

Each vector $\zeta \in \mathcal{K}_n^s$ determines an analytic function

$$\hat{\zeta}(\lambda) = \langle \zeta, u_\lambda \rangle = \sum_{k\in\mathbb{N}_0^n} c_k \lambda^k.$$

Moreover

$$|\hat{\zeta}(\lambda)| \le \|\zeta\| \, \|u_\lambda\| = \|\zeta\| \, (1 - \|\lambda\|^2)^{-1/2}.$$

Thus $\hat{\mathcal{K}}_n^s$ becomes a Hilbert space of analytic functions in which the point evaluations are continuous. To emphasize that this is an L^2 norm on these functions, we will write $\|f\|_2$ for the norm of an element f in $\hat{\mathcal{K}}_n^s$.

Let $\mathfrak{M}$ denote the WOT-closed algebra generated by the M_i. This is the WOT-closed algebra generated by the compressions of the L_i's to $\mathcal{K}_n^s$ which is the orthogonal complement of $\overline{\mathfrak{C}\mathcal{K}_n}$. Hence by Corollary 8.2, this is completely isometrically isomorphic to $\mathfrak{L}_n/\mathfrak{C}$. So we have a concrete representation of the quotient, and moreover every element of the unit ball of $\mathfrak{M}$ is in the image of the unit ball of $\mathfrak{L}_n$.

The important fact is that operators in $\mathfrak{M}$ have a nice analytic form in this functional representation. Indeed, if A is any operator in $\mathfrak{L}_n$ and $M = P_s A|_{\mathcal{K}_n^s}$ is the compression to $\mathcal{K}_n^s$, then because $A^*\nu_\lambda = \overline{\hat{A}(\lambda)}\nu_\lambda$ we obtain

$$\widehat{M\zeta}(\lambda) = \langle A\zeta, u_\lambda \rangle = \langle \zeta, A^* u_\lambda \rangle = \hat{A}(\lambda)\langle \zeta, u_\lambda \rangle = \hat{A}(\lambda)\hat{\zeta}(\lambda).$$

In particular, the operators M_i become multiplication operators by the coordinate functions z_i. The operators in $\mathfrak{M}$ are analytic multipliers on $\hat{\mathcal{K}}_n^s$, and we may write $M = M_h$ where $M = A + \mathfrak{C}$ and $h = \hat{A}$. Thus the operator norm equals the multiplier norm:

$$\|M\| = \|M_h\| = \sup\{\|hf\|_2 : \|f\|_2 \le 1\}.$$

Conversely, suppose that h is a bounded multiplier; so that M_h is a bounded operator. Then the Cesaro means $h_n = C_n(f)$ are polynomials and M_{h_n} converges to M_h in the strong operator topology. Since each M_{h_n} is a polynomial in the M_i's, it follows that M_h belongs to $\mathfrak{M}$. Thus $\mathfrak{M}$ is precisely the algebra of multipliers. So the image of the Gelfand map is realized as the algebra of multipliers for this special reproducing kernel Hilbert space.

More generally, a reproducing kernel Hilbert space (see [1]) on a set X is determined by a positive definite function $k(x,y)$ on $X \times X$, meaning that $\big[k(x_i,x_j)\big]$ is a positive definite matrix for every *finite* subset $\{x_1,\dots,x_k\}$ of X. The functions $k_x(y) = k(x,y)$ span a space of functions on X on which we define an inner product $\big\langle \sum a_i k_{x_i}, \sum b_j k_{y_j} \big\rangle = \sum a_i \overline{b_j} k(x_i,y_j)$. The completion $\mathcal{H}_k$ is called a reproducing kernel Hilbert space, and each vector ξ determines a function $\hat{\xi}(x) = \langle \xi, k_x \rangle$.

A *multiplier* is a (bounded) function h on X such that the map $\widehat{M_h\xi}(x) = h(x)\hat{\xi}(x)$ is continuous. This is equivalent to the boundedness of the adjoint map which has the nicer formulation $M_h^* k_x = \overline{h(x)} k_x$. The norm on h is just the operator norm $\|M_h\|$. One defines matrix multipliers as matrices with multipliers as coefficients. This kernel is said to have the *complete Nevanlinna–Pick property* if for every finite subset $x_1, \dots, x_s$ of X and finite set $C_1, \dots, C_k$ of $p \times p$ matrices, there is a matrix multiplier h of norm at most 1 with $\hat{h}(x_i) = C_i$ for $1 \le i \le s$ exactly when the Pick condition holds:

$$\left[(I_p - C_i C_j^*)\langle k_{x_j}, k_{x_i}\rangle\right]_{s\times s} \ge 0.$$

In our case, one has $\langle \nu_\lambda, \nu_\mu \rangle = \dfrac{1}{1 - \langle \lambda, \mu \rangle}$ which leads to the precise condition obtained above. In other words, symmetric Fock space is a complete Nevanlinna–Pick kernel.

Agler [2] reformulated the Nevanlinna–Pick problem in this way. Exactly which reproducing kernels have this property was solved by Quiggin [51] and McCullough [39]. An easy argument reduces to the irreducible case in which $k(x, y) \neq 0$ for all $x, y \in X$.

Theorem 8.5. *A necessary and sufficient for an irreducible positive definite kernel $k(x, y)$ on $X \times X$ to have the complete Nevanlinna–Pick property is that for every finite subset $x_1, \dots, x_s$ of X, the $s \times s$ matrix*

$$\left[\frac{1}{k(x_i, x_j)}\right]$$

has exactly one positive eigenvalue counting multiplicity.

In our case, this is the matrix

$$\left[1 - \langle \lambda_i, \lambda_j \rangle\right] = \mathbf{1} \cdot \mathbf{1}^* - \Lambda^* \Lambda$$

where $\mathbf{1}$ is the column vector consisting of s 1's, and $\Lambda = (\lambda_1, \dots, \lambda_s)$ is an $n \times s$ matrix with columns equal to λ_i. Now $\Lambda^*\Lambda$ is positive and norm less than s; while $\mathbf{1} \cdot \mathbf{1}^*$ is rank one and norm exactly s. It follows that $\mathbf{1} \cdot \mathbf{1}^* - \Lambda^*\Lambda$ has exactly one positive eigenvalue. So this is a complete NP kernel, which leads to a rather different proof than ours.

Agler and McCarthy [3] use this to show that (irreducible) complete Nevanlinna–Pick kernels have the rather specific form

$$k(x, y) = \frac{\overline{d(x)} d(y)}{1 - f(x, y)}$$

where d is a non-vanishing function on X and f is a positive semidefinite function on $X \times X$ taking values in the open unit disk. If n is the rank of the Hermitian form f, they then show that k is just the restriction of the kernel for symmetric Fock space $\mathcal{K}_n^s$ to some subset of the ball $\mathbb{B}_n$. In other words, the spaces $\mathcal{K}_n^s$ are in a certain sense the universal complete Nevanlinna–Pick kernels.

References

[1] G. Adams, P. McGuire and V. Paulsen, *Analytic reproducing kernels and multiplication operators*, Illinois J. Math., **36** (1992), 404–419.

[2] J. Agler, *Nevanlinna–Pick interpolation on Sobolev space*, Proc. Amer. Math. Soc., **108** (1990), 341–351.

[3] J. Agler and J.E. McCarthy, *Complete Nevanlinna–Pick kernels*, preprint, 2000.

[4] A. Arias and G. Popescu, *Factorization and reflexivity on Fock spaces*, Integral Eqtns. & Operator Theory, **23** (1995), 268–286.

[5] A. Arias and G. Popescu, *Noncommutative interpolation and Poisson transforms*, Israel J. Math., **115** (2000), 205–234.

[6] W. B. Arveson, *Subalgebras of C^*-algebras*, Acta Math., **123** (1969), 141–224.

[7] W. B. Arveson, *Subalgebras of C^*-algebras II*, Acta Math., **128** (1972), 271–308.

[8] W. B. Arveson, *Interpolation problems in nest algebras*, J. Func. Anal., **20** (1975), 208–233.

[9] W. B. Arveson, *Subalgebras of C^*-algebras III: multivariable operator theory*, Acta Math., **181** (1998), 159–228.

[10] A. Beurling, *On two problems concerning linear transformations in Hilbert space*, Acta Math., **81** (1949), 239–255.

[11] H. Bercovici, C. Foiaş and C. Pearcy, *Dual algebras with applications to invariant subspaces and dilation theory*, CBMS Notes, **56**, Amer. Math. Soc., Providence, 1985.

[12] H. Bercovici, *Hyper-reflexivity and the factorization of linear functionals*, J. Func. Anal., **158** (1998), 242–252.

[13] O. Bratteli, D. Evans and P. Jorgensen, *Compactly supported wavelets and representations of the Cuntz relations*, Appl. Comput. Harmon. Anal., **8** (2000), 166–196.

[14] O. Bratteli and P. Jorgensen, *Endomorphisms of $\mathcal{B}(\mathcal{H})$ II*, J. Func. Anal., **145** (1997), 323–373.

[15] O. Bratteli and P. Jorgensen, *Iterated function systems and permutation representations of the Cuntz algebra*, Mem. Amer. Math. Soc., **139** (1999), no. 663.

[16] O. Bratteli and P. Jorgensen, *Isometries, shifts, Cuntz algebras and miltiresolution wavelet analysis of scale N*, Integral Eqtns. & Operator Theory, **28** (1997), 382–443.

[17] O. Bratteli and P. Jorgensen, *Wavelet filters and infinite dimensional unitary groups*, preprint, 2000.

[18] O. Bratteli, P. Jorgensen, and G. L. Price, *Endomorphisms of $\mathcal{B}(\mathcal{H})$*, in "Quantization of nonlinear partial differential equations" (W. Arveson et al, Eds.), Proc. Symp. Pure Math., **59** pp. 93–138, Amer. Math. Soc., Providence, 1996.

[19] J. Bunce, *Models for n-tuples of non-commuting operators*, J. Func. Anal., **57** (1984), 21–30.

[20] E. Christensen, *Extensions of derivations II*, Math. Scand., **50** (1982), 111–112.

[21] J. Cuntz, *Simple C^*-algebras generated by isometries*, Comm. Math. Phys., **57** (1977), 173–185.

[22] K. R. Davidson, *The distance to the analytic Toeplitz operators*, Illinois J. Math., **31** (1987), 265–273.

[23] K. R. Davidson, *C^*-Algebras by Example*, Fields Institute Monograph Series, **6**, Amer. Math. Soc., Providence, RI, 1996.

[24] K. R. Davidson, E. Katsoulis and D. R. Pitts, *The Structure of Free Semigroup Algebras*, J. reine angew. Math., to appear.

[25] K. R. Davidson, D. W. Kribs and M. E. Shpigel, *Isometric Dilations of non-commuting finite rank n-tuples*, Canad. J. Math., to appear.

[26] K. R. Davidson and M. Ordower, *Some Exact Distance Constants*, Lin. Alg. Appl., **208/209** (1994), 37–55.

[27] K. R. Davidson and D. R. Pitts, *Invariant subspaces and hyper-reflexivity for free semi-group algebras*, Proc. London Math. Soc., **78** (1999), 401–430.

[28] K. R. Davidson and D. R. Pitts, *The algebraic structure of non-commutative analytic Toeplitz algebras*, Math. Ann., **311** (1998), 275–303.

[29] K. R. Davidson and D. R. Pitts, *Nevanlinna–Pick Interpolation for non-commutative analytic Toeplitz algebras*, Integral Eqtns. & Operator Theory, **31** (1998), 321–337.

[30] S.W. Drury, *A generalization of von Neumann's inequality to the complex ball*, Proc. Amer. Math. Soc., **68** (1978), 300–304.

[31] A. Frahzo, *Models for non-commuting operators*, J. Func. Anal., **48** (1982), 1–11.

[32] J. Glimm, *Type I C^*-algebras*, Ann. Math., **73** (1961), 572–612.

[33] M. Hamana, *Injective envelopes of operator systems*, Publ. Res. Inst. Math Sci., **15** (1979), 773–785.

[34] P. Jorgensen. *Minimality of the data in wavelet filters*, preprint, 2000.

[35] D.W. Kribs, *Factoring in non-commutative analytic Toeplitz algebras*, J. Operator Theory, to appear.

[36] D.W. Kribs, *Completely Positive Maps in Dilation and Wavelet Decompositions*, preprint 2000.

[37] M. Laca, *Endomorphisms of $\mathcal{B}(\mathcal{H})$ and Cuntz algebras*, J. Operator Theory, **30** (1993), 381–396.

[38] D.E. Marshall, *Blaschke products generate H^∞*, Bull. Amer. Math. Soc., **82** (1976), 494–496.

[39] S. McCullough, *The local de Branges–Rovnyak construction and complete Nevanlinna–Pick kernels*, Algebraic methods in operator theory, 15–24, Birkhauser Boston, Boston, MA, 1994.

[40] V.I. Paulsen, *Completely bounded maps and dilations*, Pitman Res. Notes Math., **146**, Longman Sci. Tech. Harlow, 1986.

[41] G. Popescu, *Models for infinite sequences of noncommuting operators*, Acta. Sci. Math. (Szeged), **53** (1989), 355–368.

[42] G. Popescu, *Isometric dilations for infinite sequences of noncommuting operators*, Trans. Amer. Math. Soc., **316** (1989), 523–536.

[43] G. Popescu, *Characteristic functions for infinite sequences of noncommuting operators*, J. Operator Theory, **22** (1989), 51–71.

[44] G. Popescu, *Multi-analytic operators and some factorization theorems*, Indiana Univ. Math. J., **38** (1989), 693-710.

[45] G. Popescu, *Von Neumann Inequality for $(\mathcal{B}(\mathcal{H})^n)_1$*, Math. Scand., **68** (1991), 292–304.

[46] G. Popescu, *Functional calculus for noncommuting operators*, Mich. J. Math., **42** (1995), 345–356.

[47] G. Popescu, *Multi-analytic operators on Fock spaces*, Math. Ann., **303** (1995), 31–46.

[48] Popescu, G., *Noncommuting disc algebras and their representations*, Proc. Amer. Math. Soc., **124** (1996), 2137–2148.

[49] G. Popescu, *Poisson transforms on some C^*-algebras generated by isometries*, J. Func. Anal., **161** (1999), 27–61.

[50] R. Powers, *An index theory for semigroups of $*$-endomorphisms of $\mathcal{B}(\mathcal{H})$ and type II factors*, Canad. J. Math., **40** (1988), 86–114.

[51] P. Quiggin, *For which reproducing kernel Hilbert spaces is Pick's theorem true?* Integral Eqtns. & Operator Theory, **16** (1993), 244–266.

[52] S. Rosenoer, *Distance estimates for von Neumann algebras*, Proc. Amer. Math. Soc., **86** (1982), 248–252.

[53] W. Rudin, *Function Theory in the Unit Ball of $\mathbb{C}^n$*, Grund. math. Wiss., **241**, Springer–Verlag, New York, 1980.

[54] B. Russo and H. Dye, *A note on unitary operators in C^*-algebras*, Duke J. Math., **33** (1966), 413–416.

[55] D. Sarason, *Generalized interpolation in H^∞*, Trans. Amer. Math. Soc., **127** (1967), 179–203.

[56] B. Sz.-Nagy and C. Foiaş, *Harmonic analysis of operators on Hilbert space*, North Holland Pub. Co., London, 1970.

[57] D. Voiculescu, *Symmetries of some reduced free product C^*-algebras*, Lect. Notes Math., **1132**, 556–588, Springer Verlag, New York, 1985.

[58] J. Wermer, *On invariant subspaces of normal operators*, Proc. Amer. Math. Soc., **3** (1952), 270–277.

2000 Mathematics Subject Classification: 47L80

Pure Math. Dept.
U. Waterloo
Waterloo, ON N2L 3G1
CANADA
E-mail address: krdavids@math.uwaterloo.ca

Some geometric invariants from resolutions of Hilbert modules

Ronald G. Douglas, Gadadhar Misra, and Cherian Varughese

1. Introduction

The model theory of Sz.-Nagy and Foias for contractions was reformulated in [15, chapter 3]. The existence of a unique minimal unitary dilation amounts to the existence of a Silov resolution for contractive Hilbert modules $\mathcal{M}$ over the disc algebra $\mathcal{A}(\mathbb{D})$ along with the fact that any two minimal Silov resolutions are isomorphic. The Silov modules appearing in the resolution may be classified using the von Neumann and Wold decomposition for isometries. This classification then allows one to identify a complete set of unitary invariants for the contractive module $\mathcal{M}$. This produces the characteristic function of Sz.-Nagy and Foias which determines the contractive module up to unitary equivalence. Using localisation techniques and Beurling's Theorem [15, p. 115], it is possible to identify this characteristic function. If we wish to adapt this method for Hilbert modules over other function algebras, then first there is the question of existence of a resolution. Secondly, given a resolution, there is the question of identifying invariants for the module from the resolution. In the following, we will not address the first question but attempt to describe some possible invariants from a resolution. Having decided to ignore the question of existence of resolutions, we must identify situations where it is easy to write down a resolution. In this paper, we consider Hilbert spaces which consist of holomorphic functions on a bounded domain $\Omega \subseteq \mathbb{C}^m$ together with the natural action of the polynomials as multiplication operators. Thus we obtain a module $\mathcal{M}$ over the algebra of polynomials in m variables. To continue our investigation, we must assume that the action is bounded which ensures that $\mathcal{M}$ is a module over the algebra $\mathcal{A}$—the closure of the polynomials with respect to supremum norm on Ω. If we consider the submodule $\mathcal{M}_0$ of all functions vanishing on a fixed subset $\mathcal{Z}$ of Ω then we obtain a short exact sequence

$$0 \longleftarrow \mathcal{M}_{\mathrm{q}} \longleftarrow \mathcal{M} \xleftarrow{X} \mathcal{M}_0 \longleftarrow 0 \tag{1.1}$$

The research of the last two authors was partly carried out with financial support from the National Board for Higher Mathematics, India.

of Hilbert modules, where X is the inclusion map and the next arrow stands for the quotient map. There is another short exact sequence which occurs, namely,

$$0 \longleftarrow \mathcal{Q} \longleftarrow \mathcal{M} \overset{X}{\longleftarrow} \mathcal{M} \longleftarrow 0, \tag{1.2}$$

where X is multiplication by some φ in $\mathcal{A}$. Let $\mathcal{S}$ be the submodule determined by taking the closure of the range of this multiplication. Then $\mathcal{Q}$ is the quotient of $\mathcal{M}$ by the submodule $\mathcal{S}$. We see that the submodule $\mathcal{S}$ is contained in the submodule $\mathcal{M}_0$ of all functions vanishing on the zero set of φ. However, whether the two submodules $\mathcal{M}_0$ and $\mathcal{S}$ coincide, when $\mathcal{Z}$ is the zero set of φ, is probably a *subtle* question.

Let $\mathcal{M}_0 \subseteq \mathcal{M}$ be a submodule and consider the resolution (1.1) of Hilbert modules over a function algebra $\mathcal{A}$. Over the past several years, we have attempted to obtain a canonical model and a complete set of unitary invariants for the quotient module $\mathcal{M}_{\mathrm{q}}$, at least in the case where $\mathcal{M}_0$ is assumed to be a submodule of functions vanishing on a hypersurface $\mathcal{Z} \subseteq \Omega$. To extract invariants from a resolution of the form (1.1), we assume that the module $\mathcal{M}$ is in the class $B_1(\Omega^*)$, where $\Omega^* = \{w \in \mathbb{C}^m : \bar{w} \in \Omega\}$. One may consult [15, p. 95] for more details about this class, some of which are also reproduced in the following section on Preliminaries. Then it is reasonable to expect that one should be able to use the existing theory to find the answer. Unfortunately, the submodule $\mathcal{M}_0$ does not necessarily belong to the same class making our quest somewhat tortuous. Here is an example.

Example 1.1. *Let $\mathcal{M} = H^2(\mathbb{D}^2)$ be the Hardy space on the bi-disc. This may be thought of as a Hilbert space of holomorphic functions defined on $\mathbb{D}^2$ determined by the reproducing kernel*

$$\hat{K}(z,w) = (1 - z_1\bar{w}_1)^{-1}(1 - z_2\bar{w}_2)^{-1},\ z = (z_1, z_2),\ w = (w_1, w_2) \in \mathbb{D}^2.$$

(We reserve the symbol $\hat{K}$ for the reproducing kernel of the Hardy space.)

Let $H_0^2(\mathbb{D}^2) = \{f \in H^2(\mathbb{D}^2) : f(0,0) = 0\}$ be the submodule $\mathcal{M}_0$ of functions vanishing at the origin. The reproducing kernel $\hat{K}_0(z,w)$ for $H_0^2(\mathbb{D}^2)$ is easily seen to be

$$\hat{K}_0(z,w) = (z_1\bar{w}_1 + z_2\bar{w}_2 - z_1z_2\bar{w}_1\bar{w}_2)\hat{K}(z,w).$$

We verify that both the functions z_1 and z_2 lie in $\cap_{\ell=1}^2 \ker PM_{z_\ell}^*$, where P is the projection on to $H_0^2(\mathbb{D}^2)$. However, at all other points $w = (w_1, w_2) \in \mathbb{D}^2$, the joint kernel is spanned by $\hat{K}_0(\cdot, w)$. Consequently, we have $\dim \cap_{\ell=1}^2 \ker P(M_{z_\ell} - w_\ell)^* = 1$. The quotient module, in this case, is the one dimensional space spanned by the constant function 1.

The assumption that $\mathcal{M}$ is in $B_1(\Omega^*)$ ensures the existence of an anti-holomorphic, hermitian line bundle on the domain Ω. The fiber of this bundle at $w \in \Omega$ is the joint kernel $\cap_{\ell=1}^m \ker(M_\ell - w_\ell)^*$, where M_ℓ is the operator induced by the module action of the generator z_ℓ in $\mathcal{A}$. It is then shown in [6] that the local equivalence of these bundles translates to unitary equivalence of elements in $B_1(\Omega^*)$. The curvature of this line bundle completely determines the unitary equivalence class. In the case of $H^2_0(\mathbb{D}^2)$, we can no longer associate a holomorphic hermitian vector bundle on all of $\mathbb{D}^2$. However, it is clear that the module is in $B_1(\mathbb{D}^2\backslash\{0\})$ and therefore there exists a line bundle on $\mathbb{D}^2\backslash\{0\}$. This is the prototypical example of our discussion in the last section, where we indicate how to enlarge the domain of this bundle by "blowing up" the singularity at the origin. The invariants for the new line bundle defined on a larger domain then provide invariants for the submodule in question.

In [12], the submodule is taken to be the maximal set of functions vanishing on an analytic hypersurface $\mathcal{Z}$ in Ω. Two descriptions are then provided for the quotient module. The first approach characterises it as a reproducing kernel Hilbert space. In the second approach, a hermitian, holomorphic line bundle on $\mathcal{Z}$ is associated with the quotient module. A complete unitary invariant for the quotient module is then the curvature of this line bundle. The submodule in [14] is taken to be the (maximal) set of functions which vanish to some given order k on the hypersurface $\mathcal{Z}$. As in the previous case, two descriptions are provided for the quotient module. A matrix-valued kernel function must now be used and, in the vector bundle picture, we have a rank k hermitian, holomorphic bundle. Some invariants for the quotient module (though not a complete set) are described. Some possible directions which might lead to other invariants are also discussed. In the section following the Preliminaries, we describe some of these results in detail along with concrete examples.

In [12] and [14], the approach chosen has been to use the resolution (1.1) of the quotient module and to construct invariants using the map X (and metric information about some vector bundles determined by $\mathcal{M}$ and $\mathcal{M}_0$). It has, however, been shown in [13] that it may be fruitful to consider other resolutions of the quotient module. Let φ be a defining function for $\mathcal{Z} \subseteq \Omega$. Assume that the closure of the range of multiplication by φ on $\mathcal{M}$ coincides with the submodule $\mathcal{M}_0$ consisting of functions that vanish on the hypersurface $\mathcal{Z}$. Then it is shown that the following is a useful resolution:

$$0 \longleftarrow \mathcal{M}_q \longleftarrow \mathcal{M} \xleftarrow{M_\varphi} \mathcal{M} \longleftarrow 0 \tag{1.3}$$

Here M_φ is the operator of multiplication by φ. If $M_\varphi(w)$, $w \in \Omega$ is the localisation of M_φ (see subsection 2.2), it is shown that $\partial\bar{\partial}\log\|M_\varphi(w)\|^2$ represents the fundamental class of $\mathcal{Z}$. We point out here that the complex

$$0 \longleftarrow \mathcal{M} \xleftarrow{M_\varphi} \mathcal{M} \longleftarrow 0$$

is the simplest example of a Koszul complex and the sequence (1.3) is an extension of this Koszul complex by the quotient module.

More general Koszul complexes are useful tools in studying the case where $\mathcal{M}_0$ is taken to be the submodule of functions vanishing on an analytic submanifold $\mathcal{Z} \subseteq \Omega$ of codimension $r > 1$. Some of the results obtained are briefly described below, the details are given in the later sections.

Assume that $\mathcal{Z}$ is a complete intersection. $\mathcal{Z}$ is then the common zero set of r defining functions $\varphi_1, \varphi_2, \cdots, \varphi_r$. It is possible to construct a useful resolution for the quotient module via a Koszul complex if we make the following further assumption on the submodule $\mathcal{M}_0$. Consider the range of the operator of multiplication by φ_j, $1 \le j \le r$, defined on $\mathcal{M}$. It is clear that this range is contained in $\mathcal{M}_0$ for each j. We assume that $\mathcal{M}_0$ is the closure of the span of these ranges. Then it is possible to construct a resolution of the quotient module of length r which allows us to identify the fundamental class of the submanifold $\mathcal{Z}$ using a generalisation of the Poincaré-Lelong equation due to Griffiths and King [17].

If φ is the defining function for the hypersurface $\mathcal{Z}$, then the closure of the range of multiplication by φ^k on $\mathcal{M}$ is a subspace of functions which vanish to order k on $\mathcal{Z}$. Whether this is the largest subspace of functions with this property, as pointed out earlier, is a subtle question. In the absence of a definite answer to this question, we make the aparently restrictive assumption of the previous paragraph. We discuss some of these details in the third section.

An alternative resolution of the quotient module is by means of a grid and it is possible to determine the fundamental class of $\mathcal{Z}$ as a product of $(1,1)$ currents (cf. [8]) obtained from this resolution. We have described the simplest case of $r = 2$ (see (4.8)) in some detail in this paper. Although working out the details of the "grid construction" in the general case is not entirely trivial, the results are similar.

2. Preliminaries

We begin by describing the standing assumptions that we make all through this paper. By a hypersurface $\mathcal{Z} \subseteq \Omega$ we mean a complex submanifold of dimension $m-1$. It follows that, given $z^{(0)} \in \mathcal{Z}$, there is a neighborhood $U \subseteq \Omega$ and a holomorphic map $\varphi : U \to \mathbb{C}$ such that $(\partial\varphi/\partial z_j)(z^{(0)}) \neq 0$ for some j, with $1 \le j \le m$ and

$$U \cap \mathcal{Z} = \{z \in U : \varphi(z) = 0\}.$$

In this case, we say that φ is a defining function for the hypersurface $\mathcal{Z}$. Whenever we discuss a zero variety, we will assume that it is the common zero set of holomorphic functions $\varphi_1, \dots, \varphi_n$ defined on Ω. It must be pointed out that our interest lies in equivalence classes of modules consisting of holomorphic functions defined on Ω. The restriction of such modules to an open subset U of Ω yield equivalent modules as pointed out in [14, p. 370]. Consequently, we can (by going to a smaller open set if necessary) ensure that the zero variety has the description given above. The only assumption we make is that if the zero variety is the intersection of a number of hypersurfaces, then it must be a complete intersection. The assumption of complete intersection allows us to get around a number of technical difficulties—none very serious. We hope to study the general case in the near future.

Let $\Omega \subseteq \mathbb{C}^m$ be a bounded, simply connected domain in $\mathbb{C}^m$. Let $\mathcal{A}(\Omega)$ denote the closure with respect to the supremum norm on Ω of functions holomorphic in a neighborhood of $\bar{\Omega}$, the closure of the domain $\Omega \subseteq \mathbb{C}^m$. Then $\mathcal{A}(\Omega)$ is a function algebra and consists of continuous functions on $\bar{\Omega}$ which are holomorphic on Ω. We assume that Ω is polynomially convex which then ensures that $\mathcal{A}(\Omega)$ is the closure of polynomials with respect to the supremum norm on Ω.

Let $\mathcal{M}$ be a Hilbert module over $\mathcal{A}(\Omega)$. We assume that the point evaluation $f \mapsto f(w)$, $f \in \mathcal{M}$ is bounded for each $w \in \Omega$. Consequently, $\mathcal{M}$ admits a reproducing kernel $K : \Omega \times \Omega \to \mathbb{C}$. We assume, in addition, that $\mathcal{M}$ consists of holomorphic functions on Ω and contains all the polynomials, and lies in the Cowen-Douglas class $B_1(\Omega^*)$, that is, $(M_1^*, \dots, M_m^*)$ is in $B_1(\Omega^*)$. Although it may not be absolutely necessary for what follows, we also assume that $\mathcal{M}$ is of rank 1.

2.1. Functional Hilbert spaces and $B_k(\Omega)$

Let E be a k-dimensional Hilbert space and $\mathcal{L}(E)$ denote the vector space of all linear transformations on E. A function $K : \Omega \times \Omega \to \mathcal{L}(E)$, satisfying

$$\sum_{i,j=1}^{n} \left\langle K(w_i, w_j)\zeta_j, \zeta_i \right\rangle_E \geq 0, \; w_1, \dots, w_n \in \Omega, \; \zeta_1, \dots, \zeta_n \in E, n \geq 1 \tag{2.1}$$

is said to be a *non negative definite (nnd) kernel* on Ω. Given such an nnd kernel K on Ω, it is easy to construct a Hilbert space $\mathcal{H}$ of functions on Ω taking values in E with the property

$$\left\langle f(w), \zeta \right\rangle_E = \left\langle f, K(\cdot, w)\zeta \right\rangle_{\mathcal{H}}, \; w \in \Omega, \; \zeta \in E, \; f \in \mathcal{H}. \tag{2.2}$$

The Hilbert space $\mathcal{H}$ is simply the completion of the linear span of all vectors of the form $K(\cdot, w)\zeta$, $w \in \Omega$, $\zeta \in E$, with inner product defined by (2.2).

Conversely, let $\mathcal{H}$ be any Hilbert space of functions on Ω taking values in E. Let $e_w : \mathcal{H} \to E$ be the evaluation functional defined by $e_w(f) = f(w)$, $w \in \Omega$, $f \in \mathcal{H}$. If e_w is bounded for each $w \in \Omega$ then it is easy to verify that the Hilbert space $\mathcal{H}$ possesses a reproducing kernel $K(z, w) = e_z e_w^*$, that is, $K(\cdot, w)\zeta \in \mathcal{H}$ for each $w \in \Omega$ and K has the reproducing property (2.2). Finally, the reproducing property (2.2) determines the reproducing kernel K uniquely.

Remark 2.1. *If we assume that $\mathcal{H}$ is a Hilbert space consisting of holomorphic functions taking values in E, then it is not hard to see that the map $w \mapsto e_w$ is weak-$*$ holomorphic. Hence it is also strongly holomorphic as a map into the dual of $\mathcal{H}$. But since $(\partial_i e_w)(h) = (\partial_i h)(w)$, for $h \in \mathcal{H}$ and $w \in \Omega$, it follows that evaluations of partial derivatives induce bounded linear functionals on $\mathcal{H}$. Therefore, we see that $\bar{\partial}_i K(\cdot, w)$ is in $\mathcal{H}$, where $\bar{\partial}_i = \frac{\partial}{\partial \bar{w}_i}$.*

Let $T = (T_1, \dots, T_m)$ be an m-tuple of commuting bounded linear operators on a separable complex Hilbert space $\mathcal{H}$. Define the operator $D_T : \mathcal{H} \to \mathcal{H} \oplus \cdots \oplus \mathcal{H}$ by $D_T(x) = (T_1 x, \dots, T_m x)$, $x \in \mathcal{H}$. For $w = (w_1, \dots, w_m) \in \Omega$, let $T - w$ denote the operator tuple $(T_1 - w_1, \dots, T_m - w_m)$. Let k be a positive integer.

Definition 2.1. *The m-tuple T is said to be in the Cowen–Douglas class $B_k(\Omega)$ if*
1. ran D_{T-w} *is closed for all* $w \in \Omega$
2. span $\{\ker D_{T-w} : w \in \Omega\}$ *is dense in* $\mathcal{H}$
3. $\dim \ker D_{T-w} = k$ *for all* $w \in \Omega$.

This class was introduced in [7]. The case of a single operator was investigated earlier in the paper [6], where it is pointed out that an operator T in $B_1(\Omega)$ is unitarily equivalent to the adjoint of the multiplication operator M on a reproducing kernel Hilbert space. Here $(Mf)(z) = zf(z)$, $f \in \mathcal{H}$. It is not very hard to see that, more generally, an m-tuple T in $B_k(\Omega)$ is unitarily equivalent to the adjoint of the m-tuple of multiplication operators $M = (M_1, \dots, M_m)$ on a reproducing kernel Hilbert space [10, Remark 2.6 a) and b)]. Conversely, Curto and Salinas [10] show that if certain conditions are imposed on the reproducing kernel, then the corresponding adjoint of the m-tuple of multiplication operators belongs to the class $B_k(\Omega)$.

Let T be an m-tuple of operators in $B_k(\Omega)$ on $\mathcal{H}$. Pick k linearly independent vectors $\gamma_1(w), \dots, \gamma_k(w)$ in $\ker D_{T-w}$, $w \in \Omega$. Define a map $\Gamma : \Omega \to \mathcal{L}(\mathbb{C}^k, \mathcal{H})$ by $\Gamma(w)\zeta = \sum_{i=0}^{k} \zeta_i \gamma_i(w)$, where $\zeta = (\zeta_1, \dots, \zeta_k) \in \mathbb{C}^k$. It is shown in [6, Proposition 1.11] and [10, Theorem 2.2] that it is possible to choose $\gamma_1(w), \dots, \gamma_k(w)$, w in some domain $\Omega_0 \subseteq \Omega$, such that Γ is holomorphic on Ω_0. Let $\mathcal{A}(\Omega, \mathbb{C}^k)$ denote the linear space of all $\mathbb{C}^k$-valued holomorphic functions on Ω. Define $U_\Gamma : \mathcal{H} \to \mathcal{A}(\Omega_0^*, \mathbb{C}^k)$ by

$$(U_\Gamma x)(w) = \Gamma(w)^* x, \; x \in \mathcal{H}, \; w \in \Omega_0. \tag{2.3}$$

Define a sesqui-linear form on $\mathcal{H}_\Gamma = \operatorname{ran} U_\Gamma$ by $< U_\Gamma f, U_\Gamma g >_\Gamma = < f, g >$, $f, g \in \mathcal{H}$. The map U_Γ is linear and injective. Hence $\mathcal{H}_\Gamma$ is a Hilbert space of $\mathbb{C}^k$-valued holomorphic functions on Ω_0^* with inner product $< \cdot, \cdot >_\Gamma$ and U_Γ is unitary. Then it is easy to verify the following (cf. [10, Remarks 2.6]).

a): $K(z, w) = \Gamma(\bar z)^* \Gamma(\bar w)$, $z, w \in \Omega_0^*$ is the reproducing kernel for the Hilbert space $\mathcal{H}_\Gamma$.

b): $M_i^* U_\Gamma = U_\Gamma T_i$, where $(M_i f)(z) = z_i f(z)$, $z \in \Omega$.

c): There exists $w_0 \in \Omega_0^*$ such that $K(z, w_0) = I$ for all $z \in \Omega_0^*$.

An nnd kernel satisfying the condition c) above is said to be normalised at w_0.

Conversely, it is possible to impose conditions on a kernel function $K : \Omega \times \Omega \to \mathcal{L}(\mathbb{C}^k)$ ensuring the boundedness of each of the multiplication operators $M_1, \dots, M_m$ on the associated reproducing kernel Hilbert space. One may impose additional conditions on K to ensure that M is in $B_k(\Omega^*)$.

To an m-tuple T in $B_k(\Omega)$, on the one hand, one may associate a holomorphic, hermitian vector bundle on Ω (cf. [6]), which is obtained as the pullback of the universal bundle on the Grassmanian $Gr(\mathcal{H}, k)$ of the Hilbert space $\mathcal{H}$ under the map $\gamma : \Omega \to Gr(\mathcal{M}, k)$, where $\gamma(w) = \ker D_{T-w}$. On the other hand, one may associate a normalised reproducing kernel (cf. [10]) on a suitable subdomain of $\Omega^* = \{w \in \mathbb{C}^m : \bar w \in \Omega\}$. It is possible to answer a number of questions regarding the m-tuple of operators T using either the vector bundle or the reproducing kernel.

2.2. Localisation

Let $\mathcal{M}$ and $\mathcal{N}$ be any two Hilbert modules over the algebra $\mathcal{A}$. Notice that there are two possible module actions on $\mathcal{M} \otimes \mathcal{N}$, i.e., the left action: $L \otimes I : (f, h \otimes k) \mapsto f \cdot h \otimes k$ and the right action: $I \otimes R : (f, h \otimes k) \mapsto h \otimes f \cdot k$. The module tensor product $\mathcal{M} \otimes_{\mathcal{A}} \mathcal{N}$ is defined to be the module obtained by identifying these two actions. Specifically, let $\mathcal{E}$ be the closed subspace of $\mathcal{M} \otimes \mathcal{N}$ generated by vectors of the form

$$\{f \cdot h \otimes k - h \otimes f \cdot k : h \in \mathcal{M}, k \in \mathcal{N} \text{ and } f \in \mathcal{A}\}.$$

Then $\mathcal{E}$ is a submodule of $\mathcal{M} \otimes \mathcal{N}$ with respect to both the left and the right actions. The module tensor product $\mathcal{M} \otimes_{\mathcal{A}} \mathcal{N}$ is defined to be $(\mathcal{M} \otimes \mathcal{N}) \ominus \mathcal{E}$ together with the compression of either the left or the right actions, which coincide on this space. For fixed $w \in \Omega$, $\mathbb{C}$ is a module over $\mathcal{A}$ with the module action

$$(f, v) \mapsto f(w)v, \;\; f \in \mathcal{A}, v \in \mathbb{C}.$$

Let $\mathbb{C}_w$ denote the one dimensional module $\mathbb{C}$ with this action. We will largely confine ourselves to the module tensor product $\mathcal{M} \otimes_{\mathcal{A}} \mathbb{C}_w$, which we denote by $\mathcal{M}(w)$.

Remark 2.2. *Let $\mathcal{M}$ be a module over the algebra $\mathcal{A}(\Omega)$. Then the subspace $\mathcal{E}$ of $\mathcal{M} \otimes \mathbb{C}_w$ is easily seen to be*

$$\mathcal{E} = \text{span } \{((f - f(w)) \cdot h) \otimes 1 : \; f \in \mathcal{A}(\Omega), \; h \in \mathcal{M}\}.$$

It is therefore clear that the localisation $\mathcal{M}(w)$ at w, which is $\mathcal{E}^\perp$ is the common eigenspace of the "adjoint action" at w, that is,

$$\mathcal{M}(w) = \cap_{f \in \mathcal{A}} \ker(M_f - f(w))^*.$$

The action of $f \in \mathcal{A}$ on $\mathcal{M}(w)$ is scalar and is determined by $f(w)$.

If we localise the submodule $H_0^2(\mathbb{D}^2)$ of Example 1.1, using the module $\mathbb{C}_w$, where $w = (w_1, w_2) \in \mathbb{D}^2$, then we find that

$$H_0^2(\mathbb{D}^2) \otimes_{\mathcal{A}(\mathbb{D}^2)} \mathbb{C}_w = \begin{cases} \mathbb{C}_0 \oplus \mathbb{C}_0 & w = (0,0) \\ \mathbb{C}_w & w \neq (0,0). \end{cases}$$

A less trivial example is obtained as follows.

Example 2.1. *Let M be the multiplication operator on the Hardy space $H^2(\mathbb{D})$ on the disc. Consider the action of the bi-disc algebra $\mathcal{A}(\mathbb{D}^2)$ on the Hardy space $H^2(\mathbb{D})$ induced by the pair (M^2, M^3). Now, the Hardy space $H^2(\mathbb{D})$ is 2-generated, as a Hilbert module over $\mathcal{A}(\mathbb{D}^2)$, by 1 and z. If we localise $H^2(\mathbb{D})$ using the module $\mathbb{C}_w$, where $w = (w_1, w_2) \in \mathbb{D}^2$, then we find that*

$$H^2(\mathbb{D}) \otimes_{\mathcal{A}(\mathbb{D}^2)} \mathbb{C}_w = \begin{cases} \mathbb{C}_0 \oplus \mathbb{C}_0 & w = (0,0) \\ \mathbb{C}_w & w = (\alpha^2, \alpha^3), \;\alpha \neq 0, \;\alpha \in \mathbb{D} \\ (0) & w \neq (\alpha^2, \alpha^3), \;\alpha \in \mathbb{D}. \end{cases}$$

If we remove the function z by considering the subspace $H_1^2(\mathbb{D})$ of $H^2(\mathbb{D})$ spanned by $\{1, z^2, \dots\}$, then $H_1^2(\mathbb{D})$ is also a contractive Hilbert module over $\mathcal{A}(\mathbb{D}^2)$, and now

$$H_1^2(\mathbb{D}) \otimes_{\mathcal{A}(\mathbb{D}^2)} \mathbb{C}_w = \begin{cases} \mathbb{C}_w & w = (\alpha^2, \alpha^3),\ \alpha \in \mathbb{D} \\ (0) & w \neq (\alpha^2, \alpha^3),\ \alpha \in \mathbb{D}. \end{cases}$$

If X is a map from a module $\mathcal{M}$ to a module $\mathcal{N}$, we denote by $X(w)$ the restriction of $X \otimes I$ (defined on $\mathcal{M} \otimes \mathbb{C}_w$) to $\mathcal{M}(w)$ followed by the projection onto $\mathcal{N}(w)$.

A Hilbert module $\mathcal{M}$ over an algebra $\mathcal{A}(\Omega)$ is said to be of *rank k* if there exists $\Gamma \subseteq \mathcal{M}$ of cardinality k such that the linear span of $\{f \cdot h : f \in \mathcal{A}(\Omega) \text{ and } h \in \Gamma\}$ is dense in $\mathcal{M}$, and no subset of $\mathcal{M}$ of smaller cardinality has this property.

If $\mathcal{M}$ is a module of rank 1, then we find (cf. [15, Lemma 5.1]) that $\dim \mathcal{M}(w)$, $w \in \Omega$, is at most 1.

Recall that $\mathcal{M}$ is said to be locally free if the sheaf determined by the localisation $\mathcal{M}(w)$ defines a finite dimensional holomorphic, hermitian vector bundle over Ω and the linear span of the fibers is dense in $\mathcal{M}$.

We point out that locally free modules over $\mathcal{A}(\Omega)$ of finite rank, say k, resemble modules in the class $B_\ell(\Omega)$, for some $\ell \leq k$.

2.3. Factorisation of the reproducing kernel for the submodule $\mathcal{M}_0$

We recall from [14, Lemma 2.6] that if $\mathcal{M}_0$ is the submodule of functions vanishing on an irreducible hypersurface $\mathcal{Z} \subseteq \Omega$, then the reproducing kernel K_0 of $\mathcal{M}_0$ admits the following factorisation:

$$K_0(z, w) = \varphi(z)\chi(z, w)\overline{\varphi(w)},\ \ z, w \in \Omega, \tag{2.4}$$

where φ is a defining function for $\mathcal{Z}$ and $\chi : \Omega \times \Omega \to \mathbb{C}$ is some function holomorphic in the first variable and anti-holomorphic in the second one. Moreover, $\chi(w, w) \neq 0$ for any $w \in \Omega$. Since φ is a defining function, we may assume, without loss of generality, that $\partial_1 \varphi(w) \neq 0$, for $w \in \Omega$. Clearly,

$$(\bar{\partial}_1 K_0)(\cdot, w) = \varphi(\cdot)\chi(\cdot, w)\overline{(\partial_1 \varphi)(w)} \text{ for } w \in \mathcal{Z}. \tag{2.5}$$

Notice that $\bar{\partial}_1 K_0(\cdot, w)$, in view of Remark 2.1, is in $\mathcal{M}_0$. Now, it follows from the reproducing property of K_0 that

$$M_f^*(\bar{\partial}_1 K_0)(\cdot, w) = \overline{(\partial_1 f)(w)} K_0(\cdot, w) + \overline{f(w)}(\bar{\partial}_1 K_0)(\cdot, w),\ \ f \in \mathcal{A}.$$

The fact that$(\bar{\partial}_1 K_0)(\cdot, w)$ is a non zero vector for $w \in \mathcal{Z}$ in $\mathcal{M}_0$ follows from equation (2.5). Therefore, we conclude that $(\bar{\partial}_1 K_0)(\cdot, w)$ is an eigenvector for M_f^* as long as $w \in \mathcal{Z}$. At this juncture, if we assume that the submodule $\mathcal{M}_0$ coincides with the range closure of the operator M_φ, then it will follow that $\mathcal{M}_0$ is of rank 1 (we are making the standing assumption that $\mathcal{M}$ is of rank 1). We then see that $\mathcal{M}_0$ is a locally free module of rank 1.

Let $\Gamma : \Omega \to \mathcal{M}_0$ be the map defined by $\Gamma(w) = \varphi(\cdot)\chi(\cdot, w)$, then using the reproducing property of K_0 and (2.5), we find that

$$< \Gamma(w), \Gamma(w') >= \chi(w', w).$$

Therefore, following the discussion in section 2.1, we see that the module $\mathcal{M}_0$ is isomorphic to a Hilbert module consisting of holomorphic functions on Ω whose reproducing kernel is χ. We have therefore proved:

Proposition 2.1. *Let $\mathcal{M}$ be a Hilbert module of rank 1 and $\mathcal{M}_0$ be a submodule of functions as above (among other things, $\mathcal{M}_0$ coincides with the range closure of the operator M_φ). Then $\mathcal{M}_0$ is locally free and is of rank 1. Moreover, the submodule $\mathcal{M}_0$ is isomorphic to a module of holomorphic functions on Ω whose reproducing kernel is χ.*

Remark 2.3. *We point out that the assumption of irreducibility of the hypersurface is superfluous for the kernel of the submodule $\mathcal{M}_0$ to admit a factorisation of the form (2.4) as long as the submodule is the closure in $\mathcal{M}$ of a principal ideal $\mathcal{I}$. We may then take φ to be the generator of this ideal and obtain a factorisation formula exactly as in (2.4) under the additional hypothesis that $\mathcal{M}_0$ equals the space of all functions vanishing on the common zero set of the ideal $\mathcal{I}$. The only difference is that, in this case, the generator φ of the ideal $\mathcal{I}$ is not necessarily irreducible and admits a factorisation $\varphi = \varphi_1 \cdots \varphi_\ell$ into irreducible factors.*

We describe an obvious example. Let $H^2_{12}(\mathbb{D}^2) \subseteq H^2(\mathbb{D}^2)$ be the submodule of functions vanishing on the set $\{(z_1, z_2) \in \mathbb{D}^2 : z_1 = 0 \text{ or } z_2 = 0\}$. Of course, this submodule is also the closure of the ideal $\mathcal{I}_{12} \subseteq H^2(\mathbb{D}^2)$ generated by z_1z_2. In this case, the reproducing kernel for the submodule is easily seen to be

$$K_0(z, w) = z_1\bar{w}_1z_2\bar{w}_2\hat{K}(z, w), \; z = (z_1, z_2), \; w = (w_1, w_2) \in \mathbb{D}^2.$$

This is a locally free module of rank 1.

We now consider the case where the submodule $\mathcal{M}_0$ consists of functions vanishing on the intersection of two hypersurfaces $\mathcal{Z}_1$ and $\mathcal{Z}_2$. Let φ_1 and φ_2 be the defining functions (on a common open set) for these two hypersurfaces. Let $\mathcal{M}_1$ and $\mathcal{M}_2$ be the two submodules of functions vanishing on these hypersurfaces and $\mathcal{M}_{12}$ be the submodule of functions vanishing on the set $\mathcal{Z}_1 \cup \mathcal{Z}_2$ which is $\mathcal{M}_1 \cap \mathcal{M}_2$. It is clear that

$$\mathcal{M}_0 = (\mathcal{M}_1 \ominus \mathcal{M}_{12}) \oplus (\mathcal{M}_2 \ominus \mathcal{M}_{12}) \oplus \mathcal{M}_{12}.$$

We have already shown (recall Proposition 2.1) that the reproducing kernels $K_\ell(z, w)$, corresponding to the submodules $\mathcal{M}_\ell$, $\ell = 1, 2$, admit factorisations. Notice that $\mathcal{M}_{12}$ consists of functions which vanish on the common zero set of the function $\varphi_1\varphi_2$. The preceding remark ensures that a similar factorisation exists for $\mathcal{M}_{12}$ as well. Therefore, we obtain the following formula for the reproducing kernel K_0 corresponding to the submodule $\mathcal{M}_0$:

$$\begin{aligned} K_0(z,w) &= (\varphi_1(z)\chi_1(z,w)\overline{\varphi_1(w)} - \varphi_1(z)\varphi_2(z)\chi_{12}(z,w)\overline{\varphi_1(w)\varphi_2(w)}) \\ &\quad +(\varphi_2(z)\chi_2(z,w)\overline{\varphi_2(w)} - \varphi_1(z)\varphi_2(z)\chi_{12}(z,w)\overline{\varphi_1(w)\varphi_2(w)}) \\ &\quad +\varphi_1(z)\varphi_2(z)\chi_{12}(z,w)\overline{\varphi_1(w)\varphi_2(w)} \\ &= \varphi_1(z)\chi_1(z,w)\overline{\varphi_1(w)} + \varphi_2(z)\chi_2(z,w)\overline{\varphi_2(w)} \\ &\quad -\varphi_1(z)\varphi_2(z)\chi_{12}(z,w)\overline{\varphi_1(w)\varphi_2(w)}, \end{aligned} \tag{2.6}$$

where $\chi_\ell(w,w) \neq 0$ for $w \in \mathcal{Z}_\ell$, $\ell = 1,2$ and $\chi_{12}(w,w) \neq 0$ for $w \in \mathcal{Z}_1 \cap \mathcal{Z}_2$. The formula for the reproducing kernel $\hat{K}_0$ which was given in Example 1.1 is an instance of the general formula given above with $\varphi_\ell(z) = z_\ell$, $\ell = 1,2$.

Remark 2.4. *In the case of a submodule $\mathcal{M}_0$ of functions vanishing on a hypersurface, although the reproducing kernel vanishes on this hypersurface, Proposition 2.1 shows how to construct an isomorphic copy of this module with a non-vanishing reproducing kernel. This amounts to dividing by φ. The resulting kernel χ may then be used to construct a bundle, via the map $w \mapsto \chi(\cdot,w)$, on all of Ω. However, in the case where the submodule consists of functions vanishing on the intersection of two hypersurfaces, as above, if we attempt to divide either by φ_1 or φ_2, then the limit as we approach the zero set may not exist. However, we can still do this as long as we are willing to enlarge the domain. The details of this "blow up" construction are discussed in the last section.*

3. The case of a principal ideal

Assuming that the zero variety $\mathcal{Z}$ is a hypersurface, we have been able to obtain complete unitary invariants for the quotient modules only under the further assumption that $\mathcal{M}_0$ consists of the maximal set of functions in $\mathcal{M}$ which vanish on $\mathcal{Z}$. Even in the case when $\mathcal{Z}$ is not assumed to be a hypersurface, the quotient module is, surprisingly, the restriction of the module $\mathcal{M}$ to $\mathcal{Z}$, as long as $\mathcal{M}_0$ continues to be the maximal set of functions in $\mathcal{M}$ which vanish on $\mathcal{Z}$. If $\mathcal{M}$ is assumed to be in $B_1(\Omega^*)$, then the quotient module can once again be described by a line bundle on $\mathcal{Z}$. The restriction of the curvature then provides a complete unitary invariant for the quotient module. Therefore, one may argue that there is no need to find any other invariant for these quotients modules. Our aim in obtaining such invariants in the case which we understand reasonably well is the hope that some analogues may be found in the case where we drop the assumption that $\mathcal{M}_0$ consists of the maximal set of functions in $\mathcal{M}$ which vanish on $\mathcal{Z}$. For instance, we have found an analogue of the alternating sum (4.1) in [14, Theorem 4.4]. However, we have not succeeded in evaluating this current. There is almost certainly a possibility of applying the methods developed by Harvey and Lawson [19] to evaluate this current which we propose to undertake during the course of our work.

We describe one instance which we have studied in some detail. Let $\mathcal{M}_0$ consist of functions vanishing to order k on a hyper-surface $\mathcal{Z}$ in Ω. Following our assumption above, $\mathcal{M}_0$ is the closure of a principal ideal in $\mathcal{M}$.

As mentioned earlier, in the case where $\mathcal{M}_0$ consists of functions in $\mathcal{M}$ that vanish to order k, a description of the quotient module is presented in [14]. A brief description of the model is presented here.

Starting with the module $\mathcal{M}$, we construct a unitarily equivalent module, denoted $J\mathcal{M}$, using the map J defined as follows:

$$J : h \mapsto \sum_{j=0}^{k-1} \partial_1^j h \otimes \epsilon_{j+1}.$$

Here $\{\epsilon_j\}_{j=1}^k$ is the standard basis in $\mathbb{C}^k$ and ∂_1 denotes the derivative with respect to z_1 if the zero set is characterised, in local coordinates, by the equation $z_1 = 0$. Hence $J\mathcal{M}$ is a vector-valued Hilbert space and the norm on it is determined by the unitarity of the map J.

The module action on $J\mathcal{M}$ is given in matrix form by $f \mapsto \mathcal{J}f$, for $f \in \mathcal{A}(\Omega)$, where

$$\mathcal{J} = \begin{pmatrix} 1 & & & & \\ & \ddots & & 0 & \\ \vdots & \binom{\ell}{j}\partial_1^{\ell-j} & 1 & & \\ & & & \ddots & \\ \partial_1^{k-1} & & \cdots & & 1 \end{pmatrix}$$

with $0 \le \ell, j \le k-1$.

We now construct a module over the algebra $\mathcal{A}(\Omega)$ as follows. The Hilbert space denoted $(J\mathcal{M})_{\text{res}}$ is obtained by restricting the elements of $J\mathcal{M}$ to the zero set, with norm given by

$$\|\mathbf{h}_0\| = \inf\{\|\mathbf{h}\| : \mathbf{h}|_{\mathcal{Z}} = \mathbf{h}_0 \text{ for } \mathbf{h} \in J\mathcal{M}\}.$$

The module action is taken to be

$$(f, \mathbf{h}|_{\mathcal{Z}}) \mapsto \mathcal{J}f|_{\mathcal{Z}} \cdot \mathbf{h}|_{\mathcal{Z}}$$

for $f \in \mathcal{A}(\Omega)$.

The reproducing kernel for $(J\mathcal{M})_{\text{res}}$ is obtained by restricting the reproducing kernel of $J\mathcal{M}$ in both arguments. We denote the resulting reproducing kernel by $(JK)_{\text{res}}$.

We reproduce below the characterisation of the quotient module [14, cf. Theorem 3.3].

Theorem 3.1. *Let $\mathcal{M}$ be a Hilbert module over the algebra $\mathcal{A}(\Omega)$ and let $\mathcal{M}_0$ be the submodule of functions h such that $\partial_1^j h$ vanishes on $\mathcal{Z}$ for $0 \le j \le k-1$. The quotient module is then equivalent to $(J\mathcal{M})_{\text{res}}$.*

The above description of the quotient module is sufficiently general and encompasses some earlier descriptions of quotient modules that were worked out in specific examples. We illustrate this by means of some examples.

Example 3.1. *Let $\mathcal{M}$ be the space $H^2(\mathbb{D}^2)$ as a module over the bi-disc algebra $\mathcal{A}(\mathbb{D}^2)$ and let $\mathcal{M}_0$ be the maximal subset of functions that vanish along the diagonal $z_1 = z_2$, where (z_1, z_2) are coordinates on $\mathbb{D}^2$. The quotient module is spanned by the (normalised) collection $\{e_k\}_{k=0}^{\infty}$ of homogeneous polynomials defined by*

$$e_k(z_1, z_2) = \frac{1}{\sqrt{k+1}} \sum_{j=0}^{k} z_1^{k-j} z_2^j.$$

The module actions of z_1 and z_2 are the same and are described by the map

$$e_k \mapsto \sqrt{\frac{k+1}{k+2}}\, e_{k+1}.$$

The general picture described above by restricting functions to the zero set yields the Bergman space on the disc. The coordinates z_1 and z_2 are now identified and the corresponding module action is the multiplication operator on the Bergman space. The unitary map that intertwines the two pictures is the one that maps e_k to $\frac{z^k}{\sqrt{k+1}}$ in the Bergman space.

We point out that some generalisations of the above example have been studied by D. N. Clark in [5]. A more general family of zero sets is studied there, of which the diagonal in $\mathbb{D}^2$ is a special case.

Example 3.2 (Continuation of Example 2.1). *The model of the quotient module as functions obtained by restricting to the zero set is a useful one. The following example further emphasises this fact. Let $\mathcal{M}$ be the module $H^2(\mathbb{D}^2)$ over the algebra $\mathcal{A}(\mathbb{D}^2)$. Let $\mathcal{M}_0$ be the submodule consisting of functions which vanish on the zero set determined by the function $z_1^3 - z_2^2$. Thus we obtain a resolution*

$$0 \longleftarrow \mathcal{M}_{\mathrm{q}} \longleftarrow \mathcal{M} \overset{X}{\longleftarrow} \mathcal{M}_0 \longleftarrow 0.$$

The zero set is one (complex) dimensional and a parametrised description of it as a subset of $\mathbb{D}^2$ is the collection of points in $\mathbb{D}^2$ of the form (α^2, α^3) with $\alpha \in \mathbb{D}$. The quotient module, which is the restriction of the original module to the zero set, therefore, consists of functions of the parameter α. Since $H^2(\mathbb{D}^2)$ is spanned by the collection of monomials $\{z_1^k z_2^\ell : k, \ell \geq 0\}$, it follows that $\mathcal{M}_{\mathrm{q}}$ is spanned by the set $\{\alpha^{2k+3\ell} : k, \ell \geq 0\}$. That is, $\mathcal{M}_{\mathrm{q}}$ is spanned by the set $\{\alpha^n : n \geq 0, n \neq 1\}$ The module action is also defined by restriction and is given by

$$p \in \mathcal{A}(\mathbb{D}^2) \mapsto p(M^2, M^3)$$

where M denotes multiplication by the coordinate α. It follows that the quotient module is indeed isomorphic to the module $H_1^2(\mathbb{D})$ discussed in Example 2.1.

Example 3.3. *We discuss an example where the submodule of functions vanish to a higher order. Now take $\mathcal{M}_0$ to be the largest collection of functions in $\mathcal{M} = H^2(\mathbb{D}^2)$ which vanish to order 2 on the diagonal $z_1 = z_2$. (See* [11] *for the details of this example). In this case the quotient module is spanned by the homogeneous polynomials $\{e_k\}$ and $\{f_k\}$, with the e_k's as in Example 3.1 and f_k's defined by*

$$\begin{aligned} f_{2m} &= \frac{1}{\alpha_m}\sum_{j=0}^{m-1}(m-j)z_1^{2m-j}z_2^j - \sum_{j=1}^{m} j z_1^{m-j} z_2^{m+j}, \ m \geq 1 \\ f_{2m+1} &= \frac{1}{\beta_m}\sum_{j=0}^{m-1}(2m+1-2j)z_1^{2m+1-j}z_2^j - \sum_{j=0}^{m}(2j+1)z_1^{m-j}z_2^{m+j+1}, \ m \geq 1 \end{aligned}$$

with

$$\begin{aligned} \alpha_m^2 &= \frac{1}{3}m(m+1)(2m+1) \\ \beta_m^2 &= \frac{2}{3}(m+1)(2m+1)(2m+3). \end{aligned}$$

The module action is described by

$$\begin{aligned} z_1 \cdot e_k &= \sqrt{\frac{k+1}{k+2}}\, e_{k+1} \\ z_1 \cdot f_k &= \sqrt{\frac{3}{(k+2)(k+3)}}\, e_{k+1} + \sqrt{\frac{k}{k+3}}\, f_{k+1}. \end{aligned}$$

The action of z_2 is similar except that the coefficient of e_{k+1} in the second equation above now has a negative sign.

Recall that the quotient module, in the general picture described above, is characterised as a module of vector-valued functions on the zero set, which is one dimensional in this case. The correspondence between this picture and the description in terms of homogeneous polynomials on $\mathbb{D}^2$ is obtained via the map

$$e_k \mapsto \begin{pmatrix} \sqrt{k+1}z^k \\ \frac{k}{2}\sqrt{k+1}z^{k-1} \end{pmatrix}, \quad f_k \mapsto \begin{pmatrix} 0 \\ \sqrt{\frac{k(k+1)(k+2)}{12}}\, z^{k-1} \end{pmatrix}.$$

The matrix-valued reproducing kernel $(JK)_{\text{res}}$ is given by (cf. [14, Lemma 2.2]*)*

$$(JK)_{\text{res}}(z,w) = \frac{1}{(1-z\bar{w})^2}\begin{pmatrix} 1 & z(1-z\bar{w})^{-1} \\ \bar{w}(1-z\bar{w})^{-1} & (1+z\bar{w})(1-z\bar{w})^{-2} \end{pmatrix}.$$

The complex geometric approach, developed in [6], is applicable to Hilbert modules which give rise to holomorphic bundles where the module action is scalar on each fiber. If two such bundles $\mathcal{E}$ and $\tilde{\mathcal{E}}$ are equivalent via the isometric bundle map $\Theta : \mathcal{E} \to \tilde{\mathcal{E}}$, then it is shown that there exists a unitary $U_\Theta : \mathcal{M} \to \widetilde{\mathcal{M}}$ which implements Θ. Since the action of $\mathcal{A}(\Omega)$ in that treatment is scalar on each fiber of the respective bundles and Θ is a bundle map, a unitary module map is obtained. Although the quotient module in the case of multiplicity k gives rise to a rank k bundle, the action of the algebra $\mathcal{A}(\Omega)$ is no longer scalar on the fiber. Hence, even if we obtain a unitary map $U : \mathcal{M}_q \to \widetilde{\mathcal{M}}_q$ using techniques from [6], we have to ensure further that this is a module map. Necessary and sufficient conditions for this to happen have not yet been obtained. (In a previous paper [12], it was assumed that $\mathcal{M}_0$ is the submodule consisting of all functions vanishing on a hyper-surface $\mathcal{Z}$. In that case, the quotient module gives rise to a line bundle on $\mathcal{Z}$ and the module action is scalar on each fiber. Hence the complex geometric approach of [6] applies.)

The methods described in [6], at least under certain conditions, provide a first step towards a solution. It is hoped that the ideas there can be augmented to provide a complete solution. The general approach is to look at the largest subalgebra of $\mathcal{A}(\Omega)$ which acts by scalars. Equivalence under this action can be examined by the results in [6]. The attempt is then to ascertain when the equivalence can be implemented via an equivalence with respect to the action of the full algebra.

There are many situations in which the quotient modules (as modules over $\mathcal{A}(\mathcal{Z})$) lie in $B_k(\mathcal{Z})$. (cf. [15, Example 5.16, p. 95] for the definition of this class and the discussion in subsection 2.1.) In these situations, we have the following possible approach to the equivalence question:

If two quotient modules are equivalent, they must be equivalent as modules over the sub-algebra $\mathcal{A}(\mathcal{Z})$. The latter then becomes a question of equivalence in $B_k(\mathcal{Z})$. This question has been studied in [6]. For a complete answer to the equivalence question, we need to determine when there is, among all the unitaries that implement the equivalence in $B_k(\mathcal{Z})$, one that intertwines the (nilpotent) action of functions depending only on the 'normal' coordinate. This question can be studied in a series of steps as follows:

Notice, firstly, that the action of z_1^p is given by a $(k-p)$-step nilpotent operator. The requirement that the unitary which describes the $B_k(\mathcal{Z})$ equivalence must intertwine these powers of z_1 translates into a sequence of conditions on the unitary. (For instance, in the case $k = 2$, where only the first power of z_1 is relevant, this requires that the unitary is upper triangular with equal entries on the diagonal. This condition, it must be emphasized, may not ensure that the unitary must also necessarily intertwine the nilpotent elements.)

We are thus led to the following vector bundle picture. If the quotient module lies in $B_k(\mathcal{Z})$, there is naturally associated a (rank k) bundle on $\mathcal{Z}$. However, this bundle now comes equipped with a collection of sub-bundles which together determine a flag on each fiber. The full equivalence of the quotient modules requires, among other things, the equivalence of these 'flag bundles'. We point out that equivalence of these flags will determine the equivalence of the modules upto "similarity". We will need the metric structure to obtain unitary invariants.

Equivalence of flag bundles, at least formally like these, is considered and characterized by Martin and Salinas [20, Theorem 4.5].

The quotient module $\mathcal{M}_q$ is described in [14]. The modules $\mathcal{M}, \mathcal{M}_0$ and $\mathcal{M}_q$ are identified with geometric objects, specifically vector bundles, though not all on the same base. The equivalence of quotient modules is then reduced to a suitable equivalence of these geometric objects.

The program will be complete only if one now obtains (hopefully complete) invariants for these geometric objects. Towards this end the work of Bott and Chern [3] and after them, Bismut, Gillet and Soulé [2], is probably relevant. These authors study invariants for chain complexes of vector bundles on a given base. However, it seems plausible that their ideas can be extended to chain complexes of more general geometric objects of the type encountered in the study of quotient modules. If these ideas can be carried to their logical end, we will have a rich geometric complement to the module theoretic results obtained so far. In fact our results produce, in many cases, natural examples to which the results of [3] as well as [19] apply.

A second line of enquiry in the geometric direction is a possible adaptation of the results of Harvey and Lawson [19] to the present situation. Let $\mathcal{E}$ and $\mathcal{E}_0$ be the vector bundles obtained, as mentioned above, by localisation, from the modules $\mathcal{M}$ and $\mathcal{M}_0$ and let ϕ be an ad-invariant polynomial (in particular, a Chern form) in the respective curvatures $\mathcal{K}$ and $\mathcal{K}_0$. Then the work of Harvey and Lawson [19] on singular connections gives a mechanism for studying these bundles since the natural connection on the bundle $\mathcal{M}_0$ is a singular one. They obtain a relation of the form

$$\phi(\mathcal{K}) - \phi(\mathcal{K}_0) = \mathrm{R}es_\phi[\mathcal{Z}] + dT_\phi,$$

where $\mathrm{R}es_\phi[\mathcal{Z}]$ is a 'residue' form related to the zero set and T_ϕ is a transgression current. Note that this incorporates a generalised Poincaré-Lelong formula which played a crucial role in the study of the quotient module in the rank one case [12].

By comparison with the rank one case, it seems likely that the current T_ϕ is directly related to the map from $\mathcal{E}_0$ to $\mathcal{E}$ induced by the inclusion map from $\mathcal{M}_0$ into $\mathcal{M}$. By studying the collection $\{\mathrm{R}es_\phi[\mathcal{Z}]\}$ for various choices of the the ad-invariant polynomial ϕ, it is hoped that a complete characterisation of the quotient module can be obtained.

4. The case of ideals which are not necessarily principal

The approaches outlined above have been in the context where $\mathcal{M}_0$ consisted of functions which vanished to some order on the hypersurface $\mathcal{Z}$. In most cases, $\mathcal{M}_0$ is the closure of a principal ideal in $\mathcal{M}$. The problem changes character completely if we allow $\mathcal{M}_0$ to be the closure of an ideal which is not necessarily principal.

Suppose that the zero variety $\mathcal{Z} \subseteq \Omega$ is a hypersurface and $\mathcal{M}_0$ is the maximal set of functions in $\mathcal{M}$ which vanish on $\mathcal{Z}$. Then using the Poincaré-Lelong formula, it follows that the alternating sum

$$\sum_{i,j=1}^{m} \frac{\partial^2}{\partial w_i \partial \bar{w}_j} \log(X(w)^* X(w)) dw_i \wedge d\bar{w}_j - \mathcal{K}_0(w) + \mathcal{K}(w) \tag{4.1}$$

represents the fundamental class $[\mathcal{Z}]$, where $\mathcal{K}$ and $\mathcal{K}_0$ are the curvatures associated with the modules $\mathcal{M}$ and $\mathcal{M}_0$ respectively and $X(w)$ is the localisation of the map X at $w \in \Omega$.

If the zero variety is not a hypersurface, there are at least two different versions of the Poincaré-Lelong formula. In each of these cases, the current of integration on $\mathcal{Z}$ is identified with a current on Ω of bi-degree $(m-r, m-r)$, where r is the codimension of the variety $\mathcal{Z}$. If $\mathcal{M}_0$ is the maximal set of functions in $\mathcal{M}$ which vanish on $\mathcal{Z}$, then we construct two different resolutions of the quotient module and find that they lead in a natural manner to the current of integration. Of course, this current is an unitary invariant for the quotient module.

As mentioned in the introduction, the complex

$$0 \longrightarrow \mathcal{M} \xrightarrow{M_\varphi} \mathcal{M} \longrightarrow 0$$

is the simplest example of a Koszul complex (discussed in the next section) and the sequence (1.3) is an extension of this Koszul complex by the quotient module.

We therefore consider the following natural extension when $\mathcal{M}_0$ is the set of functions vanishing on an analytic submanifold of higher codimension. If $\mathcal{Z}$ is defined by the vanishing of the functions $\varphi_1, \varphi_2, \cdots, \varphi_r$, we consider the multiplication operators $M_{\varphi_1}, M_{\varphi_2}, \cdots, M_{\varphi_r}$ and the associated Koszul complex (extended by the quotient module):

$$0 \longrightarrow \mathbb{H}_0 \xrightarrow{d_0} \mathbb{H}_1 \xrightarrow{d_1} \mathbb{H}_2 \xrightarrow{d_2} \cdots \xrightarrow{d_{r-1}} \mathbb{H}_r \longrightarrow \mathcal{M}_{\mathrm{q}} \longrightarrow 0 \tag{4.2}$$

We then attempt to construct invariants for the quotient module making use of the differential maps $d_0, d_1, d_2, \cdots, d_{r-1}$ of the complex. As a first step, we recover the fundamental class of $\mathcal{Z}$ using the differential maps and the application of a generalised Poincaré-Lelong formula.

It must be pointed out that although in [13] and [14] the approach has been to use a resolution of the quotient module by means of an exact sequence, we make a departure here. Our focus here has been enlarged to chain complexes which terminate in the quotient module and which may not be necessarily exact. This has been motivated by the results in [13] where it is shown that it is possible to construct invariants for the quotient module by considering complexes which are not necessarily exact.

We further assume that $\mathcal{Z}$ is an analytic submanifold of Ω defined by the vanishing of functions $\varphi_1, \varphi_2, \cdots, \varphi_r$ defined globally on Ω. These functions are assumed to be holomorphic on Ω and continuous on $\overline{\Omega}$. The module $\mathcal{M}_0$ is then taken to be the submodule of all functions in $\mathcal{M}$ which vanish on $\mathcal{Z}$ and $\mathcal{M}_{\mathrm{q}}$ is defined to be the quotient module.

4.1. The Koszul Complex

The invariant that we propose to construct uses the Koszul complex for a prescribed set of operators. We, therefore, outline the construction of the Koszul complex for an r-tuple of operators on a Hilbert space. A useful reference here is [16].

Let $\mathcal{H}$ be a Hilbert space and let $T_1, T_2, \cdots, T_r$ be a commuting r-tuple of operators on $\mathcal{H}$. Let $\wedge(\epsilon)$ be the (graded) exterior algebra on r symbols $\epsilon_1, \epsilon_2, \cdots, \epsilon_r$. Denote by $\wedge^j(\epsilon)$ the exterior space spanned by the set $\{\epsilon_{l_1} \wedge \epsilon_{l_2} \wedge \cdots \wedge \epsilon_{l_j}\}$ for $j = 0, 1, ..., r$. Let

$$\mathbb{H}_j = \mathcal{H} \otimes \wedge^j(\epsilon).$$

Define the differential $d_j : \mathbb{H}_j \to \mathbb{H}_{j+1}$ by $d_j = \sum_{i=1}^r T_i \otimes \sigma_i$ where σ_i is the operator defined by

$$\sigma_i(\xi) = \epsilon_i \wedge \xi, \ \xi \in \wedge(\epsilon).$$

The adjoint σ_i^* of σ_i is given by

$$\sigma_i^*(\xi_1 + \epsilon_i \wedge \xi_2) = \xi_2,$$

where ξ_1 and ξ_2 do not contain the symbol ϵ_i. We therefore have the following:

$$\begin{aligned} \sigma_i\sigma_j + \sigma_j\sigma_i &= 0 \\ \sigma_i\sigma_j^* + \sigma_j^*\sigma_i &= \delta_{ij} \text{ for } i,j = 1,2,...,r, \end{aligned} \tag{4.3}$$

where δ is the Kronecker symbol. It follows that $d_{j+1}d_j = 0$.

Define the Laplace operator on $\mathbb{H}_j$ by

$$\Delta_j = d_j^* d_j + d_{j-1}d_{j-1}^* \text{ for } j = 1, 2, ..., r-1.$$

Also take $\Delta_0 = d_0^* d_0$ and $\Delta_r = d_{r-1}d_{r-1}^*$.

The operator d_j^* acts on $\mathbb{H}_{j+1}$ as the adjoint of d_j and is given by

$$d_j^* = \sum_{i=1}^r T_i^* \otimes \sigma_i^*.$$

The torsion, τ, of the Koszul complex, when $\mathcal{H}$ is finite dimensional, is defined through the expression

$$\log \tau = \sum_{j=0}^r (-1)^j j \ \log \det \Delta_j.$$

Consider the case where $T_1, T_2, \cdots, T_r$ represent a family of operators which depend holomorphically on a parameter $w \in \Omega \subseteq \mathbb{C}^m$. This parametrized family of maps, $T_1, T_2, \cdots, T_r$, gives rise to a parametrized family of Koszul complexes, as above. Motivated by the definition of the torsion, we make the following formal definition which associates, under certain circumstances, an $(r-1, r-1)$ current to such a parametrized family of Koszul complexes. We note that $\det \Delta_j(w)$ is real analytic, and consequently, each term in the product below is a current. However, this does not ensure that the product is a current, in general. The results on page 161 of [17] show that the product is indeed a current in many cases. However, in our case, the hypothesis that the common zero set is a complex analytic manifold ensures the transversality of the normal directions. Hence the singular wave front sets for the factors are disjoint and the product is well defined as a current. We also point out that for the parametrized family that we obtain from the localisation of the resolution of the quotient module, this definition yields the current

$$\Theta = \frac{1}{\alpha}(\log\det(\Delta_0(w))\prod_{j=1}^{r-1}\partial\bar{\partial}\log\det(\Delta_j(w)), \tag{4.4}$$

where α is the numerical constant given by $\alpha = \prod_{j=0}^{r-1}\binom{r}{j}$.

Lemma 4.1. *Let $T_1, T_2, \cdots, T_r$ be a commuting tuple of operators as above. If $[T_i, T_j^*] = 0$ for $i, j = 1, 2, ..., r$, then Δ_j as introduced above is given by*

$$\Delta_j = \Big(\sum_{i=1}^{r} T_i T_i^*\Big) \otimes 1_j,$$

where 1_j is the identity operator on $\wedge^j(\epsilon)$.

Moreover, in this case, if $\mathcal{H}$ is finite dimensional, we have the following:

(i) $\mathrm{Tr}\ \Delta_j = \binom{r}{j}\mathrm{Tr}(\sum_{i=1}^{r} T_i T_i^*)$ *where* Tr *denotes the trace.*

(ii) $\log\det\Delta_j = \binom{r}{j}\log\ \det(\sum_{i=1}^{r} T_i T_i^*)$.

(iii) The torsion $\tau = 1$.

Proof: The first result follows from the definition of Δ_j using (4.3). The others follow from the fact that the dimension of $\wedge^j(\epsilon)$ is $\binom{r}{j}$. □

The fundamental class of $\mathcal{Z}$ is recovered by an application of the following generalised Poincaré-Lelong formula. For a proof and more details see [17].

Theorem 4.1. *Let Ω be a complex n-manifold and let*

$$\varphi_1, \varphi_2, \cdots, \varphi_r : \Omega \to \mathbb{C}$$

be holomorphic maps. Let $\mathcal{Z}$ be the common zero set of $\varphi_1, \varphi_2, \cdots, \varphi_r$. Let

$$\Theta = \log\|\varphi\|^2(\partial\bar{\partial}\ \log\|\varphi\|^2)^{r-1},$$

where $\|\varphi\|^2 = |\varphi_1|^2 + |\varphi_2|^2 + \cdots + |\varphi_r|^2$.

If $\mathcal{Z}$ has dimension $n-r$, $\partial\bar{\partial}\Theta$ defines a current of degree (r,r) and corresponds to the fundamental class of $\mathcal{Z}$. In symbols, if ω is a compactly supported $(n-r, n-r)$ form, then

$$\int_\Omega \Theta \wedge \partial\bar{\partial}\omega = \int_{\mathcal{Z}} \omega. \tag{4.5}$$

4.2. Invariants for the Quotient Module

We now have the necessary ingredients to recover the fundamental class of the zero set $\mathcal{Z}$. If $\mathcal{Z}$ is defined by the vanishing of $\varphi_1, \varphi_2, \cdots, \varphi_r$, let $M_{\varphi_1}, M_{\varphi_2}, \cdots, M_{\varphi_r}$ be the operators of multiplication by $\varphi_1, \varphi_2, \cdots, \varphi_r$. Construct the (extended) Koszul complex corresponding to the commuting operators $M_{\varphi_1}, M_{\varphi_2}, \cdots, M_{\varphi_r}$:

$$0 \longrightarrow \mathcal{M} \otimes \wedge^0(\epsilon) \xrightarrow{d_0} \mathcal{M} \otimes \wedge^1(\epsilon) \xrightarrow{d_1} \cdots \xrightarrow{d_{r-1}} \mathcal{M} \otimes \wedge^r(\epsilon) \longrightarrow \mathcal{M}_q \longrightarrow 0 \quad (4.6)$$

Since $\mathcal{M} \otimes \wedge^r(\epsilon)$ is isomorphic to $\mathcal{M}$, the map at the last stage is taken to be the quotient map. We can make $\mathcal{M} \otimes \wedge^j(\epsilon)$ a module over $\mathcal{A}$ by choosing the trivial action on the second factor. With this choice of action, it follows that

$$(\mathcal{M} \otimes \wedge^j(\epsilon)) \otimes_{\mathcal{A}} \mathbb{C}_w = (\mathcal{M} \otimes_{\mathcal{A}} \mathbb{C}_w) \otimes \wedge^j(\epsilon).$$

With the assumptions we have made on $\mathcal{M}$, it can be shown that $\mathcal{M} \otimes_{\mathcal{A}} \mathbb{C}_w$ is one dimensional. Therefore, $\dim \Big((\mathcal{M} \otimes \wedge^j(\epsilon)) \otimes_{\mathcal{A}} \mathbb{C}_w\Big) = \binom{r}{j}$.

Theorem 4.2. *Consider the localisation of the above complex (4.6) by the module $\mathbb{C}_w$, $w \in \Omega$. Let Θ be the $(r-1, r-1)$ current defined in equation (4.4), now applied to this complex. Then $\partial\bar{\partial}\Theta$ represents the fundamental class of $\mathcal{Z}$.*

Proof: Since $\varphi_1, \dots, \varphi_r \in \mathcal{A}$, it follows that the two actions $M_{\varphi_j} \otimes I$ and $I \otimes M_{\varphi_j(w)}$ are identified on $\mathcal{M} \otimes_{\mathcal{A}} \mathbb{C}_w$. Consequently the localisation of M_{φ_j}, denoted $M_{\varphi_j}(w)$, is the same as the one dimensional operator $M_{\varphi_j(w)}$. Hence localising the above extended Koszul complex is equivalent to constructing the Koszul complex for the operators $M_{\varphi_1(w)}, \dots, M_{\varphi_r(w)}$ (each of which acts on the one dimensional Hilbert space $\mathcal{H} = \mathcal{M} \otimes_{\mathcal{A}} \mathbb{C}_w$) and then extending the complex by $\mathcal{M}_q \otimes_{\mathcal{A}} \mathbb{C}_w$.

The operator $\sum_{i=1}^r T_i T_i^*$ occuring in Lemma 4.1 is the (parametrized) one dimensional operator represented by $|\varphi_1(w)|^2 + |\varphi_2(w)|^2 + \cdots + |\varphi_r(w)|^2$. Now, the result follows from an application of this Lemma and Theorem 4.1. □

Remark 4.1. *Now consider just the Koszul complex for the operators $M_{\varphi_1}, \dots, M_{\varphi_r}$:*

$$0 \longrightarrow \mathcal{M} \otimes \wedge^0(\epsilon) \xrightarrow{d_0} \mathcal{M} \otimes \wedge^1(\epsilon) \xrightarrow{d_1} \cdots \xrightarrow{d_{r-1}} \mathcal{M} \otimes \wedge^r(\epsilon) \longrightarrow 0. \quad (4.7)$$

It would be natural to ask if the torsion of the localisation of this complex is an invariant for the quotient module. If the localisation is done with the one dimensional module $\mathbb{C}_w$, it is clear that the torsion is trivial using Lemma 4.1. We could localise using higher rank modules. Lemma 4.1 would no longer be relevant (since T_i and T_j^* may not commute). However, we have noticed that, in some cases where the torsion can be computed explicitly, it still turns out be trivial. This is true, for instance, if we localize with the two dimensional module $\mathbb{C}_w^2$ using the module action described in [14]. Whether this is true in some generality is an interesting question.

Another resolution of the quotient module is obtained as a grid of short exact sequences of Hilbert modules in which each line consists of a short exact sequence. We illustrate our methods in the case where the the zero variety is a complete intersection of two hypersurfaces $\mathcal{Z}_1$ and $\mathcal{Z}_2$. Let $\mathcal{I}_1$ and $\mathcal{I}_2$ be the maximal set of functions which vanish on the hypersurfaces $\mathcal{Z}_1$ and $\mathcal{Z}_2$ respectively and let $\mathcal{I}_{12} = \mathcal{I}_1 \cap \mathcal{I}_2$. Let us consider the diagram

$$\begin{array}{ccccccccc}
 & & 0 & & 0 & & 0 & & \\
 & & \downarrow & & \downarrow & & \downarrow & & \\
0 & \longleftarrow & \mathcal{I}_2^1 & \longleftarrow & \mathcal{I}_2 & \overset{X_2}{\longleftarrow} & \mathcal{I}_{12} & \longleftarrow & 0 \\
 & & \downarrow & & \downarrow & & \downarrow X_1 & & \\
0 & \longleftarrow & \mathcal{I}^1 & \longleftarrow & \mathcal{M} & \longleftarrow & \mathcal{I}_1 & \longleftarrow & 0 \\
 & & \downarrow & & \downarrow & & \downarrow & & \\
0 & \longleftarrow & \mathcal{I}^{12} & \longleftarrow & \mathcal{I}^2 & \longleftarrow & \mathcal{I}_1^2 & \longleftarrow & 0 \\
 & & \downarrow & & \downarrow & & \downarrow & & \\
 & & 0 & & 0 & & 0 & &
\end{array} \tag{4.8}$$

where each connecting map is either an inclusion or a quotient map. We require that each row and column be a short exact sequence of Hilbert modules. Hence all the undefined symbols in the above diagram represent quotient modules at intermediate stages. Localisation of the grid above provides the necessary ingredients to obtain the current of integration on the zero variety $\mathcal{Z}_1 \cap \mathcal{Z}_2$, using the following theorem, which is a generalisation of the Poincaré-Lelong equation (cf. [8]). It is the case $p = 2$ of the theorem which is relevant to the above grid.

Theorem 4.3. *Let $\varphi = (\varphi_1, \cdots, \varphi_p) : U \to \mathbb{C}^p$ be a holomorphic mapping. If $Y_j = \varphi_j^{-1}\{0\}$ are the loci of zeros of the components of φ, one has*

$$\bar{\partial}\partial \log|\varphi_1|^2 \wedge \cdots \wedge \bar{\partial}\partial \log|\varphi_p|^2 = 2\pi i[Y_1] \wedge \cdots \wedge 2\pi i[Y_p] = (2\pi i)^p[Y]$$

where $[Y]$ is the current corresponding to the intersection $Y_1 \cap \cdots \cap Y_p$.

Notice that $\mathcal{I}^{12}$ in the grid above is the quotient module we are attempting to describe. The result above is then a natural generalisation of the approach described following the linear resolution (1.1) of the quotient module. Specifically (see discussion in subsection 2.4),

$$X_\ell(w)^* X_\ell(w) = \frac{|\varphi_1(w)\varphi_2(w)|^2 \chi_{12}(w,w)}{|\varphi_\ell(w)|^2 \chi_\ell(w,w)}, \quad \ell = 1, 2. \tag{4.9}$$

Consequently, we find that

$$\sum_{i,j=1}^m \tfrac{\partial^2}{\partial w_i \partial \bar{w}_j} \log(X_\ell(w)^* X_\ell(w)) dw_i \wedge d\bar{w}_j - \mathcal{K}_{12}(w) + \mathcal{K}_\ell(w)$$

$$= \sum_{i,j=1}^m \tfrac{\partial^2}{\partial w_i \partial \bar{w}_j} \log|\varphi_\ell(w)|^2 dw_i \wedge d\bar{w}_j, \tag{4.10}$$

for $\ell = 1, 2$. Hence Theorem 4.3 applies and we conclude that the product of the two currents described in (4.10) yield the fundamental class of the zero variety.

Even though the grid above is not a resolution of the quotient module in the usual sense, we can extract a linear exact sequence which is a resolution of the quotient. This resolution is obtained from the grid as follows.

We reproduce here the general technique of obtaining a linear complex from a two dimensional complex and then apply the procedure to the case at hand. Let

$$\begin{array}{ccc} C_{i,j+1} & \overset{\delta_{i,j}}{\longleftarrow} & C_{i,j} \\ & & d_{i,j} \downarrow \\ & & C_{i+1,j} \end{array}$$

be a planar grid for $0 \leq i, j \leq N$, where d^2 and δ^2 are zero. This grid gives rise to a linear complex in a natural manner as follows. The k-th term of the linear complex is

$$D_k = \bigoplus_{i+j=k} C_{i,j}.$$

The boundary maps $\Delta_k : D_k \to D_{k+1}$ are defined by

$$\Delta_k|_{C_{i,j}} = \delta_{i,j} + (-1)^j d_{i,j}.$$

We can apply this to the planar grid of modules above. Recall that $\mathcal{I}_k,\ k = 1, 2$ are the ideals of functions vanishing on the hypersurface $\mathcal{Z}_k = \{z \in \Omega : \varphi_k(z) = 0\}$ and that $\mathcal{I}_{12}$ is the intersection of these two ideals. This leads to the following exact sequence.

$$0 \leftarrow Q \overset{\Delta_4}{\longleftarrow} (\mathcal{M}/\mathcal{I}_1)\oplus(\mathcal{M}/\mathcal{I}_2) \overset{\Delta_3}{\longleftarrow} (\mathcal{I}_1/\mathcal{I}_{12})\oplus\mathcal{M}\oplus(\mathcal{I}_2/\mathcal{I}_{12}) \overset{\Delta_2}{\longleftarrow} \mathcal{I}_1\oplus\mathcal{I}_2 \overset{\Delta_1}{\longleftarrow} \mathcal{I}_{12} \leftarrow 0$$

The maps Δ_k are defined by

$$\begin{aligned} \Delta_1(x) &= (x, x) \\ \Delta_2(x,y) &= (P_{12}^1 x, -x + y, P_{12}^2 y) \\ \Delta_3(x,y,z) &= (-P_1 y + z, x + P_2 y) \\ \Delta_4(x,y) &= P_{11}x + P_{22}y, \end{aligned} \tag{4.11}$$

where

a) P_{12}^1 (respectively P_{12}^2) is the quotient map from $\mathcal{I}_1$ (respectively $\mathcal{I}_2$) to $\mathcal{I}_1/\mathcal{I}_{12}$ (respectively $\mathcal{I}_2/\mathcal{I}_{12}$),

b) P_1 (respectivelyP_2) is the quotient map from $\mathcal{M}$ to $\mathcal{M}/\mathcal{I}_1$ (respectively$\mathcal{M}/\mathcal{I}_2$),

c) P_{11} (respectively P_{22}) is the relevant quotient map that appears in the grid from $\mathcal{M}/\mathcal{I}_1$ (respectively $\mathcal{M}/\mathcal{I}_2$)) to the quotient module (Q).

We point out here that there is a generalisation of the planar grid to higher dimensions which becomes relevant when the codimension of the zero set is greater than two. In fact, we then need to consider a hypercubic grid in as many dimensions as the codimension. Theorem 4.3 can be used in this general case to once again determine the current of integration on the zero set. The linearisation described above also has a natural generalisation.

5. Localisation techniques and invariants for submodules

In this section, we use the localisation technique to describe some possible invariants for submodules rather than quotient modules.

Let $\mathcal{I}$ be an ideal in the polynomial ring $\mathbb{C}[z_1, \cdots, z_m]$. Let $\mathcal{M}_0 \subseteq \mathcal{M}$ be the submodule which is the closure of such an ideal. The submodule $\mathcal{M}_0$ is therefore contained in the largest set of functions in $\mathcal{M}$ which vanish on the common zero set of this ideal—we assume that ***this inclusion is indeed an equality.*** It is easy to see that the localisation (cf. [15]) of the submodule $\mathcal{M}_0$ at $w = (w_1, \cdots, w_m) \in \Omega$ is $\cap_{\ell=1}^m \ker(M_{z_\ell}|_{\mathcal{M}_0} - w_\ell)^* \subseteq \mathcal{M}_0$. These subspaces inherit a hermitian metric from the module $\mathcal{M}_0$. In case the submodule is the closure of an ideal generated by a set of polynomials, we conjecture that

$$\dim \cap_{\ell=1}^m \ker(M_{z_\ell}|_{\mathcal{M}_0} - w_\ell)^* = \begin{cases} 1 & \text{for } w \notin V(\mathcal{I}) \\ \text{codimension of } V(\mathcal{I}) & \text{for } w \in V(\mathcal{I}), \end{cases} \tag{5.1}$$

where $V(\mathcal{I})$ is the common zero set of the ideal $\mathcal{I}$ intersect Ω. It is clear that in case $\mathcal{Z} = V(\mathcal{I})$ is a hypersurface, the localisation produces a holomorphic hermitian line bundle. Consequently, the complex geometric approach developed in [6] gurantee that the curvature of this bundle is a complete invariant for the submodule $\mathcal{M}_0$. However, if $\mathcal{Z}$ is not a hypersurface, the dimension of the localisation jumps across the zero set.

If K is the kernel function for $\mathcal{M}$, then $K(\cdot, w)$ induces a map $\gamma : \Omega \to \mathcal{M}^*$ via $\gamma(w)(f) = \langle f, K(\cdot, w)\rangle$, $f \in \mathcal{M}$. Since the constant functions are in $\mathcal{M}$, it follows that $\gamma(w) \neq 0$, $w \in \Omega$. Thus γ maps into the projective space $\mathrm{Gr}(\mathcal{M}, 1)$. The bundle described in the preceding paragraph is the pull back of the canonical bundle on the projective space under this map. However, if we attempt a similar approach with $\mathcal{M}_0$, then we find that we must consider the map induced by the vectors $PK(\cdot, w)$, where P is the projection to $\mathcal{M}_0$:

The localisation $\mathcal{M}_0(w) = \mathcal{M}_0 \otimes_{\mathcal{A}} \mathbb{C}_w$ is the orthogonal complement (in $\mathcal{M}_0$) of the subspace

$$\begin{aligned} \mathcal{N}_w &= \text{span}\,\{(f \cdot h) \otimes 1 - h \otimes (f \cdot 1) : \ f \in \mathcal{A}(\Omega),\ h \in \mathcal{M}_0\} \\ &= \text{span}\,\{((f - f(w)) \cdot h) \otimes 1 : \ f \in \mathcal{A}(\Omega),\ h \in \mathcal{M}_0\} \end{aligned}$$

We clearly have $\mathcal{N}_w \subseteq \{h \in \mathcal{M}_0 : h(w) = 0\}$ for $w \notin \mathcal{Z}$. Since $\mathcal{M}_0$ is the closure of an ideal in $\mathbb{C}[z_1, \ldots, z_m] \subseteq \mathcal{M}$, it is likely that $\mathcal{N}_w = \{h \in \mathcal{M}_0 : h(w) = 0\}$, $w \notin \mathcal{Z}$. Since $\mathcal{N}_w \subseteq \mathcal{M}_0 \subseteq \mathcal{M}$, it follows that as long as $w \notin \mathcal{Z}$, and $\mathcal{N}_w \subseteq \mathcal{M}_0$ is the submodule of functions vanishing at w, we have

$$\begin{aligned}\mathcal{M}_0 \ominus \mathcal{N}_w &= (\mathcal{M} \ominus \mathcal{N}_w) \ominus (\mathcal{M} \ominus \mathcal{M}_0)\\ &= \text{clos span}\{K(\cdot, w), K(\cdot, \lambda) : \lambda \in \mathcal{Z}\} \ominus \text{clos span}\{K(\cdot, \lambda) : \lambda \in \mathcal{Z}\}\\ &= PK(\cdot, w),\end{aligned}$$

where K is the reproducing kernel of the module $\mathcal{M}$ and P is the orthogonal projection on to $\mathcal{M}_0$.

Since $K_0(\cdot, w) = PK(\cdot, w)$ is zero for $w \in V(\mathcal{I})$, we find that the associated map γ vanishes at such points. As was pointed out in Remark 2.4, it is not possible to get around this difficulty in a straightforward manner unless $\mathcal{Z}$ is a hypersurface. Nevertheless, we show how we may take limits of $PK(\cdot, w)$ along complex lines to get around this difficulty.

Example 5.1 (Continuation of Example 1.1). *Let $H^2(\mathbb{D}^2)$ be the Hardy space on the bi-disc $\mathbb{D}^2$ and let*

$$0 \leftarrow \mathcal{M}_q \leftarrow H^2(\mathbb{D}^2) \leftarrow H_0^2(\mathbb{D}^2) \leftarrow 0$$

be a short exact sequence of modules over the bi-disc algebra. Here, $H_0^2(\mathbb{D}^2)$ is the ideal generated in $H^2(\mathbb{D}^2)$ by the coordinate functions z_1, z_2 of $\mathbb{D}^2$. The quotient module in this case is 1 dimensional, the spectrum of $\mathcal{M}_q$ is the singleton set $\{0\}$. The localisation of this module $\mathcal{M}_q$ is the skyscraper sheaf at the origin. The bundle for $H_0^2(\mathbb{D}^2)$ is only defined over $\mathbb{D}^2\backslash\{0\}$. The rank of the fibre (which is 1 on $\mathbb{D}^2\backslash\{0\}$) jumps to 2 at the origin and hence we get a "hermitian" sheaf on $\mathbb{D}^2$ rather than a bundle.

A natural question to ask is how to construct invariants for such a "hermitian" sheaf so that they play the same role as the curvature does for hermitian bundles. For the example at hand, one has the following approach.

Let $w = (w_1, w_2) \in \mathbb{D}^2\backslash\{0\}$. We will compute the module $H_0^2(\mathbb{D}^2) \otimes_{\mathcal{A}} \mathbb{C}_w$, where $\mathcal{A}$ is the bi-disc algebra consisting of functions which are continuous on the closed bi-disc and are holomorphic on the bi-disc. Since $\{z_1, z_2\}$ forms a minimal set of generators for the module $H_0^2(\mathbb{D}^2)$, the set $\{z_1 \otimes 1_w, z_2 \otimes 1_w\}$ spans the whole module $H_0^2(\mathbb{D}^2) \otimes \mathbb{C}_w$. The localization $H_0^2(\mathbb{D}^2) \otimes_{\mathcal{A}} \mathbb{C}_w$ is

$$H_0^2(\mathbb{D}^2) \ominus \{h \in H_0^2(\mathbb{D}^2) : h(w) = 0\}.$$

As pointed out earlier this space is spanned by the projection of $\hat{K}(\cdot, w)$, $w = (w_1, w_2) \in \mathbb{D}^2$, to $\mathcal{M}_0$ which equals $\hat{K}(\cdot, w) - \hat{K}(\cdot, 0) = \hat{K}_0(\cdot, w)$ defined in Example 1.1. This gives a basis vector for the one dimensional space obtained by localising using $\mathbb{C}_w$ for $w \neq 0$.

From the explicit description of $\hat{K}_0$, we find that the limit of $\hat{K}_0(\cdot, w)$ as $w \to 0$ is 0. Nevertheless, the limit of $\frac{1}{\bar{w}_\ell}\hat{K}_0(\cdot, w)$ along lines through the origin as $w \to 0$ exist and is non-zero for $\ell = 1, 2$. If we parametrize the lines through the *origin by* $\bar{w}_2 = \theta_1 \bar{w}_1$ *or* $\bar{w}_1 = \theta_2 \bar{w}_2$, *we get*

$$\begin{aligned}
\lim_{\bar{w}_2=\theta_1\bar{w}_1,\, w\to 0} \frac{\hat{K}_0(z,w)}{\bar{w}_1} &= z_1 + \theta_1 z_2 \\
&= (\bar{\partial}_1 + \theta_1 \bar{\partial}_2)\hat{K}_0(z,w)|_{w=0} \\
\lim_{\bar{w}_1=\theta_2\bar{w}_2,\, w\to 0} \frac{\hat{K}_0(z,w)}{\bar{w}_2} &= \theta_2 z_1 + z_2 \\
&= (\theta_2\bar{\partial}_1 + \bar{\partial}_2)\hat{K}_0(z,w)|_{w=0}. \qquad (5.2)
\end{aligned}$$

We now construct a line bundle on the projective space $\mathbb{P}^1$ *(obtained by "blowing up the origin"). The section* s_1 *on the coordinate patch* $U = \{w \in \mathbb{D}^2 : w_1 \neq 0\}$ *is given (as an element of* $H_0^2(\mathbb{D}^2)$*) by*

$$\begin{aligned}
(s_1(\theta_1))(z) &= (\bar{\partial}_1 K_0(z,w))|_{w=0} + \theta_1(\bar{\partial}_2 K_0(z,w))|_{w=0} \\
&= z_1 + \theta_1 z_2.
\end{aligned}$$

The squared norm of this section is $1 + |\theta_1|^2$. *Consequently the curvature of the line bundle on the affine set* $\{w_1 \neq 0\}$ *is given by*

$$\begin{aligned}
\mathcal{K}(\theta_1) &= \partial_{\theta_1}\partial_{\bar{\theta}_1} \log \frac{1}{1+|\theta_1|^2} \\
&= \frac{d\theta_1 \wedge d\bar{\theta}_1}{(1+|\theta_1|^2)^2}.
\end{aligned}$$

Obviously, a similar calculation yields $\mathcal{K}$ *on the affine subset* $\{w_2 \neq 0\}$ *of* $\mathbb{P}^1$.

We point out that the above form determines a class in the cohomology group $H^2(\mathbb{P}^1, \mathbb{Z}) \sim \mathbb{Z}$, *which is the generator of* $H^2(\mathbb{P}^1, \mathbb{Z})$. *Actually, the curvature form above is induced from the Fubini–Study metric on* $\mathbb{P}^1$.

Let $\mathcal{M}_0$ be the largest submodule of functions in $\mathcal{M}$ vanishing at a fixed point $w \in \Omega \subseteq \mathbb{C}^m$. It is clear, in view of equation (2.6), that we will be able to construct a line bundle on the projective space $\mathbb{P}^{m-1}$ obtained by "blowing up" the point w.

To illustrate the ideas involved in the simplest possible terms, we assume that our domain is the bi-disc $\mathbb{D}^2$ and the submodule $\mathcal{M}_0$ is the closure of the ideal generated by the coordinate functions z_1 and z_2. Let θ_1 and θ_2 denote the usual homogeneous coordinates on the open sets U and V in the complex projective space $\mathbb{P}^1$ corresponding to the affine sets determined by $w_1 \neq 0$ and $w_2 \neq 0$ in $\mathbb{C}^2$. Let E be the holomorphic line bundle on $\mathbb{P}^1$ defined, in local coordinates, by the following sections

$$\begin{aligned}
s_1(\theta_1) &= (\bar{\partial}_1 K_0(\cdot,w))|_{w=0} + \theta_1(\bar{\partial}_2 K_0(\cdot,w))|_{w=0} \\
s_2(\theta_2) &= \theta_2(\bar{\partial}_1 K_0(\cdot,w))|_{w=0} + (\bar{\partial}_2 K_0(\cdot,w))|_{w=0}. \qquad (5.3)
\end{aligned}$$

Theorem 5.1. *Let* $\mathcal{M}_0$ *and* $\mathcal{M}_0'$ *be two modules consisting of holomorphic functions on* $\mathbb{D}^2$ *which are generated by the two coordinate functions, that is, both modules consist of functions vanishing at the origin. Assume that the localisations* $\mathcal{M}_0(w)$

and $\mathcal{M}_0'(w)$ are one dimensional away from the zero set. If the modules $\mathcal{M}_0$ and $\mathcal{M}_0'$ are isomorphic (equivalent via a unitary module map), then the corresponding bundles E and E' they determine on the projective space are equivalent.

Proof: Let $L : \mathcal{M}_0' \to \mathcal{M}_0$ be a unitary module map. By our assumption, the localisations of $\mathcal{M}_0$ and $\mathcal{M}_0'$ on $\mathbb{D}^2\backslash\{0\}$ are one dimensional and are spanned by the corresponding reproducing kernels. Recall that $M_f^* K_0'(\cdot,w) = \overline{f(w)} K_0'(\cdot,w)$ for $f \in \mathcal{A}(\mathbb{D}^2)$ and $w \neq 0$. Since L intertwines the module actions, it follows that $M_f^*(LK_0'(\cdot,w)) = \overline{f(w)} LK_0'(\cdot,w)$. Hence

$$LK_0'(\cdot,w) = \overline{g(w)} K_0(\cdot,w), \text{ for } w \neq 0. \tag{5.4}$$

It then follows that g must be holomorphic on $\mathbb{D}^2\backslash\{0\}$. Since L is unitary, we also have

$$\|K_0'(\cdot,w)\|^2 = |g(w)|^2 \|K_0(\cdot,w)\|^2, \ w \neq 0. \tag{5.5}$$

Therefore,

$$\partial\bar{\partial} \log \frac{K_0'(w,w)}{K_0(w,w)} = \partial\bar{\partial} \log |g(w)|^2 = 0 \ , w \neq 0.$$

Consider the following change of coordinates on $\mathbb{D}^2\backslash\{0\}$:

$$(w_1,w_2) \mapsto (\varrho,\theta_1), \ w_1 = \varrho, \ w_2 = \varrho\theta_1.$$

We then find that $\partial_{\theta_1}\bar{\partial}_{\theta_1} = |w_1|^2 \partial_2 \bar{\partial}_2$. Consequently, it follows that

$$\partial_{\theta_1}\bar{\partial}_{\theta_1} \log |g(\varrho,\theta_1)|^2 = 0 \text{ for } \varrho \neq 0. \tag{5.6}$$

From the form of these kernel functions given in equation (2.6), it follows that $\lim_{\varrho\to 0} |g(\varrho,\theta_1)|^2$ exists and is a real analytic function of the real and imaginary parts of θ_1. From equation (5.6), it follows that $\lim_{\varrho\to 0} g(\varrho,\theta_1)$ is holomorphic in θ_1. If s_1 and s_1' are sections of E and E' on $U \subseteq \mathbb{P}^1$, then we see that

$$Ls_1'(\theta_1) = g(0,\theta_1) s_1(\theta_1), \tag{5.7}$$

where $g(0,\theta_1) = \lim_{\varrho\to 0} g(\varrho,\theta_1)$. This is a consequence of the following observations. From Equation (5.5) and the factorisation of K_0 and K_0' described earlier, it follows that $|g(w)|$ has a finite limit at the origin. By the Riemann removable singularity theorem, it follows that g extends to a holomorphic function on all of $\mathbb{D}^2$. By performing the necessary differentiations on Equation (5.4), noting that differentiation with respect to w commutes with L and using the boundedness of g, we obtain (5.7).

From the unitarity of L it follows that

$$\|s_1'(\theta_1)\|^2 = |g(0,\theta_1)|^2 \|s_1(\theta_1)\|^2$$

and consequently the bundles determined by $\mathcal{M}_0$ and $\mathcal{M}_0'$ on $\mathbb{P}^1$ are equivalent. □

We remark that it is possible to prove a theorem analogous to the one above without making the restrictive assumption that the generators be irreducible polynomials. Indeed, in the second family of examples below, the submodules are generated by the functions z_1^2, z_2^2 and z_1z_2.

We illustrate by means of some examples the nature of the invariant we obtain from the bundle E that lives on the projective space.

As we have shown, the curvature of the bundle which lives on the projective space is an invariant for the submodule. While this curvature does provide an invariant, the family discussed in Example 5.2 consisting of holomorphic functions vanishing at 0, shows that it is not a complete invariant. However, surprisingly enough, if we consider a similar family of examples (see Example 5.3), this time consisting of functions vanishing to a higher order at 0, then it turns out that within this family it is a complete invariant. We intend to investigate the nature of this invariant, arising out of a "blow up" construction, further at a future date.

Example 5.2. *Let $\mathcal{H}_1$ be the functional Hilbert space on the disc with the reproducing kernel*

$$K_1(z_1, w_1) = \frac{1}{(1 - z_1\bar{w}_1)^\lambda}.$$

Similarly, let $\mathcal{H}_2$ be the functional Hilbert space on the disc with the reproducing kernel

$$K_2(z_2, w_2) = \frac{1}{(1 - z_2\bar{w}_2)^\mu}.$$

Let $\mathcal{H}$ be the tensor product of $\mathcal{H}_1$ and $\mathcal{H}_2$. Then $\mathcal{H}$ is a functional Hilbert space on the bi-disc $\mathbb{D}^2$. The reproducing kernel K of $\mathcal{H}$ is the product of K_1 and K_2. That is,

$$K(z_1, z_2; w_1, w_2) = \frac{1}{(1 - z_1\bar{w}_1)^\lambda(1 - z_2\bar{w}_2)^\mu}.$$

Define $\mathcal{M}^{\lambda,\mu}_{(p,q)}$ to be the subspace of functions in $\mathcal{H}$ which vanish at the point (p,q) in the bi-disc. Our goal is to decide when two modules in the set

$$\{\mathcal{M}^{\lambda,\mu}_{(p,q)} : (p,q) \in \mathbb{D}^2, \ \lambda, \mu > 0\}$$

are isomorphic. We begin by observing that

$$\dim \mathcal{M}^{\lambda,\mu}_{(p,q)}(w) = \begin{cases} 1 & \text{for } (w_1, w_2) \neq (p,q) \\ 2 & \text{for } (w_1, w_2) = (p,q). \end{cases} \tag{5.8}$$

This implies that $\mathcal{M}^{\lambda,\mu}_{(p,q)}$ is not equivalent to $\mathcal{M}^{\lambda',\mu'}_{(p',q')}$ if $(p,q) \neq (p',q')$. Therefore, we are reduced to considering submodules $\mathcal{M}^{\lambda,\mu}_{(p,q)}$ with fixed $(p,q) \in \mathbb{D}^2$.

To explicitly compute the reproducing kernel for $\mathcal{M}^{\lambda,\mu}_{(p,q)}$, we recall that for ϕ, ψ in the Möbius group, the operators

$$\begin{aligned} (U_\phi f)(z) &= (\phi'(z))^{\lambda/2} f(\phi(z)) \text{ for } f \in \mathcal{H}_1 \\ (V_\psi f)(z) &= (\psi'(z))^{\mu/2} f(\psi(z)) \text{ for } f \in \mathcal{H}_2 \end{aligned} \tag{5.9}$$

define projective unitary representations of the Möbius group on $\mathcal{H}_1$ and $\mathcal{H}_2$, respectively [22]. Choose ϕ (respectively, ψ) to be the element of the Möbius group which maps the origin to the point p (respectively, q) in the disc. Let $S_{\phi,\psi}$ denote the unitary operator $U_\phi \otimes V_\psi$. The operator $S_{\phi,\psi}$ clearly maps the subspace of functions vanishing at 0 to those vanishing at (p,q). Furthermore, if $\{v_n\}$ is an orthonormal basis for the subspace of functions in $\mathcal{H}$ which vanish at the origin, $\{S_{\phi,\psi}v_n\}$ is an orthonormal basis for the subspace of functions in $\mathcal{H}$ which vanish at the point (p,q). There is a natural basis for the subspace of functions which vanish at the origin which is constructed from the complete orthogonal set $\{z_1^\ell z_2^k : (\ell,k) \neq (0,0)\}$. This, in turn, gives us an orthonormal basis for the subspace of functions which vanish at the point (p,q), from which it is not hard to compute the reproducing kernel $K_{(p,q)}$ for $\mathcal{M}_{(p,q)}^{\lambda,\mu}$:

$$\begin{aligned} K_{(p,q)}(z,w) &= \left(\phi'(z_1)\overline{\phi'(w_1)}\right)^{\lambda/2}\left(\psi'(z_2)\overline{\psi'(w_2)}\right)^{\mu/2} \\ &\qquad \times \left(K((\phi(z_1),\psi(z_2)),(\phi(w_1),\psi(w_2))-1\right), \end{aligned}$$

where $z=(z_1,z_2)$, $w=(w_1,w_2)$ are in $\mathbb{D}^2$.

We now have a situation that is analogous to the previous example. The procedure outlined there can be applied to "blow up" the point (p,q) and to construct a line bundle on the projective space at (p,q). We can calculate the section of this line bundle using the explicit form of the reproducing kernel $K_{(p,q)}$. The section s_1 on the coordinate patch U, as an element of $\mathcal{M}_{(p,q)}^{\lambda,\mu}$, is of the form

$$\begin{aligned} (s_1(\theta_1))(z_1,z_2) &= \frac{\lambda}{1-|p|^2}\left(\frac{\phi(z_1)}{(1-\bar{p}z_1)^\lambda(1-\bar{q}z_2)^\mu}\right) \\ &\quad + \theta_1\,\frac{\mu}{1-|q|^2}\left(\frac{\psi(z_2)}{(1-\bar{p}z_1)^\lambda(1-\bar{q}z_2)^\mu}\right). \end{aligned}$$

The curvature of this line bundle turns out to be the $(1,1)$-form with coefficient

$$C_{\lambda,\mu}(\theta_1) = \frac{(1-|q|^2)^2(1-|p|^2)^2\mu\lambda}{\{(1-|q|^2)^2\lambda+(1-|p|^2)^2\mu|\theta_1|^2\}^2}.$$

In case $(p,q)=(0,0)$, we have $C_{\lambda,\mu}(\theta_1) = \frac{\lambda\mu}{(\lambda+\mu|\theta_1|^2)^2}$. We get that $C_{\lambda,\mu} = C_{\lambda',\mu'}$ if $\frac{\lambda}{\mu} = \frac{\lambda'}{\mu'}$ even though $(\lambda,\mu)\neq(\lambda',\mu')$. (Of course, from the general formula for the curvature given above, it is easy to see that we get a similar result at an arbitrary point.) In all probability the submodules corresponding to two distinct pairs (λ,μ) and (λ',μ') are not equivalent.

Theorem 5.1 was stated for the case of functions which vanish at the origin in $\mathbb{D}^2$ but can naturally be generalised to any other point. There is also a generalisation of this theorem to functions which vanish to higher order at a point. We present a brief outline for the case of functions which vanish to order 2 and illustrate the usefulness of the result by means of an example.

The sections s_1 and s_2 are now given by

$$\begin{aligned} s_1(\theta_1) &= ((\bar\partial_1 + \theta_1\bar\partial_2)^2 K_0(\cdot, w))|_{w=0} \\ s_2(\theta_2) &= ((\theta_2\bar\partial_1 + \bar\partial_2)^2 K_0(\cdot, w))|_{w=0}. \end{aligned} \tag{5.10}$$

A result analogous to Theorem 5.1 holds for the line bundle that these sections determine on the projective space $\mathbb{P}^1$. As mentioned earlier, the curvature of this line bundle gives a complete invariant for the family of modules considered in the example below.

Example 5.3. *The setting of this example is the same as that of Example 5.2 and makes use of the Hilbert spaces $\mathcal{H}_1, \mathcal{H}_2$ and $\mathcal{H}$ defined there.*

We let $\mathcal{N}^{\lambda,\mu}_{(p,q)}$ be the subspace of functions in $\mathcal{H}$ which vanish to order 2 at the point (p,q) in the bi-disc. The dimension of the localisation of $\mathcal{N}^{\lambda,\mu}_{(p,q)}$ jumps at the point (p,q) and hence $\mathcal{N}^{\lambda,\mu}_{(p,q)}$ is not equivalent to $\mathcal{N}^{\lambda',\mu'}_{(p',q')}$ if $(p,q) \neq (p',q')$

As in Example 5.2 we now consider the family of modules $\mathcal{N}^{\lambda,\mu}_{(0,0)}$. The section s_1 of the line bundle on $\mathbb{P}^1$ is given (as an element of $\mathcal{N}^{\lambda,\mu}_{(0,0)}$) by

$$(s_1(\theta_1))(z) = \lambda(\lambda+1)z_1^2 + 2\lambda\mu\theta_1 z_1 z_2 + \mu(\mu+1)\theta_1^2 z_2^2$$

The squared norm of the above section is

$$2\lambda(\lambda+1) + 4\lambda\mu|\theta_1|^2 + 2\mu(\mu+1)|\theta_1|^4$$

and consequently the curvature is

$$\mathcal{K}(\theta_1) = \frac{b + 4c|\theta_1|^2 + bc|\theta_1|^4}{(1 + b|\theta_1|^2 + c|\theta_1|^4)^2}$$

where

$$b = \frac{\mu(\mu+1)}{\lambda(\lambda+1)} \text{ and } c = \frac{\mu}{\lambda+1}.$$

It follows that two modules in this family corresponding to parameter values (λ,μ) and (λ',μ') are equivalent if and only if the corresponding values of b and c are equal. A simple computation shows that this happens if and only if $(\lambda,\mu) = (\lambda',\mu')$, showing that the curvature of the line bundle on the projective space is a complete invariant for modules in this family.

We emphasise the advantage of using the bundle that we have constructed on the projective space obtained by blowing up the zero set (the origin in our examples). It is conceivable that the inequivalence results obtained above can also be obtained by localising on $\mathbb{D}^2\backslash\{0\}$ and looking at the curvature of the corresponding line bundle as in the Cowen-Douglas theory. However, any attempt to do so, even in the simple Examples 5.2, lead to complicated calculations. Part of the complication is due to the fact that this latter bundle lives on a two dimensional base space. As illustrated above, in the Examples 5.3, it often suffices to consider only the bundle defined on the projective space.

Many of the issues that we have raised in the context of Hilbert modules have been studied in analogous algebraic settings. We have attempted to adapt some of these results and techniques. Resolutions of modules have been studied intensely during the last half of the twentieth century in both the algebraic and analytic settings. This has been extended to the context of complex geometry—in the presence of a hermitian metric—only in the last couple of decades. The work of Bismut-Gillet-Soulé and Harvey-Lawson, cited earlier, are examples of this effort, but not the only ones. The whole of what is called Arakelov theory falls in this area. The problems which are described above are, we believe, both interesting and important.

Acknowledgement: The authors would like to thank B. Bagchi and V. Pati for many hours of helpful conversations.

References

[1] N. Aronszajn, *Theory of reproducing kernels*, Trans. Amer.Math. Soc., 68 (1950), 337–404.

[2] J. M. Bismut, H. Gillet and C. Soulé, *Analytic torsion and holomorphic determinant bundles, I, II and III*, Comm. Math. Phy., 115 (1988), 49–78, 79–126, 301–351.

[3] R. Bott and S. S. Chern, *Hermitian vector bundles and the equidistribution of the zeros of their holomorphic sections*, Acta Math., 114 (1968), 71–112.

[4] X. Chen and R. G. Douglas, *Localization of Hilbert Modules*, Michigan Math. J., 39 (1992), 443–454.

[5] D. N. Clark, Restrictions of H^p functions in the polydisk, Amer. J. Math., 110 (1988), 1119–1152.

[6] M. Cowen and R. G. Douglas, *Complex geometry and operator theory*, Acta Math., 141 (1978), 187-261.

[7] M. Cowen and R. G. Douglas, *On operators possessing an open set of eigenvalues*, Memorial conf. for Féjer-Riesz, Budapest, Colloq. Math. Soc. J. Bolyai, (1980), 323 - 341.

[8] Jean-Pierre Demailly and Mikael Passare, *Courants résiduels et classe fondamentale*, Bull. Sci. Math., 119 (1995), 85–94.

[9] Jean-Pierre Demailly, *Multiplier ideal sheaves and analytic methods in algebraic geometry*, Lecture Notes, School on vanishing theorems and effective results in algebraic geometry, ICTP, Trieste, April 2000.

[10] R. E. Curto and N. Salinas, *Generalized Bergman kernels and the Cowen-Douglas theory*, Amer J. Math., 106(1984), 447–488.

[11] R. G. Douglas and G. Misra, *Some Calculations for Hilbert Modules*, J. Orissa Math. Soc., 12-15 (1993-96), 75–85.

[12] R. G. Douglas and G. Misra, *Geometric invariants for resolutions of Hilbert modules*, Operator Theory: Advances and Applications, 104 (1998), 83–112.

[13] Ronald G. Douglas and Gadadhar Misra, *On quotient modules*, Operator Theory: Advances and Applications, special volume dedicated to the memory of Bela Szokefalvi-Nagy, to appear.

[14] Ronald G. Douglas, Gadadhar Misra and Cherian Varughese, *On quotient modules—The case of arbitrary multiplicity*, J. Functional Anal., 174 (2000), 364–398.

[15] R. G. Douglas and V. I. Paulsen, *Hilbert Modules over Function Algebras*, Longman Research Notes, 217, 1989.

[16] Donggeng Gong and Joel Pincus, *Torsion invariants for finite von Neumann algebras*, Contemporary Mathematics, 185 (1995), 151–186.

[17] Phillip Griffiths and James King, *Nevanlinna theory and holomorphic mappings between algebraic varieties*, Acta Math., 130 (1973), 145–220.

[18] R. C. Gunning and H. Rossi, *Analytic functions of several complex variables*, Prentice Hall, New York, 1965.

[19] F. R. Harvey and H. B. Lawson, *A theory of characteristic currents associated with a singular connection*, Asterisque, 213 (1993), 1–268.

[20] M. Martin and N. Salinas, *Flag Manifolds and the Cowen-Douglas Theory*, J. Operator Theory, 38 (1997), 329–365.

[21] B. Sz-Nagy and C. Foias, *Harmonic Analysis of operators on Hilbert space*, North Holland, 1970.

[22] P. J. Sally, *Analytic continuation of the irreducible unitary representations of the universal covering group of $SL(2,\mathbb{R})$*, Mem. Amer. Math. Soc. 69, Providence, 1967.

[23] R. O. Wells, Jr., *Differential analysis on complex manifolds*, Springer Verlag, 1973.

Texas A&M University, College Station, Texas 77843
E-mail address, Ronald G. Douglas: `rgd@tamu.edu`

Indian Statistical Institute, R. V. College Post, Bangalore 560 059
E-mail address, Gadadhar Misra: `gm@isibang.ac.in`
E-mail address, Cherian Varughese: `cherian@isibang.ac.in`

Some remarks on spherical isometries

Jörg Eschmeier and Mihai Putinar

Abstract. Spherical isometries are subnormal tuples of commutative Hilbert space operators which generalize the notion of an isometry to the multivariable case. We review recent classification and invariant subspace results concerning spherical isometries. Then we discuss some pathological examples which show that, in general, even the simplest properties of isometries do not persist in this extension to several variables. Finally we propose a general classification scheme for commutative tuples of operators, with emphasis on mixed commutator ranges, which in particular applies to spherical isometries.

Paper dedicated to Daoxing Xia on the occasion of his seventieth birthday

1. Introduction

Let H be a separable infinite–dimensional complex Hilbert space. A bounded linear operator $V \in L(H)$ is called an *isometry* if $V^*V = I$, or equivalently, if $\|Vx\| = \|x\|$ for every vector $x \in H$. A typical example is the unilateral shift operator $U_+ \in L(l^2(\mathbb{N}))$. The celebrated theorem of Wold and von Neumann states that any isometry V can be uniquely decomposed into an orthogonal direct sum $V = U \oplus U_+^{(m)}$ of a unitary operator and an m–fold direct sum of the unilateral shift where $0 \leq m \leq \infty$ (see [SzF]). A cassical theorem of Bela Szökefalvi-Nagy asserts that any contraction $T \in L(H)$ on a Hilbert space can be realized as the compression of an isometry to a co-invariant subspace ([SzF]). On the other hand, the basic isometry U_+ can be represented as the multipli cation by the variable z on the Hardy space $H^2(\mathbb{D})$ of the unit disk in the complex plane. Thus any contractive operator comes together with a unique minimal isometric dilation, and the latter can be represented on a vector–valued function space.

This fundamental link between the geometry of Hilbert spaces and analytic function theory in the unit disk was the source for many important developments in both fields during the last half century. For details and further references the reader can consult [SzF] and [Ni].

The natural question whether a similar framework exists for commuting systems T of bounded operators has definitely a negative answer. Although a rather

Paper partially supported by the National Science Foundation Grant DMS 9800666.

refined spectral theory exists (see [T]), with many links to classical function theory, see for references [Ar], [EP], [Ru], the simplicity resulting from the Sz.–Nagy dilation theorem and the related functional model is missing in the context of several commuting operators. Instead, special classes of commuting n-tuples were studied, usually motivated by specific Hilbert–space or function–theoreti cal questions. Among them are the spherical isometries which form the object of this note.

The aim of the present paper is two fold: to review some known facts about spherical isometries, by providing complete proofs for a basic classification result, and second, to offer some pathological examples of such systems of operators. This class of commuting systems of Hilbert–space operators can be considered as one possible generalization of the notion of an isometry to the multivariable case. A simple and powerful link with the function theory in the unit ball of $\mathbb{C}^n$ exists, some rigidity results à la Wold-von Neumann are known, applications to interpolation problems and systems theory can be established, some invariant–subspace results were recently proved; but in spite of this progress, many pathologies are easily available.

Thus the present picture of the theory of spherical isometries is complex, yet unfinished.

Besides spherical isometries, a few notable several variable generalizations of isometries are: tuples of commuting isometries and the related polydisk models for operators, see [BLT], [BV], [CV], [GS]; the non-commutative framework of G. Popescu [Po1]; Cuntz algebras [Cu].

The list at the end of this note contains references to various aspects of spherical isometries, but also to more distant related subjects of function theory and operator theory. Although the bibliography is rather long, it is far from being exhaustive. We apologize for any omissions.

2. Preliminaries

2.1.

Let H be a separable infinite–dimensional complex Hilbert space and let $n \geq 1$ be a fixed integer. A system $T = (T_1, T_2, \ldots, T_n)$ of commuting bounded linear operators on H is called a *spherical isometry* if

$$T_1^*T_1 + T_2^*T_2 + \ldots + T_n^*T_n = I. \tag{1}$$

The simplest example is furnished by the Lebesgue space $H = L^2(\mu)$ of a positive Borel measure supported by the unit sphere in $\mathbb{C}^n$, and correspondingly by the n-tuple of multiplication operators by the complex variables $z = (z_1, z_2, \ldots, z_n)$, $M = (M_{z_1}, M_{z_2}, \ldots, M_{z_n})$. This system is actually a *spherical unitary*, that is, a spherical isometry $T = (T_1, \ldots, T_n)$ such that all components T_i are normal.

Let $P^2(\mu)$ be the closure of all complex polynomials in $L^2(\mu)$, and let T be the restriction of the n-tuple M to this invariant subspace. Then T is a spherical isometry, which in general is not normal. It is well known that any spherical isometry with a joint cyclic vector is of this form.

A more intricate class of examples is obtained from a system of commuting isometries $[V_i, V_j] = 0,\ V_i^* V_i = I,\ 1 \le i, j \le n$. Indeed, in this case

$$T = (\frac{V_1}{\sqrt{n}}, \dots, \frac{V_n}{\sqrt{n}})$$

is a spherical isometry.

Some recent attempts of understanding the classification of commuting isometries are contained in [BV], [DF], [GS], [BerLi]. Thus there are good reasons not to aim at a complete classification of spherical isometries up to joint unitary equivalence or joint similarity. The examples constructed in Section 3 will support and clarify this statement.

However, there are a few positive results which make the study of spherical isometries interesting. We briefly recall some of them, without proofs, in what follows.

2.2.

By using Agler's convexity and extension technique A. Athavale has proved the following basic fact, see [At2].

Theorem 2.1 (Athavale). *Any spherical isometry is subnormal.*

To be more specific, given a spherical isometry T on a Hilbert space H, there exists a Hilbert space K, containing H as a closed subspace, and a spherical unitary N on K which leaves H invariant such that $T = N|H$.

The main idea in Athavale's proof is to lift the ordered functional calculus

$$\Phi : \mathbb{C}[z, \overline{z}] \longrightarrow L(H), \quad \Phi(z^k \overline{z}^l) = T^{*l} T^k, \quad k, l \ge 0,$$

to a star homomorphism

$$\Psi : \mathbb{C}[z, \overline{z}] \longrightarrow L(K), \quad \Psi(z^k \overline{z}^l) = N^{*l} N^k, \quad k, l \ge 0.$$

The basic positivity properties of the linear map Φ, plus the isometry assumption (2.1), make this lifting possible, see [At1], [At2].

In this way, the theory of commutative subnormal tuples of operators is applicable to spherical isometries. In particular, any spherical isometry T can be uniquely decomposed as $T = N \oplus V$ into a normal tuple N with joint spectrum in the unit sphere and a *completely non-normal* or *pure* spherical isometry V.

2.3.

Let T be a spherical isometry. Taylor's joint spectrum of T will be denoted by $\sigma(T)$, while $\sigma_{ess}(T)$ will stand for the joint essential spectrum. Similarly, $\sigma_r(T)$ denotes the right spectrum and so on, see [EP] for precise definitions.

It is known that

$$\sigma(N) \subset \sigma_r(T),$$

where N is the minimal normal extension of T, see [Pu1].

Furthermore, it is elementary to check that the spectrum of T is contained in the polynomially convex hull of the spectrum of N. Indeed, Taylor's joint spectrum $\sigma(T)$ is contained in all commutative Banach algebra joint spectra $\sigma_B(T)$, $T \in B^n$.

The first non-trivial open question we encounter is: under which conditions is $\sigma(T)$ equal to the polynomial convex hull of $\sigma(N)$? For more about the operator-theoretical aspects of polynomial convexity see [AlW].

Let us consider a simplified situation. Namely assume that the finiteness restriction $\text{rank}[T_i^*, T_i] < \infty$, $1 \leq i \leq n$, is imposed.

Then the tuple T is essentially normal, and by known general theory (see for instance [EP]) the set $\sigma(T) \cap \mathbb{B} = (\sigma(T) \setminus \sigma_{ess}(T)) \cap \mathbb{B}$ is a (possibly singular) complex analytic subvariety of the unit ball of $\mathbb{C}^n$.

Although no classification of these spherical isometries with finite rank self–commutators exists, their spectrum has a simple geometric picture, first revealed in the works of D. Xia and J. Pincus [PX]. We state below one of their main results.

First some notation. For a fixed completely non-normal spherical isometry T, the subspace

$$D = (\sum_{j=1}^{n} [T_j^*, T_j]H)^-$$

will be called the *defect space* of T.

In what follows we assume that $\dim D < \infty$. Then, as we will see later, D is an invariant subspace for the operators T_i^* and $[T_i^*, T_j]$, $1 \leq i, j \leq n$. Let us denote, following [Xia1],

$$\Lambda_i = (T_i^* | D)^*, \quad C_{ij} = [T_i^*, T_j]|D, \quad 1 \leq i, j \leq n.$$

Some elementary but tedious computations carried out in [Xia1] show that the spectrum of any minimal normal extension of T is contained in the real algebraic set A on the unit sphere of $\mathbb{C}^n$ given by the system of equations

$$\det[(z_i - \Lambda_i)^*(z_j - \Lambda_j) - C_{ij}] = 0, \quad 1 \leq i, j \leq n.$$

A dimension count shows then that A is a real algebraic curve, see [PX]. Since the set $\sigma(T) \cap \partial\mathbb{B}$ "bounds" the rest of the spectrum $\sigma(T) \cap \mathbb{B}$, we find that the whole spectrum is contained in a complex algebraic curve. Hence the next theorm.

Theorem 2.2 (Pincus-Xia). *The joint spectrum of a pure spherical isometry with finite rank self-commutators is contained in the intersection of a complex algebraic curve and the closed unit ball of* $\mathbb{C}^n$.

Thus it is reasonable, although not completely proved, to expect that such a spherical isometry T admits a "uniformization" $T_j = f_j(S), 1 \leq j \leq n$, where $f_1, \ldots, f_n$ are analytic functions of a single variable and S is a single bounded linear operator. A similar spectral picture can be obtained, as for instance in [BV], via M. Livsic theory of operator colligations and their determinatal varieties.

To give a simple example, as a direct byproduct of this spectral analysis one obtains the following generalization of Morrel's theorem, cf. [Mo] and [Xia1].

Theorem 2.3 (Morrel, Xia). *Let T be a pure spherical isometry with one–dimensional defect space. Then*

$$T \equiv (a_1 U_+ + b_1, a_2 U_+ + b_2, \dots, a_n U_+ + b_n),$$

where $a = (a_1, \dots, a_n) \in \mathbb{C}^n \setminus \{0\}$, $b = (b_1, \dots, b_n) \in \mathbb{C}^n$ satisfy

$$|a|^2 + |b|^2 = 1, \qquad \langle a, b \rangle = 0,$$

and $\equiv$ is the joint unitary equivalence relation.

However, we should remark that the classification of spherical isometries having a two–dimensional defect space is not yet fully understood.

In addition to the above mentioned spectral picture, a trace formula for commutators of functions of T and T^* was established in [Xia2], [PX]. We do not expand here these interesting aspects.

2.4.

There is much analogy between the contents of the last paragraph and the recent studies of subnormal operators of finite rank self-commutators, see [McY], [Pu2], [Xia3], [Yak]. We elaborate below a single aspect: a canonical matricial decomposition of spherical isometries possessing finite–dimensional defect space.

Assume as before that the defect space D is finite dimensional, and that T is a completely non-normal spherical isometry. Let N be a minimal normal extension of T, decomposed with respect to the sum $K = H \oplus (K \ominus H)$

$$N_j = \begin{pmatrix} T_j & U_j \\ 0 & V_j \end{pmatrix}.$$

The commutator identities $[N_j^*, N_k] = [N_j, N_k] = 0$ lead to the following relations, valid for $1 \le j, k \le n$,

$$T_j U_k + U_j V_k = T_k U_j + U_k V_j,$$

$$[V_j, V_k] = 0,$$

$$[T_j^*, T_k] = U_k U_j^*,$$

$$T_j^* U_k = U_k V_j^*,$$

$$[V_j^*, V_k] = -U_j^* U_k.$$

The operators U_j turn out to be of finite rank, with ranges spanning the defect space D. The invariance of D under each operator T_j^* is therefore proved by one of the intertwining relations above.

In analogy to the theory of Jacobi matrices, we consider the finite–dimensional subspaces

$$D_m = \bigvee_{k_1 + \dots + k_n \le m} T_1^{k_1} \dots T_n^{k_n} D, \quad m \ge 0.$$

A simple recurrence argument based on the commutator inclusion $[T_j^*, T_k]H \subset D_0 = D$ shows that, for any $m \ge 0$ and $1 \le k \le n$, we have

$T_k^* D_m \subset D_m$. The complete non-normality of T implies that the Hilbert space H can be decomposed as

$$H = D_0 \oplus (D_1 \ominus D_0) \oplus (D_2 \ominus D_1) \oplus \dots .$$

With respect to this decomposition each operator T_j admits a Jacobi type block-matrix decomposition

$$T_j = \begin{pmatrix} D_1^j & 0 & 0 & 0 & \dots \\ C_1^j & D_2^j & 0 & 0 & \dots \\ 0 & C_2^j & D_3^j & 0 & \dots \\ 0 & 0 & C_3^j & D_4^j & \dots \\ \vdots & & \vdots & & \ddots \end{pmatrix}.$$

We list in addition some commutation relations deduced from $[T_j, T_k] = 0$ and the fact that $[T_j^*, T_k]$ has range in the first space D_0

$$[D_1^{j*}, D_1^k] + C_1^{j*} C_1^k = U_k U_j^* | D,$$

$$[D_{p+1}^{j*}, D_{p+1}^k] + C_{p+1}^{j*} C_{p+1}^k = C_p^k C_p^{j*},$$

$$C_{p+1}^k C_p^j = C_{p+1}^j C_p^k,$$

valid for $p \geq 1$ and $1 \leq j, k \leq n$.

Once the latter matricial decomposition is established, it is standard operator theory to prove the following classification result (see Theorem 3 in [Xia1]).

Theorem 2.4 (Xia). *Let T be a pure spherical isometry with finite-dimensional defect space D and let*

$$\Lambda_i = (T_i^* | D)^*, \qquad C_{ij} = [T_i^*, T_j]|D, \ 1 \leq i, j \leq n.$$

Then the system of matrices (Λ_i, C_{ij}), $1 \leq i \leq j \leq n$, is a complete set of unitary invariants for T.

However, the matrices (Λ_i, C_{ij}), $1 \leq i \leq j \leq n$, are not free; they are subject to certain complicated non-linear operatorial equations. In Section 4 we shall prove a more general version of the last theorem valid for arbitrary completely non-normal n-tuples.

The matricial model described in this section shows, from another perspective, the difficulties arising in the construction of spherical isometries with low dimensional defect space.

It is worth mentioning that the same matricial structure is encountered in the theory of orthogonal polynomials in several (real) variables, see [Xu1], [Xu2]. Again, this circle of ideas is not expanded in the present paper.

2.5.

Let $S \in L(H)$ be a subnormal operator. A famous theorem of Scott Brown says that the invariant-subspace lattice $\mathrm{Lat}(S)$ of S is non-trivial. Using the Scott Brown method Olin and Thomson [OT] proved that each subnormal operator $S \in L(H)$ is *reflexive*, that is, each operator in

$$\mathrm{Alg\ Lat}(S) = \{T \in L(H);\ TM \subset M \text{ for all } M \in \mathrm{Lat}(S)\}$$

belongs to the unital WOT-closed subalgebra $W_S \subset L(H)$ generated by S.

A result of K.Yan [Yan] shows that the joint invariant-subspace lattice of a subnormal n-tuple $S \in L(H)^n$ is again non-trivial, but the reflexivity question for subnormal tuples is still open, except in some special cases. For commuting families of isometries a positive answer was given by Bercovici [Ber].

Theorem 2.5 (Bercovici). *Any commuting tuple $T = (T_1, \dots, T_n) \in L(H)^n$ of isometries on a Hilbert space is reflexive.*

The question whether each spherical isometry $T \in L(H)^n$ is reflexive is open. The answer is positive if the Taylor spectrum of T is dominating for $\mathbb{B}$, that is, if

$$\|f\|_{\infty,\mathbb{B}} = \|f\|_{\infty,\mathbb{B}\cap\sigma(T)} \quad (f \in H^\infty(\mathbb{B})),$$

or if T possesses an isometric w^*-continuous $H^\infty(\mathbb{B})$-functional calculus ([E2]).

Theorem 2.6 (Eschmeier). *Each subnormal tuple $S \in L(H)^n$ with $\sigma(S) \subset \overline{\mathbb{B}}$ such that*

(i) *the Taylor spectrum of S is dominating in $\mathbb{B}$ or*
(ii) *S possesses an isometric w^*–continuous $H^\infty(\mathbb{B})$–functional calculus*

is reflexive.

It was shown by Müller and Ptak [MP] that spherical isometries T are at least *hyporeflexive*, that is, $\mathcal{W}_T = \mathrm{Alg\ Lat}(T) \cap (T)'$ where $(T)'$ denotes the commutant of T. The known reflexivity proofs for subnormal operators or tuples T depend on the fact that the w^*–closed unital subalgebra $\mathfrak{A}_T$ of $L(H)$ generated by T satisfies suitable factorization properties. Via trace duality the Banach space $\mathfrak{A}_T$ is the norm dual of the quotient space $\mathcal{C}^1(H)/{}^\perp\mathfrak{A}_T$. Hence each w^*–continuous linear functional of $\mathfrak{A}_T$ is of the form

$$L = \sum_{i=1}^{\infty} [x_i \otimes y_i]$$

where $(x_i)_{i\geq 1}$ and $(y_i)_{i\geq 1}$ are square–summable sequences in H and $[x \otimes y]$ denotes the equivalence class of the rank–one operator $H \to H$, $\xi \mapsto \langle \xi, y\rangle x$.

Theorem 2.7 (Eschmeier). *The dual algebra $\mathfrak{A}_T$ generated by a spherical isometry $T \in L(H)^n$ possesses the factorization property $(\mathbb{A}_{1,\chi_0})$, i.e., for each sequence $(L_k)_{k\geq 1}$ of w^*–continuous linear functionals $L_k : \mathfrak{A}_T \to \mathbb{C}$, there are vectors $x \in H$, $y_k \in H$ $(k \geq 1)$ such that $L_k = [x \otimes y_k]$ $(k \geq 1)$.*

A quantitative version of this result giving norm bounds for the vectors x and y_k can be found in [E3] (Corollary 1.10). In the case that $\mathfrak{A}_T \cong H^\infty(\mathbb{B})$ as dual algebras the last result can be used to prove the reflexivity of T. Necessary for the isomorphism $\mathfrak{A}_T \cong H^\infty(\mathbb{B})$ to hold is the existence of w^*–continuous characters for the dual algebra $\mathfrak{A}_T$. However, in the next section we shall see that there are spherical isometries $T \in L(H)^n$ for which the dual algebra $\mathfrak{A}_T$ has no w^*–continuous characters at all.

2.6.

There is much to say about the role of spherical isometries, or more generally of spherical contractions, in questions of bounded interpolation, corona type decompositions, commutant lifting on the unit sphere of $\mathbb{C}^n$. Similar questions are currently being investigated for commuting isometries, see [BV], [BT]. Due to space limitation, we will present elsewhere some details about these subjects.

3. Approximation on subsets of the sphere

An isometry $T \in L(H)$ with spectrum contained in the unit circle is obviously a unitary operator. A result of Izzo [I] on the failure of polynomial approximation on polynomially convex subsets of the unit sphere implies that the situation is very different in the multivariable case. For a compact set $K \subset \mathbb{C}^n$, we denote by $\hat{K}$ its polynomially convex hull in $\mathbb{C}^n$.

Theorem 3.1. *For every $n \geq 3$, there is a non–normal spherical isometry $T \in L(H)^n$ such that the polynomially convex hull of $\sigma(T)$ is contained in $\partial\mathbb{B}$.*

Proof. Fix an integer $n \geq 3$. By the cited result of Izzo [I] there is a polynomially convex compact subset K of $\partial\mathbb{B}$ such that the uniform closure $P(K)$ of the polynomials in n variables in $C(K)$ is a strict subalgebra of $C(K)$. By Hahn–Banach's theorem there is a non–zero complex measure μ on K with $\mu \in P(K)^\perp$. The tuple

$$T = M_z \in L(P^2(|\mu|))^n$$

is a spherical isometry with Taylor spectrum contained in the polynomially convex hull of $\sigma(M_z, L^2(|\mu|))$. Hence $\sigma(T) \subset K \subset \partial\mathbb{B}$.

Let us assume that T is normal. Then $P^2(|\mu|)$ would be a reducing subspace for $N = M_z \in L(L^2(|\mu|))^n$ and hence $\mathbb{C}[z, \bar{z}] \subset P^2(|\mu|)$. By the Stone–Weierstraß theorem it follows that $P^2(|\mu|) = L^2(|\mu|)$. But then, for $f \in C(K)$ arbitrary, we could choose a sequence (p_k) of polynomials with limit f in $L^2(|\mu|)$. Since

$$|\int_K f d\mu| = |\int_K (f - p_k) d\mu| \leq \int_K |f - p_k| d|\mu| \overset{(k\to\infty)}{\longrightarrow} 0$$

and since $f \in C(K)$ was arbitrary, we obtain the contradiction that $\mu = 0$. □

Since spherical isometries are subnormal, the existence of a spherical isometry as described in Theorem 3.1 is in fact equivalent to the existence of a polynomially convex compact set $K \subset \partial\mathbb{B}$ with $P(K) \neq C(K)$. Indeed, if $S \in L(H)^n$ is subnormal with minimal normal extension $N \in L(K)^n$, then S possesses the contractive functional calculus

$$A(\sigma) \to L(H), \quad f \mapsto f(N)|H,$$

where $\sigma = \sigma(S)$ and $A(\sigma)$ is the closure of $\mathcal{O}(\sigma)|\sigma$ in $C(\sigma)$. Hence, if $A(\sigma) = C(\sigma)$, then S is normal. Therefore, for T as in Theorem 3.1, the set $K = \sigma(T)^\wedge$ is a polynomially convex compact subset of $\partial\mathbb{B}$ with $P(K) \neq C(K)$.

Our next aim is to study more closely the properties of spherical isometries of the above type. The following observation is certainly well known.

Lemma 3.2. *Let $K \subset \partial\mathbb{B}$ be a polynomially convex compact set, and let $\mu \in M^+(K)$ be a measure without atoms. Then the dual algebra $P^\infty(\mu)$ has no w^*–continuous characters.*

Proof. Assume that there is a w^*–continuous character $\varkappa : P^\infty(\mu) \to \mathbb{C}$. Then there is a point $\lambda \in K$ such that $\varkappa(f) = f(\lambda)$ for all $f \in P(K)$. Since λ is a peak point for $A(\mathbb{B})$, there is a function $h \in A(\mathbb{B})$ with $h(\lambda) = 1$ and $|h(z)| < 1$ for all $z \in \overline{\mathbb{B}} \setminus \{\lambda\}$. Because of

$$\int_K \varphi h^k d\mu \overset{(k\to\infty)}{\longrightarrow} \varphi(\lambda)\mu(\{\lambda\}) = 0 \quad (\varphi \in L^1(\mu))$$

it follows that (h^k) induces a w^*–zero sequence in $P^\infty(\mu)$. But then

$$1 = \varkappa(h^k|K) \overset{(k\to\infty)}{\longrightarrow} 0.$$

This contradiction shows that the assumption was wrong. □

The last lemma can be used to modify the example given in Theorem 3.1.

Corollary 3.3. *Let $n \geq 3$. Then there is a Henkin measure $\mu \in M^+(\partial\mathbb{B})$ such that*

(i) *$S = M_z \in L(P^2(\mu))^n$ is a pure spherical isometry with $\sigma(S)^\wedge \subset \partial\mathbb{B}$,*
(ii) *$\mathfrak{A}_s \cong P^\infty(\mu)$ has no w^*–continuous characters,*
(iii) *$P^2(\mu)$ has no bounded point evaluations.*

Proof. Let $T \in L(H)^n$ be a non–normal spherical isometry such that the polynomially convex hull K of $\sigma(T)$ is contained in $\partial\mathbb{B}$. Denote by $T_0 \in L(H_0)^n$ ($H_0 \subset H$) its completely non–unitary part and fix $0 \neq h_0 \in H_0$ arbitrary. Then

$$S = T|\bigvee\{T^k h_0;\ k \in \mathbb{N}^n\}$$

is a pure cyclic spherical isometry with $\sigma(S) \subset K$. Since S is cyclic, there is a positive measure μ on K such that $S = M_z \in L(P^2(\mu))^n$ up to unitary equivalence. The measure μ is a Henkin measure because S is pure (Corollary 1.7 in [E1]). In particular, μ has no atoms and by Lemma 3.2 the dual algebra $\mathfrak{A}_s \cong P^\infty(\mu)$ has no w^*–continuous characters.

Assume that λ is a bounded point evaluation for $P^2(\mu)$. Then λ belongs to the point spectrum of S^*. In particular, $\lambda \in \partial\mathbb{B}$. Choose a function $h \in A(\mathbb{B})$ with $h(\lambda) = 1 > |h(z)|$ for all $z \in \overline{\mathbb{B}} \setminus \{\lambda\}$. By the dominated convergence theorem, (h^k) converges to zero in $P^2(\mu)$. The contradiction $1 = \lim_{k\to\infty} h^k(\lambda) = 0$ shows that bounded point evaluations cannot exist for $P^2(\mu)$. □

By the Berger–Shaw theorem each hyponormal operator which possesses a finite rationally cyclic set is essentially normal with trace–class self–commutator. It is well known that no straightforward generalization of this result to the multivariable case is possible. For instance, the commuting tuple $M_z = (M_{z_1}, \ldots, M_{z_n}) \in L(H^2(\mathbb{D}^n))^n$ consisting of the multiplication operators with the coordinate functions on the Hardy space over the unit polydisc is a cyclic subnormal tuple of isometries which is not essentially normal. In the positive direction Douglas and Yan [DY] proved that at least, for a cyclic subnormal tuple $S = (S_1, \ldots, S_n)$ with Taylor spectrum contained in an algebraic curve, all self–commutators $[S_i^*, S_j]$ $(1 \leq i, j \leq n)$ are trace–class operators.

Below we give an example of a pure cyclic spherical isometry $S = (S_1, S_2)$ such that S_2 is self–adjoint, but the pair S is not essentially normal.

Write $I = [0, 1]$ for the unit interval and define $H = H^2(\mathbb{D}) \otimes L^2(I)$. The tuple $S = (S_1, S_2)$ in $L(H)^2$ given by

$$S_1 = M_z \otimes M_\tau, \quad S_2 = I \otimes M_{\sqrt{1-\tau^2}},$$

where τ denotes the independent variable in I, is obviously a spherical isometry which is not essentially normal. A Stone–Weierstraß argument can be used to show that $1 \otimes 1$ is a cyclic vector for S. Alternatively, one can prove directly that S is unitarily equivalent to the multiplication tuple $M_z = (M_{z_1}, M_{z_2}) \in L(P^2(\mu))^2$, where $\mu = \lambda^\varphi$ is the image measure of the planar Lebesgue measure λ restricted to the closed unit disc $\overline{\mathbb{D}}$ under the topological embedding

$$\varphi : \overline{\mathbb{D}} \to \partial\mathbb{B}_2, \quad \varphi(z) = (z, \sqrt{1 - |z|^2}).$$

Since S is a polynomial in the system $T = (M_z \otimes 1,\ 1 \otimes M_\tau, 1 \otimes M_{\sqrt{1-\tau^2}})$, standard results on the spectra of tensor product systems can be used to compute the spectrum and essential spectrum of S

$$\sigma(S) = p(\sigma(T)) = p(\sigma_e(T)) = \sigma_e(S) = \{(rz, \sqrt{1-r^2});\ z \in \overline{\mathbb{D}} \text{ and } r \in I\},$$

where $p(z_1, z_2, z_3) = (z_1 \cdot z_2, z_3)$.

Since S_1 is a pure subnormal operator, the pair $S = M_z \in L(P^2(\mu))^2$ is a pure spherical isometry. Because of the self–adjoint component S_2, there can be no bounded point evaluations for $P^2(\mu)$ and the dual algebra $\mathfrak{A}_S \cong P^\infty(\mu)$ generated by S cannot possess any w^*–continuous characters. To prove these assertions it suffices to observe that S_2 has empty point spectrum and that $L^\infty(I)$ can be canonically embedded in the algebra $\mathfrak{A}_S \cong P^\infty(\mu)$. As before, since S is a pure spherical isometry, the measure μ is a Henkin measure.

Theorem 3.4. *There is a Henkin measure $\mu \in M^+(\partial\mathbb{B}_2)$ such that*

(i) $S = M_z \in L(P^2(\mu))^2$ *is a pure spherical isometry,*
(ii) S_2 *is self-adjoint, but* S *is not essentially normal,*
(iii) $\mathfrak{A}_S \cong P^\infty(\mu)$ *has no* w^**–continuous characters,*
(iv) $P^2(\mu)$ *has no bounded point evaluations.*

The above examples demonstrate some of the typical phenomena that occur when single–variable isometries are replaced by the much more general concept of a spherical isometry.

4. A decomposition theorem for commuting tuples of operators

In this section we discuss an elementary decomposition and classification of tuples of commuting operators. The decomposition resembles the Jacobi matrix decomposition of a self-adjoint operator A with a cyclic vector x, with respect to the increasing sequence of subspaces generated by succesive powers of A applied to x. In particular our matrix decomposition will apply to spherical isometries and prove Theorem 2.4 above.

Let $S = (S_1, S_2, \dots, S_n) \in L(H)^n$ be a commuting tuple of bounded linear operators on a separable Hilbert space H.

Lemma 4.1. *For any operator* $A \in L(H)$*, the space*

$$M = \bigcap_{\alpha \in \mathbb{N}^n} \ker(AS^\alpha - S^\alpha A),$$

is the largest invariant subspace for S *with the property that*

$$AS_j x = S_j Ax, \qquad x \in M, 1 \le j \le n. \tag{2}$$

Proof. For $x \in M$ and $1 \le j \le n$ we have

$$AS^\alpha(S_j x) = S^{\alpha + e_j} Ax = S^\alpha A(S_j x), \quad \alpha \in \mathbb{N}^n.$$

Conversely, if L is an invariant subspace for S such that relation (4.1) holds for every vector $x \in L$, then an elementary induction shows that $AS^\alpha x = S^\alpha Ax$ for all $x \in L$ and $\alpha \in \mathbb{N}^n$. □

Corollary 4.2. *The space*

$$H_0 = \bigcap_{\beta \in \mathbb{N}^n} \bigcap_{\alpha \in \mathbb{N}^n} \ker(S^{*\beta} S^\alpha - S^\alpha S^{*\beta}),$$

is the largest reducing subspace for S *such that* $S|_{H_0}$ *is normal.*

Proof. By applying Lemma 4.1 twice we find that H_0 is a reducing subspace for S. Obviously, H_0 is the largest reducing subspace for S such that $S|_{H_0}$ is normal. □

Lemma 4.3. *For $k \in \mathbb{N}$, the spaces*

$$M_k = \bigcap_{0\leq|\beta|\leq k}\bigcap_{\alpha\in\mathbb{N}^n} \ker(S^{*\beta}S^{\alpha} - S^{\alpha}S^{*\beta}),$$

form a decreasing sequence of invariant subspaces for S such that

$$S_j^* M_{k+1} \subset M_k, \qquad k \geq 0, 1 \leq j \leq n. \tag{3}$$

Proof. By Lemma 4.1 it is clear that the spaces M_k are invariant for S. Let $x \in M_{k+1}$ and $j \in \{1, \dots, n\}$ be given. Then

$$S^{*\beta}S^{\alpha}(S_j^* x) = S^{*(\beta+e_j)}S^{\alpha}x = S^{\alpha}S^{*\beta}(S_j^* x),$$

for all multiindices $\alpha, \beta \in \mathbb{N}^n$ satisfying $|\beta| \leq k$. □

The spaces M_k can be used to construct a canonical decomposition of the space H with respect to the tuple S. Namely, the subspaces $H_p = M_{p-1} \ominus M_p$, $p \geq 1$, satisfy the following relations.

Theorem 4.4. *Let S be a commutative n-tuple of bounded operators on H. There exists a sequence (possibly finite) of pairwise orthogonal subspaces H_p, $p \geq 0$, with the properties:*

(i) $H = \oplus_{p\geq 0} H_p$;

(ii) H_0 *is the largest reducing subspace for S such that $S|_{H_0}$ is normal;*

(iii) H_1 *is invariant under S^*;*

(iv) $S_j^* H_p \subset H_{p-1} \oplus H_p, \quad p \geq 2,\ 1 \leq j \leq n$;

(v) $S_j H_p \subset H_p \oplus H_{p+1}, \quad p \geq 1,\ 1 \leq j \leq n$;

(vi) $H_p \oplus H_{p+1} = (H_p + \sum_{j=1}^n S_j H_p)^-, \quad p \geq 1$;

(vii) $\dim H_{p+1} \leq n \dim H_p, \quad p \geq 1$.

The standard proof of this result is left to the reader.

From now on we assume that the $n-$tuple S is pure, that is, $H_0 = 0$, and hence $H = \oplus_{p=1}^{\infty} H_p$. For a fixed component $j, 1 \leq j \leq n$, condition (v) and the decomposition $H = \oplus_{p=1}^{\infty} H_p$ above imply the following matricial structure of S_j:

$$S_j = \begin{pmatrix} D_1^j & 0 & 0 & 0 & \dots \\ C_1^j & D_2^j & 0 & 0 & \dots \\ 0 & C_2^j & D_3^j & 0 & \dots \\ 0 & 0 & C_3^j & D_4^j & \dots \\ \vdots & \vdots & \vdots & \vdots & \ddots \end{pmatrix}. \tag{4}$$

Using condition (vi) in Theorem 4.4 one easily obtains that the operators

$$C_k : H_k^n \longrightarrow H_{k+1}, \quad (x_j) \mapsto \sum_{j=1}^n C_k^j x_j,$$

have dense range for $k \geq 1$.

By definition,

$$H_1^\perp = M_1 = \bigcap_{j=1}^n \bigcap_{\alpha \in \mathbb{N}^n} \ker(S_j^* S^\alpha - S^\alpha S_j^*) \subset \bigcap_{j,k=1}^n \ker[S_j^*, S_k].$$

By taking orthogonal complements this yields

$$\vee_{j,k=1}^n \mathrm{Im}[\mathrm{S}_j^*, \mathrm{S}_k] \subset \mathrm{H}_1.$$

This shows that the space H_1 is reducing for all self-commutators $[S_j^*, S_k]$, $1 \leq j, k \leq n$. Consequently, the matrix of $[S_j^*, S_k]$ with respect to the orthogonal decomposition in Theorem 4.4 contains only one non-zero term, carried by the space H_1.

If one uses the above matrix representation for S_j to compute the self-commutators $[S_j^*, S_k]$ one obtains, for all $1 \leq j, k \leq n$, the relations

$$[D_1^{j*}, D_1^k] + C_1^{j*} C_1^k = [S_j^*, S_k]|_{H_1}, \tag{5}$$

$$[D_p^{j*}, D_p^k] + C_p^{j*} C_p^k = C_{p-1}^k C_{p-1}^{j*}, \quad p \geq 2, \tag{6}$$

$$C_p^{j*} D_{p+1}^k = D_p^k C_p^{j*}, \quad p \geq 1. \tag{7}$$

Our aim is to prove that the operators

$$D_1^j = (S_j^*|_{H_1})^*, \qquad [S_j^*, S_k]|_{H_1}, \quad 1 \leq j, k \leq n,$$

form a complete set of unitary invariants for S. First we note that some of the matricial structure of S can be "straightened" into a canonical form.

Proposition 4.5. *Let $S \in L(H)^n$ be a pure commuting n-tuple. Then there are a sequence $(J_p)_{p=1}^\infty$ of Hilbert spaces and bounded operators $d_p^j : J_p \longrightarrow J_p$, $c_p : J_p^n \longrightarrow J_p^n$ satisfying*

$$J_{p+1} \subset J_p^n, \quad J_p^n \ominus J_{p+1} = \ker c_p, \quad c_p \geq 0$$

such that S is unitarily equivalent to the tuple $T = (T_1, T_2, \ldots, T_n)$ acting on $J = \oplus_{p=1}^\infty J_p$ with components

$$T_j = \begin{pmatrix} d_1^j & 0 & 0 & 0 & \ldots \\ c_1^j & d_2^j & 0 & 0 & \ldots \\ 0 & c_2^j & d_3^j & 0 & \ldots \\ 0 & 0 & c_3^j & d_4^j & \ldots \\ \vdots & \vdots & \vdots & \vdots & \ddots \end{pmatrix}, \quad 1 \leq j \leq n.$$

Here the coefficients $c_p^j : J_p \longrightarrow J_{p+1}$ are determined by the operators c_p via the rule

$$c_p(x_1, x_2, \ldots, x_n) = \sum_{j=1}^n c_p^j x_j.$$

Furthermore, one can achieve that

$$M_p(T) = J_{p+1} \oplus J_{p+2} \oplus \dots , \quad p \geq 1.$$

Proof. Let $H = \oplus_{p=1}^{\infty} H_p$ be the intrinsic decomposition obtained in Theorem 4.4. We keep the notations explained above. Define $J_1 = H_1$, $W_1 = 1_{J_1}$, and $J_2 = J_1^n \ominus \ker C_1$. Since

$$J_2 \xrightarrow{C_1} H_2$$

is injective with dense range, there exists a unitary operator

$$W_2 : J_2 \longrightarrow H_2.$$

Define $\tilde{C}_1 = W_2^* C_1 \in L(J_1^n, J_2)$ and $\tilde{D}_2^j = W_2^* D_2^j W_2 \in L(J_2)$.

Since the range of the composition

$$J_2^n \xrightarrow{\oplus W_2} H_2^n \xrightarrow{C_2} H_3,$$

is dense, there exists a unitary operator $W_3 : J_3 \longrightarrow H_3$, where $J_3 = J_2^n \ominus \ker(C_2 \circ (\oplus W_2))$. Define as before $\tilde{C}_2$ by $W_3 \circ \tilde{C}_2 = C_2 \circ (\oplus W_2)$, and set $\tilde{D}_3^j = W_3^* D_3^j W_3$, $1 \leq j \leq n$.

Continuing in this way one obtains a sequence of spaces J_p with $J_{p+1} \subset J_p^n$, $p \geq 1$ together with operators $\tilde{C}_p : J_p^n \longrightarrow J_{p+1}$, $\tilde{D}_p^j : J_p \longrightarrow J_p$, $W_p : J_p \longrightarrow H_p$, $p \geq 1, 1 \leq j \leq n$, such that, for $p \geq 1$ and $j = 1, \dots, n$ (with $\tilde{D}_1^j = D_1^j$),

(i) $W_{p+1} \tilde{C}_p = C_p(\oplus W_p)$;

(ii) each W_p is unitary;

(iii) $W_p \tilde{D}_p^j = D_p^j W_p$;

(iv) $\ker \tilde{C}_p = J_p^n \ominus J_{p+1}$;

(v) $\tilde{C}_p$ has dense range.

Note that in this way we obtain a unitary operator

$$W = \oplus W_p : J = \oplus_{p=1}^{\infty} J_p \longrightarrow H = \oplus_{p=1}^{\infty} H_p,$$

which carries the matricial decomposition of each entry of S into a matrix as in the statement, except for one condition: the maps $\tilde{C}_p : J_p^n \longrightarrow J_p^n$ need not be positive. To achieve this last requirement, we rotate again the spaces J_p with the unitaries coming from the polar decomposition of the operators $\tilde{C}_p$ and finally obtain the positive operators c_p in the statement. □

We shall say that a pure commuting n-tuple T is in *standard form* if the underlying Hilbert space has a decomposition $H = \oplus_{p=1}^{\infty} J_p$ with respect to which T admits a block-matrix representation satisfying all conditions in Proposition 4.5.

The advantage of working with tuples in standard form is illustrated by the natural proof of the next theorem.

Theorem 4.6. *Let* $T = (T_1, T_2, \dots, T_n) \in L(H)^n$ *and* $\tilde{T} = (\tilde{T}_1, \tilde{T}_2, \dots, \tilde{T}_n) \in L(\tilde{H})^n$ *be two commuting n-tuples in standard form, relative to the decompositions*

$$H = \oplus_{p=1}^{\infty} J_p, \qquad \tilde{H} = \oplus_{p=1}^{\infty} \tilde{J}_p.$$

Suppose that there exists a unitary operator $U : J_1 \longrightarrow \tilde{J}_1$ *such that*

$$U(T_j^*|_{J_1})^* = (\tilde{T}_j^*|_{\tilde{J}_1})^* U, \quad U[T_j^*, T_k] = [\tilde{T}_j^*, \tilde{T}_k]U, \quad \text{on } J_1,$$

for all $1 \le j, k \le n$.

Then T *and* $\tilde{T}$ *are unitarily equivalent.*

Proof. Let T be given by the operators

$$d_p^j : J_p \longrightarrow J_p, \quad c_p : J_p^n \longrightarrow J_p^n, \quad c_p^j : J_p \longrightarrow J_{p+1},$$

and similarly for $\tilde{T}$. Recall that $d_1^j = (T_j^*|_{J_1})^*$. Hence by equation (4.4) and the hypotheses

$$Uc_1^{j*}c_1^k = \tilde{c}_1^{j*}\tilde{c}_1^k U, \quad 1 \le j, k \le n.$$

Let $\pi_p : J_p^n \longrightarrow J_{p+1}$ be the orthogonal projection, and let $i_j : J_p \longrightarrow J_p^n$, $i_j(x) = (x\delta_j^i)_{i=1}^n$, be the canonical inclusion.

Since $c_p^j = \pi_p c_p i_j$, we obtain the relations

$$c_p^{j*}c_p^k = i_j^*(c_p\pi_p^*\pi_p c_p)i_k = i_j^* c_p^2 i_k.$$

Therefore $U^{(n)}c_1^2 = \tilde{c}_1^2 U^{(n)}$, and because $c_1, \tilde{c}_1$ are positive elements, we infer that

$$U^{(n)}c_1 = \tilde{c}_1 U^{(n)}.$$

Now, $J_2 = (\ker a_1)^\perp = (\operatorname{Im} c_1)^-$, so that $U^{(n)}J_2 = \tilde{J}_2$. Note that, due to the positivity of c_1, we have $J_1^n = (\operatorname{Im} c_1)^- \oplus \ker c_1$, and that this decomposition is carried over by the unitary map $U^{(n)}$.

Let us write $\alpha_p = (c_p^1, c_p^2, \dots, c_p^n) : J_p^n \longrightarrow J_{p+1}$ for c_p considered as a map with range space J_{p+1}. Similarly define $\tilde{\alpha}_p$.

According to relations (4.6) we obtain $\alpha_p^* d_{p+1}^k = (\oplus d_p^k)\alpha_p^*$, whence

$$\alpha_p\alpha_p^* d_{p+1}^k = \alpha_p(\oplus d_p^k)\alpha_p^*, \quad p \ge 1, 1 \le k \le n.$$

Set $U_1 = U$ and denote by $U_2 : J_2 \longrightarrow \tilde{J}_2$ the unitary map induced by $U^{(n)}$. Since $U_2\alpha_1 = \tilde{\alpha}_1 U_2$, we obtain

$$\tilde{\alpha}_1\tilde{\alpha}_1{}^* U_2 d_2^k = \tilde{\alpha}_1 U_1^{(n)}(\oplus d_1^k)\alpha_1^* = \tilde{\alpha}_1(\oplus d_1^k)U_1^{(n)}\alpha_1^* =$$

$$\tilde{\alpha}_1(\oplus d_1^k)\tilde{\alpha}_1^* U_2 = \tilde{\alpha}_1\tilde{\alpha}_1{}^*\tilde{d}_2^k U_2.$$

But $\ker \tilde{\alpha}_1\tilde{\alpha}_1^* = \ker \tilde{\alpha}_1^* = 0$, hence:

$$U_2 d_2^k = \tilde{d}_2^k U_2, \quad 1 \le k \le n. \tag{8}$$

Using the observation that

$$U_1^{(n)}c_1 i_k i_j^* c_1 = \tilde{c}_1 U_1^{(n)}(i_k i_j^*)c_1 = \tilde{c}_1(\tilde{i}_k\tilde{i}_j^*)U_1^{(n)}c_1 = \tilde{c}_1(\tilde{i}_k\tilde{i}_j^*)\tilde{c}_1 U_1^{(n)},$$

and the commutation relation $U_2\pi_1 = \tilde{\pi}_1 U_1^{(n)}$, we obtain that

$$U_2 c_1^k c_1^{j*} = U_2(\pi_1 c_1 i_k)(i_j^* c_1 \pi_1^*) = \tilde{\pi}_1 \tilde{c}_1 \tilde{i}_k \tilde{i}_j^* \tilde{c}_1 U_1^{(n)} \pi_1^* = \tilde{c}_1^k \tilde{c}_1^{j*} U_2.$$

Similarly one proves that

$$U_2 c_1^j = U_2 \pi_1 c_1 i_j = \tilde{\pi}_1 U_1^{(n)} c_1 i_j = \tilde{\pi}_1 \tilde{c}_1 U_1^{(n)} i_j = \tilde{\pi}_1 \tilde{c}_1 \tilde{i}_j U_1 = \tilde{c}_1^j U_1.$$

In conclusion, by equations (4.5) and (4.7) we find that

$$U_2 c_2^{j*} c_2^k = \tilde{c}_2^{j*} \tilde{c}_2^k U_2, \quad 1 \leq j,k \leq n.$$

Continuing in this way one obtains a sequence of unitary operators $U_p : J_p \longrightarrow \tilde{J}_p$ which intertwine the matrix elements d_p^j and c_p^j with the corresponding coefficients in the matrix of $\tilde{T}_j$.

Thus the unitary operator $U = \oplus_{p=1}^{\infty} U_p$ satisfies

$$UT_j = \tilde{T}_j U, \quad 1 \leq j \leq n,$$

and the proof is complete. □

Because every pure n-tuple is isomorphic to one in standard form, we have already proved the following result.

Corollary 4.7. *Let $S = (S_1, S_2, \ldots, S_n) \in L(H)^n$ and $\tilde{S} = (\tilde{S}_1, \tilde{S}_2, \ldots, \tilde{S}_n) \in L(\tilde{H})^n$ be two pure commuting n-tuples.*

Suppose that there exists a unitary operator $U : H_1(S) \longrightarrow H_1(\tilde{S})$ such that

$$U(S_j^*|_{H_1(S)})^* = (\tilde{S}_j^*|_{H_1(\tilde{S})})^* U, \quad U[S_j^*, S_k] = [\tilde{S}_j^*, \tilde{S}_k]U, \quad \text{on} \quad H_1(S),$$

for all $1 \leq j,k \leq n$.

Then S and $\tilde{S}$ are unitarily equivalent.

Let us turn our attention now to the special case where $S \in L(H)^n$ is a subnormal n-tuple. Then the operator

$$D = \sum_{j=1}^{n} [S_j^*, S_j]$$

is positive. The closure of its range $\mathcal{D} = \text{Im}(D)^-$ is called the *defect space* of S. We have

$$\mathcal{D}^\perp = \ker D = \bigcap_{j=1}^{n} \ker[S_j^*, S_j].$$

Let $N = (N_1, \ldots, N_n) \in L(K)^n$ be the minimal normal extension of S. Each component N_j has a matrix representation

$$N_j = \begin{pmatrix} S_j & U_j \\ 0 & V_j \end{pmatrix}$$

relative to the orthogonal decomposition $K = H \oplus H^\perp$.

Since N_j^* commutes with N_k, we obtain the relations

$$[S_j^*, S_k] = U_k U_j^*, \quad S_j^* U_k = U_k V_j^*, \quad 1 \leq j,k \leq n.$$

Thus

$$D = \sum_{j=1}^{n} U_j U_j^*,$$

and

$$\ker D = \bigcap_{j=1}^{n} \ker U_j U_j^* = \bigcap_{j=1}^{n} \ker U_j^*, \quad D = \vee_{j=1}^{n} \operatorname{Im} U_j.$$

Because of $U_j^* S_k = V_k U_j^*$ it is clear that $\ker U_j^* \in \operatorname{Lat}(S)$. Therefore

$$\ker D \in \operatorname{Lat}(S) \quad \text{and} \quad D = (\ker D)^{\perp} \in \operatorname{Lat}(S^*).$$

For $k \in \mathbb{N}$, consider the space

$$D_k = \vee\{p(S)D; \;\; p \in \mathbb{C}[z_1, z_2, \dots, z_n] \text{ with } \deg(p) \leq k\}.$$

Obviously,

$$D_k = (\sum_{|\beta| \leq k} S^{\beta} \operatorname{Im} D)^{-}, \tag{9}$$

and therefore a short computations shows that

$$D_k^{\perp} = \{x \in H; \;\; [S_j^*, S_j] S^{*\beta} x = 0 \text{ for } |\beta| \leq k \text{ and } 1 \leq j \leq n\}.$$

An elementary induction, based on (4.8) and the inclusions $\operatorname{Im}[S_j^*, S_i] \subset D$, allows one to deduce that all spaces D_k are invariant for S^*.

Coming back to the notation introduced at the beginning of this section we note the following identity

$$M_{k+1} = D_k^{\perp}, \quad k \geq 0. \tag{10}$$

Indeed, fix a vector $x \in M_{k+1}$. Then, according to Lemma 4.3, $S_j^* x \in M_k$ and

$$S_j^* S_j S^{*\beta} x = S_j^* S^{*\beta} S_j x = S^{*\beta} S_j (S_j^* x) = S_j S_j^* S^{*\beta} x,$$

for all $|\beta| \leq k$ and $1 \leq j \leq n$. Thus $M_{k+1} \subset D_k^{\perp}$.

To prove the converse inclusion we shall use induction on k. For $k = 0$, we have

$$D_0^{\perp} = \bigcap_{j=1}^{n} \ker[S_j^*, S_j] \in \operatorname{Lat}(S).$$

Hence, an n-fold application of Lemma 4.1 yields that

$$D_0^{\perp} \subset \bigcap_{j=1}^{n} \bigcap_{\alpha \in \mathbb{N}^n} \ker[S_j^*, S^{\alpha}] = M_1.$$

Suppose that $k \geq 1$ and that $D_l^{\perp} \subset M_{l+1}, \; 0 \leq l \leq k-1$. This in particular implies $D_k^{\perp} \subset D_{k-1}^{\perp} \subset M_k$. Fix a vector $x \in D_k^{\perp}$. Because of $S_j D_{k-1} \subset D_k$, we find $S_j^* x \in S_j^* D_k^{\perp} \subset D_{k-1}^{\perp} \subset M_k$ for $1 \leq j \leq n$. To see that $x \in M_{k+1}$ it is sufficient to remark that, for any β with $|\beta| = k$ and each index $j = 1, \dots, n$, we have

$$S^{*(\beta+e_j)} S^{\alpha} x = S^{*\beta} S^{\alpha} (S_j^* x) = S^{\alpha} S^{*(\beta+e_j)} x$$

for all $\alpha \in \mathbb{N}^n$.

As remarked before, for a subnormal tuple, the unitary invariants appearing in Corollary 4.7 are not free. Even for a single subnormal operator the relations among these invariants are rather complicated, see for instance [Pu2].

References

[Ag1] J. Agler, *Hypercontractions and subnormality*, J. Operator Theory **13** (1985), 201–217.

[Ag2] J. Agler, *On the representation of certain holomorphic functions defined on a polydisc*, in "Topics in Operator Theory: Ernst D. Hellinger Memorial Volume" (L. de Branges, I. Gohberg, and J. Rovnyak, Eds.), Operator Theory and Applications, Vol. 48, pp. 47–66, Birkhäuser, Basel, 1990.

[AgMc1] J. Agler and J. McCarthy, *Nevanlinna-Pick interpolation on the bidisk*, J. reine angew. Math. **506** (1999), 191–204.

[AgMc2] J. Agler and J. McCarthy, *Nevanlinna-Pick kernels and localization*, preprint.

[AD] L. A. Aizenberg and S. A. Dautov, *Holomorphic functions of several complex variables with nonnegative real part. Traces of holomorphic and pluriharmonic functions on the Shilov boundary*, Math. USSR Sbornik **28** (1976), 301–313.

[AlW] H. Alexander and J. Wermer, *Several Complex Variables and Banach Algebras*, 3-rd. edition, Springer, Berlin, 1998.

[Ar] J. Arazy, *A survey of invariant Hilbert spaces of analytic functions on bounded symmetric domains*, Contemp. Math. **185** (1995), 7–65.

[Arv] W. B. Arveson, *Subalgebras of C^*-algebras*, Acta Math. **123** (1969), 141–224; ibidem. Part III, Acta Math. **181** (1998), 159–228.

[At1] A. Athavale, *Holomorphic kernels and commuting operators*, Trans. Amer. Math. Soc. **304** (1987), 101–110.

[At2] A. Athavale, *On the intertwining of joint isometries*, J. Operator Theory **23** (1990), 339–350.

[At3] A. Athavale, *A note on joint subnormality and spherical dilations*, Rev. Roum. Math. Pures Appl. **37** (1992), 667–672.

[At4] A. Athavale, *Unitary and spherical contractions: a hypercontractive perspective*, Rev. Roum. Math. Pures Appl. **38** (1993), 387–400.

[At5] A. Athavale, *GNS decomposition and spherical dilations*, J. Funct. Analysis **154** (1998), 117–129.

[BLT] J. A. Ball, W. S. Li, D. Timotin, and T. T. Trent, *A commutant lifting theorem on the polydisc*, preprint.

[BT] J. A. Ball and T. T. Trent, *Unitary colligations, reproducing kernel Hilbert spaces, and Nevanlinna-Pick interpolation in several variables*, J. Funct. Analysis **157** (1998), 1–61.

[BV] J. A. Ball and V. Vinnikov, *Multivariable linear systems, scattering, unitary dilations, and operator model theory for row contractions*, in preparation.

[BB] F. Beatrous and J. Burbea, *Positive definiteness and its applications to interpolation problems for holomorphic functions*, Trans. Amer. Math. Soc. **284** (1984), 247–270.

[BCZ] C. A. Berger, L. A. Coburn, and K. H. Zhu, *Function theory on Cartan domains and the Berezin-Toeplitz symbol calculus*, Amer. J. Math. **110** (1988), 341–379.

[Ber] H. Bercovici, *A factorization theorem with applications to invariant subspaces and the reflexivity of isometries*, Math. Res. Lett. **1** (1994), 511–518.

[BerLi] H. Bercovici and W. S. Li, *Reflexivity of certain pairs of commuting isometries*, Topics in operator theory, operator algebras and applications (Timişoara, 1994), Romanian Acad., Bucharest, 1995, pp. 29–48.

[CLW] B. Cole, K. Lewis, and J. Wermer, *Pick conditions on a uniform algebra and von Neumann inequalities*, J. Funct. Analysis **107** (1992), 235–254.

[Cu] J. Cuntz, *Simple C^*-algebras generated by isometries*, Commun. Math. Phys. **57** (1977), 173–185.

[CV] R. E. Curto and F.-H. Vasilescu, *Standard operator models in the polidisc*, Indiana Univ. Math. Journal 42 (1993), 791–810; part II, ibidem **44** (1995), 727–746.

[D] S. W. Drury, *A generalization of von Neumann's inequality to the complex ball*, Proc. Amer. Math. Soc. 68 (1978), 300–304.

[DaP] K. R. Davidson and D. R. Pitts, *Nevanlinna-Pick interpolation for non-comutative analytic Toeplitz algebras*, Integral Eq. Operator Theory **31** (1998), 321–337.

[DF] R. G. Douglas and C. Foiaş, *Unitary invariants for commuting isometries*, preprint.

[DY] R. Douglas and K. Yan, *A multi-variable Berger–Shaw theorem*, J. Operator Theory **27** (1992), 205–217.

[E1] J. Eschmeier, *Invariant subspaces for spherical contractions*, Proc. London Math. Soc. **75** (1997), 157–176.

[E2] J. Eschmeier, *Algebras of subnormal operators on the unit ball*, J. Operator Theory **42** (1999), 37–76.

[E3] J. Eschmeier, *On the structure of spherical contractions*, to appear.

[EP] J. Eschmeier and M. Putinar, *Spectral decompositions and analytic sheaves*, London Math. Soc. Monograph, vol. 10, Clarendon Press, Oxford, 1996.

[EPP] J. Eschmeier, L. Patton, and M. Putinar, *Carathéodory-Féjer interpolation on polydisks*, Math. Res. Lett. **7** (2000), 25–34.

[FK] J. Farault and A. Koranyi, *Analysis on symmetric cones*, Clarendon Press, Oxford, 1994.

[GS] D. Gaşpar and N. Suciu, *On invariant subspaces in the bitorus*, J. Operator Theory **30** (1993), 227–241.

[Hel] J. W. Helton, *Optimization over H^∞ and the Toeplitz corona theorem*, J. Operator Theory **15** (1986), 359–375.

[I] A. J. Izzo, *Failure of polynomial approximation on polynomially convex subsets of the sphere*, Bull. London Math. Soc. **28** (1996), 393–397.

[KP] A. Koranyi and L. Pukanszky, *Holomorphic functions with positive real part on polycylinders*, Trans. Amer. Math. Soc. **108** (1963), 449–456.

[MaP] M. Martin and M. Putinar, *Lectures on hyponormal operators*, Operator Theory, Advances and Applications, 39, Birkhäuser, Basel, 1989.

[McY] J. McCarthy and L. Yang, *Subnormal operators and quadrature domains*, Adv. Math. **127** (1997), 52–72.

[Mo] B. B. Morrel, *A decomposition for some operators*, Indiana Univ. Math. J. **23** (1973), 497–511.

[MP] V. Müller and M. Ptak, *Spherical isometries are hyporeflexive*, Rocky Mountain J. Math. **29** (1999), 677–683.

[MV] V. Müller and F.-H. Vasilescu, *Standard models for some commuting multioperators*, Proc. Amer. Math. Soc. **117** (1993), 979–989.

[Ni] N. K. Nikolskii, *Treatise on the shift operator*, Springer, Berlin, 1986.

[OT] R. Olin and J. Thomson, *Algebras of subnormal operators*, J. Funct. Analysis **37** (1980), 271–301.

[Po1] G. Popescu, *Noncommutative Wold decomposition for semigroups of isometries*, Indiana Univ. Math. Journal 47 (1998), 277–296.

[Po2] G. Popescu, *Interpolation problems in several variables*, J. Mathematical Analysis Applications **227** (1998), 227–250.

[Pu1] M. Putinar, *Spectral inclusion for subnormal n-tuples*, Proc. Amer. Math. Soc. **90** (1984), 405–406.

[Pu2] M. Putinar, *Linear analysis of quadrature domains.* III, J. Math. Analysis Appl. **239** (1999), 101–117.

[PX] J. Pincus and D. Xia, *A trace formula for subnormal operator tuples*, Intergral Eq. Operator Theory 14 (1991), 390–398.

[PZ] J. Pincus and D. Zheng, *A remark on the spectral multiplicity of normal extensions of commuting subnormal operator tuples*, Integral Eq. Operator Theory 16 (1993), 145–153.

[Ro] M. Rosenblum, *A corona theorem for countable many functions*, Integral Eq. Operator Theory **3** (1980), 125–137.

[Ru] W. Rudin, *Function theory in the unit ball of* $\mathbb{C}^n$, Springer, Heidelberg 1980.

[S] F. H. Szafraniec, *On bounded holomorphic interpolation in several variables*, Mh. Math. **101** (1986), 59–66.

[SzF] B. Sz.-Nagy and C. Foaiş, *Harmonic analysis of operators on Hilbert space*, North Holland, Amsterdam, 1970.

[T] J. L. Taylor, *The analytic functional calculus for several commuting operators*, Acta Math. **125** (1970), 1–38.

[Xia1] D. Xia, *Analytic theory of subnormal n-tuples of operators*, Proc. Symp. Pure Math. **51** (1990), 617–640.

[Xia2] D. Xia, *Trace formulas for a class of subnormal tuples of operators*, Intergral Eq. Operator Theory 17 (1993), 417–439.

[Xia3] D. Xia, *On pure subnormal operators with finite rank self-commutator and related operator tuples*, Intergral Eq. Operator Theory **24** (1996), 106–125.

[Xia4] D. Xia, *Trace formulas and complete unitary invariants for some k-tuples of commuting operators*, preprint.

[Xu1] Y. Xu, *Common zeros of polynomials in several variables and higher dimensional quadrature*, Pitman Res. Notes Math. Vol. 312, Longman Sci. Techn., Harlow, 1994.

[Xu2] Y. Xu, *Orthogonal polynomials and cubature formulae on spheres and simplices*, Methods Appl. Math. **5** (1998), 169–184.

[Yak] D. Yakubovitch, *Subnormal operators of finite type*. I, Rev. Math. Iberoamericana **14** (1998), 95–115; part II, ibidem **14** (1998), 623–681.

[Yan] K. Yan, *Invariant subspaces for joint subnormal systems*, unpublished manuscript.

Fachbereich Mathematik,
Universität des Saarlandes,
D-66123 Saarbrücken,
Germany
E-mail address: eschmei@math.uni-sb.de

Depatment of Mathematics,
University of California,
Santa Barbara, CA 93106,
USA
E-mail address: mputinar@math.ucsb.edu

Higher-order Hankel forms and cohomology groups

Sarah H. Ferguson

Abstract. Higher-order Hankel forms on the standard weighted Bergman spaces are studied. A bounded projection of nth-order forms, $\mathcal{H}^{(n)}$, onto the jth-order $(0 \leq j \leq n)$ *special* forms of Janson-Peetre type is exhibited and used to give a decomposition of nth-order forms in terms of these special forms of lower or equal order. This decomposition is optimal in several ways. One, it produces these special forms in a very natural way, namely, by averaging over the *elementary* forms of some fixed order, and this process is independent of the group representation theory. It also yields a decomposition of derivations which provides estimates on certain quotient norms. This gives detailed information on the first order Hochschild groups of the disk algebra with coefficients in the nth-order Hankel forms. We also use the decomposition to show that the map on the first order groups induced by the quotient map $\mathcal{H}^{(n)} \to \mathcal{H}^{(n)}/\mathcal{H}^{(n-1)}$ is onto and the map induced by the inclusion of $\mathcal{H}^{(n-1)} \subseteq \mathcal{H}^{(n)}$ into bounded bilinear forms is the zero map.

1. Introduction

Generalized Hankel forms of higher weight were introduced by Janson and Peetre in the setting of group representations, [JP1], [JP2]. The idea is that the Mobius group acts irreducibly on the standard weighted Bergman space $A^{2,\alpha}$, $\alpha > -1$, but the induced action on bounded bilinear forms on $A^{2,\alpha} \times A^{2,\alpha}$ is not irreducible and the lowest weight component coincides with the usual Hankel forms. The higher-order components are the so-called *special* Hankel forms of higher weight, the adjective *special* meaning that these forms are orthogonal (in some dual sense) to forms of lower weight. An explicit realization of the kernels of these forms using generalized *transvectants* was given in the two papers above, as well as, the analytic theory concerning boundedness and Schatten ideal criteria. Their results were extended by Zhang, [Zh1], [Zh2], to the case of forms on $A^{2,\alpha} \times A^{2,\beta}$, $\alpha \neq \beta$ and multi-linear analogues of these forms were studied in [Ro].

Peetre and Rochberg in [PR] took an alternative approach to this theory by defining *higher-order* Hankel forms in the setting of a natural module action of

Author was supported in part by NSF grant DMS-0071514.

an algebra on a Hilbert space of holomorphic functions. For example, H^∞ (or, the disk algebra $A(\mathbb{D})$) acts on $A^{2,\alpha}$ by pointwise multiplication which induces a bi-module action on bounded bilinear forms on $A^{2,\alpha} \times A^{2,\alpha}$. It was shown that the Hilbert-Schmidt forms of order 1 which are orthogonal to forms of order 0 (order 0 = usual Hankel forms) coincided with Janson and Peetre's class. This was recently extended to all orders in [FR] using certain higher-order restriction results which are essentially dilation results and thus, have applications outside the scope of higher-order Hankel forms. See [FP] for an application to joint similarity problems.

In both approaches, it was acknowledged, although not explored, that there is some connection with cohomology. Rochberg, in [R2], made this connection explicit, primarily in an algebraic sense. The appropriate cohomology groups in this setting are the so-called *bounded* Hochschild groups of the disk algebra with values in either bounded bilinear forms, Hankel forms of order n, denoted by $\mathcal{H}^{(n)}$, or the special forms of order n, $\mathcal{H}^{(n)}/\mathcal{H}^{(n-1)}$. Bounded cohomology just means that the co-cycles are assumed to be bounded.

For the Hardy space H^2, it was shown in [CFR] that the first order Hochschild group $H^1(A(\mathbb{D}), \mathcal{H}^{(n)})$ is isomorphic to $\mathcal{H}^{(n+1)}/\mathcal{H}^{(n)}$, $n \geq 0$. The proof relied on the base case $H^1(A(\mathbb{D}), \mathcal{H}^{(0)}) \cong \mathcal{H}^{(1)}/\mathcal{H}^{(0)}$ which is just a reformulation of the results in [AP], [Bo]. The paper [F] contains a very introductory treatment of this cohomological reformulation and an alternative proof of one of the results in [Bo]. The isomorphism $H^1(A(\mathbb{D}), \mathcal{H}^{(n)}) \cong \mathcal{H}^{(n+1)}/\mathcal{H}^{(n)}$, $n \geq 1$, was then established by some algebraic trickery coupled with a new Bloch estimate on a certain $\mathcal{H}^{(0)}$-valued m-linear map, $m > 1$, defined on the m-fold product of $A(\mathbb{D})$.

For the weighted Bergman spaces $A^{2,\alpha}$, $\alpha > -1$, the isomorphism $H^1(A(\mathbb{D}), \mathcal{H}^{(0)}) \cong \mathcal{H}^{(1)}/\mathcal{H}^{(0)}$ was established in [FP] but, again, formulated in terms of the Foguel-type operators as in [AP]. The same Bloch estimate holds in the Bergman space setting (shown below) and so essentially the same proof as in [CFR] can be used to establish the isomorphism $H^1(A(\mathbb{D}), \mathcal{H}^{(n)}) \cong \mathcal{H}^{(n+1)}/\mathcal{H}^{(n)}$. However, we choose to give an alternative proof here which is structurally much more precise and gives estimates on related quotient norms which we were not able to do for the Hardy space. What is exploited here and not in [CFR] is the fact that $\mathcal{H}^{(n-1)}$ is complemented in $\mathcal{H}^{(n)}$. We will present an explicit realization of the isomorphism

$$\mathcal{H}^{(n)} \cong \mathcal{H}^{(0)} \oplus \mathcal{H}^{(1)}/\mathcal{H}^{(0)} \oplus \cdots \oplus \mathcal{H}^{(n)}/\mathcal{H}^{(n-1)}$$

which lends itself to the optimal *bounded* decomposition of $\mathcal{H}^{(n)}$.

A decomposition of a form $B \in \mathcal{H}^{(n)}$, say, $B = B_0 + B_1 + \ldots + B_n$, is said to be a bounded decomposition if the form $B_j \in \mathcal{H}^{(j)}$ and B_j has a certain elementary form (defined below). The problem of finding such a decomposition was first addressed in [R2] where an algebraic argument produced a decomposition which could be viewed as a formal Taylor series; however, it was unclear if the terms B_j were, in fact, bounded. It will follow from the work here that Rochberg's decomposition is indeed bounded. A purely formal argument (employed in [CFR] and again below) shows that every nth-order form B has $(n+1)!$ decompositions where each term, B_j, has, formally speaking, order j and is *elementary* in form. Again, the problem with any of these decompositions is that there is no apriori reason to see that the individual terms are bounded. However, results in the next section imply that any such decomposition is a bounded decomposition and thus, the decomposition in [R2] is also a bounded one.

The problem now is to find an optimal bounded decomposition; optimal in that each of the terms B_j is a bona fide special jth-order Hankel form as defined in [JP1]. The projections $Q_j^{(n)} : \mathcal{H}^{(n)} \to \mathcal{H}^{(j)}$, $j = 0, \ldots, n$, constructed below, produce forms $B_j = Q_j^{(n)} B$ satisfying $\|B_j\| \sim \|B_j + \mathcal{H}^{(j-1)}\|_{\mathcal{H}^{(j)}/\mathcal{H}^{(j-1)}}$. Furthermore, the form $Q_n^{(n)}(B)$, which is defined by a weighted average over all of the *elementary* forms of order n associated to B, is a *special* nth-order form in the Janson-Peetre setting. This is just one of the reasons these projections are very convienent. It will be shown below that

$$Q_j^{(n)}\, Q_k^{(n)} = \delta_{j,k}\, I_{\mathcal{H}^{(n)}} \quad \text{and} \quad \sum_{j=0}^{n} Q_j^{(n)} = I_{\mathcal{H}^{(n)}}.$$

Thus, these projections provide the desired decomposition which will be used to write down a corresponding bounded decomposition of derivations. This, in turn, will be used to show that the map on the 1st-order groups, induced from the quotient map $q : \mathcal{H}^{(n)} \to \mathcal{H}^{(n)}/\mathcal{H}^{(n-1)}$, is onto and that the map induced from the inclusion $\mathcal{H}^{(n-1)} \subseteq \mathcal{H}^{(n)}$ is the zero map.

1.1. Some relevant definitions

For $\alpha > -1$, μ_α denotes the probability measure on the unit disk, $\mathbb{D}$, defined by

$$d\mu_\alpha(z) = \frac{\alpha+1}{\pi}\left(1-|z|^2\right)^\alpha dA(z) \quad dA = dx\,dy. \tag{1}$$

The weighted Bergman space $A^{p,\alpha}$, $1 \le p < \infty$, is the space of holomorphic functions on $\mathbb{D}$ which are p-integrable with respect to μ_α. The norm on $A^{p,\alpha}$ is the $L^p(\mu_\alpha)$ norm. Only the spaces $A^{1,\alpha}$ and $A^{2,\alpha}$ appear here. At several points we exploit the fact that every function h in $A^{1,\alpha}$ can be written as a product of two functions $f, g \in A^{2,\alpha}$ and furthermore, there is a constant C, depending only on α, such that $h = fg$ with $\|f\|_{A^{2,\alpha}}\|g\|_{A^{2,\alpha}} \le C\|h\|_{A^{1,\alpha}}$, [Ho].

Let $\mathcal{B}_\alpha$ denote the space of bounded bilinear forms on $A^{2,\alpha}$. $\mathcal{B}_\alpha$ is a bi-module over the disk algebra, A($\mathbb{D}$), with left and right actions given by

$$a \cdot B\,(f,g) := B(af,g) \quad \text{and} \quad B \cdot a\,(f,g) := B(f,ag). \tag{2}$$

The definition of higher-order Hankel forms given here is from [PR]. A simple way to describe them is in terms of the map $\Delta : \mathcal{B}_\alpha \to \mathcal{B}_\alpha$ defined by

$$\Delta(B) = z \cdot B - B \cdot z \qquad z = \text{coordinate function}.$$

A form $B \in \mathcal{B}_\alpha$ is called an nth-order Hankel form if B is in the kernel of the map Δ^{n+1}. Let $\mathcal{H}^{(n)}$ denote the space of nth-order Hankel forms on $A^{2,\alpha}$. We suppress the parameter α since the statement of every result concerning these forms is the same for every $\alpha > -1$. Analogous results for the limiting case $\alpha = -1$ (i.e., the Hardy space) are metnioned throughout.

In symbols,

$$B \in \mathcal{H}_\alpha^{(n)} \Leftrightarrow \Delta^{n+1}(B) = 0.$$

From this definition we see that $\mathcal{H}^{(n)} \subseteq \mathcal{H}^{(n+1)}$, $n \geq 0$, and $\Delta\mathcal{H}^n \subseteq \mathcal{H}^{(n-1)}$, $n \geq 1$; in other words, the map Δ lowers order by one. The space $\mathcal{H}^{(n)}$ is weakly closed in $\mathcal{B}_\alpha$. Also,

$$\Delta(f \cdot B) = f \cdot \Delta(B) \quad \text{and} \quad \Delta(B \cdot f) = \Delta(B) \cdot f \qquad \forall\, f \in \mathrm{A}(\mathbb{D}).$$

Thus, $\mathcal{H}^{(n)}$ is invariant under the action of $\mathrm{A}(\mathbb{D})$ defined in (2).

The first problem to address is: what do these forms look like? For $n = 0$, $\mathcal{H}^{(0)}$ coincides with the usual Hankel forms on $A^{2,\alpha}$ defined by

$$B_a(f,g) := \int_{\mathbb{D}} \bar{a} f g \, d\mu_\alpha.$$

By the factorization, $A^{1,\alpha} = A^{2,\alpha} \cdot A^{2,\alpha}$ (with control on norms), the ***symbol** a*, of the form B_a, is a function necessarily in the dual of $A^{1,\alpha}$. It is well known that this dual is the Bloch class, BLOCH, which is defined to be the space of holomorphic functions b on $\mathbb{D}$ with

$$\|b\|_{\mathrm{BLOCH}} := \sup_{z\in\mathbb{D}} |b'(z)|\,(1-|z|) < \infty.$$

In other words, the form B_a is bounded on $A^{2,\alpha} \times A^{2,\alpha}$ if and only if the symbol $a \in$ BLOCH and, furthermore, this holds for all $\alpha > -1$.

We should note that even though $(A^{1,\alpha})^* =$ BLOCH for all $\alpha > -1$, the dual norm

$$\|a\|_{*,\alpha} := \sup_{\|h\|_{A^{1,\alpha}}=1} \left| \int_{\mathbb{D}} \bar{a} h \, d\mu_\alpha \right|$$

certainly depends on α. Indeed, an equivalent norm on $A^{2,\alpha}$ is

$$\|\sum_{n=0}^{\infty} b_n z^n\|^2 = \sum (n+1)^{-(1+\alpha)} |b_n|^2$$

and thus, by considering the matrix representing the Hankel form B_a with respect to the orthogonal basis $\{(n+1)^{\frac{1+\alpha}{2}} z^n\}$ of $A^{2,\alpha}$, we have

$$\|B_a\| \sim \left\| \left(\frac{(i+1)^{\frac{1+\alpha}{2}} (j+1)^{\frac{1+\alpha}{2}}}{(i+j+1)^{1+\alpha}} \, a_{i+j} \right) \right\|.$$

So, in the limiting case $\alpha = -1$, we get the norm of the Hankel matrix determined by the Taylor coefficients of the function a which is equal to the norm of the Hankel form B_a acting on the product Hardy space $H^2 \times H^2$ which, in turn, is equivalent (via the factorization $H^1 = H^2 \cdot H^2$) to the BMOA norm of the function a, [Fef]. Since BMOA is properly contained in BLOCH, the dual norm $\|a\|_{*,\alpha}$ cannot be bounded (independent of a) as $\alpha \to -1$.

Higher-order Hankel forms can be constructed as follows. Let D be the derivative operator $Df = f'$. For $n, m \geq 0$, consider the forms $D^n B_a D^m$ acting on polynomials f, g by

$$D^n B_a D^m (f, g) = \int_{\mathbb{D}} \bar{a} f^{(n)} g^{(m)} \, d\mu_\alpha.$$

The calculus between the map Δ and differentiation yields the following identities.

$$\Delta(D^n B_a D^m) = \begin{cases} nD^{n-1}B_a & n \geq 1, m = 0 \\ nD^{n-1}B_a D^m - mD^n B_a D^{m-1} & n, m \geq 1 \\ -mB_a D^{m-1} & n = 0, m \geq 1 \end{cases} \tag{3}$$

Thus, formally, the densely defined form $D^n B_a D^m$ is a Hankel form of order $n+m$. A form of this type will be called an *elementary Hankel form* of order $n+m$ even though it may not extend to a bounded form on $A^{2,\alpha} \times A^{2,\alpha}$.

A formal argument shows that every $B \in \mathcal{H}^{(n)}$ can be written as a sum of elementary Hankel forms of lower or equal order. Indeed, consider the map ρ_n on $\mathcal{B}_\alpha$ defined by

$$\rho_n(B) = \frac{1}{n!} \Delta^n (B). \tag{4}$$

If $B \in \mathcal{H}^{(n)}$ then by definition of the nth-order forms, $\rho_n(B) \in \mathcal{H}^{(0)}$. Set $B_{a_n} = \rho_n(B)$. By (3), $\rho_n(D^n B_{a_n}) = B_{a_n}$, and thus, formally, $B - D^n B_{a_n}$ has order $n-1$. Hence, $B = B_{n-1} + D^n B_{a_n}$ where B_{n-1} has order $n-1$. Applying ρ_{n-1} to the form B_{n-1} produces a function a_{n-1} such that $B_{n-1} = B_{n-2} + D^{n-1} B_{a_{n-1}}$ where B_{n-2} has order $n-2$. Continuing in this fashion yields a decomposition of B of the form

$$B = B_{a_0} + DB_{a_1} + \cdots + D^n B_{a_n}. \tag{5}$$

If, instead of $D^n B_{a_n}$ we used the form $D^{n-1} B_{a_n} D$ in the argument above (note that $\rho_n(D^{n-1} B_{a_n} D) = -B_{a_n}$), we would get a decomposition of B of the form

$$B = B_{b_0} + B_{b_1} D + DB_{b_2} D \cdots - D^{n-1} B_{a_n} D. \tag{6}$$

In particular, the symbol of the highest order elementary form in any decompostion is the symbol of $\rho_n(B) \in \mathcal{H}^{(0)}$ up to a sign factor. Also the boundedness of B implies that the functions a_j and b_j are in $A^{2,\alpha}$, $j = 0, \ldots, n-1$. However, there is no *a priori* reason that the individual elementary Hankel forms $D^j B_{a_j}$, $j = 0, \ldots, n$ are themselves bounded.

It will follow from results below that every elementary Hankel form in any of the possible $(n+1)!$ decompositions of a form $B \in \mathcal{H}^{(n)}$ is, indeed, bounded. This fails for the Hardy space H^2, [CFR]. In fact, any function a with $a^{(2)} \in$ BLOCH but, $a^{(2)}$ not in BMOA, can be used to produce a bounded second order Hankel form $B = B_{a_0} + DB_{a_1} + D^2 B_{a_2}$ for which the highest order term $D^2 B_{a_2}$ is not bounded on $H^2 \times H^2$. On the other hand, we showed that a decomposition of the form $B = B_{b_0} + B_{b_1} D - DB_{a_2} D$ is a bounded decomposition.

2. $\mathcal{H}^{(n-1)}$ is complemented in $\mathcal{H}^{(n)}$

A decomposition of forms in $\mathcal{H}^{(n)}$ which is exploited in the cohomological theory is one which establishes the isormorphism

$$\mathcal{H}^{(n)} \cong \mathcal{H}^{(0)} \oplus \mathcal{H}^{(1)}/\mathcal{H}^{(0)} \oplus \cdots \oplus \mathcal{H}^{(n)}/\mathcal{H}^{(n-1)}.$$

We will show that for each $n \geq 1$, $\mathcal{H}^{(n-1)}$ is complemented in $\mathcal{H}^{(n)}$ by producing a bounded projection of $\mathcal{H}^{(n)}$ onto $\mathcal{H}^{(n-1)}$. To do this we first need to to characterize the 0th-order Hankel forms in the image of the map Δ^n. The reason is that the map $\Delta^n|\mathcal{H}^{(n)}$ has kernel $\mathcal{H}^{(n-1)}$ and so, the quotient $\mathcal{H}^{(n)}/\mathcal{H}^{(n-1)}$ can be identified with $Im\left(\Delta^n|\mathcal{H}^{(n)}\right) = \{B_a \in \mathcal{H}^{(0)} \mid \exists\, B \in \mathcal{H}^{(n)},\ B_a = \Delta^n(B)\ \}$. The space $Im\left(\Delta^n|\mathcal{H}^{(n)}\right)$ is a subspace of $\mathcal{H}^{(0)}$ and this latter space is just the BLOCH class (via the symbol map and a standard estimate). The next result says that the symbols of the 0th-order Hankel forms in $Im\left(\Delta^n|\mathcal{H}^{(n)}\right)$ coincide with the class of holomorphic functions whose nth derivative is in BLOCH.

Theorem 2.1. *Let $a \in$ BLOCH and $n \geq 0$. There is a form $B \in \mathcal{H}^{(n)}$ such that $B_a = \Delta^n(B)$ if and only if $a^{(n)} \in$ BLOCH.*

We first prove a lemma which characterizes the class of holomorphic functions a with $a^{(n)} \in$ BLOCH in terms of boundedness criteria on elementary forms, as well as, a certain multi-linear map defined on the n-fold product of the disk algebra with values in $\mathcal{B}_\alpha$. We should point out that the equivalence of (1), (2) and (3) below is well known, [Pe]. What is new here is the addition of condition (4). Actually, the equivalence of (4) and (1) when $n = 1$ was shown in [FP] and, for the limiting case $\alpha = -1$, (i.e., the Hardy space) the result was established in [CFR] for all $n \geq 2$. It fails for $\alpha = -1$ and $n = 1$, however, in this case, replacing BLOCH with BMOA will yield the analogous result, [AP], [CFR].

Lemma 2.2. *Let $a \in$ BLOCH and $n \geq 0$. The following are equivalent.*

(1) $a^{(n)} \in$ BLOCH.
(2) *There is a $0 \leq j \leq n$ such that $\|D^j B_a D^{n-j}\|_{\mathcal{B}_\alpha} < \infty$.*
(3) *For every $0 \leq j \leq n$, $\|D^j B_a D^{n-j}\|_{\mathcal{B}_\alpha} < \infty$.*
(4) *$\exists\, C > 0$ such that $\|f_1' f_2' \cdots f_n' \cdot B_a\|_{\mathcal{B}_\alpha} \leq C \|f_1\|_\infty \cdots \|f_n\|_\infty\ \forall\, f_1, \ldots, f_n \in$ A$(\mathbb{D})$.*

Proof. Suppose (1) holds and fix $0 \leq j \leq n$. We may assume that $a(0) = a'(0) = \cdots = a^{(n)}(0) = 0$. Then standard estimates imply

$$\|D^j B_a D^{n-j}\|_{\mathcal{B}_\alpha} \sim \sup_{\|f\|_{A^{2,\alpha}} = \|g\|_{A^{2,\alpha}} = 1} \left| \int_{\mathbb{D}} \overline{a^{(n+1)}}\, f^{(j)} g^{(n-j)}\, d\mu_{\alpha+n+1} \right|.$$

By Holder's

$$\begin{aligned}&\left|\int_{\mathbb{D}} \overline{a^{(n+1)}}\, f^{(j)} g^{(n-j)}\, d\mu_{\alpha+n+1}\right|\\ &\leq \|a^{(n)}\|_{\text{BLOCH}}\, \|f^{(j)}\|_{A^{2,\alpha+2j}}\, \|g^{(n-j)}\|_{A^{2,\alpha+2(n-j)}}\\ &\preceq \|a^{(n)}\|_{\text{BLOCH}}\, \|f\|_{A^{2,\alpha}} \|g\|_{A^{2,\alpha}}.\end{aligned}$$

Thus, (1) implies (3) and clearly, (3) implies (2).

The proof that (4) implies (1) uses the same method employed for the Hardy space in [CFR]. Recall the factorization $A^{1,\alpha} = A^{2,\alpha} \cdot A^{2,\alpha}$ (with control of norms) [Ho]. If (4) holds then

$$\left|\int_{\mathbb{D}} \bar{a}\, f_1' \cdots f_n' h\, d\mu_\alpha\right| \leq \tilde{C} \|f_1\|_\infty \cdots \|f_n\|_\infty \|h\|_{A^{1,\alpha}} \quad \forall\, f_1, \dots, f_n \quad \forall\, h. \tag{7}$$

Fix $0 < \beta < n^{-1}$ and $z \in \mathbb{D}$. Then

$$\begin{aligned}|a^{(n+1)}(z)| &= c\left|\int_{\mathbb{D}} \overline{a(\lambda)} \frac{\lambda^{n+1}}{(1-\bar{z}\lambda)^{\alpha+3+n}}\, d\mu_\alpha(\lambda)\right|\\ &= c\left|\int_{\mathbb{D}} \overline{a(\lambda)} \underbrace{\left(\frac{\lambda}{1-\bar{z}\lambda}\right)^{1+\beta} \cdots \left(\frac{\lambda}{1-\bar{z}\lambda}\right)^{1+\beta}}_{n \text{ factors}} \frac{\lambda^{1-n\beta}}{(1-\bar{z}\lambda)^{\alpha+3-n\beta}}\, d\mu_\alpha(\lambda)\right|.\end{aligned}$$

Let $f \in \mathrm{A}(\mathbb{D})$ such that $f'(\lambda) = \lambda^{1+\beta}\,(1-\bar{z}\lambda)^{-1-\beta}$. Then $\|f\|_\infty \preceq (1-|z|)^{-\beta}$ and since $1 - n\beta > 0$,

$$\left\| \frac{\lambda^{1-n\beta}}{(1-\bar{z}\lambda)^{\alpha+3-n\beta}} \right\|_{A^{1,\alpha}} \preceq (1-|z|)^{n\beta-1}.$$

It follows by the estimate in (7) that $|a^{(n+1)}(z)| \leq M\left(1-|z|^2\right)^{-1}$ for all $z \in \mathbb{D}$ and thus, $a^{(n)} \in$ BLOCH.

The proof of the implication (2) $\Rightarrow$ (4) also follows along the same lines as that for the Hardy space and so, an abbreviated form of the proof will be presented here. The proof makes use of the maps $\triangle_f$, $f \in \mathrm{A}(\mathbb{D})$, acting on $\mathcal{B}_\alpha$ by

$$\triangle_f(B) = f \cdot B - B \cdot f \quad B \in \mathcal{B}_\alpha. \tag{8}$$

A fundamental identity involving these maps is

$$\triangle_{f_1} \cdots \triangle_{f_n} (D^j B_a D^{n-j}) = (-1)^{n-j}\, n!\, f_1' \cdots f_n' \cdot B_a. \tag{9}$$

This can be verified inductively from the identity $\triangle_f(DB) = f' \cdot B + D\, \triangle_f(B)$, which is just the product rule and holds for all $B \in \mathcal{B}_\alpha$, see [CFR].

Now suppose that the elementary Hankel form $D^j B_a D^{n-j}$ is bounded on $A^{2,\alpha} \times A^{2,\alpha}$ for some $0 \leq j \leq n$. By the identity (9) and the estimate $\|\triangle_f\| \leq$

$2\|f\|_\infty$,

$$\begin{aligned}\|f_1'f_2'\cdots f_n'\cdot B_a\| &= \frac{1}{n!}\|\triangle_{f_1}\cdots\triangle_{f_n}(D^jB_aD^{n-j})\| \\ &\leq 2\|D^jB_aD^{n-j}\|\|f_1\|_\infty\cdots\|f_n\|_\infty \quad \forall\, f_1,\dots,f_n\in \mathrm{A}(\mathbb{D}).\end{aligned}$$

Thus, (4) holds and the proof is now complete. □

We now prove Theorem 2.1.

Proof. Suppose that $B_a = \triangle^n(B)$ where $B\in\mathcal{H}^{(n)}$. By identity (3), $\triangle^n(D^nB_a) = n!B_a$. Hence, formally, $B - n!^{-1}D^nB_a$ has order $n-1$. It follows that on pairs of polynomials,

$$\triangle_{f_1}\cdots\triangle_{f_n}(B) = \triangle_{f_1}\cdots\triangle_{f_n}(\frac{1}{n!}D^nB_a) \quad \forall\, f_1,\dots f_n.$$

By identity (9),

$$\triangle_{f_1}\cdots\triangle_{f_n}(\frac{1}{n!}D^nB_a) = f_1'\cdots f_n'\cdot B_a \quad \forall\, f_1,\dots f_n.$$

Thus,

$$\|f_1'\cdots f_n'\cdot B_a\| \leq 2^n\|B\|\|f_1\|_\infty\cdots\|f_n\|_\infty \quad \forall\, f_1,\dots,f_n.$$

By Lemma 2.2, this inequality implies that the function $a^{(n)}\in BLOCH$.

Conversely, if $a^{(n)}\in BLOCH$, then by Lemma 2.2, the elementary Hankel form D^nB_a is bounded on $A^{2,\alpha}\times A^{2,\alpha}$ and satisfies $\triangle^n(D^nB_a) = n!B_a$. Therefore, $B_a\in Im(\triangle^n)$. □

For each $n\geq 0$ consider the map π_n, defined (formally) on $\mathcal{H}^{(0)}$ by,

$$\pi_n(B_a) = \frac{1}{2^n}\sum_{j=0}^{n}(-1)^{n-j}\binom{n}{j}D^jB_aD^{n-j}. \tag{10}$$

Let $a\in$ BLOCH with $a^{(n)}\in$ BLOCH. By Lemma 2.2, $\pi_n(B_a)$ is bounded. With respect to the orthogonal basis of $A^{2,\alpha}$ given by $\{(n+1)^{\frac{1+\alpha}{2}}z^n\}$, the matrix representing the form $\pi_n(B_a)$ is

$$\left(\left(\frac{i-j}{2}\right)^n\frac{(i+1)^{\frac{1+\alpha}{2}}(j+1)^{\frac{1+\alpha}{2}}}{(i+j+1)^{1+\alpha}}a_{i+j}\right)$$

where the a_k's are the Taylor coefficients of the function a.

By identity (3), $\triangle^n\pi_n(B_a) = n!B_a$. Thus, by Theorem 2.1 together with Lemma 2.2, $\pi_n(B_a)$ is bounded if and only if $B_a\in Im(\triangle^n)$. In particular, π_n is bounded as a map from $Im(\triangle^n)$ into $\mathcal{H}^{(n)}$. A simple computation shows that

$$\triangle\pi_n = n\,\pi_{n-1} \quad n\geq 1. \tag{11}$$

Now let $B\in\mathcal{H}^{(n)}$ and let ρ_n be the map defined in (4). Then $\rho_n(B)\in Im(\triangle^n)$ and hence, $\pi_n\rho_n(B)\in\mathcal{H}^{(n)}$. The next proposition gathers the relevant facts about the map $P_n := \pi_n\rho_n$.

Proposition 2.3. *For each $n \geq 0$, $P_n = \pi_n \rho_n$ is a bounded projection on $\mathcal{H}^{(n)}$ with* Ker $P_n = \mathcal{H}^{(n-1)}$ *and satisfying $\triangle P_n = P_{n-1} \triangle$. Furthermore,*

$$\|B + \mathcal{H}^{(n-1)}\|_{\mathcal{H}^{(n)}/\mathcal{H}^{(n-1)}} \leq \|P_n(B)\|_{\mathcal{H}^{(n)}} \leq \frac{1}{n!} \|\pi_n | Im(\Delta^n)\| \, \|B + \mathcal{H}^{(n-1)}\|_{\mathcal{H}^{(n)}/\mathcal{H}^{(n-1)}}.$$

Proof. Equip $Im(\Delta^n)$ with the quotient norm

$$\|B_a\|_{Im(\Delta^n)} = \inf \{\|B\| \mid \Delta^n(B) = B_a\} = \frac{1}{n!} \|\pi_n(B_a) + \mathcal{H}^{(n-1)}\|_{\mathcal{H}^{(n)}/\mathcal{H}^{(n-1)}}.$$

The second equality holds since $\rho_n \pi_n(B_a) = B_a$. For $B \in \mathcal{H}^{(n)}$,

$$\|\rho_n(B)\|_{Im(\Delta^n)} = \frac{1}{n!} \|B + \mathcal{H}^{(n-1)}\|_{\mathcal{H}^{(n)}/\mathcal{H}^{(n-1)}}.$$

Thus,

$$\begin{aligned} \|P_n(B)\|_{\mathcal{H}^{(n)}} &= \|\pi_n \rho_n(B)\|_{\mathcal{H}^{(n)}} \\ &\leq \|\pi_n | Im(\Delta^n)\| \|\rho_n(B)\|_{Im(\Delta^n)} \\ &= \frac{1}{n!} \|\pi_n | Im(\Delta^n)\| \, \|B + \mathcal{H}^{(n-1)}\|_{\mathcal{H}^{(n)}/\mathcal{H}^{(n-1)}}. \end{aligned}$$

Now by (11),

$$\triangle P_n = \triangle \pi_n \rho_n = n\pi_{n-1}\rho_n = \pi_{n-1}\rho_{n-1}\triangle = P_{n-1}\triangle$$

and $\rho_n \pi_n = Id_{\mathcal{H}^{(0)}}$ which implies that $P_n^2 = P_n$ and $\rho_n P_n = \rho_n$.

Since Ker $\rho_n | \mathcal{H}^{(n)} = \mathcal{H}^{(n-1)}$, we deduce that Ker $P_n = \mathcal{H}^{(n-1)}$. Therefore, $B + \mathcal{H}^{(n-1)} = P_n(B) + \mathcal{H}^{(n-1)}$ and now, the statement follows. □

What is missing from Lemma 2.2 are precise BLOCH estimates. So, we do not know at the moment if the projection P_n acts contractively on $\mathcal{H}^{(n)}$ except in the case $n = 0$ (obvious because $P_0 = Id_{\mathcal{H}^{(0)}}$), and in the case $n = 1$. The fact that P_1 is contractive can be seen as follows. Let $B \in \mathcal{H}^{(1)}$ and set $B_{a_1} = \rho_1(B)$. Then B has the form $B = B_{a_0} + DB_{a_1}$. Define the *transpose* of B by $B^T(f, g) = B(g, f)$. Clearly, $\|B^T\|_{B_\alpha} = \|B\|_{B_\alpha}$ and

$$P_1(B) = \pi_1 \rho_1(B) = \pi_1(B_{a_1}) = \frac{1}{2}(DB_{a_1} - B_{a_1}D) = \frac{1}{2}(B - B^T).$$

Thus $\|P_1(B)\| \leq \|B\|$. The argument here is so general that it cannot extend to the higher-order cases. The reason being is the fact that P_2 is not bounded on *2nd*-order forms on the Hardy space H^2, [CFR]. However, a modified definition of P_n, $n \geq 2$, will work for the Hardy space. In particular, $P_2(B) = -D\rho_2(B)D$ is bounded on 2nd-order forms on H^2, [CFR], and its easy to check that $P_2^2 = P_2$ and P_2 has kernel equal to 1st-order forms.

We can build a projection of $\mathcal{H}^{(n)}$ onto $\mathcal{H}^{(j)}$ in the obvious way. By Proposition 2.3, the map $I - P_n$ is a bounded projection of $\mathcal{H}^{(n)}$ onto $\mathcal{H}^{(n-1)}$. Thus, the map $(I - P_{n-1})(I - P_n)$ projects $\mathcal{H}^{(n)}$ onto $\mathcal{H}^{(n-2)}$ and, in general, the map

$(1-P_{j+1})\cdots(1-P_n)$ projects $\mathcal{H}^{(n)}$ onto $\mathcal{H}^{(j)}$, $j=0,\dots,n-1$. In particular, the $0th$-order forms are complemented in $\mathcal{H}^{(n)}$ for all $n \geq 1$.

Define the maps $Q_j^{(n)}$ on $\mathcal{H}^{(n)}$ by

$$Q_j^{(n)} = \begin{cases} P_j(1-P_{j+1})\cdots(1-P_n) & 0 \leq j \leq n-1 \\ P_n & j=n \end{cases}$$

Note that $Q_j^{(n)}$ maps $\mathcal{H}^{(n)}$ into $\mathcal{H}^{(j)}$ and $Q_0^{(n)}$ projects $\mathcal{H}^{(n)}$ onto $\mathcal{H}^{(0)}$.

Theorem 2.4. *The maps $Q_j^{(n)}$, $j=0,\dots,n$, satisfy*

(i) $Q_j^{(n)}\, Q_k^{(n)} = \delta_{j,k}\, Q_j^{(n)}$

(ii) $Q_0^{(n)} + \cdots + Q_n^{(n)} = Id_{\mathcal{H}^{(n)}}$

(iii) $\|Q_j^{(n)}(B)\| \leq \|P_j\|\, \|Q_j^{(n)}(B) + \mathcal{H}^{(j-1)}\|_{\mathcal{H}^{(j)}/\mathcal{H}^{(j-1)}} \quad \forall\, B \in \mathcal{H}^{(n)}$

Proof. For $k<l$, $(I-P_l)\,P_k\,|\mathcal{H}^{(k)} = P_k$. Thus, if $k<j$ then $Q_j^{(n)}\,Q_k^{(n)} = P_j Q_k^{(n)} = 0$ and if $j<k$ then, $Q_j^{(n)}\,Q_k^{(n)} = P_j\,(I-P_{j+1})\cdots(I-P_k)\,Q_k^{(n)} = 0$. If $j=k$ then $Q_j^{(n)}\,Q_j^{(n)} = P_j\,Q_j^{(n)} = Q_j^{(n)}$. To see (ii), let $B \in \mathcal{H}^{(n)}$. Note that

$$\rho_n(Q_j^{(n)}\,B) = \begin{cases} 0 & 0 \leq j \leq n-1 \\ \rho_n(P_n\,B) & j=n. \end{cases}$$

Since $B - P_n\,B \in \mathcal{H}^{(n-1)}$, $\rho_n(P_n\,B) = \rho_n(B)$ and so, $\rho_n(B - \sum_{j=0}^n Q_j^{(n)}B) = 0$. Thus, $B - \sum_{j=0}^n Q_j^{(n)}B \in \mathcal{H}^{(n-1)}$. It follows that $Q_n^{(n)}B = 0$. Now apply ρ_{n-1} to the form $B-\sum_{j=0}^n Q_j^{(n)}B = B - \sum_{j=0}^{n-1} Q_j^{(n)}B$ to see that $B - \sum_{j=0}^n Q_j^{(n)}B \in \mathcal{H}^{(n-2)}$. Continuing in this fashion yields $B = \sum_{j=0}^n Q_j^{(n)}B$ and so, (ii) holds.

Recall that $(1-P_{j+1})\cdots(1-P_n)$ projects $\mathcal{H}^{(n)}$ onto $\mathcal{H}^{(j)}$ and $P_j\,\mathcal{H}^{(j-1)} = (0)$. It follows that if $B \in \mathcal{H}^{(n)}$ and $\tilde{B} \in \mathcal{H}^{(j-1)}$ then,

$$\begin{aligned} \|Q_j^{(n)}(B)\| &= \|Q_j^{(n)}(B+\tilde{B})\| \\ &\leq \|P_j\|\, \|(1-P_{j+1})\cdots(1-P_n)\,(B+\tilde{B})\| \\ &= \|P_j\|\, \|(1-P_{j+1})\cdots(1-P_n)\,B + \tilde{B}\|. \end{aligned}$$

Thus,

$$\|Q_j^{(n)}(B)\| \leq \|P_j\|\, \|(1-P_{j+1})\cdots(1-P_n)\,B + \mathcal{H}^{(j-1)}\|_{\mathcal{H}^{(j)}/\mathcal{H}^{(j-1)}}.$$

Since the cosets $(1-P_{j+1})\cdots(1-P_n)\,B + \mathcal{H}^{(j-1)}$ and $P_j(1-P_{j+1})\cdots(1-P_n)\,B + \mathcal{H}^{(j-1)} = Q_j^{(n)}(B) + \mathcal{H}^{(j-1)}$ are equal, statement (iii) now follows. □

Theorem 2.4 provides us with the desired decomposition of $\mathcal{H}^{(n)}$, $n \geq 0$. In particular, the theorem implies that the map $B \mapsto (Q_0^{(n)}B, \cdots, Q_n^{(n)}B)$ is an isomorphism (not necessarily isometric) between $\mathcal{H}^{(n)}$ and $\oplus_{j=0}^n \mathcal{H}^{(j)}/\mathcal{H}^{(j-1)}$, $n \geq$

1. Here $\mathcal{H}^{(-1)} := (0)$. Let $B \in \mathcal{H}^{(n)}$. Using the intertwining property in Proposition 2.3 we see that

$$\rho_j (I - P_{j+1}) \dots (I - P_n) = (I - P_1) \dots (I - P_{n-j}) \rho_j \quad j = 0, \dots, n-1.$$

So we can now write

$$Q_j^{(n)}(B) = \pi_j (I - P_1) \dots (I - P_{n-j}) \rho_j \quad j = 0, \dots n.$$

For each $0 \le j \le n$, the form $(I-P_1) \dots (I-P_{n-j}) \rho_j(B)$ is a $0th$-order Hankel form which is contained in $Im(\Delta^j)$. Theorem 2.1 then provides the boundedness criteria of each piece in the decomposition in terms of the BLOCH class. Also, Proposition 2.4, part (iii), provides the estimate of the quotient norm of each piece. We record this particular bounded decomposition below. We will rely heavily on it in the next section.

Corollary 2.5. *Every $B \in \mathcal{H}^{(n)}$ has the form*

$$B = \sum_{j=0}^{n} \pi_j (B_{a_j})$$

where $a_j^{(j)} \in$ BLOCH, $j = 0, \dots, n$. Furthermore,

$$\|\pi_j (B_{a_j})\|_{\mathcal{B}_\alpha} \sim \|\pi_j (B_{a_j}) + \mathcal{H}^{(j-1)}\|_{\mathcal{H}^{(j)}/\mathcal{H}^{(j-1)}}.$$

3. The cohomology groups

Recall that the space $\mathcal{H}^{(n)}$ is a submodule of $\mathcal{B}_\alpha$, $n \ge 0$. We now consider three groups associated to the A($\mathbb{D}$)-module $\mathcal{H}^{(n)}$. One of these is the first-order Hochschild group $H^1(\mathrm{A}(\mathbb{D}), \mathcal{H}^{(n)})$. (Hochschild groups will be defined below). The inclusion $\mathcal{H}^{(n)} \subseteq \mathcal{B}_\alpha$ induces a map from $H^1(\mathrm{A}(\mathbb{D}), \mathcal{H}^{(n)})$ into $H^1(\mathrm{A}(\mathbb{D}), \mathcal{B}_\alpha)$. The range of this map is the subgroup formed by equivalence classes of derivations with values in $\mathcal{H}^{(n)}$. We will show that this map is the zero map even though the group $H^1(\mathrm{A}(\mathbb{D}), \mathcal{H}^{(n)})$ is nonzero and, in fact, isomorphic to $\mathcal{H}^{(n+1)}/\mathcal{H}^{(n)}$, $n \ge 0$. The third group is $H^1(\mathrm{A}(\mathbb{D}), \mathcal{H}^{(n)}/\mathcal{H}^{(n-1)})$. The quotient map $q_n : \mathcal{H}^{(n)} \to \mathcal{H}^{(n)}/\mathcal{H}^{(n-1)}$ induces a map of $H^1(\mathrm{A}(\mathbb{D}), \mathcal{H}^{(n)})$ into $H^1(\mathrm{A}(\mathbb{D}), \mathcal{H}^{(n)}/\mathcal{H}^{(n-1)})$. We will show that this map is onto (and 1-1.) We then have an inclusion of $2nd$-order groups $H^2(\mathrm{A}(\mathbb{D}), \mathcal{H}^{(n-1)}) \hookrightarrow H^2(\mathrm{A}(\mathbb{D}), \mathcal{H}^{(n)})$, $n \ge 1$. We remark that these results also hold for the Hardy space, [CFR].

The 1st-order Hochschild groups defined here are the *bounded* versions of the algebraic groups since we insist that the co-cycles (i.e., derivations) are bounded. A *derivation* $\delta : \mathrm{A}(\mathbb{D}) \to \mathcal{B}_\alpha$ is a bounded, linear map satisfying

$$\delta(fg) = f \cdot \delta(g) + \delta(f) \cdot g \quad \forall\, f, g \in \mathrm{A}(\mathbb{D}). \tag{12}$$

Let $\mathrm{Der}(\mathrm{A}(\mathbb{D}), \mathcal{B}_\alpha)$ denote the space of derivations. Examples of derivations are the *inner* ones defined by $\delta(f) = f \cdot B - B \cdot f$, $f \in \mathrm{A}(\mathbb{D})$ and $B \in \mathcal{B}_\alpha$. In this case, we say that δ is the inner derivation implemented by the form $B \in \mathcal{B}_\alpha$. Let $\mathrm{Inn}(\mathrm{A}(\mathbb{D}), \mathcal{B}_\alpha)$ denote the space of all inner derivations. The first order (bounded)

Hochschild group of A(𝔻) with coefficients in $\mathcal{B}_\alpha$ is defined to be the quotient $\operatorname{Der}(A(\mathbb{D}), \mathcal{B}_\alpha)/\operatorname{Inn}(A(\mathbb{D}), \mathcal{B}_\alpha)$. The additive structure on this group comes from pointwise addition of derivations. The notation for this group is $H^1(A(\mathbb{D}), \mathcal{B}_\alpha)$.

The group $H^1(A(\mathbb{D}), \mathcal{H}^{(n)})$ is defined in the same manner by restricting to derivations with values in $\mathcal{H}^{(n)}$. Let $\delta : A(\mathbb{D}) \to \mathcal{B}_\alpha$ be a derivation and let $B = \delta(z)$. By the derivation property (12) we have the formula, which holds for the monomial z^m, $m \geq 1$,

$$\delta(z^m) = \sum_{j=0}^{m-1} z^{m-j-1} \cdot B \cdot z^j. \tag{13}$$

Since $\mathcal{H}^{(n)}$ is invariant under the action of A(𝔻), the formula above shows that $B \in \mathcal{H}^{(n)}$ if and only if $\delta(p) \in \mathcal{H}^{(n)}$ for all polynomials p. Therefore, since δ is bounded and $\mathcal{H}^{(n)}$ is norm closed, δ takes on values in $\mathcal{H}^{(n)}$ if and only if $B = \delta(z) \in \mathcal{H}^{(n)}$.

If a derivation $\delta : A(\mathbb{D}) \to \mathcal{B}_\alpha$ is inner then we will be more precise by saying that δ is inner with respect to $\mathcal{B}_\alpha$ (vs $\mathcal{H}^{(n)}$). In this case, the form $B = \delta(z) = \Delta(\tilde{B})$ for some form $\tilde{B} \in \mathcal{B}_\alpha$. Now if $B \in \mathcal{H}^{(n)}$ then $\tilde{B}$ is necessarily a Hankel form of order $n+1$. In fact the map

$$B + \mathcal{H}^{(0)} \mapsto \delta(f) := f \cdot B - B \cdot f.$$

is a well-defined, one-to-one map from $\mathcal{H}^{(n+1)}/\mathcal{H}^{(0)}$ into $\operatorname{Der}(A(\mathbb{D}), \mathcal{H}^{(n)})$. We show below that this map is onto $\operatorname{Der}(A(\mathbb{D}), \mathcal{H}^{(n)})$ and use Theorem 2.5 to find an estimate on the solution to $B = \Delta(\tilde{B})$ of minimum norm.

This result for the case $n = 0$ was proved in [FP] in the language of operators. Let us a give a quick proof based on Lemma 2.2. Let $\delta : A(\mathbb{D}) \to \mathcal{H}^{(0)}$ be a derivation and set $B_a = \delta(z)$, $a \in$ BLOCH. By the the formula in (13), $\delta(z^m) = mz^{(m-1)} \cdot B_a$, $m \geq 0$. Thus, $\delta(f) = f' \cdot B_a$ for every $f \in A(\mathbb{D})$. Since δ is bounded,

$$\sup_{\|f\|_\infty = 1} \|f' \cdot B_a\| < \infty.$$

This implies, by Lemma 2.2, that $\pi_1(B_a)$ is bounded. By (9), $f \cdot \pi_1(B_a) - \pi_1(B_a) \cdot f = f' \cdot B_a$, $\forall\, f \in A(\mathbb{D})$. Thus, the derivation δ is inner and implemented by the first order Hankel form $\pi_1(B_a)$. Furthermore, since $\|P_1\| = 1$, $\|\pi_1(B_a)\| = \|\pi_1(B_a) + \mathcal{H}^{(0)}\|_{\mathcal{H}^{(1)}/\mathcal{H}^{(0)}}$.

We now proceed to prove a higher-order version of this result. First we consider the special case when $B = \pi_n(B_a)$, with $a^{(n)} \in$ BLOCH.

Theorem 3.1. *Let $n \geq 1$ and let $\delta : A(\mathbb{D}) \to \mathcal{B}_\alpha$ be a derivation with $\delta(z) = \pi_n(B_a)$. Then the form $\hat{B} = \frac{1}{n+1}\pi_{n+1}(B_a)$ is bounded and satisfies $\delta(f) = f \cdot \hat{B} - \hat{B} \cdot f$ for all $f \in A(\mathbb{D})$. Furthermore,*

$$\|\hat{B}\| \leq \|I - Q_0^{(n+1)}\| \inf\{\, \|\tilde{B}\|_{\mathcal{B}_\alpha} \mid \delta(f) = f \cdot \tilde{B} - \tilde{B} \cdot f \quad \forall\, f \in A(\mathbb{D}) \,\}$$

Proof. By formula (13) and identity (9),

$$\begin{aligned}\Delta_{f_1}\cdots\Delta_{f_n}\,\delta(z^m) &= \sum_{j=0}^{m-1} z^{m-1-j}\cdot\Delta_{f_1}\cdots\Delta_{f_n}\,\pi_n(B_a)\cdot z^j\\ &= \sum_{j=0}^{m-1} z^{m-1-j}\, n!\, f_1'\cdots f_n'\cdot B_a\cdot z^j = n!\, f_1'\cdots f_n'\cdot m\, z^{m-1}\cdot B_a.\end{aligned}$$

Thus,

$$\Delta_{f_1}\cdots\Delta_{f_n}\,\delta(f_{n+1}) = n!\, f_1'\cdots f_{n+1}'\cdot B_a \quad \forall\ f_1,\dots,f_{n+1}\in \mathrm{A}(\mathbb{D}).$$

And so,

$$\sup_{\|f_1\|_\infty=\cdots=\|f_{n+1}\|_\infty=1} \|f_1'\cdots f_{n+1}'\cdot B_a\| \le 2\|\delta\|.$$

Lemma 2.2 implies that the form $\hat{B}$ is bounded. The intertwining condition, (9), shows that $\pi_n(B_a) = \Delta(\hat{B})$ and from this, it follows that $\delta(f) = f\cdot\hat{B} - \hat{B}\cdot f$, $f\in \mathrm{A}(\mathbb{D})$.

Now the map $Q_0^{(n+1)} = (I-P_1)\cdots(I-P_{n+1})$ is the projection of $\mathcal{H}^{(n+1)}$ onto $\mathcal{H}^{(0)}$ and $Q_{n+1}^{(n+1)}\,\hat{B} = P_{n+1}\,\hat{B} = \hat{B}$. It follows by Theorem 2.4 that $Q_0^{(n+1)}\,\hat{B} = 0$ and hence,

$$\|\hat{B}\| \le \|I - Q_0^{(n+1)}\|\,\|\hat{B} + \mathcal{H}^{(0)}\|_{\mathcal{H}^{(n+1)}/\mathcal{H}^{(0)}}.$$

But,

$$\|\hat{B} + \mathcal{H}^{(0)}\|_{\mathcal{H}^{(n+1)}/\mathcal{H}^{(0)}} = \inf\{\,\|\tilde{B}\|_{\mathcal{B}_\alpha} \mid \delta(f) = f\cdot\tilde{B} - \tilde{B}\cdot f \quad \forall\ f\in \mathrm{A}(\mathbb{D})\,\}$$

which completes the proof. □

The next result says, in particular, that every derivation with values in $\mathcal{H}^{(n)}$, $n\ge 0$, is inner with respect to $\mathcal{B}_\alpha$ and implemented by a bounded Hankel form of order $n+1$. Actually, the result is more precise than this. Let $\delta : \mathrm{A}(\mathbb{D}) \to \mathcal{H}^{(n)}$ be a derivation with $\delta(z) = B$. Decompose B using Corollary 2.5, say,

$$B = \sum_{j=0}^{n} \pi_j(B_{a_j}) \quad \text{with} \quad a_j^{(j)} \in \mathrm{BLOCH},\ j = 0,\dots,n.$$

Then δ decomposes as a sum $\delta(\cdot) = \sum_{j=0}^{n}\delta_j(\cdot)$ where for each $0\le j\le n$, δ_j is the derivation defined on monomials by

$$\delta_j(z^m) = \sum_{k=0}^{m-1} z^{m-1-k}\cdot\pi_j(B_{a_j})\cdot z^j \tag{14}$$

which extends linearly to polynomials. The theorem below says that each of the derivations δ_j, $j = 0,\dots,n$, extends to a bounded derivation on $\mathrm{A}(\mathbb{D})$. This places Theorem 3.1 at our disposal and furthermore, Theorem 2.4 gives us some control on quotient norms.

Theorem 3.2. *Let $n \geq 0$ and let $\delta : \mathrm{A}(\mathbb{D}) \to \mathcal{B}_\alpha$ be a derivation with $\delta(z) = B \in \mathcal{H}^{(n)}$. Decompose $B = \sum_{j=0}^{n} \pi_j(B_{a_j})$ with $a_j^{(j)} \in$ BLOCH. Then $\hat{B} = \sum_{j=0}^{n} \frac{1}{j+1}\pi_{j+1}(B_{a_j})$ is bounded, in particular, $a_j^{(j+1)} \in$ BLOCH, and the form $\hat{B}$ satisfies $\delta(f) = f \cdot \hat{B} - \hat{B} \cdot f$ for every $f \in \mathrm{A}(\mathbb{D})$. Furthermore,*

$$\|\hat{B}\| \leq \|I - Q_0^{(n+1)}\| \inf\{\, \|\tilde{B}\| \mid \delta(f) = f \cdot \tilde{B} - \tilde{B} \cdot f \quad \forall\, f \in \mathrm{A}(\mathbb{D})\,\}.$$

Proof. We showed above that the result holds for $n = 0$. We now provide the induction step. Suppose $n \geq 1$ and that we have the result for $n-1$. Decompose $\delta = \sum_{j=0}^{n} \delta_j$ as above. Since $\delta_j(z) \in \mathcal{H}^{(j)}$, $j = 0, \ldots, n$,

$$\delta(p) - \delta_n(p) \in \mathcal{H}^{(n-1)} \quad \forall\, p \in \mathbb{C}[z].$$

Thus,

$$\Delta_{f_1} \cdots \Delta_{f_n} \delta(p) = \Delta_{f_1} \cdots \Delta_{f_n} \delta_n(p) \quad \forall\, f_1, \ldots, f_{n+1} \in \mathrm{A}(\mathbb{D}),\ \forall\, p \in \mathbb{C}[z].$$

As shown in the proof of Theorem 3.1,

$$\Delta_{f_1} \cdots \Delta_{f_n} \delta_n(p) = n!\, f_1' \cdots f_n' \cdot p' \cdot B_{a_n}$$

and hence,

$$\|f_1' \cdots f_n' \cdot p' \cdot B_{a_n}\| \leq 2\,\|\delta\| \|f_1\|_\infty \cdots \|f_n\|_\infty\, \|p\|_\infty.$$

Since this holds for all $f_1, \ldots, f_n \in \mathrm{A}(\mathbb{D})$ and for all $p \in \mathbb{C}[z]$, Lemma 2.2 implies that the form $\hat{B}_{n+1} = \frac{1}{n+1}\pi_{n+1}(B_{a_n}) \in \mathcal{H}^{(n+1)}$ and satisfies $\Delta(\hat{B}_{n+1}) = \pi_n(B_{a_n})$. Thus, δ_n is inner (and so, bounded) and implemented by $\hat{B}_{n+1}$. Therefore, the derivation $\delta - \delta_n$ is bounded and since $\delta(z) - \delta_n(z) = B - \pi_n(B_{a_n}) \in \mathcal{H}^{(n-1)}$, $\delta - \delta_n$ takes on values in $\mathcal{H}^{(n-1)}$.

The induction hypothesis tells us that the form

$$\hat{B}_n = \sum_{j=0}^{n-1} \frac{1}{j+1}\pi_{j+1}(B_{a_j})$$

is bounded and satisfies

$$\delta(f) - \delta_n(f) = f \cdot \hat{B}_n - \hat{B}_n \cdot f \quad \forall\, f \in \mathrm{A}(\mathbb{D}).$$

It follows that $\hat{B} = \hat{B}_n + \hat{B}_{n+1} = \sum_{j=0}^{n} \frac{1}{j+1}\pi_{j+1}(B_{a_j})$ is bounded and satisfies

$$\delta(f) = f \cdot \hat{B} - \hat{B} \cdot f \quad \forall\, f \in \mathrm{A}(\mathbb{D}).$$

Since $Q_0^{n+1}\hat{B} = 0$,

$$\|\hat{B}\| \leq \|I - Q_0^{n+1}\|\, \|\hat{B} + \mathcal{H}^{(0)}\|_{\mathcal{H}^{n+1}/\mathcal{H}^{(0)}}.$$

This completes the proof. □

Theorem 3.2 implies that the map

$$B \mapsto \delta_B(f) := f \cdot B - B \cdot f$$

from $\mathcal{H}^{(n+1)}$ into $\mathrm{Der}(\mathrm{A}(\mathbb{D}), \mathcal{H}^{(n)})$ is onto for all $n \geq 0$. The corollary below is now immediate.

Corollary 3.3. *The map $H^1(\mathrm{A}(\mathbb{D}), \mathcal{H}^{(n)}) \to H^1(\mathrm{A}(\mathbb{D}), \mathcal{B}_\alpha)$ induced from the containment $\mathcal{H}^{(n)} \subseteq \mathcal{B}_\alpha$ is the zero map.*

Another corollary of Theorem 3.2 is the following isomorhism.

Corollary 3.4. $H^1(\mathrm{A}(\mathbb{D}), \mathcal{H}^{(n)}) \cong \mathcal{H}^{(n+1)}/\mathcal{H}^{(n)} \quad \forall\, n \geq 0.$

Proof.

By definition

$$\mathrm{Inn}(\mathrm{A}(\mathbb{D}), \mathcal{H}^{(n)}) = \{\, \delta_B(f) := f \cdot B - B \cdot f \mid B \in \mathcal{H}^{(n)} \,\}.$$

Thus, by Theorem 3.2,

$$\mathrm{Inn}(\mathrm{A}(\mathbb{D}), \mathcal{H}^{(n)}) = \mathrm{Der}(\mathrm{A}(\mathbb{D}), \mathcal{H}^{(n-1)}) \cong \mathcal{H}^{(n)}/\mathcal{H}^{(0)}.$$

We also have, by Theorem 3.2, the isomorphism

$$\mathrm{Der}(\mathrm{A}(\mathbb{D}), \mathcal{H}^{(n)}) \cong \mathcal{H}^{(n+1)}/\mathcal{H}^{(0)}$$

and so,

$$\mathrm{Der}(\mathrm{A}(\mathbb{D}), \mathcal{H}^{(n)})/\,\mathrm{Inn}(\mathrm{A}(\mathbb{D}), \mathcal{H}^{(n)}) \cong \mathcal{H}^{(n+1)}/\mathcal{H}^{(n)}.$$

□

Our last result concerns the group $H^1(\mathrm{A}(\mathbb{D}), \mathcal{H}^{(n)}/\mathcal{H}^{(n-1)})$. The space $\mathcal{H}^{(n)}/\mathcal{H}^{(n-1)}$ is a bi-module over $\mathrm{A}(\mathbb{D})$ via the quotient action

$$f \cdot (B + \mathcal{H}^{(n-1)}) \cdot g = f \cdot B \cdot g + \mathcal{H}^{(n-1)} \quad f, g \in \mathrm{A}(\mathbb{D}),\ B \in \mathcal{H}^{(n)}.$$

The group $H^1(\mathrm{A}(\mathbb{D}), \mathcal{H}^{(n)}/\mathcal{H}^{(n-1)})$ is defined in the same way as with any $\mathrm{A}(\mathbb{D})$-bimodule , namely,

$$H^1(\mathrm{A}(\mathbb{D}), \mathcal{H}^{(n)}/\mathcal{H}^{(n-1)}) = \mathrm{Der}(\mathrm{A}(\mathbb{D}), \mathcal{H}^{(n)}/\mathcal{H}^{(n-1)})/\,\mathrm{Inn}(\mathrm{A}(\mathbb{D}), \mathcal{H}^{(n)}/\mathcal{H}^{(n-1)}).$$

But, as usual, we have the *bounded* restriction on derivations. In other words, $\mathrm{Der}(\mathrm{A}(\mathbb{D}), \mathcal{H}^{(n)}/\mathcal{H}^{(n-1)})$ is the space of derivations with values in $\mathcal{H}^{(n)}/\mathcal{H}^{(n-1)}$ which are bounded with respect to the quotient norm. The first thing to note is that $\mathrm{Inn}(\mathrm{A}(\mathbb{D}), \mathcal{H}^{(n)}/\mathcal{H}^{(n-1)}) = (0)$. To see this let $\delta \in \mathrm{Inn}(\mathrm{A}(\mathbb{D}), \mathcal{H}^{(n)}/\mathcal{H}^{(n-1)})$. Suppose δ is implemented by $B + \mathcal{H}^{(n-1)}$, $B \in \mathcal{H}^{(n)}$. Then

$$\delta(z) = z \cdot (B + \mathcal{H}^{(n-1)}) - (B + \mathcal{H}^{(n-1)}) \cdot z = \Delta(B) + \mathcal{H}^{(n-1)} = 0.$$

It follows that $\delta = 0$ and so,

$$H^1(\mathrm{A}(\mathbb{D}), \mathcal{H}^{(n)}/\mathcal{H}^{(n-1)}) = \mathrm{Der}(\mathrm{A}(\mathbb{D}), \mathcal{H}^{(n)}/\mathcal{H}^{(n-1)}).$$

Now let $\delta : \mathrm{A}(\mathbb{D}) \to \mathcal{H}^{(n)}/\mathcal{H}^{(n-1)}$ be a derivation with $\delta(z) = B + \mathcal{H}^{(n-1)}$, $B \in \mathcal{H}^{(n)}$. Decompose B according to Corollary 2.5, say,

$$B = \sum_{j=0}^{n} \pi_j(B_{a_j}) \quad a_j^{(j)} \in \mathrm{BLOCH},\ j = 0, \ldots, n$$

and let δ_j be the derivation defined on monomials by (14), extended linearly to $\mathbb{C}[z]$. Since $\delta_j(p) \in \mathcal{H}^{(j)}$, $j = 0, \dots, n$,

$$\delta(p) = \delta_n(p) + \mathcal{H}^{(n-1)} \quad \forall\, p \in \mathbb{C}[z]. \tag{15}$$

Thus, by Proposition 2.3,

$$\|\delta(p)\|_{\mathcal{H}^{(n)}/\mathcal{H}^{(n-1)}} \sim \|P_n\, \delta_n(p)\|_{B_\alpha} \quad p \in \mathbb{C}[z].$$

Claim 3.5. *The derivation δ_n is bounded.*

To prove this we note that (15) implies,

$$\triangle_{f_1} \cdots \triangle_{f_n}\, \delta(p) = \triangle_{f_1} \cdots \triangle_{f_n}\, P_n\, \delta_n(p) \qquad \forall\, f_1, \dots, f_n \in \mathrm{A}(\mathbb{D}),\ \forall\, p \in \mathbb{C}[z].$$

As shown in the proof of Theorem 3.1,

$$\triangle_{f_1} \cdots \triangle_{f_n}\, P_n\, \delta_n(p) = n!\, f_1' \cdots f_n' \cdot p' \cdot B_{a_n} \qquad \forall\, f_1, \dots, f_n \in \mathrm{A}(\mathbb{D}),\ \forall\, p \in \mathbb{C}[z].$$

The boundedness of δ implies, by Lemma 2.2, that $\pi_{n+1}(B_{a_n})$ is bounded. Since

$$\triangle\Big(\frac{\pi_{n+1}}{n+1}(B_{a_n})\Big) = \pi_n(B_{a_n}),$$

δ_n is inner, thus bounded. The claim is now proved.

We have shown that the map $\hat{q} : \mathrm{Der}(\mathrm{A}(\mathbb{D}), \mathcal{H}^{(n)}) \to H^1(\mathrm{A}(\mathbb{D}), \mathcal{H}^{(n)}/\mathcal{H}^{(n-1)})$ defined by

$$\hat{q}(\delta)\,(f) = \delta(f) + \mathcal{H}^{(n-1)}$$

is onto. Since the kernel of $\hat{q}$ is $\mathrm{Inn}(\mathrm{A}(\mathbb{D}), \mathcal{H}^{(n)})$, we have the following isomorphism of groups.

$$H^1(\mathrm{A}(\mathbb{D}), \mathcal{H}^{(n)}/\mathcal{H}^{(n-1)}) \cong H^1(\mathrm{A}(\mathbb{D}), \mathcal{H}^{(n)}) \cong \mathcal{H}^{(n+1)}/\mathcal{H}^{(n)} \quad n \geq 0.$$

References

[AP] A. B. Aleksandrov and V.V. Peller, *Hankel operators and similarity to a contraction*, Internat. Math. Res. Notices no. 6 (1996), 263–275.

[Bo] J. Bourgain, *The Similarity Problem for polynomially bounded operators on Hilbert space*, Israel Journal of Math., **54** (1986), no. 2, 227–241.

[CFR] W. S. Cohn, S. H. Ferguson and R. Rochberg, *Boundedness of higher-order Hankel forms, factorization in potential spaces and derivations*, Proc. London Math. Soc., **83** (2001), no. 3, 110–130.

[Fef] C. Fefferman, *Characterizations of bounded mean oscillation*, Bull. Amer. Math. Soc., **77** (1971), 587–588.

[F] S. H. Ferguson, *Polynomially bounded operators and Ext groups*, Proc. Amer. Math. Soc., **124** (1996), no. 9, 2779–2785.

[FP] S. H. Ferguson and S. Petrovic, *The joint similarity problem for weighted Bergman shifts*, preprint.

[FR] S. H. Ferguson and R. Rochberg, *Higher-order Hilbert-Schmidt Hankel forms on the ball*, preprint.

[Ho] C. Horowitz, *Factorization theorems for functions in the Bergman spaces*, Duke Math. J., **44** (1977), no. 1, 203–213.

[JP1] S. Janson and J. Peetre, *A new generalization of Hankel operators (the case of higher weights)*, Math. Nachr., **132** (1987), 313–328.

[JP2] S. Janson and J. Peetre, *Paracommutators—boundedness and Schatten-von Neumann properties*, Trans. Amer. Math. Soc., **305** (1988), no. 2, 467–504.

[PR] J. Peetre and R. Rochberg, *Higher order Hankel forms*, Multivariable operator theory (Seattle, WA, 1993), 283–306, Contemp. Math., 185, AMS, Providence, RI, 1995.

[Pe] V. V. Peller, *Vectorial Hankel operators and related operators of the Schatten–von Neumann class* S_p, Int. Equat. Oper. Theory, **5** (1982), 244–272.

[R1] R. Rochberg, *A Kronecker theorem for higher order Hankel forms*, Proc. Amer. Math. Soc., **123** (1995), no. 10, 3113–3118.

[R2] R. Rochberg, *Higher order Hankel forms and commutators*, Holomorphic spaces (Berkeley, CA, 1995), Axler, McCarthy, and Sarason, eds., 155–178, MSRI. Publ., 33, Cambridge Univ. Press, Cambridge, 1998.

[Ro] H. Rosengren, *Multilinear Hankel forms of higher-order and orthogonal polynomials*, Math. Scand., **82** (1998), no. 1, 53–88.

[Zh1] G. Zhang, *Ha-plitz operators between Moebius invariant subspaces*, Math. Scand., **71** (1992), 69–84.

[Zh2] G. Zhang, *Tensor products of weighted Bergman spaces and invariant Ha-plitz operators*, Math. Scand., **71** (1992), 85–95.

Department of Mathematics, Wayne State University, Detroit, MI 48202
E-mail address: `sarah@math.wayne.edu`

A Weighted Version of Almost Commutant Lifting

C. Foias, A.E. Frazho, and M.A. Kaashoek

Abstract. This paper presents a new almost commutant lifting theorem involving expansive operators which generalizes earlier results from [31] and [5]. The main theorems include a generalization for operator pencils of the Adamjan-Arov-Krein formula for the singular values. Some applications of almost commutant lifting are included.

0. Introduction

In the last ten years the commutant lifting theorem has been extended in various directions. This paper concentrates on two of these directions, namely the versions of the commutant lifting theorem involving expansive operators from [31] and [11], and the "almost" versions in which the lifting is done on appropriate subspaces of finite co-dimension; see [9, 31, 4, 15, 5]. In the present paper we combine these two directions, and establish an almost commutant lifting theorem involving expansive operators. Our main results (Theorems 4.3 and 4.4) contain as special cases all the previous results referred to above when restricted to Hilbert spaces.

The proof of our main theorems are based on the invariant subspace theorem of Krein-Langer-Iokhvidov (see the book [7]), applied to a spectral decomposition associated to a spectral gap of an appropriate self-adjoint pencil. This more general setting involves additional technical complications, but in essence our proof follows the same line of reasoning as in [31] and [5]. For the benefit of the reader we present complete proofs of all the steps, including a proof of the Krein-Langer-Iokhvidov theorem; see Theorems 2.1 and 2.5. Consequently this paper also has an expository character.

The paper consists of six sections (not counting the introduction). In the first section we present the graph space structure of the negative subspaces corresponding to the spectral decomposition associated to a gap $(0, \delta)$ in the spectrum of a self-adjoint operator Λ, and we discuss in more detail the case when zero is in the spectrum of Λ. In Section 2 we state and prove the Krein-Langer-Iokhvidov theorem. Section 3 contains new results connecting a spectral gap for one operator in a self-adjoint operator pencil to negative subspaces for another operator in the same pencil. In Section 4 we recall the weighted commutant lifting theorem from [11], state and prove an "almost" version of that result (see Theorem 4.3), and establish a generalization (Theorem 4.4) for operator pencils of the Adamjan-Arov-Krein formula for the singular numbers in [1]. In the last two sections we demonstrate how the method of almost commutant lifting (with the weight equal to the identity operator) can be used to solve a Hankel approximation problem, and an almost Nevanlinna-Pick interpolation problem. For the scalar case explicit formulas for solutions are given, some in state space form. These formulas concern "optimal" solutions when the singular number is attained (for the Hankel approximation problem) or the Pick matrix is singular (for the Nevanlinna-Pick problem). For an analysis and state space formulas for the suboptimal case, see [8] and the references therein. The last two sections are expository in character, but some new results are also presented.

Acknowledgement. The authors thank Aad Dijksma for some references and helpful comments.

1. Spectral Gaps

Let Λ be a self-adjoint operator on a separable Hilbert space $\mathcal{H}$. Assume that the open interval $(0,\delta)$ is a gap in the spectrum of Λ, that is, the set $(0,\delta)$ is in the resolvent set $\rho(\Lambda)$ of Λ. In this case, Λ admits an operator matrix representation of the form

$$\Lambda = \begin{bmatrix} -\Lambda_- & 0 \\ 0 & \Lambda_+ \end{bmatrix} \text{ on } \begin{bmatrix} \mathcal{H}_- \\ \mathcal{H}_+ \end{bmatrix} \tag{1.1}$$

where Λ_- and Λ_+ are positive operators, and Λ_+ is invertible. Throughout P_- is the orthogonal projection onto $\mathcal{H}_-$ while P_+ is the orthogonal projection onto $\mathcal{H}_+$. In the sequel the notation $h = f \oplus g$ means that f is in $\mathcal{H}_-$, the vector g is in $\mathcal{H}_+$ and $h = f + g$. If Q is a positive operator on a Hilbert space $\mathcal{K}$, then $Q^{1/2}$ denotes the positive square root of Q. Notice that $\overline{Q^{1/2}\mathcal{K}} = \overline{Q\mathcal{K}} = \mathcal{K} \ominus \ker Q$. Here ker denotes the kernel or the null space of an operator. If Q is positive and invertible, then $Q^{-1/2}$ denotes the positive square root of Q^{-1}. Finally, recall that by definition a subspace is a closed linear manifold. The following result plays a fundamental role in characterizing certain invariant subspaces.

Lemma 1.1. *Assume that Λ is a self-adjoint operator on $\mathcal{H}$, and the open interval $(0,\delta)$ is a gap in the spectrum of Λ. Let $\mathcal{M}$ be a subspace of $\mathcal{H}$. Then*

$$(\Lambda f, f) \le 0 \qquad \text{(for all } f \in \mathcal{M}) \tag{1.2}$$

if and only if the linear manifold $\mathcal{F} = P_-\mathcal{M}$ is closed in $\mathcal{H}_-$ and

$$\mathcal{M} = \left\{ \begin{bmatrix} h \\ \widehat{X}h \end{bmatrix} : h \in \mathcal{F} \right\} \tag{1.3}$$

where $\widehat{X}$ is an operator from $\mathcal{H}_-$ into $\mathcal{H}_+$ of the form $\widehat{X} = \Lambda_+^{-1/2} X \Lambda_-^{1/2}$ for some contraction X mapping $\overline{\Lambda_-^{1/2}\mathcal{H}_-}$ into $\mathcal{H}_+$. Finally, if $\overline{\Lambda_-^{1/2} P_- \mathcal{M}} = \overline{\Lambda_-^{1/2}\mathcal{H}_-}$, then the contraction X is uniquely determined by $\mathcal{M}$.

Proof. Let f be a vector in $\mathcal{H}$. Then we claim that

$$(\Lambda f, f) \leq 0 \qquad \text{if and only if} \qquad \|\Lambda_+^{1/2} P_+ f\| \leq \|\Lambda_-^{1/2} P_- f\| . \tag{1.4}$$

To see this notice that $(\Lambda f, f) \leq 0$ if and only if

$$\begin{aligned} 0 &\geq (\Lambda f, f) = (\Lambda(P_- + P_+)f, (P_- + P_+)f) \\ &= -(\Lambda_- P_- f, P_- f) + (\Lambda_+ P_+ f, P_+ f) \\ &= -\|\Lambda_-^{1/2} P_- f\|^2 + \|\Lambda_+^{1/2} P_+ f\|^2 , \end{aligned}$$

or equivalently, $\|\Lambda_-^{1/2} P_- f\| \geq \|\Lambda_+^{1/2} P_+ f\|$. Therefore (1.4) holds.

Equation (1.4) shows that if (1.2) holds, then there exists a contraction X mapping $\overline{\Lambda_-^{1/2} P_- \mathcal{M}}$ into $\mathcal{H}_+$ satisfying

$$\Lambda_+^{1/2} P_+ f = X \Lambda_-^{1/2} P_- f \qquad (f \in \mathcal{M}) . \tag{1.5}$$

Clearly, this X can be extended to a contraction, also denoted by X, mapping $\overline{\Lambda_-^{1/2}\mathcal{H}_-}$ into $\mathcal{H}_+$ and satisfying (1.5). Moreover, if $\overline{\Lambda_-^{1/2} P_- \mathcal{M}} = \overline{\Lambda_-^{1/2}\mathcal{H}_-}$, then this X is uniquely determined by (1.5). Now let $\widehat{X}$ be the operator from $\mathcal{H}_-$ into $\mathcal{H}_+$ defined by $\widehat{X} = \Lambda_+^{-1/2} X \Lambda_-^{1/2}$. Equation (1.5) implies that $P_+ f = \widehat{X} P_- f$. Hence $f = P_- f + P_+ f = P_- f \oplus \widehat{X} P_- f$. So the subspace $\mathcal{M}$ is given by the following graph

$$\mathcal{M} = \left\{ \begin{bmatrix} h \\ \widehat{X} h \end{bmatrix} : h \in P_- \mathcal{M} \right\} . \tag{1.6}$$

Setting $\mathcal{F} = P_- \mathcal{M}$, yields the form of $\mathcal{M}$ in (1.3). Because $\mathcal{M}$ is closed and represented by the graph in (1.6), it follows that $P_- \mathcal{M}$ is closed.

On the other hand, assume that $\mathcal{M}$ is given by (1.3) where $\widehat{X} = \Lambda_+^{-1/2} X \Lambda_-^{1/2}$ and X is contractive. Then any vector f in $\mathcal{M}$ admits an orthogonal decomposition of the form $f = h \oplus \widehat{X} h$ for some h in $\mathcal{F}$. Using $h = P_- f$ and $\widehat{X} h = P_+ f$, this implies that

$$\|\Lambda_+^{1/2} P_+ f\| = \|X \Lambda_-^{1/2} h\| \leq \|\Lambda_-^{1/2} h\| = \|\Lambda_-^{1/2} P_- f\| .$$

According to (1.4), the vector f satisfies $(\Lambda f, f) \leq 0$. This completes the proof.

Motivated by Lemma 1.1, let $\mathcal{C}_\Lambda$ be the closed subset of $\mathcal{H}$ defined by

$$\mathcal{C}_\Lambda = \{ f \in \mathcal{H} : (\Lambda f, f) \leq 0 \} . \tag{1.7}$$

Obviously, $\mathcal{H}_- \subset \mathcal{C}_\Lambda$. Lemma 1.1 says that a subspace $\mathcal{M}$ is contained in $\mathcal{C}_\Lambda$ if and only if $\mathcal{M}$ is given by (1.3) where $\mathcal{F} = P_- \mathcal{M}$ is closed, the operator $\widehat{X} = \Lambda_+^{-1/2} X \Lambda_-^{1/2}$ and X is a contraction mapping $\overline{\Lambda_-^{1/2}\mathcal{H}_-}$ into $\mathcal{H}_+$.

The operator $\widehat{X}$ in formula (1.3) is usually referred to as an angular operator (see [10], page 110 and the reference given there).

The following result will be used in the proofs of Theorems 5.2 and 6.2, to derive the Adamjan-Arov-Krein [1] formula for computing a finite rank Hankel approximation of a compact Hankel operator, and to solve explicitly a scalar "almost" version of the Nevanlinna-Pick interpolation problem.

Corollary 1.2. *Assume that Λ is a self-adjoint operator on $\mathcal{H}$, and the open interval $(0,\delta)$ is a gap in the spectrum of Λ. Let $\mathcal{M}$ be a subspace of $\mathcal{H}$ such that $\mathcal{M} \subset \mathcal{C}_\Lambda$ and $P_- \mathcal{M} = \mathcal{H}_-$. Then* $\ker \Lambda \subset \mathcal{M}$.

PROOF. According to Lemma 1.1,

$$\mathcal{M} = \left\{ \begin{bmatrix} h \\ \widehat{X}h \end{bmatrix} : h \in \mathcal{H}_- \right\} \tag{1.8}$$

where $\widehat{X}$ is an operator from $\mathcal{H}_-$ into $\mathcal{H}_+$. Assume that x is a vector in the kernel of Λ, that is, $\Lambda x = 0$. In particular, x is in $\mathcal{H}_-$. Set $y = x \oplus \widehat{X}x$. Then y is in $\mathcal{M}$ and $P_- y = x$. Moreover, from (1.8), it follows that y is the unique vector in $\mathcal{M}$ satisfying $P_- y = x$. Notice that $\widehat{X}x$ is in $\mathcal{H}_+$. So using the fact that $(\Lambda\widehat{X}x, \widehat{X}x)$ is positive, we obtain

$$0 \geq (\Lambda y, y) = (\Lambda x, x) + (\Lambda \widehat{X}x, \widehat{X}x) \geq (\Lambda x, x) = 0\,.$$

This readily implies that $(\Lambda\widehat{X}x, \widehat{X}x) = 0$. Since $\Lambda|\mathcal{H}_+$ is invertible, we see that $\widehat{X}x = 0$. Hence $x = y$ is a vector in $\mathcal{M}$, which completes the proof.

The following result presents a somewhat stronger version of Corollary 1.2.

Corollary 1.3. *Assume that Λ is a self-adjoint operator on $\mathcal{H}$, and the open interval $(0,\delta)$ is a gap in the spectrum of Λ. Let $\mathcal{M}$ be a subspace of $\mathcal{H}$ such that $\mathcal{M} \subset \mathcal{C}_\Lambda$ and $P_- \mathcal{M} = \mathcal{H}_-$. Let $\{x_n\}_1^\infty$ be a sequence of vectors in $\mathcal{H}_-$ such that Λx_n approaches zero as n tends to infinity. Then for each n there exists a unique vector y_n in $\mathcal{M}$ such that $P_- y_n = x_n$ and $y_n - x_n$ approaches zero as n tends to infinity.*

It is noted that when zero is contained in the spectrum of Λ, then one can always find a sequence of unit vectors $\{x_n\}_1^\infty$ in $\mathcal{H}_-$ such that Λx_n approaches zero as n tends to infinity.

PROOF OF COROLLARY 1.3. As before, Lemma 1.1 shows that $\mathcal{M}$ is given by (1.8) for some operator $\widehat{X}$ mapping $\mathcal{H}_-$ into $\mathcal{H}_+$. Set $y_n = x_n + \widehat{X}x_n$. Then y_n is in $\mathcal{M}$ and $P_- y_n = x_n$. Moreover, y_n is the only vector in $\mathcal{M}$ satisfying $P_- y_n = x_n$. Notice that

$$0 \geq (\Lambda y_n, y_n) = (\Lambda x_n, x_n) + (\Lambda \widehat{X}x_n, \widehat{X}x_n) \geq (\Lambda x_n, x_n) \to 0 \qquad (n \to \infty)\,.$$

Thus $(\Lambda\widehat{X}x_n, \widehat{X}x_n)$ approaches zero as n tends to infinity. Since $\Lambda|\mathcal{H}_+$ is invertible, $\widehat{X}x_n$ converges to zero as n tends to infinity. Therefore $y_n - x_n = \widehat{X}x_n$ approaches zero as n tends to infinity. This completes the proof.

Motivated by Lemma 1.1, let $\mathcal{B}$ denote the set of all contractions X mapping $\overline{\Lambda_-^{1/2}\mathcal{H}_-}$ into $\mathcal{H}_+$, that is,

$$\mathcal{B} = \{X \in \mathcal{L}(\overline{\Lambda_-^{1/2}\mathcal{H}_-}, \mathcal{H}_+) : \|X\| \leq 1\}. \tag{1.9}$$

(The set of all bounded linear operators mapping $\mathcal{F}$ into $\mathcal{G}$ is denoted by $\mathcal{L}(\mathcal{F}, \mathcal{G})$.) For every X in $\mathcal{B}$, we define the operator $\widehat{X}$ from $\mathcal{H}_-$ into $\mathcal{H}_+$ by $\widehat{X} = \Lambda_+^{-1/2} X \Lambda_-^{1/2}$, and let $\mathcal{M}_X$ be the subspace defined by

$$\mathcal{M}_X = \left\{ \begin{bmatrix} h \\ \widehat{X}h \end{bmatrix} : h \in \mathcal{H}_- \right\}. \tag{1.10}$$

By Lemma 1.1, we see that $\mathcal{M}_X$ is a subspace of $\mathcal{H}$ and $\mathcal{M}_X \subset \mathcal{C}_\Lambda$. Furthermore, if $\mathcal{M}$ is a subspace such that $\mathcal{M} \subset \mathcal{C}_\Lambda$, then $\mathcal{M} \subset \mathcal{M}_X$ for some X in $\mathcal{B}$. In this case, we say that X is *associated with* $\mathcal{M}$. In general there may be many different contractions X in $\mathcal{B}$ associated with $\mathcal{M}$. In fact, if $\mathcal{M} \subset \mathcal{M}_X$ and $\overline{\Lambda_-^{1/2} P_- \mathcal{M}} = \overline{\Lambda_-^{1/2}\mathcal{H}_-}$, then $\mathcal{M}$ uniquely determines X; see Lemma 1.1. Using this notation we readily obtain the following result.

Lemma 1.4. *Assume that Λ is a self-adjoint operator on $\mathcal{H}$, and the open interval $(0, \delta)$ is a gap in the spectrum of Λ. Let $\mathcal{M}$ be a subspace of $\mathcal{H}$ such that $\mathcal{M} \subset \mathcal{C}_\Lambda$. Let $\mathcal{B}_\mathcal{M}$ be the set of contractions endowed with the weak operator topology defined by*

$$\mathcal{B}_\mathcal{M} = \{X \in \mathcal{B} : \mathcal{M} \subset \mathcal{M}_X\}. \tag{1.11}$$

Then $\mathcal{B}_\mathcal{M}$ is a compact convex set.

PROOF. Clearly, the operator ball $\mathcal{B}$ defined by (1.9) is convex and compact in the weak operator topology; see Halmos [19]. Since $\mathcal{M} \subset \mathcal{C}_\Lambda$, there exists a unique contraction $X_\mathcal{M}$ mapping $\overline{\Lambda_-^{1/2} P_- \mathcal{M}}$ into $\mathcal{H}_+$ and satisfying (1.5) with $X_\mathcal{M}$ in place of X. It follows that the set $\mathcal{B}_\mathcal{M}$ is also given by

$$\mathcal{B}_\mathcal{M} = \{X \in \mathcal{B} : X | \overline{\Lambda_-^{1/2} P_- \mathcal{M}} = X_\mathcal{M}\}. \tag{1.12}$$

¿From this description for $\mathcal{B}_\mathcal{M}$ it is clear that $\mathcal{B}_\mathcal{M}$ is convex and compact in the weak operator topology, because $\mathcal{B}$ has these properties. This completes the proof.

2. Spectral Gaps and Invariant Subspaces

Throughout this section Λ is a self-adjoint operator on a separable Hilbert space $\mathcal{H}$ satisfying all the properties specified in the first paragraph of Section 1. In particular, $\mathcal{H} = \mathcal{H}_- \oplus \mathcal{H}_+$ and Λ admits an operator matrix representation of the form

$$\Lambda = \begin{bmatrix} -\Lambda_- & 0 \\ 0 & \Lambda_+ \end{bmatrix} \text{ on } \begin{bmatrix} \mathcal{H}_- \\ \mathcal{H}_+ \end{bmatrix} \tag{2.1}$$

where Λ_- and Λ_+ are positive operators, and Λ_+ is invertible. Finally, recall that $\mathcal{C}_\Lambda$ is the closed subset of $\mathcal{H}$ defined by $\mathcal{C}_\Lambda = \{f \in \mathcal{H} : (\Lambda f, f) \leq 0\}$. In this section we will prove the following formulation of the Krein-Langer-Iokhvidov invariant subspace theorem.

Theorem 2.1. *Let Λ be a self-adjoint operator of the form (2.1), where the open interval $(0, \delta)$ is a gap in the spectrum of Λ. Consider the operator*

$$T = \begin{bmatrix} T_{11} & T_{12} \\ T_{21} & T_{22} \end{bmatrix} \quad on \quad \begin{bmatrix} \mathcal{H}_- \\ \mathcal{H}_+ \end{bmatrix}, \tag{2.2}$$

where $T^*\Lambda T \leq \Lambda$ and $\Lambda_-^{1/2} T_{12}$ is compact. Then T has an invariant subspace $\mathcal{M}$ satisfying $\mathcal{M} \subset \mathcal{C}_\Lambda$ and $P_- \mathcal{M} = \mathcal{H}_-$. In this case, $\mathcal{M} = \mathcal{M}_X$ for some X in $\mathcal{B}$. Finally, $\mathcal{M}^\perp$ and $\mathcal{H}_+$ are isomorphic, and hence have the same dimension.

The condition of compactness in the above theorem was introduced by H. Langer. Motivated by Theorem 2.1, let $\mathrm{Lat}_-(T, \Lambda)$ be the set of all invariant subspaces for T given by

$$\mathrm{Lat}_-(T, \Lambda) = \{\mathcal{M} : \mathcal{M} \text{ is a } T \text{ invariant subspace, } \mathcal{M} \subset \mathcal{C}_\Lambda \text{ and } P_- \mathcal{M} = \mathcal{H}_-\}\,. \tag{2.3}$$

If $T^*\Lambda T \leq \Lambda$ and $\Lambda_-^{1/2} T_{12}$ is compact, then Theorem 2.1 shows that $\mathrm{Lat}_-(T, \Lambda)$ is not empty.

The inequality $T^*\Lambda T \leq \Lambda$ implies that T maps $\mathcal{C}_\Lambda$ into its self. Now let X be in $\mathcal{B}$, and consider the subspace $\mathcal{M}_X$ defined in (1.10). Obviously, $\mathcal{M}_X \subset \mathcal{C}_\Lambda$. Since T maps $\mathcal{C}_\Lambda$ into $\mathcal{C}_\Lambda$, the subspace $\overline{T\mathcal{M}_X}$ is also contained in $\mathcal{C}_\Lambda$. So by Lemma 1.1, there is a contraction Y in $\mathcal{B}$ associated with $\overline{T\mathcal{M}_X}$, that is, $\overline{T\mathcal{M}_X} \subset \mathcal{M}_Y$. This allows us to define a map F_T from $\mathcal{B}$ into $\mathcal{P}(\mathcal{B})$, the set of all subsets of $\mathcal{B}$. To be precise

$$F_T(X) = \{Y \in \mathcal{B} : \overline{T\mathcal{M}_X} \subset \mathcal{M}_Y\} = \mathcal{B}_{\overline{T\mathcal{M}_X}}\,. \tag{2.4}$$

Notice that $Y \in F_T(X)$ if and only if $\overline{T\mathcal{M}_X} \subset \mathcal{M}_Y$. Furthermore, $\mathcal{M}_X$ is an invariant subspace for T if and only if X is a fixed point of the map F_T, that is, $X \in F_T(X)$. Finally, it is noted that Lemma 1.4 shows that $F_T(X)$ is a compact convex subset of $\mathcal{B}$.

Notice that the set-valued map F_T has a fixed point if and only if $\mathrm{Lat}_-(T, \Lambda)$ is nonempty, that is, there exists an invariant subspace $\mathcal{M}$ for T such that

$$P_- \mathcal{M} = \mathcal{H}_- \qquad \text{and} \qquad \mathcal{M} \subset \mathcal{C}_\Lambda\,. \tag{2.5}$$

Indeed, if X is in $\mathcal{B}$ and $X \in F_T(X)$, then $\mathcal{M} = \mathcal{M}_X$ is an invariant subspace for T such that (2.5) holds. Conversely, if $\mathcal{M}$ is an invariant subspace for T such that (2.5) holds, then $\mathcal{M} = \mathcal{M}_X$ for some X in $\mathcal{B}$. Moreover, $T\mathcal{M} \subset \mathcal{M}$ implies that X is in $F_T(X)$. In particular, X is a fixed point of F_T if and only if $\mathcal{M}_X$ is in $\mathrm{Lat}_-(T, \Lambda)$. So in order to prove Theorem 2.1 it suffices to show that F_T has a fixed point. For this purpose we use the following theorem.

Theorem 2.2 (Kakutani's fixed point theorem). *Let $\mathcal{X}$ be a non-empty compact convex subset of a Hausdorff locally convex topological vector space $\mathcal{Y}$. Let G be a mapping from $\mathcal{X}$ into the set $\mathcal{P}(\mathcal{X})$ of all subsets of $\mathcal{X}$. Assume that*

(a) *the set $G(x)$ is non-empty, compact and convex for each x in $\mathcal{X}$;*
(b) *G is upper semi-continuous on $\mathcal{X}$.*

Then G has a fixed point, that is, there exists a vector x in $\mathcal{X}$ such that $x \in G(x)$.

A proof of Theorem 2.2 can be found in Chapter 10 of [21]; see also Section 15.3 in [6]. Assume that the space $\mathcal{X}$ satisfies the hypothesis of Theorem 2.2 and G is a mapping from $\mathcal{X}$ into $\mathcal{P}(\mathcal{X})$ satisfying (a). Then G is upper semi-continuous (see Theorem 10.2.4 in [21]) if and only if the set

$$\{\{x,y\} : x \in \mathcal{X},\ y \in \mathcal{X} \text{ and } y \in G(x)\}$$

is closed in $\mathcal{X} \times \mathcal{X}$. In particular, if the topology on $\mathcal{X}$ is metrizable, then G is upper semi-continuous if and only if given any two sequence $\{x_n\}_1^\infty$ and $\{y_n\}_1^\infty$ such that

$$\begin{aligned} x_n &\in \mathcal{X} & \text{and} \quad & x = \lim_{n\to\infty} x_n\,, \\ y_n &\in G(x_n) & \text{and} \quad & y = \lim_{n\to\infty} y_n\,, \end{aligned}$$

then y is in $G(x)$.

PROOF OF THEOREM 2.1. The analysis preceding Theorem 2.2 shows that there exists an invariant subspace $\mathcal{M}$ for T in $\mathrm{Lat}_-(T,\Lambda)$ if and only if the map F_T defined in (2.4) has a fixed point X. In this case, $\mathcal{M} = \mathcal{M}_X$ is one such invariant subspace. Notice that for each X in $\mathcal{B}$, we have $F_T(X) = B_{\overline{T\mathcal{M}_X}}$. By Lemma 1.4, the set $F_T(X)$ is compact and convex. Lemma 2.3 below shows that F_T is upper semi-continuous. Hence by the Kakutani fixed point theorem, F_T has a fixed point.

To complete the proof it remains to show that $\mathcal{M}^\perp$ and $\mathcal{H}_+$ are isomorphic. Using the form of $\mathcal{M}_X$ in (1.10), it follows that

$$\mathcal{M}_X^\perp = \left\{ \begin{bmatrix} -\widehat{X}^* g \\ g \end{bmatrix} : g \in \mathcal{H}_+ \right\}. \tag{2.6}$$

This readily shows that $\mathcal{H}_+$ and $\mathcal{M}^\perp$ are isomorphic, and thus have the same dimension. To verify that $\mathcal{M}_X^\perp$ is indeed given by (2.6), let $f \oplus g$ be any vector in $\mathcal{H}_- \oplus \mathcal{H}_+$ which is orthogonal to $\mathcal{M}_X$. Then $f \oplus g$ is orthogonal to $h \oplus \widehat{X}h$ for all h in $\mathcal{H}_-$, or equivalently, $f + \widehat{X}^* g$ is orthogonal to $\mathcal{H}_-$. Since f and $\widehat{X}^* g$ are vectors in $\mathcal{H}_-$, we see that $f = -\widehat{X}^* g$. Therefore $\mathcal{M}_X^\perp$ is given by (2.6). This completes the proof.

The following lemma shows that F_T is upper semi-continuous .

Lemma 2.3. *Let Λ be a self-adjoint operator of the form (2.1), where the open interval $(0,\delta)$ is a gap in the spectrum of Λ. Let T on $\mathcal{H}$ be an operator of the form (2.2) such that $T^*\Lambda T \leq \Lambda$ and $\Lambda_-^{1/2} T_{12}$ is compact. Then F_T is upper semi-continuous from $\mathcal{X}$ into $\mathcal{P}(\mathcal{X})$.*

PROOF. We begin with some preparations. Let X and Y be contractions in $\mathcal{B}$. Set

$$\widehat{X} = \Lambda_+^{-1/2} X \Lambda_-^{1/2} \quad \text{and} \quad \widehat{Y} = \Lambda_+^{-1/2} Y \Lambda_-^{1/2}. \tag{2.7}$$

We claim that $\overline{T\mathcal{M}_X} \subset \mathcal{M}_Y$ if and only if

$$T_{21} + T_{22}\widehat{X} - \widehat{Y}T_{11} - \widehat{Y}T_{12}\widehat{X} = 0. \tag{2.8}$$

Indeed, assume that $T\mathcal{M}_X \subset \mathcal{M}_Y$. Then for each h in $\mathcal{H}_-$ there exists a g in $\mathcal{H}_-$ such that

$$\begin{bmatrix} T_{11} & T_{12} \\ T_{21} & T_{22} \end{bmatrix} \begin{bmatrix} h \\ \widehat{X}h \end{bmatrix} = \begin{bmatrix} g \\ \widehat{Y}g \end{bmatrix} \qquad (h \in \mathcal{H}_-). \tag{2.9}$$

This readily implies that $g = T_{11}h + T_{12}\widehat{X}h$, and thus

$$T_{21}h + T_{22}\widehat{X}h - \widehat{Y}(T_{11}h + T_{12}\widehat{X}h) = 0. \tag{2.10}$$

Since h in (2.10) is an arbitrary element of $\mathcal{H}_-$, we obtain (2.8). To prove the reverse implication, that is, (2.8) implies $\overline{T\mathcal{M}_X} \subset \mathcal{M}_Y$, assume that (2.8) holds. Then for each h in $\mathcal{H}_-$ the equality in (2.9) holds with $g = T_{11}h + T_{12}\widehat{X}h$. Thus for each h in $\mathcal{H}_-$ we can find a g in $\mathcal{H}_-$ such that (2.9) is fulfilled. This yields $\overline{T\mathcal{M}_X} \subset \mathcal{M}_Y$. Finally, recall that $\overline{T\mathcal{M}_X} \subset \mathcal{M}_Y$ if and only if $Y \in F_T(X)$. Therefore we have also shown that $Y \in F_T(X)$ if and only if (2.8) holds.

Now let us show that F_T is upper semi-continuous. Since the space $\mathcal{H}$ is assumed to be separable, the weak operator topology on $\mathcal{B}$ is metrizable. Hence we have to prove that if

$$\begin{aligned} X_n &\in \mathcal{B} \quad &&\text{and} \quad X = \text{weak} \lim_{n\to\infty} X_n, \\ Y_n &\in F_T(X_n) \quad &&\text{and} \quad Y = \text{weak} \lim_{n\to\infty} Y_n, \end{aligned} \tag{2.11}$$

then Y is in $F_T(X)$. Here convergence is in the weak operator topology. Since Y_n is in $F_T(X_n)$ for all integers $n \geq 1$, the results of the previous paragraph show that

$$T_{21} + T_{22}\Lambda_+^{-1/2}X_n\Lambda_-^{1/2} - \Lambda_+^{-1/2}Y_n\Lambda_-^{1/2}T_{11} - \Lambda_+^{-1/2}Y_n\Lambda_-^{1/2}T_{12}\Lambda_+^{-1/2}X_n\Lambda_-^{1/2} = 0. \tag{2.12}$$

As before, let $\widehat{X}$ and $\widehat{Y}$ be defined by (2.7). By consulting (2.11), it follows that

$$T_{21} + T_{22}\widehat{X} - \widehat{Y}T_{11} = \text{weak} \lim_{n\to\infty} \left(T_{21} + T_{22}\Lambda_+^{-1/2}X_n\Lambda_-^{1/2} - \Lambda_+^{-1/2}Y_n\Lambda_-^{1/2}T_{11}\right), \tag{2.13}$$

again with convergence in the weak operator topology. According to our assumption on T the operator $\Lambda_-^{1/2}T_{12}$ is compact. This implies that

$$\Lambda_-^{1/2}T_{12}\widehat{X} = \text{strong} \lim_{n\to\infty} \Lambda_-^{1/2}T_{12}\Lambda_+^{-1/2}X_n\Lambda_-^{1/2}$$

with convergence in the strong operator topology, that is, pointwise. Using this we obtain

$$\widehat{Y}T_{12}\widehat{X} = \text{weak} \lim_{n\to\infty} \Lambda_{+}^{-1/2}Y_n\Lambda_{-}^{1/2}T_{12}\Lambda_{+}^{-1/2}X_n\Lambda_{-}^{1/2}. \tag{2.14}$$

By combining (2.13) and (2.14) with (2.12), we see that for our X and Y the identity in (2.8) is fulfilled. Therefore Y is in $F_T(X)$, which completes the proof.

Remark 2.4. *As before, assume that T and Λ satisfy the hypothesis of Theorem 2.1. The proof of Lemma 2.3 shows that $\overline{T\mathcal{M}_X} \subset \mathcal{M}_Y$ for X and Y in $\mathcal{B}$ if and only if (2.8) holds. Therefore $\mathcal{M}_X$ is an invariant subspace for T if and only if $\widehat{X}$ is a solution of the operator Riccati equation*

$$T_{21} + T_{22}\widehat{X} - \widehat{X}T_{11} - \widehat{X}T_{12}\widehat{X} = 0. \tag{2.15}$$

Moreover, Theorem 2.1 shows that there exists a solution X in $\mathcal{B}$ to this Riccati equation. In other words, the Kakutani fixed point theorem can be used to show that the Riccati equation in (2.15) has a solution.

Theorem 2.1 has the following somewhat stronger version (cf., Theorem 2.1 in [5]) which is of independent interest and is not used in this paper.

Theorem 2.5. *Let Λ be a self-adjoint operator of the form (2.1), where the open interval $(0, \delta)$ is a gap in the spectrum of Λ. Let T on $\mathcal{H}$ be an operator of the form (2.2) such that $T^*\Lambda T \leq \Lambda$ and $\Lambda_{-}^{1/2}T_{12}$ is compact. Furthermore, let $\mathcal{G}$ be a subspace of $\mathcal{H}$ satisfying $\mathcal{G} \subset \mathcal{C}_\Lambda$ and $\overline{T\mathcal{G}} = \mathcal{G}$. Then there exists a subspace $\mathcal{M}$ in $Lat_{-}(T, \Lambda)$ such that $\mathcal{G} \subset \mathcal{M}$. In this case, $\mathcal{M} = \mathcal{M}_X$ for some X in $\mathcal{B}$. Finally, $\mathcal{M}^{\perp}$ and $\mathcal{H}_{+}$ are isomorphic, and thus have the same dimension.*

By applying Theorem 2.5 with $\mathcal{G}$ being the zero subspace, one obtains Theorem 2.1 as a special case of Theorem 2.5.

PROOF OF THEOREM 2.5. As before, let

$$\mathcal{B}_\mathcal{G} = \{X \in \mathcal{B} : \mathcal{G} \subset \mathcal{M}_X\}.$$

Lemma 1.4 shows that the set $\mathcal{B}_\mathcal{G}$ is convex and compact set in the weak operator topology. We claim that F_T maps $\mathcal{B}_\mathcal{G}$ into $\mathcal{P}(\mathcal{B}_\mathcal{G})$, the set of all subsets of $\mathcal{B}_\mathcal{G}$. To see this simply observe that if X is in $\mathcal{B}_\mathcal{G}$ and Y is in $F_T(X)$, then

$$\mathcal{G} = \overline{T\mathcal{G}} \subset \overline{T\mathcal{M}_X} \subset \mathcal{M}_Y.$$

This implies that Y is in $\mathcal{B}_\mathcal{G}$. Therefore F_T maps $\mathcal{B}_\mathcal{G}$ into $\mathcal{P}(\mathcal{B}_\mathcal{G})$. Now consider the set valued function G_T mapping $\mathcal{B}_\mathcal{G}$ into $\mathcal{P}(\mathcal{B}_\mathcal{G})$ defined by $G_T = F_T|\mathcal{B}_\mathcal{G}$. Notice that $G_T(X)$ is a compact convex set for all X in $\mathcal{B}_\mathcal{G}$, because $F_T(X)$ is a compact convex set. Finally, since F_T is upper semi-continuous and $G_T = F_T|\mathcal{B}_\mathcal{G}$, it follows that G_T is also upper semi-continuous. Indeed, let X_n converge to X and Y_n converge to Y in the weak operator topology as n tends to infinity, where X_n is in $\mathcal{B}_\mathcal{G}$ and Y_n is in $G_T(X_n)$ for all integers $n \geq 1$. Then the upper semi-continuity of F_T implies that Y is in $F_T(X)$. However, X is in $\mathcal{B}_\mathcal{G}$, and thus $G_T(X) = F_T(X)$. Hence Y is in $G_T(X)$. In other words, G_T is upper semi-continuous. By the Kakutani fixed point theorem, G_T has a fixed point. This readily implies that there exists a X in $\mathcal{B}_{\mathcal{G}}$ such that $X \in G_T(X)$. Thus

$$\mathcal{G} = \overline{T\mathcal{G}} \subset \overline{T\mathcal{M}_X} \subset \mathcal{M}_X .$$

Clearly, $\mathcal{M}_X$ is in $\mathrm{Lat}_-(T, \Lambda)$. As before, equation (2.6) shows that $\mathcal{M}_X^\perp$ and $\mathcal{H}_+$ are isomorphic. This completes the proof.

3. Self-adjoint Operator Pencils and Spectral Gaps

Throughout this section Γ and Ω are positive operators on a Hilbert space $\mathcal{H}$. Furthermore, we set $\Lambda = \Gamma - \epsilon\Omega$ and $\Lambda' = \Gamma - \epsilon'\Omega$, where ϵ and ϵ' are non-negative numbers. Notice that Λ and Λ' are self-adjoint operators. As before, we assume that Λ admits an operator matrix representation of the form (2.1) where Λ_- and Λ_+ are positive operators. Finally, P_- and P'_- are respectively the spectral projections of Λ and Λ' corresponding to $(-\infty, 0]$.

Lemma 3.1. *Let $\Lambda = \Gamma - \epsilon\Omega$ and $\Lambda' = \Gamma - \epsilon'\Omega$ where Γ and Ω are positive operators on $\mathcal{H}$. Moreover, assume that the following three conditions hold.*

(i) *Zero is in the spectrum of Λ and $(0, \delta) \subset \rho(\Lambda)$ for some $\delta > 0$.*
(ii) *There exists a subspace $\mathcal{M}$ of $\mathcal{H}$ satisfying $(\Lambda' f, f) \leq 0$ for all f in $\mathcal{M}$.*
(iii) *The subspace $\overline{P_- \mathcal{M}} = \mathcal{H}_- := P_-\mathcal{H}$.*

If $\epsilon > \epsilon'$, then zero is in the spectrum of $\Gamma + \Omega$. In other words, if $\Gamma + \Omega$ is invertible, then $\epsilon' \geq \epsilon$. In particular, if Ω is invertible, then $\epsilon' \geq \epsilon$.

PROOF. Since $\epsilon > \epsilon'$, we have $\Lambda \leq \Lambda'$. Hence $(\Lambda f, f) \leq (\Lambda' f, f) \leq 0$ for all f in $\mathcal{M}$. As before, let (2.1) be the spectral decomposition of Λ corresponding to the spectral gap $(0, \delta)$. According to Lemma 1.1, the space $P_- \mathcal{M}$ is closed and there exists an operator $\widehat{X}$ from $\mathcal{H}_-$ into $\mathcal{H}_+ := (I - P_-)\mathcal{H}$ such that

$$\mathcal{M} = \left\{ \begin{bmatrix} h \\ \widehat{X}h \end{bmatrix} : h \in P_- \mathcal{M} = \mathcal{H}_- \right\} . \tag{3.1}$$

Since zero is in the boundary of the spectrum of Λ_-, there exists a sequence of unit vectors $\{x_n\}_1^\infty$ in $\mathcal{H}_-$ such that Λx_n approaches zero as n tends to infinity. Now set $y_n = x_n \oplus \widehat{X}x_n$. According to (3.1) the vectors y_n belong to $\mathcal{M}$, and hence $(\Lambda' y_n, y_n) \leq 0$. On the other hand,

$$\begin{aligned} 0 \ \geq \ & (\Lambda' y_n, y_n) = (\Lambda y_n, y_n) + (\epsilon - \epsilon')(\Omega y_n, y_n) \\ \geq \ & (\Lambda y_n, y_n) \ = (\Lambda x_n, x_n) + (\Lambda \widehat{X} x_n, \widehat{X} x_n) \\ \geq \ & (\Lambda x_n, x_n) \to 0 \qquad (\text{as } n \to \infty) . \end{aligned}$$

So we conclude that both $(\Lambda y_n, y_n)$ and $(\Omega y_n, y_n)$ approach zero as n tends to infinity. Hence $((\Gamma + \Omega)y_n, y_n)$ approaches zero as n tends to infinity. Since $\Gamma + \Omega$ is positive, we see that $(\Gamma + \Omega)y_n$ approaches zero as n tends to infinity. Notice that $\|y_n\| \geq \|x_n\| \geq 1$. Finally, let $\{z_n\}_1^\infty$ be the unit vectors defined by $z_n = y_n / \|y_n\|$

for all integers $n \geq 1$. Then $(\Gamma + \Omega)z_n$ approaches zero as n tends to infinity. In other words, zero is in the spectrum of $\Gamma + \Omega$.

We have just proved that if $\epsilon > \epsilon'$, then zero is in the spectrum of $\Gamma + \Omega$. If $\Gamma + \Omega$ is invertible, then zero is not in spectrum of $\Gamma + \Omega$, and thus $\epsilon' \geq \epsilon$. This completes the proof.

Notice that if $\mathcal{M} = P_-'\mathcal{H}$, then $(\Lambda' f, f) \leq 0$ clearly holds for all f in $\mathcal{M}$. Indeed, since P_-' is the orthogonal projection of Λ' corresponding to $(-\infty, 0]$, the operator $\Lambda'|P_-'\mathcal{H}$ is non-positive. In this case, the conclusion of Lemma 3.1 holds if condition (iii) is replaced by

$$(iii)' \qquad \overline{P_- P_-' \mathcal{H}} = P_- \mathcal{H}\,. \tag{3.2}$$

Notice that if $\Gamma + \Omega$ is not invertible, and conditions (i), (ii) and (iii) in Lemma 3.1 are satisfied, then it is possible for $\epsilon > \epsilon'$ or $\epsilon' > \epsilon$. To see this simply choose Γ to be any singular positive operator on a finite dimensional space $\mathcal{H}$, and set $\Omega = 0$. Now let $\mathcal{M} = \mathcal{H}_-$. Then all the conditions in Lemma 3.1 are satisfied independent of the choice of ϵ or ϵ'.

By applying Lemma 3.1 with $\Omega = I$, we obtain the following result.

Proposition 3.2. *Let Γ be a positive operator on $\mathcal{H}$ and P be the spectral projection of Γ corresponding to $(-\infty, \epsilon]$. Assume that ϵ is in the spectrum of Γ and $(\epsilon, \epsilon + \delta)$ in contained in the resolvent set $\rho(\Gamma)$ for some $\delta > 0$. Then*

$$\epsilon = \min\{\|\Gamma|\mathcal{M}\| : \mathcal{M} \text{ is a subspace of } \mathcal{H} \text{ satisfying } \overline{P\mathcal{M}} = P\mathcal{H}\}\,. \tag{3.3}$$

PROOF. Since ϵ is in the spectrum of Γ and $(\Gamma f, f) \leq \epsilon(f, f)$ for all f in $P\mathcal{H}$, we have $\epsilon = \|\Gamma|P\mathcal{H}\|$. Now let $\mathcal{M}$ be any subspace of $\mathcal{H}$ such that $\overline{P\mathcal{M}} = P\mathcal{H}$ and set $\epsilon' = \|\Gamma|\mathcal{M}\|$. It suffices to prove that $\epsilon' \geq \epsilon$. To this end, let us apply Lemma 3.1 with $\Omega = I$. In this case, $\Lambda = \Gamma - \epsilon I$ and $\Lambda' = \Gamma - \epsilon' I$. Clearly, zero is in the spectrum of Λ. Notice that $\Gamma + I$ is invertible. Since $\epsilon' = \|\Gamma|\mathcal{M}\|$, we have

$$(\Gamma f, f) \leq \|\Gamma f\|\|f\| \leq \epsilon'(f, f) \qquad (f \in \mathcal{M})\,.$$

Hence $(\Lambda' f, f) \leq 0$ for all f in $\mathcal{M}$. So Λ and Λ' satisfy all the conditions in Lemma 3.1. By applying Lemma 3.1 with $\Omega = I$, we arrive at $\epsilon' \geq \epsilon$. This completes the proof.

Since Γ and Ω are both positive, $\ker(\Gamma + \Omega) = \ker \Gamma \bigcap \ker \Omega$. Clearly, $\ker \Gamma \bigcap \ker \Omega$ is an invariant subspace for $\Lambda := \Gamma - \epsilon\Omega$. Because Λ is a self-adjoint operator, this implies that the closure of the range of $\Gamma + \Omega$ is an invariant subspace for Λ. Finally, recall that a self-adjoint R operator is *Fredholm* if the range of R is closed and the kernel of R is finite dimensional. The following result provides some further insight into Lemma 3.1.

Proposition 3.3. *Assume that $\Lambda = \Gamma - \epsilon\,\Omega$ and $\Lambda' = \Gamma - \epsilon'\Omega$ satisfy conditions (i) to (iii) of Lemma 3.1, the operator $\Gamma + \Omega$ is Fredholm, and let $\mathcal{R}$ be the range of $\Gamma + \Omega$. If zero belongs to the spectrum of the operator $\Lambda|\,\mathcal{R}$ on $\mathcal{R}$, then $\epsilon' \geq \epsilon$.*

PROOF. Let $P_{\mathcal{R}}$ be the orthogonal projection onto the subspace $\mathcal{R}$. Clearly, $\ker\Gamma\bigcap\ker\Omega$ is an invariant subspace for both Γ and Ω. Moreover, $\mathcal{R} = \mathcal{H}\ominus\ker\Gamma\bigcap\ker\Omega$. Hence $P_{\mathcal{R}}$ commutes with Γ, Ω, Λ, Λ', P_- and P'_-. Let us use an additional subscript $\mathcal{R}$ to denote the restriction of these operators to the subspace $\mathcal{R}$. For example, $\Gamma_{\mathcal{R}} = \Gamma|\mathcal{R}$ on $\mathcal{R}$, the operator $\Omega_{\mathcal{R}} = \Omega|\mathcal{R}$ on $\mathcal{R}$, et cetera. Thus $\Lambda_{\mathcal{R}} = \Gamma_{\mathcal{R}} - \epsilon\,\Omega_{\mathcal{R}}$ and $\Lambda'_{\mathcal{R}} = \Gamma_{\mathcal{R}} - \epsilon'\Omega_{\mathcal{R}}$. Notice that Λ and Λ' admit the following operator matrix representations:

$$\Lambda = \begin{bmatrix} \Lambda_{\mathcal{R}} & 0 \\ 0 & 0 \end{bmatrix} \text{ on } \begin{bmatrix} \mathcal{R} \\ \mathcal{R}^{\perp} \end{bmatrix} \quad\text{and}\quad \Lambda' = \begin{bmatrix} \Lambda'_{\mathcal{R}} & 0 \\ 0 & 0 \end{bmatrix} \text{ on } \begin{bmatrix} \mathcal{R} \\ \mathcal{R}^{\perp} \end{bmatrix}.$$

Because $\Gamma+\Omega$ is Fredholm, $\Gamma_{\mathcal{R}}+\Omega_{\mathcal{R}}$ is invertible. We claim that conditions (i) to (iii) of Lemma 3.1 are satisfied for $\Lambda_{\mathcal{R}}$ and $\Lambda'_{\mathcal{R}}$.

Clearly, $(0,\delta)\subset\rho(\Lambda_{\mathcal{R}})$. By assumption, zero is in the spectrum of $\Lambda_{\mathcal{R}}$. So (i) is fulfilled. Set $\mathcal{M}_{\mathcal{R}} = P_{\mathcal{R}}\mathcal{M}$. Condition (ii) of Lemma 3.1 implies that $(\Lambda'_{\mathcal{R}}g,g)\leq(g,g)$ for all g in $\mathcal{M}_{\mathcal{R}}$. Indeed, if $g = P_{\mathcal{R}}f$ for some f in $\mathcal{M}$, then

$$(\Lambda'_{\mathcal{R}}g,g) = (\Lambda' P_{\mathcal{R}}f, P_{\mathcal{R}}f) = (\Lambda' f, f)\leq 0\,.$$

Next let us show that $P_{\mathcal{R},-}\mathcal{M}_{\mathcal{R}}$ is dense in $P_{\mathcal{R},-}\mathcal{R}$, where $P_{\mathcal{R},-}$ is the spectral projection of $\Lambda_{\mathcal{R}}$ corresponding to $(-\infty,0]$. To this end, let g be any vector in $P_{\mathcal{R},-}\mathcal{R}$. Then g is in $P_-\mathcal{H}$. According to (iii) of Lemma 3.1, there exists a sequence of vectors $\{m_n\}_1^\infty$ in $\mathcal{M}$ such that P_-m_m converges to g as n tends to infinity. However, this readily implies that

$$\lim_{n\to\infty} P_{\mathcal{R},-}P_{\mathcal{R}}m_n = \lim_{n\to\infty} P_{\mathcal{R}}P_{\mathcal{R},-}P_-m_n = P_{\mathcal{R}}g = g\,.$$

Since $\mathcal{M}_{\mathcal{R}} = P_{\mathcal{R}}\mathcal{M}$, we have $\overline{P_{\mathcal{R},-}\mathcal{M}_{\mathcal{R}}} = P_{\mathcal{R},-}\mathcal{R}$. Hence condition (iii) is satisfied for $\Lambda_{\mathcal{R}}$ and $\Lambda'_{\mathcal{R}}$. Finally, because $\Gamma_{\mathcal{R}}+\Omega_{\mathcal{R}}$ is invertible, Lemma 3.1 shows that $\epsilon'\geq\epsilon$. This completes the proof.

If $\Gamma+\Omega$ is not Fredholm, then Proposition 3.3 does not necessarily hold. To see this let $\mathcal{H} = \ell^2_+$ be the Hilbert space formed by the set of all square summable unilateral infinite sequences of complex numbers. Let Γ and Ω be the diagonal operators on ℓ^2_+ defined by

$$\Gamma = \operatorname{diag}(\,1 \quad 1/4 \quad 1/6 \quad 1/8 \quad \cdots\,) \text{ and } \Omega = \operatorname{diag}(\,0 \quad 1/2 \quad 1/3 \quad 1/4 \quad \cdots\,)\,.$$

If $\epsilon = 1$ and $\epsilon' = 1/2$, then

$$\begin{aligned} \Lambda &= \Gamma-\Omega &&= \operatorname{diag}(\,1 \quad -1/4 \quad -1/6 \quad -1/8 \quad \cdots\,) \\ \Lambda' &= \Gamma-.5\Omega &&= \operatorname{diag}(\,1 \quad 0 \quad 0 \quad 0 \quad \cdots\,)\,. \end{aligned}$$

Clearly, zero is in the spectrum of Λ and $(0,1)\subset\rho(\Lambda)$. In this case, $\mathcal{H}_-$ is the subspace consisting of the set of all vectors in ℓ^2_+ whose first component is zero. To be precise, $\mathcal{H}_- = \{(f_j)_0^\infty\in\ell^2_+ : f_0 = 0\}$. Moreover, $P_- = P'_-$. So if $\mathcal{M} = \mathcal{H}_-$, then all the conditions in Lemma 3.1 are satisfied with $\epsilon' = 1/2 < 1 = \epsilon$. Notice that $\Gamma+\Omega$ is compact and thus $\Gamma+\Omega$ is not Fredholm.

4. Operator Pencils and Weighted Commutant Lifting

Recall that U' on $\mathcal{K}'$ is an isometric lifting of a contraction T' on $\mathcal{H}'$ if U' is an isometry on $\mathcal{K}'$, the space $\mathcal{H}' \subset \mathcal{K}'$ and $P'U' = T'P'$. Here P' is the orthogonal projection from $\mathcal{K}'$ onto $\mathcal{H}'$. Notice that $P'U' = T'P'$ if and only if $\mathcal{H}'$ is an invariant subspace for U'^* and $U'^*|\mathcal{H}' = T'^*$. We say that $\{T, A, T', U', \Omega\}$ is a *weighted lifting data set* if T is an operator on $\mathcal{H}$ and U' on $\mathcal{K}'$ is an isometric lifting of the contraction T' on $\mathcal{H}'$ and A is an operator intertwining T with T', that is, $T'A = AT$. Finally, Ω is a positive operator on $\mathcal{H}$ and T is Ω-expansive, that is, $\Omega \le T^*\Omega T$. This notation is used in the following result taken from [11].

Theorem 4.1. *Let $\{T, A, T', U', \Omega\}$ be a weighted lifting data set where $A^*A \le \Omega$. Then there exists an operator B mapping $\mathcal{H}$ into $\mathcal{K}'$ satisfying*

$$U'B = BT, \qquad P'B = A \qquad \text{and} \qquad B^*B \le \Omega\,. \tag{4.1}$$

If $\Omega = I$, then the weighted commutant lifting theorem, reduces to the commutant lifting theorem in Treil-Volberg [31]. If T is an isometry and $\Omega = \gamma^2 I$, then the weighted commutant lifting theorem reduces to the following form of the classical commutant lifting theorem; see [29, 30, 13, 14].

Theorem 4.2. *Let A be an operator from $\mathcal{H}$ into $\mathcal{H}'$ satisfying $\|A\| \le \gamma$ and $T'A = AT$ where T is an isometry on $\mathcal{H}$ and T' is a contraction on $\mathcal{H}'$. Let U' be an isometric lifting of T'. Then there exists an operator B mapping $\mathcal{H}$ into $\mathcal{K}'$ satisfying*

$$U'B = BT, \qquad P'B = A \qquad \text{and} \qquad \|B\| \le \gamma\,. \tag{4.2}$$

Let $\{T, A, T', U', \Omega\}$ be a weighted lifting data set. Throughout the rest of this section Γ is the positive operator on $\mathcal{H}$ defined by $\Gamma = A^*A$. In this case, $T^*\Gamma T \le \Gamma$. To see this notice that the intertwining relation $T'A = AT$ implies that

$$T^*\Gamma T = T^*A^*AT = A^*T'^*T'A \le A^*A = \Gamma\,.$$

Hence $T^*\Gamma T \le \Gamma$ and T is Γ-contractive.

Now fix an $\epsilon \ge 0$ and set $\Lambda = \Gamma - \epsilon\Omega$. Using $T^*\Gamma T \le \Gamma$ along with $\Omega \le T^*\Omega T$, it follows that $T^*\Lambda T \le \Lambda$. In the sequel we always assume that there is a gap in the spectrum of Λ, that is, $(0, \delta) \subset \rho(\Lambda)$ for some $\delta > 0$. Thus Λ admits an operator matrix representation of the form

$$\Lambda = \begin{bmatrix} -\Lambda_- & 0 \\ 0 & \Lambda_+ \end{bmatrix} \text{ on } \begin{bmatrix} \mathcal{H}_- \\ \mathcal{H}_+ \end{bmatrix} \tag{4.3}$$

where Λ_- and Λ_+ are positive operators, and Λ_+ is invertible. As before, P_- is the orthogonal projection onto $\mathcal{H}_-$ and P_+ is the orthogonal projection onto $\mathcal{H}_+$. In this setting T admits an operator matrix representation of the form

$$T = \begin{bmatrix} T_{11} & T_{12} \\ T_{21} & T_{22} \end{bmatrix} \text{ on } \begin{bmatrix} \mathcal{H}_- \\ \mathcal{H}_+ \end{bmatrix}. \tag{4.4}$$

We also assume that T satisfies the compactness condition of the Krein-Langer-Iokhvidov theorem, that is, the operator

$$\Lambda_-^{1/2}T_{12} = \Lambda_-^{1/2}P_-T|\mathcal{H}_+ : \mathcal{H}_+ \to \mathcal{H}_- \text{ is compact}. \tag{4.5}$$

According to Theorem 2.1 there exists a subspace $\mathcal{M}$ in $\mathrm{Lat}_-(T,\Lambda)$, that is, $\mathcal{M}$ is an invariant subspace for T satisfying $\mathcal{M} \subset \mathcal{C}_\Lambda$ and $P_-\mathcal{M} = \mathcal{H}_-$. In the following theorem we choose and fix such a subspace.

Theorem 4.3. *Let $\{T, A, T', U', \Omega\}$ be a weighted lifting data set. Fix $\epsilon \geq 0$, and set $\Lambda = A^*A - \epsilon\Omega$. Assume that $(0,\delta) \subset \rho(\Lambda)$ and that the operator $\Lambda_-^{1/2}T_{12}$ in (4.5) is compact. Then given $\mathcal{M}$ in $Lat_-(T,\Lambda)$, there exists an operator B mapping $\mathcal{M}$ into $\mathcal{K}'$ satisfying*

$$U'B = BT|\mathcal{M}, \qquad P'B = A|\mathcal{M} \qquad \text{and} \qquad B^*B \leq \epsilon P_\mathcal{M}\Omega|\mathcal{M}. \tag{4.6}$$

In this case, $\mathcal{M}^\perp$ and $\mathcal{H}_+$ are isomorphic, and hence have the same dimension.

PROOF. Since $\Lambda = A^*A - \epsilon\Omega$ and $\mathcal{M} \subset \mathcal{C}_\Lambda$, we have

$$P_\mathcal{M}A^*A|\mathcal{M} \leq \epsilon P_\mathcal{M}\Omega|\mathcal{M}. \tag{4.7}$$

Obviously, $T'A|\mathcal{M} = AT|\mathcal{M}$. Hence $\{T|\mathcal{M}, A|\mathcal{M}, T', U', \epsilon P_\mathcal{M}\Omega|\mathcal{M}\}$ is a weighted lifting data set satisfying (4.7). So by applying the weighted commutant lifting Theorem 4.1, we see that there exists an operator B from $\mathcal{M}$ into $\mathcal{K}'$ satisfying (4.6). Finally, Theorem 2.1 guarantees that $\mathcal{M}^\perp$ and $\mathcal{H}_+$ are isomorphic. This completes the proof.

As before, let $\{T, A, T', U', \Omega\}$ be a weighted lifting data set. Moreover, assume that $\Lambda_-^{1/2}T_{12}$ is compact and $(0,\delta) \subset \rho(\Lambda)$ where $\Lambda = A^*A - \epsilon\Omega$ for some $\epsilon \geq 0$. If $\mathcal{M}$ is an invariant subspace for T, then $\mathcal{I}(\mathcal{M}, T, A, T', U')$ denotes the set of all operators B mapping $\mathcal{M}$ into $\mathcal{K}'$ satisfying $U'B = BT|\mathcal{M}$ and $P'B = A|\mathcal{M}$. In fact, the commutant lifting theorem guarantees that there exists an operator B in $\mathcal{I}(\mathcal{M}, T, A, T', U')$ such that $\|B\| = \|A|\,\mathcal{M}\|$. Let $\mathrm{Lat}_-(T)$ denote the set of all invariant subspaces for T satisfying $P_-\mathcal{M} = \mathcal{H}_-$. For each $\mathcal{M}$ in $\mathrm{Lat}_-(T)$, let $\gamma_\mathcal{M}$ be the scalar defined by

$$\gamma_\mathcal{M} = \inf\{\eta \geq 0 : B \in \mathcal{I}(\mathcal{M}, T, A, T', U') \text{ and } B^*B \leq \eta P_\mathcal{M}\Omega|\mathcal{M}\}. \tag{4.8}$$

Notice that $\gamma_\mathcal{M}$ can be infinite. For example choose any appropriate nonzero finite rank operator A, with $\mathcal{M} = \mathcal{H}$ and $\Omega = 0$. Then B is nonzero and $\gamma_\mathcal{M}$ is infinite. If $\gamma_\mathcal{M}$ is finite, then the infimum in (4.8) is attained due to weak compactness.

Theorem 4.4. *Let $\{T, A, T', U', \Omega\}$ be a weighted lifting data set. Fix $\epsilon \geq 0$, and set $\Lambda = A^*A - \epsilon\Omega$. Assume that $(0,\delta) \subset \rho(\Lambda)$ for some $\delta > 0$, and that the operator $\Lambda_-^{1/2}P_-T|\mathcal{H}_+$ in (4.5) is compact. Finally, assume that $\Gamma + \Omega$ is Fredholm and zero is in the spectrum of $\Lambda|(\Gamma + \Omega)\mathcal{H}$. Then*

$$\epsilon = \inf\{\gamma_\mathcal{M} : \mathcal{M} \in \mathrm{Lat}_-(T)\}. \tag{4.9}$$

Moreover, the infimum is attained. In particular, if $\Gamma + \Omega$ is invertible and zero is in the spectrum of Λ, then (4.9) holds.

PROOF. According to Theorem 2.1, there exists a subspace $\mathcal{M}$ in $\mathrm{Lat}_-(T,\Lambda)$. For this subspace Theorem 4.3 shows that $\gamma_{\mathcal{M}} \leq \epsilon$. Since $\mathrm{Lat}_-(T,\Lambda)$ is contained in $\mathrm{Lat}_-(T)$, the infimum in (4.9) is finite. So to complete the proof it remains to show that

$$\epsilon \leq \gamma_{\mathcal{M}} \qquad (\text{for all } \mathcal{M} \in \mathrm{Lat}_-(T)) . \tag{4.10}$$

Assume that (4.10) does not hold, that is, for some $\mathcal{M}$ in $\mathrm{Lat}_-(T)$ we have $\gamma_{\mathcal{M}} < \epsilon$. Then there exists an operator B in $\mathcal{I}(\mathcal{M},T,A,T',U')$ and $0 \leq \gamma_{\mathcal{M}} < \eta < \epsilon$ such that $B^*B \leq \eta P_{\mathcal{M}}\Omega|\mathcal{M}$. Now set $\epsilon' = \eta$ and $\Lambda' = A^*A - \epsilon'\Omega$. Then for any f in $\mathcal{M}$, the relation $P'B = A|\,\mathcal{M}$ yields

$$(\Lambda' f, f) = ((A^*A - \epsilon'\Omega)f, f) \leq ((A^*A - B^*B)f, f) \leq 0 \qquad (f \in \mathcal{M}) .$$

Thus Λ and Λ' satisfy conditions (i) to (iii) of Lemma 3.1. By applying Proposition 3.3, we see that $\epsilon' \geq \epsilon$. This leads to a contradiction, and thus (4.10) holds. Theorem 4.3 implies that there exists an invariant subspace $\mathcal{M}$ in $\mathrm{Lat}_-(T,\Lambda)$ such that $\epsilon = \gamma_{\mathcal{M}}$. In other words, the infimum in (4.9) is attained. This completes the proof.

5. Operator approximations

Let A be a compact operator mapping $\mathcal{H}$ into $\mathcal{H}'$, and $\sigma_0 \geq \sigma_1 \geq \sigma_2 \geq \cdots$ the singular values for A. Then it is well known (see, for example [16], Theorem VI.1.5) that

$$\sigma_k = \min\{\|A - A_k\| : A_k \in \mathcal{L}(\mathcal{H},\mathcal{H}') \text{ and } \mathrm{rank}(A_k) \leq k\} . \tag{5.1}$$

Now assume that $T'A = AT$, where T is an isometry on $\mathcal{H}$ and T' is a co-isometry on $\mathcal{H}'$. Consider the following operator approximation problem:

$$\beta_k = \inf\{\|A - A_k\| : A_k \in \mathcal{L}(\mathcal{H},\mathcal{H}') ,\ T'A_k = A_kT \text{ and } \mathrm{rank}(A_k) \leq k\} . \tag{5.2}$$

Clearly, $\sigma_k \leq \beta_k$. The following result due to Adamjan, Arov and Krein [1, 2] shows that the infimum is achieved and $\sigma_k = \beta_k$; see also [31].

Theorem 5.1. *Let A be a compact operator mapping $\mathcal{H}$ into $\mathcal{H}'$ with singular values $\sigma_0 \geq \sigma_1 \geq \sigma_2 \geq \cdots$. Moreover, assume that $T'A = AT$, where T is an isometry on $\mathcal{H}$ and T' is a co-isometry on $\mathcal{H}'$. Then there exists an operator A_k in $\mathcal{L}(\mathcal{H},\mathcal{H}')$ with rank at most k such that $T'A_k = A_kT$ and*

$$\sigma_k = \|A - A_k\| . \tag{5.3}$$

In particular, this operator A_k achieves the infimum in (5.2).

PROOF. Let Λ be the operator on $\mathcal{H}$ defined by $\Lambda = A^*A - \sigma_k^2 I$. As before, let $\mathcal{H}_-$ be the subspace of $\mathcal{H}$ corresponding to the spectral projection of Λ onto $(-\infty, 0]$. In this setting, the dimension of $\mathcal{H}_+ = \mathcal{H} \ominus \mathcal{H}_-$ equals m the number of singular values strictly greater than σ_k, and thus $\dim \mathcal{H}_+ \leq k$. (The dimension of a subspace is denoted by dim.) To see this let m be the largest integer such that $\sigma_{m-1} > \sigma_k$ and $\sigma_m = \sigma_k$. Then $\mathcal{H}_+$ is the m dimensional subspace spanned by the orthogonal vectors $\{x_j\}_0^{m-1}$ where $\{x_j, y_j\}$ is the Schmidt pair corresponding to σ_j, that is, $Ax_j = \sigma_j y_j$ and $A^* y_j = \sigma_j x_j$. In particular, $\dim \mathcal{H}_+ \leq k$.

Recall that U' on $\mathcal{K}'$ is a *minimal* isometric lifting of T' if U' is an isometric lifting of T' and

$$\mathcal{K}' = \bigvee_{n=0}^{\infty} U'^n \mathcal{H}'. \tag{5.4}$$

Moreover, any contraction admits a minimal isometric lifting; see [30, 13]. So let U' on $\mathcal{K}'$ be a minimal isometric lifting of T'. Because T' is a co-isometry, U' is a unitary operator; see Remark VI.3.4 in [13] or Lemma XXVII.4.1 in [17]. Notice that $\{T, A, T', U', I\}$ is a weighted lifting data set. Since T is an isometry, we have $T^*\Lambda T \leq \Lambda$. According to Theorem 2.1 with $\epsilon = \sigma_k^2$, there exists a subspace $\mathcal{M}$ in $\text{Lat}_-(T, \Lambda)$. Using this subspace in Theorem 4.3 with $\Omega = I$, there exists an operator B mapping $\mathcal{M}$ into $\mathcal{K}'$ satisfying

$$U'B = BT|\mathcal{M}, \qquad P'B = A|\mathcal{M} \qquad \text{and} \qquad \|B\| \leq \sigma_k . \tag{5.5}$$

Here P' is the orthogonal projection of $\mathcal{K}'$ onto $\mathcal{H}'$.

Let U on $\mathcal{K}$ be any unitary extension of T, that is, U is unitary and $U|\mathcal{H} = T$. To construct a unitary extension, let U^* be any minimal isometric lifting of T^*. Since T^* is a co-isometry, this U is a unitary extension of T. Using $U'B = BT|\mathcal{M}$, it follows that there exists an operator B_k from $\mathcal{K}$ into $\mathcal{K}'$ such that $U'B_k = B_k U$, the operator $B_k|\mathcal{M} = B$ and $\|B_k\| = \|B\|$. In fact, this follows from the commutant lifting Theorem 4.2. To see this let $T_{\mathcal{M}}$ be the operator on $\mathcal{M}$ defined by $T_{\mathcal{M}} = T|\mathcal{M}$. Using $U|\mathcal{M} = T_{\mathcal{M}}$, we see that U^* is an isometric lifting of $T_{\mathcal{M}}^*$. Since U' is unitary, U'^* is an isometric lifting of U'^*. By applying the commutant lifting theorem to $T_{\mathcal{M}}^* B^* = B^* U'^*$, there exists an operator C from $\mathcal{K}'$ into $\mathcal{K}$ satisfying $U^*C = CU'^*$, with $\|C\| = \|B\|$ and $P_{\mathcal{M}} C = B^*$. (As usual, $P_{\mathcal{M}}$ is the orthogonal projection onto $\mathcal{M}$.) In particular, $C^*|\mathcal{M} = B$. By setting $B_k = C^*$ and consulting (5.5), we obtain

$$U'B_k = B_k U, \qquad P'B_k|\mathcal{M} = A|\mathcal{M} \qquad \text{and} \qquad \|B_k\| \leq \sigma_k . \tag{5.6}$$

Now let A_k be the operator from $\mathcal{H}$ into $\mathcal{H}'$ defined by $A_k = A - P'B_k|\mathcal{H}$. Notice that

$$\|A - A_k\| = \|P'B_k|\mathcal{H}\| \leq \|B_k\| \leq \sigma_k . \tag{5.7}$$

Furthermore, the second equation in (5.6) shows that $A_k \mathcal{M} = 0$. So the rank of A_k is less than or equal to the dimension of $\mathcal{M}^\perp = \mathcal{H} \ominus \mathcal{M}$. According to Theorem 2.1, the subspaces $\mathcal{M}^\perp$ and $\mathcal{H}_+$ have the same dimension. Since $\dim \mathcal{H}_+ \leq k$, we have $\operatorname{rank} A_k \leq k$. Finally, using (5.6) with h in $\mathcal{H}$, we obtain

$$\begin{aligned} T'A_k h &= T'Ah - T'P'B_k h = ATh - P'U'B_k h \\ &= ATh - P'B_k Uh = ATh - P'B_k Th = A_k Th . \end{aligned}$$

Hence $T'A_k = A_k T$. This completes the proof.

Now let us show how one can use the previous analysis to derive the explicit formula for A_k in Adamjan-Arov-Krein [1] for certain Hankel operators. To this end, let L^∞ denote the set of all Lebesgue measurable, essentially bounded functions on $[0, 2\pi]$. As usual, L^2 is the set of all Lebesgue measurable, square integrable functions on $[0, 2\pi]$. Furthermore, if g is a function in L^∞, then M_g is the multiplication operator on L^2 defined by $(M_g h)(\omega) = g(\omega)h(\omega)$ where h is a function in L^2. Throughout H^2 is the subspace of L^2 consisting of all functions whose Fourier coefficients relative to $e^{\imath n\omega}$ are zero for all integers $n < 0$. Finally, $K^2 = L^2 \ominus H^2$. Let V be the unilateral shift on L^2, that is, $(Vh)(\omega) = e^{\imath\omega}h(\omega)$ for all h in L^2. Let S be the unilateral shift on H^2, that is, $(Sh)(\omega) = e^{\imath\omega}h(\omega)$ for all h in H^2. Clearly, $S = V|L^2$. Finally, let S_1 be the backward shift on K^2, that is, $(S_1 h)(\omega) = (P_{K^2}e^{\imath\omega}h)(\omega)$. Here P_{K^2} is the orthogonal projection onto K^2. Notice that $S_1^* = V^*|K^2$. Recall that an operator A from H^2 into K^2 is *Hankel* if A intertwines S with S_1. It is well known that A is a Hankel operator if and only if there exists a function g in L^∞ such that $A = P_{K^2}M_g|H^2$; see for example [13]. Furthermore, A is a finite rank Hankel operator if and only if there exists a rational function g in L^∞ such that $A = P_{K^2}M_g|H^2$. A Hankel operator A is a compact if and only if there exists a continuous function g in L^∞ such that $A = P_{K^2}M_g|H^2$. Finally, recall that if x is a nonzero function in H^2, then $x(e^{\imath\omega}) \neq 0$ a.e. on $[0, 2\pi]$; see [19, 20]. The following result due to Adamjan, Arov and Krein [1] presents an explicit formula for certain A_k in Theorem 5.1.

Theorem 5.2. *Let $\sigma_0 \geq \sigma_1 \geq \sigma_2 \geq \cdots$ be the singular values for a compact Hankel operator A mapping H^2 into K^2. Let x be an eigenvector for A^*A with respect to the eigenvalue σ_k^2, that is, $A^*Ax = \sigma_k^2 x$ where x is a nonzero vector in H^2. Finally, let g be the function defined by*

$$g = \frac{Ax}{x}. \tag{5.8}$$

Then g is a function in L^∞ satisfying $|g(\omega)| = \sigma_k$ a.e. Moreover, $A_k = A - P_{K^2}M_g|H^2$ is a Hankel operator of rank at most k satisfying $\sigma_k = \|A - A_k\|$. In particular, for the case considered here A_k achieves the infimum in (5.2).

PROOF. Let us use the notation and operators established in the proof of Theorem 5.1. In this setting $\mathcal{H} = H^2$ and $\mathcal{H}' = K^2$ while $\mathcal{K}' = \mathcal{K} = L^2$. The operators T, T', U' and U are now given by $T = S$ and $T' = S_1$ while $U' = U = V$. Moreover, in this case, B_k is now an operator on L^2 satisfying (5.6). Using $VB_k = B_kV$, it follows that there exists a function g in L^∞ such that $B_k = M_g$. According to Corollary 1.2, the eigenvector x is in $\mathcal{M}$. Using $\|Ax\| = \sigma_k\|x\|$ in the last two equations in (5.6), we obtain

$$\sigma_k\|x\| = \|Ax\| = \|P'B_kx\| \leq \|B_kx\| \leq \sigma_k\|x\|.$$

So $\|P'B_kx\| = \|B_kx\|$. Hence $Ax = P'B_kx = B_kx = M_gx$. This readily implies that $g = Ax/x$ is given by (5.8). Since g is independent of x, the function g and the operator B_k are both uniquely determined. According to the proof of Theorem

5.1, the operator $A_k = A - P_{K^2} M_g | H^2$ is a Hankel operator of rank at most k satisfying $\sigma_k = \|A - A_k\|$.

To complete the proof it remains to show that $|g(\omega)| = \sigma_k$ a.e. The proof is taken from Adamjan-Arov-Krein [1] and given for completeness. Let $y = Ax$. Then using $S_1^n A = AS^n$ for any integer $n \geq 0$, we have

$$\begin{aligned}\int_0^{2\pi} e^{\imath n\omega} |y(\omega)|^2 \, d\omega &= (V^n Ax, Ax) = (S_1^n Ax, Ax) = (AS^n x, Ax) \\ &= (S^n x, A^* Ax) = \sigma_k^2 (S^n x, x) = \sigma_k^2 \int_0^{2\pi} e^{\imath n\omega} |x(\omega)|^2 \, d\omega \,.\end{aligned}$$

By taking the complex conjugate, we see that the previous equation holds for all integers n. Hence

$$\int_0^{2\pi} e^{\imath n\omega} \left(|y(\omega)|^2 - \sigma_k^2 |x(\omega)|^2 \right) d\omega = 0 \qquad (n = 0, \pm 1, \pm 2, \cdots). \tag{5.9}$$

So all the Fourier coefficients of $|y|^2 - \sigma_k^2 |x|^2$ are zero, and thus $|Ax| = |y| = \sigma_k |x|$ a.e. Therefore $|g| = |Ax|/|x| = \sigma_k$ a.e. This completes the proof.

6. Almost Nevanlinna-Pick interpolation

In this section we present an "almost" version of the standard tangential Nevanlinna-Pick interpolation problem from [14]. To motivate our almost interpolation problem, we begin with the "almost" version of the classical Nevanlinna-Pick interpolation problem presented in [3].

First recall that φ is a finite Blaschke product if φ is a rational function of the form

$$\varphi(\lambda) = \zeta \prod_{j=1}^{m} \frac{\lambda - \beta_j}{1 - \overline{\beta}_j \lambda}, \tag{6.1}$$

where $\{\beta_j\}_1^m$ are complex numbers in the open unit disc and ζ is a complex number of modulus one. The number of zeros including their multiplicity is called the *order* of the Blaschke product. Let $\{\alpha_j\}_1^n$ be a finite set of distinct points in the open unit disc and $\{\gamma_j\}_1^n$ a set of complex numbers. Finally, let $\gamma > 0$ be a specified tolerance and k a positive integer. Then an "almost" version of the classical Nevanlinna-Pick interpolation problem (see [3]) is to find a function f in H^∞ and a finite Blaschke product φ satisfying the following three conditions

$$\begin{aligned} &(i) \qquad f(\alpha_j) = \gamma_j \varphi(\alpha_j) \qquad (\text{for } j = 1, 2, \cdots, n); \\ &(ii) \qquad \|f\|_\infty \leq \gamma; \\ &(iii) \qquad \text{the order of } \varphi \text{ is at most } k. \end{aligned} \tag{6.2}$$

Here we will provide necessary and sufficient conditions for this problem to have a solution.

The almost Nevanlinna-Pick interpolation problem in (6.2) is a special case of a more general almost tangential Nevanlinna-Pick interpolation problem. To introduce the latter interpolation problem, we need the notion of point evaluation for operator valued functions; see [14], Section I.2. To this end, let Z be a stable operator on a finite dimensional space $\mathcal{Z}$. By stable we mean that all the eigenvalues of Z are contained in the open unit disc. Throughout this section for simplicity of presentation we will assume that $\mathcal{Z}$ is finite dimensional. Let E be an operator mapping a finite dimensional space $\mathcal{Y}$ into $\mathcal{Z}$. Finally, let F be a function in $H^\infty(\mathcal{U},\mathcal{Y})$ where $\mathcal{U}$ is also a finite dimensional space. (As usual, $H^\infty(\mathcal{U},\mathcal{Y})$ is the Banach space of all uniformly bounded analytic functions in the open unit disc with values in $\mathcal{L}(\mathcal{U},\mathcal{Y})$.) Then $(EF)(Z)_{left}$ is the operator mapping $\mathcal{U}$ into $\mathcal{Z}$ defined by

$$(EF)(Z)_{left} = \sum_{n=0}^{\infty} Z^n E F_n \,, \tag{6.3}$$

where $\sum_0^\infty \lambda^n F_n$ is the power series expansion for $F(\lambda)$ in the open unit disc. Notice that if $\mathcal{U} = \mathcal{Y} = \mathbb{C}$, then $(EF)(Z)_{left} = F(Z)E$.

Let Φ be an inner function in $H^\infty(\mathcal{U},\mathcal{U})$, that is, Φ is analytic in the open unit disc and $\Phi(e^{\imath\omega})$ is a.e. an isometry. Let T_Φ be the Toeplitz operator on $\ell_+^2(\mathcal{U})$ defined by

$$T_\Phi = \begin{bmatrix} \Phi_0 & 0 & 0 & \cdots \\ \Phi_1 & \Phi_0 & 0 & \cdots \\ \Phi_2 & \Phi_1 & \Phi_0 & \cdots \\ \vdots & \vdots & \vdots & \ddots \end{bmatrix} \quad \text{on } \ell_+^2(\mathcal{U})\,, \tag{6.4}$$

where $\Phi(\lambda) = \sum_0^\infty \lambda^n \Phi_n$ is the Taylor series expansion for Φ in the open unit disc. Because Φ is inner, it follows that T_Φ is an isometry. The degree of Φ, denoted by $\deg \Phi$, is the co-dimension of the range of T_Φ. In other words, $\deg \Phi = \dim \ker T_\Phi^*$. It is well known that the degree of Φ is finite if and only if Φ is a rational inner function. Moreover, if $\mathcal{U} = \mathbb{C}$ and φ is a finite Blaschke product, then the order of φ equals the degree of φ. So if φ is given by the finite Blaschke product in (6.1), then $m = \deg \varphi = \dim \ker T_\varphi^*$.

The data for our almost tangential Nevanlinna-Pick interpolation problem is $\{Z, E, \widetilde{E}, \gamma, k\}$ where Z on $\mathcal{Z}$ is a stable operator, E is an operator mapping $\mathcal{Y}$ into $\mathcal{Z}$ while $\widetilde{E}$ is an operator mapping $\mathcal{U}$ into $\mathcal{Z}$, the tolerance γ is a strictly positive number and k is a positive integer. Using this data our almost tangential Nevanlinna-Pick interpolation problem is to find a function F in $H^\infty(\mathcal{U},\mathcal{Y})$ and an inner function Φ in $H^\infty(\mathcal{U},\mathcal{U})$ satisfying the following three conditions

$$\begin{aligned} &(i) && (EF)(Z)_{left} = (\widetilde{E}\Phi)(Z)_{left} \\ &(ii) && \|F\|_\infty \le \gamma; \\ &(iii) && \deg \Phi \le k\,. \end{aligned} \tag{6.5}$$

If these three conditions hold, then the pair $\{F, \Phi\}$ is called an interpolant or a solution for the data $\{Z, E, \widetilde{E}, \gamma, k\}$.

To demonstrate why this interpolation problem contains the previous almost Nevanlinna-Pick interpolation problem, let $\{\alpha_j\}_1^n$ be a finite set of distinct points in the open unit disc and $\{\gamma_j\}_1^n$ a set of complex numbers. Let Z be the diagonal matrix on $\mathbb{C}^n$ defined by $Z = \mathrm{diag}[\{\alpha_j\}_1^n]$. Let E and $\widetilde{E}$ be the column vectors in $\mathbb{C}^n$ defined by

$$E = \begin{bmatrix} 1 & 1 & \cdots & 1 \end{bmatrix}^{tr} \quad \text{and} \quad \widetilde{E} = \begin{bmatrix} \gamma_1 & \gamma_2 & \cdots & \gamma_n \end{bmatrix}^{tr}. \tag{6.6}$$

Here tr denotes the transpose. Now let f be a function in H^∞ and φ a finite Blaschke product. Then the first condition in (6.2) is equivalent to $(Ef)(Z)_{left} = (\widetilde{E}\varphi)(Z)_{left}$. Recall that the order of a finite Blaschke product equals it degree. So the almost Nevanlinna-Pick interpolation problem of finding a function f in H^∞ and a finite Blaschke product φ satisfying (6.2) is a special case of our almost tangential Nevanlinna-Pick interpolation problem, that is, finding a function F in $H^\infty(\mathcal{U}, \mathcal{Y})$ and inner function Φ in $H^\infty(\mathcal{U}, \mathcal{U})$ satisfying (6.5).

As before, let Z be a stable operator on a finite dimensional space $\mathcal{Z}$ and E an operator mapping $\mathcal{Y}$ into $\mathcal{Z}$. The controllability operator for the pair $\{Z, E\}$ is the operator mapping $\ell_+^2(\mathcal{Y})$ into $\mathcal{Z}$ defined by

$$W = \begin{bmatrix} E & ZE & Z^2E & \cdots \end{bmatrix} : \ell_+^2(\mathcal{Y}) \to \mathcal{Z}. \tag{6.7}$$

(As usual, $\ell_+^2(\mathcal{Y})$ is the Hilbert space formed by the set of all square summable unilateral sequences with values in $\mathcal{Y}$.) Recall that the pair $\{Z, E\}$ is *controllable* if the operator W in (6.7) is onto $\mathcal{Z}$; see [22, 27]. The *controllability Gramian* associated with the pair $\{Z, E\}$ is the operator P on $\mathcal{Z}$ defined by $P = WW^*$. Notice that P is invertible if and only if the pair $\{Z, E\}$ is controllable. It is well known that P is the unique solution of the following Lyapunov equation

$$P = ZPZ^* + EE^*. \tag{6.8}$$

Let $\widetilde{W}$ be the controllability operator corresponding to the pair $\{Z, \widetilde{E}\}$, that is,

$$\widetilde{W} = \begin{bmatrix} \widetilde{E} & Z\widetilde{E} & Z^2\widetilde{E} & \cdots \end{bmatrix} : \ell_+^2(\mathcal{U}) \to \mathcal{Z}. \tag{6.9}$$

The controllability Gramian associated with the pair $\{Z, \widetilde{E}\}$ is the operator $\widetilde{P}$ on $\mathcal{Z}$ defined by $\widetilde{P} = \widetilde{W}\widetilde{W}^*$. Notice that $\widetilde{P}$ is the unique solution to the following Lyapunov equation

$$\widetilde{P} = Z\widetilde{P}Z^* + \widetilde{E}\widetilde{E}^*. \tag{6.10}$$

Using this notation we are ready to state the following solution to our almost tangential Nevanlinna-Pick interpolation problem.

Theorem 6.1. *Let $\{Z, E, \widetilde{E}, \gamma, k\}$ be a data set for an almost tangential Nevanlinna-Pick interpolation problem, where $\{Z, E\}$ is a stable, controllable pair. Let P and $\widetilde{P}$ be respectively the controllability Gramians for the pairs $\{Z, E\}$ and $\{Z, \widetilde{E}\}$. Then the following statements are equivalent.*

(a) *There exists a function F in $H^\infty(\mathcal{U},\mathcal{Y})$ and an inner function Φ in $H^\infty(\mathcal{U},\mathcal{U})$ satisfying the three conditions in (6.5).*
(b) *The operator $P^{-1}\widetilde{P}$ has at most k eigenvalues strictly greater that γ^2.*
(c) *The operator $\widetilde{P}-\gamma^2 P$ has at most k strictly positive eigenvalues.*

PROOF. We present the proof in six steps. The first two steps follow the approach in [14] for solving the standard tangential Nevanlinna-Pick interpolation problem.

STEP 1. Let F be a function in $H^\infty(\mathcal{U},\mathcal{Y})$. Let T_F be the Toeplitz operator mapping $\ell_+^2(\mathcal{U})$ into $\ell_+^2(\mathcal{Y})$ defined by

$$T_F = \begin{bmatrix} F_0 & 0 & 0 & \cdots \\ F_1 & F_0 & 0 & \cdots \\ F_2 & F_1 & F_0 & \cdots \\ \vdots & \vdots & \vdots & \ddots \end{bmatrix} : \ell_+^2(\mathcal{U}) \to \ell_+^2(\mathcal{Y}), \tag{6.11}$$

where $F(\lambda)=\sum_0^\infty \lambda^n F_n$ is the Taylor series expansion for F in the open unit disc. Recall that $\|T_F\| = \|F\|_\infty$. Let Φ be an inner function in $H^\infty(\mathcal{U},\mathcal{U})$ and T_Φ the Toeplitz operator defined in (6.4). By consulting (6.3), (6.4) and (6.11), we see that $(EF)(Z)_{left} = WT_F|\mathcal{U}$ and $(\widetilde{E}\Phi)(Z)_{left} = \widetilde{W}T_\Phi|\mathcal{U}$ where $\mathcal{U}$ is embedded in the first component of $\ell_+^2(\mathcal{U})$. ¿From this it readily follows that $(EF)(Z)_{left} = (\widetilde{E}\Phi)(Z)_{left}$ if and only if $WT_F = \widetilde{W}T_\Phi$. So F and Φ are functions satisfying part (a) if and only if F is a function in $H^\infty(\mathcal{U},\mathcal{Y})$ and Φ is an inner in $H^\infty(\mathcal{U},\mathcal{U})$ satisfying the following three conditions

$$WT_F = \widetilde{W}T_\Phi, \quad \|F\|_\infty \le \gamma \quad \text{and} \quad \deg\Phi \le k\,. \tag{6.12}$$

STEP 2. The next step in our proof is to put this interpolation problem in the framework of Theorem 4.3. Let $\mathcal{H}'$ be the subspace of $\ell_+^2(\mathcal{Y})$ defined by $\mathcal{H}' = (\ker W)^\perp$. Notice that $\mathcal{H}'$ also equals the range of W^*, which is closed because W is onto and $\dim\mathcal{H}' = \dim\mathcal{Z} < \infty$. Let A be the operator from $\ell_+^2(\mathcal{U})$ into $\mathcal{H}'$ uniquely determined by $WA = \widetilde{W}$. In fact, an explicit formula for A is given by

$$A = W^*P^{-1}\widetilde{W}\,. \tag{6.13}$$

To verify this simply recall that $P = WW^*$, and thus $W(W^*P^{-1}\widetilde{W}) = \widetilde{W}$. Since $W|\mathcal{H}'$ is one to one, A is the only operator from $\ell_+^2(\mathcal{U})$ into $\mathcal{H}'$ satisfying $WA = \widetilde{W}$. Hence (6.13) holds.

Let $S_\mathcal{Y}$ be the unilateral shift on $\ell_+^2(\mathcal{Y})$, that is, $S_\mathcal{Y}$ is the isometry on $\ell_+^2(\mathcal{Y})$ defined by

$$S_\mathcal{Y}\begin{bmatrix} f_0 & f_1 & f_2 & \cdots \end{bmatrix}^{tr} = \begin{bmatrix} 0 & f_0 & f_1 & \cdots \end{bmatrix}^{tr}, \tag{6.14}$$

where $\begin{bmatrix} f_0 & f_1 & f_2 & \cdots \end{bmatrix}^{tr} \in \ell_+^2(\mathcal{Y})$. Using $ZW = WS_\mathcal{Y}$, we obtain $S_\mathcal{Y}^*W^* = W^*Z^*$. So $\mathcal{H}'$ is an invariant subspace for $S_\mathcal{Y}^*$. Let T' be the operator on $\mathcal{H}'$ defined by $T' = P'S_\mathcal{Y}|\mathcal{H}'$ where P' is the orthogonal projection onto $\mathcal{H}'$. Because $S_\mathcal{Y}^*|\mathcal{H}' = T'^*$, it follows that $S_\mathcal{Y}$ is an isometric lifting of T'. Let $T = S_\mathcal{U}$ where $S_\mathcal{U}$

is the unilateral shift on $\ell^2_+(\mathcal{U})$, that is, $S_\mathcal{U}$ is the isometry defined by (6.14) where $\mathcal{U}$ replaces $\mathcal{Y}$. We claim that $T'A = AT$. To prove this notice that $T'P' = P'S_\mathcal{Y}$ and $WA = \widetilde{W}$, yield

$$WT'A = WP'S_\mathcal{Y}A = WS_\mathcal{Y}A = ZWA = Z\widetilde{W} = \widetilde{W}S_\mathcal{U} = WAT.$$

Since $W|\mathcal{H}'$ is one to one, $T'A = AT$. Hence $\{T, A, T', S_\mathcal{Y}, I\}$ is a weighted lifting data set where $\Omega = I$.

STEP 3. Since $\mathcal{H}'$ is finite dimensional, A^*A is a finite rank operator. We claim that A^*A and $P^{-1}\widetilde{P}$ have the same nonzero eigenvalues. To see this recall that if M and N are two operators acting between the appropriate spaces, then MN and NM have the same nonzero spectrum. Using $P = WW^*$ and $A = W^*P^{-1}\widetilde{W}$, we obtain $A^*A = \widetilde{W}^*P^{-1}\widetilde{W}$. Hence A^*A and $P^{-1}\widetilde{W}\widetilde{W}^* = P^{-1}\widetilde{P}$ have the same nonzero eigenvalues.

Now set $\Lambda = A^*A - \gamma^2 I$. In this setting, $\mathcal{H}_+$ is the spectral subspace spanned by the eigenvectors corresponding to the strictly positive eigenvalues of Λ. Since A^*A and $P^{-1}\widetilde{P}$ have the same nonzero eigenvalues, the dimension of $\mathcal{H}_+$ equals the number of eigenvalues (including their multiplicity) of $P^{-1}\widetilde{P}$ strictly greater than γ^2.

STEP 4. Assume that part (a) holds, that is, suppose that a function F in $H^\infty(\mathcal{U},\mathcal{Y})$ and an inner function Φ in $H^\infty(\mathcal{U},\mathcal{U})$ satisfy the three conditions in (6.5), or equivalently, (6.12). Let $\mathcal{M}$ be range of T_Φ. By definition the co-dimension of $\mathcal{M}$ equals the degree of Φ. Hence $\dim \mathcal{M}^\perp \le k$. Using $WT_F = \widetilde{W}T_\Phi$ and $WA = \widetilde{W}$, we obtain

$$WAT_\Phi = \widetilde{W}T_\Phi = WT_F = WP'T_F.$$

Recall that $\mathcal{H}' = (\ker W)^\perp$. So $P'T_F = AT_\Phi$. In particular, $\|AT_\Phi\| \le \|T_F\| = \|F\|_\infty \le \gamma$. In other words, $\|A|\mathcal{M}\| \le \gamma$. Thus $(A^*Af, f) \le \gamma^2(f,f)$ for all f in $\mathcal{M}$. Because the co-dimension of $\mathcal{M}$ is at most k, this implies that A^*A has at most k eigenvalues strictly greater than γ^2. To see this, assume that $\lambda_1 \ge \lambda_2 \ge \cdots \ge \lambda_{k+1}$ are $k+1$ eigenvalues of A^*A and $\lambda_{k+1} > \gamma^2$. Then the subspace $\mathcal{G}$ generated by the orthogonal eigenvectors for A^*A corresponding to $\{\lambda_j\}_1^{k+1}$ is a $k+1$ dimensional reducing subspace for A^*A satisfying $A^*A|\mathcal{G} \ge \lambda_{k+1}I_\mathcal{G}$. Because the co-dimension of $\mathcal{M}$ is k and the dimension of $\mathcal{G}$ is $k+1$, there exists a nonzero vector g in $\mathcal{G}\bigcap\mathcal{M}$. Thus $\lambda_{k+1}\|g\|^2 \le (A^*Ag, g) \le \gamma^2\|g\|^2$. This implies that $\lambda_{k+1} \le \gamma^2$, which contradicts the fact that $\gamma^2 < \lambda_{k+1}$. Hence A^*A has at most k eigenvalues strictly greater than γ^2. Recall that A^*A and $P^{-1}\widetilde{P}$ have the same nonzero eigenvalues. So $P^{-1}\widetilde{P}$ has at most k eigenvalues strictly greater than γ^2. Therefore part (a) implies part (b).

STEP 5. Now assume that part (b) holds, that is, $P^{-1}\widetilde{P}$ has at most k eigenvalues strictly greater than γ^2. As before, set $\Lambda = A^*A - \gamma^2 I$. Since A^*A and $P^{-1}\widetilde{P}$ have the same nonzero eigenvalues, $\dim \mathcal{H}_+ \le k$. By Theorem 2.1 there exists a subspace $\mathcal{M}$ in $\mathrm{Lat}_-(T,\Lambda)$. Moreover, $\mathcal{M}$ and $\mathcal{H}_-$ have the same co-dimension. In particular, $\dim \mathcal{M}^\perp \le k$. Since $\mathcal{M}$ is an invariant subspace for the shift $T = S_\mathcal{U}$, the Beurling-Lax-Halmos theorem implies that there exists a rational inner function Φ in $H^\infty(\mathcal{U},\mathcal{U})$ such that $\mathcal{M} = \operatorname{ran} T_\Phi$. Obviously, $\deg \Phi \leq k$. According to Theorem 4.3 with $\Omega = I$ and $\epsilon = \gamma^2$, there exists an operator B mapping $\mathcal{M}$ into $\ell_+^2(\mathcal{Y})$ satisfying

$$S_\mathcal{Y} B = BT|\mathcal{M}, \qquad P'B = A|\mathcal{M} \qquad \text{and} \qquad \|B\| \leq \gamma . \tag{6.15}$$

The first equation yields $S_\mathcal{Y} BT_\Phi = BTT_\Phi = BT_\Phi T$. In other words, BT_Φ intertwines the shift T with the shift $S_\mathcal{Y}$. So there exists a function F in $H^\infty(\mathcal{U},\mathcal{Y})$ satisfying $BT_\Phi = T_F$. Using $WA = \widetilde{W}$ along with $P'B = A|\mathcal{M}$, we obtain

$$WT_F = WBT_\Phi = WP'BT_\Phi = WAT_\Phi = \widetilde{W}T_\Phi .$$

Thus $WT_F = \widetilde{W}T_\Phi$, or equivalently, the first condition in (6.12) is satisfied. Finally, notice that

$$\|F\|_\infty = \|T_F\| = \|BT_\Phi\| \leq \|B\|\|T_\Phi\| = \|B\| \leq \gamma .$$

In other words, $\|F\|_\infty \leq \gamma$. Therefore part (b) implies part (a). So parts (a) and (b) are equivalent.

STEP 6. To complete the proof it remains to show that parts (b) and (c) are equivalent. To this end, notice that $P^{-1}\widetilde{P}$ is similar to $P^{-1/2}\widetilde{P}P^{-1/2}$. To verify this simply observe that $P^{1/2}(P^{-1}\widetilde{P})P^{-1/2} = P^{-1/2}\widetilde{P}P^{-1/2}$. This implies that $P^{-1}\widetilde{P}$ and $P^{-1/2}\widetilde{P}P^{-1/2}$ have the same eigenvalues. In particular, $P^{-1}\widetilde{P}$ has m eigenvalues strictly greater than γ^2 if and only if $P^{-1/2}\widetilde{P}P^{-1/2} - \gamma^2 I$ has m strictly positive eigenvalues. Notice that $P^{-1/2}\widetilde{P}P^{-1/2} - \gamma^2 I$ and $P^{1/2}(P^{-1/2}\widetilde{P}P^{-1/2} - \gamma^2 I)P^{1/2}$ are congruent, and thus have the same signature. Hence $P^{-1}\widetilde{P}$ has m eigenvalues strictly greater than γ^2 if and only if $\widetilde{P} - \gamma^2 P$ has m strictly positive eigenvalues. From this it readily follows that parts (b) and (c) are equivalent. This completes the proof.

To complete this section we will present an explicit solution to our almost tangential Nevanlinna-Pick interpolation problem in the scalar case.

Theorem 6.2. *Let $\{Z, E, \widetilde{E}, \gamma, k\}$ be a data set for an almost tangential Nevanlinna-Pick interpolation problem, where $\{Z, E\}$ is a stable, controllable pair and $\mathcal{U} = \mathcal{Y} = \mathbb{C}$. Assume that $\widetilde{P} - \gamma^2 P$ has at most k strictly positive eigenvalues and zero is an eigenvalue for $\widetilde{P} - \gamma^2 P$. Let u be any nonzero vector in the kernel of $\widetilde{P} - \gamma^2 P$ and let g be the rational function defined by*

$$g(\lambda) = \gamma^2 \frac{E^*(I - \lambda Z^*)^{-1} u}{\widetilde{E}^*(I - \lambda Z^*)^{-1} u} . \tag{6.16}$$

Then g admits a representation of the form $g = f/\varphi$ where f is in H^∞ and φ is a finite Blaschke product satisfying

$$f(Z)E = \varphi(Z)\widetilde{E}, \qquad \|f\|_\infty \leq \gamma \qquad \textit{and} \qquad \deg \varphi \leq k . \tag{6.17}$$

Furthermore, $\gamma^{-1}f$ *is also a finite Blaschke product and* $|g| = \gamma$ *a.e. on the unit circle. Finally,* φ *divides* $\widetilde{E}^*(I - \lambda Z^*)^{-1}u$*, that is,* $\widetilde{E}^*(I - \lambda Z^*)^{-1}u/\varphi(\lambda)$ *is in* H^∞*.*

PROOF. Let us continue to use the notation and results established in steps 1 to 3 of the proof of Theorem 6.1. In this setting, T is now the unilateral shift on ℓ_+^2 and $S_\mathcal{Y} = T$. So $\{T, A, T', T, I\}$ is a weighted lifting data set where $A = W^*P^{-1}\widetilde{W}$ and T' is the compression of T to $\mathcal{H}' = W^*\mathcal{Z}$. As before, set $\Lambda = A^*A - \gamma^2 I$. Notice that u is an eigenvector for $P^{-1}\widetilde{P}$ with eigenvalue γ^2. To see this simply observe that $(\widetilde{P} - \gamma^2 P)u = 0$ if and only if $P^{-1}\widetilde{P}u = \gamma^2 u$. Recall that A^*A and $P^{-1}\widetilde{P}$ have the same nonzero eigenvalues. In particular, γ^2 is an eigenvalue for A^*A. Because parts (*b*) and (*c*) in Theorem 6.1 are equivalent, we see that Λ has at most k strictly positive eigenvalues and zero is an eigenvalue for Λ. According to Theorem 2.1 there exists a subspace $\mathcal{M}$ in $\mathrm{Lat}_-(T, \Lambda)$. Moreover, $\mathcal{M}$ and $\mathcal{H}_-$ have the same co-dimension. In particular, $\dim \mathcal{M}^\perp \leq k$. Since $\mathcal{M}$ is an invariant subspace for the shift T, the Beurling theorem shows that there exists an inner function φ in H^∞ such that $\mathcal{M} = \mathrm{ran}\, T_\varphi$. Obviously, $\deg \varphi \leq k$. So φ is a finite Blaschke product of order less than or equal to k.

According to Theorem 4.3, there exists an operator B mapping $\mathcal{M}$ into ℓ_+^2 satisfying

$$TB = BT|\mathcal{M}, \qquad P'B = A|\mathcal{M} \qquad \text{and} \qquad \|B\| \leq \gamma. \tag{6.18}$$

The first equation yields $TBT_\varphi = BTT_\varphi = BT_\varphi T$. In other words, BT_φ commutes with the shift T. So there exists a function f in H^∞ satisfying $BT_\varphi = T_f$.

Next set $x = \widetilde{W}^*u$, where u is a nonzero vector in the kernel of $\widetilde{P} - \gamma^2 P$. We claim that x is nonzero and x is an eigenvalue for A^*A corresponding to the eigenvalue γ^2, or equivalently, x is a nonzero vector in the kernel of Λ. First notice that x is nonzero. If $x = 0$, then $0 = \widetilde{W}x = \widetilde{W}\widetilde{W}^*u = \widetilde{P}u$. Thus $0 = P^{-1}\widetilde{P}u = \gamma^2 u$ and u must be zero. (By assumption the tolerance $\gamma > 0$.) This contradicts the fact that u is nonzero. Hence x is nonzero. Using $A = W^*P^{-1}\widetilde{W}$ and $P = WW^*$, we obtain

$$A^*Ax = \widetilde{W}^*P^{-1}\widetilde{W}x = \widetilde{W}^*P^{-1}\widetilde{W}\widetilde{W}^*u = \widetilde{W}^*P^{-1}\widetilde{P}u = \gamma^2\widetilde{W}^*u = \gamma^2 x\,.$$

So x is an eigenvector for A^*A corresponding to the eigenvalue γ^2. In particular, x is in the kernel of Λ.

Because x is in the kernel of Λ, Corollary 1.2 shows that x is in $\mathcal{M}$. Using $\|Ax\| = \gamma\|x\|$ in the last two equations in (6.18), we obtain

$$\gamma\|x\| = \|Ax\| = \|P'Bx\| \leq \|Bx\| \leq \gamma\|x\|\,.$$

So $\|P'Bx\| = \|Bx\|$. Hence $P'Bx = Bx$. Notice that $T_\varphi T_\varphi^*$ is the orthogonal projection onto $\mathcal{M}$. Therefore $Ax = P'Bx = Bx = BT_\varphi T_\varphi^* x = T_f T_\varphi^* x$. In particular, this implies that

$$Ax = T_f T_\varphi^* x \qquad \text{(when } x \in \ker \Lambda\text{)}\,. \tag{6.19}$$

Now let $\mathcal{F}_+$ be the Fourier transform mapping ℓ_+^2 onto the Hardy space H^2. To be precise, $\mathcal{F}_+(h_n)_0^\infty = \sum_0^\infty \lambda^n h_n$ where $(h_n)_0^\infty$ is a vector in ℓ_+^2 and λ is in the open unit disc. Because x is in $\mathcal{M} = \operatorname{ran} T_\varphi$, it follows that φ divides $\mathcal{F}_+ x$. By taking the Fourier transform of both sides in (6.19), we arrive at $(\mathcal{F}_+ Ax) = f\overline{\varphi}(\mathcal{F}_+ x)$ a.e. on the unit circle. Because x is nonzero, the H^2 function $\mathcal{F}_+ x$ is a.e. nonzero on the unit circle; see [19, 20]. So if g is the function defined by $g = \mathcal{F}_+ Ax/\mathcal{F}_+ x$, then

$$g(\lambda) := \frac{(\mathcal{F}_+ Ax)(\lambda)}{(\mathcal{F}_+ x)(\lambda)} = \frac{f(\lambda)}{\varphi(\lambda)}\,. \tag{6.20}$$

In the next paragraph we will show that g is a rational function.

Now let us compute the Fourier transform of x. By consulting the form of $\widetilde{W}$ in (6.9) with $\mathcal{U} = \mathbb{C}$ and λ in the open unit disc, we have

$$\begin{aligned}(\mathcal{F}_+ x)(\lambda) &= (\mathcal{F}_+ \widetilde{W}^* u)(\lambda) = \sum_{n=0}^{\infty} \lambda^n \widetilde{E}^* Z^{*n} u \\ &= \widetilde{E}^* \sum_{n=0}^{\infty} \lambda^n Z^{*n} u = \widetilde{E}^*(I - \lambda Z^*)^{-1} u\,.\end{aligned} \tag{6.21}$$

In other words, $(\mathcal{F}_+ x)(\lambda) = \widetilde{E}^*(I - \lambda Z^*)^{-1} u$. A similar calculation shows that $(\mathcal{F}_+ W^* u)(\lambda) = E^*(I - \lambda Z^*)^{-1} u$. Using $A = W^* P^{-1} \widetilde{W}$ and $x = \widetilde{W}^* u$, we obtain

$$\begin{aligned}(\mathcal{F}_+ Ax)(\lambda) &= (\mathcal{F}_+ W^* P^{-1} \widetilde{W}\widetilde{W}^* u)(\lambda) = (\mathcal{F}_+ W^* P^{-1} \widetilde{P} u)(\lambda) \\ &= \gamma^2 (\mathcal{F}_+ W^* u)(\lambda) = \gamma^2 E^*(I - \lambda Z^*)^{-1} u\,.\end{aligned} \tag{6.22}$$

Equations (6.21) and (6.22), yield

$$(\mathcal{F}_+ Ax)(\lambda) = \gamma^2 E^*(I - \lambda Z^*)^{-1} u \quad \text{and} \quad (\mathcal{F}_+ x)(\lambda) = \widetilde{E}^*(I - \lambda Z^*)^{-1} u\,. \tag{6.23}$$

By consulting (6.20), we arrive at the formula for g in (6.16). In particular, g is a rational function.

To complete the proof it remains to show that $\gamma^{-1} f$ is a finite Blaschke product. Let $p = \mathcal{F}_+ Ax$ and $q = \mathcal{F}_+ x$. Notice that $(I - \lambda Z^*)^{-1} = \operatorname{adj}(I - \lambda Z^*)/\det(I - \lambda Z^*)$ where adj is the algebraic adjoint. Because Z is stable, equation (6.23) shows that p and q are rational function in H^∞. Clearly, $g = p/q$ is also rational. Following some of the ideas from Adamjan-Arov-Krein [1], as in the proof of Theorem 5.2, consider any integer $n \geq 0$. Then using $P'T = T'P'$ and $T'A = AT$, we obtain

$$\begin{aligned}\int_0^{2\pi} e^{in\omega} |p(\omega)|^2\, d\omega &= (T^n Ax, Ax) = (P'T^n Ax, Ax) = (T'^n Ax, Ax) \\ &= (AT^n x, Ax) = (T^n x, A^* Ax) = \gamma^2 (T^n x, x) \\ &= \gamma^2 \int_0^{2\pi} e^{in\omega} |q(\omega)|^2\, d\omega\,.\end{aligned}$$

By taking the complex conjugate, we see that

$$\int_0^{2\pi} e^{\imath n\omega} \left(|p(\omega)|^2 - \gamma^2|q(\omega)|^2\right) d\omega = 0 \qquad (n = 0, \pm 1, \pm 2, \cdots). \tag{6.24}$$

So all the Fourier coefficients of $|p|^2 - \gamma^2|q|^2$ are zero, and thus $|p| = \gamma|q|$ on the unit circle. Therefore $|g| = |p|/|q| = \gamma$ on the unit circle. Recall that $g = f/\varphi$ and g is rational. So $f = g\varphi$ is also a rational function. Because $|f| = |g\varphi| = \gamma$ on the unit circle and f is in H^∞, we see that $\gamma^{-1}f$ is a finite Blaschke product. This completes the proof.

Let us now return to the "almost" version of the classical Nevanlinna-Pick interpolation problem associated with (6.2). In this problem, $\{\alpha_j\}_1^n$ is a finite set of distinct points in the open unit disc and $\{\gamma_j\}_1^n$ a set of complex numbers. Let Z be the diagonal matrix on $\mathbb{C}^n$ defined by $Z = \mathrm{diag}[\{\alpha_j\}_1^n]$. Obviously, Z is stable. Let E and $\widetilde{E}$ be the column vectors in $\mathbb{C}^n$ defined by (6.6). Notice that the pair $\{Z, E\}$ is controllable. In this setting, it is well known (see Section I.2 in [14]) that $\widetilde{P} - \gamma^2 P$ is the Pick matrix given by

$$(\widetilde{P} - \gamma^2 P)_{ij} = \frac{\gamma_i \overline{\gamma}_j - \gamma^2}{1 - \alpha_i \overline{\alpha}_j} \qquad (\text{for } i, j = 1, 2, \cdots, n). \tag{6.25}$$

Moreover, if $u = [u_1, u_2, \cdots, u_n]^{tr}$ is a nonzero vector in the kernel of $\widetilde{P} - \gamma^2 P$, then the rational function g in (6.16) is given by

$$g(\lambda) = \frac{\sum_{k=1}^n u_k(1 - \overline{\alpha}_k\lambda)^{-1}}{\sum_{k=1}^n \overline{\gamma}_k u_k(1 - \overline{\alpha}_k\lambda)^{-1}}. \tag{6.26}$$

So if $\widetilde{P} - \gamma^2 P$ has at most k strictly positive eigenvalues and the kernel of $\widetilde{P} - \gamma^2 P$ is nonzero, then the function g in (6.26) admits a representation of the form $g = f/\varphi$ where $\gamma^{-1}f$ and φ are finite Blaschke products satisfying (6.2). A parametrization of all solutions of the classical scalar almost Nevanlinna-Pick problem with the Pick matrix invertible (and with the unit disk replaced by the upper half plane) can be found in [12]. Notice that in our case the Pick matrix is singular.

In the next paragraph and the following corollary we deal with the case when the representation $g = f/\varphi$ in Theorem 6.2 is a co-prime representation. We begin by defining the latter notion.

Let g be a scalar rational function. Then g is in L^∞ if and only if g has no poles on the unit circle. If g is a rational function in L^∞, then g admits a representation of the form $g = a/b$ where a is a function in H^∞ and b is a finite Blaschke product. One can compute a and b by collecting the appropriate poles and zeros of g inside and outside the unit disc. Notice that a and b are not unique. For example, if ψ is any finite Blaschke product, then clearly $g = a\psi/(b\psi)$ where $a\psi$ is in H^∞ and $b\psi$ is a finite Blaschke product. By canceling out the common finite Blaschke factors between a and b, we see that $g = h/\phi$ where $h \in H^\infty$ and the finite Blaschke product ϕ have no common zeros. Motivated by this we say that $g = h/\phi$ is a *co-prime representation* of g if h is a rational function in H^∞ and ϕ is a finite Blaschke product such that h and ϕ have no common zeros.

Clearly, any rational function in L^∞ admits a co-prime representation. Moreover, a co-prime representation is unique up to a constant of modulus one. To be precise, if $g = h/\phi$ and $g = h_1/\phi_1$ are two co-prime representations, then $h = \zeta h_1$ and $\phi = \zeta\phi_1$ where ζ is a constant of modulus one. Recall that the function g in (6.16) of Theorem 6.2, admits a representation of the form $g = f/\varphi$ where $f \in H^\infty$ and the finite Blaschke product φ satisfy the three conditions in (6.17). (In this case, $\gamma^{-1}f$ is also a finite Blaschke product.) However, the representation $g = f/\varphi$ may not be co-prime, that is, f and φ may have a non-trivial common finite Blaschke factor. Obviously, one can eliminate the common finite Blaschke factor and obtain a co-prime representation $g = h/\phi$. However, the functions h and ϕ may not be a solution to the corresponding almost tangential Nevanlinna-Pick interpolation problem; see the example after the proof of Corollary 6.3. The following result addresses a general case when a co-prime representation of g in (6.16) yields a solution to our interpolation problem.

Corollary 6.3. *Let $\{Z, E, \widetilde{E}, \gamma, k\}$ be a data set for an almost tangential Nevanlinna-Pick interpolation problem satisfying the hypothesis of Theorem 6.2. Moreover, assume that the set of all eigenvalues of Z and the set of zeros of $\widetilde{E}^*(I - \lambda Z^*)^{-1}u$ have no points in common. Let $g = h/\phi$ be a co-prime representation of the rational function g defined in (6.16). Then the functions h and ϕ satisfy the following three conditions*

$$h(Z)E = \phi(Z)\widetilde{E}, \qquad \|h\|_\infty \leq \gamma \qquad \text{and} \qquad \deg\phi \leq k\,. \tag{6.27}$$

Furthermore, $\gamma^{-1}h$ is a finite Blaschke product. Finally, the set of all eigenvalues of Z and the set of zeros of ϕ have no points in common.

PROOF. According to Theorem 6.2, the function g admits a representation of the form $g = f/\varphi$, where $f \in H^\infty$ and the finite Blaschke product φ satisfy the three conditions in (6.17). Let $f = \psi h$ and $\varphi = \psi\phi$ where ψ is a finite Blaschke product such that h and ϕ have no common zeros. By construction $g = h/\phi$ is a co-prime representation of g. By consulting the proof of Theorem 6.2, we see that $x = \widetilde{W}^*u$ is in $\mathcal{M} = \operatorname{ran} T_\varphi$. Since $\widetilde{E}^*(I - \lambda Z^*)^{-1}u$ is the Fourier transform of x, it follows that φ divides the inner part of $\widetilde{E}^*(I - \lambda Z^*)^{-1}u$. So the zeros of φ do not contain any eigenvalues of Z. In particular, $\psi(Z)$ is invertible. By employing the first equation in (6.17), we obtain

$$\psi(Z)h(Z)E = f(Z)E = \varphi(Z)\widetilde{E} = \psi(Z)\phi(Z)\widetilde{E}\,.$$

Since $\psi(Z)$ is invertible, $h(Z)E = \phi(Z)\widetilde{E}$. Clearly, $\|h\|_\infty = \|f\|_\infty \leq \gamma$ and $\deg\phi \leq \deg\varphi \leq k$. In other words, the three conditions in (6.27) hold. Because $\gamma^{-1}f$ is a finite Blaschke product, $\gamma^{-1}h$ is also a finite Blaschke product. This completes the proof is complete.

It is possible for an eigenvalue of Z to be a zero of the rational function $\widetilde{E}^*(I - \lambda Z^*)^{-1}u$. In this case, the conclusion of Corollary 6.3 may not hold. For example, take

$$Z = \begin{bmatrix} 1/4 & 0 & 0 \\ 0 & 1/2 & 0 \\ 0 & 0 & 3/4 \end{bmatrix}, \qquad E = \begin{bmatrix} 1 \\ 1 \\ 1 \end{bmatrix} \qquad \text{and} \qquad \widetilde{E} = \begin{bmatrix} 1 \\ 1 \\ 0 \end{bmatrix}. \tag{6.28}$$

Set $\gamma = 1$. Then a standard computation shows that

$$\widetilde{P} - P = \begin{bmatrix} 0 & 0 & -16/13 \\ 0 & 0 & -8/5 \\ -16/13 & -8/5 & -16/7 \end{bmatrix}. \tag{6.29}$$

Clearly, the kernel of $\widetilde{P} - P$ is nonzero. Moreover, $\widetilde{P} - P$ has one strictly positive eigenvalue. (In fact, -3.46, 1.18 and 0 are the eigenvalues of $\widetilde{P} - P$.) So we set $k = 1$. Notice that $u = [13, -10, 0]^{tr}$ is in the kernel of $\widetilde{P} - P$. Using this u, we obtain

$$\widetilde{E}^*(I - \lambda Z^*)^{-1}u = \gamma^2 E^*(I - \lambda Z^*)^{-1}u = \frac{(\lambda - 0.75)}{(1 - 0.75\lambda)} \frac{3(\lambda - 4/3)}{(1 - 0.25\lambda)(1 - 0.5\lambda)} \tag{6.30}$$

The first equality in (6.30) is a simple consequence of the fact that the last component of u is zero. Obviously, $\varphi = (\lambda - 0.75)/(1 - 0.75\lambda)$ is the inner part of $\widetilde{E}^*(I - \lambda Z^*)^{-1}u$. Furthermore, if we set $f = \varphi$, then it is easy to verify that all three conditions in (6.17) hold. Notice that 0.75 is a zero of $\widetilde{E}^*(I - \lambda Z^*)^{-1}u$ and an eigenvalue of Z. So we cannot use Corollary 6.3 to compute a solution to this almost Nevanlinna-Pick interpolation problem. According to equation (6.30), the function $g = 1$. Clearly, $g = h/\phi$ with $h = \phi = 1$ is a co-prime representation of g. However, $h(Z)E = E \neq \widetilde{E} = \phi(Z)\widetilde{E}$.

To complete this section let us make some connections to the Takagi interpolation problem. For some state space solutions and a more detailed discussion of the Takagi interpolation problem see [8]. Assume that $\{Z, E, \widetilde{E}, \gamma, k\}$ is a data set for an almost tangential Nevanlinna-Pick interpolation problem satisfying the hypothesis of Corollary 6.3. Then one Takagi interpolation problem is to find a rational function g in L^∞ with at most k poles inside the open unit disc such that

$$g(Z)E = \widetilde{E} \qquad \text{and} \qquad \|g\|_\infty \leq \gamma. \tag{6.31}$$

Here we also assume that no eigenvalue of Z is pole of g, and thus $g(Z)$ is a well defined operator. We claim that the function g defined in (6.16) is a solution to the previous Takagi interpolation problem. To verify this, let $g = h/\phi$ be a co-prime representation of g. According to Corollary 6.3, the three conditions in (6.27) hold. In particular, $\|g\|_\infty \leq \gamma$. Moreover, no eigenvalue of Z is zero of ϕ. So $\phi(Z)$ is invertible. Since $h(Z)E = \phi(Z)\widetilde{E}$, we have $\phi(Z)^{-1}h(Z)E = \widetilde{E}$. In other words, $g(Z)E = \widetilde{E}$. Hence (6.31) holds. Because $\deg\phi \leq k$, the function g has at most k poles inside the open unit disc. Therefore the function g defined in (6.16) is a solution to the previous Takagi interpolation problem.

References

[1] V. A. Adamjan, D. Z. Arov and M.G. Krein, *Analytic properties of Schmidt pairs for a Hankel operator and the generalized Schur-Takagi problem*, Math USSR Sbornick, **15** (1971), 31–73.

[2] V. A. Adamjan, D. Z. Arov and M.G. Krein, *Infinite Hankel block matrices and related extension problems*, Izv. Akad. Nauk. Armjan SSR, Matematika, **6** (1971), 87–112 (English Translation Amer. Math. Soc. Transl., **III** (1978), 133–156.)

[3] D. Alpay, T. Constantinescu, A. Dijksma and J. Rovnyak, *On Nudelḿan's problem and interpolation in the generalized Schur and Nevanlinna classes*, Preprint.

[4] R. Arocena, T. Ya. Azizov, A. Dijksma and S.A.M. Marcantognini, *On commutant lifting with finite defect*, J. Operator Theory, **35** (1996), 117–132.

[5] R. Arocena, T. Ya. Azizov, A. Dijksma and S.A.M. Marcantognini, *On commutant lifting with finite defect, II*, J. Funct. Analysis, **144** (1997), 105–116.

[6] J.P. Aubin, *Applied Functional Analysis*, Wiley, New York, 1979.

[7] T. Ya. Azizov and I. S. Iokhvidov, *Linear Operators in Spaces with an Indefinite Metric*, Wiley, New York, 1989.

[8] J. A. Ball, I. Gohberg and L. Rodman, *Interpolation for Rational Matrix Functions*, Operator Theory: Advances and Applications, **45**, Birkhäuser-Verlag, Basel, 1990.

[9] J. A. Ball and J. W. Helton, *A Beurling-Lax Theorem for the Lie group U(m,n) which contains most classical interpolation theory*, J. Operator Theory, **9** (1983), 107–142.

[10] H. Bart, I. Gohberg and M.A. Kaashoek, *Minimal factorization of matrix and operator functions*, Operator Theory: Advances and Applications, **1**, Birkhäuser Verlag, Basel, 1979.

[11] A. Biswas, C. Foias and A. E. Frazho, *Weighted Commutant Lifting*, Acta Sci. Math. (Szeged), **65** (1999), 657–686.

[12] A. Dijksma and H. Langer, *Notes on a Nevanlinna-Pick interpolation problem for generalized Nevanlinna functions*, in: Topics in Interpolation Theory (eds. H. Dym, B. Fritzsche, V. Katsnelson, B. Kristein), Operator Theory: Advances and Applications, **95**, Birkhäuser-Verlag, Basel, 1997, pp. 69–91.

[13] C. Foias and A. E. Frazho, *The Commutant Lifting Approach to Interpolation Problems*, Operator Theory: Advances and Applications, **44**, Birkhäuser-Verlag, Basel, 1990.

[14] C. Foias, A.E. Frazho, I. Gohberg and M. A. Kaashoek, *Metric Constrained Interpolation, Commutant Lifting and Systems*, Operator Theory: Advances and Applications, **100**, Birkhäuser-Verlag, Basel, 1998.

[15] A. Gheondea, *Contractive intertwining dilations for quasi-contractions*, Z. Anal. Anwendung., **16** (1996), 31–44.

[16] I. Gohberg, S. Goldberg and M. A. Kaashoek *Classes of Linear Operators I*, Operator Theory: Advances and Applications, **49**, Birkhäuser-Verlag, Basel, 1990.

[17] I. Gohberg, S. Goldberg and M. A. Kaashoek *Classes of Linear Operators II*, Operator Theory: Advances and Applications, **63**, Birkhäuser-Verlag, Basel, 1993.

[18] I. Gohberg, M. A. Kaashoek and H. J. Woerdeman, *The band method for positive and strictly contractive extension problems: an alternative version and new applications*, Integral Equations and Operator Theory, **12** (1989), 343–382.

[19] P.R. Halmos, *A Hilbert Space Problem Book,* Springer-Verlag, New York, 1982.

[20] K. Hoffman, *Banach Spaces of Analytic Functions,* Engelwood Cliffs, Prentice Hall, New Jersey, 1962.

[21] V.I. Istratescu, *Fixed Point Theory,* Reidel Publishing Co., Dordrecht, 1981.

[22] T. Kailath, *Linear Systems,* Englewood Cliffs, Prentice-Hall, New Jersey, 1980.

[23] N.K. Nikolskii, *Treatise of the shift operator,* Springer-Verlag, New York, 1985.

[24] S. Parrott, *On a quotient norm and the Sz.-Nagy-Foias lifting theorem,* J. Funct. Analysis, **30** (1978), 311-328.

[25] S.C. Power, *Hankel Operators on Hilbert Space,* Research notes in Math. **64**, Pitman, Boston, 1982.

[26] F. Riesz and B. Sz.-Nagy, *Leçons d' Analyse Fonctionnelle,* Akadémia Kaidó, Budapest, 1952 (2nd ed. 1955); latest english edition: *Functional Analysis*, Dover Publications, New York, 1990.

[27] W.J. Rugh, *Linear System Theory,* Prentice-Hall, New Jersey, 1993.

[28] D. Sarason, *Generalized interpolating in* H^∞, Trans. American Math. Soc., **127** (1967), 179–203.

[29] B. Sz.-Nagy and C. Foias, *Dilation des commutants d'opérateurs,* C. R. Acad. Sci. Paris, série A, **266** (1968), 493–495.

[30] B. Sz.-Nagy and C. Foias, *Harmonic Analysis of Operators on Hilbert Space,* North Holland Publishing Co., Amsterdam-Budapest, 1970.

[31] S. Treil and A. Volberg, *A fixed point approach to Nehari's problem and its applications, Operator Theory: Advances and Applications,* **71**, Birkhäuser-Verlag, Basel, 1994, 165–186.

C. Foias, Department of Mathematics, Indiana University,
Bloomington, IN 47405-5701, U.S.A.
E-mail address: `foias@indiana.edu`

A.E. Frazho, School of Aeronautics and Astronautics, Purdue University
West Lafayette, IN 47907-1282, U.S.A.
E-mail address: `frazho@ecn.purdue.edu`

M.A. Kaashoek, Faculteit Wiskunde en Informatica, Vrije Universiteit,
De Boelelaan 1081a, 1081 HV Amsterdam, The Netherlands
E-mail address: `kaash@cs.vu.nl`

Hyperbolicity of semigroups and Fourier multipliers

Yuri Latushkin and Roman Shvydkoy

Abstract. We present a characterization of hyperbolicity for strongly continuous semigroups on Banach spaces in terms of Fourier multiplier properties of the resolvent of the generator. Hyperbolicity with respect to classical solutions is also considered. Our approach unifies and simplifies the M. Kaashoek–S. Verduyn Lunel theory and multiplier-type results previously obtained by S. Clark, M. Hieber, S. Montgomery-Smith, F. Räbiger, T. Randolph, and L. Weis.

1. Introduction

Suppose X is a complex Banach space and $\mathbf{T} = (T_t)_{t\geq 0}$ is a strongly continuous semigroup of operators on X. Let A denote its infinitesimal generator.

An autonomous version of a well-known result that goes back to O. Perron says the following: a homogeneous differential equation $\dot{u} = Au$ admits exponential dichotomy on $\mathbb{R}$ if and only if the inhomogeneous equation $\dot{u} = Au + f$ has a unique mild solution $u \in F(\mathbb{R}; X)$ for each $f \in F(\mathbb{R}; X)$, see [DK] or [LZ], and [CL], and the literature therein. Here $F(\mathbb{R}; X)$ is a space of X-valued functions, for instance, $F(\mathbb{R}; X) = L_p(\mathbb{R}; X)$, $1 \leq p < \infty$. The exponential dichotomy for $\dot{u} = Au$ means that the semigroup generated by A is hyperbolic, that is, condition $\sigma(T_t) \cap \{|z| = 1\} = \emptyset$, $t \neq 0$, holds for the spectrum $\sigma(\cdot)$.

Passing, formally, to the Fourier transforms in the equation $\dot{u} = Au + f$ we have that the solution u is given by $u = Mf$, where $M : f \mapsto [R(i\cdot; A)\hat{f}]^{\vee}$, $R(\lambda; A)$ is the resolvent operator, and $\wedge, \vee$ are the Fourier transforms. Thus, heuristically, the above-mentioned Perron-type theorem could be reformulated to state that the hyperbolicity of the semigroup is equivalent to the fact that the function $s \mapsto R(is; A)$ is a Fourier multiplier on $L_p(\mathbb{R}; X)$, $1 \leq p < \infty$, see, e.g., [A, H1] for the definition of Fourier multipliers. One of the objectives of the current paper is to systematically study the connections of hyperbolicity and L_p-Fourier multiplier properties of the resolvent.

The use of Fourier multipliers for stability and hyperbolicity for strongly continuous semigroups has a fairly long history. To put our paper in this context, we briefly review relevant results. Probably, the first Fourier multiplier type result was obtained in the important paper [KVL] by M. Kaashoek and S. Verduyn Lunel.

These authors used scalar functions ("matrix elements" of the resolvent) defined by

$$r_\rho(s,x,x^*) = \langle x^*, R(\rho+is;A)x\rangle, \quad \rho\in\mathbb{R}, \quad s\in\mathbb{R}, \quad x\in X, \quad x\in X^*.$$

They proved that $\mathbf{T}$ is hyperbolic if and only if the following two conditions holds:

(i) $|\langle r_\rho, \Phi\rangle| \le K\|x\|\|x^*\|\|\check{\Phi}\|_{L^1}$ for some $K>0$, $\rho_0>0$ and all ρ with $|\rho|<\rho_0$ and all $\Phi\in\mathcal{S}$, the Schwartz class of scalar functions on $\mathbb{R}$;

(ii) the Césaro integral

$$\begin{aligned} G_0x &= \frac{1}{2\pi}(C,1)\int_{\mathbb{R}} R(is;A)x\,ds \\ &= \frac{1}{2\pi}\lim_{N\to\infty}\frac{1}{N}\int_0^N\int_{-l}^{l} R(is,A)x\,ds\,dl \end{aligned}$$

converges for all $x\in X$.

Remark, that one of the results of the current paper (Theorem 2.7) shows that condition (ii), in fact, follows from (i).

L. Weis in [W2] used Fourier multiplier properties of the resolvent on Besov spaces to give an alternative proof of the fact that the supremum $\omega_\alpha(\mathbf{T})$ of the growth bounds of "α–smooth" solutions T_tx are majorated by the boundedness abscissa $s_0(A)$ of the resolvent. Remark, that in Section 3 of the present paper we derive a formula (Theorem 3.6) for $\omega_\alpha(\mathbf{T})$ in terms of Fourier multipliers on L_p. Moreover, in Sections 4 and 5 we use Fourier multipliers to study an analogue of dichotomy (hyperbolicity) for the smooth solutions.

A similar formula for $\omega_0(\mathbf{T})$ in terms of the resolvent of the generator was obtained in [CLRM], see also [LMS] and formula (5.20) in [CL]. Formally, Fourier multipliers have not been used in [LMS] and [CLRM]. The hyperbolicity of $\mathbf{T}$ was characterized in [LMS] and [CLRM], see also [CL], in terms of the invertibility of generator Γ of the evolution semigroup $\{E^t\}$ defined on $L_p(\mathbb{R};X)$ as $(E^tf)(\tau) = e^{tA}f(\tau-t)$. However, a simple calculation (see Remark 2.2 below) shows that $\Gamma^{-1} = -M$. Thus, formula (5.20) in [CL] for the growth bound of $\mathbf{T}$ is, in fact, a Fourier multiplier result that is generalized in Theorem 3.6 below.

Via completely different approach based on an explicit use of Fourier multipliers, M. Hieber [H2] gave a characterization of uniform stability for $\mathbf{T}$ in terms of Fourier multiplier properties of the resolvent. Also, he proved a formula for $\omega_0(\mathbf{T})$ that is contained in Theorem 3.6 when $\alpha=0$. An important ingredient of his proof was the use of well-known Datko-van Neerven Theorem saying that $\mathbf{T}$ is uniformly stable if and only if the convolution with $\mathbf{T}$ is a bounded operator on $L_p(\mathbb{R};X)$. Since the resolvent is the Fourier transform of $\mathbf{T}$, the Fourier multipliers characterization of uniform stability follows.

Among other things, this result with a different proof was given in [LR], where Datko-van Neerven Theorem was also used. In fact, Theorem 3.6 was proved in [LR] for $\alpha=0$ or $\alpha=1$. Also, a spectral mapping theorem from [LMS] was explained in [LR] using Fourier multipliers instead of evolution semigroups. In addition, a particular case of Theorem 4.1 of the current paper (with a different proof) was established in [LR]. Thus, in the present paper we use new technique to "tie the ends", and give a universal treatment for the results in [KVL, CLRM, H2, LR] in a more general context.

Acknowledgment. Yuri Latushkin was supported by the Summer Research Fellowship and by the Research Board of the University of Missouri. He thanks S. Verduyn Lunel for fruitful discussions during his visit to Amsterdam; without these discussions this paper would has not been written. Roman Shvydkoy was partially supported by the NSF grant DMS-9870027.

2. Characterization of hyperbolicity

Let us fix some notation:

- $\mathbf{T} = (T_t)_{t\geq 0}$ is a strongly continuous semigroup on a Banach space X with the generator A;
- $\mathcal{L}(X)$ – the set of bounded linear operators on X;
- $R(\lambda, A) = R(\lambda)$ is the resolvent of A;
- $\omega_0 = \omega_0(\mathbf{T})$ denotes the growth bound of $\mathbf{T}$, i.e. $\omega_0(\mathbf{T}) = \inf\{\omega \in \mathbb{R} : \|T_t\| \leq M_\omega e^{\omega t}\}$;
- $s_0(A)$ denotes the abscissa of uniform boundedness of the resolvent, i.e. $s_0(A) = \inf\{s \in \mathbb{R} : \sup\{\|R(\lambda)\| : \operatorname{Re}\lambda > s\} < \infty\}$;
- $r_\rho(s, x, x^*) = r_\rho(s) = \langle x^*, R(is + \rho)x\rangle$; $s \in \mathbb{R}$, $x \in X$, $x^* \in X^*$, $\rho \in \mathbb{R}$;
- $\widehat{f}(t) = \int_{\mathbb{R}} f(s)e^{-ist}ds$; $\check{f}(t) = \frac{1}{2\pi}\int_{\mathbb{R}} f(s)e^{ist}ds$;
- $\mathcal{S}$ stands for the class of Schwartz functions;
- $\langle r, \Phi\rangle$ denotes the value of a distribution r on $\Phi \in \mathcal{S}$.

Definition 2.1. *We say that the semigroup* $\mathbf{T}$ *is hyperbolic if there is a bounded projection* P *on* X*, called splitting, such that* $PT_t = T_tP$ *for all* $t > 0$ *and there exist positive numbers* ω *and* M *such that*

1. $\|T_tx\| \leq Ke^{-\omega t}\|x\|$, *for all* $t > 0$ *and* $x \in \operatorname{Im} P$,
2. $\|T_tx\| \geq Ke^{\omega t}\|x\|$, *for all* $t < 0$ *and* $x \in \operatorname{Ker} P$.

The semigroup $\mathbf{T}$ *is called uniformly exponentially stable if* $P = I$.

In other words, conditions 1 and 2 say that $(T_t)_{t\geq 0}$ is uniformly exponentially stable on $\operatorname{Im} P$; all the T_t's are invertible on $\operatorname{Ker} P$ and the semigroup $(T_{-t})_{t\geq 0}$ is uniformly exponentially stable there.

Definition 2.2. *The function*

$$G(t) = \begin{cases} T_tP, & t > 0 \\ -T_t(I - P), & t < 0 \end{cases}$$

is called the Green's function corresponding to the hyperbolic semigroup $\mathbf{T}$.

Definition 2.1 allows an equivalent reformulation in terms of spectral properties of $\mathbf{T}$. Namely, $\mathbf{T}$ is hyperbolic if and only if the unit circle $\mathbb{T}$ lies in the resolvent set of T_t for one/all t (see [EN, Proposition V.1.15]).

Let us recall the following inversion result.

Lemma 2.3. *Suppose $\rho > s_0(A)$ and $x \in X$, then*

$$F_t(x) = \frac{1}{2\pi i}(C,1)\int_{\mathrm{Re}\,\lambda=\rho} e^{\lambda t}R(\lambda)x d\lambda, \qquad t \in \mathbb{R},$$

where F_t is defined as

$$F_t(x) = \begin{cases} T_t x, & t > 0 \\ \frac{1}{2}x, & t = 0 \\ 0, & t < 0 \end{cases}.$$

In particular, $\check{r}_\rho(t,x,x^*) = e^{-\rho t}\langle x^*, F_t(x)\rangle$.

The proof can be found in [vN, Theorem 1.3.3]. See also Corollary 3.5.

Below we establish some algebraic properties of the distributions $\check{r}_\rho$ for small $|\rho|$ without any additional assumptions on $s_0(A)$. The reader will easily recognize the semigroup properties, in the case $s_0(A) < 0$.

In order to be able to threat r_ρ's as distributions and to justify some computations, we assume that the function $s \mapsto \|R(is)\|$ is bounded on $\mathbb{R}$, though the proofs below require merely that this function grows not faster then a power of $|s|$.

Lemma 2.4. *If $\tau > 0$, then*

$$\check{r}_0(t-\tau, T_\tau x, x^*) = \check{r}_0(t,x,x^*) - \langle x^*, T_t x\rangle \chi_{[0,\tau]}(t), \quad t \in \mathbb{R}.$$

Proof. Let us take arbitrary $\Phi \in \mathcal{S}$. Then

$$\begin{aligned}\langle \check{r}_0(\cdot-\tau, T_\tau x, x^*), \Phi\rangle &= \langle \check{r}_0(\cdot, T_\tau x, x^*), \Phi(\cdot+\tau)\rangle \\ &= \langle r_0(\cdot, T_\tau x, x^*), e^{-i\tau\cdot}\check{\Phi}\rangle = \int_{-\infty}^{+\infty} \langle x^*, R(is)e^{-is\tau}T_\tau x\rangle \check{\Phi}(s)ds.\end{aligned}$$

Note that

$$e^{-is\tau}R(is)T_\tau x = R(is)x - \int_0^\tau T_r x \cdot e^{-isr}dr. \tag{1}$$

Continuing the line of equalities, we obtain:

$$\begin{aligned}\langle \check{r}_0(\cdot-\tau, T_\tau x, x^*), \Phi\rangle &= \int_{-\infty}^{+\infty} \langle x^*, R(is)x\rangle \check{\Phi}(s)ds \\ &- \int_{-\infty}^{+\infty} \left\langle x^*, \int_0^\tau T_r e^{-isr}xdr\right\rangle \check{\Phi}(s)ds = \langle \check{r}_0, \Phi\rangle \\ &- \int_0^\tau \langle x^*, T_r x\rangle \Phi(r)dr = \langle \check{r}_0 - \langle x^*, T_\cdot x\rangle \chi_{[0,\tau]}(\cdot), \Phi\rangle.\end{aligned}$$

□

Lemma 2.5. $\check{r}_0(t, T_\tau x, x^*) = \check{r}_0(t, x, T_\tau^* x^*)$, $\tau > 0$, $t \in \mathbb{R}$.

The proof is obvious.

Lemma 2.6. $\check{r}_0(t) = e^{\rho t}\check{r}_\rho(t)$ *for all* ρ *with* $|\rho| < \rho_0$ *and* $t \in \mathbb{R}$.

Proof. We choose ρ_0 such that $\sup\{\|R(is+\rho)\| : s \in \mathbb{R}, |\rho| < \rho_0\}$ is finite. Suppose $\Phi \in \mathcal{S}$ has compact support. Then $\check{\Phi}$ is an entire function. Moreover,

$$\lim_{\alpha\to\infty} \check{\Phi}(\alpha + i\beta) = 0, \tag{2}$$

uniformly for all β from some finite interval $[a, b]$. It is an immediate consequence of the following equality:

$$\check{\Phi}(\alpha + i\beta) = \int_{\mathbb{R}} e^{\beta x}\Phi(x)e^{i\alpha x}dx = -\frac{1}{i\alpha}\int_{\mathbb{R}}[\beta e^{\beta x}\Phi(x) + e^{\beta x}\Phi'(x)]e^{i\alpha x}dx.$$

Now using Cauchy's theorem and (2) we get

$$\begin{aligned}
\langle \check{r}_\rho, \Phi\rangle &= \langle r_\rho, \check{\Phi}\rangle = \int_{\mathbb{R}}\langle x^*, R(is+\rho)x\rangle\check{\Phi}(s)ds \\
&= \int_{\mathbb{R}}\langle x^*, R(i(s-i\rho))x\rangle\check{\Phi}(s)ds = \int_{\mathbb{R}-i\rho}\langle x^*, R(i\lambda)x\rangle\check{\Phi}(\lambda+i\rho)d\lambda \\
&= \int_{\mathbb{R}}\langle x^*, R(is)x\rangle\check{\Phi}(s+i\rho)ds \\
&\quad + 2i\lim_{\alpha\to\pm\infty}\int_0^\rho \langle x^*, R(i(\alpha+i\beta))x\rangle\check{\Phi}(\alpha+i(\beta+\rho))d\beta \\
&= \int_{\mathbb{R}}\langle x^*, R(is)x\rangle\check{\Phi}(s+i\rho)ds = \langle r_0, \check{\Phi}(\cdot+i\rho)\rangle \\
&= \langle \check{r}_0, e^{-\rho\cdot}\Phi\rangle,
\end{aligned}$$

and the result follows. □

Now we are in a position to prove our main theorem. Let us denote by M_ρ the operator acting by the rule

$$M_\rho : f \mapsto [R(i\cdot+\rho)\hat{f}]^\vee.$$

Recall that a function $m \in L_\infty(\mathbb{R}; \mathcal{L}(X))$ is called a *Fourier multiplier* on $L_p(\mathbb{R}; X)$ if the operator $M : f \mapsto [m(\cdot)\hat{f}]^\vee$ is a bounded operator on $L_p(\mathbb{R}; X)$. Let $L_{1,\infty}(\mathbb{R}; X)$ denote the weak-L_1 space with values in X (see, e.g., [T, 1.18.6]), that is, the set of all X-valued strongly continuous functions f with the finite norm

$$\|f\|_{L_{1,\infty}} := \sup_{\sigma>0}\left\{\sigma\text{mes}\left(\{s \in \mathbb{R} : \|f(s)\| \geq \sigma\}\right) < \infty\right\}.$$

Note that $L_{1,\infty}(\mathbb{R}; X) \subset L_1(\mathbb{R}; X)$.

Theorem 2.7. *For a strongly continuous semigroup* $\mathbf{T}$ *on* X *the following conditions are equivalent:*

1) $\mathbf{T}$ *is hyperbolic;*
2) $R(i\cdot)$ *is a Fourier multiplier on* $L_p(\mathbb{R}, X)$ *for some/all* p, $1 \leq p < \infty$*;*
3) *There exists a* $\rho_0 > 0$ *such that for all* ρ *with* $|\rho| < \rho_0$, M_ρ *maps* $L_1(\mathbb{R}, X)$ *into* $L_{1,\infty}(\mathbb{R}, X)$*;*
4) *There exists a* $\rho_0 > 0$ *such that for all* ρ *with* $|\rho| < \rho_0$ *and all* $\Phi \in \mathcal{S}$ *we have* $|\langle r_\rho, \Phi\rangle| \leq K_\rho \|x\|\|x^*\|\|\check{\Phi}\|_1$.

Furthermore, if one of these properties holds, then for every $t \in \mathbb{R}$ *and* $x \in X$ *the integral*

$$G(t)x = \frac{1}{2\pi}(C, 1) \int_{\mathbb{R}} R(is)xe^{ist}ds$$

converges and represents the Green's function of $\mathbf{T}$. *Moreover,* $M_0 f = G * f$ *for* $f \in L_1(\mathbb{R}, X)$, *and the splitting projection is given by the formula*

$$P = \frac{1}{2}I + G(0). \tag{3}$$

Proof. 1)⇒4). This is a part of Theorem 0.2 from [KVL].

4)⇒2). It follows from 4) that $\check{r}_\rho \in L_\infty, |\rho| < \rho_0$ and $\|\check{r}_\rho\|_\infty \leq K_\rho\|x\|\|x^*\|$. By Lemma 2.6, $\check{r}_0(t) = e^{-\rho t}\check{r}_{-\rho}(t)$ a.e. and $\check{r}_0(t) = e^{\rho t}\check{r}_\rho(t)$ a.e. for some $\rho > 0$. So, $|\check{r}_0(t)| \leq e^{-\rho|t|}K\|x\|\|x^*\|$ a.e. for every $x \in X$ and $x^* \in X^*$, where $K = \max\{K_\rho, K_{-\rho}\}$. Now let us fix p, $1 \leq p < \infty$, and consider a function $\Phi = \sum_{k=1}^n x_k \otimes \Phi_k$, where $\Phi_k \in \mathcal{S}$ and $\{\Phi_k\}$ have disjoint supports. Then $\|\Phi\|_{L_p}^p = \sum_{k=1}^n \|x_k\|^p\|\Phi_k\|_{L_p}^p$. So, we get the following estimates:

$$\begin{aligned}
\|M_0(\Phi)\|_{L_p}^p &= \int_{\mathbb{R}} \|M_0(\Phi)(t)\|^p dt = \frac{1}{2\pi}\int_{\mathbb{R}}\left\|\int_{\mathbb{R}} R(is)\hat{\Phi}(s)e^{ist}ds\right\|^p dt \\
&= \frac{1}{2\pi}\int_{\mathbb{R}} \sup_{\|x^*\|\leq 1}\left|\sum_{k=1}^n \int_{\mathbb{R}} r_0(s, x_k, x^*)\hat{\Phi}_k(s)e^{ist}ds\right|^p dt \\
&= \frac{1}{2\pi}\int_{\mathbb{R}} \sup_{\|x^*\|\leq 1}\left|\sum_{k=1}^n \int_{\mathbb{R}} \check{r}_0(\tau, x_k, x^*)\Phi_k(t-\tau)d\tau\right|^p dt \\
&\leq K^p \int_{\mathbb{R}}\left(\sum_{k=1}^n \int_{\mathbb{R}} e^{-\rho|\tau|}\|x_k\||\Phi_k(t-\tau)|d\tau\right)^p dt \\
&= K^p \int_{\mathbb{R}}\left(\int_{\mathbb{R}} e^{-\rho|\tau|}\left(\sum_{k=1}^n \|x_k\||\Phi_k(t-\tau)|\right)d\tau\right)^p dt \\
&\leq C_\rho K^p \int_{\mathbb{R}}\int_{\mathbb{R}} e^{-\rho|\tau|}\left(\sum_{k=1}^n \|x_k\||\Phi_k(t-\tau)|\right)^p d\tau dt
\end{aligned}$$

$$= C_\rho K^p \int_{\mathbb{R}} e^{-\rho|\tau|} \int_{\mathbb{R}} \left(\sum_{k=1}^{n} \|x_k\| |\Phi_k(t-\tau)| \right)^p dt d\tau$$

$$= C_\rho K^p \int_{\mathbb{R}} e^{-\rho|\tau|} \sum_{k=1}^{n} \|x_k\|^p \|\Phi_k\|_{L_p}^p d\tau = C'_\rho K^p \|\Phi\|_{L_p}^p.$$

Since the functions Φ are dense in $L_p(\mathbb{R}, X)$, the proof of 4)⇒2) is finished.

2)⇒1). Suppose 2) holds for some p, $1 \le p < \infty$. Then, by the transference principle (see, for example, [SW, Thm. VII.3.8]), $\{R(ik + i\xi)\}_{k\in\mathbb{Z}}$ is a multiplier in $L_p(\mathbb{T}, X)$ for all $\xi \in \mathbb{R}$, where $\mathbb{T}$ is the unit circle. So, using results from [LMS, Theorem 2.3] or [LR, Theorem 1], we conclude that $e^{2\pi i\xi} \in \rho(T_{2\pi})$ for all $\xi \in \mathbb{R}$. Thus, $\mathbb{T} \subset \rho(T_{2\pi})$ and hence $\mathbf{T}$ is hyperbolic.

This completes the proof of 1)⇔2)⇔4).

2)⇒3). It is easy to see using the resolvent identity, that there exists a $\rho_0 > 0$ such that $R(i\cdot +\rho)$ is a $L_1(\mathbb{R}, X)$-multiplier for all ρ such that $|\rho| < \rho_0$.

3)⇒4). Without loss of generality, assume $\rho = 0$. Denote

$$\mu = \sup_{0\le\tau\le1} \|T_\tau\|$$

and fix $x \in X$, $x^* \in X^*$, $\|x\| = \|x^*\| = 1$. Let us take a function $\Phi \in \mathcal{S}$. By condition 3) we have

$$\|M_0(\check{\Phi} \otimes x)\|_{1,\infty} \le K\|\hat{\Phi}\|_1.$$

So, $\mathrm{mes}\{\tau : \|M_0(\hat{\Phi} \otimes x)(\tau)\| > 2K\|\hat{\Phi}\|_1\} \le \frac{1}{2}$. This implies that there is a τ, $-1 < \tau < 0$, such that

$$\|M_0(\check{\Phi} \otimes x)(\tau)\| \le 2K\|\hat{\Phi}\|_1.$$

Let us apply the functional $T^*_{-\tau}x^*$ to the left-hand side of the inequality. Then we have:

$$\left|\frac{1}{2\pi} < T^*_{-\tau}x^*, [R\cdot\Phi\otimes x]^\vee(\tau) > \right| \le 2\mu KC\|\hat{\Phi}\|_1.$$

By Lemma 2.4 and 2.5 the expression under the absolute value sign is equal to

$$\check{r}_0(\cdot, x, T^*_{-\tau}x^*) * \check{\Phi}(\tau) = \langle \check{r}_0(\cdot, T_{-\tau}x, x^*), \check{\Phi}(\tau - \cdot)\rangle$$

$$= \langle \check{r}_0(\cdot + \tau, T_{-\tau}x, x^*), \hat{\Phi}\rangle = \langle \check{r}_0(\cdot, x, x^*), \hat{\Phi}\rangle - \int_0^{-\tau} \langle x^*, T_t x\rangle \hat{\Phi}(t) dt.$$

By the triangle inequality, we have

$$|\langle r_0(\cdot, x, x^*), \Phi\rangle| \le 2K\mu\|\hat{\Phi}\|_1 + \mu\|\hat{\Phi}\|_1 \le 3K\mu\|\check{\Phi}\|_1,$$

which is what we wanted.

Now we turn to the second part of the theorem. First, we prove an auxiliary Fejér-type lemma (probably, well-known).

Lemma 2.8. *If $f \in L_1(\mathbb{R}, X)$, then the integral*

$$\frac{1}{2\pi}(C,1)\int_{\mathbb{R}} \hat{f}(s)e^{ist}ds$$

converges to $f(t)$ a.e. Moreover,

$$f = \frac{1}{2\pi}L_1 - \lim_{N\to\infty}\frac{1}{N}\int_0^N\int_{-\ell}^{\ell}\hat{f}(s)e^{is\cdot}ds.$$

Proof.

$$\begin{aligned}\frac{1}{2\pi}\frac{1}{N}\int_0^N\int_{-\ell}^{\ell}\hat{f}(s)e^{ist}ds &= \frac{1}{2\pi}\int_{-N}^{N}\hat{f}(s)e^{ist}\left(1-\frac{|s|}{N}\right)ds\\ &= \frac{1}{2\pi}\int_{-N}^{N}\int_{-\infty}^{+\infty}f(r)e^{isr}dr\cdot e^{ist}\left(1-\frac{|s|}{N}\right)ds\\ &= \int_{-\infty}^{+\infty}f(r)\frac{1}{2\pi}\int_{-N}^{N}e^{is(t-r)}\left(1-\frac{|s|}{N}\right)dsdr.\end{aligned}$$

The inner integral is equal to $K_N(t-r) = \frac{1}{\pi N(t-r)^2}[1-\cos N(t-r)]$. One can easily check that K_N is a positive kernel in L_1, that is, $(K_N * f)(\cdot)$ tends to $f(\cdot)$ a.e. and in L_1 as $N \to \infty$. □

Suppose $f \in L_1(\mathbb{R})$. Then by 2) we have that $M_0(f \otimes x) \in L_1(\mathbb{R}, X)$. By Lemma 2.8, there is a $\tau \in (-1, 0)$ such that

$$M_0(f\otimes x)(\tau) = \frac{1}{2\pi}(C,1)\int_{\mathbb{R}} R(is)x\hat{f}(s)e^{is\tau}ds.$$

Let us apply the operator $T_{-\tau}$. Then using (1) we obtain:

$$\begin{aligned}T_{-\tau}([R\hat{f}\otimes x]^{\vee}(\tau)) &= \frac{1}{2\pi}(C,1)\int_{\mathbb{R}} R(is)T_{-\tau}xe^{is\tau}\hat{f}(s)ds\\ &= \frac{1}{2\pi}(C,1)\int_{\mathbb{R}}\Big[R(is)x\hat{f}(s)\\ &\qquad - \int_0^{-\tau}T_r xe^{-isr}dr\cdot\hat{f}(s)\Big]ds.\end{aligned}$$

Since

$$f(-r) = L_1 - \lim_{N\to\infty}\frac{1}{2\pi}\frac{1}{N}\int_0^N\int_{-\ell}^{\ell}\hat{f}(s)e^{-isr}dsd\ell$$

and $V\varphi = \int_0^{-\tau} T_r x\cdot\varphi(r)dr$ is a bounded linear operator from $L_1(\mathbb{R})$ to X, we conclude that the $(C,1)$-integral of the second summand converges and equals $\int_0^{-\tau} T_r x\cdot f(-r)dr$. This means, in particular, that $\frac{1}{2\pi}(C,1)\int_{\mathbb{R}} R(is)x\hat{f}(s)ds$ converges. Let us denote it by $G(0,f)$. Also let

$$G(t,f) = G(0, f(\cdot - t)) = \frac{1}{2\pi}(C,1)\int_{\mathbb{R}} R(is)x\hat{f}(s)e^{ist}ds$$

for $f \in L_1$, $t \in \mathbb{R}$, $x \in X$. Now we introduce the following operators:

$$S_N^t(f,x) = \frac{1}{2\pi}\frac{1}{N}\int_0^N \int_{-\ell}^{\ell} R(is)x\hat{f}(s)e^{ist}dsd\ell;$$

$$I_N^t(x) = \frac{1}{2\pi}\frac{1}{N}\int_0^N \int_{-\ell}^{\ell} R(is)xe^{ist}dsd\ell. \tag{4}$$

It is easy to see that $\|S_N^t(f,x)\| \le C_N\|f\|_{L_1}\|x\|$. On the other hand, we have just proved that $G(t,f)x = \lim_{N\to\infty} S_N^t(f,x)$ exists for all $f \in L_1$, $x \in X$. So, by the boundedness principle for bilinear operators, $\|S_N^t\| \le C$, where C does not depend on N and t.

Let f_ϵ, $\epsilon > 0$, be a kernel in $L_1(\mathbb{R})$, that is, $f_\epsilon * \Phi \to \Phi$ as $\epsilon \to 0$ for each $\Phi \in L_1(\mathbb{R})$. Then $I_N^t(x) = \lim_{\epsilon\to 0} S_N^t(f_\epsilon, x)$ and hence, $\|I_N^t\| \le C$.

Let us show that $G(t)x = \lim_{N\to\infty} I_N^t(x)$ exists for all $x \in D(A^2)$. This will be enough to prove that $G(t)x$ exists for all $x \in X$. Fix $x \in D(A^2)$ and notice that

$$\begin{aligned}
I_N^t(x) &= \frac{1}{2\pi}\frac{1}{N}\int_0^N\int_{-\ell}^{\ell} R(is)xe^{ist}dsd\ell \\
&= \frac{1}{2\pi}\frac{1}{N}\int_0^1\int_{-\ell}^{\ell} R(is)xe^{ist}dsd\ell + \frac{1}{2\pi}\frac{1}{N}\int_1^N\int_{|s|\le 1} R(is)xe^{ist}dsd\ell \\
&\quad + \frac{1}{2\pi}\frac{1}{N}\int_1^N\int_{1\le|s|\le\ell} R(is)xe^{ist}dsd\ell \\
&= \frac{1}{2\pi}\frac{1}{N}\int_0^1\int_{-\ell}^{\ell} R(is)xe^{ist}dsd\ell + \frac{1}{2\pi}\frac{N-1}{N}\int_{|s|\le 1} R(is)xe^{ist}ds \\
&\quad + \frac{1}{2\pi}\frac{1}{N}\int_1^N\int_{1\le|s|\le\ell}\left[-\frac{R(is)A^2x}{s^2} + \frac{x}{is} - \frac{Ax}{s^2}\right]e^{ist}dsd\ell.
\end{aligned}$$

So,

$$\begin{aligned}
\lim_{N\to\infty} I_N^t(x) &= \frac{1}{2\pi}\int_{|s|\le 1} R(is)xe^{ist}ds \\
&\quad - \frac{1}{2\pi}\int_{|s|\ge 1}\left[\frac{Ax}{s^2} + \frac{R(is)A^2x}{s^2}\right]e^{ist}ds \\
&\quad + \frac{x}{2\pi i}\int_t^{+\infty}\frac{sin(s)}{s}ds \cdot \chi_{\mathbb{R}\setminus\{0\}}(t).
\end{aligned} \tag{5}$$

Finally, it is only left to verify that $G(t)$ is indeed the Green's function. Let us prove the first equality in Definition 2.2, the second one being analogous. We have

$$\begin{aligned}T_\tau Px &= \frac{1}{2}T_\tau x + \frac{1}{2\pi}(C,1)\int_{\mathbb{R}} R(is)xT_\tau x ds \\ &= \frac{1}{2}T_\tau x + \frac{1}{2\pi}(C,1)\int_{\mathbb{R}} R(is)xe^{ist} ds \\ &\quad - \frac{1}{2\pi}(C,1)\int_{\mathbb{R}}\int_0^\tau T_r xe^{-isr}dr \cdot e^{is\tau}ds \\ &= G(\tau)x + \frac{1}{2}T_\tau x - \frac{1}{2}T_\tau x = G(\tau)x,\end{aligned}$$

where we use the ordinary Fejér's theorem.

It follows from the above that $G(\tau)$ is an exponentially decaying function. So, $f \mapsto G * f$ is a bounded operator on L_1. On the other hand, $M_0\Phi = G * \Phi$ for all $\Phi \in \mathcal{S}$. Hence, $M_0 f = G * f$ for all $f \in L_1$.

The proof of (3) can be found in [KVL]. □

Corollary 2.9 ([G, H]). *Suppose X is a Hilbert space. Then the semigroup $\mathbf{T}$ is hyperbolic if and only if the resolvent $R(\lambda, A)$ is bounded in some strip containing the imaginary axes.*

Remark 2.10. *Condition 3) can be considerably weakened in the following way. Suppose F is a space of functions on $\mathbb{R}$ with the following property: for any $f \in F$ there is a $t \in [-1,0]$ such that $|f(t)| \leq c\|f\|_F$. Many quasi-normed function spaces have this property, for example, $L_{q,r}(\mathbb{R})$, $H_p(\mathbb{R})$, $C_0(\mathbb{R})$, or any function lattice with $\|\chi_{[-1,0]}\| \neq 0$. Denote by $F(X)$ the space of all strongly measurable functions f with values in X such that $\|f(\cdot)\| \in F$. Our proof shows that it is enough to require that M_ρ maps $L_1(\mathbb{R}, X)$ into $F(X)$ (see also the proof of Theorem 3.1).*

Remark 2.11. *Recall that the generator Γ of the evolution semigroup $(E^t)_{t\geq 0}$, defined on $L_p(\mathbb{R}, X)$ by $(E^t f)(s) = T_t f(s-t)$, is the closure of the operator $-d/dt + A$ on the domain $D(-d/dt) \cap D(A)$. It is known, see* [CL, Thm.2.39], *that $\mathbf{T}$ is hyperbolic if and only if the operator Γ is invertible on one/all $L_p(\mathbb{R}, X)$, $1 \leq p < \infty$. This result immediately implies that conditions 1) and 2) in Theorem 2.7 are equivalent. Indeed, if $x \in D(A)$ and $\Phi \in \mathcal{S}$, then $\Gamma(\Phi \otimes x) = -\Phi' \otimes x + \Phi \otimes Ax$. Using elementary properties of Fourier transform, we have that*

$$M_0\Gamma(\Phi \otimes x) = -\Phi \otimes x, \quad x \in D(A)$$

and

$$\Gamma M_0(\Phi \otimes x) = -\Phi \otimes x, \quad x \in X,$$

and the result follows.

It is worth noting that in the special case $s_0(A) < 0$, by Lemma 2.3, the splitting projection turns into the identity and our theorem gives the characterization of uniform exponential stability observed in [LR].

There is a Mikhlin-type sufficient condition due to M. Hieber [H1] for an operator-valued symbol to be L_1-multiplier. Applied to the resolvent it yields the following: if there exists a $\delta > \frac{3}{4}$ such that $\sup\{|s|^{\delta}\|R(is)\|\} < \infty$, then $R(i\cdot)$ is a multiplier.

Yet another condition for operator-valued symbol to be a multiplier is recently developed in [W1]. It works if X is a UMD-space and says that if the families $\{R(is)\}_{s\in\mathbb{R}}$ and $\{sR^2(is)\}_{s\in\mathbb{R}}$ are R-bounded, then $R(i\cdot)$ is a multiplier.

3. Extension to the case $\alpha > 0$

It turns out that many arguments from Section 2 work in a more general situation, when the resolvent multiplier is restricted to $L_p(\mathbb{R}, X_\alpha)$, where X_α is the domain of the fractional power $(A-\omega)^\alpha$, endowed with the norm $\|x\|_\alpha = \|(A-\omega)^\alpha x\|$. In this section we show that $R(i\cdot+\rho)$ is a multiplier from $L_p(\mathbb{R}, X_\alpha)$ to $L_p(\mathbb{R}, X)$ for small values of ρ if and only if the following modified Kaashoek - Verduyn Lunel inequality holds: $|\langle r_\rho, \Phi\rangle| \le K\|x\|_\alpha\|x^*\|\|\hat{\Phi}\|_{L_1}$. Also in this case $G(t)x$ exists for all $x \in X_\alpha$ and is exponentially decaying as $|t| \to \infty$. As a by-product of this results we obtain the following relationship between the fractional growth bound $\omega_\alpha(\mathbf{T})$ and its spectral analogue $s_\alpha(A)$ (see (10) for the definitions): $\omega_\alpha(\mathbf{T})$ is the infimum of all $\omega > s_\alpha(A)$ such that $R(i\cdot+\omega)$ is a multiplier from $L_p(\mathbb{R}, X_\alpha)$ to $L_p(\mathbb{R}, X)$. In the particular case, when X is a Hilbert space, the latter condition will be shown to hold for all $\omega > s_\alpha(A)$. So, $s_\alpha(A) = \omega_\alpha$, which gives a different proof of G. Weiss's [Ws] result for arbitrary $\alpha \ge 0$, also obtained by L. Weis and V. Wrobel in [WW]. The main result in this section is an extension of Theorem 2.7 to the case of arbitrary $\alpha > 0$. To be more precise, we treat only conditions 2)-4), as hyperbolicity is ambiguous in this situation and therefore it is postponed to the next section.

One can notice that most of the proof of Theorem 2.7 work for all $\alpha > 0$ if one replaces all X-norms by X_α-norms. However, the "some/all" part of condition 2), being an easy consequence of results in [LMS] and the spectral characterization of hyperbolicity in case $\alpha = 0$, requires some additional duality argument.

Before we state our main theorem, let us recall the notion of fractional power of A. Suppose $\omega > \max\{\omega_0+3, 3\}$. Denote $A-\omega$ by A_ω. Let γ be the path consisting of two rays $\Gamma_1 = \{-1+te^{i\theta} : t \in [0,+\infty)\}$ and $\Gamma_2 = \{-1-te^{i\theta} : t \in [0,+\infty)\}$ going upwards. We assume that θ, $\theta < \frac{\pi}{6}$, is small enough to ensure the inequality $\|R(\mu+\omega)\| \le C\frac{1}{1+|\mu|}$ in the sector generated by γ. For any $\alpha > 0$ we define A_ω^α as the inverse to the operator $A_\omega^{-\alpha}$ acting on X by the rule

$$A_\omega^{-\alpha}(x) = \frac{1}{2\pi i}\int_\gamma \mu^{-\alpha}R(\mu+\omega)x\,d\mu.$$

Let us denote by X_α the domain of A_ω^α endowed with the norm $\|x\|_\alpha = \|(A-\omega)^\alpha x\|$. Then X_α is a Banach space and it does not depend on the particular choice of ω, $\omega > \omega_0$, see [EN] for more information concerning fractional powers.

Theorem 3.1. *Assume that there exists a $\rho_0 > 0$ such that*

$$\sup\left\{\frac{\|R(\lambda)\|}{1+|\lambda|^\alpha} : |\operatorname{Re}\lambda| < \rho_0\right\} < \infty. \tag{6}$$

Then the following conditions are equivalent:

1) *$R(i\cdot+\rho)$ is a multiplier from $L_p(\mathbb{R},X_\alpha)$ to $L_p(\mathbb{R},X)$, for some/all p, $1 \le p < \infty$, and all ρ, $|\rho| < \rho_0$;*
2) *$R(i\cdot+\rho)$ is a multiplier from $L_p(\mathbb{R},X_\alpha)$ to $E(X)$, for some p, $1 \le p < \infty$, and all ρ, $|\rho| < \rho_0$, where E is a rearrangement invariant quasi-Banach lattice;*
3) *$R(i\cdot+\rho)$ is a multiplier from $L_1(\mathbb{R},X_\alpha)$ to $F(X)$ for all ρ, $|\rho| < \rho_0$, where F is some rearrangement invariant quasi-Banach lattice;*
4) *$|\langle r_\rho, \Phi\rangle| \le K\|x\|_\alpha\|x^*\|\|\hat{\Phi}\|_{L_1}$ for all $\Phi \in \mathcal{S}$, $x \in X_\alpha$, $x^* \in X^*$ and $|\rho| < \rho_0$.*

If one of these conditions holds, then the integral

$$G(t)x = \frac{1}{2\pi}(C,1)\int_{\mathbb{R}} R(is)xe^{ist}ds$$

converges for all $x \in X_\alpha$, and $\|G(t)x\| \le K\|x\|_\alpha e^{-\rho|t|}$ for all ρ, $0 < \rho < \rho_0$. Moreover, $M_0 f = G * f$ for $f \in L_1(\mathbb{R},X)$.

Proof. 1)⇒2) is evident.

2)⇒3). Assume for simplicity that $\rho = 0$. First we claim that M_0 maps $L_p(\mathbb{R},X_\alpha)$ into $L_\infty(\mathbb{R},X)$. To prove this, let us take an arbitrary function Φ of the form $\sum_{i=1}^n \Phi_i x_i$, where $x_i \in X_\alpha$ and $\Phi_i \in \mathcal{S}$. Then, by condition 2),

$$\|M_0(\Phi)\|_{E(X)} \le K\|\Phi\|_{L_p(\mathbb{R},X_\alpha)}.$$

It implies that for every $n \in \mathbb{Z}$ there exists a $t \in [n, n+1]$ such that

$$\|[R\hat{\Phi}]^\vee(t)\|_X \le \frac{2K}{\varphi(1)}\|\Phi\|_{L_p(\mathbb{R},X_\alpha)},$$

where φ is the characteristic function of E. For any fixed $\tau \in [0,2]$, let us apply the operator T_τ to the right-hand side of this inequality. Then we get

$$\|\frac{1}{2\pi}\int_{\mathbb{R}} T_\tau R(is)\hat{\Phi}(s)e^{ist}ds\| \le C\|\Phi\|_{L_p(\mathbb{R},X_\alpha)}.$$

Now using equality (1) we obtain the following

$$\begin{aligned}\frac{1}{2\pi}\int_{\mathbb{R}} T_\tau R(is)\hat{\Phi}(s)e^{ist}ds &= \frac{1}{2\pi}\int_{\mathbb{R}} R(is)\hat{\Phi}(s)e^{is(t+\tau)}ds \\ &\quad -\frac{1}{2\pi}\int_{\mathbb{R}}\int_0^\tau T_r\hat{\Phi}(s)e^{is(\tau-r)}drds \\ &= M_0(\Phi)(t+\tau) - \frac{1}{2\pi}\int_0^\tau T_r\Phi(\tau-r)dr.\end{aligned}$$

Thus, $\|M_0(\Phi)(t+\tau)\|_X \le \tilde{C}\|\Phi\|_{L_p(\mathbb{R},X_\alpha)}$. By the choice of τ and t we have the same inequality on the whole real line. Since τ was chosen arbitrary, the claim is proved.

Let us observe that the boundedness of M_0 is equivalent to the fact that $R(i\cdot)A_\omega^{-\alpha}$ is an $L_p(\mathbb{R},X) - L_\infty(\mathbb{R},X)$ multiplier.

Denote by $X^\odot$ the sun dual to X on which the dual semigroup is strongly continuous (see [EN]). One can easily check, by duality, that for a test function $\Phi = \sum_{i=1}^n \Phi_i x_i^\odot$ one has

$$\left\|[R^\odot(A^\odot)^{-\alpha}\hat{\Phi}]^\vee\right\|_{L_q(\mathbb{R},X^\odot)} \le \tilde{C}\|\Phi\|_{L_1(\mathbb{R},X^\odot)},$$

where $\frac{1}{p}+\frac{1}{q}=1$ and $A^\odot$ is the generator of the sun dual semigroup. In other words, $M_0^\odot$ maps $L_1(\mathbb{R},X_\alpha^\odot)$ into $L_q(\mathbb{R},X^\odot)$.

By what we just proved, $M_0^\odot$ is bounded from $L_1(\mathbb{R},X_\alpha^\odot)$ to $L_\infty(\mathbb{R},X^\odot)$, and again by duality, M_0 maps $L_1(\mathbb{R},X_\alpha)$ into $L_\infty(\mathbb{R},X)$, which proves condition 3) with $F=L_\infty$.

The proofs of all other implications are completely analogous to those of Theorem 2.7.

Let us now turn to the second part of our theorem. Although its proof is also essentially the same, some comments will be in order. By Lemma 3.2, proved below, assumption (6) is equivalent to $\sup\{\|R(\lambda)\|_{X_\alpha\to X} : |\operatorname{Re}\lambda| < \rho_0\} < \infty$. So, the operators S_N^t, introduced in (4), are bounded from $X_\alpha \times L_1(\mathbb{R})$ to X. Uniform boundedness follows from the fact that $\lim\limits_{N\to\infty} S_N^t(x,f)$ exists for all $x\in X_\alpha$ and $f\in L_1(\mathbb{R})$ by Lemma 2.8. Consequently, $\|I_N^t\|_{X_\alpha\to X}\le C$. Formula (5) still makes sense for all $x\in X_{\alpha+2}$, because then $A^2x\in X_\alpha$ and all the integrals converge absolutely. So, $G(t)x$ exists for all $x\in X_\alpha$, and it is continuous in t, $t\ne 0$.

Since $\langle x^*, G(t)x\rangle = \check{r}_0(t,x,x^*)$, by condition 3) and Lemma 2.6 we have that $|\langle x^*, G(t)x\rangle| \le Ke^{-\rho|t|}\|x\|_\alpha\|x^*\|$ almost everywhere and hence, by the continuity of $G(t)x$, for all $t\in\mathbb{R}$. Thus, $\|G(t)x\|_X \le Ke^{-\rho|t|}\|x\|_\alpha$ and the proof is finished. □

Lemma 3.2. *Let $S=\{\lambda\in\mathbb{C} : a<\operatorname{Re}\lambda<b\}$, $a,b\in\mathbb{R}$, be a subset of $\rho(A)$, where $a\in\mathbb{R}$, $b\in\mathbb{R}$. Then conditions*

$$\sup\left\{\frac{\|R(\lambda)\|}{1+|\lambda|^\alpha} : \lambda\in S\right\} < \infty \quad \textit{and} \quad \sup\left\{\|R(\lambda)A_\omega^{-\alpha}\| : \lambda\in S\right\}<\infty$$

are equivalent.

Proof. Since b is finite, there are constants $c>0$ and φ_0, $0<\varphi_0<\pi$ such that $|\mu - e^{i\varphi}| > |\mu| + c$ for all $\mu\in\gamma$ and $\varphi_0<|\varphi|<\pi-\varphi_0$. Pick $N>1$ large enough to satisfy $\frac{\omega}{N}<\frac{c}{2}$ and such that whenever $\lambda\in S$ and $|\lambda|>N$, then $\varphi_0<|\arg\lambda|<\pi-\varphi_0$ and λ does not belong to the sector bounded by the contour $|\lambda|\gamma$. For all such λ we have

$$|\mu + \frac{\omega}{|\lambda|} - e^{i\arg\lambda}| > |\mu| + \frac{c}{2}. \tag{7}$$

Let us consider the following integral:

$$I_\lambda = \int_\gamma \frac{\mu^{-\alpha}}{\mu+\omega-\lambda} d\mu, \quad \lambda \in S, \quad |\lambda| > N.$$

By the choice of N, the integrand does not have singular points between γ and $|\lambda|\gamma$. By the Cauchy Theorem, we have

$$I_\lambda = \int_{|\lambda|\gamma} \frac{\mu^{-\alpha}}{\mu+\omega-\lambda} d\mu = \frac{1}{|\lambda|^\alpha} \int_\gamma \frac{\mu^{-\alpha}}{\mu + \frac{\omega}{|\lambda|} - e^{i \arg \lambda}} d\mu.$$

Inequality (7) implies that the absolute value of the last integral is bounded from above by a constant that does not depend on λ, whenever $\lambda \in S$, $|\lambda| > N$. The analogous estimate from below follows from geometric considerations. Thus,

$$\frac{d_1}{|\lambda|^\alpha} \le |I_\lambda| \le \frac{d_2}{|\lambda|^\alpha}, \tag{8}$$

for some positive d_1 and d_2.

Suppose $x \in X$. Then

$$\begin{aligned} R(\lambda)A_\omega^{-\alpha}x &= \frac{1}{2\pi i}\int_\gamma \mu^{-\alpha}R(\lambda)R(\mu+\omega)x d\mu \\ &= \frac{1}{2\pi i} I_\lambda R(\lambda)x - \frac{1}{2\pi i}\int_\gamma \frac{\mu^{-\alpha}}{\mu+\omega-\lambda} R(\mu+\omega)x d\mu. \end{aligned}$$

Let us notice that $|\mu+\omega-\lambda| \ge K(|\mu|+1)$ for some $K>0$ and all $\mu \in \gamma$, $\lambda \in S$, $|\lambda| > N$, whereas $\|R(\mu+\omega)\| \le C\frac{1}{1+|\mu|}$. Consequently,

$$\left\| \int_\gamma \frac{\mu^{-\alpha}}{\mu+\omega-\lambda} R(\mu+\omega)x d\mu \right\| \le K_\alpha \|x\|.$$

In combination with (8) this gives the following estimates:

$$\begin{aligned} \|R(\lambda)A_\omega^{-\alpha}x\| &\ge d_1 \frac{\|R(\lambda)x\|}{|\lambda|} - K_\alpha\|x\|, \\ \|R(\lambda)A_\omega^{-\alpha}x\| &\le d_2 \frac{\|R(\lambda)x\|}{|\lambda|} + K_\alpha\|x\|, \end{aligned}$$

for all $\lambda \in S$, $|\lambda| > N$ and $x \in X$, which proves the lemma. □

Remark 3.3. *In view of Lemma 3.2, assumption* (6) *in Theorem 3.1, in fact, follows from condition 1) or 3).*

Remark 3.4. *Just as in the proof of Theorem 2.7 one can show the following identities:*

$$\begin{aligned} G(t) &= T_t P, \quad t > 0 \\ G(t)T_{-t} &= -(I-P), \quad t < 0 \end{aligned} \tag{9}$$

on X_α, where P is defined as $\frac{1}{2}I + G(0)$.

Let us now recall the definition of the *fractional growth bound* $\omega_\alpha(\mathbf{T})$ and its spectral counterpart $s_\alpha(A)$: $\omega_\alpha(\mathbf{T})$ is the infimum of all $\omega \in \mathbb{R}$ such that $\|T_t x\| \le M_\omega e^{\omega t}\|x\|_\alpha$, for some $M_\omega > 0$ and all $x \in X_\alpha$ and $t \ge 0$, and

$$s_\alpha(A) = \inf\left\{ s : \sup\left\{ \frac{\|R(\lambda)\|}{1+|\operatorname{Im}\lambda|^\alpha} : \operatorname{Re}\lambda > s \right\} < \infty \right\}. \tag{10}$$

As another consequence of Lemma 3.2 we get the following inversion formula (see [vN] for the case $\alpha = 0$).

Corollary 3.5. *Let $x \in X_\alpha$ and $h > s_\alpha(A)$. If F_t is defined as in Lemma 2.3, then*

$$F_t(x) = \frac{1}{2\pi i}(C,1)\int_{\operatorname{Re}\lambda = h} e^{\lambda t} R(\lambda) x \, d\lambda$$

for all $t \in \mathbb{R}$.

Proof. If $h \ge \omega$, then our statement is the ordinary inversion formula (see Lemma 2.3). Otherwise, by the resolvent identity, we have

$$R(u+iv)x = (1-(u-\omega)R(u+iv))A_\omega^{-\alpha} R(\omega+iv)A_\omega^\alpha x,$$

for all u, $h \le u \le \omega$. So, in view of Lemma 3.2, $\lim_{v\to\infty} R(u+iv)x = 0$ uniformly in $u \in [h,\omega]$. Then, by the Cauchy Theorem, we get

$$\begin{aligned} \frac{1}{2\pi i}(C,1)\int_{\operatorname{Re}\lambda = h} e^{\lambda t} R(\lambda) x \, d\lambda &= \frac{1}{2\pi i}(C,1)\int_{\operatorname{Re}\lambda = \omega} e^{\lambda t} R(\lambda) x \, d\lambda \\ &= F_t(x). \end{aligned}$$

□

Let us recall the inequality $\omega_\alpha(\mathbf{T}) \ge s_\alpha(A)$ (see [Ws]). Now suppose $\omega > s_\alpha(A)$ and $R(i\cdot+\omega)$ is a multiplier from $L_p(\mathbb{R}, X_\alpha)$ to $L_p(\mathbb{R}, X)$. One can easily notice that the implication 1)⇒4) of Theorem 3.1 was proved individually for every ρ. Thus,

$$|\langle r_\omega, \Phi\rangle| \le |\langle \check{r}_\omega, \hat{\Phi}\rangle| \le K\|x\|_\alpha \|x^*\| \|\hat{\Phi}\|_{L_1}.$$

However, by Corollary 3.5, $\check{r}_\omega(t) = e^{-\omega t}\langle x^*, T_t x\rangle$, $t > 0$, which implies $\|T_t x\| \le e^{\omega t}\|x\|_\alpha$. So, $\omega_\alpha(\mathbf{T}) \le \omega$.

On the other hand, if $\omega > \omega_\alpha(\mathbf{T})$, then $\|e^{-\omega t}T_t\|_{X_\alpha \to X}$ is exponentially decaying. Consequently, the operator M_ω, being a convolution with the kernel $e^{-\omega t}T_t$, maps $L_p(\mathbb{R}, X_\alpha)$ into $L_p(\mathbb{R}, X)$ as a bounded operator. Thus, we have proved the following result.

Theorem 3.6. *For any C_0-semigroup $\mathbf{T}$ on a Banach space X, $\omega_\alpha(\mathbf{T})$ is the infimum over all $\omega > s_\alpha(A)$ such that $R(i\cdot+\omega)$ is a multiplier from $L_p(\mathbb{R}, X_\alpha)$ to $L_p(\mathbb{R}, X)$, for some p, $1 \le p < \infty$.*

Corollary 3.7 ([Ws, WW]). *If X is a Hilbert space, then $\omega_\alpha(\mathbf{T}) = s_\alpha(A)$ for any strongly continuous semigroup $\mathbf{T}$ and $\alpha \ge 0$.*

There are many results about properties of the constants $\omega_\alpha(\mathbf{T})$, $s_\alpha(A)$ and relations between them. We refer the reader to paper [WW] for a detailed exposition of the subject.

We conclude this section by proving an α-analogue of Perron's Theorem, cf. [LR]. Let us recall the classical result: a C_0-semigroup $\mathbf{T}$ with generator A is hyperbolic if and only if for every $g \in L_p(\mathbb{R}, X)$ the following integral equation

$$u(\theta) = T_{\theta-\tau}u(\tau) + \int_\tau^\theta T_{\theta-s}g(s)ds, \quad \theta \geq \tau, \tag{11}$$

has unique solution in $L_p(\mathbb{R}, X)$ (see, e.g. [CL, Theorem 4.33]).

In case of arbitrary $\alpha \geq 0$, we are looking for a necessary and sufficient condition on $\mathbf{T}$, which provides existence and uniqueness of solution to (11) in $L_p(\mathbb{R}, X)$ for any given $g \in L_p(\mathbb{R}, X_\alpha)$. It turns out that the multiplier property of $R(is)$ is the condition we need.

Theorem 3.8. *Suppose $i\mathbb{R} \subset \rho(A)$. Then the following assertions are equivalent:*

1) *$R(i\cdot)$ is a multiplier from $L_p(\mathbb{R}, X_\alpha)$ to $L_p(\mathbb{R}, X)$;*
2) *for every $g \in L_p(\mathbb{R}, X_\alpha)$ there exists a unique solution of (11) belonging to $L_p(\mathbb{R}, X)$.*

Before we prove the theorem let us state one auxiliary fact, see [MRS] or [CL, Prop.4.32].

Lemma 3.9. *A function u is a solution of (11) if and only if $u \in D(\Gamma)$ and $\Gamma u = -g$, where Γ is the generator of the associated evolution semigroup.*

Proof. 2) $\Rightarrow$ 1). Denote by L the linear operator that maps $g \in L_p(\mathbb{R}, X_\alpha)$ to the corresponding solution of (11). By the Closed Graph Theorem, L is bounded. We prove that actually $L = M_0$. Indeed, by Lemma 3.9, $Lg \in D(\Gamma)$ and $\Gamma Lg = -g$, for every $g \in L_p(\mathbb{R}, X_\alpha)$. On the other hand, a straightforward computation shows that if g is a C^∞-function with compact support, then $M_0 g \in D(\Gamma)$ and $\Gamma M_0 g = -g$. Thus, $\Gamma(M_0 g - Lg) = 0$. However, if $\Gamma u = 0$ for some $u \in D(\Gamma)$, then again by Lemma 3.9, u is a solution of (11) corresponding to $g = 0$. By the uniqueness, we get $u = 0$. So, $M_0 g = Lg$ on a dense subspace of $L_p(\mathbb{R}, X_\alpha)$ and boundedness of M_0 is proved.

1) $\Rightarrow$ 2). Suppose M_0 is bounded from $L_p(\mathbb{R}, X_\alpha)$ to $L_p(\mathbb{R}, X)$. For a fixed C^∞-function g having compact support, we show that $u = M_0 g$ solves (11). Indeed, using (1), we get

$$\begin{aligned} u(\theta) - T_{\theta-\tau}u(\tau) &= \int_{\mathbb{R}} R(is)\hat{g}(s)e^{is\theta}ds - \int_{\mathbb{R}} R(is)T_{\theta-\tau}\hat{g}(s)e^{is\tau}ds \\ &= \int_0^{\theta-\tau} T_r \int_{\mathbb{R}} e^{is(\theta-r)}\hat{g}(s)dsdr \\ &= \int_0^{\theta-\tau} T_r g(\theta-r)dr = \int_\tau^\theta T_{\theta-r}g(r)dr, \end{aligned}$$

which is precisely (11).

Now suppose g is an arbitrary function from $L_p(\mathbb{R}, X_\alpha)$. Let us approximate g by functions (g_n) of considered type. Then $u_n = M_0 g_n$ converge to $u = M_0 g$ in $L_p(\mathbb{R}, X)$ and, without loss of generality, pointwise on a set $E \subset \mathbb{R}$ with $mes\{\mathbb{R}\backslash E\} = 0$. Thus, (11) is true for u, g and all θ and τ from E. To get (11) for all θ and τ, we will modify u on the set $\mathbb{R}\backslash E$. To this end, let us take a decreasing sequence $(\tau_n) \subset E$ such that $\lim \tau_n = -\infty$. Observe that the functions $f_n(\theta) = T_{\theta-\tau_n} u(\tau_n) + \int_{\tau_n}^{\theta} T_{\theta-s} g(s)ds$ defined for $\theta \geq \tau_n$ are continuous. Since $u = f_n = f_m$ on $(+\infty, \max(\tau_n, \tau_m)] \cap E$, we get $f_n = f_m$ everywhere in the half-line $(+\infty, \max(\tau_n, \tau_m)]$. Put $\tilde{u}$ to be f_n on $(+\infty, \tau_n]$. By the above, $\tilde{u}$ is a well-defined function on all $\mathbb{R}$. Obviously, $u = \tilde{u}$ on E. Let us show that $\tilde{u}$ satisfies (11). Indeed, for any $\theta \geq \tau$ and $\tau > \tau_n$ we have

$$\begin{aligned} T_{\theta-\tau}\tilde{u}(\tau) + \int_{\tau}^{\theta} T_{\theta-s} g(s)ds &= T_{\theta-\tau}[T_{\tau-\tau_n} u(\tau_n) + \int_{\tau_n}^{\tau} T_{\tau-s} g(s)ds] \\ &\quad + \int_{\tau}^{\theta} T_{\theta-s} g(s)ds \\ &= T_{\theta-\tau_n} u(\tau_n) + \int_{\tau_n}^{\theta} T_{\tau-s} g(s)ds = \tilde{u}(\theta). \end{aligned}$$

□

Clearly, assertion 1) in Theorem 3.8 is weaker than condition 1) in Theorem 3.1. We do not know if they are equivalent. In case $\alpha = 0$, though, we can apply the resolvent identity to argue that if $R(i\cdot)$ is a multiplier, then $R(i \cdot +\rho)$ is also a multiplier for small values of ρ. So, by Theorem 2.7, this is equivalent to hyperbolicity of the semigroup $\mathbf{T}$, and our statement turns into classical Perron's Theorem.

4. An α-analogue of hyperbolicity

We begin with a discrete version of Theorem 3.1 in the spirit of [LR, Theorem 5]. Denote by $\operatorname{Rg} T$ the range of an operator T.

Theorem 4.1. *Suppose $i\mathbb{Z} \subset \rho(A)$. Then the following conditions are equivalent:*

1) $X_\alpha \subset \operatorname{Rg}(I - T_{2\pi})$;
2) *The sum* $(C, 1) \sum_{k\in\mathbb{Z}} R(ik)x$ *exists in X-norm for all $x \in X_\alpha$;*
3) $\{R(ik)\}_{k\in\mathbb{Z}}$ *is a multiplier from $L_p(\mathbb{T}, X_\alpha)$ to $L_p(\mathbb{T}, X)$ for some/all* $1 \leq p < \infty$;
4) $\{R(ik)\}_{k\in\mathbb{Z}}$ *is a multiplier from $L_1(\mathbb{T}, X_\alpha)$ to $F(\mathbb{T}, X)$, where F is some quasi-normed function lattice;*
5) *There exists a constant $K > 0$ such that*

$$|\langle r_0, \Phi\rangle| = |\sum_{k\in\mathbb{Z}} r_0(k, x, x^*)\Phi(k)| \leq K\|x\|_\alpha \|x^*\| \|\check{\Phi}\|_{L_1(\mathbb{T})}$$

holds for all $x \in X_\alpha$, $x^* \in X^*$, *and* $\Phi \in C_\infty(\mathbb{T})$.

Proof. 1)⇔2). Note that

$$\frac{1}{2\pi}R(ik)(I - T_{2\pi})x = \frac{1}{2\pi}\int_0^{2\pi} e^{-ikt}T_t x dt,$$

for all $x \in X$. So,

$$\frac{1}{2\pi}(C,1)\sum_{k\in\mathbb{Z}} R(ik)(I - T_{2\pi})x = \frac{1}{2}(I + T_{2\pi})x.$$

Thus, 1) implies 2).

Now assume 1). Denote $S = \frac{1}{2\pi}(C,1)\sum_{k\in\mathbb{Z}} R(ik)x$. Then

$$(\frac{1}{2}I + S)(I - T_{2\pi})x = (I - T_{2\pi})(\frac{1}{2}I + S)x = x, \tag{12}$$

for all $x \in X_\alpha$, and 1) follows.

Following [LR] we denote by K the operator of convolution with the semigroup, i.e.

$$Kf(t) = \int_0^{2\pi} T_s f((t - s)[\bmod 2\pi])ds.$$

Clearly, K is bounded on $L_p(\mathbb{T}, X)$ for all $1 \le p < \infty$ and $\alpha > 0$. Now we define the discrete multiplier operator L by the rule

$$Lf(\theta) = \sum_{k\in\mathbb{Z}} R(ik)\hat{f}(k)e^{ik\theta}, \quad \theta \in [0, 2\pi],$$

where f is a trigonometric polynomial. One can check the identity

$$K = L(I - T_{2\pi}) = (I - T_{2\pi})L. \tag{13}$$

By the assumption and the spectral mapping theorem for the point spectrum the operator $(I - T_{2\pi})$ is one-to-one. Suppose 1) holds. Then $(I - T_{2\pi})$ has the left inverse $U_1 := (I - T_{2\pi})^{-1}_{left}$ defined on X_α. By the Closed Graph Theorem U_1 is bounded as an operator from X_α to X. Then (13) says that $KU_1 = L$ on trigonometric polynomials with values in X_α. So, L maps $L_p(\mathbb{T}, X_\alpha)$ into $L_p(\mathbb{T}, X)$ for all $1 \le p < \infty$, which is what is stated in 3).

If the assertion in 3) is true only for some p, then as in the proof of Theorem 3.1, $LA_\omega^{-\alpha}$ maps $L_p(\mathbb{T}, X)$ into $L_\infty(\mathbb{T}, X)$. By duality, $(LA_\omega^{-\alpha})^*$ maps $L_1(\mathbb{T}, X^\odot)$ into $L_q(\mathbb{T}, X^\odot)$ and hence into $L_\infty(\mathbb{T}, X^\odot)$. So, $LA_\omega^{-\alpha}$ is a bounded operator from $L_1(\mathbb{T}, X)$ to $L_\infty(\mathbb{T}, X)$, which proves 4) with $F = L_\infty(\mathbb{T}, X)$.

Assume 4). Then for every $f \in C_\infty(\mathbb{T}, X)$ there is a $\theta \in [0, 2\pi]$ such that

$$\left\| \sum_{k\in\mathbb{Z}} R(ik)\hat{f}(k)e^{ik\theta} \right\| \le K\|f\|_1.$$

Applying T_θ in the above inequality and using (1) we have:

$$\left\| \sum_{k\in\mathbb{Z}} R(ik)\hat{f}(k) \right\| \leq K'\|f\|_1.$$

In particular, for $f = \check{\Phi} \otimes x$ the last inequality yields 5).

If 5) holds, then taking $\Phi = \sum_{n=1}^{N} \Phi_n \otimes x_n$, with $\Phi_n \in C_\infty(\mathbb{T})$ having disjoint supports we get the following estimates

$$\begin{aligned} \|L\Phi\|_1 &= \frac{1}{2\pi}\int_0^{2\pi} \sup_{\|x^*\|=1} \left| \sum_{n=1}^{N}\sum_{k\in\mathbb{Z}} r_0(k, x_n, x^*)\hat{\Phi}_n(k)e^{ik\theta} \right| d\theta \\ &\leq \sum_{n=1}^{N} \frac{1}{2\pi}\int_0^{2\pi} \sup_{\|x^*\|=1} |\langle r_0, \widehat{\Phi(\cdot+\theta)}\rangle| d\theta \\ &\leq \sum_{n=1}^{N} \|x_n\|_\alpha \|\Phi_n\|_1 = \|\Phi\|_1. \end{aligned}$$

So, we proved 3).

Finally, similarly to the proof of convergence of the $(C,1)$-integral in Theorem 3.1, we can show that 2) follows from L_1-boundedness of $LA_\omega^{-\alpha}$. □

In Theorem 2.7 we have proved that conditions 2) through 4), involving multipliers, are equivalent to the hyperbolicity of the semigroup, that is, to a spectral property of $\mathbf{T}$. A natural question is to see if the multipliers-type conditions 2) through 4) in Theorem 3.1 are equivalent to a spectral property that could be formulated in terms of $\mathbf{T}$ acting on the Banach space X, and *not* in terms of a space of X-valued functions. Theorem 4.1 suggests that each of the conditions 2) through 4) in Theorem 3.1 implies that the inclusion $X_\alpha \subset \mathrm{Rg}(zI - T_{2\pi})$ holds for all z from some annulus $\mathbb{A}$ containing $\mathbb{T}$. It turns out that this inclusion alone is not equivalent to any of the conditions in Theorem 3.1. Below we will find the needed complement, but let us first make some observations.

Assume that $\mathcal{S}$ is some strip containing the imaginary axes and $\mathcal{S} \subset \rho(A)$. Suppose also that $X_\alpha \subset \mathrm{Rg}(zI - T_{2\pi})$ for all z from some annulus $\mathbb{A}$ containing $\mathbb{T}$. By the Point Spectrum Mapping Theorem $(zI - T_{2\pi})$ is one-to-one. Thus, the left inverse operator $U_z = (zI - T_{2\pi})^{-1}_{left} : X_\alpha \to X$ exists and is bounded by the Closed Graph Theorem. The family $\mathbb{U} = \{U_z\}_{z\in\mathbb{A}}$ obeys the resolvent identity on vectors from $X_{2\alpha}$. However, to prove analyticity, first of all one needs uniform boundedness of $\mathbb{U}$. And that is the condition we are looking for.

Theorem 4.2. *Suppose there is a strip $\mathcal{S}$ such that $i\mathbb{R} \subset \mathcal{S} \subset \rho(A)$. Then any of the equivalent conditions of Theorem 3.1 holds if and only if there exists an annulus $\mathbb{A}$ containing $\mathbb{T}$ such that $X_\alpha \subset \mathrm{Rg}(zI - T_{2\pi})$ for all $z \in \mathbb{A}$, and* $\sup\{\|U_z\|_{X_\alpha\to X} : z \in \mathbb{A}\} < \infty$.

Proof. Suppose that M_ρ maps $L_p(\mathbb{R}, X_\alpha)$ into $L_p(\mathbb{R}, X)$ for $|\rho| < 2\rho_0$. By the Uniform Boundedness Principle M_ρ are uniformly bounded for all $|\rho| < \rho_0$. Then, by transference, $\{R(i(k+\xi)+\rho\}_{k\in\mathbb{Z}}$ is a multiplier uniformly in $\xi \in [0,1)$ and $|\rho| < \rho_0$. In view of just proved Theorem 4.1 we get $X_\alpha \subset Rg(zI - T_{2\pi})$ for all z from some open annulus $\mathbb{A}$ containing $\mathbb{T}$. In order to show uniform boundedness of $\mathbb{U}$, let us look at identity (12) first. It shows, in particular, that $U_1 = \frac{1}{2} + S$. Just like in the second part of the proof of Theorem 2.7, one can estimate $\|S\|_{X_\alpha \to X}$ by the multiplier norm of $\{R(ik)\}$. Rescaling gives the same conclusion for all U_z. Since norms of the corresponding multipliers are uniformly bounded, the desired result is proved.

Now let us prove the converse statement.

Clearly, the family $\mathbb{U} = \{U_z\}_{z\in\mathbb{A}}$ obeys the resolvent identity on vectors from $X_{2\alpha}$. Since, in addition, it is bounded, the mapping $z \to U_z$ is strongly continuous on vectors from $X_{2\alpha}$ and, hence, on all X_α. Again by the resolvent identity, U_z is strongly analytic on $X_{2\alpha}$. Since for any $x \in X_\alpha$, $U_z x$ is the uniform limit of a sequence $U_z x_n$ with $x_n \in X_{2\alpha}$, $U_z x$ is analytic.

It suffices to show that the integral $G(t)x = (C,1)\int_{\mathbb{R}} R(is)xe^{ist}ds$ converges for all $x \in X_\alpha$ and there exists a $\beta > 0$ such that $\|G(t)x\| \le Ke^{-\beta|t|}\|x\|_\alpha$.

So, let us fix $x \in X_\alpha$ and $t \in \mathbb{R}$. Then for any $s \in \mathbb{R}$, $x = (e^{2\pi is} - T_{2\pi})U_{e^{2\pi is}}x$. Thus

$$
\begin{aligned}
R(is)xe^{ist} &= R(is)(I - e^{-2\pi is}T_{2\pi})U_{e^{2\pi is}}xe^{ist}e^{2\pi is} \\
&= e^{(2\pi+t)is}\int_0^{2\pi} e^{-ris}T_r U_{e^{2\pi is}}x dr.
\end{aligned}
$$

¿From this we get

$$
\begin{aligned}
G(t)x &= \lim_{N\to\infty}\int_0^{2\pi} T_r \int_{-N}^{N} U_{e^{2\pi is}}xe^{(2\pi+t-r)is}\left(1 - \frac{|s|}{N}\right)dsdr \\
&= \lim_{N\to\infty}\int_0^{2\pi} T_r \int_0^1 U_{e^{2\pi is}}xe^{(2\pi+t-r)is}\sum_{n=-N}^{N} e^{(t-r)in}\left(1 - \frac{|s+n|}{N}\right)dsdr \\
&= \lim_{N\to\infty}\int_0^{2\pi} T_r \int_0^1 U_{e^{2\pi is}}xe^{(2\pi+t-r)is}\sum_{n=-N}^{N} e^{(t-r)in}\left(1 - \frac{|n|}{N}\right)dsdr \\
&= \lim_{N\to\infty}\int_0^{2\pi} T_r \int_0^1 U_{e^{2\pi is}}xe^{(2\pi+t-r)is}\mathcal{F}_N(-r + t \bmod 2\pi)dsdr,
\end{aligned}
$$

where $\mathcal{F}_N$ is the Fejér kernel. Passing to limit inside the integral we get

$$
\begin{aligned}
G(t) &= \frac{1}{2}\left[T_{2\pi}\int_0^1 U_{e^{2\pi is}}xe^{ist}ds + \int_0^1 U_{e^{2\pi is}}xe^{is(2\pi+t)}ds\right] \\
&= \frac{1}{2}\left[T_{2\pi}\frac{1}{2\pi i}\int_{\mathbb{T}} z^t U_z x dz + \frac{1}{2\pi i}\int_{\mathbb{T}} z^{2\pi+t}U_z x dz\right],
\end{aligned}
$$

if $t = 0 \bmod 2\pi$. And

$$\begin{aligned} G(t) &= T_{t \bmod 2\pi} \int_0^1 U_{e^{2\pi i s}} x e^{(2\pi + t - t \bmod 2\pi) is} ds \\ &= T_{t \bmod 2\pi} \frac{1}{2\pi i} \int_{\mathbb{T}} z^{2\pi + t - t \bmod 2\pi} U_z x dz, \end{aligned}$$

otherwise. In either case, replacing $\mathbb{T}$ by $(1+\varepsilon)\mathbb{T}$, if $t < 0$, or by $(1-\varepsilon)\mathbb{T}$, otherwise, we get the desired exponential decay. □

5. Strong α-hyperbolicity

In this section we introduce yet another notion of α-hyperbolicity for strongly continuous semigroups. The spectral property we considered in the previous section, though strong enough, fails to produce any splitting projection, which is so natural in the case $\alpha > 0$. Therefore, we investigate a notion of strong α-hyperbolicity, in which we force such a projection to exist.

Definition 5.1. *A C_0-semigroup $\mathbf{T} = (T_t)_{t \geq 0}$ is said to be strongly α-hyperbolic if there exists a projection P on X, called splitting, such that $PT_t = T_tP$, $t \geq 0$ and the following two conditions hold:*

1. *$\omega_\alpha(\mathbf{T}|_{\operatorname{Im} P}) < 0$;*
2. *the restriction of $\mathbf{T}$ on $\operatorname{Ker} P$ is a group, and $\omega_\alpha(\mathbf{T}^{-1}|_{\operatorname{Ker} P}) < 0$, where $\mathbf{T}^{-1} = (T_{-t}|_{\operatorname{Ker} P})_{t \geq 0}$.*

The function $G(t)$ defined as in Definition 2.2 is called the Green's function corresponding to the α-hyperbolic semigroup $\mathbf{T}$.

It is an immediate consequence of the definition that Green's function exponentially decays at infinity on vectors from X_α.

Now we prove an analogue of Theorem 2.7 for α-hyperbolic semigroups.

Theorem 5.2. *A semigroup $\mathbf{T}$ is α-hyperbolic if and only if one of the equivalent conditions of Theorem 3.1 is satisfied and the operator*

$$G(t)x = \frac{1}{2\pi}(C, 1) \int_{\mathbb{R}} R(is) x e^{ist} ds$$

has a continuous extension to all of X for each $t \in \mathbb{R}$.

If this is the case, $G(t)$ represents the Green's function. Furthermore, the splitting projection is unique and given by

$$P = \frac{1}{2} I + G(0). \tag{14}$$

Proof. Let us prove necessity.

If $\mathbf{T}$ is α-hyperbolic, then there is a splitting projection P. Suppose $x \in (\operatorname{Im} P)_\alpha$. Then by Corollary 3.5 applied to the semigroup $\mathbf{T}|_{\operatorname{Im} P}$, we have $F_t(x) = G(t)x$. In particular, $Px = x = \frac{1}{2}x + G(0)x$. On the other hand, if $x \in (\operatorname{Ker} P)_\alpha$, then by the same reason, $\tilde{F}_t(x) = \frac{1}{2\pi}(C,1)\int_{\mathbb{R}} R(is,-A)xe^{ist}ds = -G(-t)x$, where $\tilde{F}_t(x)$ is defined by

$$\tilde{F}_t(x) = \begin{cases} T_{-t}x, & t > 0 \\ \frac{1}{2}x, & t = 0 \\ 0, & t < 0 \end{cases}.$$

So, $Px = 0 = \frac{1}{2}x + G(0)x$. Since $X_\alpha = (\operatorname{Im} P)_\alpha + (\operatorname{Ker} P)_\alpha$ is dense in X, this shows that $G(t)$ continuously extends to all of X and equality (14) is true. The uniqueness of P follows automatically from (14).

Since the $(X_\alpha \to X)$-norm of $G(t)$ is exponentially decreasing and $M_0(\Phi) = G * \Phi$ for all $\Phi \in \mathcal{S}$, M_0 is bounded from $L_1(\mathbb{R}, X_\alpha)$ to $L_1(\mathbb{R}, X)$. To show boundedness of M_ρ, it is enough to notice that if $\mathbf{T}$ is α-hyperbolic, then the scaled semigroup $e^{\rho\cdot}\mathbf{T}$ is also α-hyperbolic, for small values of ρ.

Now we prove sufficiency.

Let us introduce the operator $P = \frac{1}{2}I + G(0)$. Since Theorem 2.7 is valid, and hence formulas (9) in Remark 3.4 are true, the norm of T_t on $P(X_\alpha)$ is exponentially decaying. Consequently, by the ordinary inversion formula for Laplace transform, we get $G(0)x = \frac{1}{2}x$, for all $x \in P(X_\alpha)$. This implies $P^2 = P$ on all X , in view of the continuity of P. So, P is a projection.

Obviously, $PT_t = T_tP$. On the other hand, since $P(X_\alpha) = (\operatorname{Im} P)_\alpha$, we have $\omega_\alpha(\mathbf{T}|_{\operatorname{Im} P}) < 0$ and condition 1 of Definition 5.1 is proved.

To show invertibility of T_t on $\operatorname{Im}(I - P)$, we apply formula (9). It implies that $\|G(t)\|\|T_{-t}x\| \geq \|x\|$, for x in $\operatorname{Im}(I-P)$, and hence, $T_{-t}|_{\operatorname{Im}(I-P)}$ is invertible. Another application of (9) and the second part of Theorem 2.7 proves condition 2 in Definition 5.1. □

References

[A] H. Amann, *Operator-valued Fourier multipliers, vector-valued Besov spaces, and applications,* Math. Nachr., **186** (1997), 5–56.

[CL] C. Chicone and Y. Latushkin, *Evolution Semigroups in Dynamical Systems and Differential Equations,* Math. Surv. Monogr., **70**, AMS, Providence, 1999.

[CLRM] S. Clark, Y. Latushkin, T. Randolph, and S. Montgomery-Smith, *Stability radius and internal versus external stability in Banach spaces: an evolution semigroups approach,* SIAM J. Control and Optimization, to appear.

[DK] J. Daleckij and M. Krein, *Stability of differential equations in Banach space,* Amer. Math. Soc., Providence, RI, 1974.

[EN] K. Engel and R. Nagel, *One-parameter semigroups for linear evolution equations,* Springer-Verlag, New York, 2000.

[G] L. Gearhart, *Spectral theory for contraction semigroups on Hilbert space,* Trans. Amer. Math. Soc., **236** (1978), 385–394.

[H] I. Herbst, *The spectrum of Hilbert space semigroups,* J. Operator Theory, **10** (1983), 87–94.

[H1] M. Hieber, *Operator valued Fourier multipliers,* Progress in Nonlinear Differential Equations and Their Applications, **35**, (1999), 363–380.

[H2] M. Hieber, *A characterization of the growth bound of a semigroup via Fourier multipliers*, Preprint, 1999.

[KVL] M. A. Kaashoek and S. M. Verduyn Lunel, *An integrability condition on the resolvent for hyperbolicity of the semigroup*, J. Diff. Eqns., **112** (1994), 374–406.

[LMS] Y. Latushkin and S. Montgomery-Smith, *Evolutionary semigroups and Lyapunov theorems in Banach spaces*, J. Funct. Anal., **127** (1995), 173–197.

[LR] Y. Latushkin and F. Räbiger, *Fourier multipliers in stability and control theory,* Preprint.

[LZ] B. M. Levitan and V. V. Zhikov, *Almost periodic functions and differential equations*, Cambridge Univ. Press, Cambridge, 1982.

[MRS] Nguyen Van Minh, F. Räbiger, and R. Schnaubelt, *Exponential stability, exponential expansiveness, and exponential dichotomy of evolution equations on the half-line*, Integr. Eqns. Oper. Theor., **32** (1998), 332–353.

[SW] E. Stein and G. Weiss, *Introduction to Fourier analysis on Euclidean spaces*, Princeton, 1971.

[T] H. Triebel, *Interpolation theory, function spaces, differential operators*, Berlin, 1978.

[vN] J. M. A. M. van Neerven, *The asymptotic behavior of semigroups of linear operators*, Oper. Theory Adv. Appl., **88**, Birkhäuser-Verlag, 1996.

[Ws] G. Weiss, *The resolvent growth assumption for semigroup on Hilbert spaces,* J. Math. Anal. Appl., **145** (1990), 154–171.

[W1] L. Weis, *Operator-valued Fourier multiplier theorems and maximal L_p-regularity*, preprint.

[W2] L. Weis, *Stability theorems for semigroups via multiplier theorems*, in *Differential equations, asymptotic analysis, and mathematical physics (Potsdam, 1996)* Math. Res., **100** 1997, pp. 407–411.

[WW] L. Weis, V. Wrobel, *Asymptotic behavior of C_0-semigroups in Banach spaces*, Proc. Amer. Math. Soc., **124** (1996), 3663–3671.

Department of Mathematics,
University of Missouri,
Columbia, MO 65211, USA
E-mail address: `yuri@math.missouri.edu`, `RIS1DB@mizzou.edu`

Relaxations of Quadratic Programs in Operator Theory and System Analysis

Alexandre Megretski

Abstract. The paper describes a class of mathematical problems at an intersection of operator theory and combinatorics, and discusses their application in complex system analysis. The main object of study is *duality gap bounds* in *quadratic programming* which deals with problems of maximizing quadratic functionals subject to quadratic constraints. Such optimization is known to be universal, in the sense that many computationally hard questions can be reduced to quadratic programming. On the other hand, it is conjectured that an efficient algorithm of solving general non-convex quadratic programs exactly does not exist.

A specific technique of "relaxation", which essentially replaces deterministic decision parameters by random variables, is known experimentally to yield high quality approximate solutions in some non-convex quadratic programs arising in engineering applications. However, proving good error bounds for a particular relaxation scheme is usually a challenging mathematical problem. In this paper relaxation techniques of dynamical system analysis will be described. It will be shown how operator theoretic methods can be used to give error bounds for these techniques or to provide counterexamples. On the other hand, it will be demonstrated that some difficult problems of operator theory have equivalent formulations in terms of relaxation bounds in quadratic programming, and can be approached using the insights from combinatorics and system theory.

1. Introduction

It is always intriguing to discover that interest in a specific class of difficult mathematical problems is shared by seemingly distant fields of pure and applied research. Usually there is also some hope that exchanging the ideas and open questions between the fields will eventually lead to new results. This paper is written exactly with this idea in mind. It contains few new theorems, and is instead concentrated on showing similarity between different problems.

The main object of study in this paper is *duality gap bounds* in *quadratic programming.* Quadratic programming, which appears naturally in numerous applications, deals with problems of maximizing quadratic functionals subject to quadratic constraints. Such optimization is known to be universal (even very special settings lead to NP complete problems, [4]), which guarantees that many computationally hard questions can be reduced to quadratic programming. On the other hand, it is conjectured that an efficient algorithm of solving general non-convex quadratic programs exactly does not exist (P≠NP).

A specific technique of "relaxation", which essentially replaces deterministic decision parameters by random variables, is known experimentally to yield high quality approximate solutions in some non-convex quadratic programs arising in engineering applications. The error induced by such relaxation can be interpreted as duality gap in non-convex optimization. Proving good duality gap bounds for a particular relaxation scheme (equivalent quadratic programs may lead to non-equivalent relaxations) is usually a challenging mathematical problem. An important point to be made by this paper is that some well known problems from operator theory can be formulated exactly as problems of proving some particular relaxation error bounds.

As an application motivation, this paper describes relaxation techniques of dynamical system analysis and design. It will be shown how the search for a Lyapunov function, or for another type of stability certificate, leads naturally to non-convex quadratic programming. In fact, a number of modern techniques in system theory are primarily relaxation methods. Ideas from operator theory can be used to give error bounds for these techniques or to provide counterexamples.

The paper is organized as follows. First, a general description of quadratic programs and relaxation techniques is given, and some well known duality gap estimation problems are formulated. Next, a more detailed account of the use of quadratic programming and relaxations in system analysis is presented. Finally, some proofs and counterexamples are given for special duality gap bounds.

Notation and Terminology

In this paper matrices with real entries are used to describe finite dimensional vectors and their linear transformations. $\mathbf{R}^n$ denotes the set of single-column real matrices of length n. S_n denotes the set of symmetric n-by-n matrices with real entries. The prime $'$ means transposition or Hermitian conjugation (for complex matrices), so that $\sigma(x) = x'Qx$ defines a real-valued quadratic form on $\mathbf{R}^n = \{x\}$ for any $Q \in S_n$. For $Q \in S_n$, $Q \geq 0$ means that $x'Qx \geq 0$ for all $x \in \mathbf{R}^n$. A *quadratic functional* on $\mathbf{R}^n$ is any function $\sigma : \mathbf{R}^n \to \mathbf{R}$ defined by

$$\sigma(x) = \begin{bmatrix} x \\ 1 \end{bmatrix}' \Sigma \begin{bmatrix} x \\ 1 \end{bmatrix},$$

where $\Sigma \in S_{n+1}$ is a constant matrix. A *quadratic S_m-valued functional* on $\mathbf{R}^n$ is any function $\alpha : \mathbf{R}^n \to S_m$ such that $\sigma_v(x) = v'\alpha(x)v$ is a quadratic functional on $\mathbf{R}^n$ for any fixed $v \in \mathbf{R}^m$.

Vector-valued random variables will be used in this paper. $\mathbf{E}\eta$ denotes the expected value of an integrable vector random variable η.

2. The Duality Gap Bound Problem

In this section, a general description of quadratic programs and relaxation techniques is given, and some well known duality gap estimation problems are formulated.

2.1. Relaxation and Duality Gap

This subsection gives specific definitions of quadratic programs, relaxed quadratic programs, and duality gaps in quadratic programming.

2.1.1. QUADRATIC PROGRAMS. Let α, β be two symmetric S_m-valued quadratic functions on $\mathbf{R}^n$ such that $\alpha(x) \geq 0$ and $\alpha(x) \neq 0$ for all x. Any such pair (α, β) defines a *quadratic program* – the problem of finding the supremum of the functional

$$J(x) = \sup\{\lambda : \ \beta(x) - \lambda\alpha(x) \geq 0\} \to \sup. \tag{1}$$

The variable x with respect to which the optimization is performed is called *decision variable*. The supremum of J is denoted by J_*.

Example 2.1. *The problem of finding the maximum of $x_1x_2 + x_2x_3 - x_1x_3$ where the real variables x_1, x_2, x_3 range over the interval $[-1, 1]$ can be viewed as a special case of quadratic program (1) with $n = 3$, $m = 4$,*

$$\alpha = \begin{bmatrix} 1 & 0 & 0 & 0 \\ 0 & 0 & 0 & 0 \\ 0 & 0 & 0 & 0 \\ 0 & 0 & 0 & 0 \end{bmatrix}, \quad \beta = \begin{bmatrix} x_1x_2 + x_2x_3 - x_1x_3 & 0 & 0 & 0 \\ 0 & 1 - x_1^2 & 0 & 0 \\ 0 & 0 & 1 - x_2^2 & 0 \\ 0 & 0 & 0 & 1 - x_3^2 \end{bmatrix}.$$

In this case the supremum J_* of $J(x)$ equals 1.

2.1.2. QUADRATIC OPTIMIZATION ON A HYPERCUBE. Let $u_1, u_2, \ldots, u_n$ be given vectors in a real Hilbert space. How large and how small can be the length of the sum

$$u = x_1u_1 + x_2u_2 + \cdots + x_nu_n, \quad x_k = \pm 1$$

when the "sign" coefficients $x_k \in \{-1, 1\}$ can be selected arbitrarily? Note that already the special case of this problem when u_k are finite dimensional vectors with integer coefficients, is known to be NP-complete.

Both questions (maximization and minimization) can be reduced to the same problem of finding maximum of a quadratic form over the set of vertices of a hypercube:

$$\sigma(x) = x'Qx \to \max, \quad \text{subject to } x \in \{-1, 1\}^n, \tag{2}$$

where $Q \in S_n$ is a given symmetric matrix. Here $Q_{ij} = \langle u_i, u_j \rangle$ when the length of u is maximized, and $Q_{ij} = -\langle u_i, u_j \rangle$ when the length of u is minimized.

This problem is frequently referred to as the MAX-CUT problem, because of the following interpretation. Let the indexes from 1 to n correspond to nodes on a graph. Let $-2Q_{ij}$ be the "value of the benefit" of "cutting off" the edge connecting nodes i and j. Any decision vector $x \in \{-1,1\}^n$ defines a partition of the graph's nodes into two sets N_+ and N_-, the i-th node being in N_+ if and only if $x_i = 1$. If any edge connecting a node from N_+ and a node from N_- must be "cut off", maximizing $x'Qx$ maximizes the total benefit of dividing the graph's nodes into two groups and cutting the edges between the groups.

Let $\bar{Q}$ denote the matrix obtained from Q by replacing all its diagonal terms with zeros. Then

$$x'Qx = x'\bar{Q}x + \mathrm{tr}[Q] \quad \forall\, x \in \{-1,1\}^n.$$

Therefore, the maximums of $x'Qx$ and $x'\bar{Q}x$ on $\{-1,1\}^n$ are achieved on the same vectors x, and their difference is a known constant.

It is frequently more convenient to consider the equivalent problem of maximizing $x'\bar{Q}x$ rather than that of optimizing $x'Qx$, due to the following simple observation.

Lemma 2.2. *Any argument of maximum of $x'\bar{Q}x$ over the set $\{-1,1\}^n$ of vertices of the hypercube $[-1,1]^n$ is also an argument of maximum in the optimization problem*

$$\sigma(x) = x'\bar{Q}x \to \max, \quad \text{subject to} \quad x \in [-1,1]^n, \tag{3}$$

i.e. that of maximizing $x'\bar{Q}x$ over the whole hypercube $[-1,1]^n$. In particular, the maximum of $x'\bar{Q}x$ over $\{-1,1\}^n$ is always non-negative.

Proof. Since the diagonal elements of $\bar{Q}$ are zero, $x'\bar{Q}x$ is *linear* with respect to any single component x_i of x. Hence maximum of $x'\bar{Q}x$ over $[-1,1]^n$ can be achieved with all $x_i \in \{-1,1\}$. □

Just as in Example 2.1, problem (3) can be shown to be equivalent to a quadratic program (1) with

$$\alpha(x) = \begin{bmatrix} 1 & 0 & 0 & 0 \\ 0 & 0 & 0 & 0 \\ 0 & 0 & \ddots & 0 \\ 0 & 0 & 0 & 0 \end{bmatrix}, \quad \beta(x) = \begin{bmatrix} x'\bar{Q}x & 0 & 0 & 0 \\ 0 & 1-x_1^2 & 0 & 0 \\ 0 & 0 & \ddots & 0 \\ 0 & 0 & 0 & 1-x_n^2 \end{bmatrix}.$$

2.1.3. SEMIDEFINITE PROGRAMMING. In the special case when $\alpha(x), \beta(x)$ are *affine* functions of x problem (1) is called a *semidefinite program*. Semidefinite programs are much easier to solve than general quadratic programs. An informal explanation is that in this case the function J is *quasi-convex*, i.e. its level sets are convex. In general, one would expect convex optimization problems to have relatively simple solutions. At least, this turns out to be the case for semidefinite programs, which admit polynomial time solution algorithms [7, 1].

2.1.4. RELAXATION. Due to their universality (NP completeness means that virtually any optimization problem can be re-written as an equivalent quadratic program) it is very unlikely that a "smart" way of solving finite dimensional quadratic programs both accurately and efficiently can ever be found. For most quadratic programs, efficient solution algorithms are not available. However, there exists a simple way of finding efficiently an upper bound of $J = J(x)$. This method is based on replacing vector x by a random variable w with a finite second moment, i.e. such that $\mathbf{E}|w|^2 < \infty$, and also replacing quadratic matrix-valued functionals $\alpha(x), \beta(x)$ by their expected values:

$$\hat{J}(w) = \sup\{\lambda : \ \mathbf{E}\beta(w) - \lambda\mathbf{E}\alpha(w) \geq 0\} \to \sup . \tag{4}$$

Since w can be a random variable which takes a single value $w = x$ with probability 1, where $x \in \mathbf{R}^n$ can be chosen arbitrarily, the supremum $\hat{J}_*$ of $\hat{J}$ cannot be smaller than the supremum J_* of J.

For an arbitrary square integrable random n-vector w let its *covariance and mean value matrix* W be defined by

$$W = \mathbf{E}\begin{bmatrix} w \\ 1 \end{bmatrix}\begin{bmatrix} w \\ 1 \end{bmatrix}'.$$

Note that any $W \in S_{n+1}$ is a covariance and mean value matrix of some random n-vector w if and only if the conditions

$$W \geq 0, \quad W_{n+1,n+1} = 1 \tag{5}$$

are satisfied. The expected value of a quadratic functional of w is a linear function of W, according to

$$\mathbf{E}\begin{bmatrix} w \\ 1 \end{bmatrix}'\Sigma\begin{bmatrix} w \\ 1 \end{bmatrix} = \mathrm{tr}[\Sigma W].$$

Hence the matrices $\mathbf{E}\alpha(w), \mathbf{E}\beta(w)$ depends *linearly* on W, and the relaxed quadratic program (4) is a *semidefinite program* with respect to W.

Example 2.3.

$$x_1x_2 + x_2x_3 - x_1x_3 \to \sup, \quad \text{subject to} \quad x_k^2 \leq 1$$

will be "relaxed" to

$$W_{12} + W_{23} - W_{13} \to \sup, \quad \text{subject to} \quad W_{kk} \leq 1, \quad W \geq 0,$$

where $W = (W_{ij})_{i,j=1}^4$ *is a symmetric matrix such that* $W_{44} = 1$. *In fact, the last column and the last row of* W *do not have any effect on the relaxed problem formulation, and hence* W *can be assumed to range over* S_3. *The supremum in the relaxed problem equals* $3/2$.

2.1.5. Non-Uniqueness of Relaxations. An important observation is that *equivalent* quadratic programs frequently define non-equivalent relaxations. As a rule, equivalent quadratic programs are obtained by adding redundant constraints to a quadratic program.

Example 2.4. *Let $Q \in S_n$ be a symmetric matrix. Consider the problem of maximizing $J(x) = x'Qx$ over the unit cube $\{x\} = [-1,1]^n$. One way to pose this problem as a quadratic program, as shown before, is by using diagonal $\alpha(x), \beta(x) \in S_{n+1}$, where $\alpha_{11} = 1$, $\alpha_{ii} = 0$ for $i > 1$, $\beta_{11}(x) = x'Qx$, and $\beta_{ii}(x) = 1 - x_{i-1}^2$ for $i > 1$. The standard relaxation of this setup is equivalent to maximizing $\mathrm{tr}[QW]$ over those $W \in S_n$ for which $W \geq 0$ and $W_{ii} \leq 1$.*

On the other hand, simple inspection shows that $x_ix_j + x_jx_k - x_ix_k \in [-1,1]$ for all $x_i, x_j, x_k \in [-1,1]$. Let us define $\bar{\alpha}(x), \bar{\beta}(x)$ as diagonal matrices from S_N, where $N = n+1+n(n-1)(n-2)/2$, such that the first $n+1$ diagonal elements of $\bar{\alpha}(x), \bar{\beta}(x)$ are same as those of $\alpha, \beta(x)$, and the remaining $n(n-1)(n-2)/2$ diagonal elements equal zero for $\bar{\alpha}$, and equal $1 - x_ix_j - x_jx_k + x_ix_k$ for $\bar{\beta}(x)$, where (i,j,k) ranges over the set of all ordered triples with $i > j$ and $k \notin \{i,j\}$ (a total of $n(n-1)(n-2)/2$ triples).

By construction, the quadratic program defined by $\bar{\alpha}(x)$ and $\bar{\beta}(x)$ is equivalent to the quadratic program defined by $\alpha(x)$ and $\beta(x)$. However, it is easy to see that the standard relaxation of the latter is equivalent to maximizing $\mathrm{tr}[QW]$ over those $W \in S_n$ for which $W \geq 0$, $W_{ii} \leq 1$, and $W_{ij} + W_{jk} - W_{ik} \leq 1$. As it is shown by Example 2.3, the maximum in the new relaxation can be strictly less than the maximum in the original relaxation.

Given a quadratic program, usually there are numerous ways to improve the quality of relaxation-based optimization by formulating equivalent quadratic programs with redundant constraints. As a rule, such improvement comes at a cost of solving larger semidefinite programs.

2.1.6. The Duality Gap. Another approach to finding an upper bound for the supremum in (1) is based on introducing Lagrange multipliers and solving the corresponding dual problem defined as

$$\bar{J}(Z) = \inf\{\bar{\lambda} : \mathrm{tr}[Z(\bar{\lambda}\alpha(x) - \beta(x))] > 0 \ \ \forall\, x \in \mathbf{R}^n\} \to \inf_{Z \in S_m, Z \geq 0}. \qquad (6)$$

Lemma 2.5. *Assume there exists a bounded optimizing sequence in problem (1), i.e. $x_i \in \mathbf{R}^n$, $\sup|x_i| \leq r < \infty$, and $J(x_i) \to J_* = \inf J$. Then the infimum in (6) equals the supremum in (4).*

Proof. Let $Z \in S_m$, $Z \geq 0$ and $\bar{\lambda} \in \mathbf{R}$ be such that $\mathrm{tr}[Z(\bar{\lambda}\alpha(x) - \beta(x))] > 0$ for all x. Then for any random vector w and $\lambda \in \mathbf{R}$ such that $\mathbf{E}\beta(w) \geq \lambda\mathbf{E}\alpha(w)$ it follows

$$\lambda\mathbf{E}\mathrm{tr}[Z\alpha(w)] \leq \mathbf{E}\mathrm{tr}[Z\beta(w)] < \bar{\lambda}\mathbf{E}\mathrm{tr}[Z\alpha(w)].$$

Since $\mathbf{E}\mathrm{tr}[Z\alpha(w)] \geq 0$, this implies $\mathbf{E}\mathrm{tr}[Z\alpha(w)] > 0$ and hence $\lambda < \bar{\lambda}$. Therefore the infimum in (6) is not smaller than the supremum in (4). To prove the opposite inequality, assume that the inequality $\mathbf{E}\beta(w) \geq \lambda_*\mathbf{E}\alpha(w)$ is impossible for some

$\lambda_* \in \mathbf{R}$. Note that the set B of all expected values $b = \mathbf{E}(\beta(w) - \lambda_*\alpha(w))$, where w is a square integrable random vector and $\mathbf{E}|w|^2 \le r^2$, is convex and compact. Hence there exists a hyperplane in S_m which separates B from the convex cone of all positive semidefinite matrices. In other words, there exists $Z \in S_m$, $Z \ge 0$, $Z \ne 0$ such that $\mathrm{tr}[Zb] < 0$ for any $b \in B$. Hence $\mathrm{tr}[Z(\beta(x) - \lambda_*\alpha(x))] < 0$ for any $x \in \mathbf{R}^n$, $|x| \le r$, i.e. $\bar{J}(Z) \ge \lambda_*$. Hence the infimum in (6) is not larger than the supremum in (4). □

As a consequence of Lemma 2.5, the difference between the maxima in relaxed and non-relaxed versions of the same quadratic program can be interpreted as a *duality gap* in the original non-convex optimization problem.

2.1.7. The Duality Gap Estimation Problem. While the use of quadratic programs and their relaxations is a critical issue in many applications, a large class of difficult *mathematical questions* is associated with finding general a-priory bounds for the relaxation gap. This paper is devoted to the problem, called the *duality gap estimation problem*: *find good upper bounds of the relaxation gap in special classes of quadratic programs.*

Let J_* and $\hat{J}_*$ be the maxima in a quadratic program and its semidefinite program relaxation respectively. In a typical application, J_* is unknown, while $\hat{J}_*$ is known and serves as an *upper* bound of J_*. An estimate of the duality gap usually has the form of an inequality

$$J_* \ge \rho(\hat{J}_*, \alpha, \beta),$$

where $\rho(\cdot)$ is an *easily computable* function of its arguments (of course, J_* itself is a function of α, β, but it is presumed that J_* is difficult to compute).

When J_* is known to be non-negative, the duality gap bound frequently has the form

$$J_* \ge \rho^{-1}\hat{J}_*$$

where ρ is a constant which is the same for a class of quadratic programs. By the definition, ρ is a measure of success of the relaxation technique, applied to a particular quadratic program. When ρ equals 1, the quadratic program is essentially equivalent to its relaxed form. When $\rho \gg 1$, the relaxation technique is not accurate.

Let $N = n + m$ denote the "size" of quadratic program (1). An *informal* evaluation of the difficulty of obtaining duality gap bounds $\rho = \rho(N)$ is as follows:

- **"No gap".** The case of no gap bound ($\rho(N) = 1$) is very rare, but usually not difficult to prove, if true.
- **"Bounded gap".** The cases when $\rho(N) < c$ where c does not depend on n are rare. The corresponding duality gap estimates are usually difficult to prove.
- **"Logarithmic gap".** The case when a logarithmic growth bound, such as $\rho(N) < c\log(N)$, is available, is of interest when unboundedness of $\rho(N)$ can be proven.

- **"Power gap".** As a rule, a power law growth bound $\rho(N) < c \cdot N^r, r > 0$ can be established easily for many quadratic programs.

2.2. Polynomially Bounded Operators

A linear operator $A: H \to H$ on a Hilbert space H is called *polynomially bounded* if there exists a constant M such that

$$\|p(A)\| \leq M\|p\|_\infty$$

for any scalar polynomial p, where

$$\|p\|_\infty = \max\{|p(z)| : |z| \leq 1\}$$

is the so-called infinity norm. If A is *similar to a contraction*, i.e. can be represented in the form $A = STS^{-1}$ where $\|T\| \leq 1$ and $\|S\|, \|S^{-1}\| < \infty$, then, due to a von Neumann theorem, A is polynomially bounded with $M \leq \|S\| \cdot \|S^{-1}\|$. The problem posed by P. Halmos in [3] and finally solved by G. Pisier in [9] asks whether *any* polynomially bounded operator A is similar to a contraction.

The intention of this subsection is to show that a finite dimensional version of the problem is equivalent to verifying boundedness of a relaxation gap in a particular class of quadratic programs.

2.2.1. Finite Dimensional Formulation. Let A be an n-by-n matrix. A is called *marginally stable* if the norms of A^k, $k > 0$, are bounded. While it appears that any alternative definition of marginal stability for finite dimensional matrices will lead to the same condition, there is a number of definitions of numerical *degree of stability* which are far from being equivalent when $n \to \infty$.

The following measures of marginal stability will be studied in this subsection:

$$\gamma_{pb}(A) = \sup\{\|p(A)\| : p \in \mathbf{P}, \|p\|_\infty \leq 1\},$$

where $\mathbf{P}$ is the set of scalar polynomials,

$$\gamma_{cb}(A) = \sup\{\|p(A)\| : p \in \mathbf{P}_*, \|p\|_\infty \leq 1\},$$

where $\mathbf{P}_*$ is the set of matrix polynomials,

$$\gamma_{sc}(A) = \inf\{\|S^{-1}\| : \|SAS^{-1}\| \leq 1, \|S\| \leq 1\}.$$

It is easy to show that all three quantities $\gamma_{pb}(A)$, $\gamma_{cb}(A)$ and $\gamma_{sc}(A)$ are finite for all marginally stable n-by-n matrices A.

Let the function $\phi = \phi(M, n)$ be defined for $M \geq 1$ and $n = 1, 2, \ldots$ by

$$\phi(M, n) = \sup\{\gamma_{sc}(A) : \gamma_{pb}(A) \leq M\}, \tag{7}$$

where the supremum is taken over all n-by-n marginally stable matrices A.

The finite dimensional version of the Halmos problem studied here asks whether $\phi(M, n)$ is bounded as $n \to \infty$ for any fixed $M > 1$.

2.2.2. SCHUR MATRICES AND LYAPUNOV EQUATIONS. Note that it is sufficient to consider the case when A in (7) is a ***strictly stable*** (Shur) matrix, i.e. all eigenvalues lie strictly within the unit disc. Indeed, if A is marginally stable but not strictly stable then for any $r \in (0,1)$ the matrix $A_r = rA$ is strictly stable and polynomially bounded. Hence $\|S_r A_r S_r^{-1}\| \leq 1$ for some matrices S_r such that $\|S_r\| = 1$, $\|S_r^{-1}\| \leq \phi(M,n)$. As $r \to 1$, S_r will converge to a matrix S such that $\|SAS^{-1}\| \leq 1$, $\|S\| \leq 1$, and $\|S^{-1}\| \leq \phi(M,n)$.

Given an n-by-n Schur matrix A, for any $X \in S_n$ there exists the *unique* solution $Q = \mathbf{L}_A(X)$ of the *Lyapunov equation*

$$Q - AQA' = X.$$

An alternative way to define Q is by

$$\mathbf{L}_A(X) = \sum_{k=0}^{\infty} A^k X (A')^k.$$

In particular, $Q \geq 0$ whenever $X \geq 0$. The opposite is not true: there exist matrices X which are not positive semidefinite such that $\mathbf{L}_A(X) \geq 0$ for *some* Schur matrices A.

2.2.3. POLYNOMIAL BOUNDEDNESS AND LYAPUNOV EQUATIONS. The following statement relates positive semidefiniteness of solutions of Lyapunov equations to polynomial boundedness and complete boundedness.

Lemma 2.6. *Let A be a strictly stable n-by-n matrix, $w_1, \ldots, w_k$, $v_1, \ldots v_m \in \mathbf{R}^n$. Let $\bar{v} \in \mathbf{R}^{mn}$ be the column vector obtained by stacking v_i's one over the other. Let $\bar{w} \in \mathbf{R}^{kn}$ be constructed similarly from w_i's:*

$$\bar{v} = \begin{bmatrix} v_1 \\ \vdots \\ v_m \end{bmatrix}, \quad \bar{w} = \begin{bmatrix} w_1 \\ \vdots \\ w_k \end{bmatrix}.$$

A sequence $\{p_i\}$ of polynomial k-by-m matrices $p_i \in \mathbf{P}_*$ such that $\|p_i\|_\infty \leq 1$ and $p_i(A)\bar{v} \to \bar{w}$ as $i \to \infty$ exists if and only if $Q \geq 0$ where Q is the solution of the Lyapunov equation

$$Q - AQA' = V - W, \tag{8}$$

with

$$V = \sum_{i=1}^{m} v_i v_i', \quad W = \sum_{i=1}^{k} w_i w_i'. \tag{9}$$

Proof. Let $l_+^2(\mathbf{R}^r)$ denote the standard Hilbert space of one-sided sequences $(x_i)_{i=0}^{\infty}$ of vectors $x_i \in \mathbf{R}^r$. Let τ denote the backward shift operator on $l_+^2(\mathbf{R}^r)$. Let H_v

be the (finite dimensional) τ-invariant linear subspace of $l_+^2(\mathbf{R}^m)$:

$$H_v = \left\{ x_v = x_v(z) = (x_i) : \ x_i = \begin{bmatrix} v_1'(A')^i z \\ \vdots \\ v_m'(A')^i z \end{bmatrix}, \ z \in \mathbf{R}^n \right\}.$$

Consider the map $L_w : \ H_v \to l_+^2(\mathbf{R}^k)$ defined by $L_w x_v(z) = x_w(z)$ where $x_w(z)$ is defined similarly to x_v using vectors w_j.

If $\bar{w} = p(A)\bar{v}$ where $p \in \mathbf{P}_*$ and $\|p\|_\infty \leq 1$ then $L_w = p(\tau)|_{H_v}$ is correctly defined and contractive. Hence

$$\|x_v(z)\|^2 \geq \|x_w(z)\|^2 \ \ \forall \ z \in \mathbf{R}^n. \tag{10}$$

Since by construction

$$\|x_v(z)\|^2 - \|x_w(z)\|^2 = z'Qz,$$

the inequality $Q \geq 0$ follows.

If $Q \geq 0$ then (10) holds, and hence L_w is a correctly defined contraction which commutes with τ. Hence L_w can be extended to the whole $l_+^2(\mathbf{R}^m)$ preserving contractiveness and commutativity. The extension $\bar{L}_w$ will have the form $x \to h(\tau)x$ where h is a rational matrix function with $\|h\|_\infty \leq 1$. Now p_i can be defined as converging polynomial approximations of h. □

Lemma 2.6 allows one to express polynomial boundedness and complete boundedness in terms of solutions of Lyapunov equations. According to Lemma 2.6 with $k = m = 1$, the number $\gamma_{pb}(A)$ is the maximum in the quadratic program

$$\gamma_{pb}(A) = \sup\{|w|^2 : \ \mathbf{L}_A(vv' - ww') \geq 0, \ |v|^2 \leq 1\}. \tag{11}$$

Using Lemma 2.6 with arbitrary k, m shows that $\gamma_{cb}(A)$ is the maximum in the standard relaxation of (11) obtained by replacing the rank one matrices vv' and ww' with arbitrary positive semidefinite matrices V, W:

$$\gamma_{cb}(A) = \sup\{\mathrm{tr}[W] : \ V \geq 0, \ W \geq 0, \ \mathbf{L}_A(V - W) \geq 0, \ \mathrm{tr}[V] \leq 1\}. \tag{12}$$

2.2.4. Similarity to a Contraction and Lyapunov Equations. The following statement relates positive semidefiniteness of solutions of Lyapunov equations to similarity to a contraction.

Lemma 2.7. *Let A be a stable n-by-n matrix, $r > 0$. A matrix S such that*

$$\|SAS^{-1}\| \leq 1, \ \|S^{-1}\| \leq 1, \ \|S\| < r$$

does not exist if and only if there exist matrices $V, W \in S_n$ such that

$$V \geq 0, \ W \geq 0, \ \mathrm{tr}[V] = 1, \ \mathrm{tr}[W] \geq r^2, \ \ \mathbf{L}_A(V - W) \geq 0. \tag{13}$$

Proof. In terms of $R = S'S$ the conditions imposed on S have the form

$$R - A'RA \geq 0, \ R - I \geq 0, \ (r^2 - \epsilon)I - R \geq 0, \tag{14}$$

where $\epsilon > 0$ is sufficiently small. If V, W satisfy (13) and $Q = \mathbf{L}_A(V - W) \geq 0$ then

$$\begin{aligned} 0 &\leq \operatorname{tr}[Q(R - A'RA) + V((r^2 - \epsilon)I - R) + W(R - I)] \\ &= \operatorname{tr}[(Q - AQA' - V + W)R + r^2 V - \epsilon V - W] \\ &\leq 0 + r^2 - \epsilon - r^2 = -\epsilon, \end{aligned}$$

which proves that (13) and (14) cannot hold simultaneously. On the other hand, if conditions (14) cannot be satisfied then a separation argument similar to the one used in the proof of Lemma 2.5 proves existence of V, W satisfying conditions (13). □

Lemma 2.7 together with Lemma 2.6 prove that $\gamma_{cb}(A) = \gamma_{sc}(A)$. Perhaps more importantly, it is now shown that $\gamma_{sc}(A)$ equals maximum in the standard relaxation of the non-convex quadratic program (11) in which $\gamma_{pb}(A)$ is the maximum. Therefore, the question of whether polynomial boundedness implies similarity to a contraction can be viewed as a question about boundedness of the standard relaxation gap in a specific non-convex quadratic program.

Due to the recent result by G. Pisier [9] the relaxation gap in (11) is not bounded. Earlier, J. Bourgain has shown that the relaxation gap grows not faster than $\log(n)$, according to

$$\gamma_{sc}(A) \leq \text{const} \cdot \gamma_{pb}(A)^4 \log(n+1).$$

2.3. Quadratic Optimization on a Hypercube

Recall that the problem of quadratic optimization on a hypercube is that of maximizing $J(x) = x'Qx$ over $x \in \{-1, 1\}^n$, where $Q \in S_n$ is a given symmetric matrix. When the diagonal entries Q_{ii} of Q are non-negative, an equivalent form is

$$J_*(Q) = \max\{J(x) = x'Qx : \ x \in [-1, 1]^n\} = ?$$

The corresponding standard relaxed semidefinite program is

$$\hat{J}_*(Q) = \max\{\hat{J}(X) = \operatorname{tr}[QX] : \ X = X' \geq 0, \ X_{kk} \leq 1\} = ?$$

It is possible to formulate a number of hypotheses about the gap between $\hat{J}_*$ and J_*.

2.3.1. Uniform Bounds of Duality Gap.

Which conditions should be imposed on Q to guarantee that

$$J_*(Q) \geq c\hat{J}_*(Q) \tag{15}$$

where c is a constant not depending on Q (in particular, on its size)?

For an arbitrary symmetric matrix $Q \in S_n$ (possibly such that some $Q_{ii} < 0$), conjecture (15) turns out to be false, though it can be shown that, as a function of n, c would grow not faster than $\log(n)$. On the other hand, it was shown in

[8] that(15) holds for any positive semidefinite $Q \geq 0$ with $c = 2/\pi$. For the case when Q is diagonally dominant, i.e. when

$$Q_{ii} \geq \sum_{j \neq i} (|Q_{ij}| + |Q_{ji}|) \ \ \forall\, i,$$

it was shown in [2] that (15) holds with a better constant $c \approx 0.88$. The so-called Grothendieck inequality (see [5]) states that (15) is true with $c \approx 0.56$ whenever Q has the block form

$$Q = \begin{bmatrix} Q_{11} & Q_{12} \\ Q_{21} & Q_{22} \end{bmatrix},$$

where Q_{11} and Q_{22} are *square* zero matrices. The answer to the question of whether there exists $c > 0$ such that (15) is true for all matrices Q with $Q_{ii} = 0$ is unknown to the author.

Inequality (15) will also be discussed later.

2.3.2. VECTOR SUMS WITH BOUNDED COEFFICIENTS. Let $\{u_k\}_{k=0}^{\infty}$ be a sequence of elements of a Hilbert space such that

$$J(x) = \left\| \sum u_k x_k \right\| \leq \gamma < \infty \text{ whenever } \sup_k |x_k| \leq 1. \tag{16}$$

It is well known and easy to show that (16) implies

$$\sum \|u_k\|^2 \leq \gamma,$$

which corresponds to the trivial case $c = 0$ of inequality (15).

Indeed, if $x_k = 0$ for $k > n$ then

$$J(x) = \bar{x}_n' Q^n \bar{x}_n$$

where $\bar{x}_n$ is the vector with components $x_0, \ldots, x_n$, and Q^n is the Gramm matrix of $u_0, \ldots, u_n$. Since $Q_{kk}^n \geq 0$, the maximum of $\bar{x}_n' Q^n \bar{x}_n$ subject to $|x_k| \leq 1$ is achieved at the extremal points $x_k \in \{-1, 1\}$. Therefore

$$\max\{\bar{x}_n' Q^n \bar{x}_n : \ x_k \in [-1,1]\} = \mathrm{tr}[Q^n] + \max\{\bar{x}_n' Q^n \bar{x}_n : \ x_k \in \{-1,1\}\} \geq \mathrm{tr}[Q^n].$$

Similarly, the result of [8] can be interpreted in the following way: (16) implies existence of a sequence $\{d_k\}$ such that

$$\sum d_k \leq \frac{\pi}{2}\gamma, \tag{17}$$

and

$$\|\sum u_k x_k\|^2 \leq \sum d_k |x_k|^2 \tag{18}$$

for *any* x_k (not only for those bounded by 1). Proving (15) for matrices Q with all $Q_{ii} = 0$ would be equivalent to showing that (17) can be replaced by

$$\sum d_k + \frac{1-c}{c} \sum \|u_k\|^2 \leq \frac{\gamma}{c}.$$

2.3.3. Gramm Matrices of Bounded Functions. A dual interpretation of questions associated with duality gap estimation for quadratic optimization on a hypercube is given in terms of Gramm matrices. Let $V = \{v_k\}_{k=1}^{\infty}$ be a sequence of (real) scalar functions $v_k = v_k(t) \in L^2(0,1)$ which is uniformly bounded by 1, i.e.

$$|v_k(t)| \leq 1 \;\; \forall \;\; k, t.$$

Which matrices W belong to the set $\mathbf{W} = \{W\}$ of Gramm matrices of such a sequence? Obviously W must be a Gramm matrix of some set of vectors of length not exceeding 1. However, this is not enough. The result of [8] is equivalent to saying that for any Gramm matrix R with $R_{ii} \leq 1$ there exists $W \in \mathbf{W}$ such that

$$R \leq \frac{\pi}{2} W.$$

Similarly, the Grothendieck inequality implies that if any such R is decomposed as

$$R = \begin{bmatrix} R_{11} & R_{12} \\ R_{21} & R_{22} \end{bmatrix}$$

where R_{11} and R_{22} are square matrices then cR_{12} is equal to the corresponding block element W_{12} of some $W \in \mathbf{W}$, where $c \approx 0.56$. Similarly, showing that for any Gramm matrix R with $R_{ii} \leq 1$ there exists $W \in \mathbf{W}$ such that

$$cR + (1 - c)I = W$$

would be equivalent to proving (15) for all Q with $Q_{ii} = 0$.

3. Relaxations in Dynamical System Analysis

Many questions in analysis and design of dynamical systems come in the form of non-convex quadratic programs. As a consequence, techniques of system theory are frequently based on relaxation and duality gap estimation.

3.1. Stability of Systems with Structured Uncertainty

Analysis of linear time invariant (LTI) systems with structured uncertainty is a prominent example of the use of quadratic programming and its relaxations in system analysis.

3.1.1. Models of Stable LTI Systems and their Interconnections. A stable LTI system S with n inputs and n outputs can be represented mathematically by a function H in the Hardy space $H_n^{\infty}(\mathbf{T})$ of n-by-n matrix-valued analytical functions (H would be called *transfer matrix* of the system). For a pair S_1, S_2 of such stable LTI systems with transfer matrices H_1, H_2, it is said that the *feedback interconnection* S_{12} of S_1 and S_2 is stable if there exists

$$H_{12} = (I - H_1 H_2)^{-1} \in H_n^{\infty}(\mathbf{T}),$$

in which case H_{12} can be regarded as the transfer matrix of a system S_{12} defined by the interconnection of S_1 and S_2.

3.1.2. UNCERTAIN MODELS AND ROBUST STABILITY. Feedback interconnections can be used to represent systems with models which are not known precisely, or *uncertain systems.* A rich class of such systems can be modeled as feedback interconnections of a stable LTI system with a known transfer matrix $H \in H_n^\infty(\mathbf{T})$ and a stable LTI system with an unknown transfer matrix Δ, assumed to belong to the set $\bar{\Delta}$ of all *diagonal* matrix functions in $H_n^\infty(\mathbf{T})$, with H^∞ norm not exceeding 1.

Such uncertain model $U = U(H)$ (completely defined by $H \in H_n^\infty(\mathbf{T})$) is called ***robustly stable*** if there exists a constant c such that the inverse $(I - H\Delta)^{-1}$ exists, belongs to the class $H_n^\infty(\mathbf{T})$, and satisfies the inequality

$$\|(I - H\Delta)^{-1}\|_\infty \le c$$

for all $\Delta \in \bar{\Delta}$.

3.1.3. CRITERIA OF ROBUST STABILITY. It is easy to show that uncertain model $U(H)$ with $H \in H_n^\infty(\mathbf{T})$ is robustly stable if and only if there exists a constant c such that

$$\|(I - MZ)^{-1}\| \le c \quad \forall\, Z \in \bar{Z} \tag{19}$$

for any essential value M of H where $\bar{Z}$ is the set of all diagonal complex n-by-n matrices with norm not exceeding 1. In most applications, H is continuous and a bound of its derivative is available. A common practice is to sample the values $M = M_k$ of H on a "dense" (in a practical, not mathematical, sense) grid in T and check that each of the samples satisfies (19).

The main difficulty, however, lies in the fact that checking (19) for a given n-by-n matrix M is computationally challenging. In particular, the problem is known to be NP-hard [10]. As a result, one is forced to look for an easier-to-verify condition on M which would imply (19), without being equivalent to it. The frequently used sufficient condition for (19) is existence of a diagonal matrix $D > 0$, $D \le I$ and $\epsilon > 0$ such that

$$D - M'DM \ge \epsilon I, \tag{20}$$

which implies (19) with $c = c(\epsilon)$.

Indeed, let $R = D^{1/2}$, $T = RMR^{-1}$. Then $D \le I$ implies $\|R\| \le 1$ while (20) implies $D \ge \epsilon I$, which means $\|R^{-1}\| \le \epsilon^{-1/2}$. In addition, multiplying (20) by R^{-1} on both sides yields $I - T'T \ge \epsilon D^{-1} \ge \epsilon I$, hence $\|T\| \le (1-\epsilon)^{1/2} \le 1 - \epsilon/2$. Therefore

$$\|(I - MZ)^{-1}\| = \|(R^{-1}(I - TZ)R)^{-1}\| \le 2\epsilon^{-3/2}.$$

Since (20) can be considered as a semidefinite program with respect to D and ϵ (with an objective to maximize ϵ to check whether it can be made larger than zero), (20) is relatively inexpensive to verify.

In the field of robustness analysis, the notion of ***structured singular value*** $\mu(M)$ of M is used to represent condition (19): the non-negative number $\mu = \mu(M)$ is defined in such a way that

(a) (19) holds for some c if and only if $\mu(M) < 1$;
(b) $\mu(rM) = |r|\mu(M)$ for any $r \in \mathbf{R}$.

The standard *upper bound* $\hat{\mu}(M)$ *of* $\mu(M)$ is defined in such way that

(a) (20) holds for some $\epsilon > 0$ if and only if $\hat{\mu}(M) < 1$;
(b) $\hat{\mu}(rM) = |r|\hat{\mu}(M)$ for any $r \in \mathbf{R}$.

Thus, robust stability of $U(H)$ is equivalent to the inequality $\operatorname{ess\,sup} \mu(H) < 1$ and is implied by the easier-to-verify inequality $\operatorname{ess\,sup} \hat{\mu}(H) < 1$.

3.1.4. QUALITY OF THE UPPER BOUND OF μ. The question about quality of the upper bound $\hat{\mu}(M)$ of $\mu(M)$ was posed about 20 years ago. It is known for a long time that $\hat{\mu}(M) = \mu(M)$ for any matrix M of size 3-by-3, and less. In addition, the equality $\hat{\mu}(M) = \mu(M)$ holds for any matrix M of rank 1. It is very difficult to find matrices M for which the ratio $\mu(M)/\hat{\mu}(M)$ is small. It was conjectured that

$$\mu(M) \geq c\hat{\mu}(M) \tag{21}$$

where the constant c does not depend on M (in particular, on the size of M).

It turns out that (21) is equivalent to existence of a constant duality gap bound in an appropriately constructed non-convex quadratic program. Indeed, for $x \in \mathbf{R}^n$ let $\alpha(x), \beta(x)$ be the diagonal $(n+1)$-by-$(n+1)$ matrices

$$\alpha = \begin{bmatrix} |e_1'x|^2 & & & \\ & \ddots & & \\ & & |e_n'x|^2 & \\ & & & 1 \end{bmatrix}, \quad \beta = \begin{bmatrix} |e_1'Mx|^2 & & & \\ & \ddots & & \\ & & |e_n'Mx|^2 & \\ & & & |x|^2 \end{bmatrix}, \tag{22}$$

where $\{e_i\}$ is the standard basis in $\mathbf{C}^n$. By construction, $J_* = \sup J(x) = \mu(M)^2$ and $\hat{J}_* = \sup \hat{J}(w) = \hat{\mu}(M)^2$ for the quadratic program (1) and its relaxation (4) with α, β defined by (22).

3.1.5. AN ANALOG IN OPERATOR THEORY. The "μ-gap" question of whether the ratio $\hat{\mu}(M)/\mu(M)$ has a finite upper bound remained unanswered until recently Treil [11] used an operator-theoretic analog of the original finite dimensional problem to show that the answer is negative. His construction is as follows.

Let A be a bounded operator on $L^2(\nu)$. For any $\phi \in L^\infty(\nu)$ let M_ϕ be the operator of multiplication by ϕ, acting on $L^2(\nu)$. Does there exist a $\rho > 0$ such that, for any bounded operator A on $L^2(\nu)$, we have

$$\|M_\psi A M_{1/\psi}\| \leq \rho \text{ for some } \psi, 1/\psi \in L^\infty(\nu)$$

whenever

$$I - M_\phi A \text{ is invertible for all } \|\phi\|_\infty < 1?$$

When A is defined by the singular integral

$$(Af)(t) = \frac{1}{\pi}\int_\Gamma \frac{f(s)}{s-t}\,ds$$

where $d\nu = ds$ is the arclength on a smooth curve Γ with a large Ahlfors constant, ρ is unbounded. Therefore, the infinite dimensional analog of the μ-gap problem has a negative answer. A convergence argument is then used in [11] to show that the original finite dimensional version must have a negative answer as well.

3.2. Analysis of Nonlinear Systems

The dominant idea of rigorous analysis of nonlinear systems is the use of the so-called system invariants (for example, Lyapunov functions). However, finding invariants of generic nonlinear system models is a difficult task. The use of relaxations in non-convex quadratic programming allows one to make substantial progress in this direction.

3.2.1. System Analysis and System Invariants. Consider a nonlinear dynamical system model of the form

$$x_{t+1} = f(x_t, v_t),\ \ z_t = g(x_t, v_t),\ \ t = 0, 1, 2, \ldots, \tag{23}$$

where x_t, v_t, z_t are "hybrid" signal vectors, which means that the components of x, v, z can be logical (i.e. ranging over the set $\{-1, 1\}$) and analog (i.e. ranging over $[-1, 1]$ or $\mathbf{R}$) elements. Here x_t is the system state, v_t is the disturbance input, modeling the environment, and z_t is the performance output. System (23) is said to meet the performance expectations if for any admissible initial condition x_0 there exists a constant $\gamma = \gamma(x_0)$ such that

$$\sum_{t=0}^{\infty} g(x_t, v_t) \geq -\gamma(x_0), \tag{24}$$

for all possible inputs $v = v_t$.

It can be shown that many important system specifications, including those representing stability, safety, and efficiency, can be expressed in terms of performance expectations with an appropriate selection of function g in (23).

Proving rigorously that a given complex system meets performance expectations is usually very difficult. Chances for successful analysis are greatly improved when a ***system invariant***, or ***quasi-Lyapunov function*** $V : \{x\} \to \mathbf{R}$ is available, defined by the condition

$$V(x_{t+1}) - V(x_t) \leq g(x_t, v_t) \tag{25}$$

for all possible trajectories of (23). Designing efficient algorithms for finding such functions V is a major problem in system analysis.

3.2.2. Analysis Using Partial Invariants. An approach to *automatic* search for system invariants can be based on the notion of *partial Lyapunov functions*. Assume that an auxiliary variable w_t is defined by $w_t = \phi(x_t, v_t)$ in such a way that the total range of possible values of w_t is a subset of

$$\Omega = \{(v, u) \in \mathbf{R}^p \times \{-1, 1\}^q : \ |h_i v| \leq 1\}, \tag{26}$$

where $\{h_i\}$ is a finite set of linear functionals on $\mathbf{R}^p$. In other words, Ω is a direct product of a polytope in $\mathbf{R}^p$ (symmetric with respect to the origin) and a complete set of vertices of a cube. A function $V_k : \{x\} \to \mathbf{R}$ is called a partial Lyapunov function (for system (23), with respect to the auxiliary variable w) if there exists a quadratic functional

$$\sigma_k(w) = w'Q_k w + 2L_k w + C_k$$

such that

$$V_k(x_{t+1}) - V_k(x_t) \le \sigma_k(w_t) \tag{27}$$

for any possible trajectory of (23). The inequality (27) in that case is called an *Integral Quadratic Constraint* (IQC).

Typically, partial Lyapunov functions can be derived for the simple building blocks (elementary logic, delay elements, integrators, uncertainty bounds etc.) in terms of which the whole system (23) can be described. Usually, more useful partial Lyapunov functions can be obtained after the state of the system is re-defined to include some past history (for example, by concatenating the original state x_t with a stable LTI transformation of the past states $x_{t-\tau}$).

Once available, partial Lyapunov functions can be used in an automatic search for a quasi-Lyapunov function in the form

$$V(x_t) = V_0(x_t) + \sum_{k=1}^{N} \tau_k V_k(x_t), \quad \tau_k \ge 0, \tag{28}$$

provided that V_0 satisfies the inequality

$$V_0(x_{t+1}) - V_0(x_t) \le \sigma_0(w_t) + g(x_t, v_t), \tag{29}$$

where σ_0 is a quadratic functional, for any possible trajectory of (23).

The set of the coefficients $\tau = \tau_k$ for which (28) yields a function V satisfying (25) is convex. Therefore, an efficient algorithm for finding τ will be available, as soon as one can find an acceptable "feasibility oracle", i.e. an algorithm which quickly checks the inequality in (25) for a particular V from (28) (and, in particular, presents t, x_t, v_t such that (25) is not valid, if this is the case). This is where partial Lyapunov functions can help, because, due to (27), the inequality in (28) will be implied whenever

$$\sigma(w) \le 0 \ \ \forall\, w \in \Omega, \tag{30}$$

for

$$\sigma(w) = \sigma_0(w) + \sum \tau_k \sigma_k(w).$$

In contrast with (25), the inequality in (30) is relatively easy to verify. In the standard theory of IQC analysis [6] this is done by simply checking that $\sigma(w) \ge 0$ for *all* w (i.e. not only those from Ω). This is good enough when absolute stability and performance of nonlinear and time-varying perturbations of linear systems are analyzed, as long as the general behavior remains similar to the linear one. However, analysis of essentially nonlinear behavior of hybrid systems requires a more accurate treatment of (30).

3.2.3. RELAXATION. The task of checking (30) is essentially equivalent to verifying that maximum of the quadratic functional

$$\tilde{\sigma}(w) = w'Qw + 2L'w,$$

on Ω does not exceed $-C$ where

$$Q = Q_0 + \sum_{k=1}^{N} \tau_k Q_k, \; L = L_0 + \sum_{k=1}^{N} \tau_k L_k, \; C = C_0 + \sum_{k=1}^{N} \tau_k C_k.$$

As usually, one can assume without loss of generality that the quadratic form $w'Qw$ does not contain any terms $q_{kk}u_k^2$, where u_k are those components of $w = (v, u)$ that range over the binary set $u_k \in \{-1, 1\}$. Indeed, any such term can be replaced by the constant q_{kk}. When the coefficients at u_k^2 are zero, allowing u_k to range over $[-1, 1]$, rather than $\{-1, 1\}$, will not change the maximum of $\sigma(w)$. Therefore, optimization of $\tilde{\sigma}(w)$ over Ω is equivalent to optimization of $\bar{\sigma}(w, \theta)$ over

$$\bar{\Omega} = \{(v, u, \theta) \in \mathbf{R}^p \times [-1, 1]^q \times [-1, 1] : \; |h_k v| \leq 1\},$$

where

$$\bar{\sigma}(w, \theta) = w'Qw + 2\theta L'w.$$

Finally, we can conclude that verifying (30) reduces to finding the maximum in the MAX-CUT quadratic program (generally non-convex):

$$J(x) = x'Px \to \max \text{ subject to } |a_k' x| \leq 1 \; (k = 1, \ldots, n). \tag{31}$$

The optimization problem in (31) can be "relaxed" to a semidefinite program in several ways, the simplest (and presumably the roughest) of which is

$$\hat{J}(X) = \mathrm{tr}(PX) \to \max \text{ subject to } a_k' X a_k \leq 1, \; X = X' \geq 0. \tag{32}$$

In order to construct a quasi-lyapunov function V in (25), it is be sufficient to find $\tau_k \geq 0$ such that

$$\mathrm{tr}\left(X \left(\begin{bmatrix} Q_0 & L_0 \\ L_0' & 0 \end{bmatrix} + \sum_{k=1}^{N} \tau_k \begin{bmatrix} Q_k & L_k \\ L_k' & 0 \end{bmatrix} \right) \right) \leq C_0 - \sum_{k=1}^{N} \tau_k C_k \; \forall \, X \in \Omega_r, \tag{33}$$

where

$$W_r = \left\{ X = X' \geq 0 : \; \begin{bmatrix} h_k \\ 0 \end{bmatrix}' X \begin{bmatrix} h_k \\ 0 \end{bmatrix} \leq 1, \begin{bmatrix} 0 \\ e_k \end{bmatrix}' X \begin{bmatrix} 0 \\ e_k \end{bmatrix} \leq 1 \right\}. \tag{34}$$

(Here e_k, $k = 1, \ldots, q$ is the coordinate basis in $\mathbf{R}^q$.)

3.2.4. QUASI-LINEAR MODELS. A major obstacle in using partial invariants for system analysis is derivation of a sufficiently large and representative set of partial invariants. While partial invariants can be derived for simple subsystems, and are "inherited" when such subsystem is included in a complex interconnection, the success of the analysis may well depend on the partial invariants describing interaction between subsystem. At this point, the number of partial invariants to consider begins growing exponentially, which makes the whole analysis process inefficient.

A different approach to automatic search for system invariants can be proposed, based on the notion of a *quasi-linear model.* For simplicity, consider system (23) where

$$(x_t, f_t) \in \Omega_0 = [-1,1]^p \times \{-1,1\}^q$$

is a quasi-cube. System (23) is said to be *quasi-linear* if there exist matrices D, F, L of appropriate size, a quadratic functional

$$\sigma_0(w) = \begin{bmatrix} w \\ 1 \end{bmatrix}' \Sigma_0 \begin{bmatrix} w \\ 1 \end{bmatrix},$$

and a quasi-cube

$$\Omega = [-1,1]^a \times \{-1,1\}^b,$$

such that

$$\{(w_0, f(w_0), g(w_0)) : \ w_0 \in \Omega_0\} = \{(Dw, Fw, \sigma_0(w)) : \ Lw = 0, \ w \in \Omega\}. \quad (35)$$

In other words, a quasi-linear model is defined by a set of *linear* equations and a quadratic "cost" on a quasi-cube.

When $b = 0$ in the definition of Ω (i.e. when Ω is a hypercube in the usual sense), the only *functions* f, g that can be defined by (35) are respectively linear and quadratic (it is also possible to produce some *multi-valued* mappings f, g). However, it is well known that, when some components of $w \in \Omega$ are allowed to be "bits" (modeled as elements of $\{-1,1\}$ in this case), the set of functions (f, g) which one can define by (35) is dense in the uniform metric. In other words, an arbitrary system model (23) defined by *continuous* functions f, g on a quasi-cube can be approximated arbitrarily well by quasi-linear models.

For example, the operation of *quantization* of a scalar $x_0 \in [-1,1]$ can be defined by the linear equations

$$x_{k-1} = 0.5(x_k + w_k), \quad k = 1, 2, \ldots,$$

where the variables x_k range over the interval $[-1,1]$, while the variables w_k (which represent a binary expansion of x_0) range over $\{-1,1\}$. Similarly, the logical operation "AND" (i.e. $w_3 = \phi(w_1, w_2) = 1$ if and only if $w_1 = w_2 = 1$) can be defined by the single equation

$$w_1 + w_2 - 2w_3 + w_4 = 1,$$

where $w_i \in \{-1,1\}$ for $i = 1,\dots,4$. Using the operations of quantization, logical operations, and digital-to-analog conversion (which is simply a weighted summation of the corresponding bits), one can define a rich set of transformations.

3.2.5. QUADRATIC STABILITY OF QUASI-LINEAR MODELS. Let (23) be a quasi-linear system defined by (35). A quadratic functional $\sigma : \Omega \to \mathbf{R}$ is a system invariant if there exists $r > 0$ such that

$$\rho(\sigma, r) = \max_{w\in\Omega}\{\sigma(Fw) - \sigma(Dw) - \sigma_0(w) - r|Lw|^2\} \le 0. \tag{36}$$

To find σ, r such that (36) holds, one can try to minimize ρ as a function of σ, r. Note that $\rho(\cdot)$ is a maximum of a family of linear functionals, and hence is a convex function of its arguments. Therefore, minimization of ρ can be performed efficiently, assuming that $\rho(\sigma, r)$ can be evaluated for any fixed σ, r. The task of calculating the maximum in (36) is equivalent to the MAX-CUT optimization problem.

The procedure can be used to search automatically for quadratic invariants of systems with quasi-linear models. Moreover, while the invariants searched for are limited to quadratic functions of the "extended" system state $w \in \Omega$, the resulting system invariant will in general be non-quadratic as a function of w_0. Still, it is not clear that a system invariant in this specific form can be found whenever a system invariant exists.

To increase flexibility of the approach, the search for a quadratic invariant can be applied to a "lifted" model of system (35). Indeed, if (23) can be represented in the form of (35) then the "lifted" system with

$$\bar{x}_k = \begin{bmatrix} x_{mk+1} \\ x_{mk+2} \\ \vdots \\ x_{m(k+1)} \end{bmatrix}, \quad \bar{v}_k = \begin{bmatrix} v_{mk+1} \\ v_{mk+2} \\ \vdots \\ v_{m(k+1)} \end{bmatrix}$$

can be represented in the "lifted" form as well, according to

$$\bar{w} = \begin{bmatrix} w_1 \\ \vdots \\ w_m \end{bmatrix} \in \Omega^m, \quad \bar{D}\bar{w} = Dw_1, \quad \bar{F}\bar{w} = Fw_m, \quad \bar{L}\bar{w} = \begin{bmatrix} Lw_1 \\ \vdots \\ Lw_m \\ Fw_1 - Dw_2 \\ \vdots \\ Fw_{m-1} - Dw_m \end{bmatrix}$$

In general, one can expect that a combination of quasi-linear modeling with the use of partial invariants on a lifted system model will be most convenient for automatic generation of system invariants.

4. Gaussian Proofs of Duality Gap Bounds

In this section, the technique of Gaussian randomization is used to prove some old and new duality gap bounds for the problem of quadratic optimization over the hypercube.

4.1. Using Signs of Gaussian Vectors

Several duality gap bounds in the MAX-CUT problem can be proven by working with the signs of Gaussian vectors.

4.1.1. The Idea of Goemans and Williamson. The idea of working with the signs $v_i = \text{sgn}(w_i)$ of a set $\{w_i\}$ of jointly Gaussian random variables was originally used by Goemans and Williamson in [2] to prove the following uniform bound of the duality gap in the MAX-CUT problem.

Theorem 4.1. *(Goemans, Williamson [2]) Let $Q \in S_n$ be a matrix such that*

$$Q_{ii} \geq \sum_{j \neq i} (|Q_{ij}| + |Q_{ji}|).$$

Then $J_* \geq c\hat{J}_*$ where

$$J_* = \max\{J(x) = x'Qx : \quad x \in \{-1, 1\}^n\},$$

$$\hat{J}_* = \max\{\hat{J}(W) = \text{tr}[QW] : \quad W \in S_n, \ W \geq 0, \ W_{ii} = 1\},$$

$$c = \min\left\{\frac{2}{\pi}\frac{t}{1-\cos(t)} : \quad t \in (0, \pi)\right\} \approx 0.8786.$$

Moreover, if w is a Gaussian random vector such that $\mathbf{E}w = 0$ and $\mathbf{E}ww' = \bar{W}$ is the argument of the maximum of $\hat{J}(W)$ (i.e. $\hat{J}(\bar{W}) = \hat{J}_*$, $\bar{W} = \bar{W}' \geq 0$, $\bar{W}_{ii} = 1$) then

$$\mathbf{E}\text{sgn}(w)'Q\text{sgn}(w) \geq c\hat{J}_*,$$

where in $\text{sgn}(w)$ the sign function is applied to w component-wise.

Proof. Following [2] let us prove first that if f and g are two jointly Gaussian random variables such that

$$\mathbf{E}f = \mathbf{E}g = 0, \quad \mathbf{E}|f|^2 = \mathbf{E}|g|^2 = 1, \quad \mathbf{E}fg = \cos(t)$$

then

$$\mathbf{E}\text{sgn}(f)\text{sgn}(g) = \phi(t) = 1 - \frac{2t}{\pi}. \tag{37}$$

Indeed, let $f - g\cos(t) = h\sin(t)$ where $\mathbf{E}h = 0$ and $\mathbf{E}|h|^2 = 1$. Then $\mathbf{E}fh = 0$ and hence the random variables f and h are independent. Therefore the vector random variable

$$z = \begin{bmatrix} z_1 \\ z_2 \end{bmatrix} = \frac{1}{|f|^2 + |h|^2}\begin{bmatrix} f \\ h \end{bmatrix}$$

is uniformly distributed on the unit circle in $\mathbf{R}^2$, which implies that

$$\mathbf{E}\mathrm{sgn}(f)\mathrm{sgn}(g) = \mathbf{E}\mathrm{sgn}(z_1)\mathrm{sgn}(z_1\cos(t) + z_2\sin(t)) = \phi(t)$$

is an affine function of t. Since $\phi(0) = 1$ and $\phi(\pi) = -1$, the conclusion is that $\phi(t) = 1 - 2t/\pi$.

Now let $v_i = \mathrm{sgn}(w_i)$. Since $\mathbf{E}|v_i|^2 = 1$, (37) shows that

$$\mathbf{E}v_iv_j = 1 - \frac{2}{\pi}\arccos(W_{ij}).$$

Hence, for $t_{ij} = \arccos(W_{ij}) \in [0,\pi]$,

$$1 - \mathbf{E}v_iv_j = \frac{2t_{ij}}{\pi} \geq c(1-\cos(t_{ij})) = c(1-\mathbf{E}w_iw_j).$$

Since

$$\min_{t\in(0,\pi)} \frac{2}{\pi}\frac{t}{1-\cos(t)} = \min_{t\in(0,\pi)} \frac{2}{\pi}\frac{\pi - t}{1+\cos(t)},$$

it is also true that

$$1 + \mathbf{E}v_iv_j = \frac{2}{\pi}(\pi - t_{ij}) \geq c(1+\cos(t_{ij})) = c(1+\mathbf{E}w_iw_j).$$

Therefore, for

$$r = \mathrm{tr}[Q] - \sum_{j\neq i}(|Q_{ij}| + |Q_{ji}|),$$

it follows that

$$\begin{aligned}\mathbf{E}v'Qv &= r + \sum_{i\neq j}|Q_{ij}|(1+\mathrm{sgn}(Q_{ij})\mathbf{E}v_iv_j)\\ &\geq cr + c\sum_{i\neq j}|Q_{ij}|(1+\mathrm{sgn}(Q_{ij})\mathbf{E}w_iw_j)\\ &= c\,\mathrm{tr}[QW].\end{aligned}$$

Since $J_* \geq v'Qv$, this proves the theorem. □

4.1.2. A REMARK ON ORTHOGONALITY. Other duality gap bounds can be proven using an orthogonality feature of the sign of a Gaussian variable. The constant

$$\nu = \left(\frac{2}{\pi}\right)^{1/2},$$

which equals $\mathbf{E}|w|$ for a Gaussian random variable w with zero mean and unit variance, will play an important role.

Lemma 4.2. *Let f, g be two jointly Gaussian random variables with zero mean. Let $v = (\mathbf{E}g^2)^{1/2}\mathrm{sgn}(g)$. Then $\mathbf{E}f(\nu g - v) = 0$.*

Proof. A direct calculation shows that $\mathbf{E}g(\nu g - v) = 0$. On the other hand, f can be written in the form $f = f_0 + ag$ where $a \in \mathbf{R}$ and f_0 is independent of g. Hence f_0 and $\nu g - v$ are independent as well, and therefore $\mathbf{E}f_0(\nu g - v) = 0$. □

4.1.3. THE NESTEROV-YE THEOREM. An immediate consequence of Lemma 4.2 is the following theorem proven by Nesterov [8] and Ye [12].

Theorem 4.3. *For $Q \in S_n$ let*

$$J_+ = \max\{J(x) = x'Qx : \quad x \in [-1,1]^n\},$$

$$\hat{J}_+ = \max\{\hat{J}(W) = \mathrm{tr}[QW] : \quad W \in S_n, \ W \geq 0, \ W_{ii} \leq 1\},$$

$$\hat{J}_- = \min\{\hat{J}(W) = \mathrm{tr}[QW] : \quad W \in S_n, \ W \geq 0, \ W_{ii} \leq 1\}.$$

Then

$$J_+ \geq \frac{2}{\pi}\hat{J}_+ + (1 - \frac{2}{\pi})\hat{J}_-.$$

Moreover, if $\bar{W}$ is the argument of maximum of $\hat{J}(W)$ subject to $W = W' \geq 0$ and $W_{ii} \leq 1$ then

$$\mathbf{E}v'Qv \geq \nu^2 \hat{J}_+ + (1 - \nu^2)\hat{J}_-, \tag{38}$$

where

$$v_i = (\mathbf{E}w_i^2)^{1/2}\mathrm{sgn}(w_i)$$

and w is a Gaussian random vector with $\mathbf{E}w = 0$, $\mathbf{E}ww' = \bar{W}$.

Proof. By Lemma 4.2

$$\mathbf{E}w'F(\nu w - v) = 0$$

for *any* constant matrix F. Note also that since

$$\mathbf{E}(\nu w_i - v_i)^2 = (1 - \nu^2)\mathbf{E}w_i^2 \leq 1 - \nu^2,$$

for all i, the inequality

$$\mathbf{E}(\nu w - v)'Q(\nu w - v) \geq (1 - \nu^2)\hat{J}_-$$

takes place. Hence

$$\begin{aligned} \mathbf{E}v'Qv &= \mathbf{E}(\nu w + (v - \nu w))'Q(\nu w + (v - \nu w)) \\ &= \nu^2 \mathbf{E}w'Qw + \mathbf{E}(\nu w - v)'Q(\nu w - v) \\ &\geq \nu^2 \hat{J}_+ + (1 - \nu^2)J_-. \end{aligned}$$

□

Theorem 4.3 was proven in [8] for the case $Q \geq 0$ and in [12] for the general case. In [12], the following implication of (38) is formulated as the main result:

$$\frac{J_+ - \mathbf{E}v'Qv}{J_+ - J_-} \leq \frac{\pi}{2} - 1, \tag{39}$$

where

$$J_- = \min\{J(x) = x'Qx : \quad x \in [-1,1]^n\}.$$

To derive (39) from Theorem 4.3, note first that applying the Theorem with Q replaced by $-Q$ yields

$$J_- \le \nu^2 \hat{J}_- + (1-\nu^2)\hat{J}_+.$$

Hence

$$\begin{aligned}\frac{J_+ - \mathbf{E}v'Qv}{J_+ - J_-} &\le \frac{J_+ - (\nu^2\hat{J}_+ + (1-\nu^2)\hat{J}_-)}{J_+ - (\nu^2\hat{J}_- + (1-\nu^2)\hat{J}_+)} \\ &\le \frac{\hat{J}_+ - (\nu^2\hat{J}_+ + (1-\nu^2)\hat{J}_-)}{\hat{J}_+ - (\nu^2\hat{J}_- + (1-\nu^2)\hat{J}_+)} \\ &= \nu^{-2} - 1.\end{aligned}$$

Note that the second inequality is valid since the function

$$x \mapsto \frac{x - (\nu^2\hat{J}_+ + (1-\nu^2)\hat{J}_-)}{x - (\nu^2\hat{J}_- + (1-\nu^2)\hat{J}_+)}$$

is non-decreasing for $x \ge \nu^2\hat{J}_+ + (1-\nu^2)\hat{J}_-$, and $J_+ \le \hat{J}_+$.

4.1.4. SUBTRACTION OF EVEN-CYCLIC MATRICES. Inequality (38) implies

$$J_+ \ge (4/\pi - 1)\hat{J}_+$$

whenever it can be proven that $\hat{J}_- \ge -\hat{J}_+$. An interesting case when this can be done easily is associated with matrices with even cycles only.

Definition 4.4. *Let us call a symmetric matrix $C \in S_n$* even-cyclic *if its diagonal entries equal zero and the graph Γ_C formed by connecting those nodes $i, j \in \{1, \dots, n\}$ for which $C_{ij} \ne 0$ contains no cycles of odd length.*

Theorem 4.5. *If $Q - C \ge 0$ for some even-cyclic matrix C then $J_+ \ge (4/\pi - 1)\hat{J}_+$.*

Proof. Without loss of generality assume that the graph Γ_C is connected. For any n-vector x and an even-cyclic matrix C let $z = x_C$ be the vector defined as follows: $z_i = x_i$ whenever node i of Γ_C is connected to node 1 through an even number of edges, and $z_i = -x_i$ otherwise. Since Γ_C is even-cyclic, the definition of $z = x_C$ is correct. By construction, $x'Cx = -x_C'Cx_C$ for any x. Hence $\hat{J}_- \ge -\hat{J}_+$ for any $Q \ge C$. □

4.1.5. THE GROTHENDIECK INEQUALITY. Theorem 4.5 can be considered as a "generalized" Grothendieck inequality. However, for the original Grothendieck inequality, the resulting constant $4/\pi - 1$ is far from being the best known.

Theorem 4.6. *The inequality*

$$J_* \ge (4/\pi - 1)\hat{J}_*$$

holds for all matrices

$$Q = \begin{bmatrix} 0 & Q_{12} \\ Q_{12}' & 0 \end{bmatrix},$$

where the diagonal blocks are of square form.

Proof. Q is an even-cyclic matrix. □

4.2. Slowly Increasing Duality Gap

Some slowly increasing duality gap bounds can be derived using the fast decay of the Gaussian density away from the origin.

4.2.1. A LOGARITHMIC UPPER BOUND FOR DUALITY. Taking into account difficulty of finding good duality gap bounds, it is surprising to see that the following theorem is true.

Theorem 4.7. *For* $n > 60$, $\hat{J}_* \leq 2\log(n)J_*$ *for any problem data* P, a_k, n *in (31),(32).*

Proof. Let $X = X' \geq 0$ be any matrix satisfying the constraints $a_k' X a_k \leq 1$. Let us show that then there exists a vector x satisfying the conditions $|a_k' x| \leq 1$ such that $2\log(n)x'Px \geq \mathrm{tr}[XP]$.

Indeed, let w will be a Gaussian random vector with $\mathbf{E}[w] = 0$ and $\mathbf{E}[ww'] = X$. Let $q = \max_k\{|a_k' w|\}$, $v = w/q$. Then v is a random variable satisfying the constraints $|a_k' v| \leq 1$ with probability 1. Moreover, we have

$$\begin{aligned}\mathrm{tr}[XP] &= \mathbf{E}[w'Pw] \\ &= \mathbf{E}[q^2(v'Pv)] \\ &\leq \mathbf{E}[v'Pv]\mathbf{E}[q^2].\end{aligned}$$

Hence, in order to complete the proof, it is sufficient to show that $\mathbf{E}[q^2] \leq 2\log(n)$.

For $r \geq 0$ let $f(r)$ be the probability that $q^2 \geq r$. Since $f(r)$ does not exceed the sum of probabilities $f_k(r)$ that $|a_k' w|^2 \geq r$, and since $a_k' w$ are Gaussian random variables with zero mean and variance not exceeding 1,

$$f(r) \leq n\nu \int_{\sqrt{r}}^{\infty} e^{-t^2/2} dt. \tag{40}$$

Since $f(r) \leq 1$ for all r, for any $\lambda > 0$ we have

$$\int_0^{\infty} f(r)dr \leq \lambda + \int_{\lambda}^{\infty} f(r)dr.$$

Hence

$$\mathbf{E}[q^2] \leq \lambda + n\nu \int_{\lambda}^{\infty} f(r)dr = \lambda + n\nu \int_{\lambda}^{\infty} \int_{\sqrt{r}}^{\infty} e^{-t^2/2} dt dr.$$

Changing the order of integration and integrating over r yields

$$\int_{\lambda}^{\infty} \int_{\sqrt{r}}^{\infty} e^{-t^2/2} dt dr = \int_{\sqrt{\lambda}}^{\infty} (t^2 - \lambda) e^{-t^2/2} dt.$$

Substituting $t = (s + \lambda)^{1/2}$ into the last integral yields

$$\int_{\sqrt{\lambda}}^{\infty} (t^2 - \lambda) e^{-t^2/2} dt = \frac{1}{2} \int_0^{\infty} s e^{-(s+\lambda)/2} (s + \lambda)^{-1/2} ds.$$

Replacing $(s+\lambda)^{-1/2}$ with its upper bound $\lambda^{-1/2}$ yields

$$\int_0^\infty se^{-(s+\lambda)/2}(s+\lambda)^{1/2}ds \leq \lambda^{-1/2}e^{-\lambda/2}\int_0^\infty se^{-s/2}ds = 4\lambda^{-1/2}e^{-\lambda/2}.$$

Hence

$$\mathbf{E}[q^2] \leq \lambda + 4n\sqrt{\frac{1}{2\pi}}\exp(-\lambda/2)\lambda^{-1/2}.$$

Taking $\lambda = 2\log(n) - 1$ yields $\mathbf{E}[q^2] \leq 2\log(n)$ for $n > 60$. □

4.2.2. A Lower Bound for the Duality Gap. The following theorem shows that it is not possible to replace the logarithmic relaxation gap bound of Theorem 4.7 with a constant bound which does not grow with n.

Theorem 4.8. *There exists a sequence of matrices $P = P_n$ of size $n \times n$, $n \to \infty$, such that, for $a_i = e_i$ being the standard basis vectors, $\hat{J}_*(P_n)/J_*(P_n) \to \infty$.*

Proof. For a fixed positive integer m and an arbitrarily small $\epsilon > 0$ let $\mathbf{U} = \{u_k\}_{k=1}^N \subset \mathbf{R}^m$ be an ϵ-dense subset of the unit sphere, i.e. $|u_k| = 1$ for all k and

$$\max_k |u_k'x|^2 \geq (1-\epsilon)|x|^2 \quad \forall\, x \in \mathbf{R}^m. \tag{41}$$

Let U be the m-by-N matrix whose k-th column is u_k. Let $X = (u_i'u_j)_{ij} = U'U$ be the Gramm matrix of $\mathbf{U}$.

First, let us show that X does not belong to the set Λ of all convex combinations of rank one matrices $\theta_i\theta_i'$, where $\|\theta_i\|_\infty < \sqrt{m(1-\epsilon)}$.

Indeed, if

$$X = U'U = \sum c_i\theta_i\theta_i', \quad c_i > 0, \quad \sum c_i \leq 1, \tag{42}$$

then all θ_i satisfy the inequality $c_i^{-1}U'U \geq \theta_i\theta_i'$, and hence belong to the image of U', i.e. $\theta_i = U'v_i$ for some $v_i \in \mathbf{R}^m$. According to (41)

$$|v_i|^2 \leq (1-\epsilon)^{-1}\max|u_k'v_i|^2 = (1-\epsilon)^{-1}\|\theta_i\|_\infty^2 < m.$$

On the other hand, since U' is left invertible, identity (42) implies

$$I_m = \sum c_iv_iv_i'.$$

Comparing traces on both sides yields a contradiction

$$m = \sum c_i|v_i|^2 < m.$$

Now let $P = P' \neq 0$ define the linear functional separating X from Λ. By construction, $\hat{J}_*(P) \geq m(1-\epsilon)J_*(P)$. □

4.2.3. DISCUSSION. Existence of a logarithmic gap bound for the relaxation of (31) into (32) appears to be a consequence of convexity of all constraints in (31). In contrast, quadratic programs derived in most of the classical robustness analysis techniques typically have the relaxation gap growth according to a power law. Thus, boundedness of signals subject to linear constraints is expected to be easier to exploit in the analysis than the usual "sector inequalities". Therefore, modeling of nonlinear systems in terms of signals subject to linear equations and amplitude constraints, though more difficult than the canonical approach of the "gain bounds", should be considered as a promising research direction.

Acknowledgement

This work was supported by NSF, AFOSR, DARPA, and the Esther and Harold E. Edgerton Associate Professorship at MIT. The author is grateful to Alexander Borichev for many useful remarks.

References

[1] F. Alizadeh, *Interior point methods in semidefinite programming with applications to combinatorial optimization*, SIAM Journal on Optimization, **5** (1995), 13–51.

[2] M.X. Goemans and D.P. Williamson, *Improved approximation algorithms for maximum cut and satisfiability problems using semidefinite programming*, J. ACM, **42** (1995), 1115–1145.

[3] P. Halmos, *Ten problems in Hilbert space*, Bull. Amer. math. Soc., **76** (1970), 887–933.

[4] J. Hastad, *Some optimal inapproximability results*, Proc. of the 29th ACM Symp. on Theory Comput., 1997.

[5] J.-L. Krivine, *Constantes de Grothendieck et fonctions de type positif sur les spheres*, Adv. in Math., **31** (1979), 16–30.

[6] A. Megretski and A. Rantzer, *System Analysis via Integral Quadratic Constraints*, IEEE Transactions on Automatic Control, volume 47, no. 6, pp. 819–830, June 1997.

[7] Y. Nesterov and A. Nemirovski, *Interior Point Polynomial Methods in Convex Programming*, SIAM, Philadelphia, PA, 1994.

[8] Y. Nesterov, *Quality of semidefinite relaxation for nonconvex quadratic optimization*, Manuscript, 1997.

[9] G. Pisier, *A polynomially bounded operator on Hilbert space which is not similar to a contraction*, Journal Amer. Math. Soc., **10** (1997), no. 2, 351–369.

[10] O. Toker and H. Ozbay, *On the complexity of purely complex μ computation and related problems in multidimensional systems*, IEEE Trans. Aut. Contr., **43** (1998), no. 3, 409–414.

[11] S. Treil, *The gap between complex structured singular value μ and its upper bound is infinite*, To appear in IEEE Trans. Aut. Contr.

[12] Yinyu Ye, *Approximating quadratic programming with bound constraints*, Manuscript, 1997.

Department of Electrical Engineering and Computer Science,
MIT,
35-415 Cambridge 02139, USA
E-mail address: ameg@mit.edu

Bellman function in stochastic control and harmonic analysis

F. Nazarov, S. Treil, and A. Volberg

Abstract. The stochastic optimal control uses the differential equation of Bellman and its solution—the Bellman function. We show how the homonym function in harmonic analysis is (and how it is not) the same stochastic optimal control Bellman function. Then we present several creatures from Bellman's Zoo: a function that proves the inverse Hölder inequality, as well as several other harmonic analysis Bellman functions and their corresponding Bellman PDE's. Finally we translate the approach of Burkholder to the language of "our" Bellman function.

The goal of this paper is almost entirely methodological: we relate the ideas between each other, rather than presenting the new ones.

CONTENTS

All authors are supported by the NSF grant DMS 9970395.

1. General approach: Bellman equation for stochastic optimal control

1.1. Stochastic optimal control

The goal of this paper is methodological. We want to relate the equation of Bellman from optimal control of stochastic processes to the concept of Bellman function that appeared in the first preprint version of [10] in 1995, was developed extensively in [8], and proved to be very useful in [9], [5], [11], [12], [13], [17], [14]. We want also to connect the ideas behind the construction of Burkholder's function $b(x,y) = (|x|-(p-1)|y|)(|x|+|y|)^{p-1}$ (which gives the exact estimates in certain martingale inequalities) to the concept of the Bellman equation.

Let us start with the problem of control of rather general stochastic process. We will borrow the pieces of exposition from the book [6].

Let we have a stochastic process x^t in $\mathbb{R}^d$, satisfying the following integral equation

$$x^t = x + \int_0^t \sigma(\alpha^s, x^s)\, dw^s + \int_0^t b(\alpha^s, x^s) ds \tag{1}$$

Here t is the time, w^t is a d_1-dimensional Wiener process (white noise), $\sigma = \sigma(\alpha, y)$ is a $d \times d_1$ matrix, and b is a d-dimensional vector. The process α^t is supposed to be a control that we have to choose. We denote by $A \subset \mathbb{R}^{d_2}$ the set of admissible controls, i. e. the domain where the vector of control parameters α runs.

The choice of stochastic process α^s (in our cases it will be also d-dimensional, $d_2 = d$, except for the case in Section 3.3) gives us different "motions"—different solutions of (1). Of course, the questions about existence and uniqueness of the solution immediately arise, but in this paper we just assume that the solution exists and unique.

From the practical point of view it is reasonable to consider that the values of the control process α^s at the time $s > 0$ are chosen on the basis of observation of the process x^t up to the moment s, so we think that (for each individual trajectory, i. e. for each point ω of the probability space Ω) α^s is the function of the trajectory $\{(t, x^t) : 0 \le t \le s\}$, that is $\alpha^s = \alpha^s(x^{[0,s]})$.

Suppose we are given the profit function $f^\alpha(y)$; on the trajectory x^t for the time interval $[t, t+\Delta t]$ the profit is $f^{\alpha^t}(x^t)\Delta t + o(\Delta t)$.

Therefore, on the whole trajectory we get $\int_0^\infty f^{\alpha^t}(x^t)\, dt$.

We want to choose the control $\alpha = \{\alpha^s(x^{[0,s]})\}$ to *maximize* the average profit

$$v^\alpha(x) = \mathbb{E} \int_0^\infty f^{\alpha^t}(x^t)\, dt + \lim_{t\to\infty} \mathbb{E}(F(x^t)) \tag{2}$$

for the process x^t starting at x. Here F is a bonus function—one gets it when one retires.

The optimal average gain is what is called the Bellman function for stochastic control:

$$v(x) = \sup_{\alpha} v^{\alpha}(x) \tag{3}$$

here *supremum* is taken over all admissible controls α; $x \in \mathbb{R}^d$ is the starting point of the process: it is the same x as in (1).

It satisfies a well known Bellman differential equation, which we are going to explain now. The deduction of the Bellman equation for the stochastic optimal control is based on two things:

1. Bellman's principle,
2. Ito's formula.

We want to review them to show what changes have to be made to obtain our Bellman function of harmonic analysis pedigree.

Bellman's principle states that

$$v(x) = \sup_{\alpha^s} \mathbb{E}\left[\int_0^t f^{\alpha^s}(x^s)\,ds + v(x^t)\right] \tag{4}$$

for each $t \geq 0$. Here the supremum is taken over all admissible control processes α^s.

To explain it, let us fix some time $t > 0$, and let us consider an individual trajectory of the system. The profit for the interval $[0, t]$ is given by

$$\int_0^t f^{\alpha^s}(x^s) ds.$$

Suppose that the trajectory of the process has reached the point, say y, at the moment t. Then the maximal average profit we can gain starting at the moment t from the point y is exactly $v(y)$. Indeed, since the increments of w^s for $s \geq t$ do not depend on $w^{\tau}, \tau < t$, and they behave as corresponding increments after time 0, and the equations are time invariant, there is no difference between starting at time 0 or at time t. Applying now full probability formula to take into account all possible endpoints x^t, we get exactly the Bellman Principle (4).

Ito's formula gives us a representation of $v(x^t)$ as a stochastic integral (one can also say that it is a formula for the stochastic differential of $v(x^t)$). We then going to hit this formula with the averaging $\mathbb{E}$ over the probability.

For a reader who is not familiar with the stochastic differential/integral calculus, we are going to explain the version of Ito's formula we need.

Let us fix a moment of time s, and a small increment Δs. We want to estimate the difference $v(x^{s+\Delta s}) - v(x^s)$. Recall that $w^s = (w_1^s, w_2^s, \dots, w_{d_1}^s)^T \in \mathbb{R}^{d_1}$, $x = (x_1, x_2, \dots x_d)^T \in \mathbb{R}^d$, and denote $\Delta w^s := w^{s+\Delta s} - w^s$, $\Delta w_k^s := w_k^{s+\Delta s} - w_k^s$, etc.

Using Taylor's formula (we think that v is smooth, which might not be the case, but we do not care now—we are heading towards formal deduction of the

Bellman equation), we will have to consider the term

$$\sum_{k=1}^{d} \frac{\partial v}{\partial x_k}(x^s) \sum_{j=1}^{d_1} \sigma_{k,j}(\alpha^s, x^s) \Delta w_j^s + \sum_{k=1}^{d} \frac{\partial v}{\partial x_k}(x^s) b_k(\alpha^s, x^s) \Delta s.$$

Since we are going to average over the probability, the first term in the sum will vanish—the terms Δw_k^s are independent of x^s and have zero averages. The second term can be rewritten as $\big(\mathcal{L}_1^{\alpha^s}(x^s)v\big)(x^s)\Delta s$, where the first order differential operator $\mathcal{L}_1^\alpha(x)$ is given by

$$\mathcal{L}_1^\alpha(x) := \sum_{k=1}^{d} b_k(\alpha, x) \frac{\partial}{\partial x_k}$$

The main difference between the stochastic calculus and the regular one is that in the former we need to consider higher order terms. The next term in the Taylor formula will be

$$\frac{1}{2} \sum_{i,j} \frac{\partial^2 v}{\partial x_i \partial x_j} \Big(\Sigma_k \sigma^{jk} \Delta w_k^s + b_j(\alpha^s, x^s)\Delta s\Big) \cdot \Big(\Sigma_k \sigma^{ik} \Delta w_k^s + b_i(\alpha^s, x^s)\Delta s\Big) \tag{5}$$

Let us see what happens with this term after averaging over the probability: the terms involving $\Delta w_k^s \Delta s$ disappear again, and we can disregard the terms involving $(\Delta s)^2$, for they are $o(\Delta s)$. Since

$$\mathbb{E}\Delta w_k^s \Delta w_m^s = \begin{cases} 0, & k \neq m \\ \Delta s & k = m \end{cases},$$

the averaging of the (5) gives

$$\mathbb{E}\Big[\Big(\mathcal{L}_2^{\alpha^s}(x^s)\, v\Big)(x^s)\Big]\ \Delta s, \tag{6}$$

where the second order differential operator $\mathcal{L}_2^\alpha(x)$ is given by

$$\mathcal{L}_2^\alpha(x) := \sum_{i,j=1}^{d} a^{ij}(\alpha, x) \frac{\partial^2}{\partial x_i \partial x_j},$$

and

$$a^{ij}(\alpha, x) := \frac{1}{2} \sum_{k=1}^{d_1} \sigma^{ik}(\alpha, x) \sigma^{jk}(\alpha, x).$$

So, we have got an extra term with Δs!

The higher order Taylor terms other than (5) will give us Δs to powers greater than 1, and, therefore, can be omitted by an obvious reason.

Gathering all together and integrating with respect to ds we get

$$\mathbb{E}(v(x^t)) = v(x) + \mathbb{E} \int_0^t \mathcal{L}^{\alpha^s}(x^s) v(x^s)\, ds \tag{7}$$

where $\mathcal{L}^\alpha(x) := \mathcal{L}_1^\alpha(x) + \mathcal{L}_2^\alpha(x)$. That is exactly the application of Ito's formula we need. [1]

Putting (7) into the Bellman Principle (4) one gets:

$$0 = \sup_{\alpha^s} \mathbb{E}\left[\int_0^t f^{\alpha^s}(x^s)\,ds + \int_0^t \mathcal{L}^{\alpha^s}(x^s)v(x^s)\,ds\right]$$

Dividing by t and letting t tend to zero, one gets Bellman's equation on Bellman function v:

$$\sup_{\alpha\in A}\,[\mathcal{L}^\alpha(x)v(x) + f^\alpha(x)] = 0 \tag{8}$$

(Of course, to justify taking the limit one has to make some assumptions, so the above presentation is not a complete proof, but just a general scheme, explaining how the Bellman equation was derived).

Notice that in the above Bellman equation (8) the supremum is taken not over all control processes α^s, but over all admissible values $\alpha \in A \subset \mathbb{R}^{d_2}$.

Let us also notice, that the bonus function F is not included in the equation: it usually figures in the boundary conditions and (or) inequalities that v has so satisfy. We will see examples later, see Section 3.2 below.

Note, that it is possible to reverse the above reasonings, namely to show that if (under some additional assumptions) a function v satisfies the Bellman equation (8) with appropriate boundary conditions, then it has to be the Bellman function (3).

In applications to the Harmonic Analysis we will be more interested in *supersolutions* of the Bellman equation, i. e. functions V satisfying

$$\sup_{\alpha\in A}\,[\mathcal{L}^\alpha(x)V(x) + f^\alpha(x)] \le 0. \tag{9}$$

Suppose the bonus function $F \equiv 0$ and the profit density $f^\alpha(x) \ge 0$. Then clearly the Bellman function v from (3) is nonnegative. Then we claim that any $V \ge 0$ satisfying the "Bellman inequality" (9) majorates v, $V \ge v$.

Indeed, (9) implies that $-\mathcal{L}^\alpha(x)V(x) \ge f^\alpha(x)$. Using this inequality in the Ito's formula (7) for V we get

$$\begin{aligned} V(x) &= \mathbb{E}V(x^t) - \mathbb{E}\int_0^t (\mathcal{L}^{\alpha^s}(x^s)V)(x^s)ds \\ &\ge \mathbb{E}V(x^t) + \mathbb{E}\int_0^t f^{\alpha^s}(x^s)ds \ge \mathbb{E}\int_0^t f^{\alpha^s}(x^s)ds \end{aligned}$$

(the last inequality holds because $V \ge 0$). Taking limit as $t \to \infty$ and then supremum over all controls α, we get the desired inequality $V(x) \ge v(x)$.

One more remark: if $\sigma(\alpha, x) \equiv 0$, i. e. the stochastic integral (with respect to the Wiener process w) disappears from the equation (1) and we have only integral with ds, then the problem becomes essentially deterministic. Namely, to

[1] Ito's formula gives us an integral representation of $v(x^t)$ *before* we average it. Its proof is more complicated than the reasoning presented above.

find the optimal control α one just needs to find it for each trajectory separately, i. e. to solve a deterministic problem. In the Bellman equation (8) the second order term $\mathcal{L}_2$ disappears, and we get as a result a first order PDE, which is exactly the classical (deterministic) Bellman–Jacobi equation from optimal control. The second order operator $\mathcal{L}_2$ reflects the specifics of the stochastic case.

2. Harmonic analysis Bellman function

2.1. Harmonic analysis problems

By "harmonic analysis problems" we always will mean here some *dyadic* problem, dealing with averages over dyadic intervals. It is often possible to pass from a dyadic problem to a problem with analytic or harmonic function by using some kind of Green's formula, see [9].

By dyadic lattice $\mathcal{D}$ we mean the collection of all dyadic interval, i. e. the collection of all intervals of form $[j \cdot 2^k, (j+1) \cdot 2^k)$, $j, k \in \mathbb{Z}$.

If I is an interval, we denote by $|I|$ its length, and by χ_I its characteristic function, $\chi_I(x) = 1$ if $x \in I$, and $\chi_I(x) = 0$ if $x \notin I$. Symbols I_+ and I_- denote right and left halves of the interval I respectively.

And finally, given a function f we denote by $\langle f \rangle_I$ its average over the interval I,

$$\langle f \rangle_I := |I|^{-1} \int_I f(x)dx$$

For an interval I define the Haar function $h_I := |I|^{-1/2}(\chi_{I_+} - \chi_{I_-})$. The *Haar system* $\{h_I : I \in \mathcal{D}\}$ is an orthonormal basis in $L^2(\mathbb{R})$.

2.2. General remarks

In "our" Bellman function deduction we must have several changes in the scheme. The most important change is that our time will be discrete. Let us postpone this modification. So, our time is still continuous.

The second change is that in most of our problems the corresponding Wiener process will be one-dimensional ($d_1 = 1$). So, it simplifies the things a bit.

The matrix $\sigma(\alpha, x)$ will be so just a column $(\sigma_1, \sigma_2, \dots, \sigma_d)^T$. Very often it will be very simple, just $\sigma_k = \alpha_k$, $k = 1, 2, \dots, d$.

The first order operator $\mathcal{L}_1^\alpha$ will remain the same, but the second order part will be much simpler,

$$\mathcal{L}_2^\alpha(x) = \frac{1}{2}\sum_{i,j} \sigma_i(\alpha, x)\sigma_j(\alpha, x)\frac{\partial^2}{\partial x_i \partial x_j}$$

$$\Big(= \frac{1}{2}\sum_{i,j} \alpha_i\alpha_j \frac{\partial^2}{\partial x_i \partial x_j} \quad \text{if } \sigma_k = \alpha_k,\ k = 1, 2, \dots, d. \Big)$$

Usually, the solutions of Bellman equation (8) are quite difficult to find, so in most cases we will be looking for ***supersolutions*** (9), when the equality is replaced by $\leq$

sign. The Bellman inequality (9) in our case usually mean some kind of convexity. For example, if there is no linear term $\mathcal{L}_1$, and the profit density $f^\alpha \equiv 0$, then, in the case where $\sigma_k = \alpha_k$ for every k, the Bellman inequality (9) becomes

$$\frac{1}{2} \sup_\alpha \sum_{i,j=1}^{d} \alpha_i \alpha_j \frac{\partial^2 v}{\partial x_i \partial x_j} \leq 0,$$

i. e. exactly the concavity condition for v.

But let us consider a simple example to illustrate what we were saying.

2.3. An example: A_∞ weights and their associated Carleson measures

We call a positive function w on the line $\mathbb{R}$ an A_∞ weight (dyadic A_∞ weight, actually) if

$$\langle w \rangle_J \leq c_1\, e^{\langle \log w \rangle_J}, \quad \forall J \in \mathcal{D} \tag{10}$$

Here $\mathcal{D}$ denotes a dyadic lattice on $\mathbb{R}$, and $\langle \cdot \rangle_J$ denotes the averaging over J.

We are going to use our Bellman function technique to prove the following result of Buckley [1] that can be found (along with it "continuous analogs) also in the paper of Fefferman–Kenig–Pipher [4].

Theorem 2.1. *Let $w \in A_\infty$. Then*

$$\forall I \in \mathcal{D}, \quad \frac{1}{|I|} \sum_{\ell \subset I} \left(\frac{\langle w \rangle_{\ell_+} - \langle w \rangle_{\ell_-}}{\langle w \rangle_\ell} \right)^2 |\ell| \leq c_2, \tag{11}$$

where c_2 depends only on c_1 in (10).

The meaning of (11) is that a certain measure associated with $w \in A_\infty$ must be a Carleson measure.

2.3.1. "Common sense" approach.

We present first a "naïve" approach, based essentially just on common sense, and then show how it correspond to the general scheme presented above.

Let us fix the averages $\mathbf{w} := \langle w \rangle_I$, $\mathbf{u} := \langle \log w \rangle_I$.

Define a function $\mathcal{B} = \mathcal{B}_I = \mathcal{B}_I(\mathbf{u}, \mathbf{w})$ of two real variables $\mathbf{u}, \mathbf{w}$ as

$$\mathcal{B}_I(\mathbf{u}, \mathbf{w}) := \sup \frac{1}{|I|} \sum_{J \subset I} \left(\frac{\langle w \rangle_{J_+} - \langle w \rangle_{J_-}}{\langle w \rangle_J} \right)^2 |J|,$$

where the supremum is taken over all weights w on I satisfying the A_∞ condition (10) and such, that $\langle w \rangle_I = \mathbf{w}$, $\langle \log w \rangle_I = \mathbf{u}$. We use the letter $\mathcal{B}$ here for *Bellman.*

Note, that the function $\mathcal{B}_I$ does not depend on the interval I: it clearly doesn't change if one shifts the interval I, and the scaling is chosen in such a way, that $\mathcal{B}$ remains the same if one changes the size of I. Therefore, in what follows, we just skip the subscript I.

Clearly, the above Theorem 2.1 holds if and only if $\mathcal{B} \leq c_2$. Let us try to investigate the properties of the function $\mathcal{B}$. Clearly, the following properties hold

1. Domain: the function $\mathcal{B}$ is defined on the domain D $e^{\mathbf{u}} \le \mathbf{w} \le c_1 e^{\mathbf{u}}$.

   The first inequality here is just Jensen inequality $e^{\langle \log w \rangle_I} \le \langle w \rangle_I$ (geometric mean is at most the arithmetic mean), and the second one is the A_∞ condition (10).

   It takes some effort to prove that for any pair $\mathbf{u}, \mathbf{w}$ from the domain one can find an A_∞ weight w such that $\langle \log w \rangle_I = \mathbf{u}$, $\langle w \rangle_I = \mathbf{w}$, but we really do not need this fact to prove the theorem: we need it only to be sure that we did not miss anything in our Bellman function approach.
2. $0 \le \mathcal{B} \le c_2$; the first inequality just follows from the definition, and the second one is just our belief that the theorem is true;
3. The main inequality: for any three pairs $(\mathbf{u}, \mathbf{w})$, $(\mathbf{u}_+, \mathbf{w}_+)$, $(\mathbf{u}_-, \mathbf{w}_-)$ in the domain satisfying

$$\mathbf{u} = \frac{1}{2}(\mathbf{u}_+ + \mathbf{u}_-), \qquad \mathbf{w} = \frac{1}{2}(\mathbf{w}_+ + \mathbf{w}_-)$$

   the following inequality

$$\mathcal{B}(\mathbf{u}, \mathbf{w}) \ge \left(\frac{\mathbf{w}_+ + \mathbf{w}_-}{\mathbf{w}} \right)^2 + \frac{1}{2}(\mathcal{B}(\mathbf{u}_+, \mathbf{w}_+) + \mathcal{B}(\mathbf{u}_-, \mathbf{w}_-)) \tag{12}$$

   holds.

The explanation for this main inequality is rather simple: consider all A_∞ weights w such that $\langle \log w \rangle_I = \mathbf{u}$, $\langle w \rangle_I = \mathbf{w}$, and the corresponding averages over subintervals I_+ and I_- are $\mathbf{u}_+$, $\mathbf{w}_+$, and $\mathbf{u}_-$, $\mathbf{w}_-$ respectively. Then

$$\frac{1}{|I|} \sum_{J \subset I} \left(\frac{\langle w \rangle_{J_+} - \langle w \rangle_{J_-}}{\langle w \rangle_J} \right)^2 |J| = \left(\frac{\mathbf{w}_+ + \mathbf{w}_-}{\mathbf{w}} \right)^2 + \frac{1}{|I|} \sum_{J \subset I_+} \dots + \frac{1}{|I|} \sum_{J \subset I_-} \dots$$

Taking the supremum of the left side over all such w we get at most $\mathcal{B}(\mathbf{u}, \mathbf{w})$. On the other hand, since one can define the function w independently on the intervals I_+ and I_-, the supremum of the right side gives us $\left(\frac{\mathbf{w}_+ + \mathbf{w}_-}{\mathbf{w}} \right)^2 + \frac{1}{2}(\mathcal{B}(\mathbf{u}_+, \mathbf{w}_+) + \mathcal{B}(\mathbf{u}_-, \mathbf{w}_-))$. The main inequality is proved. □

2.3.2. Connections with stochastic control. Now, let us see how does it fit in the general scheme of stochastic control. First of all, the time here is discrete. Instead of the Wiener process, we consider its simplest discrete time analogue—the process $\sum_{k=0}^{n} \xi^k$, $n = 0, 1, 2, \dots$, where ξ^k are independent coin tosses, i. e. the independent random quantities taking values 1 and -1 with probabilities 1/2. Now we have to decide what is the state space variable x and what is the control α. For x it is easy to deicide: it is two-dimensional vector of averages, $x = (\mathbf{u}, \mathbf{w})$.

The equation of the process is

$$x^{n+1} = x^n + \alpha^n \xi^n, \qquad x^0 = (\mathbf{u}, \mathbf{w})^T$$

The interpretation is the following. We start from the interval I and going down to its subintervals, so at the moment n we are on a subinterval J of lengths $2^{-n}|I|$.

The position of the interval J is determined by the sequence of coin tosses ξ^k: we move from the interval J to J_+ if $\xi^n = 1$, and to J_- if $\xi^n = -1$.

If we are given a weight w, it completely defines the process, namely, the initial state $(\mathbf{u}, \mathbf{w})$ (just the averages over the interval I) as well as the control α. Namely, the vector $x^n = (\mathbf{u}^n, \mathbf{w}^n)$ is just the vector of averages over an appropriate interval J, and if $x^n = (\mathbf{u}^n, \mathbf{w}^n) = (\langle \log w \rangle_J, \langle w \rangle_J)$, then $\alpha^n = (x^{n+1} - x^n)/\xi^n$, so

$$\alpha_1^n = \frac{1}{2}\left(\langle \log w \rangle_{J_+} - \langle \log w \rangle_{J_-}\right), \qquad \alpha_2^n = \frac{1}{2}\left(\langle w \rangle_{J_+} - \langle w \rangle_{J_-}\right).$$

On the other hand, if we have a process (i. e. the initial state $(\mathbf{u}, \mathbf{w})$ and the control process α) such that $\lim_{n\to\infty} e^{\mathbf{u}^n} = \lim_{n\to\infty} \mathbf{w}^n$ with probability 1, then it uniquely defines an A_∞ weight w.

Indeed, the process defines the "averages" $\mathbf{u}_J$, $\mathbf{w}_J$ for all dyadic subintervals J of I, and we just put

$$w = \lim_{n\to\infty} \sum_{J\subset I:|J|=2^{-n}|I|} \mathbf{w}_J \chi_J.$$

Since we always stay in the domain, the resulting weight w clearly satisfies the A_∞ condition (10).

Since we established one to one correspondence between weights and controlled processes, in what follows we will switch freely from one language to another.

The profit density in our case is $4(\alpha_2/x_2)^2 = 4(\alpha_2/\mathbf{w})^2$, so we need to maximize the average gain

$$\mathbb{E}\sum_{n=0}^{\infty} 4\cdot\left(\frac{\alpha_2^n}{\mathbf{w}^n}\right)^2 = \frac{1}{|I|}\sum_{J\subset I}\left(\frac{\langle w \rangle_{J_+} - \langle w \rangle_{J_-}}{\langle w \rangle_J}\right)^2 |J|.$$

(the term $|J|/|I|$ in the right hand sum is the probability of getting to the interval J).

The Bellman principle for the discrete process (for time $t = 1$) reads

$$\mathcal{B}(\mathbf{u}, \mathbf{w}) = \sup \mathbb{E}\left\{4\cdot\left(\frac{\alpha_2^0}{\mathbf{w}}\right)^2 + \mathcal{B}(\mathbf{u}^1, \mathbf{w}^1)\right\}$$

where the supremum is taken over all admissible control processes α. The explanation of the Bellman principle is exactly the same as in the case of continuous time.

Let us now deduce the Bellman equation. For a given control process α let us compute the expectation

$$\mathbb{E}\left\{4\cdot\left(\frac{\alpha_2^0}{\mathbf{w}}\right)^2 + \mathcal{B}(\mathbf{u}^1, \mathbf{w}^1)\right\} = 4\cdot\left(\frac{\alpha_2^0}{\mathbf{w}}\right)^2 + \frac{1}{2}\left(\mathcal{B}(x+\alpha^0) + \mathcal{B}(x-\alpha^0)\right)$$

If we now denote $\mathbf{u}_\pm := \mathbf{u} \pm \alpha_1^0$, $\mathbf{w}_\pm := \mathbf{w} \pm \alpha_2^0$, then the above identity immediately implies the main inequality (12) (property 3 from Section 2.3.1).

One can be tempted to take the supremum over all possible values of α^0 to get the Bellman equation

$$\mathcal{B}(\mathbf{u},\mathbf{w}) = \sup\left\{4\cdot\left(\frac{\alpha_2^0}{\mathbf{w}}\right)^2 + \frac{1}{2}\left(\mathcal{B}(\mathbf{u}+\alpha_1^0,\mathbf{w}+\alpha_2^0) + \mathcal{B}(\mathbf{u}-\alpha_1^0,\mathbf{w}-\alpha_2^0)\right)\right\}, \tag{13}$$

but unfortunately this leads nowhere: the obvious choice $\alpha^0 = 0$ gives us trivial identity $\mathcal{B}(x) = \mathcal{B}(x)$.

So, what we have now, is an inequality for a ***supersolution.***

2.3.3. FROM SUPERSOLUTION TO THE THEOREM. As we already have discussed above in Section 1.1, a ***supersolution*** majorates the Bellman function, so if we find *any* functions satisfying the conditions 1–3 from Section 2.3.1 we prove Theorem 2.1. For the sake of completeness, let us present the reasoning in this discrete case.

Take an arbitrary A_∞ weight w, and for a dyadic subinterval J of I let us denote $\mathbf{u}_J := \langle \log w\rangle_J$, $\mathbf{w}_J := \langle w\rangle_J$. The main inequality 3 (Bellman inequality) implies

$$|I|\cdot\mathcal{B}(\mathbf{u}_I,\mathbf{w}_I) \geq |I|\cdot\left(\frac{\mathbf{w}_{I_+}-\mathbf{w}_{I_-}}{\mathbf{w}_I}\right)^2 + |I_+|\cdot\mathcal{B}(\mathbf{u}_{I_+},\mathbf{w}_{I_+}) + |I_-|\cdot\mathcal{B}(\mathbf{u}_{I_-},\mathbf{w}_{I_-})$$

Writing similar inequalities for I_+ and I_-, then for the intervals of the next generation, etc, and them summing everything up to the intervals of the generation $n+1$, we get

$$\begin{aligned}
|I|\cdot c_2 &\geq |I|\cdot\mathcal{B}(\mathbf{u}_I,\mathbf{w}_I)\\
&\geq \sum_{\substack{J\subset I\\ 2^{-n}|I|<|J|\leq|I|}} |J|\cdot\left(\frac{\mathbf{w}_{J_+}-\mathbf{w}_{J_-}}{\mathbf{w}_J}\right)^2 + \sum_{\substack{J\subset I\\ |J|=2^{-n}|I|}} |J_-|\cdot\mathcal{B}(\mathbf{u}_{J_-},\mathbf{w}_{J_-})\\
&\geq \sum_{\substack{J\subset I\\ 2^{-n}|I|<|J|\leq|I|}} |J|\cdot\left(\frac{\mathbf{w}_{J_+}-\mathbf{w}_{J_-}}{\mathbf{w}_J}\right)^2
\end{aligned}$$

(the last inequality holds because $\mathcal{B}\geq 0$). Taking limit as $n\to\infty$, we get exactly the conclusion of the theorem.

2.3.4. CONTINUOUS (INFINITESIMAL) VERSION. To solve the above Bellman equation (inequality), we first move to its continuous (infinitesimal) version.

Let $x = (\mathbf{u},\mathbf{w})$ in the interior of the domain of $\mathcal{B}$ and $\alpha = (\alpha_1,\alpha_2) = (d\mathbf{u}, d\mathbf{w})$ be fixed, and let $x_\pm = (\mathbf{u}_\pm,\mathbf{w}_\pm) = x\pm\alpha$. (Here we are skipping the superscript 0 at α_k from (13)). Multiplying α by a small constant, we can always assume that the whole interval $[x_-,x_+]$ belongs to the domain.

Consider functions φ, Φ of the scalar argument t,

$$\varphi(t) = \mathcal{B}(x+\alpha t),\ \Phi(t) = \mathcal{B}(x) - \frac{1}{2}\Big(\mathcal{B}(x+\alpha t)+\mathcal{B}(x-\alpha t)\Big) - \left(\frac{2\alpha_2 t}{\mathbf{w}}\right)^2, \quad -1\leq t\leq 1.$$

If the function $\mathcal{B}$ is C^2-smooth

$$\lim_{t\to 0}\frac{\varphi(t)+\varphi(-t)-2\varphi(0)}{t^2}=\varphi''(0)=\sum_{j,k=1}^{2}\frac{\partial^2\mathcal{B}(x)}{\partial x_j\partial x_k}\alpha_j\alpha_k \tag{14}$$

The Bellman Inequality (12) implies that $\Phi(t)\geq 0\ \forall t\in[-1,1]$. Dividing this inequality by t^2 and taking limit as $t\to 0$, we get (taking into account (14))

$$-\frac{1}{2}\sum_{j,k=1}^{2}\frac{\partial^2\mathcal{B}(x)}{\partial x_j\partial x_k}\alpha_j\alpha_k\geq 4\cdot\left(\frac{\alpha_2}{\mathbf{w}}\right)^2, \tag{15}$$

or, equivalently $-d^2\mathcal{B}\geq 8(d\mathbf{w}/\mathbf{w})^2$. We have proved the above inequality only for small α, but because of the homogeneity, it holds for all α.

Note, that if we consider a continuous time controlled process

$$x^t=\int_0^t\alpha^s dw^s,\qquad x^s=(x_1^s,x_2^s),\quad \alpha^s=(\alpha_1^s,\alpha_2^s),$$

(here w^s is one dimensional Wiener process, and not a weight) with profit density $f^\alpha(x)=4(\alpha_2/x_2)$, then the Bellman equation for this process will be

$$\sup_\alpha\left\{\frac{1}{2}\sum_{j,k=1}^{2}\frac{\partial^2\mathcal{B}(x)}{\partial x_j\partial x_k}\alpha_j\alpha_k+4\cdot\left(\frac{\alpha_2}{x_2}\right)^2\right\}=0. \tag{16}$$

The inequality (15) ($-d^2\mathcal{B}\geq 8(d\mathbf{w}/\mathbf{w})^2$), which we have just derived from the discrete Bellman inequality, is simply the inequality for a supersolution of the continuous time Bellman equation (16).

This is not a coincidence: we get the infinitesimal version of the discrete Bellman equation if we assume that all changes of state variables are infinitely small. Mathematically it can be expressed (at least formally) by considering the process

$$x^{n+1}=x^n+\alpha^n\xi^n,$$

where ξ^n are independent random variables taking values $-1/\sqrt{N}$ and $1/\sqrt{N}$ with probabilities 1/2, and maximizing the average $\mathbb{E}\sum_{n=0}^{\infty}\left(\frac{\alpha_2}{x_2}\right)^2\frac{1}{N}$.

Taking the limit as $N\to\infty$ (assuming that the time elapsed from n to $n+1$ is $1/N$), we get (at least formally) the continuous time process.

We leave all the details as an exercise for the reader. A detailed proof of this equivalence with justification of all limiting procedures, is beyond the scope and goals of this paper.

We will call equation (16) the *Bellman equation* for Theorem 2.1. Note, that formally this equation is not well defined, namely, any function satisfying $-d^2\mathcal{B}\geq 8(d\mathbf{w}/bw)^2$ automatically satisfies (16) (supremum is attained at $\alpha=0$). To get the equation one has to require, for example that ***supremum*** is attained at some $\alpha\neq 0$.

It is possible to show (under some additional assumption) that if $\mathcal{B}$ is C^2-smooth, then the Bellman function *must* satisfy this equation. However, in most situations it is rather hard to show that the Bellman function is smooth (moreover, it is often not the case), so it does not make a lot of sense proving that the Bellman equation is necessary.

Besides, very often we are not able to solve the equation, and guessing a supersolution is the best we can do. And in rare cases when we can solve the equation, it is usually possible to show that the solution is indeed the Bellman function.

2.3.5. SOLUTION OF THE BELLMAN EQUATION. In this case we are lucky—we can just guess a solution of the Bellman equation (16). Namely,

$$\mathcal{B}(\mathbf{u},\mathbf{w}) := 8\cdot(\log\mathbf{w}-\mathbf{u})$$

clearly satisfies the equation. What is more important, it is indeed the Bellman function! Note, that equation (16) does not have a unique solution: for example, one can get infinitely many solutions adding arbitrary linear functions.

Let us first show, that the function $\mathcal{B}$ satisfies conditions 1–3 from Section 2.3.1. Clearly, $\mathcal{B}(\mathbf{u},\mathbf{w})\geq 0$ if $e^{\mathbf{u}}\leq\mathbf{w}$, and $\mathcal{B}(\mathbf{u},\mathbf{w})\leq c_2 = 8\log c_1$ if $\mathbf{w}\leq c_1e^{\mathbf{u}}$. So, conditions 1 and 2 are satisfied.

Let us prove the main inequality (12). Denote $x=(\mathbf{u},\mathbf{w})$, $x_\pm=(\mathbf{u}_\pm,\mathbf{w}_\pm)=(\mathbf{u}\pm\alpha_1,\mathbf{w}\pm\alpha_2)$. Then

$$\begin{aligned}\mathcal{B}(x)-\frac{1}{2}\Big(\mathcal{B}(x_+)+\mathcal{B}(x_-)\Big) &= 8\cdot\Big(\log\mathbf{w}-\frac{1}{2}(\log(\mathbf{w}+\alpha_2)+\log(\mathbf{w}-\alpha_2))\Big)\\ &= -4\log(1-\left(\frac{\alpha_2}{\mathbf{w}}\right)^2)\geq 4\cdot\left(\frac{\alpha_2}{\mathbf{w}}\right)^2=\left(\frac{\mathbf{w}_+-\mathbf{w}_-}{\mathbf{w}}\right)^2.\end{aligned}$$

The main inequality (12), and so the theorem are proved.

Let us show now, that the function $\mathcal{B}$ we constructed is indeed the Bellman function. Note first, that $\langle\log w\rangle_I=\log(\langle w\rangle_I)$ if and only if $w\equiv$ Const, so the Bellman function is 0 if $\mathbf{u}=\log\mathbf{w}$. Our function $\mathcal{B}$ satisfies this property, so this is a good indication.

To prove that the function $\mathcal{B}(\mathbf{u},\mathbf{w}) := 8\cdot(\log\mathbf{w}-\mathbf{u})$ is the Bellman function let us fix a point $x=(\mathbf{u},\mathbf{w})$ in the domain, and given $\varepsilon>0$ construct a process (choose "control" α)

$$x^{n+1}=x^n+\alpha^n\xi^n$$

in the domain, starting at x, such that $\lim_{n\to\infty}e^{\mathbf{u}^n}=\lim_{n\to\infty}\mathbf{w}^n$ with probability 1. Recall, that ξ^n, $n=0,1,2,\ldots$ is the sequence of independent coin tosses, each ξ^n takes values ±1 with probability $1/2$.

As we had discussed above, such processes correspond to A_∞ weights. So, if we are able, given $\varepsilon>0$ to pick control $\alpha=\{\alpha^n\}_{n=0}^\infty$ such, that

$$\mathbb{E}\sum_{n=0}^\infty\left(\frac{\alpha^n}{\mathbf{w}^n}\right)^2\geq(1-\varepsilon)\mathcal{B}(x),$$

then we are done (constructed an ε-optimal weight).

To construct such a process, let us fix in the domain a line segment $L=[A,B]$ containing x and with both endpoints A and B on the lower bound $\mathbf{w}=e^{\mathbf{u}}$ of the

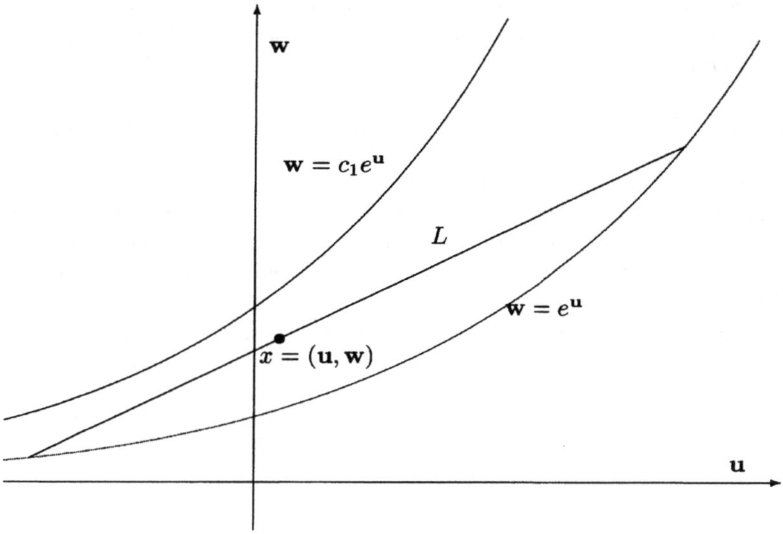

FIGURE 1. Constructing suboptimal random walk.

domain, see Fig. 1. Our process will be just a discrete random walk on the line L with small steps. Namely, at each step we will move by on the line L by δ, picking one of two possible directions with probability $1/2$.

Formally, it will be written as follows. Let $\boldsymbol{\ell}$ be a unit vector parallel to L. Let $\delta > 0$ be a small number to be picked later. We put $\alpha^n := \delta\boldsymbol{\ell}$ if $\delta \leq d := \mathrm{dist}(x, \{A, B\})$, and put $\alpha^n := d\boldsymbol{\ell}$ otherwise. The latter simply means that we have to decrease the step near the boundary, to avoid going out of the domain Ω. It also means, that when we hit the boundary, we just stay there forever.

It is clear, that with probability 1 all trajectories eventually hit the boundary. Applying the Taylor's formula and using compactness of L we can find $\delta > 0$ such, that

$$\mathcal{B}(x) - \frac{1}{2}\Big(\mathcal{B}(x+\alpha) + \mathcal{B}(x-\alpha)\Big) \leq (1+\varepsilon)\cdot 4 \cdot \left(\frac{\alpha_2}{\mathbf{w}}\right)^2.$$

Since with probability 1 all trajectories end on the lower boundary $\mathbf{w} = e^{\mathbf{u}}$ and on this boundary $\mathcal{B}(x) = 0$, we can conclude that

$$\mathbb{E}\sum_{n=0}^{\infty}\left(\frac{\alpha^n}{\mathbf{w}^n}\right)^2 \geq \frac{1}{(1+\varepsilon)}\mathcal{B}(x) \geq (1-\varepsilon)\mathcal{B}(x),$$

and that is exactly what we need.

2.4. Concluding remarks

The above example illustrates pretty much main ideas of our Bellman function method: determine variables (usually averages), find discrete Bellman inequality, then its infinitesimal version, and finally pick a supersolution. There are few points we want to make

- Domain of $\mathcal{B}$ is essential, to emphasize this we even included it as a property. For example, it is impossible to find a *bounded* solution of the Bellman inequality (12) (or (15)) on the whole plane.
- Infinitesimal Bellman equation (inequality) can be obtained (at least formally) as the Bellman equation (inequality) for the corresponding continuous time process. This formalism sometimes can be easier than investigation of the discrete case

This feature of our Bellman function theory—guessing the supersolution of Bellman equation—will eventually be replaced hopefully by some intelligent method. But so far no such method is available—so the guessing of simple blocks and putting them together as in the lego game is the only approach now. The excuse is that our Bellman equation is usually very complicated.

3. New creatures from Bellman's Zoo

There are many of them to be found in [8]. Here we have a few more.

3.1. A two-weight inequality

Theorem 3.1. *Let u, v be two positive functions such that for any dyadic interval I,*

$$\langle u\rangle_J \langle v\rangle_J \le 1$$

Then for any dyadic interval I

$$\frac{1}{|I|}\sum_{J\in\mathcal{D}, J\subset I} |\langle u\rangle_{J_+} - \langle u\rangle_{J_-}| \cdot |\langle v\rangle_{J_+} - \langle v\rangle_{J_-}| \cdot |J| \le 16\, \langle u\rangle_I^{1/2} \langle v\rangle_I^{1/2} \tag{17}$$

This inequality is a particular case of two-weight inequality for martingale transform. The general two-weight inequality for martingale transform was found in [10] (and it has far reaching consequences in [17], [14], [5], [11], [12], [13]). But this particular case (17) was our first Bellman function inequality, and it appeared in the first preprint version of [10].

Let us deduce the Bellman equation (inequality) for this problem. The choice of variables is obvious, we take the averages $\mathbf{u} = \langle u\rangle_I$, $\mathbf{v} = \langle v\rangle_I$. Let us fix a dyadic interval I and define the function $\mathcal{B} = \mathcal{B}_I$ as follows:

$$\mathcal{B}(\mathbf{u}, \mathbf{v}) := \sup\Big\{\frac{1}{|I|}\sum_{J\in\mathcal{D}, J\subset I} |\langle u\rangle_{J_+} - \langle u\rangle_{J_-}| \cdot |\langle v\rangle_{J_+} - \langle v\rangle_{J_-}| \cdot |J|\Big\}$$

where the supremum is taken over all non-negative functions u and v satisfying $\langle u\rangle_J \langle v\rangle_J \leq 1$ for all dyadic intervals J, and such, that $\langle u\rangle_I = \mathbf{u}$, $\langle v\rangle_I = \mathbf{v}$. Note, that again, because of right scaling, the function $\mathcal{B}$ does not depend on the interval I, so we will skip it.

The Bellman function $\mathcal{B}$ satisfies the following properties:

1. Domain Ω: $\mathbf{u}, \mathbf{v} \geq 0$, $\mathbf{uv} \leq 1$;
2. Range: $0 \leq \mathcal{B} \leq 16\sqrt{\mathbf{uv}}$;
3. Main inequality: for any three pairs $x = (\mathbf{u}, \mathbf{v})$, $x_+ = (\mathbf{u}_+, \mathbf{v}_+)$, $x_- = (\mathbf{u}_-, \mathbf{v}_+)$ belonging to the domain Ω and satisfying $x = (x_+ + x_-)/2$, the inequality

$$\mathcal{B}(x) \geq \frac{1}{2}\Big(\mathcal{B}(x_+) + \mathcal{B}(x_-)\Big) + |\mathbf{u}_+ - \mathbf{u}_-| \cdot |\mathbf{v}_+ - \mathbf{v}_-| \tag{18}$$

holds.

The inequality $\mathcal{B} \geq 0$ follows immediately from the definition of $\mathcal{B}$, and the inequality $\mathcal{B}(\mathbf{u}, \mathbf{v}) \leq 32\sqrt{\mathbf{uv}}$ is our belief that the theorem is true.

The explanation for the main inequality (18) is also simple. Let us consider all weights u, v satisfying $\langle u\rangle_J \langle v\rangle_J \leq 1$ for all dyadic intervals J, and such that the averages over the intervals I, I_+, I_- are fixed,

$$\langle u\rangle_I = \mathbf{u}, \quad \langle v\rangle_I = \mathbf{v}, \qquad \langle u\rangle_{I_\pm} = \mathbf{u}_\pm, \quad \langle v\rangle_{I_\pm} = \mathbf{v}_\pm$$

Then, for any such u, v

$$\begin{aligned}\mathcal{B}(\mathbf{u}, \mathbf{v}) &\geq \frac{1}{|I|} \sum_{J \in \mathcal{D}, J \subset I} |\langle u\rangle_{J_+} - \langle u\rangle_{J_-}| \cdot |\langle v\rangle_{J_+} - \langle v\rangle_{J_-}| \cdot |J| \\ &= |\mathbf{u}_+ - \mathbf{u}_-| \cdot |\mathbf{v}_+ - \mathbf{v}_-| + \frac{1}{2}\Big(\sum_{J \subset I_+} \cdots + \sum_{J \subset I_-} \cdots\Big),\end{aligned}$$

and taking the supremum over all such u, v we get the main inequality (18).

Again, if we find any function satisfying the conditions 1–3 above, the theorem is proved. Indeed, for any weights u, v from the theorem the main inequality (18) implies

$$\begin{aligned}&|\langle u\rangle_{I_+} - \langle u\rangle_{I_-}| \cdot |\langle v\rangle_{I_+} - \langle v\rangle_{I_-}| \cdot |I| \\ &\qquad \leq |I|\mathcal{B}(\langle u\rangle_I, \langle v\rangle_I) - |I_+|\mathcal{B}(\langle u\rangle_{I_+}, \langle v\rangle_{I_+}) - |I_-|\mathcal{B}(\langle u\rangle_{I_-}, \langle v\rangle_{I_-})\end{aligned}$$

Writing similar inequalities for I_+ and I_-, then for the intervals of the next generation and summing everything up to the intervals of the generation $n+1$, we get

$$\sum_{\substack{J\subset I\\ 2^{-n}|I|<|J|\le|I|}} |\langle u\rangle_{J_+} - \langle u\rangle_{J_-}|\cdot|\langle v\rangle_{J_+} - \langle v\rangle_{J_-}|\cdot|J|$$
$$\le |I|\mathcal{B}(\langle u\rangle_I,\langle v\rangle_I) - \sum_{\substack{J\subset I\\ |J|=2^{-n}|I|}} \mathcal{B}(\langle u\rangle_J,\langle v\rangle_J)$$
$$\le |I|\mathcal{B}(\langle u\rangle_I,\langle v\rangle_I) \le 16|I|\sqrt{\langle u\rangle_I\langle v\rangle_I}.$$

Taking limit as $n\to\infty$ we get the conclusion of the theorem.

The stochastic control interpretation is as follows. We have a controlled process

$$x^{n+1} = x^n + \alpha^n\xi^n$$

where ξ^n is the sequence of independent coin tosses, and we want to maximize the average profit $\mathbb{E}(\sum 4|\alpha_1^n|\cdot|\alpha_2^n|)$.

Here again the weights u, v determine the process, namely, the initial state $x=(\mathbf{u},\mathbf{v})$ (it is just the averages over the interval I) and the control $\alpha=(\alpha_1,\alpha_2)$. Namely, as it was discussed in Section 2.3.2, the sequence of coin tosses ξ^n determines the interval we are on: we move from an interval J to J_+ if $\xi^n=1$ and to J_- if $\xi^1=-1$. If at the time n we are on an interval J, then the state space vector x is just the vector of averages over J,

$$x^n = (\mathbf{u}^n,\mathbf{v}^n) = (\langle u\rangle_J,\langle v\rangle_J)$$

and the control $\alpha^n=(\alpha_1^n,\alpha_2^n)$ is determined by

$$\alpha_1^n = \frac{1}{2}\Big(\langle u\rangle_{J_+} - \langle u\rangle_{J_-}\Big),\qquad \alpha_1^n = \frac{1}{2}\Big(\langle v\rangle_{J_+} - \langle v\rangle_{J_-}\Big).$$

In this situation we have that the average profit $\mathbb{E}(\sum 4|\alpha_1^n|\cdot|\alpha_2^n|)$ is exactly the sum in the left side of (17).

Again, as in Section 2.3.4 we can find that the main inequality (18) implies its infinitesimal version

$$-\frac{1}{2}\sum_{j,k=1}^{2}\frac{\partial^2\mathcal{B}(x)}{\partial x_j\partial x_k}\alpha_j\alpha_k \ge 4|\alpha_1|\cdot|\alpha_2|, \tag{19}$$

or equivalently, $-d^2\mathcal{B}\ge 8|d\mathbf{u}|\cdot|d\mathbf{v}|$.

This inequality also can be formally obtained by considering a continuous time analogue of our process and writing corresponding Bellman equation for it. The continuous time version of our process is

$$x^t = \int_0^t \alpha^s dw^s$$

where w^t is one dimensional Wiener process, $x^t=(x_1^t,x_2^t)$, $\alpha^t=(\alpha_1^t,\alpha_2^t)$, and we want to maximize $\mathbb{E}(\int_0^\infty 4|\alpha_1^t\alpha_2^t|dt)$, so the profit density $f^\alpha=|\alpha_1\alpha_2|$. The corresponding Bellman equation (see Section 1.1) is

$$\sup_{\alpha}\left\{\frac{1}{2}\sum_{j,k=1}^{2}\frac{\partial^2\mathcal{B}(x)}{\partial x_j\partial x_k}\alpha_j\alpha_k+4|\alpha_1|\cdot|\alpha_2|\right\}=0$$

so the partial differential inequality (19) is simply the inequality for ***supersolution.***

We do not know how to solve the equation, but it is really not too hard to *guess* a supersolution. It is an easy exercise in multivariable calculus to compute the Hessian (the matrix of second derivatives) and to verify that for the function

$$\mathcal{B}(\mathbf{u},\mathbf{v}):=4\cdot(4\sqrt{\mathbf{uv}}-\mathbf{uv})$$

the partial differential inequality (19) holds.

We are almost done. We are saying "almost", because to prove the theorem we need the discrete main inequality (18). If the domain Ω were convex, the discrete inequality (18) would immediately follow from the infinitesimal version (19). Unfortunately, the domain is not convex, so additional work is needed. Let, according to our notational convention $x_\pm=(\mathbf{u}_\pm,\mathbf{v}_\pm)=x\pm\alpha=(\mathbf{u},\mathbf{v})\pm(\alpha_1,\alpha_2)$. Then, if $\alpha_1\alpha_2\geq 0$, and the points x, x_+ and x_- are in the domain Ω, then the whole interval $[x_-,x_+]$ is in the domain. Therefore, for such α_1,α_2 we can apply Taylor's formula to get the main inequality (18) from it infinitesimal version (19). There is also an elementary way to get to that conclusion. Since

$$\mathbf{uv}-\frac{1}{2}\Big((\mathbf{u}+\alpha_1)(\mathbf{v}+\alpha_2)+(\mathbf{u}-\alpha_1)(\mathbf{v}-\alpha_2)\Big)=-\alpha_1\alpha_2 \tag{20}$$

and the function $\sqrt{\mathbf{uv}}$ is concave, the main inequality (18) follows immediately if $\alpha_1\alpha_2>0$.

To treat the case $\alpha_1\alpha_2<0$ we note, that the function $t\mapsto\sqrt{t}$ is concave, so

$$\begin{aligned}\sqrt{\mathbf{uv}}-\frac{1}{2}\Big(\sqrt{\mathbf{u}_+\mathbf{v}_+}+\sqrt{\mathbf{u}_-\mathbf{v}_-}\Big)&\geq\sqrt{\mathbf{uv}}-\sqrt{\frac{\mathbf{u}_+\mathbf{v}_++\mathbf{u}_-\mathbf{v}_-}{2}}\\&=\sqrt{\mathbf{uv}}-\sqrt{\mathbf{uv}+\alpha_1\alpha_2}\geq-\frac{1}{2}\frac{\alpha_1\alpha_2}{\sqrt{\mathbf{uv}}}\geq\frac{1}{2}|\alpha_1\alpha_2|.\end{aligned}$$

The term $-\mathbf{uv}$ works against us in the case $\alpha_1\alpha_2<0$, see (20), but it gets overpowered by $4\sqrt{\mathbf{uv}}$, so the main inequality (18) holds in this case as well.

Our "first harmonic analysis Bellman function"

$$\mathcal{B}(\mathbf{u},\mathbf{v}):=4\cdot(4\sqrt{\mathbf{uv}}-\mathbf{uv})$$

looks rather simple. A much more complicated sibling of this Bellman function helped us to solve the two-weight Martingale transform in [10]. But, actually, even this "simple" one has a challenging minus sign. However, this minus sign is not too difficult to explain: it serves the goal to make $\mathcal{B}$ strictly concave (it must be quite concave—see (19)) exactly where $\sqrt{\mathbf{uv}}$ looses concavity (near the diagonal $\mathbf{u}=\mathbf{v}$).

3.2. Inverse Hölder inequality

Let us recall that a weight (a nonnegative locally L^1 function) w is called a dyadic A_∞ weight if

$$\langle w\rangle_J \leq c\exp(\langle \log w\rangle_J), \qquad \text{for any dyadic interval } J \tag{21}$$

The best constant c is called the A_∞-norm of w. Of course, the A_∞ norm is not a real norm, but we will use the term anyway.

Theorem 3.2. *If w is in dyadic A_∞, then there exists $p>1$ (depending only on A_∞-norm of w) such that for any dyadic interval I*

$$\langle w^p\rangle_I \leq C\,\langle w\rangle_I^p \tag{22}$$

To prove the theorem let us fix the averages

$$x_1=\mathbf{w}:=\langle w\rangle_I, \qquad x_2=\mathbf{u}:=\langle \log w\rangle_I, \tag{23}$$

and define the Bellman function $\mathcal{B}$ by

$$\mathcal{B}(x)=\mathcal{B}(\mathbf{w},\mathbf{u}):=|I|^{-1}\sup\Big\{\int_I w^p dx\ :\ w \text{ satisfying (21), (23) }\Big\}$$

(clearly it does not depend on the interval I).

In terms of stochastic processes the equation can be written as

$$x^{n+1}=x^n+\begin{pmatrix}\alpha_1^n\\ \alpha_2^n\end{pmatrix}\xi^n, \qquad x^0=(\mathbf{w},\mathbf{u})^T.$$

and we want to maximize

$$\lim_{n\to\infty}(E|x_1^n|^p).$$

The variables are easy to read: if at the moment of time n we are at an interval I, then $x^n=(x_1^n,x_2^n)^T=(\mathbf{w}^n,\mathbf{u}^n)^T=(\langle w\rangle_I,\langle \log w\rangle_I)^T$, and

$$\alpha_1^n=(\langle w\rangle_{I_+}-\langle w\rangle_{I_-})/2, \qquad \alpha_2^n=(\langle \log w\rangle_{I_+}-\langle \log w\rangle_{I_-})/2,$$

There is no profit function to optimize, $f^\alpha(x)=0$, only the retirement bonus $F(x)=|x_1|^p$.

The properties of $\mathcal{B}$:

1. Domain: $\Omega=\Omega_c=\{\mathbf{w},\mathbf{u}:1\leq \mathbf{w}\,e^{-\mathbf{u}}\leq c\}$
2. Range: $\mathbf{w}^p\leq \mathcal{B}(x)\leq C\mathbf{w}^p$ (the first inequality is just the Hölder inequality $\langle w\rangle_I^p\leq\langle w^p\rangle_I$, and the second one should hold if the theorem is true).
3. Main inequality: for any $x,x_+,x_-\in\Omega$ and satisfying $x=(x_++x_-)/2$, the inequality $\mathcal{B}(x)\geq(\mathcal{B}(x_+)+\mathcal{B}(x_-))/2$ holds.

Again, if we find a function $\mathcal{B}$ satisfying the above properties 1–3, the theorem is proved. Indeed, fix an A_∞ weight w, and for an interval I denote

$$x_I=(\mathbf{w}_I,\mathbf{u}_I)=(\langle w\rangle_I,\langle \log w\rangle_I).$$

Main inequality 3 implies

$$|I_+|\cdot\mathcal{B}(x_{I_+})+|I_-|\cdot\mathcal{B}(x_{I_-})\leq|I|\cdot\mathcal{B}(x_I).$$

Applying the same inequality to I_+ and I_-, and then to their subintervals, we get after going n generations down

$$\sum_{\substack{J\subset I\\ |I|=2^{-n}|I|}} \langle w\rangle_J^p |J| \le \sum_{\substack{J\subset I\\ |I|=2^{-n}|I|}} \mathcal{B}(x_J)|J| \le |I|\cdot\mathcal{B}(x_I) \le C\cdot|I|\langle w\rangle_I^p.$$

Taking limit as $n\to\infty$ we get

$$\int_I w^p \le C\cdot|I|\langle w\rangle_I^p,$$

which is exactly the conclusion of the theorem. □

The infinitesimal version of the main inequality is simply the concavity condition $d^2\mathcal{B}\le 0$. It can be also obtained by writing the formal Bellman equation for the continuous analogue of our process.

So, we need to find a concave function $\mathcal{B}$ in Ω, satisfying the inequalities 2. We actually should prove that there exists $p=p(c)>1$, for which such a concave function exists in Ω_c.

And indeed such a function can be found in the form $\mathcal{B}(x)=x_1^p\varphi(x_1\,e^{-x_2})$. The reader can find φ on $[1,c]$ in such a way that the above aggregate is concave in Ω_c, g being bounded and $\varphi\ge 1$.

Looks like we are done, but a careful reader can notice that there remains an issue to be resolved. Namely, because the domain $\Omega=\Omega_c$ is not convex, the concavity condition $d^2\mathcal{B}\le 0$ does not imply main inequality 3.

To overcome this difficulty, let us notice, that if all 3 points, x, x_+ and x_- are in the domain Ω_c, then the whole interval $[x_-,x_+]$ is in the domain Ω_{c^2}. This would immediately solve all our problems, since a concave function in the domain Ω_{c^2} satisfies the main inequality 3 in the smaller domain Ω_c.

To show that the interval $[x_-,x_+]$ lies in Ω_{c^2}, define $\mathbf{w}_\tau:=\mathbf{w}+(\mathbf{w}_+-\mathbf{w}_-)\tau/2$, $\mathbf{u}_\tau:=\mathbf{u}+(\mathbf{u}_+-\mathbf{u}_-)\tau/2$, so $\mathbf{w}_{\pm1}=\mathbf{w}_\pm$, $\mathbf{u}_{\pm1}=\mathbf{u}_\pm$. Define $g(\tau)=\log\mathbf{w}_\tau-\mathbf{u}_\tau$. Note that $g(\tau)$ is concave ($g''\le 0$) non-negative function on $[-1,1]$ and $g(0)\le\log c$. Therefore, the whole graph of g lies within the "butterfly" on Figure 2, and we get $g(\tau)\le 2g(0)\le 2\log c$, which means $\mathbf{w}_\tau e^{-\mathbf{u}_\tau}\le C^2$.

3.3. Sharp constant in the (dyadic) Carleson embedding theorem

In this section we are going to prove the following theorem.

Theorem 3.3. *If the values μ_I, $I\in\mathcal{D}$ satisfy the Carleson measure condition, that means*

$$\sum_{\substack{J \text{ is dyadic}\\ J\subset I}} \mu_J \le |I| \qquad \textit{for any dyadic interval } I,$$

then for any $f\in L^2$

$$\sum_{I \text{ is dyadic}} \mu_I|\langle f\rangle_I|^2 \le 4\|f\|_{L^2}^2,$$

and, moreover, the constant 4 is sharp (cannot be replaced by a smaller one)

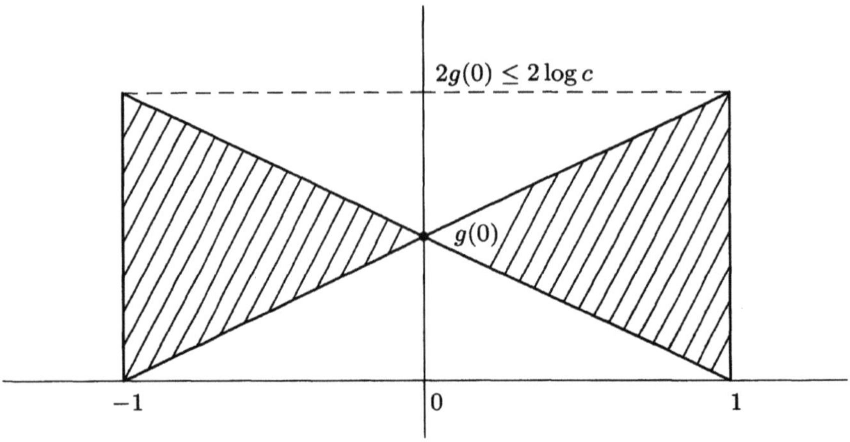

FIGURE 2. The graph of $g(\tau) = \log \mathbf{w}_\tau - \mathbf{u}_\tau$ has to be inside the "butterfly".

It is the well known Carleson embedding theorem (dyadic version). There exist several different proofs of it. For example, it was proved in [8] (with the same constant) by the Bellman function method. The new fact here is that the constant is sharp.

Let us prove the theorem. Let us first define the Bellman function. Fix a dyadic interval I and real numbers F, $\mathbf{f}$, M. Consider all sequences $\{\mu_J\}_{J\in\mathcal{D}}$ satisfying the Carleson condition

$$\frac{1}{|J|}\sum_{J'\subset J}\mu_{J'} \le 1 \qquad \forall J \in \mathcal{D} \tag{24}$$

and such that

$$\frac{1}{|I|}\sum_{J\subset I}\mu_J = M\,. \tag{25}$$

Consider also all functions $f \in L^2$ for which the quantities

$$\langle f^2\rangle_I = F\,, \qquad \langle f\rangle_I = \mathbf{f} \tag{26}$$

are fixed, and define the Bellman function $\mathcal{B}$ as

$$\mathcal{B}(F,\mathbf{f},M) = \frac{1}{|I|}\sup\Big\{\sum_{J\subset I}\mu_J\cdot\langle f\rangle_J^2 \;:\; f,\ \{\mu_J\}_{J\in\mathcal{D}} \text{ satisfy (24), (25), (26)}\Big\}.$$

In terms of stochastic processes, we have a process

$$x^{n+1} = x^n + \begin{pmatrix} \alpha_1^n \\ \alpha_2^n \\ \alpha_3^n \end{pmatrix} \xi^n + \begin{pmatrix} 0 \\ 0 \\ \alpha_4^n \end{pmatrix}, \qquad x^0 = (F, \mathbf{f}, M)^T,$$

and we want to maximize $\mathbb{E}(\sum_n (\mathbf{f}^n)^2 \alpha_4^n)$. Here the state space vector x is just the vector of averages $(F, \mathbf{f}, M)^T$, that is if at the moment n we are at the interval J, then

$$x^n = (F^n, \mathbf{f}^n, M^n)^T = \Big(\langle |f|^2 \rangle_J, \langle f \rangle_J, |J|^{-1} \sum_{J' \subset J} \mu_J \Big)^T.$$

The vector α of controls is determined by

$$\alpha_1^n = (F_{J_+} - F_{J_-})/2, \quad \alpha_2^n = (\mathbf{f}_{J_+} - \mathbf{f}_{J_-})/2 \quad \alpha_3^n = (M_{J_+} - M_{J_-})/2 \quad \alpha_4^n = |J|^{-1} \mu_J$$

Note that $\mathcal{B}(F, \mathbf{f}, M)$ does not depend on the choice of an interval I.

3.3.1. PROPERTIES OF $\mathcal{B}(F, \mathbf{f}, M)$.

1. Domain: $\mathbf{f}^2 \leq F$, $0 \leq M \leq 1$ (Cauchy inequality and the Carleson condition);
2. Range: $0 \leq \mathcal{B}(F, \mathbf{f}, M) \leq CF$ (our belief that the theorem is true);
3. Key property: for all $0 \leq \Delta M \leq M$

$$\mathcal{B}(F, \mathbf{f}, M) \geq \Delta M f^2 + \frac{1}{2} \Big\{ \mathcal{B}(F_+, \mathbf{f}_+, M_+) + \mathcal{B}(F_-, \mathbf{f}_-, M_-) \Big\},$$

whenever all triples $(F, \mathbf{f}, M)$, $(F_\pm, \mathbf{f}_\pm, M_\pm)$ belong to the domain and

$$F = \frac{1}{2}(F_+ + F_-), \qquad \mathbf{f} = \frac{1}{2}(\mathbf{f}_+ + \mathbf{f}_-), \quad \text{and} \quad M = \frac{1}{2}(M_+ + M_-) + \Delta M.$$

The main property 3 is simply discrete Bellman inequality. It was explained in details how to get it in [8]; one can also just apply the technique from Sections 1, 2 to get the discrete Bellman inequality.

It is also easy to show, and was shown in [8] that existence of a function $\mathcal{B}$ satisfying the above properties 1–3 implies the estimate

$$\sum_{I \text{ is dyadic}} \mu_I |\langle f \rangle_I|^2 \leq C \|f\|_{L^2}^2,$$

where C is the same as in in the property 2. So we need to show that $C = 4$ is the best possible constant in 2.

We need to get a continuous version of the main property 3. It again was done in [8], and here we present an alternative explanation. Namely, let us just write the Bellman equation for the continuous analogue of our process. It will be

$$\sup_\alpha \Big\{ \frac{1}{2} \sum_{j,k=1}^{3} \frac{\partial^2 \mathcal{B}}{\partial x_j \partial x_k} \alpha_j \alpha_k + \frac{\partial \mathcal{B}}{\partial M} \alpha_4 \Big\} = 0. \tag{27}$$

This immediately implies the following two inequalities

3′. $d^2\mathcal{B} \le 0$;

3″. $\dfrac{\partial \mathcal{B}}{\partial M} \ge \mathbf{f}^2$,

which, as it was shown in [8], constitute the infinitesimal version of the main property 3 (the derivatives here can be understood in the sense of distributions). Inequalities 3′, 3″ clearly follows from **3** (just consider infinitely small changes of variables). Since the domain is convex, it is not difficult to show, that the infinitesimal version implies, in turn, the discrete version **3**, see again [8].

To "solve" the Bellman equation, let us eliminate one variable, using a trick of Burkholder, see [2]. We will also use this trick in the next section.

Namely, let us suppose the "real" Bellman function (i. e. obtained as sup) $\mathcal{B}$ satisfies the inequality $\mathcal{B} \le cF$. Define

$$u(\mathbf{f}, M) := \sup_F \{\mathcal{B}(F, \mathbf{f}, M) - C \cdot F\}.$$

We claim, that the function u is concave, and satisfies the inequalities $\partial u/\partial M \ge \mathbf{f}^2$ and $-C\mathbf{f}^2 \le u(\mathbf{f}, M) \le 0$.

The last inequality is trivial: we simply take supremum in the inequality

$$-CF \le \mathcal{B}(F, \mathbf{f}, M) - CF \le 0.$$

The concavity is not completely trivial. We know, that the *infimum* of concave functions is concave, but that is not true for *supremum*. However, we have not just a family of concave functions, but the function that is concave in all variables, and the following simple lemma takes care of everything.

Lemma 3.4. *Let $\varphi(x,y)$ be a concave function, and let $\Phi(x) := \sup_y \varphi(x,y)$. Then Φ is concave.*

The proof of this lemma is a simple exercise in convex analysis, and we omit it.

The inequality $\partial u/\partial M \ge \mathbf{f}^2$ is also not difficult, and we leave its proof to the reader.

Summarizing, if the Carleson Embedding Theorem holds with constant C, there exists a convex function $u(\mathbf{f}, M)$ satisfying $\partial u/\partial M \ge \mathbf{f}^2$ and such that $-C\mathbf{f}^2 \le u(\mathbf{f}, M) \le 0$. On the other hand, if such u exists, we can define $\mathcal{B}(F, \mathbf{f}, M) := u(\mathbf{f}, M) + CF$, so the Carleson Embedding Theorem holds with the same constant C. Hence, the best constant in the Carleson Embedding Theorem is exactly the best constant for which the function u exists.

Now we use the homogeneity of $\mathcal{B}$ and u to eliminate one more variable. Namely, the "real" Bellman function $\mathcal{B}$ (defined as the *supremum*) satisfies the following property:

$$\mathcal{B}(a^2F, a\mathbf{f}, M) = a^2\mathcal{B}(F, \mathbf{f}, M) \qquad \forall a \in \mathbb{R}.$$

Therefore

$$u(a\mathbf{f}, M) = a^2 u(\mathbf{f}, M), \tag{28}$$

i. e. u can be represented as

$$u(\mathbf{f}, M) = \mathbf{f}^2\varphi(M). \tag{29}$$

Note, that to get the homogeneity condition (28), one does not have to know anything about the "real" Bellman function. Namely, if we have a function $\widetilde{u}$ satisfying all the inequalities, then any its "rescaling" $u_a := a^{-2}\widetilde{u}(a\mathbf{f}, M)$ satisfies all the inequalities too (because all the inequalities have correct homogeneity). So, if we define $u(\mathbf{f}, M) := \inf_{a\neq 0} a^{-2}\widetilde{u}(a\mathbf{f}, M)$, it satisfies (28). It is clear that u is concave (as *infimum* of concave functions), and it is easy to show that $\partial u/\partial M \geq \mathbf{f}^2$ (in the sense of distributions).

For the function u defined by (29) the Hessian (matrix of second derivatives) equals to

$$\begin{pmatrix} 2\varphi(M) & 2\mathbf{f}\varphi'(M) \\ 2\mathbf{f}\varphi'(M) & \mathbf{f}^2\varphi''(M) \end{pmatrix},$$

so the concavity of u is equivalent to the following two inequalities:

$$\varphi \leq 0, \qquad \varphi\varphi'' - 2(\varphi')^2 \geq 0 \qquad \text{for } M \in [0,1]$$

The inequality $\partial u/\partial M \geq \mathbf{f}^2$ means simply that $\varphi' \geq 1$, and φ also has to satisfy $-C \leq \varphi \leq 0$.

Now it is an easy task to find φ. The inequality $\varphi\varphi'' - 2(\varphi')^2 \geq 0$ can be rewritten as

$$\varphi^4 \left(\varphi'/\varphi^2\right)' \geq 0.$$

The critical case $(= 0)$ gives rise to the differential equation

$$\left(\varphi'/\varphi^2\right)' = 0$$

which has general solution

$$\varphi(M) = 1/(C_1 + C_2 M).$$

Let us consider only solutions that are non-positive on $[0,1]$ and satisfy $\varphi' \geq 1$ there. Among all such solutions function

$$\varphi(M) = -4/(1+M)$$

has the best lower bound on $[0,1]$, $\inf\{\varphi(M) : M \in [0,1]\} = -4$.

It is not difficult to show that if we consider on $[0,1]$ all non-negative solutions of the inequality $\varphi\varphi'' - 2(\varphi')^2 \geq 0$ which satisfy $\varphi' \geq 1$, we do not improve the lower bound.

So, $C = 4$ is the best constant in the Carleson Embedding Theorem, and it is given by the function $u(\mathbf{f}, M) = -4\mathbf{f}^2/(1+M)$.

Note that

$$\mathcal{B}(F, \mathbf{f}, M) = u(\mathbf{f}, M) + 4F = 4(F - \mathbf{f}^2/(1+M))$$

is exactly the Bellman function that was obtained in [8]. It means that the best estimate in the Carleson Embedding Theorem was already obtained in [8], although we did not realize that until now.

Note also, that the function $\mathcal{B}$ defined above is not the "real" Bellman function: in particular, it does not satisfy the Bellman equation (27) (recall, that "satisfies" means that supremum is attained at some non-zero α; it is always attained at $\alpha = 0$.) Of course, this is only an indication, not a proof, for we did not prove that the Bellman Equation in this sense (supremum is attained at non-zero α) is necessary. But it is easy to see that on the boundary $F = \mathbf{f}^2$ the "real" Bellman function must satisfy the boundary condition $\mathcal{B}(F, \mathbf{f}, M) = M\mathbf{f}^2 = MF$, and the function we constructed does not satisfy this condition.

So, our Bellman function only touches the real one along some set.

4. Burkholder meets Bellman

Let us recall a famous result of Burkholder [2] about basis constant for the Haar system in L^p. Symbol h_I here denotes the normalized Haar function for the interval I.

Theorem 4.1. *Let $1 < p < \infty$, and let $f = \sum_{I\in\mathcal{D}} c_I h_I$ (we can always assume that the sum is finite) and $g = \sum_{I\in\mathcal{D}} \varepsilon_I c_I h_I$, where $\varepsilon_I = \pm 1$. Then for any choice of the sequence $\varepsilon_I (= \pm 1)$*

$$\|g\|_p \le C_p \|f\|_p;$$

here $C_p = p^* - 1$, $p^* = \max\{p, p'\}$, where p' is the conjugate exponent, $1/p + 1/p' = 1$.

Moreover, the constant C_p here is sharp.

Let us translate Burkholder's approach to the Bellman function language.

But first, few trivial observations. First of all, by duality, it is sufficient to prove the theorem only for $p > 2$ (case $p = 2$ is trivial because of orthogonality). So, from now on we assume that $2 < p < \infty$, so $p^* = p$. The second observation is that the theorem holds true if ε_I are arbitrary numbers in the interval $[-1, 1]$, because any such sequence belongs to the convex hall of ± 1 sequences (it is sufficient to consider only sequences with finitely many non-zero terms).

4.1. Bellman equation for Burkholder's theorem

To construct the Bellman function, let us fix a dyadic interval I, and consider all functions $f = \sum_{J\in\mathcal{D}} \varsigma_J h_J$, $g = \sum_{J\in\mathcal{D}} \varepsilon_J \varsigma_J h_J$, $\varepsilon_J = \pm 1$, such that the averages over the interval I are fixed:

$$\langle f \rangle_I = \mathbf{f}, \qquad \langle g \rangle_I = \mathbf{g}, \qquad \langle |f|^p \rangle_I = F. \tag{30}$$

Define function $\mathcal{B}(\mathbf{f}, \mathbf{g}, F)$ of three real variables (all functions are assumed to be real-valued) as $\mathcal{B}(\mathbf{f}, \mathbf{g}, F) := \sup\{\langle |g|^p \rangle_I \ : \ f, g \text{ satisfying } (30)\}$. Clearly, it does not depend on the interval I, due to the right scaling.

In terms of stochastic processes, the equation of our motion is

$$x^{n+1} = x^n + \alpha^n \xi^n, \qquad x^0 = (\mathbf{f}, \mathbf{g}, F)^T,$$

where ξ^n are independent (scalar) "coin tosses", and we want to maximize $\lim_{n\to\infty}\mathbb{E}(|x_2^n|^p)$ (only "retirement bonus" here).

All variables and controls α here are easy to understand: if at the moment n we are at an interval J, then

$$x^n = (x_1^n, x_2^n, x_3^n)^T = (\langle f\rangle_J, \langle g\rangle_J, \langle |f|^p\rangle_J)^T,$$

and

$$\alpha^n = (\alpha_1^n, \alpha_2^n, \alpha_3^n)^T = \Big(\frac{\langle f\rangle_{J_+} - \langle f\rangle_{J_-}}{2}, \frac{\langle g\rangle_{J_+} - \langle g\rangle_{J_-}}{2}, \frac{\langle |f|^p\rangle_{J_+} - \langle |f|^p\rangle_{J_-}}{2}\Big)^T.$$

Let us check the properties of the function $\mathcal{B}$: it is defined on the (convex) domain $|\mathbf{f}|^p \le F$ (Hölder inequality), and it has to satisfy the inequalities

$$\mathcal{B}(\mathbf{f}, \mathbf{g}, F) \ge |\mathbf{g}|^p, \tag{31}$$

$$\mathcal{B}(0, 0, F) \le C_p^p F \tag{32}$$

The last inequality (32) here is our belief that the theorem is true: if we consider $f = \sum_{J\subset I} c_J h_J$, then $\langle f\rangle_I = \langle g\rangle_I = 0$.

What about the main property—Bellman equation? Since there is no profit function, only the "retirement bonus", it looks like the Bellman equation (inequality) is simply the concavity condition. However, one can easily see that it is impossible to find a concave function $\mathcal{B}$ in the domain $|\mathbf{f}|^p \le F$, satisfying (31) and (32). So, what is wrong?

The answer is simple: we did not take into account that the controls α are not arbitrary, they satisfy some restrictions. Namely, since $f = \sum_J c_J h_J$ and $g = \sum_J \varepsilon_J c_J h_J$, $\varepsilon_J = \pm 1$, we have that for any interval I,

$$|\langle f\rangle_{I_+} - \langle f\rangle_{I_-}| = |\langle g\rangle_{I_+} - \langle g\rangle_{I_-}|,$$

so $|\alpha_1| = |\alpha_2|$. So, the Bellman inequality is not the concavity, but *restricted concavity*

$$\mathcal{B}(\mathbf{f}, \mathbf{g}, F) \ge \Big(\mathcal{B}(\mathbf{f} + \alpha_1, \mathbf{g} + \alpha_2, F + \alpha_3) + \mathcal{B}(\mathbf{f} - \alpha_1, \mathbf{g} - \alpha_2, F - \alpha_3)\Big)/2 \tag{33}$$

whenever $|\alpha_1| = |\alpha_2|$ (and all the arguments belong to the domain).

Note, that if we construct *any* function $\mathcal{B} = \mathcal{B}(\mathbf{f}, \mathbf{g}, F)$ on the domain $|\mathbf{f}|^p \le F$, satisfying the Bellman inequality (33) and the estimates (31), (32), then we prove the theorem.

Indeed, consider f supported by a dyadic interval I, $f = \sum_{J\subset I} c_J h_J$. Then the Bellman inequality (33) implies that

$$\begin{aligned}|I|\cdot\mathcal{B}(0,0,\langle |f|^p\rangle_I) &= |I|\cdot\mathcal{B}(\langle f\rangle_I, \langle g\rangle_I, \langle |f|^p\rangle_I)\\ &\ge |I_+|\cdot\mathcal{B}(\langle f\rangle_{I_+}, \langle g\rangle_{I_+}, \langle |f|^p\rangle_{I_+}) + |I_-|\cdot\mathcal{B}(\langle f\rangle_{I_-}, \langle g\rangle_{I_-}, \langle |f|^p\rangle_{I_-})\end{aligned}$$

Writing the Bellman inequalities to I_+ and I_-, then to their halves, we get after going n generations down:

$$|I|\cdot\mathcal{B}(0,0,\langle|f|^p\rangle_I) \geq \sum_{J\subset I:|J|=2^{-n}|I|} |J|\cdot\mathcal{B}(\langle f\rangle_J,\langle g\rangle_J,\langle|f|^p\rangle_J)$$
$$\geq \sum_{J\subset I:|J|=2^{-n}|I|} |J|\cdot\langle|g|^p\rangle_J$$

(the last inequality here is just (31)).

Taking limit as $n\to\infty$ and using (32) we get

$$C_p^p\int_I|f|^p = C_p^p|I|\langle|f|^p\rangle_I \geq |I|\cdot\mathcal{B}(0,0,\langle|f|^p\rangle_I) \geq \int_I|g|^p.$$

To complete the proof it remains to notice that any finite sum $f=\sum c_J h_J$ is supported by the union of two disjoint dyadic intervals.

4.2. Finding $\mathcal{B}$: biconcave function

To find $\mathcal{B}$, let us first understand what does the restricted concavity (33) mean.

To do that let us do a simple change of variables. Let us introduce new variables, $\mathbf{x} := \mathbf{g}+\mathbf{f}$, $\mathbf{y} := \mathbf{g}-\mathbf{f}$, so

$$\mathbf{f} = (\mathbf{x}-\mathbf{y})/2, \qquad \mathbf{g} = (\mathbf{x}+\mathbf{y})/2.$$

Define $\widetilde{\mathcal{B}}(\mathbf{x},\mathbf{y},F) := \mathcal{B}(\mathbf{f},\mathbf{g},F) = \mathcal{B}((\mathbf{x}-\mathbf{y})/2,(\mathbf{x}+\mathbf{y})/2,F)$. Then the restricted concavity (33) of $\mathcal{B}$ just mean that the function $\widetilde{\mathcal{B}}$ for any fixed $\mathbf{y}$ is concave in arguments $\mathbf{x}$, F, and for any fixed $\mathbf{x}$ is concave in arguments $\mathbf{y}$, F. The inequalities (31), (32) can be translated into

$$\widetilde{\mathcal{B}}(\mathbf{x},\mathbf{y},F) \geq \left|\frac{\mathbf{x}+\mathbf{y}}{2}\right|^p$$
$$\widetilde{\mathcal{B}}(0,0,F) \leq C_p^pF.$$

Domain of $\widetilde{\mathcal{B}}$ will be $\left|\frac{\mathbf{x}-\mathbf{y}}{2}\right|^p \leq F$.

Burkholder in [2, 3] did the above change of variables in the original problem, using the so-called "zigzag martingales".

To find $\widetilde{\mathcal{B}}$, one can use the homogeneity of the "real" Bellman function $\mathcal{B}$ to get the condition

$$\widetilde{\mathcal{B}}(a\mathbf{x},a\mathbf{y},|a|^pF) = |a|^p\widetilde{\mathcal{B}}(\mathbf{x},\mathbf{y},F).$$

This allows us to reduce the number of variable to two, and the partial differential equations appearing from the concavity conditions can be solved (by Burkholder, not by us).

It is really amazing, that Burkholder, see [3] was able to solve these PDE's: they are really complicated, and solving them definitely goes beyond our abilities.

However, there is an easier method, also due to Burkholder. This method gives us a *supersolution* with the best constant, as well as the proof that the constant is best. We essentially following Burkholder [2] here, and we refer the reader to [2, 3] for all the details.

Suppose we found a function $\mathcal{B}$ satisfying $\mathcal{B}(0,0,F) \leq C_p^p F$ with some C_p, as well as all other necessary inequalities. Changing variables to $\mathbf{x}$, $\mathbf{y}$ we get the function $\widetilde{\mathcal{B}}(\mathbf{x},\mathbf{y},F)$. Let us now eliminate variable F, using the same trick as in the previous section with the Carleson Embedding Theorem. Define

$$u(\mathbf{x},\mathbf{y}) := \sup_{F \geq \left|\frac{\mathbf{x}-\mathbf{y}}{2}\right|^p} \left\{\widetilde{\mathcal{B}}(\mathbf{x},\mathbf{y},F) - C_p^p F\right\}.$$

Using Lemma 3.4 we get that the function u is *biconcave*, that means the functions $u(\mathbf{x},\,\cdot\,)$ and $u(\,\cdot\,,\mathbf{y})$ are concave. The function u also satisfies the inequalities

$$u(0,0) \leq 0, \qquad u(\mathbf{x},\mathbf{y}) \geq \left|\frac{\mathbf{x}+\mathbf{y}}{2}\right|^p - C_p^p \cdot \left|\frac{\mathbf{x}-\mathbf{y}}{2}\right|^p. \tag{34}$$

The first one is trivial, because $\widetilde{\mathcal{B}}(0,0,F) - C_p^p F \leq 0$, and the second one follows from the inequality

$$\widetilde{\mathcal{B}}(\mathbf{x},\mathbf{y},F) - C_p^p F \geq \left|\frac{\mathbf{x}+\mathbf{y}}{2}\right|^p - C_p^p F$$

and the fact that $\sup\left\{-F \,:\, F \geq \left|\frac{\mathbf{x}-\mathbf{y}}{2}\right|^p\right\} = -\left|\frac{\mathbf{x}-\mathbf{y}}{2}\right|^p$.

We have proved, that if a function $\mathcal{B}$ exists for some C_p, then a function u exists for the same constant C_p. The converse is also true: if a function u exists for some C_p, then we can define $\widetilde{\mathcal{B}}(\mathbf{x},\mathbf{y},F) := u(\mathbf{x},\mathbf{y}) + C_p^p F$, and then changing variables $\mathbf{x}$, $\mathbf{y}$ back to $\mathbf{f}$ and $\mathbf{g}$ we get $\mathcal{B}$. Thus the estimate in the Burkholder's theorem is equivalent to the existence of such biconvex u, and the best constant is the best C_p for which such u exists.

Note, that without loss of generality one can assume that u is symmetric and homogeneous of degree p:

$$u(\mathbf{x},\mathbf{y}) = u(\mathbf{y},\mathbf{x}), \qquad u(a\mathbf{x},a\mathbf{y}) = |a|^p u(\mathbf{x},\mathbf{y}), \quad a \in \mathbb{R}. \tag{35}$$

For a function u obtained from the "real" Bellman function it can be deduced from the symmetry and homogeneity of $\mathcal{B}$, but there is an alternative, very simple explanation. Namely, if a function $u = u(\mathbf{x},\mathbf{y})$ satisfies inequalities (34), then the functions $|a|^{-p}u(a\mathbf{x},a\mathbf{y})$ and $u(\mathbf{y},\mathbf{x})$ satisfy them as well. So, if we have some biconcave u_0, then we can define a symmetric and homogeneous biconcave u by

$$u(\mathbf{x},\mathbf{y}) = \inf_{a\neq 0} |a|^{-p}\big(u_0(a\mathbf{x},a\mathbf{y}) + u_0(a\mathbf{y},a\mathbf{x})\big)/2$$

(infimum of concave functions is concave). Note, that $u(0,0) = 0$ even if $u_0(0,0) > 0$, so we do not have to worry about first inequality in (34).

The fact that u is homogeneous means that it is (almost) completely defined by $w(\mathbf{x}) := u(\mathbf{x},1)$, namely $u(\mathbf{x},\mathbf{y}) = |\mathbf{y}|^p w(\mathbf{x}/\mathbf{y})$ if $y \neq 0$.

If we denote

$$v(\mathbf{x},\mathbf{y}) = \left|\frac{\mathbf{x}+\mathbf{y}}{2}\right|^p - C_p^p \cdot \left|\frac{\mathbf{x}-\mathbf{y}}{2}\right|^p$$

then clearly $w(\mathbf{x}) \geq v(\mathbf{x}, 1)$. The function w is clearly concave. Note, that a p-homogeneous u satisfies $u(\mathbf{x}, \mathbf{y}) = u(\mathbf{y}, \mathbf{x})$ if and only if

$$w(\mathbf{x}) = |\mathbf{x}|^p w(1/\mathbf{x}). \tag{36}$$

Indeed,

$$u(\mathbf{x}, \mathbf{y}) = |\mathbf{y}|^p w(\mathbf{x}/\mathbf{y}) = |\mathbf{y}|^p |\mathbf{x}/\mathbf{y}|^p w(\mathbf{y}/\mathbf{x}) = |\mathbf{x}|^p w(\mathbf{y}/\mathbf{x}) = u(\mathbf{y}, \mathbf{x}).$$

In this case u can be completely restored from w by putting $u(\mathbf{x}, 0) = |x|^p w(0)$. Note, that if w is concave (and satisfies (36)), then the restored function u is biconcave: indeed, for any fixed $\mathbf{y}$ the function $u(\,\cdot\,, \mathbf{y})$ is clearly concave, and the symmetry condition $u(\mathbf{x}, \mathbf{y}) = u(\mathbf{y}, \mathbf{x})$ implies that $u(\mathbf{x}, \,\cdot\,)$ is concave as well.

Therefore, given $C_p \geq 1$, the existence of the biconcave u (and so the Burkholder's Theorem) is equivalent to the existence of a concave w satisfying (36) and such, that $w(\mathbf{x}) \geq v(\mathbf{x}, 1)$.

Let $\mathbf{x}_0$ be the solution of the equation $v(\mathbf{x}, 1) = 0$ on the interval $[-1, 1]$, $\mathbf{x}_0 = (C_p - 1)/(C_p + 1)$.

Lemma 4.2. *A concave function w as above exists if and only if $C_p \geq p - 1$ (we are considering only the case $p \geq 2$ here). Moreover, if $C_p = p - 1$, then w is uniquely defined on the interval $[\mathbf{x}_0, 1/\mathbf{x}_0]$, and it is an affine function on $[\mathbf{x}_0, 1]$.*

Proof. If

$$L := \lim_{\mathbf{x} \to 1-} \frac{w(\mathbf{x}) - w(1)}{\mathbf{x} - 1}, \qquad R := \lim_{\mathbf{x} \to 1+} \frac{w(\mathbf{x}) - w(1)}{\mathbf{x} - 1}$$

then (36) implies that $L = pw(1) - R$. Since w is concave, we have

$$pw(1) = L + R \leq 2L \leq 2\frac{w(1) - w(\mathbf{x}_0)}{1 - \mathbf{x}_0} = (C_p + 1)(w(1) - w(\mathbf{x}_0)) \leq (C_p + 1)w(1);$$

the last inequality here follows from the fact that $w(\mathbf{x}_0) \geq v(\mathbf{x}_0, 1) = 0$.

So, $pw(1) \leq (C_p + 1)w(1)$, and since $w(1) \geq v(1, 1) > 0$, we get $C_p \geq p - 1$.

If $C_p = p - 1$, then all the inequalities in the above chain become equalities, and that means on the interval $[\mathbf{x}_0, 1]$ ($\mathbf{x}_0 = 1 - 2/p$ in this case) the function w must be an affine function satisfying $w(\mathbf{x}_0) = 0$. Since w is a concave majorant of $v(\mathbf{x}, 1)$, the slope must be equal to the derivative $\frac{d}{dx} v(\mathbf{x}, 1)$ at $\mathbf{x} = \mathbf{x}_0$. So, indeed, the function w (if it exists) is uniquely defined on $[\mathbf{x}_0, 1]$ (and so on $[\mathbf{x}_0, 1/\mathbf{x}_0]$).

To prove that the condition $C_p \geq p - 1$ is sufficient it is enough to construct w for the critical case $C_p = p - 1$. The function w is uniquely defined on the interval $[\mathbf{x}_0, 1/\mathbf{x}_0]$, and outside of this interval we can just put $w(\mathbf{x}) = v(\mathbf{x}, 1)$. Of course, it remains to check that w is a concave majorant of $v(\mathbf{x}, 1)$, and it is nor hard to do.

Note, that function w is not unique, there are other choices possible. For example, we can choose w to be an affine function on $[-1, 1]$ (extension of the affine function on $[\mathbf{x}_0, 1]$), and then extend it to $\mathbb{R}$ according to (36). Again, one has to check that w majorates $v(\mathbf{x}, 1)$ and that it is concave (outside of $[-1, 1]$). The resulting function is given by

$$w(\mathbf{x}) = \alpha_p \left\{ \left|\frac{\mathbf{x}+1}{2}\right| - (p^* - 1)\left|\frac{\mathbf{x}-1}{2}\right| \right\} \left\{ \left|\frac{\mathbf{x}+1}{2}\right| + \left|\frac{\mathbf{x}-1}{2}\right| \right\}^{p-1} \tag{37}$$

where $\alpha_p = p \cdot (1 - 1/p^*)^{p-1}$. Although we only considered the case $p > 2$, the resulting function works for $p < 2$ as well, and that is why we are using $p^* := \max\{p, p'\}$, $1/p + 1/p' = 1$ here. □

Formula (37) gives rise to

$$u(\mathbf{x}, \mathbf{y}) = \left\{ \left|\frac{\mathbf{x}+\mathbf{y}}{2}\right| - (p^* - 1)\left|\frac{\mathbf{x}-\mathbf{y}}{2}\right| \right\} \left\{ \left|\frac{\mathbf{x}+\mathbf{y}}{2}\right| + \left|\frac{\mathbf{x}-\mathbf{y}}{2}\right| \right\}^{p-1}.$$

Going back to the original variables $\mathbf{f}$, $\mathbf{g}$ we get a Bellman function

$$\mathcal{B}(\mathbf{f}, \mathbf{g}, F) = \alpha_p \{|\mathbf{g}| - (p^* - 1)|\mathbf{f}|\} \{|\mathbf{f}| + |\mathbf{g}|\}^{p-1} + (p^* - 1)F.$$

5. Conclusions

- Our Bellman equation in many cases (in all cases we considered here) can be interpreted as a particular case of the classical Bellman equation for stochastic control (see [6]). Recall that the deterministic Bellman equation is a first order PDE. The second order terms in our case are due to the probabilistic nature of the problem (Ito's formula).
- Every (dyadic) harmonic analysis problem, which scales correctly, has the Bellman equation assigned to it.
- The process that correspond to a dyadic harmonic analysis problem is a discrete time process, so the Bellman equation is in fact a system of discrete (finite difference) inequalities. One can consider their continuous (infinitesimal) version by taking infinitely small changes of variables. This infinitesimal version can be also obtained (at least formally) by writing the Bellman equation for a continuous time analogue of our discrete time process.
- Moving from a continuous (infinitesimal) Bellman equation (function) is usually pretty easy if the domain is convex. If the domain is not convex, some extra work is needed, and it can be rather tricky to get from the continuous version to the discrete one.
- The Bellman equation is best to interpret as a system of inequalities. There are several reasons for that. First of all, if we just formally write the equation (8), in our situation any ***supersolution***, i. e. any function satisfying the inequalities satisfies the equation as well (supremum is attained and equal 0 at $\alpha = 0$). So something else is needed to get the equation: for example one can require that the supremum is attained at some non-zero α.

Moreover, if we find a way to interpret the equation, its solvability is not equivalent to our original problem. We do not know if it is necessary, because we do not know *apriori* that $\mathcal{B}$ is smooth, and we do not know if it is sufficient.

So, while thinking about equation can provide some useful heuristics, it is better to *work* with it as with a system of inequalities.

- The Bellman equations, appearing in harmonic analysis are quite different from the classical PDEs: we do not know any existence and uniqueness results (see also above item about interpretation). The boundary values are not essential in our case (although in some situations they can be written), but some bounds on $\mathcal{B}$ are very important.
- The domain is very important: it can mean the difference between true and false statements. For example, it is well known that the Muckenhoupt condition is sufficient for the boundedness of Haar multipliers in $L^2(w)$, see [16], but its two-weight analogue is not, see [10]. From the point of view of Bellman function, the difference is not in the equation, but in the domain.
- In classical stochastic control, the "retirement bonus" F usually does not appear, but in harmonic analysis it is very often an important part. Unlike *profit function* f^α it does not figure in the equation, but gives rise to bounds on the Bellman function.
- It is very difficult usually (if at all possible) to find the exact solution of the corresponding Bellman equation. Problem of finding a *supersolution* is often much more tractable.
- Harmonic analysis Bellman equations (inequalities) not always can be written in infinitesimal (continuous) form, see, for example [7].

References

[1] St. Buckley, *Summation conditions on weights*, Mich. Math. J., **40** (1993), 153–170.

[2] D. L. Burkholder, *Explorations in martingale theory and its applications*, Ecole d'Eté de Probabilité de Saint-Flour XIX, 1989, 1–66, Lecture Notes in Mathematics, **1464**, Springer, Berlin, 1991.

[3] D. L. Burkholder, *Boundary value problems and sharp inequalities for martingale transforms,* Ann. Prob., **12** (1984), 647–802.

[4] R. Fefferman, C. Kenig, J. Pipher, *The theory of weights and the Dirichlet problem for elliptic equations*, Annals of Math., **134** (1991), 65–124.

[5] S. Hucovic, S. Treil, A. Volberg, *The Bellman functions and the sharp square estimates for square functions*, Operator Theory: Advances and Applications, the volume in memory of S. A. Vinogradov, v. 113, Birkhauser Verlag, 2000.

[6] N. Krylov, *Optimal Control of Diffusion Processes*, Springer-Verlag, New York, 1980, 308 p.

[7] F. Nazarov, *The Beurling lemma via the Bellman function*, Preprint, MSU, 1996.

[8] F. Nazarov, S. Treil, *The hunt for Bellman function: applications to estimates of singular integral operators and to other classical problems in harmonic analysis,* Algebra i Analysis, **8** (1997), no. 5, 32–162.

[9] F. Nazarov, S. Treil, *Hunt–Muckenhoupt–Wheeden inequality is now trivial,* C. R. Acad. Sci. Paris, Ser. I, **323** (1996), 717–722.

[10] F. Nazarov, S. Treil, and A. Volberg, *The Bellman functions and two-weght inequalities for Haar multipliers,* J. of the Amer. Math. Soc., **12** (1999), no. 4, 909–928.

[11] F. Nazarov, A. Volberg, *Heating of the Beurling operator and estimates of its norm,* Preprint, MSU, 2000, p. 1–12.

[12] F. Nazarov, A. Volberg, *Bellman function and $T1$ conditions for two-weight Hilbert transform with application to Carleson imbedding theorem for spaces K_θ*, Preprint, MSU, 2000, p. 1–28.

[13] St. Petermichl, A. Volberg, *Heating of the Beurling operator: weakly quasiregular maps on the plane are quasiregular,* Preprint, MSU, 2000, p. 1–25.

[14] St. Petermichl, J. Wittwer, *A sharp weighted estimate on the norm of Hilbert transform via invariant A_2 characteristic of the weight,* Preprint, MSU, 2000.

[15] E. Stein, *Harmonic Analysis: Real-variable Methods, Orthogonality, And Oscillatory Integrals,* Princeton Math. Series, 43, Monographs in Harmonic Analysis, Princeton Univ. Press, Princeton, NJ, 1993.

[16] S. Treil, A. Volberg, *Wavelets and the angle between past and future,* J. Funct. Analysis, **143** (1997), no. 2, 269–308.

[17] J. Wittwer, Ph. D. Thesis, University of Chicago, 2000.

Department of Mathematics
Michigan State University
East Lansing, MI 48823
USA

Department of Mathematics
Brown University
151 Thayer Street, Box 1917
Providence, RI 02912
USA

Department of Mathematics
Michigan State University
East Lansing, MI 48823
USA

Integral representations and uniqueness sets for star-invariant subspaces.

Alexei Poltoratski

Abstract. We discuss properties of pseudocontinuable functions in the unit disk of the complex plane in the context of operator theory.

We start with a review of recent developments in the study of the boundary behavior of the Cauchy integrals and their role in rank-one perturbation problems. After that we obtain new results on integral representations, uniqueness sets and separating measures for the model spaces K^p_θ.

1. Introduction

This note is devoted to pseudocontinuable functions in the unit disk $\mathbb{D} = \{|z| < 1\}$ and their applications in operator theory. We say that an analytic function f in $\mathbb{D}$ has a pseudocontinuation if there exists a certain meromorphic function $f^{\#}$ from the Nevanlinna class in $\mathbb{C} \setminus \bar{\mathbb{D}}$ satisfying

$$\lim_{r\to 1-} f(r\xi) = \lim_{r\to 1+} (f^{\#})(r\xi)$$

for almost every ξ on the unit circle $\mathbb{T} = \{|z| = 1\}$. Here we assume that f has boundary values a. e. on the circle, which is a common feature for all "reasonable" analytic functions in $\mathbb{D}$.

The importance of pseudocontinuable functions was established by the famous theorem of R. Douglas, H. Shapiro and A. Shields [9]. It says that a function from the Hardy space H^2 in the unit disk is a cyclic vector for the backward shift operator $S^* : f \mapsto \frac{f-f(0)}{z}$ if and only if it does not posess a pseudocontinuation. Hence, all proper invariant subspaces of S^* are filled with pseudocontinuable functions. Moreover, as was shown by S. Nagy and C. Foias, "most" of the contractive operators in Hilbert spaces can be realized as compressions of the shift operator $S : f \mapsto zf$ on such subspaces of H^2. This fact lies in the foundation of the operator model theory, see [11].

To give a precise definition of these subspaces of pseudocontinuable functions let us first consider the case $p = 2$. For any inner function θ we denote by K^2_θ the orthogonal difference $H^2 \ominus \theta H^2$. The space K^2_θ is a Hilbert space. It possesses a

The author is supported in part by N.S.F. grant DMS 9970151 .

system of reproducing kernels k_λ with the property that $f(\lambda) = < f, k_\lambda >$ for any $\lambda \in \mathbb{D}$ and any $f \in K^2_\theta$. The kernels are given by the formula

$$k_\lambda(z) = \frac{1 - \overline{\theta(\lambda)}\theta(z)}{1 - \bar{\lambda} z}. \tag{1}$$

The space K^p_θ, $p > 0$ can be defined as the closed span of k_λ's in the topology of the Hardy space H^p. When $p \geq 1$ the spaces K^p_θ are the only closed invariant subspaces of the backward shift operator S^*, see [11] for $p > 1$ and [6] for $p = 1$. (When $p < 1$ the spaces K^p_θ are not the only invariant subspaces of S^*, see [6].)

A necessary and sufficient condition for a function f to belong to K^p_θ, $p \geq 1$ is that the function $\tilde{f}$ defined on $\mathbb{T}$ as $\bar{f}\theta$ belongs to H^p and $\tilde{f}(0) = 0$. Now the property that any $f \in K^p_\theta$ admits a pseudocontinuation can be easily seen from the fact that $\bar{\tilde{f}}\theta$ is meromorphic outside $\mathbb{D}$ and equals f a. e. on $\mathbb{T}$.

In this paper we will focus on the following aspect of the operator model theory discovered by D. Clark in [7]. Let P_θ denote the orthogonal projector from H^2 onto K^2_θ. Then the classical model operator S_θ is defined as a compression of S onto K^2_θ: $S_\theta : K^2_\theta \to K^2_\theta$, $S_\theta = P_\theta S$. The operator S_θ is a contraction with deficiency indices (1,1). Hence it can be made unitary by a rank-one perturbation. Namely, if $\alpha \in \mathbb{T} = \{|z| = 1\}$ and $\theta(0) = 0$ (this restriction is not important but simplifies most formulas) one could define the operator U_α on K^2_θ as

$$U_\alpha f = z\left(f - (f, \frac{\theta}{z})\frac{\theta}{z}\right) + \alpha(f, \frac{\theta}{z}). \tag{2}$$

It is not difficult to check that U_α is unitary for any $\alpha \in \mathbb{T}$. The relation

$$U_\alpha = S_\theta + \alpha(\cdot, \frac{\theta}{z}) \tag{3}$$

shows that it is a rank-one perturbation of S_θ. Moreover, the family $\{U_\alpha\}_{\alpha \in \mathbb{T}}$ contains *all* possible rank one unitary perturbations of the model contraction S_θ. The resulting operators U_α bear all the information about the model contraction S_θ but are much easier to study, as they are *unitary*.

Perhaps the most remarkable part of Clark's construction is the explicit formula for the spectral measures of the operators U_α. To unveil these measures one simply has to notice that, since θ is inner, the function $(\alpha+\theta)(\alpha-\theta)^{-1}$ has positive real part in $\mathbb{D}$ for any $\alpha \in \mathbb{T}$. Hence, there is a unique finite positive Borel measure σ_α on $\mathbb{T}$ satisfying

$$\operatorname{Re} \frac{\alpha + \theta}{\alpha - \theta} = P\sigma_\alpha \tag{4}$$

where $P\sigma_\alpha$ denotes the Poisson integral

$$P\sigma_\alpha = \int_{\mathbb{T}} \frac{1 - |z|^2}{|z - \xi|^2} d\sigma_\alpha(\xi). \tag{5}$$

It turns out that σ_α is the spectral measure for U_α. We will denote the family of measures satisfying (4) by $M_\theta = \{\sigma_\alpha\}_{\alpha \in \mathbb{T}}$.

Since $(\alpha+\theta)(\alpha-\theta)^{-1}$ is imaginary a. e. on $\mathbb{T}$, the measure σ_α is singular. It is not difficult to see from (4) that σ_α is concentrated on a set where $\theta=\alpha$. Thus one can say that the spectral measures of rank one unitary perturbations of S_θ are weighted level sets of θ. In particular, σ_α has a point mass at $\xi\in\mathbb{T}$ iff $\theta(\xi)=\alpha$ and θ has a non-tangential derivative at ξ, i. e. $\frac{\theta(z)-\alpha}{z-\xi}$ has a finite non-tangential limit at ξ (see [17]).

It follows that the measures σ_α are not only singular with respect to the Lebesgue measure, but also pairwise singular. Moreover, using Rokhlin's terminology, σ_α's constitute a "partition of unity", i. e. if put together they produce the normalized Lebesgue measure m on $\mathbb{T}$:

$$\int_{\mathbb{T}}\sigma_\alpha dm(\alpha)=m, \tag{6}$$

see [1].

Clark's construction builds a bridge between the comlex analytic operator theory, in the style of the Havin-Nikolski seminar or Nikolski's book [11], on one side and Gohberg-Krein-Livshits perturbation theory and mathematial physics on the other. To see this connection let us look at the same construction from the other direction: If U_1 is an arbitrary unitary operator with simple spectrum and φ, $||\varphi||=1$ is its cyclic vector then one can consider the standard family of rank one unitary perturbations of U_1 generated by φ:

$$U_\alpha=U_1+(1-\alpha)(\cdot,U_1^{-1}\varphi)\varphi,\quad \alpha\in\mathbb{T}. \tag{7}$$

Let θ be the characteristic function for the contraction $U_1-(\cdot,U_1^{-1}\varphi)\varphi$ defined as in the operator model theory. Then the spectral measures of U_α are immediately produced by (4). If we replace the unitary U_1 with a self-adjoint A_0 and consider the family of self-adjoint rank one perturbations

$$A_\lambda=A_0+\lambda(\cdot,\varphi)\varphi\quad \lambda\in\mathbb{R},$$

which is a more traditional setting for the perturbation theory, then the same scheme will give the spectral measures of A_λ: one will just have to write down the upper half-plane analogue of (4), see [14, 15]. From the rank one situation one can, with some additional effort, pass to the case of finite rank and trace class perturbations.

Now let us consider the family of unitary operators U_α defined in K_θ^2 by the formula (3). If σ_α is the spectral measure of U_α then by the spectral theorem U_α is unitarily equivalent to M_α^z , the operator of multiplication by z in $L^2(\sigma_\alpha)$. Thus there exists the generalized Fourier transform Φ_α – the unitary operator from K_θ^2 to $L^2(\sigma_\alpha)$ satisfying

$$U_\alpha=\Phi_\alpha^* M_\alpha^z\Phi_\alpha. \tag{8}$$

As was shown in [12], the Fourier transform Φ_α acts in the most natural way: it sends every function into its boundary values. Namely, for each $\alpha\in\mathbb{T}$ every function f from K_θ^2 has non-tangential boundary values almost everywhere with respect to the (singular) measure σ_α and these boundary values are equal to $\Phi_\alpha f$ almost everywhere with respect to σ_α , see Section 2.

To write down the formula for the inverse operator Φ^*_α we need to introduce the so-called normalized Cauchy transform. Denote by M the space of all finite complex Borel measures on $\mathbb{T}$, by M_+ the subset consisting of positive measures. If $\mu \in M$ we can consider its Cauchy integral $K\mu$ in the unit disk:

$$K\mu(z) = \int_{\mathbb{T}} \frac{d\mu(\xi)}{1 - \bar{\xi} z}.$$

The normalized Cauchy transform C_μ is defined for functions $f \in L^1(|\mu|)$ as

$$C_\mu f(z) = \frac{K f\mu(z)}{K\mu(z)}.$$

It turns out that the inverse Fourier transform Φ^*_α can be defined as

$$\Phi^*_\alpha f = C_{\sigma_\alpha} f.$$

This paper is devoted to the discussion of the transforms Φ_α, Φ^*_α and the related results on pseudocontinuable functions. Both the direct operator Φ_α and its inverse Φ^*_α have many intriguing function theoretic properties. These properties were studied in a number of papers, see for instance [2], [5], [12] and [16]. In Section 2 we give a brief survey of this area.

Sections 3 and 4 contain new results on integral representaions and uniqueness in K^p_θ. In Section 4 we study the following question: When does the generalized Fourier transform Φ_α have a non-trivilal kernel in K^p_θ, $p \neq 2$? This question can be considered within the general framework of the uncertainity principle in harmonic analysis as formulated in [10].

In order to prove our uniqueness results, in Section 3 we have to solve the following elementary problem that may be of independent interest to some readers: When does a ratio of two Cauchy integrals of singular measures belong to H^p, $p \geq 1$? As we find out in Section 4, all the pseudocontinuable functions from K^p_θ can be represented as such ratios.

Acknowledgement. The author would like to thank A. B. Aleksandrov for useful discussions and A. Borichev for important comments and corrections.

2. The Fourier transforms Φ_α and Φ^*_α

Let θ be an inner function. Consider the operators U_α on K^2_θ defined by (3). Then, as we mentioned in the introduction, their spectral measures σ_α are given by (4). Recall that Φ_α is defined as the unitary operator intertwining U_α and M^z_α , the operator of multiplicaton by z in $L^2(\sigma_\alpha)$, see (8).

As was shown by D. Clark in [7], Φ_α can be defined in the following way. Recall that θ has non-tangential boundary limits equal to α σ_α-a. e. Hence any reproducing kernel

$$k_\lambda(z) = \frac{1-\overline{\theta(\lambda)}\theta(z)}{1-\bar{\lambda}z}, \ \lambda \in \mathbb{D},$$

has non-tangential boundary limits equal to $\frac{1-\overline{\theta(\lambda)}\alpha}{1-\bar{\lambda}z}$ σ_α-a. e. Put

$$\Phi_\alpha k_\lambda = \frac{1-\overline{\theta(\lambda)}\alpha}{1-\bar{\lambda}z} \in L^2(\sigma_\alpha). \tag{9}$$

It turns out that Φ_α preserves the norms of finite linear combinations of reproducing kernels. Since linear combinations of k_λ's are dense in K_θ^2 as well as their images are in $L^2(\sigma_\alpha)$, Φ_α can be extended to a unitary operator from K_θ^2 to $L^2(\sigma_\alpha)$. It is not difficult to check that (8) is satisfied on reproducing kernels and therefore on the whole K_θ^2.

After Clark's paper appeared, D. Sarason rose the natural question about the action of Φ_α on functions in K_θ^2 other than finite linear combinations of reproducing kernels. The problem here is that, unlike k_λ, an arbitrary H^p-function does not have to have non-tangential boundary limits σ_α-a. e. This question was answered by the following

Theorem 2.1 ([12]). *Let θ be an inner function, $M_\theta = \{\sigma_\alpha\}_{\alpha\in\mathbb{T}}$ be the associated family of measures on $\mathbb{T}$ defined by (4). For any $f \in K_\theta^2$ and any $\alpha \in \mathbb{T}$ the non-tangential boundary limit $f^*(\xi)$ exists for σ_α-almost every $\xi \in \mathbb{T}$. The unitary operator $\Phi_\alpha : K_\theta^2 \to L^2(\sigma_\alpha)$ satisfying (8) can be defined as*

$$(\Phi_\alpha f)(\xi) = f^*(\xi) \ \ for \ \ \sigma_\alpha - a.e. \ \ \xi.$$

In particular, the boundary values of f belong to $L^2(\sigma_\alpha)$ for any α and have the same norm as f (an analog of Plancherel's formula).

The following result on convergence in average is much easier to prove than Theorem 2.1. We are able to supply the proof here. We keep the notations of the last statement.

Theorem 2.2 ([12]). *Let $f \in K_\theta^2$, $f(z) = \sum_{n=0}^{\infty} a_n z^n$. Then for any $\alpha \in \mathbb{T}$ the partial sums $\sum_{n=0}^{k} a_n z^n$ converge to f^* (i. e. to $\Phi_\alpha f$) in $L^2(\sigma_\alpha)$ as $k \to \infty$.*

Proof. It is not difficult to show that

$$\sum_{n=0}^{k} a_n z^n = \Phi_\alpha f - z^k \Phi_\alpha (S^*)^k f \ \text{ for } \ \sigma_\alpha - a.e. \ \ \xi.$$

The equality can first be verified for $f = k_\lambda$ and then extended to the whole space in the standard way.

Since $||(S^*)^k f||_{K_\theta^2} \to 0$ and Φ_α is unitary, $||\Phi_\alpha (S^*)^k f||_{L^2(\sigma_\alpha)} \to 0$. □

If one applies the Abel-Poisson summation method to the series of f one immediately obtaines the following

Corollary 2.3 ([12]). *The functions f_r^*, $f_r(z) = f(rz)$ converge to f^* $(\Phi_\alpha f)$ in $L^2(\sigma_\alpha)$ as $r \to 1-$.*

We will return to Φ_α and discuss its uniqueness properties in Section 4. As for now, let us move on to the inverse transform Φ_α^*.

To write down an explicit formula for $\Phi_\alpha^* : L^2(\sigma_\alpha) \to K_\theta^2$ let us start with the reproducing kernels. As usual we assume that $\theta(0) = 0$. Denote $l_\lambda^\alpha = \frac{1-\overline{\theta(\lambda)}\alpha}{1-\bar{\lambda}z} \in L^2(\sigma_\alpha)$. Let us recall that by (9)

$$\Phi_\alpha^* l_\lambda^\alpha = k_\lambda.$$

Since $K\sigma_\alpha = 1/(1-\bar{\alpha}\theta)$, it is not difficult to verify that

$$k_\lambda = \frac{K l_\lambda^\alpha \sigma_\alpha}{K\sigma_\alpha} = C_{\sigma_\alpha} l_\lambda^\alpha$$

($K\mu$ and C_μ are as defined in the introduction). Now, if f is an arbitrary function from $L^2(\sigma_\alpha)$ we can consider a sequence f_n of finite linear combinations of l_λ^α's tending to f in $L^2(\sigma_\alpha)$. Since Φ^* is unitary, $\Phi^* f_n$ will tend to $\Phi^* f$ in H^2. At the same time, $C_{\sigma_\alpha} f_n$ will tend to $C_{\sigma_\alpha} f$ pointwise in the disk. Hence, we can extend from l_λ^α's to the whole $L^2(\sigma_\alpha)$ to conclude that Φ_α^* is the normalized Cauchy transform associated with σ_α:

$$\Phi_\alpha^* : f \mapsto C_{\sigma_\alpha} f. \tag{10}$$

Let $\mu \in M_+$ be a singular measure. Then there exists an inner function satisfying the equation

$$P\mu = \operatorname{Re}\frac{1-\theta}{1+\theta}.$$

By our definitions this means that $\mu = \sigma_1 \in M_\theta$ and $C_\mu = \Phi_1^*$. Hence, any transform C_μ associated with singular positive μ can be represented as $\Phi_\alpha^* : L^2(\mu) \to K_\theta^2$ for a suitably chosen θ. This allows us to switch to the discussion of the normalized Cauchy transform C_μ. Some of the results below do not require μ to be singular or real.

Together with Theorem 2.1, (10) gives the following result:

Theorem 2.4 ([12]). *Let $\mu \in M$ and $f \in L^1(|\mu|)$. Then the function $C_\mu f$ has finite non-tangential boundary values μ-a. e. These values are equal to f μ^s-a. e., where μ^s is the singular component of μ.*

The case of singular positive μ follows directly from Theorem 2.1. The general case requires a little additional effort, see [12]. One can also formulate this statement in a slightly more general form:

Theorem 2.5. *Let $\mu, \nu \in M$. Then the meromorphic function $\frac{K\mu}{K\nu}$ has non-tangential boundary limits ν-a. e. on $\mathbb{T}$.*

If $\mu = f\nu + \gamma$ where $f \in L^1(|\nu|)$ and $\gamma \in M$, $\gamma \perp \nu$ then these non-tangential limits are equal to f ν^s-a. e.

We suggest it to the reader as an exercise to deduce Theorem 2.5 from Theorem 2.4.

In Section 3 we will need the following simple corollary of the last result:

Corollary 2.6. *Let $\mu, \nu \in M$. Suppose that ν is singular and $\mu \perp \nu$. Then $\frac{K\mu}{K\nu}$ has non-tangential boundary limits equal to 0 ν-a. e. on $\mathbb{T}$.*

Remark 2.7. *If ϕ is an arbitrary (not necessarily inner) function from the unit ball of H^∞ one can still define the family of measures M_ϕ on $\mathbb{T}$ satisfying (4). In the case of non-inner ϕ the measures will not be singular. Nevertheless, the whole construction works in this case as well. The role of K^2_θ will now be played by the de Branges-Rovnyak space N_ϕ (see* [17]*, where it was denoted by M_ϕ). The measures σ_α will, once again, present a family of spectral measures of rank-one perturbations of a unitary operator (whose spectral measure is σ_1), see* [14]*. The operator C_{σ_α} will map $L^2(\sigma_\alpha)$ onto N_ϕ. In Theorems 2.1, 2.2 and Corollary 2.3 K^2_θ can be replaced by N_ϕ. Theorem 2.4 says that any $f \in N_\phi$ has boundary values σ_α-a. e. for any $\alpha \in \mathbb{T}$.*

This extension allows us to include not only singular but all cyclic unitary families of rank-one perturbations in our model. It also provides motivation to study the operator C_μ for arbitrary positive μ.

The rest of this section is devoted to the L^p-estimates for C_μ and the associated maximal non-tangential operator.

For any $\mu \in M_+$ C_μ is bounded as an operator from $L^2(\mu)$ to H^2 (or to $L^2(m)$). The situation becomes more interesting for $p \neq 2$:

Theorem 2.8 ([2]). *For any $\mu \in M_+$ the operator C_μ is bounded as an operator from $L^p(\mu)$ to H^p (to $L^p(m)$) for $1 < p \leq 2$. The operator C_μ is unbounded, in general, for $p > 2$. In particular, if μ is singular and $C_\mu : L^p(\mu) \mapsto H^p$ $(L^p(m))$ is bounded for some $p > 2$ then μ is discrete.*

Theorem 2.4 shows that for any $f \in L^p(\mu)$, $p \geq 1$ the function $C_\mu f$ is defined μ-a. e. on $\mathbb{T}$. Hence it makes sense to ask if $C_\mu f \in L^p(\mu)$ i. e. to study the boundedness of C_μ as a transform in $L^p(\mu)$. The question appears to be even more interesting (and more difficult) if one considers the associated non-tangential maximal operator. The results in this direction can be used to estimate the asymptotics of the resolvents of our operators inside the disk (upper half-plane in the case of self-adjoint operators). These problems can also be viewed as a generalisation of the Hunt-Muckenhoupt-Weeden theory to the case of arbitrary weights, see [16].

Let us start with C_μ. The problem of its boundedness as a transform in $L^p(\mu)$ is solved by Theorem 2.4 for all $p \geq 1$ in the most difficult case of singular μ. For such measures C_μ is not only bounded, it is *identical.* This reflects the fact that for singular μ the measure $f\mu$ is never "antianalytic." For arbitrary μ the situation is not as nice. The same "forces" that make Theorem 2.8 fail for $p > 2$ play a role here. All in all, we have the following

Theorem 2.9 ([16]). *For any $\mu \in M_+(\mathbb{T})$ the operator C_μ is bounded in $L^p(\mu)$ for $1 < p \leq 2$.*

The operator is unbounded (in general) for $p > 2$, see Example 1 in [16]. It is not necessarily bounded when $p = 1$ either: the simplest counterexample is $\mu = m$.

Now let us discuss the action of the non-tangential maximal operator $MC_\mu f$ in $L^p(\mu)$. For any function g in $\mathbb{D}$ we define its non-tangential maximal function Mg as

$$Mg(\xi) = \sup_{z \in \Gamma_\xi} |g(z)|$$

where $\xi \in \mathbb{T}$ and Γ_ξ is the sector $\{|z - \xi| \leq \sqrt{2}\,\mathrm{Re}(1 - \bar{\xi} z), |z| > \frac{1}{\sqrt{2}}\}$. If $\mu \in M_+$ then $MC_\mu f$ is finite μ-a.e. on $\mathbb{T}$ for any $f \in L^1(\mu)$ by Theorem 2.4. That allows us to ask if the maximal operator $MC_\mu : f \mapsto MC_\mu f$ is bounded in $L^p(\mu)$.

As usual, we say that an operator has the weak type (p, p) if it acts from $L^p(\mu)$ into the "weak $L^p(\mu)$", defined as

$$L^{p,\infty}(\mu) = \{f |\ \mu(\{|f| > t\}) < \frac{C}{t^p},\ \text{for some } C < \infty\}$$

(with the inf of such C raised into the power $1/p$ viewed as a norm).

One of the results of [12] implies that MC_μ has the weak type (2,2). For $p \neq 2$ we have the following:

Theorem 2.10 ([16]). *For any $\mu \in M_+(\mathbb{T})$ the maximal operator MC_μ is bounded in $L^p(\mu)$ for $1 < p < 2$.*

The operator MC_μ is, generally, unbounded for $p > 2$. An example from [16] presents a singular μ and $f \in L^\infty(\mu)$ such that $MC_\mu f \notin L^p(\mu)$ for any $p > 2$. The operator is obviously unbounded for $p = 1$ as well. Hence, the remaining question is weather the weak type (2,2) can be improved to the strong type. It seems that the answer is negative, but a counterexample is yet to be found. Apart from the boundedness, it is also unknown if the operator has the weak type (1,1).

In the classical theory of Cauchy transforms in L^p spaces with absolutely continuous weights the maximal operator always follows the pattern of the Cauchy transform itself: they are bounded under the same conditions. For arbitrary weights and the normalized Cauchy transform the situation is different. As we saw above, if we restrict our attention to the case of singular measures μ, then the boundedness of C_μ will follow trivially from Theorem 2.4 for *all* $p \geq 1$. The maximal operator however will still be, generally, unbounded for $p > 2$ as follows from the example from [16] mentioned above.

3. Ratios of Cauchy integrals belonging to H^p

In this section we first answer the following basic question: Let $\mu, \nu \in M_+$ be two singular measures. Suppose that $\mu \perp \nu$. Can $K\mu/K\nu$ belong to H^p?

The answer is, trivially, "yes" for $p < 1$, since $|K\nu| \geq (1/2)||\nu||$ and $K\mu \in H^p$. The following theorem shows that it is also possible for $1 \leq p < 2$, but not for $p \geq 2$.

Theorem 3.1.

I) Let $\mu \in M_+$ be a singular measure. Then there exists singular $\nu \in M_+$ such that $\mu \perp \nu$ and $K\mu/K\nu \in H^p$ for every $0 < p < 2$.

II) Let $\nu \in M_+$ be a singular measure with non-trivial continuous part. Then there exists a singular $\mu \in M_+$ such that $\mu \perp \nu$ and $K\mu/K\nu \in H^p$ for every $0 < p < 2$.

III) For any singular measures $\mu \in M$ and $\sigma \in M_+$ such that $\mu \perp \sigma$ the ratio $K\mu/K\sigma$ is not in H^2.

Remark 3.2. *The measure μ in part I) does not have to be positive. The proof would require an additional technical step. Here we only discuss the positive case for the sake of brevity.*

Proof.

I) We first consider an important particular case when μ is δ_1 (the unit point mass at 1). Then $K\mu = (1-z)^{-1}$. Choose singular $\eta \in M_+$ such that $(1-z)^{-1}$ belongs to any $L^p(\eta), p < 2$ but does not belong to $L^2(\eta)$. Let θ be the inner function defined by $K\eta = (1+\theta)^{-1}$. Consider the corresponding family $M_\theta = \{\eta_\alpha\}$ (then $\eta = \eta_{-1} \in M_\theta$). Now we can put $\nu = \eta_1$.

Indeed, it is well known (see for instance [17], pages 51-53) that since $(1-z)^{-1} \notin L^2(\eta)$, no other measure from M_θ can have a point mass at 1. Hence $\nu \perp \mu$. Also, since $(1-z)^{-1} \in L^p(\eta)$ for any $p < 2$ the function $C_\eta(1-z)^{-1}$ belongs to H^p for any $p < 2$ by Theorem 2.8. But, since $1-\theta = 2$ η-a. e., the boundary values of $\frac{1-\theta}{1-z}$ are equal to $2/(1-z)$ η-a. e. and therefore $\Phi^*(1-z)^{-1} = C_\eta(1-z)^{-1} = \frac{1}{2}\frac{1-\theta}{1-z}$ by Theorem 2.1. Hence $\frac{1-\theta}{1-z}$ belongs to H^p for any $p < 2$. It is left to recall that $\frac{1-\theta}{1-z} = K\mu/K\nu$.

If μ is an arbitrary positive singular measure, consider once again the corresponding inner function I, $K\mu = (1-I)^{-1}$. WLOG $||\mu|| = 1$. Then $K\mu(z) = K\delta_1(I(z))$. By the previous argument $K\delta_1/K\sigma \in H^p$, $p < 2$ for some singular $\sigma \in M_+$, $\sigma \perp \delta_1$. Therefore, $K\delta_1(I(z))/K\sigma(I(z)) \in H^p$, $p < 2$. It is left to notice that $K\sigma(I(z))$ is a Cauchy integral of a positive singular measure ν, $\mu \perp \nu$. Namely, if we put $\nu = \int \mu_\alpha d\sigma(\alpha)$ where $\mu_\alpha \in M_I$ then

$$\frac{K\mu}{K\nu} = \frac{K\delta_1(I(z))}{K\sigma(I(z))} \in H^p,\ p < 2.$$

It is not difficult to show that such ν and μ are mutually singular.

II) Let $\Sigma \subset \mathbb{T}$ be a closed set such that $|\Sigma| = 0$, $\nu_{p.p.}(\Sigma) = 0$ and $\nu_c(\Sigma) > 0$ ($\nu_{p.p.}$ and ν_c are the pure point and continuous parts of ν respectively). Denote by η the restriction of ν on Σ. Then η is continuous. Let $E_1, E_2, ...$ be a sequence of open sets on $\mathbb{T}$ containing Σ such that $E_1 \supset E_2 \supset ...$ and $|E_n| = 2^{-n}$. Denote by d_n the distance between Σ and $\bar{E}_n$ ($\bar{E}_n = \mathbb{T} \setminus E_n$).

Construction of μ. We construct μ, $\mu \perp \nu$ using a Cantor-type process.

Step 1. Let $E_1 = \cup I^{1,n}$ where $I^{1,n}$, $n = 1, 2, ...$ are disjoint open arcs. Inside each arc $I^{1,n}$ such that $\eta(I^{1,n}) > 0$ cover η-almost all Σ with dijoint open arcs

$J_1^{1,n}, J_2^{1,n}, ..., J_{2k}^{1,n}$, $k \geq 1$ satisfying $\eta(J_1^{1,n}) = \eta(J_2^{1,n}) = ... = \eta(J_{2k}^{1,n})$ and $|J_i^{1,n}| < (d_1)^2$.

After that we define the function f_1 to equal 2 on $J_1^{1,n}, J_3^{1,n}, ..., J_{2k-1}^{1,n}$ for $n = 1, 2, ...$ and 0 on the rest of $\mathbb{T}$. Then we put $\mu_1 = f_1\eta$.

During this step we obtained a countable collection of arcs $J_i^{1,n}$. To simplify further notations we reenumerate them as $J^{1,k}$, $k = 1, 2, ...$. The particular order of the new enumeration is not important.

Step m, $m \geq 2$. Inside each arc $J^{m-1,n}$ such that $\mu_{m-1}(J^{m-1,n}) > 0$ cover μ_{m-1}-almost all Σ with disjoint open arcs $J_1^{m,n}, J_2^{m,n}, ..., J_{(m+1)k}^{m,n}$, $k \geq 1$ satisfying $\mu_{m-1}(J_1^{m,n}) = \mu_{m-1}(J_2^{m,n}) = ... = \mu_{m-1}(J_{(m+1)k}^{m,n})$ and $|J_i^{m,n}| < (d_m)^2$. (Note that the last condition provides that all $J_i^{m,n} \subset E_m$.)

After that we define function f_m to equal 0 on $J_{m+1}^{m,n}, J_{2m+2}^{m,n}, ..., J_{km+k}^{m,n}$, to $(m+1)/m$ on other arcs $J_i^{m,n}$ for $n = 1, 2, ...$ and to 0 outside of $\cup_{i=1}^{(m+1)k} J_i^{m,n}$. Then we put $\mu_m = f_m\mu_{m-1}$. Before proceeding to step $m+1$ we again redenote $J_i^{m,n}$ as $J^{m,k}$, $k \in \mathbb{N}$.

The obtained sequence μ_m converges in the $*$-weak topology of M. We denote its limit by μ. It is not difficult to see that μ is singular. By the standard Cantor argument, $\mu \perp \eta$: it is concentrated on a set $\Sigma' \subset \Sigma$ such that $\eta(\Sigma') = \prod(1 - \frac{1}{n}) = 0$. Therefore $\mu \perp \nu_c$. Since $\nu_{p.p.}(\Sigma) = 0$, $\mu \perp \nu$.

Estimate of $K\mu/K\nu$. Note that after the first step of the above construction we only change our measure inside the arcs $J_i^{1,n} \subset E_1$ leaving the total mass of each arc unchanged. The condition $|J_i^{1,n}| < (d_1)^2$ means that each arc is rather far from $\bar{E}_1$. Hence the Cauchy integral $K\mu$ can be estimated through $K\mu_1 = Kf_1\eta$ on $\bar{E}_1$. More precisely, it is not difficult to see that, since $d_1 \leq 1/2$, the condition $|J_i^{1,n}| < (d_1)^2$ implies

$$|K\mu| < |Kf_1\eta| + ||\mu|| \leq |Kf_1\eta| + ||\nu_c||$$

on $\bar{E}_1$. WLOG $||\nu|| = 1$. Since $|K\nu| \geq 1/2$ and $f_1\eta = f_1\nu$ the last inequality gives

$$||K\mu/K\nu||_{L^2(\bar{E}_1)} \leq C_1||Kf_1\nu/K\nu||_{H^2} + C_2$$

for some $C_1, C_2 > 0$. By our construction $||f_1||_{L^2(\nu)} \leq \sqrt{2}$. Note that $||Kf_1\nu/K\nu||_{H^2} = ||f_1||_{L^2(\nu)}$ because $C_\nu = \Phi_1^*$ where $\Phi_1 : K_\theta^2 \to L^2(\nu)$ is the Fourier transform discussed in the introduction corresponding to θ satisfying $K\nu = (1-\theta)^{-1}$. Hence,

$$||K\mu/K\nu||_{L^2(\bar{E}_1)} \leq C_1||Kf_1\nu/K\nu||_{H^2} + C_2 \leq \sqrt{2}C_1 + C_2.$$

In a similar way, on $E_n \setminus E_{n+1}$

$$||K\mu/K\nu||_{L^2(E_n\setminus E_{n+1})} \leq C_1||\frac{K(\prod_1^{n+1} f_k)\nu}{K\nu}||_{H^2} + C_2 \leq C_1||\prod_1^{n+1} f_k||_{L^2(\nu)} + C_2.$$

By our construction the product $\prod_1^{n+1} f_k$ is equal to $n+1$ on the set of measure $1/(n+1)$ and to 0 elsewhere. Hence

$$||K\mu/K\nu||_{L^2(E_n\setminus E_{n+1})} < C_3\sqrt{n+1}$$

for some $C_3 > 0$.

If $0 < p < 2$ then by Hölder's inequality

$$||K\mu/K\nu||^p_{L^p(E_n\setminus E_{n+1})} \le \left[||K\mu/K\nu||^2_{L^2(E_n\setminus E_{n+1})}\right]^{p/2} [m(E_n \setminus E_{n+1})]^{\frac{2-p}{2}}$$

$$\le C_3^p(n+1)^{p/2}[2^{-n-1}]^{\frac{2-p}{2}}.$$

Hence, $||K\mu/K\nu||^p_{L^p(\mathbb{T})} < \infty$.

III) Let $\mu \in M$, $\sigma \in M_+$, $\mu \perp \sigma$ be two singular measures such that $f = K\mu/K\sigma \in H^2$. Then $f \in K^2_\theta$ where inner θ is defined by $(1-\theta)^{-1} = K\sigma$ (see Lemma 4.2 below). Hence by Clark's Theorem $f = KF\sigma/K\sigma$ for some $F \in L^2(\sigma)$. Therefore, $\mu = F\sigma$ which contradicts the condition $\mu \perp \sigma$. □

Remark 3.3. *Another way to construct an inner θ such that $\frac{1-\theta}{1-z}$ belongs to any H^p, $p < 2$ but not to H^2 (as in part I above) is suggested by A. B. Aleksandrov. Theorem 3.3 from* [2] *says that any Carleson Blaschke product with zeros a_n satisfying $\sum \frac{1-|a_n|^2}{|1-a_n|^p} < \infty$ for all $p < 2$ but $\sum \frac{1-|a_n|^2}{|1-a_n|^2} = \infty$ will have the desired properties.*

In the rest of this section we will try to understand how "close" the measures μ and ν should be to each other in order for $K\mu/K\nu$ to be in H^p. If $\mu \in M$ is singular we denote by $\operatorname{Hol}\mu$ the holomorphic support of μ:

$$\operatorname{Hol}\mu = \{\xi \in \mathbb{T} |\ \lim_{z \underset{\sphericalangle}{\longrightarrow} \xi} K\mu(z) = \infty\}.$$

The notation $\operatorname{Hol}^\# \mu$ stands for a slightly larger set:

$$\operatorname{Hol}^\# \mu = \{\xi \in \mathbb{T} |\ \limsup_{z \underset{\sphericalangle}{\longrightarrow} \xi} |K\mu(z)| = \infty\}.$$

We define the space $H^{1,\infty}$ ("weak" H^1) as the subspace of H^p, $p < 1$ consisting of functions f whose boundary values f^* satisfy

$$\sup_t t\, m(\{|f^*| > t\}) < \infty.$$

By a theorem of Kolmogorov all Cauchy integrals belong to $H^{1,\infty}$ (it also follows, for instance, from Theorem 3.5 below). Any quotient $K\nu/K\mu$ where $\mu \in M_+$ will also belong to $H^{1,\infty}$ since $|K\mu| \ge 1/2||\mu||$.

We denote by $H^{o(1,\infty)}$ the subset of $H^{1,\infty}$ consisting of functions satisfying

$$\lim_{t\to\infty} t\, m(\{|f^*| > t\}) = 0.$$

We have the following result:

Theorem 3.4. *Let $\mu \in M$, $\nu \in M_+$, $\mu \perp \nu$ be two singular measures.*

I. If $K\mu/K\nu \in H^1$ then $\operatorname{Hol}\nu$ supports μ.

II. If $K\mu/K\nu \in H^{o(1,\infty)}$ then $\operatorname{Hol}^{\#}\nu$ supports μ.

Before we prove Theorem 3.4 we need some prepapation. To simplify notations, in what follows we write $K\mu$ instead of $K\mu^*$ to denote the non-tangential boundary values of $K\mu$ and $\chi_{\{|K\mu|>t\}}$ denotes the characteristic function of the set $\{|K\mu| > t\} \subset \mathbb{T}$. As usual, μ_s stands for the singular component of μ.

Theorem 3.5 ([13]). *Let $\mu \in M$. Then*

$$\pi t \chi_{\{|K\mu|>t\}} m \xrightarrow[t\to\infty]{*-weakly} |\mu_s|.$$

We will also need the following

Lemma 3.6. *Let $\mu \in M$, $\nu \in M_+$, $\mu \perp \nu$ be two singular measures. The property $K\mu/K\nu \in H^{o(1,\infty)}$ is spectral for μ, i. e. if $K\mu/K\nu \in H^{o(1,\infty)}$ then $Kf\mu/K\nu \in H^{o(1,\infty)}$ for any $f \in L^1(|\mu|)$.*

Proof. First, let μ be positive.

To show that $Kf\mu/K\nu \in H^{o(1,\infty)}$ approximate f in $L^1(\mu)$ by a smooth function g. It is not difficult to check that for any smooth g the function $Kg\mu/K\mu$ is bounded. Hence $Kg\mu/K\nu \in H^{o(1,\infty)}$. Since the $H^{1,\infty}$-norm of $K(f-g)\mu/K\nu$ can be made arbitrarily small, we obtain the statement.

Now, let μ be an arbitrary real measure, $\mu = \mu_+ - \mu_-$, where $\mu_\pm$ are mutually singular positive measures. To reduce this case to the previous one, we need to demonstrate that if $K\mu/K\nu \in H^{o(1,\infty)}$ then $K\mu_\pm/K\nu \in H^{o(1,\infty)}$.

Let E, F be disjoint open sets such that $\mu_+(E) > ||\mu_+|| - \epsilon$ and $\mu_-(F) > ||\mu_-|| - \epsilon$. By Theorem 3.5

$$\pi t m(\{|K\mu_+| > t\} \cap E) \to \mu_+(E) > ||\mu_+|| - \epsilon \text{ as } t \to \infty.$$

Hence

$$\pi t m(\{|K\mu_+| > t\} \cap E) > ||\mu_+|| - \epsilon - o(1) \quad \text{as} \quad t \to \infty. \tag{11}$$

In the same way

$$\pi t m(\{|K\mu_-| > t\} \cap E) < \epsilon + o(1) \quad \text{as} \quad t \to \infty. \tag{12}$$

Since $|K\nu| \geq 1/2$, $\{|K\mu_+/K\nu| > t\} \subset \{|K\mu_+| > t/2\}$. Hence by (11)

$$m(\{|K\mu_+/K\nu| > t\} \setminus E) < \frac{2\epsilon}{\pi t} + o(1/t)$$

and, similarly, by (12)

$$m(\{|K\mu_-/K\nu| > t\} \cap E) < \frac{2\epsilon}{\pi t} + o(1/t).$$

Since $K\mu/K\nu \in H^{o(1,\infty)}$, this implies

$$m(\{|K\mu_+/K\nu| > t\}) < m(\{|K\mu_+/K\nu| > t\} \cap E) + \frac{2\epsilon}{\pi t} + o(1/t)$$

$$\leq m\left(\left[\{|K\mu/K\nu| > t/2\} \cup \{|K\mu_-/K\nu| > t/2\}\right] \cap E\right) + \frac{2\epsilon}{\pi t} + o(1/t)$$

$$\leq m(\{|K\mu/K\nu| > t/2\} \cap E) + m(\{|K\mu_-/K\nu| > t/2\} \cap E) + \frac{2\epsilon}{\pi t} + o(1/t)$$

$$< \frac{6\epsilon}{\pi t} + o(1/t) \quad \text{as} \quad t \to \infty.$$

Therefore $K\mu_+/K\nu \in H^{o(1,\infty)}$. The measure μ_- can be treated in the same way.

The case of arbitrary complex μ can be proved by separating its real and imaginary parts. □

We say that a point $\xi \in \mathbb{T}$ is a density point of a set $E \subset \mathbb{T}$ if

$$\frac{|I_\varepsilon \cap E|}{\varepsilon} \to 1 \quad \text{as} \quad \varepsilon \to 0,$$

where I_ε is the arc of size ε centered at ξ.

Proof of Theorem 3.4. Denote $f = K\mu/K\nu$.

I. Since f is summable on $\mathbb{T}$, $f = Kf$. Suppose that there exists a set E, $E \cap \operatorname{Hol}\nu = \emptyset$ such that $|\mu|(E) > 0$. Then $1/K\nu = Kf/K\mu$ does not tend to 0 as $z \underset{\sphericalangle}{\longrightarrow} \xi$ for μ-a. e. ξ in E. But this contradicts Corollary 2.6 since $fm \perp \mu$.

II. By Lemma 3.6 we can assume that there exists $C > 0$ such that for μ-a. e. ξ

$$\sup_{z=r\xi,\ 0<r<1} |K\nu(z)| < C < \infty. \tag{13}$$

(Otherwise just multiply μ by

$$\phi_C = \chi_{\{\xi \mid \sup_{z=r\xi,\ 0<r<1} |K\nu(z)|<C\}},$$

prove the statement for $\phi_C\mu$ instead of μ and then consider larger and larger C'c.)

Denote $L_n = \{\xi \in \mathbb{T} \mid |K\mu(\xi)| > n\}$. If $\xi \in L_n$ is a density point for L_n, consider I_ξ – an open arc containing ξ satisfying

$$|L_n \cap I_\xi| > 3/4|I_\xi|. \tag{14}$$

Such an I_ξ can always be chosen to satisfy $\mu(I_\xi) > 0$. Indeed, since $I_\xi \cap L_n \neq \emptyset$, if both endpoints of I_ξ are not in L_n then automatically $\mu(I_\xi) > 0$. If one or both endpoints are in L_n one can consider a slightly larger arc such that (14) still holds but the endpoints are no longer in L_n. By the standard argument, we can choose a collection $I_1^n, I_2^n, \ldots$ of such open arcs I_ξ that covers almost all L_n. Moreover, I_i^n can be chosen to form a covering of rank 2 (each point is covered by at most 2 arcs).

Recall that, since μ is a singular measure, Theorem 3.5 implies $|L_n| > D\frac{\|\mu\|}{n}$ for some $D > 0$. Notice that since $f \in H^{o(1,\infty)}$ there exists a sequence $A_n \to \infty$ such that

$$|L_n \cap \{|K\nu| > A_n\}| > 3/4|L_n|.$$

Since I_i^n form a covering of rank 2, there exists I_k^n such that

$$|I_k^n \cap \{|K\nu| > A_n\}| > 3/8|I_k^n|. \tag{15}$$

Since $K\nu$ is an outer function, we can consider the function $G = \sqrt{K\nu}$. The function $\operatorname{Re} G$ is positive and is at least $\sqrt{A_n/2}$ on $\{|K\nu| > A_n\}$. Together with (15), the properties of the Poisson kernel now imply that $\sup_{0<r<1} \operatorname{Re} G(r\xi) > c\sqrt{A_n}$ for any $\xi \in I_k^n$ for some $c > 0$. But since $\mu(I_k^n) > 0$ and A_n can be arbitrarily large (recall that $A_n \to \infty$) this contradicts (13). □

4. Uniqueness and integral representations of pseudocontinuable functions

In this section we first discuss for what θ and p the generalized Fourier transform Φ_α (defined as in the introduction) may have a non-trivial kernel in K_θ^p. One can also look at this problem from the following function theoretic viewpoint.

If X is a linear space of functions in $\mathbb{D}$ we say that $\Sigma \subset \mathbb{T}$ is a uniqueness set for X if the only $f \in X$ such that

$$\lim_{z \underset{\sphericalangle}{\longrightarrow} \xi} f(z) = 0 \quad \text{for all} \quad \xi \in \Sigma$$

is the zero function.

Note that the existence of boundary values on Σ for all functions in X is not required. If Σ is a uniqueness set for X and $f, g \in X$ are two different functions, our definition means that f and g behave differently near Σ. Namely, the limit

$$\lim_{z \underset{\sphericalangle}{\longrightarrow} \xi} f - g$$

either does not exist or is non-zero for at least one point $\xi \in \Sigma$.

Also, if μ is a finite Borel measure on $\mathbb{T}$ we say that μ separates points in X (μ is a separating measure for X) if the only $f \in X$ satisfying

$$\lim_{z \underset{\sphericalangle}{\longrightarrow} \xi} f(z) = 0 \quad \text{for } \mu\text{-a. e.} \quad \xi \in \mathbb{T}$$

is the zero function.

The classical theory implies that Σ is a uniqueness set for H^p iff $m(\Sigma) > 0$. Similarly, $\mu \in M$, $\mu \neq 0$ is a separating measure for H^p iff μ has a nontrivial absolutely continuous part. The situation is very different for the spaces K_θ^p.

It is hardly possible to discribe all uniqueness sets for all K_θ^p in any reasonable terms. The main difficulty here is that the structure of uniqueness sets depends too much on the concrete properties of θ, e. g. [8]. For instance, if θ is a finite Blaschke product of degree n then K_θ^p is n-dimensional and any n points on $\mathbb{T}$ constitute a uniqueness set. Nothing like this takes place if we consider an infinite Blaschke product or a singular inner function. In that case, however, other "uncommon" uniqueness sets may appear if the spectrum of θ contains an arc or if θ is a so-called single-component function (see [2] for the definition).

As was announced before, in this note we are only going to discuss for what θ and p the measures σ_α (defined by (4)) are separating and when their natural supports, the level sets $\{\theta = \alpha\}$, are uniqueness sets. (We denote by $\{\theta = \alpha\}$ the subset of $\mathbb{T}$ consisting of points where the non-tangential boundary limit of θ is equal to α.)

If θ is inner we denote by $\sigma(\theta)$ the spectrum of θ:

$$\sigma(\theta) = \{\xi \in \mathbb{T} | \liminf_{z \to \xi} |\theta(z)| = 0\}.$$

Theorem 4.1.

I. If $p \geq 2$ then every $\sigma_\alpha \in M_\theta$ is a separating measure for K^p_θ for any inner θ (and therefore any set $\{\theta = \alpha\}$, $\alpha \in \mathbb{T}$ is a uniqueness set);

II. If $1 \leq p < 2$ then there exists θ such that none of the measures $\sigma_\alpha \in M_\theta$ is a separating measure for K^p_θ;

III. If $p < 1$ then there may not be any uniqueness sets of Lebesgue measure 0 (singular separating measures) for the space K^p_θ. In particular, if $\sigma(\theta) \supset \mathbb{T}$ there are no uniqueness sets of Lebesgue measure 0 (singular separating measures).

For $1 \leq p < 2$ it is unknown wether the set $\{\theta = \alpha\}$ is a uniqueness set for K^p_θ. Although this seems highly unlikely, we were unable to find a counterexample. Instead we prove that $\{\theta = \alpha\}$ is a uniqueness set for a large subset of K^p_θ consisting of functions admitting certain integral representations, see Theorem 4.5 below.

To prove Theorem 4.1 we now turn to the discussion of integral representations for pseudocontinuable functions.

For $\mu \in M$ the Cauchy integral $K\mu$ is naturally defined in $\mathbb{C} \setminus \mathbb{T}$. Moreover, if μ is singular then by Privalov's theorem the two parts of $K\mu$ defined inside and outside the disk have equal boundary values on the circle at a. e. point. Hence $K\mu$ presents an example of a pseudocontinuable function in $\mathbb{D}$. Now, let $\mu \in M$ and $\nu \in M_+$ be two singular measures. Then, once again, $\frac{K\mu}{K\nu}$ is an example of a pseudocontinuable function in $\mathbb{D}$.

For any singular μ $K\mu$ belongs to any H^p, $p < 1$ but not to H^1. However, as we will demonstrate shortly, our second example $\frac{K\mu}{K\nu}$ covers all pseudocontinuable functions in H^1: for any function $f \in K^p_\theta$, $p \geq 1$ there exist $\mu \in M$ and $\nu \in M_+$ such that $f = \frac{K\mu}{K\nu}$. We will also show how these measures are related to θ.

Lemma 4.2. *Let $f = \frac{K\mu}{K\nu}$ where $\mu \in M$, $\nu \in M_+$ are singular. If $f \in H^p$, $p \geq 1$ then $f \in K^p_\theta$, where the inner function θ is uniquely defined by the equation*

$$K\nu = \frac{1}{1-\theta}. \tag{16}$$

Proof. It is easy to check that for any singular positive ν there is a unique θ satisfying (16). Let us notice that $\bar{\theta} f = (\bar{\theta} - 1)K\mu$. Since $K\mu(\infty) = 0$, the antianalytic function $g(z) = K\mu(1/\bar{z})$ satisfies $g(0) = 0$. Hence $\bar{\theta} f = (\bar{\theta} - 1)g \in \bar{H}^p_0$. □

As we mentioned before, the relation (16) between ν and θ means simply that $\nu = \sigma_1 \in M_\theta$. Along with the representaion $f = \frac{K\mu}{K\sigma_1}$ our function always posesses representations involving other σ_α's. In fact, the definition of the inverse Fourier transform Φ^*_α (see formula (10)) implies that if $f \in K^2_\theta$ then for any $\alpha \in \mathbb{T}$ there is a unique measure $\mu \in M$ such that $f = \frac{K\mu}{K\sigma_\alpha}$. By Theorem 2.1 the measure μ satisfies $\mu = f^*\sigma_\alpha$ where $f^* \in L^2(\sigma_\alpha)$ is the non-tangential boundary values of f. For $p < 2$ we have

Lemma 4.3. *Let $f \in K^p_\theta$, $p \geq 1$. Then for almost any $\alpha \in \mathbb{T}$ $f = \frac{Kf^*\sigma_\alpha}{K\sigma_\alpha}$, where $f^* \in L^p(\sigma_\alpha)$ is the non-tangential boundary values of f.*

Proof. Consider a sequence f_n of K^2_θ-functions tending to f in H^p. By (6) the boundary values $f^*_n \to f^*$ in $L^p(\sigma_\alpha)$ for almost every α. If $f^*_n \to f^*$ in $L^p(\sigma_\alpha)$, then $Kf^*_n\sigma_\alpha \to Kf^*\sigma_\alpha$ pointwise in the disk. Hence, pointwise in the disk,

$$f = \lim_{n\to\infty} f_n = \lim_{n\to\infty} \frac{Kf^*_n\sigma_\alpha}{K\sigma_\alpha} = \frac{Kf^*\sigma_\alpha}{K\sigma_\alpha}.$$

□

We will say that the function $f \in K^p_\theta$ admits an α-representation if $f = \frac{K\mu}{K\nu}$ for some $\mu \in M$ and $\nu = \sigma_\alpha \in M_\theta$. We will call a representation $f = \frac{K\mu}{K\nu}$ regular if $\mu << \nu$.

Using this terminology we can say that any $f \in K^2_\theta$ has regular α-representations for all α. Lemma 4.3 says that for $1 \leq p < 2$ f has regular α-representations for almost all α. Let us also make the following observations:

Remark 4.4.

I) Each $f \in K^p_\theta$, $0 < p \leq \infty$ admits at most one α-representaion.

II) By Theorem 3.1 and Lemma 4.2 functions from K^2_θ admit only regular representaions, whereas functions from $K^p_\theta, 1 \leq p < 2$ may have both regular and irregular ones. The same function $f \in K^p_\theta$, $1 \leq p < 2$ may have regular α-representations for some α and irregular ones for other α. By Lemma 4.3 any f admits almost all α-representations and they are regular. Hence, f can only have irregular α-representaions for a zero-set of α's.

III) By Theorem 3.4 if $f \in K^p_\theta$, $1 \leq p < 2$, $f = \frac{K\mu}{K\sigma_\alpha}$ then μ is supported on the set $\{\theta = \alpha\}$. If $f \in K^{o(1,\infty)}_\theta$ $(= K^q_\theta \cap H^{o(1,\infty)}$, $0 < q < 1)$ then μ is supported on the set

$$\{\xi \in \mathbb{T} \mid \lim \theta(z_n) = \alpha \textit{ for some sequence } z_n \underset{\sphericalangle}{\longrightarrow} \xi\}.$$

IV) In general, a K^p_θ-function may not admit an α-representaion for some α if $1 \leq p < 2$. In [2] it was shown that a function $f \in K^p_\theta$, $1 \leq p < 2$ may have no non-tangential boundary values on a set E, $\sigma_\alpha(E) > 0$. At the same time, it follows from Theorem 2.5 that if a function has an α-representaion then it must have non-tangential boundary values σ_α-a. e. Thus, such an f does not have an α-representaion (neither a regular nor an irregular one).

V) Denote

$$\left(K^p_\theta\right)_\alpha = \{f \in K^p_\theta \mid f \textit{ admits an } \alpha\textit{-representation } \}.$$

Note that $(K^p_\theta)_\alpha$, $1 \le p$, $p \ne 2$ *contains* $\Phi^*_\alpha(L^p(\sigma_\alpha))$ *but is rarely equal to that set. The set* $\Phi^*_\alpha(L^p(\sigma_\alpha))$ *consists of functions admitting regular* α*-representations* $Kf\sigma_\alpha/K\sigma_\alpha$ *with* $f \in L^p(\sigma_\alpha)$. *If* $p < 2$ *an* α*-representaion may not be regular. If* $p > 2$ *the representation is always regular by Theorem 3.1, but* f *may not be in* $L^p(\sigma_\alpha)$, *as follows from the results of* [2].

VI) We have $\left(K^2_\theta\right)_\alpha = K^2_\theta$ *for any* α. *For* $1 \le p < 2$ *Lemma 4.3 implies*

$$\bigcup_{\alpha\in\Omega} (K^p_\theta)_\alpha = K^p_\theta$$

where Ω *is any subset of* $\mathbb{T}$ *of non-zero Lebesgue measure.*

VII) As was shown in [5] *for any* $1 \le p$ *and any* α, $\beta \in \mathbb{T}$, $\alpha \ne \beta$

$$\Phi^*_\alpha(L^p(\sigma_\alpha)) + \Phi^*_\beta(L^p(\sigma_\beta)) = K^p_\theta$$

i. e. for any $f \in K^p_\theta$ *there exist* $h \in L^p(\sigma_\alpha)$ *and* $g \in L^p(\sigma_\beta)$ *such that* $f = C_{\sigma_\alpha}h + C_{\sigma_\beta}g$.

Proof of Theorem 4.1. I) Follows from the fact that any $f \in K^p_\theta$, $p \ge 2$ has the regular α-representation $f = Kf^*\sigma_\alpha/K\sigma_\alpha$ and from Theorem 2.5.

II) If θ is such that any $\sigma_\alpha \in M_\theta$ has a non-trivial continuous part then by Theorem 3.1 for any α there exists a non-zero $\mu \in M$, $\mu \perp \sigma_\alpha$ such that $\frac{K\mu}{K\sigma_\alpha} \in H^p$. By Lemma 4.2 $f = \frac{K\mu}{K\sigma_\alpha} \in K^p_\theta$. By Theorem 2.5 the boundary values of f are 0 σ_α-a.e.

To find such a θ take, for instance, any inner function that does not have angular derivatives at any point on $\mathbb{T}$. Then all σ_α will be continuous (see [17], pages 51-53).

III) It is well-known (see [6]) that the necessary and sufficient condition for a function f to belong to K^p_θ, $0 < p < 1$ is to satisfy $\bar{f}\theta \in H^p$ and to admit an analytic continuation through $\mathbb{T}$ into the domain of θ in $\mathbb{C}$. If $\sigma(\theta) = \mathbb{T}$ then the second requirement disappears. Also if $f = K\mu$ for some singular $\mu \in M$ then $f - ||\mu||/2$ is imaginary on $\mathbb{T}$ and therefore $\bar{f}\theta = (-f + ||\mu||)\theta \in H^p$. Hence, K^p_θ, $p < 1$ contains all Cauchy integrals of singular measures. Let $\Sigma \subset \mathbb{T}$ be a uniqueness set of Lebesgue measure 0. Then there exists a singular measure μ such that $P\mu(\xi) = \infty$ at every $\xi \in \Sigma$. Consider the Herglotz integral of μ

$$H\mu = \int_{\mathbb{T}} \frac{1 + \bar{\xi}z}{1 - \bar{\xi}z} d\mu(\xi).$$

Then the function $(H\mu)^{-1}$ has positive real part in $\mathbb{D}$ and is real at 0. Hence it is a Herglotz integral of another measure ν. Since $(H\mu)^{-1}$ is imaginary a.e. on $\mathbb{T}$, ν is singular. To obtain a contradiction it is left to notice that $K\nu(\xi) \in K^p_\theta$ but $K\nu(\xi) = 0$ at any $\xi \in \Sigma$. □

As we mentioned before, it is unclear whether or not $\{\theta = \alpha\}$ is a uniqueness set for K^p_θ for every θ when $1 \le p < 2$. We can prove, however, that $(K^p_\theta)_\alpha$, the set of functions admitting an α-representation, satisfies the following

Theorem 4.5. *For any inner* θ, $\alpha \in \mathbb{T}$ *and* $p \geq 1$ *the set* $\{\theta = \alpha\}$ *is the uniqueness set for* $(K_\theta^p)_\alpha$.

Proof. Let $f \in (K_\theta^p)_\alpha$, $f = \frac{K\mu}{K\sigma_\alpha}$, $\mu \neq 0$. By Theorem 2.5 the ratio $\frac{K\sigma_\alpha}{K\mu}$ has finite non-tangential boundary limits μ-a. e. Hence f has non-zero limits μ-a. e. But by Theorem 3.4 μ is supported on $\{\theta = \alpha\}$. □

References

[1] A. B. Aleksandrov, *Multiplicity of boundary values of inner functions,* Izv. Acad. Nauk. Arm. SSR, Matematica 22, **5** (1987), 490–503.

[2] A. B. Aleksandrov, *Inner functions and related spaces of pseudocontinuable,* Zap. Nauchn. Sem. Leningrad. Otdel. Mat. Inst. Steklov. (LOMI), **170** (1989), Russian, 7–33.

[3] A. B. Aleksandrov, *On the existance of angular boundary values for pseudocontinuable functions,* Zap. Nauchn. Sem. Leningrad. Otdel. Mat. Inst. Steklov. (LOMI), **222** (1995), Russian.

[4] A. B. Aleksandrov, *Izometric embeddings of coinvariant subspaces of the shift operator,* Zap. Nauchn. Sem. Leningrad. Otdel. Mat. Inst. Steklov. (LOMI), **232** (1996), Russian.

[5] A. B. Aleksandrov, *On the maximum principle for pseudocontinuable functions,* Zap. Nauchn. Sem. Leningrad. Otdel. Mat. Inst. Steklov. (LOMI), **217** (1994), Russian.

[6] A. B. Aleksandrov, *Invariant subspaces of the backward shift operator in the space* H^p ($p \in (0, 1)$), Zap. Nauchn. Sem. Leningrad. Otdel. Mat. Inst. Steklov. (LOMI), **92** (1979), Russian.

[7] D. Clark, *One dimensional perturbations of restricted shifts,* J. d'Anal. Math., **25** (1972), 169–191.

[8] K. Dyakonov, *Zero sets and multiplier theorems for star-invariant subspaces,* Preprint.

[9] R. Douglas, H. Shapiro and A. Shields, *Cyclic vectors and invariant subspaces for the backward shift operator operator,* Ann. Inst. Fourier, **20** (1970), 37–76.

[10] V. P. Havin and B. Jöricke, *The Uncertainity Principle in Harmonic Analysis,* Springer–Verlag, Berlin, 1994.

[11] N. K. Nikolski, *Treatise on the shift operator,* Grundlehren der mathematischen Wissenschaften [Fundamental Principles of Mathematical Sciences], **273** (1986), Springer–Verlag, Berlin–New York.

[12] A. Poltoratski, *On the boundary behavior of pseudocontinuable functions,* St. Petersburg Math. J., **5** (1994), 389–406.

[13] A. Poltoratski, *On the distributions of boundary values of Cauchy integrals,* St. Petersburg Math. J., **124** (1996), 2455–2463.

[14] A. Poltoratski, *The Krein spectral shift and rank one perturbations of spectra,* Algebra i Analiz, **10** (1998) No 5, 143–183, Russian; English translation to appear in St. Petersburg Math. J.

[15] A. Poltoratski, *Finite rank perturbations of singular spectra,* Internat. Math. Res. Notices, **9** (1997), 421–436.

[16] A. Poltoratski, *The Maximal properties of the normalized Cauchy transform,* Preprint (1999).

[17] D. Sarason, *Sub-Hardy Hilbert Spaces in the unit disk,* The University of Arkansas lecture notes in the mathematical sciences, **10** (1994), J. Wiley and Sons, New York.

Key words and phrases: Cauchy integrals, pseudocontinuable functions, rank-one perturbations
2000 Mathematics Subject Classification: 30E20

Dept. of Math., Texas A&M University, College Station, TX 77843
E-mail address: alexei@math.tamu.edu

Well-Posed Linear Systems, Lax–Phillips Scattering, and L^p-Multipliers

Olof J. Staffans

Abstract. We discuss the connection between Lax–Phillips scattering theory and the theory of well-posed linear systems, and show that the latter theory is a natural extension of the former. As a consequence of this, there is a close connection between the Lax–Phillips generator and the generators of the corresponding well-posed linear system. All the essential information about these two systems is contained in the system operator $S_\Sigma = \left[\begin{smallmatrix} A & B \\ C\&D \end{smallmatrix}\right]$, where A is the generator of the (central) semigroup, B is the control operator, and $C\&D$ is the combined observation/feedthrough operator. In the important Hilbert space case this system operator can be written in the more familiar form $S_\Sigma = \left[\begin{smallmatrix} A & B \\ \overline{C} & D \end{smallmatrix}\right]$, where $\overline{C}$ is a (not necessarily uniquely determined) observation operator and D is the corresponding (generalized) feedthrough operator. The system operator is closed and densely defined. In the reflexive case the adjoint of S_Σ is the system operator of the dual system. We give formulas for the Lax–Phillips generator and resolvent in terms of the system operator. By applying the Hille–Yoshida theorem to the Lax–Phillips semigroup we get necessary and sufficient conditions for the L^p-admissibility or joint L^p-admissibility of a control operator B and an observation operator C. This leads to a criterion for an H^∞-function to be an L^p-multiplier.

1. Background

This review of the relationship between the Lax–Phillips scattering theory on one hand and the theory of well-posed linear systems on the other hand has a very definite *date of conception*: the talk on '*Passive Linear Systems and Scattering Theory*' by Prof. *D. Z. Arov* given at MTNS in Padova in 1998. He said:

'In the connection with Lax–Phillips scattering scheme *Yu Smulijan [1986] proposed the following definition* of a linear continuous time-invariant system. It is *a little bit different from the one proposed in control theory by D. Salamon.*'

This gave me the motivation to take a closer look at the scattering theory by Lax and Phillips to find out how the theory developed by Arov and Smulijan differs from the one developed by Salamon. After studying [4] and [14, 15] for some time

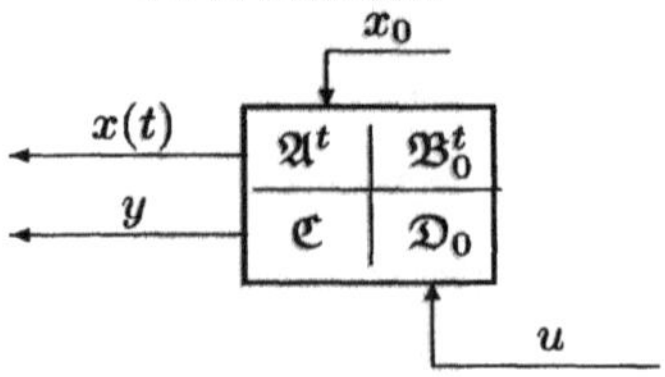

FIGURE 1. Input/State/Output Diagram of Σ

I found the rather surprising answer: there is *no real difference* (although different people tend to *emphasize* different aspects of the theory.)

Thus, this is a presentation of some of the basic notions of the general theory of well-posed linear systems developed by, among others, Adamjan, Arov, Lax, Helton, Nudelman, Ober, Phillips, Salamon, Smulijan, and Weiss. To me *Arov's notation* which he (at least partially) inherited from Lax and Phillips felt quite cumbersome, since I am used to a control theory type notation. Therefore I use a set of notations which is an extension of the standard control theory type notations (and which resemble those used by Salamon and by Weiss).

I apologize for the fact that I do not in all instances know *which results should be credited to whom*. Many of these results have been *discovered and then rediscovered*, maybe even several times.

A preliminary version of this review was presented in [24]. Details and proofs are given in the paper [26] by George Weiss and myself, and in the book manuscript [25] available (in postscript form) at http://www.abo.fi/~staffans/.

2. Introduction

Many infinite-dimensional systems can be described by the equations

$$\begin{aligned} x'(t) &= Ax(t) + Bu(t), \\ y(t) &= \overline{C}x(t) + Du(t), \quad t \geq 0, \\ x(0) &= x_0, \end{aligned} \tag{1}$$

on a triple of Banach spaces, namely, the input space U, the state space X, and the output space Y. We have $u(t) \in U$, $x(t) \in X$ and $y(t) \in Y$. The operator A is the generator of a strongly continuous semigroup, and it is usually unbounded. Also B and $\overline{C}$ are usually unbounded, whereas D is bounded.

Because of the presence of the unboudned operators A, B, and $\overline{C}$ it is often convenient to use the 'integral' representation of the system, which consists of the four operators from the initial state x_0 and the input function u to the final state $x(t)$ and the output function y:

$$\begin{aligned} x(t) &= \mathfrak{A}^t x_0 + \mathfrak{B}^t_0 u, \\ y &= \mathfrak{C} x_0 + \mathfrak{D}_0 u. \end{aligned} \tag{2}$$

Here, $\mathfrak{A}^t$ is the semigroup generated by A (which maps the initial state x_0 into the final state $x(t)$), $\mathfrak{B}^t_0$ is the map from the input u (restricted to the interval $[0,t]$) to the final state $x(t)$, $\mathfrak{C}$ is the map from the initial state x_0 to the output y, and $\mathfrak{D}_0$ is the input-output map from u (restricted to $[0,\infty)$) to y. If the operators B, $\overline{C}$, and D in (1) are bounded, then we get formulas for the corresponding operators $\mathfrak{B}_0^t$, $\mathfrak{C}$, and $\mathfrak{D}_0$ in (2) by using the standard 'variation of parameters formula' (recall that $\mathfrak{A}^t$ is the semigroup generated by A):

$$\begin{aligned} \mathfrak{B}_0^t u &= \int_0^t \mathfrak{A}^{t-v} Bu(v)\,dv, && t \geq 0, \\ (\mathfrak{C}x_0)(t) &= \overline{C}\mathfrak{A}^t x_0, && t \geq 0, \\ (\mathfrak{D}_0 u)(t) &= \overline{C}\int_0^t \mathfrak{A}^{t-v} Bu(v)\,dv + Du(t), && t \geq 0. \end{aligned} \tag{3}$$

As we shall see later, these formulas remain valid also for certain classes of unbounded operators B and $\overline{C}$.

For the moment, let us ignore (1) and instead focus on the *well-posedness* of the system (2). The standard well-posedness assumption is that (2) behaves well in an L^p-setting, where $1 \leq p < \infty$, i.e., $x(t) \in X$ and $y \in L^p_{\text{loc}}(\mathbf{R}^+;Y)$ depend continuously on $x_0 \in X$ and on $u \in L^p_{\text{loc}}(\mathbf{R}^+;U)$. If this is the case, we call the operators $\left[\begin{smallmatrix} \mathfrak{A} & \mathfrak{B} \\ \mathfrak{C} & \mathfrak{D} \end{smallmatrix}\right]$ a *L^p-well-posed linear system*, where

$$\mathfrak{B}u = \lim_{t\to\infty} \mathfrak{B}_0^t \tau^{-t} u, \qquad \mathfrak{D}u = \lim_{t\to\infty} \tau^t \mathfrak{D}_0 \tau^{-t} u,$$

each defined for those $u \in L^p_{\text{loc}}(\mathbf{R};U)$ for which the respective limit exists; here $(\tau^t u)(s) = u(s+t)$, $-\infty < s,\ t < \infty$, is the bilateral left shift by t. In the case where (2) is induced by the system (1) with bounded B, $\overline{C}$, and D, we have

$$\begin{aligned} \mathfrak{B}u &= \int_{-\infty}^0 \mathfrak{A}^{-v} Bu(v)\,dv, && \\ (\mathfrak{C}x_0)(t) &= \overline{C}\mathfrak{A}^t x_0, && t \geq 0, \\ \mathfrak{D}u &= \overline{C}\int_{-\infty}^t \mathfrak{A}^{t-v} Bu(v)\,dv + Du(t), && t \in \mathbf{R}, \end{aligned} \tag{4}$$

at least for those u whose support is bounded to the left.

As we shall see in Section 3, it is possible to define a well-posed linear system $\Sigma = \left[\begin{smallmatrix} \mathfrak{A} & \mathfrak{B} \\ \mathfrak{C} & \mathfrak{D} \end{smallmatrix}\right]$ without any reference to the system of equations (1).

The classical *Lax–Phillips model* was developed by Lax and Phillips in [14] (conservative systems) and [15] (dissipative systems) to provide a mathematical description of a scattering process where an incoming wave hits an obstacle and is scattered into an outgoing wave. It was soon realized (see [1], [4], and [10]) that it is possible to extend the classical Lax–Phillips model into a more general model of a well-posed infinite-dimensional system by relaxing some of the original assumptions on the incoming and outgoing subspaces, and by replacing the standard dissipativity assumption by a well-posedness assumption. In this extended formulation the Lax–Phillips model is a semigroup with a particular structure: it acts as an exponentially weighted incoming shift on the incoming subspace, as an exponentially weighted outgoing shift on the outgoing subspace, and the central part of the semigroup describes 'a generalized scattering process'. As we shall see in Section 4, this central part can be taken to be an arbitrary well-posed linear system. Thus there is a one-to-one correspondence between the class of all well-posed linear systems and all extended Lax–Phillips models.

We begin by presenting the most basic results about a L^p-well-posed linear system (Section 3) and the corresponding Lax–Phillips model (Section 4). We proceed in Section 5 to show that there is a close connection between the Lax–Phillips generator and the generators of the corresponding well-posed linear system. All the essential information about these two systems is contained in the ***system operator*** $S_\Sigma = \left[\begin{smallmatrix} A & B \\ C\&D \end{smallmatrix}\right]$ where A is the generator of the (central) semigroup, B is the control operator, and $C\&D$ is the combined observation/feedthrough operator. In the important case where X and U are Hilbert spaces this system operator can be written in the more familiar form $S_\Sigma = \left[\begin{smallmatrix} A & B \\ \overline{C} & D \end{smallmatrix}\right]$, where $\overline{C}$ is a (not necessarily unique) observation operator and D is the corresponding (generalized) feedthrough operator. We system operator is closed and densely defined from $X \times U$ to $X \times Y$. In the reflexive case the adjoint of S_Σ is the system operator of the dual system. We give formulas for the Lax–Phillips generator and resolvent in terms of the system operator. Finally, in the last section we apply the Hille–Yoshida theorem to the Lax–Phillips semigroup and get necessary and sufficient conditions for the admissibility or joint admissibility of a control operator B and an observation operator C. This leads to a criterion for an H^∞-function to be an L^p-multiplier.

3. Well-posed linear systems

As already outlined in Section 2, it is possible to define a well-posed linear system $\Sigma = \left[\begin{smallmatrix} \mathfrak{A} & \mathfrak{B} \\ \mathfrak{C} & \mathfrak{D} \end{smallmatrix}\right]$ without any reference to the system of equations (1). For this, we have to introduce some spaces and some simple operators. We denote $\mathbf{R} = (-\infty, \infty)$, $\mathbf{R}^+ = [0, \infty)$, $\mathbf{R}^- = (-\infty, 0)$,

$$(\pi_J u)(s) = \begin{cases} u(s), & s \in J, \\ 0, & s \notin J, \end{cases} \qquad \text{for all } J \subset \mathbf{R},$$

$$(\pi_+ u)(s) = \pi_{\mathbf{R}^+} u = \begin{cases} u(s), & s \in \mathbf{R}^+, \\ 0, & s \in \mathbf{R}^-, \end{cases},$$

$$(\pi_- u)(s) = \pi_{\mathbf{R}^-} u = \begin{cases} 0, & s \in \mathbf{R}^+, \\ u(s), & s \in \mathbf{R}^-, \end{cases},$$

$$(\tau^t u)(s) = u(t+s), \quad s,\ t \in \mathbf{R}.$$

Thus π_+ is 'restriction to $\mathbf{R}^+$', π_- is the 'restriction to $\mathbf{R}^-$', and τ^t shift to the left for $t > 0$ and to the right for $t < 0$. Moreover, we define

$$\begin{aligned} \tau_+^t &= \pi_+\tau^t, \qquad & t \geq 0, \\ \tau_-^t &= \tau^t\pi_-, \qquad & t \geq 0, \end{aligned}$$

so that τ_+^t is the left shift by $t \geq 0$ on $\mathbf{R}^+$ and τ_-^t is the left shift by $t \geq 0$ on $\mathbf{R}^-$. Thus, τ_+ is an 'incoming left shift' and τ_- is an 'outgoing left shift'.

The space $L^p_c(\mathbf{R}^-;U)$ consists of all the functions $u \in L^p(\mathbf{R}^-;U)$ with a bounded support. The space $L^p_{c,\mathrm{loc}}(\mathbf{R};U)$ consists of all the functions $u : \mathbf{R} \to U$ that are locally in L^p and whose support is bounded to the left. We interpret $L^p_c(\mathbf{R}^-;U)$ as the subspace of functions in $L^p_{c,\mathrm{loc}}(\mathbf{R};U)$ which vanish on $\mathbf{R}^+$, and $L^p_{\mathrm{loc}}(\mathbf{R}^+;U)$ as the subspace of functions in $L^p_{c,\mathrm{loc}}(\mathbf{R};U)$ which vanish on $\mathbf{R}^-$. A sequence of functions u_n converges in $L^p_{c,\mathrm{loc}}(\mathbf{R};U)$ to a function u if the common support of all the functions u_n is bounded to the left and u_n converges to u in the L^p sense on every bounded time interval. The continuity of $\mathfrak{B}$, $\mathfrak{C}$ and $\mathfrak{D}$ in the following definition is with respect to this convergence.

Definition 3.1. *Let U, X, and Y be Banach spaces, and let $1 \leq p < \infty$. An L^p-*well-posed linear system Σ *on (Y, X, U) is a quadruple $\Sigma = \left[\begin{smallmatrix} \mathfrak{A} & \mathfrak{B} \\ \mathfrak{C} & \mathfrak{D} \end{smallmatrix}\right]$ of continuous linear operators satisfying the following conditions:*

(i) *$t \mapsto \mathfrak{A}^t$ is a strongly continuous semigroup of operators on X;*

(ii) *$\mathfrak{B}\colon L^p_c(\mathbf{R}^-;U) \to X$ satisfies $\mathfrak{A}^t\mathfrak{B}u = \mathfrak{B}\tau_-^t u$, for all $u \in L^p_c(\mathbf{R}^-;U)$ and all $t \in \mathbf{R}^+$;*

(iii) *$\mathfrak{C}\colon X \to L^p_{\mathrm{loc}}(\mathbf{R};Y)$ satisfies $\mathfrak{C}\mathfrak{A}^t x = \tau_+^t\mathfrak{C}x$, for all $x \in X$ and all $t \in \mathbf{R}^+$;*

(iv) *$\mathfrak{D}\colon L^p_{c,\mathrm{loc}}(\mathbf{R};U) \to L^p_{c,\mathrm{loc}}(\mathbf{R};Y)$ satisfies $\tau^t\mathfrak{D}u = \mathfrak{D}\tau^t u$, $\pi_-\mathfrak{D}\pi_+u = 0$, and $\pi_+\mathfrak{D}\pi_-u = \mathfrak{C}\mathfrak{B}u$, for all $u \in L^p_{c,\mathrm{loc}}(\mathbf{R};U)$ and all $t \in \mathbf{R}$.*

The different components of Σ are called as follows: U is the input space, X is the state space, Y is the output space, $\mathfrak{A}$ is the semigroup, $\mathfrak{B}$ is the input (or reachability, or controllability) map, $\mathfrak{C}$ is the output (or observability) map, and $\mathfrak{D}$ is the input-output map. The state $x(t) \in X$ at time $t \in \mathbf{R}^+$ and the output $y \in L^p_{\mathrm{loc}}(\mathbf{R}^+;Y)$ of Σ with initial time zero, initial state $x_0 \in X$ and input function $u \in L^p_{\mathrm{loc}}(\mathbf{R}^+;U)$ are given by (2) *with $\mathfrak{B}_0^t = \mathfrak{B}\tau^t\pi_{[0,t]}u$ and $\mathfrak{D}_0 u = \mathfrak{D}\pi_+u$.*

It is easy to see that the operators defined in (4) (with bounded B, C, and D) satisfy these conditions. Moreover, we only have to integrate over a finite interval since the support of u is bounded to the left. (There is also a similar theory for the case $p = \infty$; see [25].)

Every well-posed linear system has as a finite exponential *growth bound*. By the growth bound $\omega_{\mathfrak{A}}$ of a system $\Sigma = \left[\begin{smallmatrix} \mathfrak{A} & \mathfrak{B} \\ \mathfrak{C} & \mathfrak{D} \end{smallmatrix}\right]$ we understand the growth bound of its semigroup $\mathfrak{A}$:

$$\omega_{\mathfrak{A}} = \lim_{t\to\infty} \frac{\log(\|\mathfrak{A}^t\|)}{t} = \inf_{t>0} \frac{\log(\|\mathfrak{A}^t\|)}{t}.$$

As is well known, we always have $\omega_{\mathfrak{A}} < \infty$, but possibly $\omega_{\mathfrak{A}} = -\infty$. To explain in which sense the other operators $\mathfrak{B}$, $\mathfrak{C}$, and $\mathfrak{D}$ are exponentially bounded we introduce exponentially weighted L^p-spaces of the following type: we let $L^p_\omega(\mathbf{R}^+;U)$ represent the space of functions $u\colon \mathbf{R}^+ \to U$ for which $t \mapsto e^{-\omega t}u(t)$ belongs to $L^p(\mathbf{R}^+;U)$.

Theorem 3.2. *Let* $\Sigma = \left[\begin{smallmatrix}\mathfrak{A} & \mathfrak{B}\\ \mathfrak{C} & \mathfrak{D}\end{smallmatrix}\right]$ *be a* L^p*-well-posed linear system,* $1 \le p < \infty$*, on* (Y,X,U) *with growth bound* $\omega_{\mathfrak{A}}$*, and let* $\omega > \omega_{\mathfrak{A}}$*. Then* $\mathfrak{B}$ *has a unique extension to an bounded linear operator from* $L^p_\omega(\mathbf{R}^-;U)$ *to* X*,* $\mathfrak{C}$ *is a bounded linear operator from* X *to* $L^p_\omega(\mathbf{R}^+;Y)$*, and* $\mathfrak{D}$ *has a unique extension to a bounded linear operator from* $L^p_\omega(\mathbf{R};U)$ *to* $L^p_\omega(\mathbf{R};Y)$*.*

Every well-posed linear system also has a *transfer function*:

Theorem 3.3. *Let* $\Sigma = \left[\begin{smallmatrix}\mathfrak{A} & \mathfrak{B}\\ \mathfrak{C} & \mathfrak{D}\end{smallmatrix}\right]$ *be a* L^p*-well-posed linear system,* $1 \le p < \infty$*, on* (Y,X,U) *with growth bound* $\omega_{\mathfrak{A}}$*. Then there is a unique analytic* $\mathcal{L}(U;Y)$*-valued* transfer function $\widehat{\mathfrak{D}}$ *defined (at least) on* $\Re z > \omega_{\mathfrak{A}}$ *determined by the fact that the Laplace transform* $\widehat{\mathfrak{D}_0 u}$ *of the input-output term* $\mathfrak{D}_0 u$ *in* (2) *is given by, for all* $u \in L^p_{\omega_{\mathfrak{A}}}(\mathbf{R}^+;U)$*,*

$$\widehat{\mathfrak{D}_0 u} = \widehat{\mathfrak{D}}(z)\hat{u}(z), \quad \Re z > \omega_{\mathfrak{A}},$$

where $\hat{u}$ *is the Laplace transform of* u*.*

Thus, $\widehat{\mathfrak{D}}$ can be interpreted as an '$L^p_\omega(U;Y)$-multiplier' for every $\omega > \omega_{\mathfrak{A}}$.

The following theorem gives us a first connection between an arbitrary L^p-well-posed linear system and a system of equations of the type (1):

Theorem 3.4. *Every* L^p*-well-posed linear system* $\Sigma = \left[\begin{smallmatrix}\mathfrak{A} & \mathfrak{B}\\ \mathfrak{C} & \mathfrak{D}\end{smallmatrix}\right]$ *with* $1 \le p < \infty$ *has a unique closed (unbounded) densly defined* system operator

$$S_\Sigma\colon X \times U \supset \mathcal{D}(S_\Sigma) \to X \times Y$$

with the following properties. If $x_0 \in X$*,* $u \in W^{1,p}_{\mathrm{loc}}(\mathbf{R}^+;U)$ *and* $\left[\begin{smallmatrix}x_0\\ u(0)\end{smallmatrix}\right] \in \mathcal{D}(S_\Sigma)$*, then the state* $x(t)$ *and the output* $y(t)$ *of* Σ *with initial state* x_0*, and input* u *satisfies* $\left[\begin{smallmatrix}x(t)\\ u(t)\end{smallmatrix}\right] \in \mathcal{D}(S_\Sigma)$ *for all* $t \ge 0$*, and*

$$\begin{aligned}\begin{bmatrix}x'(t)\\ y(t)\end{bmatrix} &= S_\Sigma \begin{bmatrix}x(t)\\ u(t)\end{bmatrix}, \qquad t \ge 0,\\ x(0) &= x_0.\end{aligned} \tag{5}$$

The proof of this theorem is given in [26] (and also in [25]).

Note that (5) reduces to (1) for *smooth input functions* and *compatible initial conditions* provided S_Σ can be written in the form $S_\Sigma = \left[\begin{smallmatrix}A & B\\ C & D\end{smallmatrix}\right]$. Is this always possible?

Before giving a (partial) answer to this question we need to introduce two auxiliary spaces X_1 and X_{-1}. Choose any γ in the resolvent set of the generator A of $\mathfrak{A}$. We let $X_1 = \mathcal{D}(A)$, with the norm $\|x\|_{X_1} = \|(\gamma I - A)x\|_X$, and X_{-1} is the completion of X with the norm $\|x\|_{X_{-1}} = \|(\gamma I - A)^{-1}x\|_X$. We have $X_1 \subset X \subset X_{-1}$ with continuous and dense imbeddings. The semigroup $\mathfrak{A}$ can be restricted to to a strongly continuous semigroup on X_1 and extended to a strongly continuous semigroup on X_{-1} (which still we denote by the same symbol). We denote the space of bounded linear operators from U to Y by $\mathcal{L}(U;Y)$.

Theorem 3.5. *Every L^p-well-posed linear system $\Sigma = \left[\begin{smallmatrix}\mathfrak{A} & \mathfrak{B} \\ \mathfrak{C} & \mathfrak{D}\end{smallmatrix}\right]$ with $1 \le p < \infty$ has a unique* control operator $B \in \mathcal{L}(U;X_{-1})$ *and a unique combined* observation/feedthrough operator $C\&D\colon \mathcal{D}(S_\Sigma) \to X \times Y$, *such that S_Σ can be written in the form*

$$S_\Sigma \begin{bmatrix} x \\ u \end{bmatrix} = \begin{bmatrix} A & B \\ C\&D \end{bmatrix} \begin{bmatrix} x \\ u \end{bmatrix}, \qquad \begin{bmatrix} x \\ u \end{bmatrix} \in \mathcal{D}(S_\Sigma).$$

Thus, the state $x(t)$ and the output $y(t)$ of Σ in Theorem 3.4 satisfy

$$\begin{aligned} x'(t) &= Ax(t) + Bu(t), \\ y(t) &= C\&D \begin{bmatrix} x(t) \\ u(t) \end{bmatrix}, \qquad t \ge 0, \\ x(0) &= x_0, \end{aligned}$$

where the equation for x' is valid in X_{-1}. Moreover, $\mathcal{D}(S_\Sigma)$ is given by

$$\mathcal{D}(S_\Sigma) = \left\{ \left[\begin{smallmatrix} x \\ u \end{smallmatrix}\right] \in X \times U \;\middle|\; Ax + Bu \in X \right\}.$$

In particular, if $x \in X_1$, then $\left[\begin{smallmatrix} x \\ 0 \end{smallmatrix}\right] \in \mathcal{D}(S_\Sigma)$, and we can define the observation operator $C \in \mathcal{L}(X_1;Y)$ *by*

$$Cx = C\&D \begin{bmatrix} x \\ 0 \end{bmatrix}, \qquad x \in X_1.$$

This theorem is actually older than Theorem 3.4; see [4], [5], [17, 18], and [27, 28] (or [25]) for the proof. In [4] the combined observation/feedthrough operator is denoted by N. The control operator B is said to be *bounded* if the range of B lies in X, in which case $B \in \mathcal{L}(U;X)$. The observation operator C is said to be *bounded* if it can be extended to an operator in $\mathcal{L}(X;Y)$.

There is an simple connection between the transfer function introduced in Theorem 3.3 and the operators introduced in Theorem 3.5.

Theorem 3.6. *With the notation of Theorems 3.3 and 3.5, the transfer function $\widehat{\mathfrak{D}}$ of Σ is given by*

$$\widehat{\mathfrak{D}}(z) = (C\&D) \begin{bmatrix} (zI - A)^{-1}B \\ I \end{bmatrix}, \qquad \Re z > \omega_{\mathfrak{A}}.$$

Conversely, for all $z \in \mathbf{C}$ with $\Re z > \omega_{\mathfrak{A}}$ and for all $\left[\begin{smallmatrix} x \\ u \end{smallmatrix}\right] \in \mathcal{D}(S_\Sigma)$ we have

$$C\&D \begin{bmatrix} x \\ u \end{bmatrix} = C[x + (zI - A)^{-1}Bu] + \widehat{\mathfrak{D}}(z)u.$$

For more details, explanations and examples we refer the reader to [1], [2, 3], [4], [5], [17, 18], [19, 20, 22, 21, 23, 25], [27, 28, 29, 30, 32, 33], [36] (and the references therein). Most of the available literature deals with Hilbert spaces and $p = 2$.

Let us now return to the question of the possibility to split S_Σ even further into $S_\Sigma = \left[\begin{smallmatrix} A & B \\ C & D \end{smallmatrix}\right]$. For the purpose of the following discussion, let us temporarily split S_Σ into $S_\Sigma = \left[\begin{smallmatrix} A\&B \\ C\&D \end{smallmatrix}\right]$, where, with the notation of Theorem 3.4, $A\&B\colon \mathcal{D}(S_\Sigma) \to X$ maps $\left[\begin{smallmatrix} x \\ u \end{smallmatrix}\right]$ into x' and (as in Theorem 3.5) $C\&D\colon \mathcal{D}(S_\Sigma) \to Y$ maps $\left[\begin{smallmatrix} x \\ u \end{smallmatrix}\right]$ into y. According to Theorem 3.5, it is always possible to extend the domain of $A\&B$ to all of $X \times U$ by allowing the values of $A\&B$ to belong to the larger space X_{-1}. This extension is unique since $\mathcal{D}(A\&B) = \mathcal{D}(S_\Sigma)$ is dense in $X \times U$. If we denote the extended operator by $\overline{A\&B}$, then

$$\overline{A\&B}\begin{bmatrix} x \\ u \end{bmatrix} = \begin{bmatrix} A & B \end{bmatrix}\begin{bmatrix} x \\ u \end{bmatrix} = Ax + Bu, \qquad x \in X, \qquad u \in U,$$

where

$$Ax = \overline{A\&B}\begin{bmatrix} x \\ 0 \end{bmatrix}, \qquad Bu = \overline{A\&B}\begin{bmatrix} 0 \\ u \end{bmatrix};$$

here A represents the extension of the original generator A to an operator $X \to X_{-1}$.

In order to get a similar splitting of $C\&D$ into $C\&D = \begin{bmatrix} \overline{C} & D \end{bmatrix}$ we need to extend $C\&D$ in a similar fashion. This extension is more difficult since we cannot, in general, replace the original range space Y of $C\&D$ be a larger space Y_{-1}. For example, if Y is finite-dimensional, then there is no natural candidate for the space Y_{-1}. The smallest possible domain of the extended operator $\overline{C\&D}$ is $Z \times U$, where Z is defined as follows. We choose any γ in the resolvent set of A, and let

$$Z = \big\{ z \in X \mid z = (\gamma I - A)^{-1}(x + Bu) \\ \text{for some } x \in X \text{ and } u \in U \big\}. \tag{6}$$

This is a Banach space with the norm

$$|z|_Z = \inf_{(\gamma I - A)^{-1}(x+Bu)=z} \big(|x|_X^2 + |u|_U^2 \big)^{1/2},$$

satisfying $X_1 \subset Z \subset X$, and it is a Hilbert space if both X and U are Hilbert spaces. It is easy to see that $\mathcal{D}(S_\Sigma) \subset Z \times U$, but the embedding $\mathcal{D}(S_\Sigma) \subset Z \times U$ need not be dense.

Definition 3.7. *The well-posed linear system* $\Sigma = \left[\begin{smallmatrix} \mathfrak{A} & \mathfrak{B} \\ \mathfrak{C} & \mathfrak{D} \end{smallmatrix}\right]$ *is* compatible *if its combined observation/feedthrough operator $C\&D$ can be extended to an operator $\overline{C\&D} \in \mathcal{L}(Z \times U; Y)$. We define the corresponding extended observation operator*

$\overline{C} \in \mathcal{L}(Z;Y)$ *and feedthrough operator* $D \in \mathcal{L}(U;Y)$ *by*

$$\begin{aligned} \overline{C} &= \overline{C\&D}\begin{bmatrix} x \\ 0 \end{bmatrix}, \quad x \in Z, \\ Du &= \overline{C\&D}\begin{bmatrix} 0 \\ u \end{bmatrix}, \quad u \in U. \end{aligned} \tag{7}$$

The extension of $C\&D$ to $Z \times U$ need not be unique since $\mathcal{D}(S_\Sigma)$ not be dense in $Z \times U$. This means that $\overline{C}$ and D need not be unique either. However, there is a one-to-one correspondence between $\overline{C\&D}$, $\overline{C}$ and D, i.e., any one of these three operators determine the other two uniquely.

In spite of the possible non-uniqueness of the extended observation operator $\overline{C}$ and the corresponding feedthrough operator D, independently of how these operators are chosen, it is still true that the formula for the output y in Theorem 3.5 simplifies into

$$y(t) = \overline{C}x(t) + Du(t), \quad t \geq 0,$$

and the formula for the transfer function given in Theorem 3.3 simplifies into

$$\widehat{\mathfrak{D}}(z) = \overline{C}(zI - A)^{-1}B + D. \qquad \Re z > \omega_{\mathfrak{A}}.$$

In particular, the formula (3) holds whenever $u \in W^{1,p}([0,t];U)$ and $\left[\begin{smallmatrix} x_0 \\ u(0) \end{smallmatrix}\right] \in \mathcal{D}(S_\Sigma)$.

It has for some time been considered an open question among specialists whether every L^p-well-posed linear system is compatible. Recently it was discovered that the answer to this question is positive, at least in the Hilbert space case.

Theorem 3.8. *Let* Σ *be a* L^p*-well-posed linear system on* (Y, X, U). *Then* Σ *is compatible in (at least) the following cases:*

(i) X *and* U *are Hilbert spaces;*

(ii) *At least one of the spaces* X, U, *or* Y *is finite-dimensional.*

The more difficult part (i) of this theorem was proved in [26], and (the easy) part (ii) in [25].

4. The Lax–Phillips Scattering Model

Instead of using a L^p-well-posed linear system to formalize the idea of having an output and state at time $t > 0$ which depend continuously on an input and the initial state we can proceed in a different way which leads to a generalized *Lax–Phillips scattering model.* This is a semigroup $\mathbf{T}$ defined on $\mathcal{Y} \times X \times \mathcal{U} = L^p_\omega(\mathbf{R}^-;Y) \times X \times L^p_\omega(\mathbf{R}^+;U)$ with certain additional properties. (Here $L^p_\omega(\mathbf{R}^-;Y)$ consists of all the functions $y\colon \mathbf{R}^- \to Y$ for which $t \mapsto e^{-\omega t}y(t)$ belongs to $L^p(\mathbf{R}^-;Y)$ and similarly for $L^p_\omega(\mathbf{R}^+;U)$.) We call $\mathcal{U}$ the *incoming subspace*, X the *central state space*, and $\mathcal{Y}$ the *outgoing subspace.* In the classical cases treated in [14, 15] ω is taken to be zero and $\mathbb{T}$ is required to be unitary (the conservative case) or a contraction semigroup (the nonconservative case).

We claim that there is a one-to-one correspondence between the class of all well-posed linear systems and the class of all Lax–Phillips models. The parameter $\omega \in \mathbf{R}$ can be chosen in an arbitrary way (the best choice depends on the particular application).

Let $\Sigma = \left[\begin{smallmatrix} \mathfrak{A} & \mathfrak{B} \\ \mathfrak{C} & \mathfrak{D} \end{smallmatrix}\right]$ be a given L^p-well-posed linear system. To each such system we construct a Lax–Phillips model $\mathbb{T}$ on $\mathcal{Y} \times X \times \mathcal{U}$ as follows. The initial data consists of the initial incoming state $u_0 \in \mathcal{U}$ representing the future values of the input, the initial central state $x_0 \in X$ is identical to the initial state of Σ, and the initial outgoing state $y_0 \in \mathcal{Y}$ represents the past values of the output. At time $t \geq 0$ the incoming state u_t is the left-shifted input $\tau_+^t u_0$ (the unused part of the input). The central state x_t at time t is equal to the state $x(t) = \mathfrak{A}^t x_0 + \mathfrak{B}_0^t u$ of Σ at time t with initial time zero, initial state x_0, and input u_0 (it depends only on x_0 and on the discarded part $\pi_{[0,t]}u$ of u). The outgoing state y_t at time t consists of two parts: it is the sum of $\tau_-^t y_0$ (the left-shifted original outgoing state) and $\tau^t \pi_{[0,t]}(\mathfrak{C}x_0 + \mathfrak{D}_0 u_0)$ (the restriction of the output $\mathfrak{C}x_0 + \mathfrak{D}_0 u_0$ of Σ to the interval $[0, t]$ shifted to the left by τ^t so that the support of the shifted and truncated output is $(-t, 0)$). Formalizing this idea we get the following theorem, where we use the notations

$$\mathfrak{B}_0^t = \mathfrak{B}\tau^t \pi_{[0,t]}, \qquad \mathfrak{C}_0^t = \pi_{[0,t]}\mathfrak{C}, \qquad \mathfrak{D}_0^t = \pi_{[0,t]}\mathfrak{D}\pi_{[0,t]}.$$

Theorem 4.1. *Let $\omega \in \mathbf{R}$, $\mathcal{Y} = L^2_\omega(\mathbf{R}^-; Y)$ and $\mathcal{U} = L^2_\omega(\mathbf{R}^+; U)$. For all $t \geq 0$ we define on $\mathcal{Y} \times X \times \mathcal{U}$ the operator $\mathbb{T}^t$ by*

$$\mathbb{T}^t = \begin{bmatrix} \tau^t & 0 & 0 \\ 0 & I & 0 \\ 0 & 0 & \tau_+^t \end{bmatrix} \begin{bmatrix} I & \mathfrak{C}_0^t & \mathfrak{D}_0^t \\ 0 & \mathfrak{A}^t & \mathfrak{B}_0^t \\ 0 & 0 & I \end{bmatrix}.$$

Then $\mathbb{T}$ is a strongly continuous semigroup. If x and y are the state trajectory and the output function of Σ corresponding to the initial state $x_0 \in X$ and the input function $u_0 \in \mathcal{U}$, and if we define $y(t) = y_0(t)$ for $t < 0$, then for all $t \geq 0$,

$$\begin{bmatrix} \pi_{(-\infty,t]}y \\ x(t) \\ \pi_{[t,\infty)}u_0 \end{bmatrix} = \begin{bmatrix} \tau^{-t} & 0 & 0 \\ 0 & I & 0 \\ 0 & 0 & \tau^{-t} \end{bmatrix} \mathbb{T}^t \begin{bmatrix} y_0 \\ x_0 \\ u_0 \end{bmatrix}. \tag{8}$$

Formula (8) shows that at any time $t \geq 0$, the *first component* of $\mathbb{T}_t \left[\begin{smallmatrix} y_0 \\ x_0 \\ u_0 \end{smallmatrix}\right]$ represents the *past output*, the *second component* represents the *present state* and the *third component* represents the *future input.*

Here the strong continuity of $\mathbb{T}$ is obvious, and so is the property $\mathbb{T}(0) = I$. The proof of the semigroup property $\mathbb{T}(s + t) = \mathbb{T}(s)\mathbb{T}(t)$ for s, $t \geq 0$ is a short algebraic computation based on Definition 3.1 (see [26] or [25] for details).

The semigroup $\mathbb{T}$ in Theorem 4.1 has an additional 'causality' property, which in the Hilbert space case where $p = 2$ and U, X, and Y are Hilbert spaces can be described as follows: for all $t \geq 0$, the images of the central and incoming states under $\mathbb{T}^t$ are orthogonal to the image of the outgoing state, and the null space of $\mathbb{T}^t$ projected onto the central and outgoing spaces is orthogonal to the null space of $\mathbb{T}^t$ projected onto the incoming space. In the general case these properties can easiest be characterized in the following way.

Definition 4.2. *A* Lax–Phillips model *of type L^p_ω is a semigroup on $\mathcal{Y} \times X \times \mathcal{U} = L^p_\omega(\mathbf{R}^-; Y) \times X \times L^p_\omega(\mathbf{R}^+; U)$ with the structure*

$$\mathbb{T}^t = \begin{bmatrix} \tau^t_- & \mathbb{C}^t & \mathbb{D}^t \\ 0 & \mathbf{A}^t & \mathbb{B}^t \\ 0 & 0 & \tau^t_+ \end{bmatrix}, \tag{9}$$

where $\mathbf{A}$ is strongly continuous and $\mathbb{B}^t$, $\mathbb{C}^t$, and $\mathbb{D}^t$ satisfy the causality conditions

$$\begin{aligned} \mathbb{C}^t &= \pi_{(-t,0)}\mathbb{C}^t, & \mathbb{D}^t &= \pi_{(-t,0)}\mathbb{D}^t, \\ \mathbb{D}^t &= \mathbb{D}^t\pi_{[0,t]}, & \mathbb{B}^t &= \mathbb{B}^t\pi_{[0,t]}. \end{aligned} \tag{10}$$

This set of conditions is a rewritten version of conditions (1.2) in [15]. Helton [10] uses the name *inertness* for this additional causality property.

Corollary 4.3. *The semigroup $\mathbb{T}$ constructed in Theorem 4.1 is a Lax–Phillips model of type L^p_ω.*

This is immediate from Theorem 4.1 and Definition 4.2. We call the semigroup $\mathbb{T}$ in Theorem 4.1 the *Lax–Phillips model (of type L^p_ω) induced by* Σ.

It is only slightly more difficult to prove a converse to Corollary 4.3: To every Lax–Phillips model there corresponds a well-posed linear system which induces this Lax–Phillips model:

Theorem 4.4. *Let $\mathbb{T}$ be a Lax–Phillips model of type L^p_ω. With the notations of Definition 4.2, let*

$$\begin{aligned} \mathfrak{A} &= \mathbf{A}, & \mathfrak{B} &= \lim_{s\to\infty} \mathbb{B}^s\tau^{-s}, \\ \mathfrak{C} &= \lim_{t\to\infty} \tau^{-t}\mathbb{C}^t, & \mathfrak{D} &= \lim_{\substack{t\to\infty \\ s\to\infty}} \tau^{-s}\mathbb{D}^{s+t}\tau^{-t}. \end{aligned} \tag{11}$$

Then $\Sigma = \left[\begin{array}{c|c} \mathfrak{A} & \mathfrak{B} \\ \hline \mathfrak{C} & \mathfrak{D} \end{array}\right]$ is an L^p-well-posed linear system on (Y, X, U), and $\mathbb{T}$ is the Lax–Phillips model induced by this system.

The proof of Theorem 4.4 is another algebraic computation given in [25].

Corollary 4.5. *For each $\omega \in \mathbf{R}$ and $1 \le p < \infty$, there is a one-to-one correspondence between the class of all L^p-well-posed linear systems and all Lax–Phillips models of type L^p_ω: every L^p-well-posed linear system Σ induces a unique Lax–Phillips model $\mathbb{T}$ of type L^p_ω, and conversely, every Lax–Phillips model $\mathbb{T}$ of type L^p_ω induces a unique L^p-well-posed linear system Σ.*

This is a union of Corollary 4.3 and Theorem 4.4. Parts of this corollary (where either the input operator or output operator vanishes) were proved by

Grabowski and Callier [7] and by Engel [6]. It is also (implicitly) contained in [4] and mentioned in [3].

Our next theorem describes the generator of the Lax–Phillips semigroup:

Theorem 4.6. *Let $1 \le p < \infty$ and $\omega \in \mathbf{R}$, let $\Sigma = \left[\begin{smallmatrix} \mathfrak{A} & \mathfrak{B} \\ \mathfrak{C} & \mathfrak{D} \end{smallmatrix}\right]$ be a L^p-well-posed linear system on (Y, X, U) with system operator $S_\Sigma = \left[\begin{smallmatrix} A & B \\ C\&D \end{smallmatrix}\right]$, and let T be the generator of the corresponding Lax–Phillips model $\mathbb{T}$ of type L^p_ω defined in Definition 4.2.*

(i) *The domain of T consists of all the vectors $\left[\begin{smallmatrix} y_0 \\ x_0 \\ u_0 \end{smallmatrix}\right] \in W^{1,p}_\omega(\mathbf{R}^-; Y) \times X \times W^{1,p}_\omega(\mathbf{R}^+; U)$ which satisfy $\left[\begin{smallmatrix} x_0 \\ u_0(0) \end{smallmatrix}\right] \in \mathcal{D}(S_\Sigma)$ and $y_0(0) = C\&D \left[\begin{smallmatrix} x_0 \\ u_0(0) \end{smallmatrix}\right]$, and on its domain T is given by*

$$T \begin{bmatrix} y_0 \\ x_0 \\ u_0 \end{bmatrix} = \begin{bmatrix} y_0' \\ Ax_0 + Bu_0(0) \\ u_0' \end{bmatrix}.$$

Thus, the following three conditions are equivalent (here $\hat{u}$ and $\hat{u}_0$ are the Laplace transforms of u and u_0, and $\hat{y}$ and $\hat{y}_0$ are the left-sided Laplace transforms of y and y_0):

(a) $\left[\begin{smallmatrix} y_0 \\ x_0 \\ u_0 \end{smallmatrix}\right] \in \mathcal{D}(T)$ *and* $\left[\begin{smallmatrix} y \\ x \\ u \end{smallmatrix}\right] = T \left[\begin{smallmatrix} y_0 \\ x_0 \\ u_0 \end{smallmatrix}\right]$;

(b) $y_0 \in W^{1,p}_\omega(\mathbf{R}^-; Y)$, $x_0 \in X$, $u_0 \in W^{1,p}_\omega(\mathbf{R}^+; U)$, $\left[\begin{smallmatrix} x_0 \\ u_0(0) \end{smallmatrix}\right] \in \mathcal{D}(S_\Sigma)$ *and*

$$\begin{bmatrix} x \\ y_0(0) \end{bmatrix} = S_\Sigma \begin{bmatrix} x_0 \\ u_0(0) \end{bmatrix}, \qquad \begin{bmatrix} y \\ u \end{bmatrix} = \begin{bmatrix} y_0' \\ u_0' \end{bmatrix}.$$

(c) $y_0 \in W^{1,p}_\omega(\mathbf{R}^-; Y)$, $x_0 \in X$, $u_0 \in W^{1,p}_\omega(\mathbf{R}^+; U)$, $\left[\begin{smallmatrix} x_0 \\ u_0(0) \end{smallmatrix}\right] \in \mathcal{D}(S_\Sigma)$ *and*

$$\begin{aligned} \begin{bmatrix} x \\ y_0(0) \end{bmatrix} &= S_\Sigma \begin{bmatrix} x_0 \\ u_0(0) \end{bmatrix}, \\ \hat{y}(z) &= z\hat{y}_0(z) - y_0(0), \qquad && \Re z < \omega, \\ \hat{u}(z) &= z\hat{u}_0(z) + u_0(0), \qquad && \Re z > \omega. \end{aligned}$$

(ii) *The spectrum of T contains the vertical line $\{\Re\alpha = \omega\}$. A point α with $\Re\alpha > \omega$ belongs to the spectrum of T iff it belongs to the spectrum of A, and a point α whith $\Re\alpha < \omega$ belongs to the spectrum of T iff $\left[\begin{smallmatrix} \alpha I - A & -B \\ -C\&D \end{smallmatrix}\right]$ is not invertible.*

(iii) *Let $\alpha \in \rho(T)$ with $\Re\alpha > \omega$ and let $\left[\begin{smallmatrix} y \\ x \\ u \end{smallmatrix}\right] \in L^2_\omega(\mathbf{R}^-; Y) \times X \times L^2_\omega(\mathbf{R}^+; U)$. Denote $\widehat{\mathfrak{D}}(\alpha) = C\&D \left[\begin{smallmatrix} (\alpha I - A)^{-1} B \\ I \end{smallmatrix}\right]$. Then the following three conditions are equivalent:*

(a) $\left[\begin{smallmatrix} y_0 \\ x_0 \\ u_0 \end{smallmatrix}\right] = (\alpha I - T)^{-1} \left[\begin{smallmatrix} y \\ x \\ u \end{smallmatrix}\right]$;

(b) $$\begin{cases} \begin{bmatrix} x_0 \\ y_0(0) \end{bmatrix} = \begin{bmatrix} (\alpha I - A)^{-1} & (\alpha I - A)^{-1}B \\ C(\alpha I - A)^{-1} & \widehat{\mathfrak{D}}(\alpha) \end{bmatrix} \begin{bmatrix} x \\ \hat{u}(\alpha) \end{bmatrix}, \\ y_0(t) = e^{\alpha t} y_0(0) + \int_t^0 e^{\alpha(t-s)} y(s)\, ds, \qquad t \le 0, \\ u_0(t) = \int_t^\infty e^{\alpha(t-s)} u(s)\, ds, \qquad t \ge 0. \end{cases}$$

(c) $$\begin{cases} \begin{bmatrix} x_0 \\ y_0(0) \end{bmatrix} = \begin{bmatrix} (\alpha I - A)^{-1} & (\alpha I - A)^{-1}B \\ C(\alpha I - A)^{-1} & \widehat{\mathfrak{D}}(\alpha) \end{bmatrix} \begin{bmatrix} x \\ \hat{u}(\alpha) \end{bmatrix}, \\ \hat{y}_0(z) = \dfrac{\hat{y}(z) + y_0(0)}{\alpha - z}, \qquad \Re z < \omega, \\ \hat{u}_0(z) = \dfrac{\hat{u}(z) - \hat{u}(\alpha)}{\alpha - z}, \qquad \Re z > \omega. \end{cases}$$

(iv) *Let $\alpha \in \rho(T)$ with $\Re\alpha < \omega$ and let $\left[\begin{smallmatrix} y \\ x \\ u \end{smallmatrix}\right] \in L^2_\omega(\mathbf{R}^-;Y) \times X \times L^2_\omega(\mathbf{R}^+;U)$. Then the following three conditions are equivalent:*

(a) $\left[\begin{smallmatrix} y_0 \\ x_0 \\ u_0 \end{smallmatrix}\right] = (\alpha I - T)^{-1} \left[\begin{smallmatrix} y \\ x \\ u \end{smallmatrix}\right]$;

(b) $$\begin{cases} \begin{bmatrix} x_0 \\ u_0(0) \end{bmatrix} = \begin{bmatrix} \alpha I - A & -B \\ & -C\&D \end{bmatrix}^{-1} \begin{bmatrix} x \\ \hat{y}(\alpha) \end{bmatrix}, \\ y_0(t) = -\int_{-\infty}^t e^{\alpha(t-s)} y(s)\, ds, \qquad t \le 0, \\ u_0(t) = e^{\alpha t} u_0(0) - \int_0^t e^{\alpha(t-s)} u(s)\, ds, \qquad t \ge 0. \end{cases}$$

(c) $$\begin{cases} \begin{bmatrix} x_0 \\ u_0(0) \end{bmatrix} = \begin{bmatrix} \alpha I - A & -B \\ & -C\&D \end{bmatrix}^{-1} \begin{bmatrix} x \\ \hat{y}(\alpha) \end{bmatrix}, \\ \hat{y}_0(z) = \dfrac{\hat{y}(z) - \hat{y}(\alpha)}{\alpha - z}, \qquad \Re z < \omega, \\ \hat{u}_0(z) = \dfrac{\hat{u}(z) + u_0(0)}{\alpha - z}, \qquad \Re z > \omega. \end{cases}$$

The proof of this theorem is given in [26] (and also in [25]).

There are a number of important ingredients in the Lax–Phillips scattering theory, such as the backward and forward wave operators, the scattering operator, and the scattering matrix. All of these have natural analogies in the theory of well-posed linear systems. In the discussion below we choose $\omega > \omega_{\mathfrak{A}}$, where $\omega_{\mathfrak{A}}$ is the growth rate of $\mathfrak{A}$.

The *backward wave operator* W_- (denoted by W_2 in [15, Theorem 1.2]) is the limit of the last column of $\mathbf{T}_{\mathcal{T}}^{-t}$ as $t \to \infty$. It maps $L^p_\omega(\mathbf{R};U)$ into $L^p_\omega(\mathbf{R}^-;Y) \times X \times L^p_\omega(\mathbf{R}^+;U)$, and it is given by (cf. Theorem 4.4)

$$W_- u = \begin{bmatrix} \pi_-\mathfrak{D} \\ \mathfrak{B} \\ \pi_+ \end{bmatrix} u. \tag{12}$$

Thus, it keeps the future input $\pi_+ u$ intact, and maps the past input $\pi_- u$ into the past output $\pi_- \mathfrak{D} u$ and the present central state $\mathfrak{B} u$.

The *forward wave operator* W_+ (denoted by W_1 in [15, Theorem 1.2]) is the limit of the first row of $\tau^{-t}\mathbb{T}$ as $t \to \infty$. It maps $L^p_\omega(\mathbf{R}^-; Y) \times X \times L^p_\omega(\mathbf{R}^+; U)$ into $L^p_\omega(\mathbf{R}; Y)$, and it is given by (cf. Theorem 4.4)

$$W_+ \begin{bmatrix} y \\ x_0 \\ u \end{bmatrix} = \begin{bmatrix} \pi_- & \mathfrak{C} & \mathfrak{D}\pi_+ \end{bmatrix} \begin{bmatrix} y \\ x_0 \\ u \end{bmatrix}. \tag{13}$$

Thus, it keeps the past output $\pi_- y$ intact, and maps the present central state x_0 and the future input $\pi_+ u$ into the future output $\mathfrak{C} x_0 + \mathfrak{D}\pi_+ u$.

The *scattering operator* in Lax–Phillips theory is the product $W_+ W_-$, and it is given by

$$W_+ W_- = \begin{bmatrix} \pi_- & \mathfrak{C} & \mathfrak{D}\pi_+ \end{bmatrix} \begin{bmatrix} \pi_- \mathfrak{D} \\ \mathfrak{B} \\ \pi_+ \end{bmatrix} = \pi_- \mathfrak{D} + \mathfrak{C}\mathfrak{B} + \pi_+ \mathfrak{D} = \mathfrak{D}. \tag{14}$$

Thus, the scattering operator is nothing but the (bilaterally shift-invariant) input-output map $\mathfrak{D}$ of the corresponding well-posed linear system.

To get the *scattering matrix* of the Lax–Phillips system we apply the scattering operator $\mathfrak{D}$ to an input of the form $u(t) = e^{zt} u_0$, where $z \in \mathbf{C}$ has a sufficiently large real part and $u_0 \in U$ is fixed; see [15, pp. 187–188]. Because of the shift-invariance of $\mathfrak{D}$, the resulting output is of the type $y(t) = e^{zt} y_0$ for some $y_0 \in Y$. The scattering matrix (evaluated at z) is defined to be the operator that maps $u_0 \in U$ into $y_0 \in Y$. It follows from [30, p. 194] that the scattering matrix of a Lax–Phillips system is equal to the transfer function $\widehat{\mathfrak{D}}$ of the corresponding well-posed linear system.

In their study of the *conservative* case, Lax and Phillips [14] assume some additional *controllability* and *observability* properties of the system:

(i) The *image of the incoming subspace* $\mathcal{U}$ under $\mathbf{T}^t$, $0 \le t < \infty$, is *dense* in the state space $\mathcal{Y} \times X \times \mathcal{U}$.
(ii) If the *projection* of a trajectory onto the *outgoing subspace* $\mathcal{Y}$ *vanishes*, then the trajectory is *identically zero.*

These additional controllability and observability assumptions imply the following additional conclusions (in the conservative case):

(i) Both $\mathfrak{A}^t$ and $\mathfrak{A}^{*t}$ *tend strongly to zero* as $t \to \infty$,
(ii) $\widehat{\mathfrak{D}}$ is *inner from both sides*, i.e., $\widehat{\mathfrak{D}}(i\omega)$ is unitary for almost all real ω,
(iii) Both the backward and the forward *wave operators are unitary*,
(iv) The system is *exactly controllable and exactly observable* in infinite time.

Without the additional controllability and observability assumptions none of the additional conclusions listed above need hold.

According to [4], *every contractive H^∞-function over $\mathbf{C}_0$* (= the right half-plane) is the *transfer function of some unitary system* (which need not be controllable or observable). It is also the transfer function of some *controllable and observable dissipative system.* For more details on conservative and dissipative systems, see [4], [14, 15], and [35]

5. Admissibility

According to Corollary 4.5, there is a one-to-one correspondence between the class of all L^p-well-posed linear systems and all Lax–Phillips scattering models of type L^p. This means that we can reduce the study of the generators of a well-posed linear system to the study of the generators the Lax–Phillips semigroup. This way we can obtain necessary and sufficient conditions for the admissibility or joint admissibility of a control operator B and an observation operator C. These notions are defined as follows.

As always we let U, X, and Y be Banach spaces and let $1 \le p < \infty$. We let $\mathfrak{A}$ be a strongly continuous semigroup on the Banach space X with generator A, and define the spaces X_1 and X_{-1} as in Section 2. This time we specify, in addition, some $\omega \in \mathbf{R}$, and suppose that $\mathfrak{A}$ is ω-bounded, i.e., $\sup_{t>0} \mathrm{e}^{-\omega t}\|\mathfrak{A}^t\| < \infty$.

An operator $B \in \mathcal{L}(U; X_{-1})$ is an *L^p-admissible control operator for A* if for some $t > 0$ (hence for all $t > 0$) the operator

$$\mathfrak{B}_0^t u = \int_0^t \mathfrak{A}^{t-s} B u(s)\, ds, \qquad u \in L^p([0,t]; U), \tag{15}$$

maps $L^p([0,t];U)$ into X (instead of X_{-1}). (This operator is then bounded with values in X). We call B *ω-bounded* if the resulting input map

$$\mathfrak{B}u = \lim_{v \to -\infty} \int_v^0 \mathfrak{A}^{-s} B u(s)\, ds, \qquad u \in L^p_\omega(\mathbf{R}^-; U) \tag{16}$$

is ω-bounded, i.e., it defines a bounded linear operator from $L^p_\omega(\mathbf{R}^-;U)$ to X.

The operator $C \in \mathcal{L}(X_1; Y)$ is an *L^p-admissible observation operator for A* if the map

$$(\mathfrak{C}x)(t) = C\mathfrak{A}^t x, \qquad x \in X_1, \qquad t \ge 0, \tag{17}$$

can be extended to a bounded linear operator $X \to L^p_{\mathrm{loc}}(\mathbf{R}^+;Y)$, and it is ω-bounded if the resulting output map $\mathfrak{C}$ is ω-bounded, i.e., it maps X into $L^p_\omega(\mathbf{R}^+;Y)$.

The operators $B \in \mathcal{L}(U; X_{-1})$ and $C \in \mathcal{L}(X_1; Y)$ are *jointly L^p-admissible for A* if B is an L^p-admissible control operator for A, C is an L^p-admissible observation operator for A, and the operator $\mathfrak{D}\colon W^{1,p}_{c,\mathrm{loc}}(\mathbf{R};U) \to C_c(\mathbf{R};Y)$ defined by

$$(\mathfrak{D}u)(t) = C\big[\mathfrak{B}\tau^t u - (\alpha I - A)^{-1} B u(t)\big] + D_\alpha u(t), \qquad t \in \mathbf{R}, \tag{18}$$

can be extended to a continuous operator $L^p_{c,\mathrm{loc}}(\mathbf{R};U) \to L^p_{c,\mathrm{loc}}(\mathbf{R};Y)$. Here $\alpha \in \rho(A)$ and $D_\alpha \in \mathcal{L}(U;Y)$ can be chosen in an arbitrary way. By introducing the combined observation/feedthrough operator

$$C\&D \begin{bmatrix} x \\ u \end{bmatrix} = C\big[x - (\alpha I - A)^{-1}Bu\big] + D_\alpha u. \tag{19}$$

we can simplify the formula for $(\mathfrak{D}u)(t)$ into

$$(\mathfrak{D}u)(t) = C\&D \begin{bmatrix} \mathfrak{B}\tau^t u \\ u(t) \end{bmatrix}, \qquad t \in \mathbf{R}. \tag{20}$$

We call B and C *jointly ω-bounded* if both B and C are ω-bounded and, in addition, the operator $\mathfrak{D}$ can be extended to a bounded linear operator from $L^p_\omega(\mathbf{R};U)$ to $L^p_\omega(\mathbf{R};Y)$. If (and only if) B and C are jointly admissible, then the four operator $\mathfrak{A}$, $\mathfrak{B}$, $\mathfrak{C}$, and $\mathfrak{D}$ can be combined into a L^p-well-posed linear system $\left[\begin{smallmatrix} \mathfrak{A} & \mathfrak{B} \\ \mathfrak{C} & \mathfrak{D} \end{smallmatrix}\right]$ with system operator $\left[\begin{smallmatrix} A & B \\ \multicolumn{2}{c}{C\&D} \end{smallmatrix}\right]$. (Here $\mathfrak{D}$ is determined by A, B, and C only modulo a constant static term.)

Before looking at the general case of L^p-admissibility, let us treat the important special case where $p = 2$ and U, X, and Y are Hilbert spaces. In this case there is a very simple characterization of the class of all L^2-well-posed ω-bounded transfer function:

Proposition 5.1. *Let U and Y be Hilbert spaces. A $\mathcal{L}(U;Y)$-valued function $\widehat{\mathfrak{D}}$ defined on $\Re\lambda > \omega$ is the transfer function of an ω-bounded L^2-well-posed linear system if and only if it is analytic and bounded on $\Re\lambda > \omega$ (i.e., it belongs to H^∞).*

This was proved independently by (at least) Salamon [18] and Curtain and Weiss [5].

The admissibility of a control operator B or an observation operator C is much more delicate in this case. In 1990 George Weiss [31] made the following conjecture:

Conjecture 5.2. *Let U, X, and Y be Hilbert spaces, and let A generate a C_0 semigroup on X. Then*

(i) *$B \in \mathcal{L}(U;X_{-1})$ is an L^2-admissible ω-bounded control operator for A if and only if there is a constant $K > 0$ such that*

$$\|(\lambda I - A)^{-1}B\| \le \frac{K}{\sqrt{\Re\lambda - \omega}}, \qquad \Re\lambda > \omega.$$

(ii) *$C \in \mathcal{L}(X_1;Y)$ is an L^2-admissible ω-bounded observation operator for A if and only if there is a constant $K > 0$ such that*

$$\|C(\lambda I - A)^{-1}\| \le \frac{K}{\sqrt{\Re\lambda - \omega}}, \qquad \Re\lambda > \omega.$$

It is easy to see that the given conditions are necessary. These two conjectures are dual of each other, so it suffices to prove or disprove one of them.

It was discovered recently by Zwart and Jacob [37] that Weiss' conjecture is false in general. It is not true even if we restrict the dimensions of U and Y to be one (see [13]) or if we require the semigroup to be a contraction semigroup (see [12]). However, it is true in several special cases. For example, the second conjecture about the observation operator is known to be true in the following special cases (here we denote the semigroup generated by A by $t \mapsto \mathfrak{A}^t$ and take $\omega = 0$):

(i) Y is finite-dimensional and $\mathfrak{A}$ is normal [8], [9], [31], [34],
(ii) Y is finite-dimensional and $\mathfrak{A}$ is the right-shift on $L^2(\mathbf{R}^+)$ [16],
(iii) Y is finite-dimensional and $\mathfrak{A}$ is a contraction semigroup [11],
(iv) $\mathfrak{A}$ is exponentially stable and $\mathfrak{A}^t$ is right-invertible for some (hence all) $t > 0$ [31] [34].

Let us now return to the general case of L^p-admissibility and Banach spaces. By applying the Hille–Yoshida theorem to the semigroup in Corollary 4.5 we get the following necessary and sufficient conditions for admissibility:

Theorem 5.3. *Let $\omega \in \mathbf{R}$, $1 \le p < \infty$, and let A be the generator of an ω-bounded C_0 semigroup on X.*

(i) *$B \in \mathcal{L}(U; X_{-1})$ is an L^p-admissible ω-bounded control operator for A if and only if there is a constant $M > 0$ such that, for all $u \in L^p_\omega(\mathbf{R}^+; U)$, $\lambda > \omega$, and $n = 0, 1, 2, \ldots$,*

$$\left| \frac{\partial^n}{\partial \lambda^n} (\lambda I - A)^{-1} B \hat{u}(\lambda) \right|_X \le \frac{M n!}{(\lambda - \omega)^{n+1}} \|u\|_{L^p_\omega(\mathbf{R}^+;U)}. \tag{21}$$

(ii) *$C \in \mathcal{L}(X_1; Y)$ is an L^p-admissible ω-bounded observation operator for A if and only if there is a constant $M > 0$ such that, for all $x_0 \in X$, $\lambda > \omega$, and $n = 0, 1, 2, \ldots$,*

$$\left(\int_0^\infty \left| \frac{\partial^n}{\partial \lambda^n} \mathrm{e}^{-(\lambda-\omega)t} C(\lambda I - A)^{-1} x_0 \right|_Y^p dt \right)^{1/p} \le \frac{M n!}{(\lambda - \omega)^{n+1}} |x_0|_X. \tag{22}$$

(iii) *The operators $B \in \mathcal{L}(U; X_{-1})$ and $C \in \mathcal{L}(X_1; U)$ are jointly L^p admissible and ω-bounded iff B is an L^p-admissible ω-bounded control operator for A (cf. (i)), C is an admissible ω-bounded observation operator for A (cf. (ii)) and there is a constant $M > 0$ such that, for all $u \in L^p_\omega(\mathbf{R}^+; U)$, $\lambda > \omega$, and $n = 0, 1, 2, \ldots$,*

$$\left(\int_0^\infty \left| \frac{\partial^n}{\partial \lambda^n} \mathrm{e}^{-(\lambda-\omega)t} \widehat{\mathfrak{D}}(\lambda) \hat{u}(\lambda) \right|_Y^p dt \right)^{1/p} \le \frac{M n!}{(\lambda - \omega)^{n+1}} \|u\|_{L^p_\omega(\mathbf{R}^+;U)}, \tag{23}$$

where

$$\widehat{\mathfrak{D}}(\lambda) = (\alpha - \lambda) C(\lambda I - A)^{-1} (\alpha I - A)^{-1} B + D_\alpha; \tag{24}$$

here α with $\Re\alpha > \omega$ and $D_\alpha \in \mathcal{L}(U; Y)$ can be chosen in an arbitrary manner.

Part (ii) of this theorem was proved by Grabowski and Callier [7] in the exponentially stable Hilbert space case (i.e., $\mathfrak{A}$ is exponentially stable, $\omega = 0$, $p = 2$, and U, X, and Y are Hilbert spaces), and the corresponding case of part (i) can be derived from (ii) by duality. The general case of parts (i) and (ii) was proved by Engel [6]. Part (iii) may (or may not) be new. A proof of the full theorem is given in [25].

Condition (23) does not depend on the particular *realization* $\left[\begin{smallmatrix}\mathfrak{A} & \mathfrak{B}\\ \mathfrak{C} & \mathfrak{D}\end{smallmatrix}\right]$ of $\mathfrak{D}$, i.e., it does not contain any direct references to $\mathfrak{A}$, $\mathfrak{B}$, and $\mathfrak{C}$, but only to $\widehat{\mathfrak{D}}$ which is completely determined by $\mathfrak{D}$. This indicates that the following conjecture may be true:

Conjecture 5.4. *An analytic $\mathcal{L}(U;Y)$-valued function $\widehat{\mathfrak{D}}$ on $\Re\lambda > \omega$ is the transfer function of an ω-bounded L^p-well-posed linear system if and only if there is a constant $M > 0$ such that* (23) *holds for all $u \in L^p_\omega(\mathbf{R}^+;U)$, $\lambda > \omega$, and $n = 0, 1, 2, \ldots$.*

Clearly, by Theorem 5.3, condition (23) is necessary for $\widehat{\mathfrak{D}}$ to be a L^p-well-posed ω-bounded transfer function, and we conjecture that it is also sufficient. This would give us a necessary and sufficient condition for an H^∞ function to be an L^p-multiplier.

References

[1] A. Adamajan and Damir Z. Arov, *On unitary couplings of semiunitary operators*, In *Eleven Papers in Analysis*, American Mathematical Society Translations, Vol. 95, pp. 75–129, American Mathematical Society, Providence, R.I., 1970.

[2] Damir Z. Arov, *Passive linear stationary dynamic systems*, Siberian Math. J., **20** (1979), 149–162.

[3] Damir Z. Arov, *Passive linear systems and scattering theory*, In *Dynamical Systems, Control Coding, Computer Vision*, Progress in Systems and Control Theory, Vol. 25, pp. 27–44, Birkhäuser Verlag, Basel, Boston, Berlin, 1999.

[4] Damir Z. Arov and Michael A. Nudelman, *Passive linear stationary dynamical scattering systems with continuous time*, Integral Equations Operator Theory, **24** (1996), 1–45.

[5] Ruth F. Curtain and George Weiss, *Well posedness of triples of operators (in the sense of linear systems theory)*, In *Control and Optimization of Distributed Parameter Systems*, International Series of Numerical Mathematics, Vol. 91, pp. 41–59, Birkhäuser-Verlag, Basel, 1989.

[6] Klaus-J. Engel, *On the characterization of admissible control and observation operators*, Systems Control Lett., **34** (1998), 225–227.

[7] Piotr Grabowski and Frank M. Callier, *Admissible observation operators. Semigroup criteria of admissibility*, Integral Equations Operator Theory, **25** (1996), 182–198.

[8] Scott Hansen and George Weiss, *The operator Carleson measure criterion for admissibility of control operators for diagonal semigroups on l^2*, Systems Control Lett., **16** (1991), 219–227.

[9] Scott Hansen and George Weiss, *New results on the operator Carleson measure criterion*, IMA J. Math. Control Inform., **14** (1997), 3–32.

[10] J. William Helton, *Systems with infinite-dimensional state space: the Hilbert space approach*, Proceedings of the IEEE, **64** (1976), 145–160.

[11] Birgit Jacob and Jonathan R. Partington, *The Weiss conjecture on admissibility of observation operators for contraction semigroups*, To appear in Integral Equations and Operator Theory, 2000.

[12] Birgit Jacob, Jonathan R. Partington, and Sandra Pott, *Admissible and weakly admissible observation operators for the right shift semigroup*, Manuscript, 2000.

[13] Birgit Jacob and Hans Zwart, *Disproof of two conjectures of George Weiss* Memorandum 1546, Faculty of Mathematical Sciences, University of Twente, The Netherlands, 2000.

[14] P. D. Lax and R. S. Phillips, *Scattering Theory*. Academic Press, New York, 1967.

[15] P. D. Lax and R. S. Phillips, *Scattering theory for dissipative hyperbolic systems*, J. Funct. Anal., **14** (1973), 172–235.

[16] Jonathan R. Partington and George Weiss, *Admissible observation operators for the right shift semigroup*, Math. Control Signals Systems, **13** (2000), 179–192.

[17] Dietmar Salamon, *Infinite dimensional linear systems with unbounded control and observation: a functional analytic approach*, Trans. Amer. Math. Soc., **300** (1987), 383–431.

[18] Dietmar Salamon, *Realization theory in Hilbert space*, Math. Systems Theory, **21** (1989), 147–164.

[19] Olof J. Staffans, *Quadratic optimal control of stable well-posed linear systems*, Trans. Amer. Math. Soc., **349** (1997), 3679–3715.

[20] Olof J. Staffans, *Coprime factorizations and well-posed linear systems*, SIAM J. Control Optim., **36** (1998a), 1268–1292.

[21] Olof J. Staffans, *Quadratic optimal control of well-posed linear systems*, SIAM J. Control Optim., **37** (1998b), 131–164.

[22] Olof J. Staffans, *Feedback representations of critical controls for well-posed linear systems*, Internat. J. Robust Nonlinear Control, **8** (1998c), 1189–1217.

[23] Olof J. Staffans, *Admissible factorizations of Hankel operators induce well-posed linear systems*, Systems Control Lett., **37** (1999).

[24] Olof J. Staffans, *Lax–Phillips scattering and well-posed linear systems*, Proceedings of the 7th IEEE Mediterranean Conference on Control and Systems, CD-ROM, Haifa, Israel, July 28–30, 1999.

[25] Olof J. Staffans, *Well-Posed Linear Systems*, Book manuscript, 2000.

[26] Olof J. Staffans and George Weiss, *Transfer functions of regular linear systems. Part II: the system operator and the Lax-Phillips semigroup*, Manuscript, 2000.

[27] George Weiss, *Admissibility of unbounded control operators*, SIAM J. Control Optim., **27** (1989a), 527–545.

[28] George Weiss, *Admissible observation operators for linear semigroups*, Israel J. Math., **65** (1989b), 17–43.

[29] George Weiss, *The representation of regular linear systems on Hilbert spaces*, In *Control and Optimization of Distributed Parameter Systems*, International Series of Numerical Mathematics, Vol. 91, pp. 401–416, Birkhäuser-Verlag, Basel, 1989c.

[30] George Weiss, *Representations of shift-invariant operators on L^2 by H^∞ transfer functions: an elementary proof, a generalization to L^p, and a counterexample for* L^∞, Math. Control Signals Systems, **4** (1991), 193–203.

[31] George Weiss, *Two conjectures on the admissibility of control operators*, Control and Optimization of Distributed Parameter Systems, pp. 367–378, Birkhäuser-Verlag, Basel, 1991.

[32] George Weiss, *Transfer functions of regular linear systems. Part I: characterizations of regularity*, Trans. Amer. Math. Soc., **342** (1994a), 827–854.

[33] George Weiss, *Regular linear systems with feedback*, Math. Control Signals Systems, **7** (1994b), 23–57.

[34] George Weiss, *A powerful generalization of the Carleson measure theorem?*, Open Problems in Mathematical Systems and Control Theory, pp. 267–272, Spinger-Verlag, London, 1999.

[35] George Weiss, Olof J. Staffans, and Marius Tuchnak, *Well-posed linear systems—a survey with emphasis on conservative systems*, Manuscript, 2000.

[36] Martin Weiss and George Weiss, *Optimal control of stable weakly regular linear systems*, Math. Control Signals Systems, **10** (1997), 287–330.

[37] Hans Zwart and Birgit Jacob, *Disproof of an admissibility conjecture of Weiss*, Memorandum 1539, Faculty of Mathematical Sciences, University of Twente, The Netherlands, 2000.

Åbo Akademi University
Department of Mathematics
FIN-20500 Åbo, Finland
http://www.abo.fi/~staffans/
E-mail address: `Olof.Staffans@abo.fi`

Spectral theory for delay equations

Sjoerd M. Verduyn Lunel

Abstract. For dynamical systems governed by feedback laws, time delays arise naturally in the feedback loop to represent effects due to communication, transmission, transportation or inertia effects. The introduction of time delays in a system of differential equations results in an infinite dimensional state space. The solution operator associated with a differential delay equation is a nonself-adjoint operator defined on a Banach space. This implies that general abstract theorems cannot directly be applied. In this paper we discuss the spectral properties of autonomous and periodic differential delay equations, series expansions of solutions, completeness of eigenvectors and generalized eigenvectors, and solutions of delay equations that decay faster than any exponential.

Introduction

The aim of this paper is to present the spectral theory for autonomous and periodic differential delay equations. We start with an introduction to the theory of delay equations. The basic topics include the very definition, the state space approach and the solution operator. After this introduction, we turn to an analysis of the spectral properties of autonomous and periodic differential delay equations. Not only are differential delay equations a rich class of nonself-adjoint problems with very interesting spectral issues which serve as motivation for a general theory (cf. [29, 31, 25, 10]). They are also important in applications (cf. [7, 13] and the references given there). As motivation for the theory that we develop in this paper, we briefly discuss some examples.

In the implementation of any feedback control system, e.g., the control of partial differential equations through the application of forces on the boundary, it is very likely that time delays will occur. Therefore, it is of importance to understand the sensitivity of the control system with respect to the introduction of small delays in the feedback loop. For some systems, small delays lead to destabilization while other systems are robust with respect to small time delays. In [14] a first attempt was made for a unifying theory which explains the underlying mechanisms in terms of spectral properties of the solution operator.

The research was supported by an N.W.O. PIONIER-grant under 600-61-410.

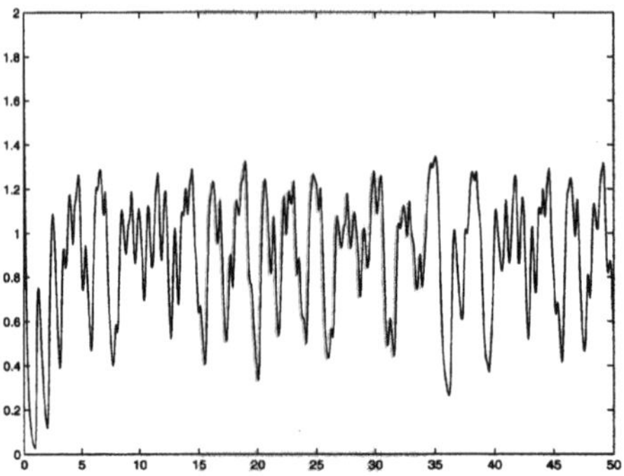

FIGURE 1. Chaotic behavior of the solutions of (1) for the initial data $\varphi(\theta) = 2$, $-1 \leq \theta \leq 0$.

The second example concerns the identifiability of unknown parameters that appear in special classes of nonself-adjoint evolutionary systems. Parameter identifiability is concerned with the question whether the parameters of a specific model can be identified from knowledge about certain solutions of the model, assuming perfect data. Completeness of the eigenvectors and generalized eigenvectors of the solution operator plays a crucial role in these results ([32]).

Another motivating example are the so-called Lang-Kobayashi equations describing a laser with delayed optical feedback. These are a three-dimensional system of differential equations with one point delay (see [19] for the mathematical theory and a description of the model).

To illustrate how complicated the behavior of scalar differential delay equations can become, we plot the solution of

$$\dot{x}(t) = -5x(t) + 10\frac{x(t-1)}{1+x(t-1)^8}, \qquad x_0 = \varphi \tag{1}$$

with initial data $\varphi(\theta) = 2$ for $-1 \leq t \leq 0$ in Fig. 1. Equation (1) was introduced and studied by Mackey [8] as a model to describe different periodic diseases. Note that the solution intersects the steady state $x(t) = 1$, but this turns out not to be a contradiction to the uniqueness of solutions because the states of the solutions are different (see Section 1).

We conclude this introduction with an outline of the paper. In Section 1 we define the state of a solution and discuss existence and uniqueness of solutions. In Section 2 we discuss a functional analytic approach toward differential delay equations based on semiflows which is essential in the modern treatment of differential delay equations. In Section 3 we discuss the spectral theory for autonomous delay equations. For autonomous equations the theory is well developed, see [7], but since we restrict our attention to the special case of one point delay – relevant in applications as we have seen – it is possible to present a novel simpler approach based on resolvent estimates for the characteristic matrix and series expansions of solutions. We discuss completeness and non-completeness results, the existence of solutions that decay faster than any exponential, and the convergence of series of spectral projections. Our results extend earlier results in [3, 21, 2, 26]. In Section 4 we extend the main results for autonomous equations to periodic differential delay equations by studying the monodromy operator associated with the equation. Basic results regarding Floquet multipliers were obtained in [12, 22, 34, 31], see also [13, 7], but many questions about the fine structure of the spectrum of the monodromy operator remain open. In the same spirit as for autonomous equations we prove new results concerning the convergence of series of spectral projections. We present results in the scalar case and refer to work in progress [25] for the complete picture.

1. Basic theory for delay equations

In this section we present a short introduction to the basic theory for differential delay equations. We start to discuss the existence and uniqueness of solutions and then continue with a discussion of linear autonomous delay equations.

1.1. Existence and uniqueness of solutions

Since all our examples are of the form

$$\dot{x}(t) = F(x(t), x(t-\tau)), \qquad t \geq 0, \tag{2}$$

where $F : \mathbb{R}^n \times \mathbb{R}^n \to \mathbb{R}^n$ is a Lipschitz continuous vector field and the time delay $\tau \geq 0$ is a fixed real number, we shall restrict the introduction to the basic theory of equations of this type.

The solution of (2) is defined to be a vector-valued function $x : [-\tau, \infty) \to \mathbb{R}^n$ satisfying (2) for positive time. A moment of reflection indicates that, in order for (2) to have a unique solution, the minimal amount of information of initial data is given by a continuous function on the interval $[-\tau, 0]$.

In fact, if $x(t) = \varphi(t)$ for $-\tau \leq t \leq 0$, where φ is a given continuous function, then the solution $x(t)$ for $0 \leq t \leq \tau$ satisfies the non-autonomous ordinary differential equation

$$\dot{x}(t) = F(x(t), \varphi(t-\tau)) \qquad \text{for } 0 \leq t \leq \tau, \quad x(0) = \varphi(0). \tag{3}$$

This equation has a unique solution and the solution of (3) on $[0, \tau]$ coincides with the solution of (2) on $[0, \tau]$ with initial data φ on $[-\tau, 0]$. Once the solution x is known on $[0, \tau]$, we can repeat the same procedure, starting with the solution on $[0, \tau]$, to find the solution $x(t)$ for $\tau \leq t \leq 2\tau$, etc. This process is called ***the method of steps*** and yields a unique, globally defined, solution $x(\,\cdot\,; \varphi)$ of (2), given an initial condition φ on $[-\tau, 0]$. Note that the solution becomes smoother in t as t increases. If we start with a continuous function on $[-\tau, 0]$, then the solution

$x(\cdot\,;\varphi)$ is continuously differentiable on $(0,\tau)$, twice continuously differentiable on $(\tau, 2\tau)$, etc.

From the method of steps, it also follows that, in general, solutions of (2) are only defined on $[-\tau,\infty)$. Backward continuation of solutions of differential delay equations for negative time requires additional smoothness of the initial function φ on $[-\tau, 0]$. Furthermore, there is the additional problem that, if there exists a backward continuation, then this backward continuation is not necessarily unique, see Theorem 3.21.

Since there is only one time delay in equation (2), we can rescale time and assume that $\tau = 1$. To understand the large time behavior of solutions of (2), one begins with an analysis of the behavior of solutions of (2) near steady states and periodic solutions. A steady state of (2) is a solution of the form $x(t) = \overline{x}$, where $\overline{x}$ is a solution of the algebraic equation $F(\overline{x},\overline{x}) = 0$. To understand the behavior of solutions near a steady state, one has to linearize around the steady state $\overline{x}$. This results in an autonomous linear delay equation

$$\dot{y}(t) = B_0 y(t) + B_1 y(t-1), \qquad t \geq 0 \tag{4}$$

with $B_0 = D_1F(\overline{x},\overline{x})$ and $B_1 = D_2F(\overline{x},\overline{x})$. Here D_i denotes the partial derivative of the function F with respect to the ith variable.

The linearization of (2) around an ω-periodic solution $x(t) = p(t)$ of (2), yields a ω-periodic linear delay equation

$$\dot{y}(t) = B_0(t)y(t) + B_1(t)y(t-1), \qquad t \geq s \tag{5}$$

with $s \in \mathbb{R}$, $B_0(t) = D_1F(p(t), p(t-1))$ and $B_1 = D_2F(p(t), p(t-1))$.

In the following section we present the basic theory for linear autonomous delay equations. In Section 2.2 we return to periodic equations.

1.2. Linear autonomous equations

For autonomous linear delay equations the theory is well-developed, using principles from the theory of Laplace transformation. If we define $\mathcal{L}(y)$ to be the Laplace transform of the function y, i.e.,

$$\mathcal{L}(y)(z) = \int_0^\infty e^{-zt} y(t)\, dt$$

and apply the Laplace transform to the autonomous equation (4) with initial data $y(\theta) = \varphi(\theta)$ for $-1 \leq \theta \leq 0$, we obtain

$$\Delta(z)\mathcal{L}(y)(z) = \varphi(0) + B_1 \int_0^1 e^{-zt}\varphi(t-1)\, dt, \tag{6}$$

where $\Delta(z)$ denotes the characteristic matrix of (4) and is given by

$$\Delta(z) = zI - B_0 - B_1 e^{-z}. \tag{7}$$

The idea is to obtain an explicit representation of y by using the inverse Laplace transform, the Cauchy theorem and a residue calculus. In order to follow this approach, we need good estimates for $\Delta(z)^{-1}$ when z approaches infinity.

Since the zeros of the determinant of $\Delta(z)$, i.e., the roots of the equation

$$\det \Delta(z) = \det\left[zI - B_0 - B_1 e^{-z}\right] = 0 \tag{8}$$

satisfy a transcendental equation, there are, in general, infinitely many zeros. Characteristic equations of type (8) are well-studied (cf. [3]). We state here some basic results and refer to Chapter I and XI of [7] for further details and references.

Lemma 1.1. *The roots of the transcendental equation* (8) *have the following properties*

(*i*) *There exists a right half-plane* $\{z \in \mathbb{C} \mid \operatorname{Re} z > \gamma_0\}$ *without roots of* (8).
(*ii*) *The number of roots of* (8) *in a given strip* $\{z \in \mathbb{C} \mid \gamma_- < \operatorname{Re} z \leq \gamma_+\}$ *is finite.*
(*iii*) *The roots of* (8) *in the left half-plane, necessarily satisfy*

$$|\operatorname{Im} z| \leq C e^{-\operatorname{Re} z},$$

where C is a constant determined by B_0 and B_1.

The asymptotic behavior of the solutions of (4) as t tends to infinity is completely controlled by the behavior of the roots of the characteristic equation (8).

Theorem 1.2. *Let y be a solution of* (4) *corresponding to some initial function φ. For any $\gamma \in \mathbb{R}$ such that* (8) *has no roots on the line* $\operatorname{Re} z = \gamma$, *we have the asymptotic expansion of the solution*

$$y(t) = \sum_{j=1}^{m} p_j(t) e^{\lambda_j t} + o(e^{\gamma t}) \quad \textit{for } t \to \infty, \tag{9}$$

where $\lambda_1, \ldots, \lambda_m$ are the finitely many roots of (8) *with real part exceeding γ and where $p_1(t), \ldots, p_m(t)$ are polynomials in t.*

Corollary 1.3. *All solutions of* (4) *converge to zero exponentially as $t \to \infty$ if and only if* (8) *has no roots in the right half-plane* $\{z \mid \operatorname{Re} z \geq 0\}$.

Therefore questions about exponential stability of steady states of (2) can be reduced to questions about the location of the zeros of the entire function $\det \Delta(z)$ where $\Delta(z)$ is given by (7).

The proof of Theorem 1.2 is based on the representation of the solution y of (4) using the inverse Laplace transform. From (6) and Lemma 1.1, it follows that

$$y(t) = \frac{1}{2\pi i} \int_{\gamma_0 - i\infty}^{\gamma_0 + i\infty} e^{zt} \Delta(z)^{-1} \left[\varphi(0) + B_1 \int_0^1 e^{-zt} \varphi(t-1)\, dt\right] dz. \tag{10}$$

Next the idea is to shift the line of integration to the left, while keeping track of the residues corresponding to the singularities of $\Delta(z)^{-1}$ that we pass (see [7] for details).

What happens when we let $\gamma \to -\infty$ in the series expansion (9)? Do we get a convergent infinite series and if so, is it a faithful representation of $y(t)$ or does $y(t)$ contain a component which goes to zero (as $t \to \infty$) faster than any exponential? A solution which does go to zero faster than any exponential is called a *small solution.* How do we recognize from the characteristic matrix $\Delta(z)$ whether or not small solutions exist? And how do we recognize whether the infinite expansion in polynomial-exponential functions is convergent and represents the solution for arbitrary initial data or, otherwise, how can we characterize those initial data for which it does?

For autonomous functional differential equations the importance of these questions have been addressed in [2, 5, 6, 24], and it was one of the motivations to develop a functional analytic approach toward delay equations (see [7]).

In the next section we shall outline this approach which is based on analyzing solution operators acting on function spaces of initial data rather than analyzing $\mathbb{C}^n$-valued solutions. We explain the natural connection between the questions above and very interesting questions about the completeness of systems of eigenvectors and generalized eigenvectors of nonself-adjoint operators.

2. A functional analytic approach

The dynamical system approach to differential delay equations is to associate with (2) a semiflow acting on the space of initial data, defined by the time evolution of segments of solutions (see Fig. 2).

Let $\mathcal{C}$ denote the Banach space of continuous functions defined on the interval $[-1, 0]$ with values in $\mathbb{C}^n$, provided with the sup-norm $\|\varphi\| := \sup_{-1\leq\theta\leq 0} |\varphi(\theta)|$ for $\varphi \in \mathcal{C}$. If the solution of (2) with initial data $x(\theta) = \varphi(\theta)$, $-1 \leq \theta \leq 0$, is denoted by $x(\,\cdot\,;\varphi)$ and the state of the solution $x(\,\cdot\,;\varphi)$ is defined by

$$x_t(\theta;\varphi) = x(t+\theta;\varphi), \qquad -1 \leq \theta \leq 0,$$

then the semiflow $\Sigma(t;\,\cdot\,) : \mathcal{C} \to \mathcal{C}$ is defined by

$$\Sigma(t;\varphi) = x_t(\,\cdot\,;\varphi), \qquad t \geq 0.$$

2.1. Linear autonomous equations

The linearization of the semiflow $\Sigma(t;\,\cdot\,)$ around an equilibrium leads to a strongly continuous semigroup $T(t)$ defined by translation along the solutions of (4), i.e., $T(t)\varphi = x_t$, where x the solution of

$$\dot{x}(t) = B_0x(t) + B_1x(t-1), \qquad x_0 = \varphi, \quad \varphi \in \mathcal{C}. \tag{11}$$

An important observation is that the evolution of the state x_t is given by an abstract ordinary differential equation in the infinite dimensional state space $\mathcal{C}$. Given equation (11) this abstract ordinary differential equation can be computed explicitly and is given by

$$\frac{du}{dt} = Au, \qquad u(0) = \varphi,\ \varphi \in \mathcal{C}, \tag{12}$$

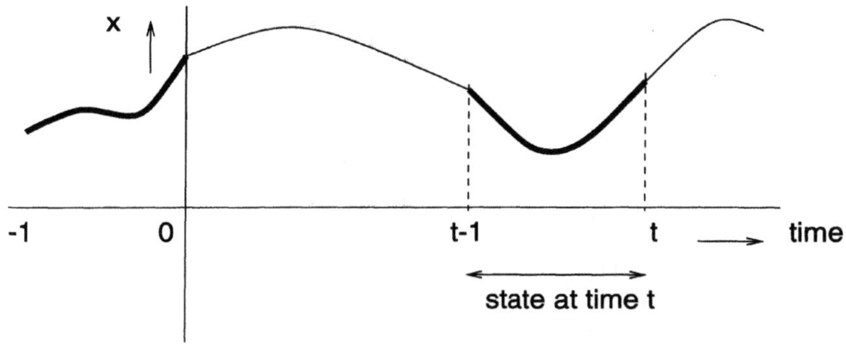

FIGURE 2. The state space approach

where $A(\mathcal{C} \to \mathcal{C})$ is an unbounded operator defined by

$$\begin{cases} A\varphi &= \frac{d\varphi}{d\theta} \\ \mathcal{D}(A) &= \{\varphi \in \mathcal{C} \mid A\varphi \in \mathcal{C} \text{ and } \frac{d\varphi}{d\theta}(0) = B_0\varphi(0) + B_1\varphi(-1)\}. \end{cases} \tag{13}$$

Thus, the system of differential equations (11) with time delays can be viewed as a transport equation with nonlocal boundary conditions. Furthermore, the solutions of (11) are in one-to-one correspondence with the solutions of the infinite dimensional ordinary differential equation (12) and this correspondence is given by

$$u(t)(\theta) = x(t+\theta).$$

This observation originated with Krasovskii and has been crucial in the development of the qualitative theory of differential delay equations.

The semigroup $T(t)$ satisfies the following basic properties

(i) $T(0) = I$;
(ii) $T(t+s) = T(t)T(s)$ for $t, s \geq 0$;
(iii) for every $\varphi \in C$, the maps $t \to T(t)\varphi$ are continuous.

Furthermore, the operator A defined by (13) is the infinitesimal generator of the semigroup $T(t)$ and is formally defined by

$$A\varphi := \lim_{t \downarrow 0} \frac{1}{t}\left[T(t)\varphi - \varphi\right] \quad \text{for every } \varphi \in \mathcal{D}(A), \tag{14}$$

where $\mathcal{D}(A)$ consists of all $\varphi \in \mathcal{C}$ for which the limit in (14) exists.

The growth bound of $T(t)$ is defined by

$$\omega_0 := \lim_{t \to \infty} \frac{1}{t} \log \|T(t)\|.$$

It is known that $\omega_0 < \infty$ and that, for $\operatorname{Re} z > \omega_0$, the resolvent of A equals the Laplace transform of the semigroup $T(t)$

$$(zI - A)^{-1} = \int_0^\infty e^{-zt} T(t)\, dt \quad \text{for } \operatorname{Re} z > \omega_0. \tag{15}$$

Let $\lambda \in \sigma(A)$ be an eigenvalue of A, the kernel $\mathcal{N}(\lambda I - A)$ is called the eigenspace at λ and its dimension d_λ the *geometric multiplicity*. The generalized eigenspace $\mathcal{M}_\lambda$ is the smallest closed subspace that contains all $\mathcal{N}((\lambda I - A)^j)$, $j = 1, 2, \dots$ and its dimension is called the *algebraic multiplicity*. It is known that there is a close connection between the spectral properties of the infinitesimal generator A and the characteristic matrix $\Delta(z)$ associated with (11) (cf. [7] and [17]). In particular, the geometric multiplicity d_λ equals the dimension of the null space of $\Delta(\lambda)$ at λ and the algebraic multiplicity is equal to the multiplicity of $z = \lambda$ as a zero of $\det \Delta(z)$. Furthermore, the generalized eigenspace at λ is given by

$$\mathcal{M}_\lambda = \mathcal{N}((\lambda I - A)^{k_\lambda}),$$

where k_λ is the order of $z = \lambda$ as a pole of $\Delta(z)^{-1}$. The eigenvalue λ is called *simple* if $m_\lambda = 1$.

Let $\{\varphi_1, \dots, \varphi_{m_\lambda}\}$ be a basis of eigenvectors and generalized eigenvectors of A at λ. Define the row m_λ-vector $\Phi = \{\varphi_1, \dots, \varphi_{m_\lambda}\}$. Since $\mathcal{M}_\lambda$ is invariant under A, there exists a $m_\lambda \times m_\lambda$ matrix M such that

$$A\Phi = \Phi M.$$

The action of the semigroup $T(t)$ on Φ is given by $T(t)\Phi = \Phi e^{tM}$. In particular, it follows that $\Phi(\theta) = \Phi(0)e^{\theta M}$, $-1 \le \theta \le 0$. So, in order to obtain a complete description of the subspace $\mathcal{M}_\lambda$, we have to compute $\Phi(0)$.

From standard spectral theory it follows that the spectral projection onto $\mathcal{M}_\lambda$ along $\mathcal{R}((\lambda I - A)^{k_\lambda})$ can be represented by a Dunford integral

$$P_\lambda = \frac{1}{2\pi i} \int_{\Gamma_\lambda} (zI - A)^{-1}\, dz, \tag{16}$$

where Γ_λ is a small circle such that λ is the only singularity of $(zI - A)^{-1}$ inside Γ_λ. A direct computation yields the following representation for the resolvent of A

$$(zI - A)^{-1}\varphi = \frac{1}{\det \Delta(z)} P(z)\varphi, \tag{17}$$

where

$$\begin{aligned} (P(z)\varphi)(\theta) &= e^{z\theta}\Big\{ \operatorname{adj} \Delta(z)[\varphi(0) + B_1 \int_0^1 e^{-z\sigma}\varphi(\sigma - 1)\, d\sigma] \\ &\qquad + \det \Delta(z) \int_\theta^0 e^{-z\sigma}\varphi(\sigma)\, d\sigma \Big\}. \end{aligned} \tag{18}$$

For example, if λ is a simple eigenvalue, this representation together with formula (16) can be used to obtain an explicit representation for the spectral projection onto the eigenspace corresponding to a simple eigenvalue

$$\begin{aligned} P_\lambda \varphi &= \operatorname*{Res}_{z=\lambda} (zI - A)^{-1}\varphi \\ &= e^{\lambda \cdot} H(\lambda)\big(\varphi(0) + B_1 \int_0^1 e^{-\lambda\sigma}\varphi(\sigma - 1)\, d\sigma\big). \end{aligned} \tag{19}$$

where $H(\lambda) = \lim_{z\to\lambda}(z-\lambda)\Delta(z)^{-1}$ is the residue of $\Delta(z)^{-1}$ at $z = \lambda$. Equivalently, we can write

$$P_\lambda\varphi = \langle \phi_\lambda^*, \varphi\rangle \phi_\lambda,$$

where ϕ_λ is an eigenvector at λ for A, ϕ_λ^* is an eigenvector at λ for A^* and $\langle \phi_\lambda^*, \phi_\lambda\rangle = 1$ (here $\langle \cdot, \cdot \rangle$ denotes the duality pairing between $\mathcal{C}$ and the dual space $\mathcal{C}^*$).

Thus $\Phi(0)$ can be found by providing a basis for the range of P_λ. In [17] and Section IV.4 of [7] a systematic procedure has been developed to construct a canonical basis using Jordan chains. We shall here describe the underlying idea.

Definition 2.1. *A sequence of vectors $\{\phi_0, \phi_1, \dots, \phi_{r-1}\}$ in C with $\phi_0 \neq 0$ is called a Jordan chain of A at λ if*

$$A\phi_0 = \lambda\phi_0, \qquad A\phi_j = \lambda\phi_j + \phi_{j-1},$$

for $j = 1, 2, \dots, r-1$. The integer r is called the rank of the Jordan chain. Note that ϕ_0 is an eigenvector of A, the vectors $\phi_1, \dots, \phi_{r-1}$ are called generalized eigenvectors at λ.

One can organize the Jordan chains according to the following procedure [11]. Choose an eigenvector, say $\phi_\lambda^{1,0}$, with maximal rank, say r_1. Next, choose a Jordan chain

$$\{\phi_\lambda^{1,0}, \dots, \phi_\lambda^{1,r_1-1}\}$$

of length r_1 and let N_1 be the complement in $\mathcal{N}(\lambda I - A)$ of the subspace spanned by $\phi_\lambda^{1,0}$. In N_1 we choose an eigenvector $\phi_\lambda^{2,0}$ of maximal rank, say r_2, and let

$$\{\phi_\lambda^{2,0}, \dots, \phi_\lambda^{2,r_2-1}\}$$

be a corresponding Jordan chain of length r_2. Next let N_2 be the complement in N_1 of the subspace spanned by $\phi_\lambda^{2,0}$ and replace N_1 by N_2 in the above-described procedure.

This construction yields a basis $\{\phi_\lambda^{1,0}, \dots, \phi_\lambda^{d_\lambda,0}\}$ for $\mathcal{N}(\lambda I - A)$ and a corresponding canonical system of Jordan chains

$$\phi_\lambda^{1,0}, \dots, \phi_\lambda^{1,r_1-1}, \dots, \phi_\lambda^{d_\lambda,0}, \dots, \phi_\lambda^{d_\lambda, r_{d_\lambda}-1}.$$

Note that $\max\{r_j \mid j = 1, \dots, d_\lambda\} = k_\lambda$ and that

$$\sum_{j=1}^{d_\lambda} r_j = m_\lambda.$$

Because of the construction, the canonical system of Jordan chains is a basis for the generalized eigenspace $\mathcal{M}_\lambda$. With respect to this basis, the matrix M has Jordan normal form. We shall call such a basis a canonical basis of eigenvectors and generalized eigenvectors for A at λ.

To apply these ideas to delay equations, we shall first restrict ourselves to the case that $z = \lambda$ is a simple pole of $\Delta(z)^{-1}$, i.e., $\Delta(z)^{-1} = (z-\lambda)^{-1}H(z)$ with H analytic at $z = \lambda$.

Lemma 2.2. *If $z = \lambda$ is a simple pole of $\Delta(z)^{-1}$, then the generalized eigenspace at λ, $\mathcal{M}_\lambda = \mathcal{R}(P_\lambda)$, is given by*

$$\mathcal{M}_\lambda = \{\theta \mapsto e^{\lambda\theta}v \mid v \in \mathcal{N}(\Delta(\lambda))\}$$

and equals the eigenspace at λ (i.e., the geometric and the algebraic multiplicity of λ are equal).

Proof. First note that

$$\mathcal{R}(H(\lambda)) = \mathcal{N}(\Delta(\lambda)).$$

Indeed, since $\Delta(z)H(z) = H(z)\Delta(z) = (z-\lambda)I$. It follows that $\Delta(\lambda)H(\lambda) = 0$. Assume that $v \in \mathbb{C}^n$ is such that

$$\Delta(\lambda)v = 0$$

and define

$$w = \lim_{z\to\lambda} \frac{1}{z-\lambda}\Delta(z)v = \Delta'(\lambda)v.$$

Then

$$H(\lambda)w = \lim_{z\to\lambda} \frac{1}{z-\lambda}H(z)\Delta(z)v = \lim_{z\to\lambda} \frac{z-\lambda}{z-\lambda}v = v.$$

So $H(\lambda)w = v$. The proof of the lemma is now easily completed using the representation (19) for P_λ. □

The characterization of $\mathcal{M}_\lambda$ for simple poles of $\Delta(z)^{-1}$ implies that

$$\Phi(0) = (v_1, \dots, v_m),$$

where $\{v_j\}_{j=1}^m$ is a basis for $\mathcal{N}(\Delta(\lambda))$.

To describe the procedure in general, one first has to extend the notion of a Jordan chain. An ordered set $(x_0, x_1, \dots, x_{k-1})$ of vectors in $\mathbb{C}^n$ is called a Jordan chain for $\Delta(z)$ at $z = \lambda$ if $x_0 \neq 0$ and

$$\Delta(z)[x_0 + (z-\lambda)x_1 + \dots + (z-\lambda)^{k-1}x_{k-1}] = O((z-\lambda)^k)$$

for $|z-\lambda| \to 0$. The number k is called the length of the chain and the maximal length of a chain starting with x_0 is called the rank of x_0.

We now have the following theorem (see [17] for a proof).

Theorem 2.3. *The spectrum of the generator A consists of eigenvalues of finite type only,*

$$\sigma(A) = \{\lambda \mid \det \Delta(\lambda) = 0\}.$$

For $\lambda \in \sigma(A)$, the geometric multiplicity of λ equals the dimension of the null space of $\Delta(\lambda)$, the algebraic multiplicity of λ equals the order of λ as a zero of $\det \Delta(z)$ *and the ascent of λ equals the order of λ as a pole of $\Delta^{-1}(z)$. Furthermore, a canonical basis of eigenvectors and generalized eigenvectors for A at λ can be obtained in the following way: If*

$$\{(\gamma_{i,0}, \dots, \gamma_{i,k_i-1}) \mid i = 1, \dots, d_\lambda\}$$

is a canonical system of Jordan chains for $\Delta(z)$ at $z = \lambda \in \Omega$, then

$$\{\chi_{i,0}, \dots, \chi_{i,k_i-1} \mid i = 1, \dots, d_\lambda\},$$

where

$$\chi_{i,\nu}(\theta) = e^{\lambda\theta} \sum_{l=0}^{\nu} \gamma_{i,\nu-l} \frac{\theta^l}{l!}$$

is a canonical basis for A at λ.

The generalized eigenspace $\mathcal{M}$ is the smallest subspace that contains all $\mathcal{M}_\lambda$, $\lambda \in \sigma(A)$. If $\mathcal{M}$ is dense in $\mathcal{C}$, we call the system of eigenvectors and generalized eigenvectors of A *complete.* A Jordan chain of A gives rise to a special solution of (11). If $\{\varphi_0, \varphi_1, \dots, \varphi_{r-1}\}$ is a Jordan chain of A at λ and

$$p(t) = \sum_{\nu=0}^{r-1} \frac{1}{\nu!} t^\nu \varphi_{r-1-\nu}(0),$$

then

$$x(t) = p(t)e^{\lambda t}, \tag{20}$$

is a solution of (11) for all t. This solution corresponds to the special solutions in Theorem 1.2. So, if the linear space spanned by all eigenvectors and generalized eigenvectors of the infinitesimal generator is dense in $\mathcal{C}$, then each solution of (11) can be approximated by a linear combination of solutions of elementary solutions of the form (20).

In Section 3 we shall return to these issues, but first we discuss the basic theory for periodic equations.

2.2. Periodic linear equations

We continue our geometric approach to include linear periodic delay equations. Consider

$$\dot{y}(t) = B_0(t)y(t) + B_1(t)y(t-1), \qquad t \geq s, \tag{21}$$

where the coefficients B_0 and B_1 are real continuous matrix-valued periodic functions with minimal period ω, i.e., $B_0(t+\omega) = B_0(t)$ and $B_1(t+\omega) = B_1(t)$ for all $t \in \mathbb{R}$. We have seen that, if $y(s+\theta) = \varphi(\theta)$, $-1 \leq \theta \leq 0$, is given, then (21) has a unique solution $y(t; s, \varphi)$ for $t \geq s$.

A transformation of variables allows us to reduce (21) even further by assuming that the coefficient B_0 in (21) is identically zero. Define for $t \geq s-1$

$$z(t) = y(t)e^{-\int_{s-1}^{t} B_0(\sigma)\,d\sigma} \quad \text{and} \quad \widehat{B}_1(t) = B_1(t)e^{-\int_{t-1}^{t} B_0(\sigma)\,d\sigma}, \tag{22}$$

then $\widehat{B}_1(t)$ is a real continuous periodic function of period ω. A straightforward computation yields that $z(t)$ satisfies the differential delay equation

$$\dot{z}(t) = \widehat{B}_1(t)z(t-1), \qquad t \geq s$$

with initial data

$$z(s+\theta) = \varphi(\theta)e^{-\int_{-1}^{\theta} B_0(s+\sigma)\,d\sigma}, \qquad -1 \leq \theta \leq 0.$$

Therefore, we can assume in (21) that $B_0 = 0$ and, for future reference, we state this reduced equation explicitly

$$\dot{x}(t) = B_1(t)x(t-1), \qquad t \geq s. \tag{23}$$

The evolutionary system associated with (23) is again given by translation along the solution

$$T(t,s)\varphi = x_t(s,\varphi), \tag{24}$$

where $x_t(s,\varphi)(\theta) = x(t+\theta; s,\varphi)$ for $-1 \leq \theta \leq 0$. Since B_1 is a periodic function of period ω, the property $T(t,\tau)T(\tau,s) = T(t,s)$ for $t \geq \tau \geq s$ implies that

$$T(t+\omega,s) = T(t,s)T(s+\omega,s) \qquad \text{for } t \geq s. \tag{25}$$

The periodicity property (25) allows us to define the monodromy map or period map $\Pi(s) : C \to C$ associated with (23) as follows

$$\Pi(s)\varphi = T(s+\omega,s)\varphi, \qquad \varphi \in C. \tag{26}$$

From the general theory for differential delay equations (see [7, 13]), it follows that $\Pi(s)$ is a compact operator, i.e., $\Pi(s)$ is a bounded operator with the property that the closure of the image of the unit ball in C is compact. Hence, the spectrum $\sigma(\Pi(s))$ of $\Pi(s)$ is at most countable with the only possible accumulation point being zero. If $\mu \neq 0$ belongs to $\sigma(\Pi(s))$, then μ is in the point spectrum of Π, i.e., there exists a $\varphi \in C$, $\varphi \neq 0$, such that $\Pi(s)\varphi = \mu\varphi$.

If μ belongs to the nonzero point spectrum of $\Pi(s)$, then μ is called a *characteristic multiplier* of (23) and λ for which $\mu = e^{\lambda\omega}$ (unique up to multiples of $2\pi i$) is called a *characteristic exponent* of (23).

The characteristic multipliers are in fact independent of s and the generalized eigenspace $\mathcal{M}_\mu(s)$ of $\Pi(s)$ at μ is defined to be $\mathcal{N}\big(\mu I - \Pi(s))^{k_\lambda}\big)$, where k_λ is the smallest integer such that

$$\mathcal{M}_\mu(s) = \mathcal{N}\big((\mu I - \Pi(s))^{k_\lambda}\big) = \mathcal{N}\big((\mu I - \Pi(s))^{k_\lambda+1}\big).$$

Since $\Pi(s)$ is compact, there are two closed subspaces $\mathcal{M}_\mu(s)$ and $\mathcal{Q}_\mu(s)$ of C such that the following properties hold:

(i) $\mathcal{C} = \mathcal{M}_\mu(s) \oplus \mathcal{Q}_\mu(s)$.
(ii) $m_\mu = \dim \mathcal{M}_\mu(s) < \infty$.
(iii) $\mathcal{M}_\mu(s)$ and $\mathcal{Q}_\mu(s)$ are $\Pi(s)$-invariant.
(iv) $\sigma(\Pi(s)|\mathcal{M}_\mu(s)) = \{\mu\}$ and $\sigma(\Pi(s)|\mathcal{Q}_\mu(s)) = \sigma(\Pi(s)) \setminus \{\mu\}$.

Let $\varphi_1(s), \dots, \varphi_{m_\mu}(s)$ be a basis of eigenvectors and generalized eigenvectors of $\Pi(s)$ at μ. Define the row m_λ-vector $\Phi(s) = \{\varphi_1(s), \dots, \varphi_{m_\lambda}(s)\}$. Since $\mathcal{M}_\lambda(s)$ is invariant under $\Pi(s)$, there exists a $m_\lambda \times m_\lambda$ matrix $M(s)$ such that

$$\Pi(s)\Phi(s) = \Phi(s)M(s)$$

and Property (iv) implies that the only eigenvalue of $M(s)$ is $\mu \neq 0$. Therefore, there is a $m_\mu \times m_\mu$ matrix $B(s)$ such that $B(s) = \log M(s)$.

Define the vector $P(t)$ with elements in $\mathcal{C}$ by

$$P(t) = T(t,0)\Phi e^{-Bt}$$

and let $\Phi = \Phi(0)$, $M = M(0)$ and $B = B(0)$. Then, for $t \geq 0$,

$$\begin{aligned} P(t+\omega) &= T(t+\omega,0)\Phi e^{-B(t+\omega)} = T(t,0)T(\omega,0)\Phi e^{-B}e^{-Bt} \\ &= T(t,0)\Pi(s)\Phi e^{-B}e^{-Bt} = T(t,0)\Phi M e^{-B}e^{-Bt} \\ &= P(t). \end{aligned}$$

Since $P(t)$ can be extended periodically for $t \in \mathbb{R}$, we conclude that

$$T(t,0)\Phi = P(t)e^{Bt} \qquad \text{for} \quad t \in \mathbb{R}.$$

As for the case $s = 0$, one can define $T(t,s)\Phi(s)$ for all $t \in \mathbb{R}$ and, for any real number τ and $\mu \in \sigma(\Pi(s)) \setminus \{0\}$, we have

$$\begin{aligned} \Pi(\tau)T(\tau,s)\Phi(s) &= T(\tau+\omega,\tau)T(\tau,s)\Phi(s) \\ &= T(\tau+\omega,s)\Phi(s) \\ &= T(\tau,s)T(s+\omega,s)\Phi(s) \\ &= T(\tau,s)\Phi(s)M(s). \end{aligned}$$

Therefore,

$$[\mu I - \Pi(\tau)]T(\tau,s)\Phi(s) = T(\tau,s)\Phi(s)(\mu I - M(s))$$

and it follows that $\mu \in \sigma(\Pi(\tau))$. Thus the dimension of $\mathcal{M}_\mu(\tau)$ is at least as large as the dimension of $\mathcal{M}_\mu(s)$. Since one can reverse the role of s and τ, we obtain that the characteristic multipliers of (23) are independent of the starting time and if $\Phi(s)$ is a basis for $\mathcal{M}_\mu(s)$, then $T(t,s)\Phi(s)$ is a basis for $\mathcal{M}_\mu(t)$ for any $t \in \mathbb{R}$. In particular, the subspaces $\mathcal{M}_\mu(s)$ and $\mathcal{M}_\mu(t)$ are diffeomorphic for all s and t.

Solutions of (23) with initial value in $\mathcal{M}_\mu(s)$ are of the Floquet type, namely of the form

$$x(t) = e^{\lambda t}p(t), \tag{27}$$

where $\mu = e^\lambda$ and $p(t) = p(t+\omega)$. In other words, on each generalized eigenspace $\mathcal{M}_\mu(s)$ there exists a periodic transformation that transforms the periodic equation to an autonomous equation (corresponding to the Floquet theory for ordinary differential equations [1]).

The spectral projection onto $\mathcal{M}_\mu(s)$ along $\mathcal{Q}_\mu(s)$ can again be represented by a Dunford integral

$$P_\mu(s) = \frac{1}{2\pi i}\int_{\Gamma_\mu} (zI - \Pi(s))^{-1}\, dz, \tag{28}$$

where Γ_μ is a small circle such that μ is the only singularity of $(zI - \Pi(s))^{-1}$ inside Γ_μ.

Let

$$\mathcal{M}(s) = \oplus_{\mu\in\sigma(\Pi(0))\setminus\{0\}}\mathcal{M}_\mu(s)$$

denote the generalized eigenspace of $\Pi(s)$. Similarly as for autonomous equations, we can relate the Floquet solutions, corresponding to initial data $\varphi \in \mathcal{M}(s)$, to an arbitrary solution $x_t(s;\varphi)$.

Theorem 2.4. *Let μ_j, $j = 1, 2, \dots$, denote the nonzero eigenvalues of $\Pi(s)$ ordered by decreasing modulus and $\varphi \in C$. If γ is an arbitrary real number, then there are positive constants ϵ and M such that for $t \geq s$*

$$\Big\|x_t(s;\varphi) - \sum_{|\mu_n|\geq \exp\gamma\omega} P_{\mu_n}(s)x_t(s;\varphi)\Big\| \leq Me^{(\gamma-\epsilon)(t-s)}\|\varphi\|.$$

Proof. Let $k = k(\gamma)$ be the integer such that $|\mu_k| \geq e^{\gamma\omega}$ and $|\mu_{k+1}| < e^{\gamma\omega}$. Set $\Sigma = \{\mu_1, \mu_2, \dots, \mu_k\}$ and define

$$\mathcal{M}_\Sigma(s) = \oplus_{\mu\in\Sigma}\mathcal{M}_\mu(s) \quad \text{and} \quad \mathcal{Q}_\Sigma(s) = \bigcap_{\mu\in\Sigma}\mathcal{Q}_\mu(s).$$

Then, one has $C = \mathcal{M}_\Sigma(s) \oplus \mathcal{Q}_\Sigma(s)$. To prove the exponential estimate, set

$$R_k(s)\varphi = \varphi - \sum_{j=1}^{k} P_{\mu_j}(s)\varphi. \tag{29}$$

It follows that $R_k(s)\varphi \in \mathcal{Q}_\Sigma(s)$ and $T(t,s)R_k(s)\varphi \in \mathcal{Q}_\Sigma(s)$ for all $\varphi \in C$. Furthermore, there exists a constant M_0 such that $\|R_k(s)\varphi\| \leq M_0\|\varphi\|$. If ϵ is such that $e^{(\gamma-2\epsilon)\omega} = |\mu_{k+1}|$, then the spectral radius of

$$\widehat{\Pi}(s) \stackrel{\text{def}}{=} \Pi(s)|\mathcal{Q}_\Sigma(s)$$

is $e^{(\gamma-2\epsilon)\omega}$. Therefore, $\lim_{n\to\infty}\|\widehat{\Pi}(s)^n\|^{1/n} = e^{(\gamma-2\epsilon)\omega}$ and this implies that for some $k > 0$

$$\|\widehat{\Pi}(s)^k R_k(s)\varphi\| \leq e^{(\gamma-\epsilon)k\omega}\|R_k(s)\varphi\|.$$

Since $\widehat{T}(t,s) = T(t,s)|\mathcal{Q}_\Sigma(s)$, $s \leq t \leq s+\omega$, is pointwise bounded, the uniform boundedness principle yields a constant M_1 such that $\|\widehat{T}(t,s)\| \leq M_1$ for $s \leq t \leq s+\omega$. Set $M_2 = M_1 \max_{j=0,1,\dots,k-1}\|\widehat{\Pi}(s)^j\|$ and let $t \geq s$ be given. If m_t is the largest integer so that $s + m_t k\omega \leq t$, then

$$\|\widehat{T}(t,s)R_k(s)\varphi\| \leq M_2\|\widehat{\Pi}(s)^k R_k(s)\varphi\|^{m_t} \leq Me^{(\gamma-\epsilon)(t-s)},$$

where $M = M_0 M_2$. This proves the exponential estimate for the remainder term. □

The solution $x_t(s, P_{\mu_n}\varphi)$ corresponds to the special solution (27). So, if the linear space spanned by all eigenvectors and generalized eigenvectors of $\Pi(s)$ is dense in $\mathcal{C}$, then each solution of (23) can be approximated by a linear combination of solutions of Floquet solutions of the form (27). In Section 4 we return to these issues, but first we present an illustration of the general theory and compute the monodromy operator in the scalar case.

Consider the scalar periodic delay equation

$$\dot{x}(t) = b(t)x(t-1), \qquad t \geq s, \tag{30}$$

where $b(t)$ is a scalar-valued periodic function with minimal period ω. If the period $\omega = 1$, then the monodromy operator $\Pi(s)$ is explicitly given by

$$(\Pi(s)\varphi)(\theta) = \varphi(0) + \int_{-1}^{\theta} b(\tau + s)\varphi(\tau)\,d\tau, \qquad -1 \leq \theta \leq 0. \tag{31}$$

To compute the eigenvalues of $\Pi(s)$, we can either use the explicit representation (31) for the resolvent (see Section 4) or use the theory of ordinary differential equations. To illustrate the second approach, assume first that $\omega = q^{-1}$, i.e., the time delay is an integer multiple of the period of the coefficient. In this case the eigenvalues and eigenvectors of $\Pi(s)$ can be easily computed using the fact that the eigenvectors and generalized eigenvectors of $\Pi(s)$ are exactly the initial data of the Floquet solutions $x(t) = e^{\lambda t}p(t)$, where $p(t+\omega) = p(t)$. Substitution of such a solution into equation (30) yields the following equation for $p(t)$

$$\dot{p}(t) + (\lambda - b(t)e^{-\lambda})p(t) = 0, \qquad p(0) = p(\omega). \tag{32}$$

Equation (32) has an ω-periodic solution if and only if

$$\frac{1}{\omega}\int_0^{\omega} (\lambda - b(s)e^{-\lambda})\,ds = 0$$

and the solution is given by

$$p(t) = p(0)e^{-\int_0^t (\lambda - b(s)e^{-\lambda})\,ds}.$$

Therefore μ is an eigenvalue of $\Pi(s)$ if and only if $\mu = e^{\lambda}$, where λ is a root of the equation

$$\lambda - [b]e^{-\lambda} = 0, \qquad [b] = \frac{1}{\omega}\int_0^{\omega} b(s)\,ds.$$

The corresponding eigenvector is given by

$$\begin{aligned} \phi_\mu(\theta) &= e^{\lambda\theta}e^{-\int_0^{\theta}(\lambda - b(s)e^{-\lambda})\,ds}\phi(0) \\ &= e^{e^{-\lambda}\int_0^{\theta} b(s)ds}\phi(0). \end{aligned}$$

Thus, it follows, that the characteristic exponents of (30) coincide with the spectrum corresponding to the autonomous differential delay equation

$$\dot{y}(t) = [b]y(t-1), \tag{33}$$

but the eigenvectors corresponding to (23) and (33) are different (see Section 4 for further information).

Finally note, as the abstract theory predicts, that the spectrum of $\Pi(s)$ is independent of s and hence, that the characteristic multiplies of (30) are independent of s. As an example, consider the case that $b(t) = \sin(2\pi t)$, then $[b] = 0$ and the nonzero point spectrum of $\Pi(s)$ consists of the single point $\mu = 1$. Therefore, the eigenvectors and generalized eigenvalues of $\Pi(s)$ corresponding to the equation

$$\dot{x}(t) = \sin(2\pi t)x(t-1),$$

are not dense in the space $\mathcal{C}$.

The case that ω is an integer multiple of the delay is more difficult. This is a consequence of the fact that for $\omega = 1/q$, equation (32) for $p(t)$ becomes a delay equation which can be reduced to a system of differential equations. Therefore, there no longer exists a simple criterion for the existence of periodic solutions. For example, if $\omega = 2$, then substitution of a solution $x(t) = e^{\lambda t}p(t)$ into equation (30) yields the following equation for $p(t)$

$$\dot{p}(t) + \lambda p(t) - b(t)e^{-\lambda}p(t-1) = 0.$$

If we set $p_1(t) = p(t)$ and $p_2(t) = p(t-1)$, then

$$\begin{cases} \dot{p}_1(t) = -\lambda p_1(t) + b(t)e^{-\lambda}p_2(t), \\ \dot{p}_2(t) = -\lambda p_2(t) + b(t+1)e^{-\lambda}p_1(t). \end{cases} \tag{34}$$

Although there is no simple criterion for the existence of periodic solutions, the system of ordinary differential equations (34) can still be used to describe the asymptotics of the eigenvalues of $\Pi(s)$, we refer to [23] or [22, 25] for a different approach based on the analysis of the resolvent of the monodromy operator. Finally, we remark that the more irrational the period ω becomes, the larger the dimension of the corresponding system of ordinary differential equations. Consequently, in case ω is irrational not much is known about the spectral data of $\Pi(s)$.

3. Spectral theory for autonomous equations

In this section we are interested in the question whether we can obtain a convergent series by letting $\gamma \to -\infty$ in (9). In Section 2 we have seen that the initial data corresponding to solutions $y(t) = p(t)e^{\lambda t}$ that arise in (9) are precisely the eigenvectors and generalized eigenvectors of the infinitesimal generator A defined in (13). Therefore, the question of whether we obtain a convergent series by letting

$\gamma \to -\infty$ in (9) can be rephrased as a question concerning the convergence of the spectral projections of the infinitesimal generator A

$$T(t)\varphi = \sum_{\lambda\in\sigma(A)} T(t)P_\lambda\varphi, \tag{35}$$

where the convergence is in the state space $\mathcal{C}$. The initial question is contained in this problem. Indeed, every solution of (11) has a convergent series expansion if and only if (35) holds for every $t > 1$.

To analyze the behavior of sums of spectral projections, we shall use the Riesz projection and the Cauchy theorem on residues. In order to do so, we need good estimates for the resolvent of A near infinity. The explicit representation for the resolvent of A given in (17) allows us to obtain these estimates using the theory of entire functions.

3.1. Exponential type calculus

Let X be a complex Banach space. An entire function $F : \mathbb{C} \to X$ is of *order* ρ if and only if

$$\limsup_{r\to\infty} \frac{\log\log M(r)}{\log r} = \rho,$$

where

$$M(r) = \max_{0\le\theta\le 2\pi} \left\{\|F(re^{i\theta})\|\right\}.$$

An entire function of order at most 1 is of *exponential type* if and only if

$$\limsup_{r\to\infty} \frac{\log M(r)}{r} = \mathrm{E}(F),$$

where $0 \le \mathrm{E}(F) < \infty$. In that case, $\mathrm{E}(F)$ is called the exponential type of F.

We start with a special case of the Paley-Wiener theorem (see Theorem 6.9.1 of [4]).

Theorem 3.1. *Let $F : \mathbb{C} \to \mathbb{C}$ be an entire function which is uniformly bounded in the closed right half-plane* $\mathrm{Re}\, z \ge 0$. *Then F is of exponential type α and L^2-integrable along the imaginary axis if and only if*

$$F(z) = \int_0^\alpha e^{-z\tau}\varphi(\tau)\,d\tau,$$

where $\varphi \in L^2([0,\alpha],\mathbb{C})$ and φ does not vanish a.e. in any neighbourhood of α.

For a large class of entire functions of exponential type, one can describe quite precisely the behavior at infinity. The first result is known as the Ahlfors-Heins theorem (see Theorem 7.26 of [4]).

Theorem 3.2. *Let $F : \mathbb{C} \to \mathbb{C}$ be an entire function of exponential type. If*

$$\int_1^\infty \frac{\log|F(\pm iy)|}{y^2}\,dy \quad \text{exist}, \tag{36}$$

then for almost all $\theta \in (\frac{\pi}{2}, \frac{3\pi}{2})$

$$\lim_{r\to\infty} \frac{\log|F(re^{i\theta})|}{r} = -\mathrm{E}(F)\cos\theta. \tag{37}$$

For every fixed $\theta_0 \in (0, \frac{\pi}{2})$, *there exists a sequence* r_j, *such that* $r_j \to \infty$ *and*

$$\lim_{j\to\infty} \frac{\log|F(r_j e^{i\theta})|}{r_j} = -\mathrm{E}(F)\cos\theta,$$

uniformly in $\frac{\pi}{2} + \theta_0 \le \theta \le \frac{3\pi}{2} - \theta_0$.

As a corollary, we have a class of entire functions of exponential type for which there is a simple calculus of exponential types.

Corollary 3.3. *Let* $F_1 : \mathbb{C} \to \mathbb{C}$ *and* $F_2 : \mathbb{C} \to \mathbb{C}$ *be entire functions of exponential type. If* F_1 *and* F_2 *are polynomially bounded in the right half-plane* $\operatorname{Re} z \ge 0$, *then*

$$\mathrm{E}(F_1F_2) = \mathrm{E}(F_1) + \mathrm{E}(F_2). \tag{38}$$

If, in addition, F_1/F_2 *is entire, then*

$$\mathrm{E}(F_1/F_2) = \mathrm{E}(F_1) - \mathrm{E}(F_2).$$

If $\Delta(z)$ is defined by (7) then $\det\Delta(z)$ and the entries of $\operatorname{adj}\Delta(z)$ are entire functions of exponential type which are polynomially bounded in the right half-plane. Therefore, Corollary 3.3 applies and we derive

$$\mathrm{E}(\det\Delta(z)) \le n$$

and

$$\mathrm{E}(\operatorname{adj}\Delta(z)) = \max_{1\le i,j\le n} \mathrm{E}\big((\operatorname{adj}\Delta(z))_{ij}\big) \le n-1.$$

The following corollary of the Paley-Wiener theorem is needed in the sequel.

Corollary 3.4. *Let* $F : \mathbb{C} \to \mathbb{C}$ *be an entire function of exponential type zero. If there exists a* γ *such that* $|F(z)|$ *is uniformly bounded in* $\{z \mid \operatorname{Re} z \ge \gamma\}$ *and* $|F(z)| \to 0$ *as* $\operatorname{Re} z \to \infty$, *then* $F = 0$.

Proof. Set

$$G(z) = \frac{F(z+\gamma) - F(\gamma)}{z}.$$

Then G is an entire function of exponential type which is L^2-integrable along the imaginary axis and bounded in the half-plane $\operatorname{Re} z \ge 0$. Since G has exponential type zero, Theorem 3.1 implies that $G = 0$. Therefore, $F(z)$ is constant, but $|F(z)| \to 0$ as $\operatorname{Re} z \to \infty$. Thus $F = 0$. □

3.2. Basic estimates

In this section we estimate the inverse of the characteristic matrix in the complex plane outside small circles centered around the zeros of $\det \Delta(z)$. These estimates provide the basic tool to estimate the resolvent of the generator near infinity.

Consider the characteristic matrix (8)

$$\Delta(z) = zI - B_0 - B_1 e^{-z},$$

where B_0 and B_1 are constant $(n \times n)$-matrices. In order to estimate $|\Delta(z)^{-1}|$, we first analyze the behavior of $|\Delta_1(z)^{-1}|$, where

$$\Delta_1(z) = zI - B_1 e^{-z}. \tag{39}$$

Since $e^z \Delta_1(z) = ze^z - B_1$, estimates for $|\Delta_1(z)^{-1}|$ follow from estimates for the resolvent $R(w) := (wI - B_1)^{-1}$ of B_1.

From the Neumann series, we have

$$|R(w)| \leq (|w| - |B_1|)^{-1} \qquad \text{for} \quad |w| > |B_1|. \tag{40}$$

Therefore, for $z \in \mathbb{C}$ with $|ze^z| > |B_1|$,

$$|\Delta_1(z)^{-1}| = |e^z R(ze^z)| \leq |e^z|(|ze^z| - |B_1|)^{-1}$$

and hence, for $z \in \mathbb{C}$ with $|ze^z| \geq C_0^{-1}|B_1|$ and $C_0 < 1$, we find

$$|\Delta_1(z)^{-1}| \leq \frac{1}{|z|} \frac{|ze^z|}{|ze^z| - |B_1|} \leq \frac{1}{|z|}(1 - C_0)^{-1}. \tag{41}$$

In particular, it follows that $\det \Delta_1(z) \neq 0$ for $z \in \mathbb{C}$ in the region with $|ze^z| \geq C_0^{-1}|B_1|$.

To estimate $|\Delta_1(z)^{-1}|$ for $z \in \mathbb{C}$ with $|ze^z|$ close to zero, we need further information about the behavior $R(w)$, the resolvent of B_1, near the poles of $R(w)$. The poles of $R(w)$ are exactly the eigenvalues of B_1 and on the resolvent set $\rho(B_1) = \mathbb{C} \backslash \sigma(B_1)$, the function $w \mapsto R(w)$ is holomorphic. To extend the estimate (40) away from the poles of $R(w)$, it remains to estimate $|R(w)|$ on the compact set K, where

$$K = \{w \in \mathbb{C} \mid |w| \leq |B_1| \text{ and } d(w, \sigma(B_1)) \geq \epsilon\}$$

and $d(w, H) = \inf\{d(w, h) \mid h \in H\}$ denotes the distance from a point $w \in \mathbb{C}$ to the set $H \subset \mathbb{C}$.

Since $w \mapsto R(w)$ is continuous on K, we conclude that given $\epsilon > 0$ and $w \in \mathbb{C}$ with $d(w, \sigma(B_1)) \geq \epsilon$, there exists a constant $M = M(\epsilon)$ such that $|R(w)| \leq M$. Therefore, for $z \in \mathbb{C}$ with $d(ze^z, \sigma(B_1)) \geq \epsilon$, we find

$$|\Delta_1(z)^{-1}| \leq |e^z| M. \tag{42}$$

Furthermore, to each nonzero eigenvalue μ_j, $j = 1, 2, \ldots, m$, of B_1, there corresponds a chain of zeros z_{jk}, $k = 1, 2, \ldots$, of $\det \Delta_1(z)$ defined by

$$z_{jk} e^{z_{jk}} = \mu_j.$$

Continuity of the complex function $z \mapsto ze^z$ implies that, given $\epsilon > 0$ and $z \in \mathbb{C}$ with $d(ze^z, \sigma(B_1)) \geq \epsilon$, there exists a $\delta > 0$ and a $C > 0$ such that, if $|z| > C$, then z is outside circles of radius δ centered at the zeros of $\det \Delta_1(z)$.

If B_1 is invertible, then $0 \notin \sigma(B_1)$ and we obtain that, given $\epsilon > 0$, there exists a positive constant C_0 such that, for $z \in \mathbb{C}$ with $|ze^z| \leq C_0$, we have that $d(ze^z, \sigma(B_1)) \geq \epsilon$. Hence, if B_1 is invertible, then for $z \in \mathbb{C}$ and $|ze^z| \leq C_0$, we have the following estimate

$$|\Delta_1(z)^{-1}| \leq |e^z| M. \tag{43}$$

In particular, it follows that $\det \Delta_1(z) \neq 0$ for $z \in \mathbb{C}$ with $|ze^z| \leq C_0$. Furthermore, it follows from (42) that, for $z \in \mathbb{C}$ and $C_0 < |ze^z| < C_0^{-1}$, the same estimate (43) holds as long as $d(ze^z, \sigma(B_1)) > \epsilon$. Or, equivalently, as long as $z \in \mathbb{C}$ is outside circles of radius δ centered around the zeros of $\det \Delta_1(z)$.

Summarizing we have proved the following lemma.

Lemma 3.5. *If $\Delta_1(z) = zI - B_1 e^{-z}$ and B_1 is invertible, then, for every $\delta > 0$, there exists a constant $M = M(\delta)$ such that*

$$|\Delta_1(z)^{-1}| \leq M \min\{\frac{1}{|z|}, |e^z|\}, \tag{44}$$

for $z \in \mathbb{C}$ outside circles of radius δ centered around the zeros of $\det \Delta_1(z)$. *Furthermore, there exists a constant C_0, $0 < C_0 < 1$, such that the zeros of* $\det \Delta_1(z)$ *are inside the set $V(C_0) = \{z \in \mathbb{C} \mid C_0 < |ze^z| < C_0^{-1}\}$.*

Next we use estimate (44) for $|\Delta_1(z)^{-1}|$ to find an estimate for $|\Delta(z)^{-1}|$. Fix $\delta > 0$ and let $z \in \mathbb{C}$ outside circles of radius δ centered around the zeros of $\det \Delta_1(z)$. Since $\det \Delta_1(z) \neq 0$, we have

$$\Delta(z) = \Delta_1(z)\big(I - \Delta_1(z)^{-1} B_0\big) \tag{45}$$

and it follows that

$$\Delta(z)^{-1} = (I - \Delta_1(z)^{-1} B_0)^{-1} \Delta_1(z)^{-1}.$$

So, it remains to estimate $|(I - \Delta_1(z)^{-1} B_0)^{-1}|$. From the Neumann series, it follows that it suffices to estimate $|\Delta_1(z)^{-1} B_0|$ and, from Lemma 3.5, we obtain that, for $|z| > 2M|B_0|$,

$$\begin{aligned} \left|\left(I - \Delta_1(z)^{-1} B_0\right)^{-1}\right| &\leq \left(1 - |\Delta_1(z)^{-1} B_0|\right)^{-1} \\ &\leq 2. \end{aligned}$$

Thus, for $|z| > 2M|B_0|$, we have that $\Delta(z)$ is invertible and

$$|\Delta(z)^{-1}| \leq 2M \min\{\frac{1}{|z|}, |e^z|\}, \tag{46}$$

for $z \in \mathbb{C}$ outside circles of radius δ centered around the zeros of $\det \Delta_1(z)$. In particular, for $|z| > 2M|B_0|$, the zeros of $\det \Delta(z)$ are inside circles of radius δ centered around the zeros of $\det \Delta_1(z)$. In particular, inside the set $V(C_0)$ and (46) holds for $z \in \mathbb{C}$ outside circles of radius 2δ centered around the zeros of $\det \Delta(z)$.

We summarize the results of the discussion in a lemma.

Lemma 3.6. *If $\Delta(z) = zI - B_0 - B_1 e^{-z}$ and B_1 is invertible, then, for every $\delta > 0$, there exists a constants C and $M = M(\delta)$ such that*

$$|\Delta(z)^{-1}| \leq M \min\{\frac{1}{|z|}, |e^z|\},$$

for $z \in \mathbb{C}$ with $|z| > C$ outside circles of radius δ centered around the zeros of $\det \Delta(z)$. *Furthermore, there exists a constant C_0, $0 < C_0 < 1$, such that the zeros of* $\det \Delta(z)$ *are inside the set $V(C_0) = \{z \in \mathbb{C} \mid C_0 < |ze^z| < C_0^{-1}\}$.*

Actually, we can use Lemma 3.5 to prove estimates for $|\Delta(z)^{-1}|$ for much more general characteristic matrices $\Delta(z)$. In fact, the same proof shows that Lemma 3.6 remains true for the characteristic matrix

$$\Delta(z) = \Delta_1(z) - \int_0^1 e^{-z\theta} d\zeta_0(\theta),$$

where ζ_0 is a matrix-valued function of bounded variation that is continuous at $\theta = 1$. See [31] for a proof of this fact using different arguments.

Next we consider the case that $B_1 \neq 0$ is not invertible. In this case we need more information about the singularity of the resolvent $R(w)$ of B_1 at $w = 0$. This information can be obtained from the Laurent series of $R(w)$ at $w = 0$

$$R(w) = \sum_{n=-k_0}^{\infty} w^n A_n.$$

The coefficients A_n are given by

$$A_n = \frac{1}{2\pi i} \int_\Gamma z^{-n-1} R(z)\, dz,$$

where Γ is a positively-oriented small circle of radius δ with center at $w = 0$ excluding any other eigenvalue of B_1. The coefficient A_{-1} is exactly the spectral projection onto the generalized eigenspace of B_1 at $\lambda = 0$ (see [18]).

Therefore, for $w \in \mathbb{C}$ with $0 < |w| \leq \delta$, there exists a constant $M_1 = M_1(\delta)$ such that

$$|R(w)| \leq M_1 |w|^{-k_0},$$

where k_0 is the order of the pole of $R(w)$ at $w = 0$. Therefore, for $z \in \mathbb{C}$ with $0 < |ze^z| \leq \epsilon$, we find

$$|\Delta_1(z)^{-1}| \leq M |z|^{-k_0} |e^{(1-k_0)z}|.$$

Thus, in the general case, Lemma 3.5 becomes

Lemma 3.7. *If $\Delta_1(z) = zI - B_1 e^{-z}$ and k_0 is the order of pole of the resolvent of $B_1 \neq 0$ at zero, then there is a constant C_0, $0 < C_0 < 1$, such that, for every $\delta > 0$, there exists a constant $M = M(\delta)$ and*

$$|\Delta_1(z)^{-1}| \leq M \min\{\frac{1}{|z|}, |e^z|\},$$

for $z \in \mathbb{C}$ *with* $|ze^z| > C_0$ *and* z *outside circles of radius* δ *centered around the zeros of* $\det \Delta_1(z)$. *If* $0 < |ze^z| \le C_0$, *then* $\Delta_1(z)$ *is invertible and*

$$|\Delta_1(z)^{-1}| \le M|z|^{-k_0}|e^{(1-k_0)z}|.$$

Furthermore, the nonzero zeros of $\det \Delta_1(z)$ *are inside the set* $V(C_0) = \{z \in \mathbb{C} \mid C_0 < |ze^z| < C_0^{-1}\}$.

As before, we use Lemma 3.7 and (45) to derive estimates for $|\Delta(z)^{-1}|$. In the general case, however, we have that $|\Delta_1(z)^{-1}|$ can grow in the region

$$W = \{z \in \mathbb{C} \mid |z| \ge C \text{ and } |ze^z| \le C_0\}.$$

Therefore, we cannot use the Neumann series to estimate $|(I - \Delta_1(z)^{-1}B_0)^{-1}|$. To provide the estimate for $|\Delta(z)^{-1}|$ in the region W, we have to use a different approach. First we claim that $\Delta(z)$ is invertible for $z \in W$. To prove this claim, note that Lemma 3.7 implies that, for $z \in W$, the matrix $\Delta_1(z)$ is invertible. Furthermore, for C_0 sufficiently small, we have the inequality

$$\frac{1}{2}|B_1| \le |ze^z I - B_1| \le 2|B_1|$$

and hence

$$\frac{1}{2}|B_1| \le |e^z\Delta_1(z)| \le 2|B_1|.$$

Since $e^z\Delta(z) = e^z\Delta_1(z) - e^zB_0$ and $|z| > 2C_0$

$$|e^zB_0| \le C_0|z|^{-1}|B_0| \le \frac{1}{2}|B_1|,$$

it follows that, for C large, C_0 small and $z \in W$, the matrix $\Delta(z)$ is invertible. Secondly, we claim that, for $z \in W$, the spectrum of $\Delta_1(z)^{-1}B_0$ is outside a circle of radius $r = r(C, C_0)$ centered at 1. To prove this claim, we assume that the statement is not true. Then there exists a $\mu \in \mathbb{C}$ with $|\mu - 1| < r$ and a vector $v \in \mathbb{C}^n$ with $v \ne 0$ such that $\Delta_1(z)^{-1}B_0v = \mu v$. But then $B_0v = \mu\Delta_1(z)v$ and

$$[\Delta(z) + (\mu - 1)\Delta_1(z)]v = 0.$$

Since $\Delta(z)$ is invertible, it suffices to show that, for $|\mu - 1|$ close to zero, we have $|\Delta(z)| > |\mu - 1||\Delta_1(z)|$. Indeed, this estimate implies that $\Delta(z) + (\mu - 1)\Delta_1(z)$ is invertible and hence $v = 0$ which contradicts the assumption. Note that if $z \in W$, then we can choose $C_0 > 0$ small and $C > 0$ large such that

$$|e^z\Delta(z)| = |ze^zI - e^zB_0 - B_1| > \frac{1}{2}|B_1|$$

and

$$|e^z\Delta_1(z)| = |ze^zI - B_1| \le 2|B_1|.$$

Thus if we choose $r < 1/4$ and $\mu \in \mathbb{C}$ with $|\mu - 1| < r$, then $|\Delta(z)| > |\mu - 1||\Delta_1(z)|$ and this proves the claim.

This observation allows us to use the Laurent series to conclude that there exists a constant $M_2 = M_2(r)$ such that

$$\left|\left(I - \Delta_1(z)^{-1}B_0\right)^{-1}\right| \leq M_2.$$

So representation (45) together with Lemma 3.7 yields an estimate for $|\Delta(z)^{-1}|$ when $z \in W$.

Thus, in the general case, we arrive at the following estimates for $|\Delta(z)^{-1}|$.

Lemma 3.8. *If $\Delta(z) = zI - B_0 - B_1e^{-z}$ and k_0 is the order of pole of the resolvent of $B_1 \neq 0$ at zero, then there is a constant C_0, $0 < C_0 < 1$, such that, for every $\delta > 0$, there exists a constant $M = M(\delta)$ and*

$$|\Delta(z)^{-1}| \leq M \min\{\frac{1}{|z|}, |e^z|\},$$

for $z \in \mathbb{C}$ with $|ze^z| > C_0$ and z outside circles of radius δ centered around the zeros of $\det \Delta(z)$. If $0 < |ze^z| \leq C_0$ and $|z| > C$, then $\Delta(z)$ is invertible and

$$|\Delta(z)^{-1}| \leq M|z|^{-k_0}|e^{(1-k_0)z}|.$$

Furthermore, there exist positive constants C_0 and C such that the zeros of $\det \Delta(z)$ with $|z| > C$ are inside the set $V(C_0) = \{z \in \mathbb{C} \mid C_0 < |ze^z| < C_0^{-1}\}$.

Remark that Lemma 3.8 does not immediately generalize to more general perturbations of $\Delta_1(z)$ and the precise estimates for $|\Delta(z)^{-1}|$ now do depend on the lower order terms in $\Delta(z)$.

To illustrate this fact consider the characteristic matrix

$$\Delta(z) = zI - B_0e^{-z/2} - B_1e^{-z}.$$

Then

$$e^{-z}\Delta(z)^{-1} = \left(I - \Delta_1(z)B_0e^{z/2}\right)^{-1}\Delta_1(z)^{-1}.$$

Note that in the region where $\Delta_1(z)$ is invertible, we do not necessary have that $\Delta(z)$ is invertible. Indeed, $|\Delta_1(z)B_0e^{z/2}|$ is close to 1 for $|ze^{z/2}|$ close to 1. This leads to chains of zeros of $\det \Delta(z)$ with different asymptotics then the zeros of $\det \Delta_1(z)$ and the behavior of $\Delta(z)^{-1}$ is not completely controlled by the behavior of $\Delta_1(z)^{-1}$. See [28] and Chapter V of [7] for a different approach based on an analysis of $\det \Delta(z)$ directly.

Instead of proving estimates for the norm of $\Delta(z)^{-1}$, we can also prove estimates for the norm of the vectors $\Delta(z)^{-1}v$ when v is restricted to belong to a subspace of $\mathbb{C}^n$. An inspection of the proofs of Lemma 3.6 and Lemma 3.8 yields the following lemma which will be useful in the sequel.

Lemma 3.9. *If $\Delta(z) = zI - B_0 - B_1e^{-z}$ and $v \in \mathbb{C}^n$ satisfies*

$$\frac{1}{2\pi i}\int_{\Gamma_0}(wI - B_1)^{-1}v\,dw = 0,$$

where Γ_0 is a small circle centered at zero enclosing no other eigenvalues of B_1, then, for every $\delta > 0$, there exists a constants C and $M = M(\delta)$ such that

$$|\Delta(z)^{-1}v| \leq M \min\{\frac{1}{|z|}, |e^z|\},$$

for $z \in \mathbb{C}$ with $|z| > C$ outside circles of radius δ centered around the zeros of $\det \Delta(z)$.

We end this subsection with the introduction of a sequence of contours that will be used to compute the complex line integral in the next section.

Using the estimates for $\Delta(z)^{-1}$ in Lemma 3.6 and Lemma 3.8, we can construct a sequence of real numbers ρ_l, such that $\rho_l \to \infty$, and a sequence of closed contours Γ_l, $l = 0, 1, \ldots$, such that for some positive constants k, ϵ and δ:

(i) Γ_l is contained in the interior of Γ_{l+1} and there are at most k zeros between Γ_l and Γ_{l+1};
(ii) the contours have at least distance $\epsilon > 0$ from the set of zeros of $\det \Delta(z)$;
(iii) the contour Γ_l lies along the circle $|z| = \rho_l$ outside $V(C_0)$; inside $V(C_0)$, the contour lies between the circle $|z| = \rho_l - \delta$ and the circle $|z| = \rho_l + \delta$;
(iv) the length of the portion of Γ_l within $V(C_0)$ is bounded for $l \to \infty$.

For any real γ we denote the part of the sequence of contours Γ_l contained in the left half-plane $\{z \mid \operatorname{Re} z \leq \gamma\}$ by $\Gamma_l^-(\gamma)$.

We end this section with an auxiliary result from complex analysis (see, for example, Lemma V.5.10 of [7]).

Lemma 3.10. *For any real number γ*

$$\lim_{l\to\infty} \Big| \int_{\Gamma_l^-(\gamma)} e^{zt} \min(\frac{|e^{-z}|}{|z|}, 1)\, dz \Big| = 0, \qquad \textit{for } t > 0.$$

The convergence is uniform for t in any interval $0 < t_0 < t < t_1 < \infty$.

3.3. Series expansions for autonomous equations

In this section we use the estimates for $|\Delta(z)^{-1}|$ to analyze the behavior of the series of spectral projections. As explained before, the idea is to obtain an explicit representation for $T(t)\varphi$ itself, using the inverse of the Laplace transform. The starting point is the following inversion formula, see Theorem 11.6.1 of [16],

$$T(t)\varphi = \lim_{\omega\to\infty} \frac{1}{2\pi i} \int_{\gamma - i\omega}^{\gamma + i\omega} e^{zt}(zI - A)^{-1}\varphi\, dz, \qquad t > 0, \tag{47}$$

for $\gamma > \omega(A)$ and for every $\varphi \in \mathcal{D}(A)$.

In order to use the contours Γ_l introduced in the previous subsection to compute the integral in (47), we need the following observation. If $\lambda \in \sigma(A)$ is a pole of $z \mapsto e^{zt}(zI - A)^{-1}\varphi$, then

$$\operatorname*{Res}_{z=\lambda} e^{zt}(zI - A)^{-1}\varphi = T(t)P_\lambda\varphi,$$

where P_λ denotes the spectral projection onto $\mathcal{M}_\lambda$. (This identity follows from (47), (16) and the resolvent equation.)

We are now ready to prove the following theorem.

Theorem 3.11. *Let $T(t)$ denote the semigroup associated with*

$$\dot{x}(t) = B_0 x(t) + B_1 x(t-1),$$

where B_1 is a nonsingular matrix. If $A(\mathcal{C} \to \mathcal{C})$ denotes the generator of $T(t)$, then, for every $\varphi \in \mathcal{D}(A)$, we have

$$\lim_{N\to\infty} \|T(t)\varphi - \sum_{j=0}^{N} T(t)P_{\lambda_j}\varphi\| = 0 \qquad \text{for } t > 0$$

uniformly on compact t-sets. Here λ_j, $j = 0, 1, \ldots$ denote the eigenvalues of A ordered by increasing modulus and P_{λ_j} denotes the spectral projection given by (16).

Proof. The Cauchy theorem implies that

$$T(t)\varphi = \lim_{l\to\infty} \Big\{ \sum_{j=0}^{m(l)} T(t)P_{\lambda_j}\varphi - \frac{1}{2\pi i}\int_{\Gamma_l^-} e^{zt}(zI - A)^{-1}\varphi\, dz \Big\},$$

where $\lambda_0, \ldots, \lambda_{m(l)}$ are the zeros of $\det \Delta$ inside the area enclosed by the line $\operatorname{Re} z = \gamma$ and the contour Γ_l^-. And it suffices to prove that, for every $\varphi \in \mathcal{D}(A)$,

$$\lim_{l\to\infty} \|\frac{1}{2\pi i}\int_{\Gamma_l^-} e^{zt}(zI - A)^{-1}\varphi\, dz\| = 0. \tag{48}$$

To analyze this limit, we shall use the representation for the resolvent of A given in (17) which we rewrite as follows

$$(zI - A)^{-1}\varphi = \Delta(z)^{-1} C(z)\varphi, \tag{49}$$

where

$$\begin{aligned}\big(C(z)\varphi\big)(\theta) &= e^{z\theta}\{\varphi(0) + B_1\int_0^1 e^{-z\sigma}\varphi(\sigma - 1)\, d\sigma + \Delta(z)\int_\theta^0 e^{-z\sigma}\varphi(\sigma)\, d\sigma\} \\ &= e^{z\theta}\varphi(0) + B_1\int_0^{\theta+1} e^{-z\sigma}\varphi(\sigma - 1)\, d\sigma \\ &\qquad + (zI - B_0)\int_0^{-\theta} e^{-z\sigma}\varphi(\sigma + \theta)\, d\sigma. \end{aligned} \tag{50}$$

Note that

$$\|C(z)\varphi\| = \sup_{-1\le\theta\le 0} \big|\big(C(z)\varphi\big)(\theta)\big| \le M|z|\,|e^{-z}|\,\|\varphi\|. \tag{51}$$

Together with the identity

$$(zI - A)^{-1}\varphi = \frac{1}{z}(zI - A)^{-1}A\varphi + \frac{1}{z}\varphi, \qquad \varphi \in \mathcal{D}(A),$$

it follows that, for $\varphi \in \mathcal{D}(A)$,

$$\|(zI-A)^{-1}\varphi\| \le C_1 \left|e^{-z}\Delta(z)^{-1}\right| \|A\varphi\| + C_2|z|^{-1}\|\varphi\|.$$

Because of the estimates for $|\Delta(z)^{-1}|$ and the definition of the contours Γ_l, there exists a constant l_0 such that, for $z \in \Gamma_l$, $l \ge l_0$, we have

$$\left|e^{-z}\Delta(z)^{-1}\right| \le K \min\{\frac{|e^{-z}|}{|z|}, 1\}. \tag{52}$$

Therefore, we obtain the following estimate for the resolvent of A

$$\|(zI-A)^{-1}\varphi\| \le C_1 K \min\{\frac{|e^{-z}|}{|z|}, 1\}\|A\varphi\| + C_2\frac{1}{|z|}\|\varphi\|$$

and an application of Lemma 3.10 shows (48). □

Since $\mathcal{R}(T(t)) \subset \mathcal{D}(A)$ for $t > 1$, the next result follows immediately, see [30] for details and further results.

Corollary 3.12. *For every $\varphi \in C$, the solution $x(\cdot\,;\varphi)$ of the differential delay equation*

$$\dot{x}(t) = B_0x(t) + B_1x(t-1),$$

where B_1 is a nonsingular matrix has a convergent series expansion

$$x(t) = \sum_{j=0}^{\infty} e^{\lambda_j t}p_j(t), \qquad t > 0.$$

The next corollary is a completeness result. See [28, 29, 30, 33] for much more general results.

Corollary 3.13. *Let $T(t)$ denote the semigroup associated with*

$$\dot{x}(t) = B_0x(t) + B_1x(t-1),$$

where B_1 is a nonsingular matrix. If $A(C \to C)$ denotes the generator of $T(t)$, then the system of eigenvectors and generalized eigenvectors of A is complete in C.

Proof. Let $\mathcal{M}$ denote the linear space spanned by the eigenvectors and generalized eigenvectors of A. To prove the statement we have to show that $\overline{\mathcal{M}} = C$. Let $\varphi \in C$. Since $\mathcal{D}(A)$ is dense in C, we can choose a sequence $\varphi_j \in \mathcal{D}(A)$ such that $\varphi_j \to \varphi$ in C. Then $T(t)\varphi_j \to T(t)\varphi$ uniformly in t on compact subsets of $[0,\infty)$. Because of Theorem 3.11, for every φ_j and $\epsilon > 0$, we have for $t \ge 0$

$$\lim_{N\to\infty} \|T(t+\epsilon)\varphi - \sum_{j=0}^{N} T(t+\epsilon)P_{\lambda_j}\varphi\| = 0.$$

Define $\varphi_{j,\epsilon} = T(\epsilon)\varphi_j$ and recall from the C_0-semigroup property of $T(t)$ that

$$\|\varphi_{j,\epsilon} - \varphi_j\| \to 0 \qquad \text{as } \epsilon \downarrow 0.$$

Therefore, we can construct a subsequence $\{\hat{\varphi}_j\}$ of $\{\varphi_{j,\epsilon}\}$, such that $T(t)\hat{\varphi}_j$ has a convergent spectral projection series uniformly in t on compact subsets of $\mathbb{R}_+$ and

$\hat{\varphi}_j$ converges to φ in C as $j \to \infty$. So, we have proved $\hat{\varphi}_j \in \overline{\mathcal{M}}$ and hence $\varphi \in \overline{\mathcal{M}}$. Thus $\overline{\mathcal{M}} = C$ and this proves the corollary. □

To illustrate the corollary we consider the following example

$$\begin{cases} \dot{x}_1(t) = x_1(t) + x_2(t-1), \\ \dot{x}_2(t) = x_1(t-1). \end{cases} \tag{53}$$

Since B_1 is invertible, the system of eigenvectors and generalized eigenvectors of the generator of the semigroup associated with (58) is complete.

Next we consider the case that B_1 is not invertible. From an inspection of the proof of Theorem 3.11, we can make the following observation. If, for $\varphi \in \mathcal{D}(A)$, there exist positive constants τ, l_0 and K_4 such that, for $l \geq l_0$ and $z \in \Gamma_l^-(\gamma)$,

$$\|e^{z\tau}(zI - A)^{-1}\varphi\| \leq K_4 \min\{\frac{\max\{|e^{-z}|, 1\}}{|z|}, 1\} \, (\|\varphi\| + \|A\varphi\|), \tag{54}$$

then the conclusion of Theorem 3.11 holds. Therefore, for $t > \tau$,

$$T(t)\varphi = \lim_{N\to\infty} \sum_{j=0}^{N} T(t)P_{\lambda_j}\varphi.$$

To investigate condition (54) further, we use (49) and the basic estimates for $|\Delta(z)^{-1}|$. First note that, it follows from Lemma 3.8, that, for $z \in \mathbb{C}$ and $|ze^z| > C_0$ for some C_0, $0 < C_0 < 1$, the estimate for $|\Delta(z)^{-1}|$ is the same as in the case that B_1 is invertible. Thus to investigate (54), we can assume that $|ze^z| \leq C_0$. In this case, Lemma 3.8 yields the following estimate

$$\left|e^{-k_0 z}\Delta(z)^{-1}C(z)\varphi\right| \leq M_1,$$

where k_0 is the order of the pole of the resolvent of B_1 at zero.

This proves the following generalization of Theorem 3.11 and Corollary 3.12.

Theorem 3.14. *Let $T(t)$ denote the semigroup associated with*

$$\dot{x}(t) = B_0 x(t) + B_1 x(t-1).$$

If $\varphi \in \mathcal{D}(A)$, then

$$T(t)\varphi = \sum_{j=1}^{\infty} T(t)P_{\lambda_j}\varphi, \qquad \text{for } t > k_0,$$

where k_0 is the order of the pole of the resolvent of B_1 at zero.

Corollary 3.15. *For every initial condition $\varphi \in C$, the solution $x(\,\cdot\,;\varphi)$ of the differential delay equation*

$$\dot{x}(t) = B_0 x(t) + B_1 x(t-1)$$

has a convergent series expansion for $t > k_0$.

Instead of shifting in time to achieve convergence of the spectral series expansion, we can also restrict φ to a smaller subspace of $\mathcal{C}$ and use Lemma 3.10 to prove (54). For sake of simplicity, we assume in the next discussion that $B_0 = 0$. The argument can easily be extended to hold for $B_0 \neq 0$ using (45) or by using a transformation of variables in the differential delay equation.

If the function $w \mapsto (wI - B_1)^{-1}C(z)\varphi$ is analytic at $w = 0$, then, for every $\epsilon > 0$, there exists a constant $M_1 = M_1(\epsilon)$ such that for $|w| \leq \epsilon$

$$\|(wI - B_1)^{-1}C(z)\varphi\| \leq M_1\|C(z)\varphi\|.$$

Thus, for $z \in \mathbb{C}$ with $|ze^z| \leq \epsilon$,

$$\begin{aligned}\|\Delta_1(z)^{-1}C(z)\varphi\| &\leq \|e^z(ze^z - B_1)^{-1}C(z)\varphi\| \\ &\leq M_1|e^z|\|C(z)\varphi\|.\end{aligned}$$

Recall that, for $\varphi \in \mathcal{D}(A)$, we can use integration by parts to obtain that, for every $\varphi \in \mathcal{D}(A)$, there exists a positive constant M_2 such that, for $z \in \mathbb{C}$,

$$\|C(z)\varphi\| \leq M_2 \max\{1, |e^{-z}|\}\big(\|\varphi\| + \|A\varphi\|\big), \tag{55}$$

where $C(z)$ is defined in (50). So if, for every $z \in \mathbb{C}$ and $-1 \leq \theta \leq 0$, the vector $\big(C(z)\varphi\big)(\theta)$ has no component in the generalized eigenspace of B_1 at 0, i.e.,

$$\frac{1}{2\pi i}\int_{\Gamma_0}(wI - B_1)^{-1}\big(C(z)\varphi\big)(\theta)\,dw = 0, \tag{56}$$

where Γ_0 is a small circle centered at zero enclosing no other eigenvalues of B_1, then Lemma 3.10 and the estimate for $C(z)\varphi$ in (55) show that (54) holds.

This proves the following theorem.

Theorem 3.16. *Let $A(\mathcal{C} \to \mathcal{C})$ denote the generator of the semigroup $T(t)$ associated with*

$$\dot{x}(t) = B_0x(t) + B_1x(t-1).$$

If, for $\varphi \in \mathcal{D}(A)$, condition (56) *holds, then*

$$T(t)\varphi = \lim_{N\to\infty}\sum_{j=0}^{N} T(t)P_{\lambda_j}\varphi \qquad \text{for } t > 0$$

uniformly on compact t-sets.

Using the exponential type of an entire function (see Section 3.1), we can further simplify condition (56). Suppose that $w \mapsto (wI - B_1)^{-1}C(z)\varphi$ has a pole at $w = 0$. From the Laurent series

$$(wI - B_1)^{-1}C(z)\varphi = \sum_{n=-k_1}^{\infty} w^nA_n(wI - B_1)^{-1}C(z)\varphi,$$

where $k_1 \leq k_0$ is the order of the pole at $w = 0$. Hence, for $|ze^z|$ small, we have that

$$\|\Delta_1(z)^{-1}C(z)\varphi\| \sim M_2\frac{|e^{-k_1z}|}{|z|^{k_1}}|e^z|\|C(z)\varphi\|$$

and, consequently, the norm of $\Delta_1(z)^{-1}C(z)\varphi$ grows exponentially in the region of the complex plane where $|ze^z|$ is small.

This leads us to the following definition

$$\mathcal{E} = \{\varphi \in \mathcal{C} \mid \mathrm{E}(\operatorname{adj}\Delta(z)C(z)\varphi) \le \mathrm{E}(\det\Delta(z))\}. \tag{57}$$

From the discussion above, it follows that, if $\varphi \in \mathcal{E}$, then $w \mapsto (wI - B_1)^{-1}C(z)\varphi$ is analytic at $w = 0$. Therefore, Theorem 3.16 shows that, for $\varphi \in \mathcal{E} \cap \mathcal{D}(A)$, the solution $T(t)\varphi$ has a convergent spectral series for $t > 0$. So, an argument similar as given in the proof of Corollary 3.13 yields $\mathcal{E} \subset \overline{\mathcal{M}}$.

Actually, we have the following theorem.

Theorem 3.17. *The closure of the generalized eigenspace of the generator A given by* (13) *satisfies*

$$\overline{\mathcal{M}} = \mathcal{E}.$$

Proof. We already know that $\mathcal{E} \subset \overline{\mathcal{M}}$ and it suffices to prove the inclusion $\overline{\mathcal{M}} \subset \mathcal{E}$. The proof consists of two parts. First we prove that $\mathcal{M} \subset \mathcal{E}$ and then, we show that $\mathcal{E}$ is actually a closed subspace of $\mathcal{C}$.

Basic properties of the exponential type (see Section 3.1) show that $\mathcal{E}$ is a subspace of $\mathcal{C}$. So, in order to show that $\mathcal{M} \subset \mathcal{E}$, it suffices to prove that the elements of a basis for $\mathcal{M}$ belong to $\mathcal{E}$. We have seen, in Theorem 2.3, that a basis for $\mathcal{M}$ consists of the canonical system of Jordan chains. For an element $\phi_\lambda^{i,j} \in \mathcal{M}_\lambda$, $\lambda \in \sigma(A)$, of this basis, we have that

$$(zI - A)^{-1}\phi_\lambda^{i,j} = \sum_{k=1}^{j} (z-\lambda)^{-k}\phi_{j-k}.$$

Therefore, for $\varphi \in \mathcal{M}$, the function $z \mapsto (zI - A)^{-1}\varphi$ is rational and hence $\mathrm{E}(\operatorname{adj}\Delta(z)C(z)\varphi) = \mathrm{E}(\det\Delta(z))$. So $\varphi \in \mathcal{E}$ and it remains to prove that $\mathcal{E}$ is closed.

From the definitions of $\operatorname{adj}\Delta(z)$ and $C(z)\varphi_j$, it follows that we can use the Paley-Wiener theorem to write

$$\operatorname{adj}\Delta(z)C(z)\varphi_j = p_j(z) + z^{n+1}\int_0^\sigma e^{-zs}\chi_j(s)\,ds.$$

Here p_j is a polynomial of degree n, $\chi_j \in L^\infty[0,n]$ and $\sigma \le n$ equals the exponential type of $\operatorname{adj}\Delta(z)C(z)\varphi_j$. If φ_j converges to φ in $\mathcal{C}$, then χ_j converges to χ in $L^\infty[0,n]$. Let $\tau = \mathrm{E}(\det\Delta(z))$. If φ_j belongs to $\mathcal{E}$ for $j = 1, 2, \dots$, then $\operatorname{supp}(\chi_j) \subset [0,\tau]$ for $j = 1, 2, \dots$. Since $\chi_j \to \chi$ in $L^\infty[0,n]$, it follows that $\operatorname{supp}(\chi) \subset [0,\tau]$ and hence $\varphi \in \mathcal{E}$. This shows that $\mathcal{E}$ is closed and completes the proof of the theorem. □

The characterization of the generalized eigenspace of A allows us to prove a noncompleteness result. If B_1 is not invertible, then we have seen that, there exist $\varphi \in \mathcal{C}$ such that $\Delta(z)^{-1}C(z)\varphi$ grows exponentially for $|ze^z|$ sufficiently small. Hence $\mathcal{E} \ne \mathcal{C}$ and we have the following generalization of Corollary 3.13.

Corollary 3.18. *If $A(\mathcal{C} \to \mathcal{C})$ denotes the generator given by* (13)*, then the system of eigenvectors and generalized eigenvectors of A is complete in $\mathcal{C}$ if and only if B_1 is invertible.*

Another interesting application of the characterization is the following observation. Suppose that $\varphi \in \overline{\mathcal{M}}$ is such that $z \mapsto (zI - A)^{-1}\varphi$ is an entire function, i.e., all the spectral projections of φ are identically zero, then it follows from Corollary 3.4 that $\varphi = 0$. This observation can be used to characterize the complement of $\overline{\mathcal{M}}$ in $\mathcal{C}$.

The following result provides the answer (see [30] for the details).

Theorem 3.19. *Let*

$$\mathcal{S} = \{\varphi \in \mathcal{C} \mid z \mapsto (zI - A)^{-1}\varphi \quad \textit{is entire}\}.$$

Then

$$\mathcal{C} = \overline{\overline{\mathcal{M}} \oplus \mathcal{S}}.$$

Initial data in the subspace $\mathcal{S}$ play a special role and correspond to so-called small solutions of the differential delay equation (11). First a definition.

Definition 3.20. *A solution x of* (11) *is called a small solution if*

$$\lim_{t\to\infty} x(t)e^{kt} = 0 \qquad \textit{for every } k \in \mathbb{R}.$$

Small solutions that are not identically zero are called nontrivial.

Let $\varphi \in \mathcal{C}$ be such that $x(\,\cdot\,;\varphi)$ is a small solution. Then

$$\lim_{t\to\infty} e^{kt}\|T(t)\varphi\| = 0 \qquad \text{for every } k \in \mathbb{R}$$

and hence

$$z \mapsto \int_0^\infty e^{-zt}T(t)\varphi\,dt$$

defines an entire function. But the Laplace transform of $T(t)\varphi$ equals the resolvent of the infinitesimal generator, see (15). Therefore, by analytic continuation, we have $\varphi \in \mathcal{S}$. Thus, the small solutions correspond to solutions with initial data belonging to $\mathcal{S}$.

The following theorem is a consequence of the Paley-Wiener theorem and the characterization of $\overline{\mathcal{M}}$ given in Theorem 3.17. It yields a further characterization of $\mathcal{S}$ and $\overline{\mathcal{M}}$ and, in particular, shows that, the small solutions of a linear autonomous delay equation are identically zero after finite time [15, 27].

Theorem 3.21. *If k_0 denotes the order of the pole of the resolvent of B_1 at zero and if $T(t)$ denotes the solution semigroup generated by* (13)*, then for $t \geq k_0$*

$$\mathcal{S} = \mathcal{N}\big(T(t)\big) \quad \textit{and} \quad \overline{\mathcal{M}} = \overline{\mathcal{R}\big(T(t)\big)}$$

and these relations do not hold for any t smaller than k_0.

Corollary 3.22. *If $T(t)$ denotes the solution semigroup generated by (13), then $\mathcal{N}\big(T(t)\big) = \{0\}$ if and only if B_1 is singular. Furthermore, the system of eigenvectors and generalized eigenvectors of A is complete if and only if there are no nontrivial small solutions.*

To illustrate the corollary we consider the following example

$$\begin{cases} \dot{x}_1(t) = x_1(t) + x_2(t-1), \\ \dot{x}_2(t) = 2x_1(t) + x_2(t-1). \end{cases} \tag{58}$$

The small solutions are identical zero for $t \geq 0$ and given by the initial data $x_0 = \varphi$ with $\varphi = (\varphi_1, \varphi_2)^\top \in C$ and $\varphi_1(0) = 0$ and $\varphi_2 = 0$.

The results in Theorem 3.17, Theorem 3.19 and Theorem 3.21 hold, under fairly general assumptions, for a large class of semigroups [29, 31, 10]. In the next section we show how to extend these results to linear periodic delay equations.

4. Spectral theory for periodic equations

In this section we extend the spectral theory for autonomous equations to periodic equations. As we have seen in Section 2, notably in Theorem 2.4, this requires detailed information about the spectral data of the monodromy operator. In particular, to understand whether we obtain a convergent series, by letting $\gamma \to -\infty$ in Theorem 2.4, we have to analyze the behavior of sums of spectral projections. In order to do so, we use the same approach as for autonomous equations. This means that we need good estimates for the resolvent $(wI - \Pi)^{-1}$ of the monodromy operator Π near $w = 0$. We provide these estimates using the theory of entire functions and an explicit representation for the resolvent of Π. Unfortunately, this approach puts restrictions on the possible period ω that is allowed in our analysis and we have to assume that ω is rational.

4.1. The resolvent of the monodromy operator

We study the monodromy operator $\Pi_\omega = \Pi_\omega(0)$ associated with

$$\dot{x}(t) = B(t)x(t-1), \qquad t \geq s, \tag{59}$$

where $B(t)$ is a periodic function with minimal period ω.

Suppose, at first, that $\omega = 1/q$, where q is a positive integer (the delay is an integer multiple of the period). We then have, by definition, the following representation for Π_ω

$$(\Pi_\omega \varphi)(\theta) = \begin{cases} \varphi(0) + \int_{-\omega}^{\theta} B(s)\varphi(s)\,ds, & -\omega \leq \theta \leq 0, \\ \varphi(\theta + \omega), & -1 \leq \theta \leq -\omega. \end{cases} \tag{60}$$

In particular, it follows that, the operator Π_ω^q is explicitly given by $\Pi_\omega^q = \Pi_1$, where

$$(\Pi_1 \varphi)(\theta) = \varphi(0) + \int_{-1}^{\theta} B(s)\varphi(s)\,ds, \qquad -1 \leq \theta \leq 0. \tag{61}$$

Thus, if $\omega = 1/q$, then $\Pi_\omega^q = \Pi_1$ and it suffices to analyze the monodromy operator Π_1.

To compute the resolvent of Π_1, we set $\varphi = (e^\lambda - \Pi_1)^{-1}\psi$ or, equivalently,

$$\psi = e^\lambda \varphi - \Pi_1 \varphi$$

and assume first that ψ is differentiable. If we differentiate this equation with respect to θ and use representation (61) for Π_1, we arrive at the following nonlocal boundary value problem

$$\varphi' = e^{-\lambda} B\varphi + e^{-\lambda}\psi', \qquad \psi(-1) = e^\lambda \varphi(-1) - \varphi(0). \tag{62}$$

Let $\Omega_s^t(\lambda)$ denote the fundamental matrix solution of the homogeneous equation $\varphi' = e^{-\lambda}B\varphi$. The solution to (62) is given by

$$\varphi(\theta) = \Omega_{-1}^\theta(\lambda)\varphi(-1) + e^{-\lambda}\int_{-1}^\theta \Omega_s^\theta(\lambda)\psi'(s)\,ds. \tag{63}$$

Next we solve $\varphi(-1)$ from the boundary condition in (62). Substitution of the solution (63) into the boundary condition yields

$$\psi(-1) = e^\lambda\varphi(-1) - \Omega_{-1}^0(\lambda)\varphi(-1) - e^{-\lambda}\int_{-1}^0 \Omega_s^0(\lambda)\psi'(s)\,ds.$$

Hence

$$\varphi(-1) = \left[e^\lambda I - \Omega_{-1}^0(\lambda)\right]^{-1}\big(\psi(-1) + e^{-\lambda}\int_{-1}^0 \Omega_s^0(\lambda)\psi'(s)\,ds\big), \tag{64}$$

whenever

$$\det\left[e^\lambda I - \Omega_{-1}^0(\lambda)\right] \neq 0.$$

Thus, the solution to the nonlocal boundary value problem (62) is given by

$$\begin{aligned}\varphi(\theta) &= \Omega_{-1}^\theta(\lambda)\left[e^\lambda I - \Omega_{-1}^0(\lambda)\right]^{-1}\Big(\psi(-1) + e^{-\lambda}\int_{-1}^0 \Omega_s^0(\lambda)\psi'(s)\,ds \\ &\qquad (e^\lambda I - \Omega_{-1}^0(\lambda))\Omega_\theta^{-1}(\lambda)e^{-\lambda}\int_{-1}^\theta \Omega_s^\theta(\lambda)\psi'(s)\,ds\Big).\end{aligned}$$

A further simplification, using basic properties of $\Omega_s^t(\lambda)$, yields the following representation for the resolvent of Π_1

$$\begin{aligned}\left((e^\lambda I - \Pi_1)^{-1}\psi\right)(\theta) &= \Omega_{-1}^\theta(\lambda)\left[e^\lambda I - \Omega_{-1}^0(\lambda)\right]^{-1}\Big(\psi(-1) \\ &\qquad + e^{-\lambda}\int_\theta^0 \Omega_s^0(\lambda)\psi'(s)ds + \int_{-1}^\theta \Omega_s^{-1}(\lambda)\psi'(s)\,ds\Big).\end{aligned} \tag{65}$$

Here we have assumed that ψ is differentiable and thus, representation (65) holds on a dense set of $\mathcal{C}$. Integration by parts in (65) can be used to extend the formula to hold for arbitrary $\psi \in \mathcal{C}$, but, in the sequel, we only use the formula in case ψ is differentiable.

4.2. Basic estimates

From representation (65) for the resolvent of Π_1, it follows that resolvent estimates for Π_1 can be reduced to estimates for the matrix valued function $\Omega_s^t(\lambda)$ for e^λ near zero. Note that, we have the following identity

$$\Omega_{-1}^\theta(\lambda)\left[e^\lambda I - \Omega_{-1}^0(\lambda)\right]^{-1} = \Omega_0^\theta(\lambda)\left[e^\lambda \Omega_0^{-1}(\lambda) - I\right]^{-1} \tag{66}$$

and, in order to estimate the resolvent of Π_1, it suffices to estimate the expressions

$$\left|e^{-\lambda}\Omega_s^\theta(\lambda)\left(I - e^\lambda\Omega_0^{-1}(\lambda)\right)^{-1}\right| \quad \text{for} \quad -1 \le \theta \le s \le 0 \tag{67}$$

and

$$\left|\Omega_s^\theta(\lambda)\Omega_0^{-1}(\lambda)\left(I - e^\lambda\Omega_0^{-1}(\lambda)\right)^{-1}\right| \quad \text{for} \quad -1 \le s \le \theta \le 0. \tag{68}$$

Since, in general, there is no formula for $\Omega_s^t(\lambda)$, estimates for (67) and (68) are not easy to obtain. In the sequel, we restrict ourselves to the scalar case, but we remark that work in progress [25] shows that, for e^λ small, using the theory for ordinary differential equations with a small complex parameter and the theory of exponential dichotomies, analogous estimates can be made in the matrix case.

In the scalar case, the fundamental solution $\Omega_s^t(\lambda)$ has a simple representation

$$\Omega_s^t(\lambda) = e^{e^{-\lambda}\int_s^t b(\sigma)d\sigma}, \tag{69}$$

where we have written $b(t)$ for $B(t)$ in (59) to stress the scalar case. In particular, we have

$$\left|\Omega_s^t(\lambda)\right| = e^{e^{-\operatorname{Re}\lambda}\cos\operatorname{Im}\lambda\int_s^t b(\sigma)d\sigma}. \tag{70}$$

Furthermore, we have the identity

$$e^\lambda\Omega_0^{-1}(\lambda) = e^{\lambda - [b]e^{-\lambda}},$$

where $[b] = \int_{-1}^0 b(\sigma)\,d\sigma$. Thus, estimates for (67) and (68) reduce to lower bounds for $\lambda - [b]e^{-\lambda}$ away from the zeros of $\lambda - [b]e^{-\lambda}$. Following the same approach as in Section 3.2, we arrive at the following estimate. For every $\epsilon > 0$, there exists a constant M such that, for every λ outside circles of radius ϵ centered at the zeros of $\lambda - [b]e^{-\lambda}$, we have

$$\left|\left(1 - e^\lambda\Omega_0^{-1}(\lambda)\right)^{-1}\right| \le M\min\left\{1, e^{-\operatorname{Re}\lambda + [b]e^{-\operatorname{Re}\lambda}\cos\operatorname{Im}\lambda}\right\} \tag{71}$$

and (67) can be estimated by

$$me^{-\lambda}H(\lambda) \le \left|e^{-\lambda}\Omega_s^\theta(\lambda)\left(1 - e^\lambda\Omega_0^{-1}(\lambda)\right)^{-1}\right| \le Me^{-\lambda}H(\lambda), \tag{72}$$

where m and M are positive constants and the function $H(\lambda)$ is defined by

$$H(\lambda) = \min\left\{e^{e^{-\operatorname{Re}\lambda}\cos\operatorname{Im}\lambda\int_s^\theta b(\sigma)d\sigma}, e^{-\operatorname{Re}\lambda + e^{-\operatorname{Re}\lambda}\cos\operatorname{Im}\lambda([b] - \int_\theta^s b(\sigma)d\sigma)}\right\}.$$

Next we analyze the behavior of the function $H(\lambda)$. Since $H(\lambda)$ is bounded in any right half plane, it suffices to analyze $H(\lambda)$ for $\operatorname{Re}\lambda \le -\gamma$ with γ sufficiently large.

If $\int_\theta^s b(\sigma)\,d\sigma \geq 0$, then $H(\lambda)$ is bounded for λ with $\cos \operatorname{Im} \lambda \geq 0$ and, if $\cos \operatorname{Im} \lambda < 0$, then $H(\lambda)$ is bounded when

$$0 \leq \int_\theta^s b(\sigma)\,d\sigma \leq [b]. \tag{73}$$

If $\int_\theta^s b(\sigma)\,d\sigma \leq 0$, then $H(\lambda)$ is bounded for λ with $\cos \operatorname{Im} \lambda \leq 0$ and, if $\cos \operatorname{Im} \lambda > 0$, then $H(\lambda)$ is bounded when

$$0 \geq \int_\theta^s b(\sigma)\,d\sigma \geq [b]. \tag{74}$$

Thus, in conclusion, $H(\lambda)$ is bounded if and only if the function $b(t)$ has no sign change, i.e., either $b(t) \geq 0$ for every t or $b(t) \leq 0$ for every t. Furthermore, if $b(t)$ changes sign, then $\left|\Omega_s^\theta(\lambda)\left(1 - e^\lambda \Omega_0^{-1}(\lambda)\right)^{-1}\right|$ grows exponentially.

Similarly, (68) can be estimated by

$$mK(\lambda) \leq \left|\Omega_s^\theta(\lambda)\Omega_0^{-1}(\lambda)\left(I - e^\lambda \Omega_0^{-1}(\lambda)\right)^{-1}\right| \leq MK(\lambda), \tag{75}$$

where m and M are positive constants and the function $K(\lambda)$ is defined by

$$K(\lambda) = \min\left\{e^{e^{-\operatorname{Re}\lambda} \cos \operatorname{Im} \lambda \,(\int_s^\theta b(\sigma) d\sigma - [b])}, e^{-\operatorname{Re}\lambda + e^{-\operatorname{Re}\lambda} \cos \operatorname{Im} \lambda \int_\theta^s b(\sigma) d\sigma}\right\}.$$

The analysis of the behavior of the function $K(\lambda)$ is analogous to the analysis of the function $H(\lambda)$. And, again, $K(\lambda)$ is bounded if and only if the function $b(t)$ has no sign change. If $b(t)$ changes sign, then (68) grows exponentially.

Summarizing we have proved the following theorem.

Theorem 4.1. *Let* Π_1 *denote the monodromy operator* (61) *associated with the periodic differential delay equation* (59). *If* $B(t) = b(t)$ *is a real scalar-valued periodic function with period* 1 *and has constant sign, then, for every* $\epsilon > 0$, *there exists a positive constant* M *such that, for* λ *outside circles of radius* ϵ *centered at the zeros of* $\lambda - [b]e^{-\lambda}$, *we have*

$$\|(e^\lambda I - \Pi_1)^{-1}\Pi_1\varphi\| \leq Me^{-\lambda} \sup_{-1 \leq \theta \leq 0} |b(\theta)\varphi(\theta)|.$$

The resolvent estimate allows us to present convergence results for the series of spectral projections introduced in Theorem 2.4. It also allows us to answer the question whether the eigenvectors and generalized eigenvectors of the monodromy operator Π_1 are dense in the space $\mathcal{C}$. But first we use Theorem 4.1 to derive estimates for $(I - z\Pi_1)^{-1}$. If we set $z = e^{-\lambda}$ in (65), then

$$(e^\lambda I - \Pi_1)^{-1} = z(I - z\Pi_1)^{-1}$$

and hence,

$$\|(I - z\Pi_1)^{-1}\Pi_1\varphi\| \leq M \sup_{-1 \leq \theta \leq 0} |b(\theta)\varphi(\theta)|, \tag{76}$$

for $z \in \mathbb{C}$ outside circles of radius ϵ centered at the zeros of $z - [b]e^{-z}$.

Furthermore, from (65) and representation (69) for $\Omega_s^t(\lambda)$, we obtain that, for $\psi \in \mathcal{R}(\Pi_1)$,

$$(I - z\Pi_1)^{-1}\psi = e^{z\int_0^\theta b(\sigma)\,d\sigma}\left[e^{-z[b]} - z\right]^{-1}\left(\psi(-1) + z\int_\theta^0 e^{z\int_s^0 b(\sigma)d\sigma}\psi'(s)\,ds + \int_{-1}^\theta e^{z\int_s^{-1} b(\sigma)d\sigma}\psi'(s)\,ds\right). \quad (77)$$

To analyze the existence of nontrivial small solutions for (59), we also need the following corollary.

Corollary 4.2. *Let Π_1 denote the monodromy operator (61) associated with the periodic differential delay equation (59). Suppose that $\varphi \in C$ is given such that the function $z \mapsto (I - z\Pi_1)^{-1}\varphi$ is an entire function. If $B(t) = b(t)$ is a nonzero real scalar-valued periodic function with period 1 which has constant sign, then $\varphi \in \mathcal{N}(\Pi_1^2)$.*

Proof. If $z \mapsto (I - z\Pi_1)^{-1}\varphi$ is an entire function, then $z \mapsto (I - z\Pi_1)^{-1}\Pi_1\varphi$ is an entire function. Thus, estimate (76) shows that the function is bounded and, hence, by Liouville's theorem, equal to a constant. From the Neumann series, we have

$$(I - z\Pi_1)^{-1}\Pi_1\varphi = \Pi_1\varphi + z\Pi_1^2\varphi + \cdots$$

and, hence, $\Pi_1^2\varphi = 0$. So $\varphi \in \mathcal{N}(\Pi_1^2)$. □

4.3. Series expansions for periodic equations

Consider a solution $x(\cdot, s; \varphi)$ of the periodic differential delay equation (59). In this section we use the basic estimates for the resolvent of the monodromy operator to study the series of spectral projections corresponding to the solution $x(\cdot, s; \varphi)$ introduced in Theorem 2.4 for scalar equations. Since the spectrum of $\Pi_1(s)$ is independent of s and $x_t(s; \varphi) = x_{t+s}(0; \varphi)$, we can restrict to the case $\Pi_1 = \Pi_1(0)$.

We shall prove the following theorem.

Theorem 4.3. *Suppose Π_1 denotes the monodromy operator associated with*

$$\dot{x}(t) = b(t)x(t-1), \qquad t \geq 0, \quad (78)$$

where $b(t)$ is a nonzero real scalar-valued periodic function with period 1. Let μ_j, $j = 1, 2, \ldots$, denote the nonzero eigenvalues of Π_1 ordered by decreasing modulus and $\varphi \in \mathcal{R}(\Pi_1^2)$. If $b(t)$ has constant sign, then the function φ has a convergent series expansion

$$\varphi = \lim_{\gamma \to -\infty} \sum_{|\mu_n| \geq \exp\gamma} P_{\mu_n}\varphi.$$

Proof. We use the notation from the proof of Theorem 2.4. Since the spectrum of Π_1 is a compact set, there exists a real R large so that $\sigma(\Pi_1)$ is contained in a ball with radius R. For a positive real number ρ, let Γ_ρ denote a circle with radius ρ centered at zero. Let γ be a negative real number and $\psi \in C$ be such that $\varphi = \Pi_1^2\psi$. If $k = k(\gamma)$ is the integer satisfying $|\mu_k| \geq e^\gamma$ and $|\mu_{k+1}| < e^\gamma$, then

$$\frac{1}{2\pi i}\int_{\Gamma_R}(wI - \Pi_1)^{-1}\Pi_1^2\varphi\, dw = \sum_{j=1}^{k} P_{\mu_j}\varphi + \frac{1}{2\pi i}\int_{\Gamma_{e^\gamma}}(wI - \Pi_1)^{-1}\Pi_1^2\varphi\, dw.$$

We also have that

$$\frac{1}{2\pi i}\int_{\Gamma_{e^\gamma}}(wI - \Pi_1)^{-1}\Pi_1^2\varphi\, dw = \frac{1}{2\pi i}\int_{\Gamma_{e^\gamma}} w(wI - \Pi_1)^{-1}\Pi_1\varphi\, dw.$$

Therefore, the series of spectral projections converges, as γ tends to $-\infty$, if and only if

$$\lim_{k\to\infty}\frac{1}{2\pi i}\int_{\Gamma_{e^{\gamma_k}}} w(wI - \Pi_1)^{-1}\Pi_1\varphi\, dw \quad \text{exists,} \tag{79}$$

where $\gamma_k \to -\infty$ as $k \to \infty$. Since, the eigenvalues μ_k can be represented by e^{λ_k}, where λ_k is a zero of $\lambda - [b]e^{-\lambda}$, we make the transformation of variables $w = z^{-1}$ to arrive at

$$\frac{1}{2\pi i}\int_{\Gamma_{e^{\gamma_k}}} w(wI - \Pi_1)^{-1}\Pi_1\varphi\, dw = \frac{1}{2\pi i}\int_{\Gamma_{e^{-\gamma_k}}}(I - z\Pi_1)^{-1}\Pi_1\varphi\,\frac{dz}{z^2}.$$

The contours $\Gamma_{e^{-\gamma_k}}$ can be assumed to be approximately circles at distance at least ϵ from the zeros of $\lambda - [b]e^{-\lambda}$. From estimate (76), it follows that

$$\left|\frac{1}{2\pi i}\int_{\Gamma_{e^{-\gamma_k}}}(I - z\Pi_1)^{-1}\Pi_1\varphi\,\frac{dz}{z^2}\right| \leq Me^{\gamma_k}$$

and the limit (79) exists. □

If the zeros of the coefficient $b(t)$ are isolated, then the operator Π_1 has a trivial kernel. This means that we can improve Corollary 4.2.

Corollary 4.4. *If $b(t)$ is a nonzero real scalar-valued periodic function with period 1 which has constant sign and isolated zeros, then the periodic differential delay equation*

$$\dot{x}(t) = b(t)x(t-1), \qquad t \geq 0,$$

has no nontrivial small solutions.

Proof. Let φ be the initial function of a nontrivial small solution. If we define $\psi = \Pi_1^2\varphi$, then ψ belongs to the range of Π_1^2 and also is the initial function of a nontrivial small solution. According to Theorem 4.3,

$$\psi = \sum_{j=1}^{\infty} P_{\mu_j}\psi.$$

The solution corresponding to the initial data $P_{\mu_j}\psi$ is given by $e^{\lambda_j t}p(t)$, where p is a periodic function of period 1 and $e^{\lambda_j} = \mu_j$. In particular, if the solution with initial data ψ decays faster than any exponential, then all the spectral projections

$P_{\mu_j}\psi$ are identical zero. Therefore, the functions $\mu \mapsto (\mu I - \Pi_1)^{-1}\psi$ and $z \mapsto z(I - z\Pi_1)^{-1}\psi$ are entire functions. Thus Corollary 4.2 implies that $\Pi_1^2\psi = 0$, hence $\psi = 0$ and also $\varphi = 0$. □

If we can prove that $\mathcal{R}(\Pi_1^2)$ is dense in $\mathcal{C}$, then Theorem 4.3 would also give us a completeness result. See [12, 35, 31] for related results. To prove that $\mathcal{R}(\Pi_1^2)$ is dense, it suffices to prove that Π_1^* has a trivial kernel. A direct computation yields that the adjoint of Π_1 is given by

$$(\Pi_1^* f)(s) = \begin{cases} 0 & \text{for } s = 0, \\ f(-1) + \int_s^0 b(\tau) f(\tau)\, d\tau & \text{for } -1 \le s < 0. \end{cases}$$

Therefore, if the zeros of b are isolated, then Π_1^* is one-to-one and $\mathcal{R}(\Pi_1)$ is dense in $\mathcal{C}$. An argument similar as given in the proof of Corollary 3.13 yields the following completeness result.

Corollary 4.5. *If $b(t)$ is a nonzero real scalar-valued periodic function with period 1 which has constant sign and isolated zeros, then the monodromy operator Π_1 defined by (61) has a complete span of eigenvectors and generalized eigenvectors.*

Observe from (63) and (64) that, if we can find $\psi \in \mathcal{C}$ such that

$$\psi(-1) + \int_{-1}^{0} \Omega_s^0(\lambda)\psi'(s)\, ds = 0, \tag{80}$$

then

$$\big((e^\lambda I - \Pi_1)^{-1}\psi\big)(\theta) = \int_{-1}^{0} \Omega_s^\theta(\lambda)\psi'(s)\, ds$$

is an entire function. Therefore, all the spectral projections $P_{\mu_j}\psi$ are identical to zero and Theorem 2.4 implies that ψ is the initial function of a nontrivial small solution.

To construct a function ψ that satisfies (80), we proceed as follows. If $b(t)$ changes sign, then there exist θ_1 and θ_2 such that $-1 \le \theta_1 \le \theta_2 \le 0$ and $\int_{\theta_1}^{\theta_2} b(\sigma)\, d\sigma = 0$. Define

$$\psi(\theta) = \begin{cases} 0 & \text{for } -1 \le \theta \le \theta_1, \\ \int_{\theta_1}^{\theta} b(\sigma)\, d\sigma & \text{for } \theta_1 \le \theta \le \theta_2, \\ 0 & \text{for } \theta_2 \le \theta \le 0, \end{cases} \tag{81}$$

then $\psi \in \mathcal{C}$ and $\psi'(\theta) = b(\theta)\chi_{[\theta_1,\theta_2]}$, where $\chi_{[\theta_1,\theta_2]}$ denotes the characteristic function of the interval $[\theta_1, \theta_2]$. Substitution of (81) into (80) yields

$$\begin{aligned}\int_{\theta_1}^{\theta_2} e^{e^{-\lambda}\int_s^0 b(\sigma)\,d\sigma} b(s)\,ds &= -e^{\lambda}\big[e^{e^{-\lambda}\int_{\theta_2}^0 b(\sigma)\,d\sigma} - e^{e^{-\lambda}\int_{\theta_1}^0 b(\sigma)\,d\sigma}\big] \\ &= -e^{\lambda}e^{e^{-\lambda}\int_{\theta_2}^0 b(\sigma)\,d\sigma}\big[1 - e^{e^{-\lambda}\int_{\theta_1}^{\theta_2} b(\sigma)\,d\sigma}\big] \\ &= 0 \qquad (\text{since } \int_{\theta_1}^{\theta_2} b(\sigma)\,d\sigma = 0).\end{aligned}$$

This proves that if b changes sign, then there exist nontrivial small solutions. One can actually compute the small solution corresponding to the initial data ψ given by (81), by iterating the period map Π_1 acting on ψ. If ψ is given by (81), then

$$(\Pi_1\psi)(\theta) = \begin{cases} 0 & \text{for } -1 \le \theta \le \theta_1, \\ \int_{\theta_1}^{\theta} b(s) \int_{\theta_1}^{s} b(\sigma)\,d\sigma\,ds & \text{for } \theta_1 \le \theta \le \theta_2, \\ \int_{\theta_1}^{\theta_2} b(s) \int_{\theta_1}^{s} b(\sigma)\,d\sigma\,ds & \text{for } \theta_2 \le \theta \le 0. \end{cases}$$

This yields

$$(\Pi_1\psi)(\theta) = \begin{cases} 0 & \text{for } -1 \le \theta \le \theta_1, \\ 1/2\big[\int_{\theta_1}^{\theta} b(\sigma)\,d\sigma\big]^2 & \text{for } \theta_1 \le \theta \le \theta_2, \\ 0 & \text{for } \theta_2 \le \theta \le 0 \end{cases}$$

and, by induction,

$$(\Pi_1^n\psi)(\theta) = \begin{cases} 0 & \text{for } -1 \le \theta \le \theta_1, \\ 1/(n+1)!\big[\int_{\theta_1}^{\theta} b(\sigma)\,d\sigma\big]^n & \text{for } \theta_1 \le \theta \le \theta_2, \\ 0 & \text{for } \theta_2 \le \theta \le 0. \end{cases} \tag{82}$$

Since $x(n+\theta;\psi) = (\Pi_1^n\psi)(\theta)$, this is an explicit representation for the small solution with initial condition ψ given by (81).

Note that for periodic delay equations, small solutions are no longer necessarily identically zero after finite time. In general, not much is known about the exact decay rate of a nontrivial small solution of a periodic differential delay equation. We can make, however, the following observation. If ψ is the initial data corresponding to a small solution, then we have seen that $z \mapsto (I - z\Pi_1)^{-1}\psi$ is an entire function, since

$$(I - z\Pi_1)^{-1}\psi = \sum_{n=1}^{\infty} z^n \Pi_1^n \psi$$

is an entire function of finite exponential type, it follows that

$$\|x_n(\psi)\| = \|\Pi_1^n\psi\| \le M\frac{c^n}{n!}.$$

The small solution constructed above decays at this rate, but we do not know if there exist nontrivial small solutions of periodic delay equations that can actually decay even faster.

Let $\mathcal{M}$ denote the linear space spanned by the eigenvectors and generalized eigenvectors of Π_1 defined by (61). Since

$$\overline{\mathcal{M}}^{\perp} = \bigcap_{\mu\in\sigma(\Pi_1^*)\setminus\{0\}} \mathcal{N}(P_\mu^*) = \{ f \mid z \mapsto (I - z\Pi_1^*)^{-1}\varphi \text{ is entire } \},$$

a similar argument can be used to show that, if $b(t)$ changes sign, then the system of eigenvectors of Π_1 is not complete.

As we have seen for autonomous equations, completeness of the system of eigenvectors and generalized eigenvectors of the monodromy operator can also be studied directly, using an explicit characterization of $\overline{\mathcal{M}}$ and the exponential type calculus. For the present operator Π_1, this is an illustration of a very general approach, which can be used to analyze completeness questions for many different compact operators (see [10] for details and general results).

To apply the exponential type calculus introduced in Section 3.1, it is easier to work with the function $z \mapsto (I - z\Pi_1)^{-1}\psi$. Recall, from representation (77), that $(I - z\Pi_1)^{-1}\psi$ is a quotient of entire functions of finite exponential type. Similar as before, define

$$\mathcal{E} = \{ \psi \in \mathcal{C} \mid \mathrm{E}\big((e^{-z[b]} - z)(I - z\Pi_1)^{-1}\psi\big) \le [b] \}.$$

Exactly as for autonomous equations, it follows that $\mathcal{E}$ is a closed subspace of $\mathcal{C}$ which contains $\overline{\mathcal{M}}$. Furthermore, if $\psi \in \mathcal{E} \cap \mathcal{R}(\Pi_1^2)$, then the function

$$\lambda \mapsto (e^{\lambda} I - \Pi_1)^{-1}\psi$$

is uniformly bounded away from the poles. An inspection of the proof of Theorem 4.3 shows that ψ has a convergent spectral projection series and $\overline{\mathcal{M}} = \mathcal{E}$. Furthermore, it follows from (77) that $\mathcal{E} = \mathcal{C}$ when $b(t)$ has constant sign. If $b(t)$ changes sign, we have constructed an initial function ψ such that $(I - z\Pi_1)^{-1}\psi$ is an entire function of positive exponential type and, hence, $\mathcal{E} \neq \mathcal{C}$.

Actually, we have the following theorem (see [31] and [10] for a general result for a large class of compact operators).

Theorem 4.6. *Let $b(t)$ be a nonzero real scalar-valued periodic function with period 1 and let Π_1 be the monodromy operator associated with*

$$\dot{x}(t) = b(t)x(t-1), \qquad t \ge 0.$$

If $\mathcal{M}$ denotes the linear span of the eigenvectors of Π_1 and

$$\mathcal{S} = \{ \varphi \in \mathcal{C} \mid z \mapsto (I - z\Pi_1)^{-1}\varphi \text{ is entire } \},$$

then

$$\mathcal{C} = \overline{\overline{\mathcal{M}} \oplus \mathcal{S}}.$$

We like to remark that our results can be extended to periodic differential delay equations

$$\dot{x}(t) = b(t)x(t-1), \qquad t \geq 0,$$

where b is a scalar-valued complex periodic function of period 1. An inspection of the proof in the real case, see (73) and (74), shows that these conditions have to be satisfied for both the real and imaginary part of $b(t)$. Therefore, the behavior of the curve $\zeta : [0,1] \to \mathbb{C}$ given by

$$\zeta(t) = \int_{-1}^{t} b(s)\, ds$$

determines whether the monodromy operator Π_1 has a complete set of eigenvectors. In fact, we have seen that completeness fails if there exist $-1 \leq \theta_1 < \theta_2 \leq 0$ such that $\int_{\theta_1}^{\theta_2} b(s)\, ds = 0$. Or, equivalently, if the curve $\zeta(t)$ has a self intersection. More general, the monodromy operator Π_1 has a complete set of eigenvectors if and only if the curve $\zeta(t)$ is an invertible function.

We state the result as a theorem (see [25] for details and further generalizations).

Theorem 4.7. *If $b(t)$ is a nonzero complex scalar-valued periodic function with period 1 such that the real and imaginary part of $b(t)$ have constant sign, then the monodromy operator Π_1 associated with*

$$\dot{x}(t) = b(t)x(t-1), \qquad t \geq 0,$$

has a complete span of eigenvectors.

To conclude this paper, we remark that we can extend our results to the case that the period ω is rational. By considering an appropriate iterate of the monodromy operator, we can actually assume that the period an integer multiple of the delay, i.e.,

$$\dot{x}(t) = b(t)x(t-1), \tag{83}$$

where $b(t) = b(t+m)$, for some positive integer m. The proof in this general setting reduces to estimates for the fundamental matrix $\Omega_s^t(\lambda)$ of the ordinary differential equation

$$\dot{y}(t) = e^{-\lambda} B y(t),$$

where the matrix B is given by

$$B = \begin{bmatrix} 0 & \cdots & \cdots & 0 & b(t) \\ b(t+1) & 0 & \cdots & \cdots & 0 \\ 0 & b(t+2) & 0 & \cdots & 0 \\ \vdots & \ddots & \ddots & \ddots & \vdots \\ 0 & \cdots & 0 & b(t+(m-1)) & 0 \end{bmatrix}$$

Again a sign change of the coefficient b in (83) determines whether or not the system of eigenvectors and generalized eigenvectors of the monodromy operator Π is complete in $\mathcal{C}$ (see [22, 25]).

The more irrational the period ω becomes, the larger the size of the matrix B and the spectral data of the monodromy operator Π associated with (83) is not yet understood for ω irrational.

References

[1] H. Amann, *Gewöhnliche Differentialgleichungen*, de Gruyter, Berlin, 1983.

[2] H. T. Banks and A. Manitius, *Projection series for retarded functional differential equations with applications to optimal control problems*, J. Diff. Eqn., **18** (1975), 296–332.

[3] R. Bellman and K. L. Cooke, *Differential-Difference Equations*, Academic Press, New York, 1963.

[4] R. Boas, *Entire Functions*, Academic Press, New York, 1954.

[5] M. C. Delfour and A. Manitius, *The structural operator F and its role in the theory of retarded systems I*, J. Math. Anal. Appl., **73** (1980), 466–490.

[6] M. C. Delfour and A. Manitius, *The structural operator F and its role in the theory of retarded systems II*, J. Math. Anal. Appl., **74** (1980), 359–381.

[7] O. Diekmann, S. A. van Gils, S. M. Verduyn Lunel, and H. O. Walther, *Delay Equations: Functional-, Complex-, and Nonlinear Analysis*, Springer-Verlag, New York, Applied Mathematical Sciences Vol. 110, 1995.

[8] L. Glass and M. C. Mackey, *Pathological conditions resulting from instabilities in physiological control systems*, Ann. N.Y. Acad. Sci., **316** (1979), 214–235.

[9] I. Gohberg, S. Goldberg, and M.A. Kaashoek, *Classes of Linear Operators I*, Birkhäuser Verlag, Basel, 1990.

[10] I. Gohberg, M. A. Kaashoek, and S.M. Verduyn Lunel, *New completeness theorems for compact operators and applications*, In preparation.

[11] I. C. Gohberg and E. I. Sigal, *An operator generalization of the logarithmic residue theorem and the theorem of Rouché*, Mat. Sb., **84** (1971), 609–629. (Russian) (Math. USSR-Sb., **13** (1971), 603–625.)

[12] W. Hahn, *On difference differential equations with periodic coefficients*, J. Math. Anal. Appl., **3** (1961), 70–101.

[13] J. K. Hale and S. M. Verduyn Lunel, *Introduction to Functional Differential Equations*, Springer-Verlag, New York, Applied Mathematical Sciences Vol. 99, 1993.

[14] J.K. Hale and S. M. Verduyn Lunel, *Effects of small delays on stability and control*, Operator Theory and Analysis, The M. A. Kaashoek Anniversary Volume (eds. H. Bart, I. Gohberg and A. C. M. Ran), Operator Theory: Advances and Applications, Vol. 122, pp. 275–301, Birkhäuser, 2001.

[15] D. Henry, *Small solutions of linear autonomous functional differential equations*, J. Differential Eqns., **8** (1970), 494–501.

[16] E. Hille and R. Phillips, *Functional Analysis and Semigroups*, American Mathematical Society, Providence, RI, 1957.

[17] M.A. Kaashoek and S.M. Verduyn Lunel, *Characteristic matrices and spectral properties of evolutionary systems*, Trans. Amer. Math. Soc., **334** (1992), 479–517.

[18] T. Kato, *Perturbation Theory for Linear Operators (2nd edn.)*, Springer-Verlag, Berlin, 1976.

[19] B. Krauskopf and D. Lenstra, *Fundamental Issues of Nonlinear Laser Dynamics*, AIP Conference Proceedings, Volume 548, American Institute of Physics, Melville, New York, 2000.

[20] B. Ja. Levin, *Distribution of Zeros of Entire Function*, Amer. Math. Soc., Providence, 1972.

[21] N. Levinson and C. McCalla, *Completeness and independence of the exponential solutions of some functional differential equations*, Studies in Appl. Math., **53** (1974), 1–15.

[22] Yu. I. Lyubich and V. A. Tkachenko, *Floquet's theory for equations with retarded argument*, Diff. Uravneniya, **5** (1969), 648–656.

[23] J. Mallet-Paret and S. M. Verduyn Lunel, *Asymptotics of the Floquet multipliers for periodic delay equations with rational period*, In preparation.

[24] A. Manitius, *Completeness and F-completeness of eigenfunctions associated with retarded functional differential equations*, J. Differential Eqns., **35** (1980), 1–29.

[25] V. Matsaev and S. M. Verduyn Lunel, *Spectral theory for periodic differential delay equations*, In preparation.

[26] D. Salamon, *Control and Observation of Neutral Systems*, Research Notes on Mathematics 91, Pitman, London, 1984.

[27] S. M. Verduyn Lunel, *A sharp version of Henry's theorem on small solutions*, J. Differential Eqns., **62** (1986), 266–274.

[28] S. M. Verduyn Lunel, *Series expansions and small solutions for Volterra equations of convolution type*, J. Differential Eqns., **85** (1990), 17–53.

[29] S. M. Verduyn Lunel, *The closure of the generalized eigenspace of a class of infinitesimal generators*, Proc. Roy. Soc. Edinburgh Sect. A, **117** (1991), 171–192.

[30] S. M. Verduyn Lunel, *Series expansions for functional differential equations*, Integral Equations and Operator Theory, **22** (1995), 93–123.

[31] S. M. Verduyn Lunel, *About completeness for a class of unbounded operators*, J. Differential Eqns., **120** (1995), 108–132.

[32] S. M. Verduyn Lunel, *Inverse problems for nonself-adjoint evolutionary systems*, Topics in Functional Differential and Difference Equations (eds. T. Faria and P. Freitas), Fields Institute Communications, Vol. 29, pp. 321–347, American Mathematical Society, Providence, 2001.

[33] S. M. Verduyn Lunel and D. V. Yakubovich, *A functional model approach to linear neutral functional differential equations*, Integral Equations and Operator Theory, **27** (1997), 347–378.

[34] H.-O. Walther, *Bifurcation from periodic solutions in functional differential equations*, Math. Z., **182** (1983), 269–289.

[35] A. M. Zverkin, *The completeness of a system of Floquet type solutions for equations with retardations*, Differential Equations, **4** (1968), 249–251.

2000 Mathematics Subject Classification: Primary 47N20; Secondary 34K05

Mathematisch Instituut,
Universiteit Leiden,
P.O. Box 9512,
2300 RA Leiden, The Netherlands
E-mail address: verduyn@math.leidenuniv.nl

On a singular perturbation of a multiplication operator

D. Yafaev

Abstract. We consider a model in perturbation theory where the Hamiltonian $\mathbf{H}$ is the sum of the operator $\mathbf{H}_0$ of multiplication by $|x|^{2l}$ in the space $L_2(\mathbb{R}^d)$ and of the integral operator $\mathbf{V}$ with kernel $v(\langle x, y\rangle)$. The real function v is supposed to be periodic with the zero mean value. We find necessary and sufficient conditions for the negative spectrum of the operator $\mathbf{H}$ to be finite and calculate explicitly the total number $\mathbf{N}$ of its negative eigenvalues. It turns out that $\mathbf{N}$ (if it is finite) is determined by the set of p such that $v^{(p)}(0) < 0$.

1.

Let H_0 be a self-adjoint positive operator in the Hilbert space $\mathcal{H}$ and let a symmetric operator V be in some sense compact relative to H_0. Then the essential spectra of the operators H_0 and $H = H_0 + V$ coincide. Our goal here is to discuss additional assumptions which imply that the negative spectrum of the operator H is finite and to calculate the total number N of its negative eigenvalues. In the abstract framework the answer is given by the Birman-Schwinger principle which, roughly speaking, means the following. Let $V \leq 0$. Suppose that the form $(H_0^{-1/2} V H_0^{-1/2} u, u)$, after removal of its "singular" part (of dimension $r < \infty$), equals (Au, u) where the operator A is well-defined in the space $\mathcal{H}$. Denote by M the total number of eigenvalues of A lying below the point -1. Then $N = M + r$. In particular, if A is compact, then the negative spectrum of H is finite. If A is only bounded, then the operator $H_\gamma = H_0 + \gamma V$ has exactly r negative eigenvalues for sufficiently small coupling constants γ. The precise formulation of these results can be found in the original paper [3] (see also the book [7]).

We present here a new formulation of the Birman-Schwinger principle adapted to the case where V is not sign-definite and illustrate it on the example of the model mentioned in the abstract. We emphasize that our definition of the singular part of the form $(H_0^{-1/2} V H_0^{-1/2} u, u)$ is technically different from that of [3]. Let us start however with a discussion of some well-known results.

A perturbation of a multiplication operator H_0 by an integral operator V is called the Friedrichs or Friedrichs-Faddeev model. To be more precise, let $\mathcal{G} =$

$L_2(\mathbb{R}_+;\mathfrak{N})$ where $\mathfrak{N}$ is an auxiliary Hilbert space. We set $(H_0 f)(\lambda) = \lambda f(\lambda)$ and

$$(Vf)(\lambda) = \int_0^\infty g(\lambda,\mu) f(\mu) d\mu. \tag{1}$$

The "kernel" $g(\lambda,\mu)$ is assumed to be a compact operator in $\mathfrak{N}$ for all $\lambda,\mu \in \mathbb{R}_+$ and to satisfy the estimates

$$|g(\lambda,\mu)| \leq C(1+\lambda+\mu)^{-\theta}, \quad \theta > 1/2, \tag{2}$$

and

$$|g(\lambda',\mu') - g(\lambda,\mu)| \leq C(|\lambda'-\lambda|^\alpha + |\mu'-\mu|^\alpha)(1+\lambda+\mu)^{-\theta} \tag{3}$$

for some $\alpha > 1/2$. Moreover, for all $\lambda,\mu \in \mathbb{R}_+$

$$g(\lambda,0) = g(0,\mu) = 0. \tag{4}$$

The kernel $g(\lambda,\mu) = g^*(\mu,\lambda)$ is Hermitian, so that the perturbation V is symmetric and the operator H is self-adjoint on domain $\mathcal{D}(H) = \mathcal{D}(H_0)$. Under these assumptions it is shown in [5] that the wave operators $W_\pm(H,H_0)$ for the pair H_0, $H = H_0 + V$ exist and are complete. Moreover, the operator H does not have the singular continuous spectrum, and its point spectrum is finite. Of course, all these results can be generalized to the case where the role of $\mathbb{R}_+$ is played by an arbitary (possibly, bounded) interval.

Actually, as shown in [9], the wave operators $W_\pm(H,H_0)$ exist and are complete under the assumption $\alpha > 0$. On the other hand, if $\alpha < 1/2$ the singular continuous spectrum of H might appear (see [6]) even for a one-dimensional perturbation when $g(\lambda,\mu) = g_0(\lambda)\overline{g_0(\mu)}$. Our main concern here is a study of the discrete spectrum. In particular, we show in Proposition 4.3 that for an arbitrary $\alpha > 0$ the discrete spectrum of the operator H is finite and the discrete spectrum of the operator $H_\gamma = H_0 + \gamma V$ is empty if γ is small enough.

As another example, we consider the Schrödinger operator $H = H_0 + V$ in the space $\mathcal{H} = L_2(\mathbb{R}^d)$ where $H_0 = -\Delta$ and V is multiplication by a real bounded function satisfying the condition $v(x) = o(|x|^{-2})$ as $|x| \to \infty$. In this case the corresponding operator A is compact and the deficiency dimension r equals 1 for $d = 1,2$ and $r = 0$ for $d \geq 3$. Therefore the operator H has only finite negative spectrum. Moreover, the operators $H_\gamma = H_0 + \gamma V$ are nonnegative for sufficiently small γ if $d \geq 3$. If $d = 1,2$ and $V \leq 0$, $V \neq 0$, then the operators $H_\gamma = H_0 + \gamma V$ have exactly one negative eigenvalue for small $\gamma > 0$. Similar results are true for the perturbed polyharmonic operator $H = H_0 + V$ where $H_0 = (-\Delta)^l$ and $l > 0$ is not necessarily integer. This operator has only finite discrete spectrum if

$$v(x) = o(|x|^{-2l}), \quad |x| \to \infty, \tag{5}$$

and the operators $H_\gamma = H_0 + \gamma V$ are nonnegative for sufficiently small γ if $2l < d$.

The condition (5) is in some sense optimal. If $v(x)$ decays as $v_0|x|^{-2l}$ at infinity, then the corresponding operator A has continuous spectrum and hence for sufficiently large negative v_0 the negative spectrum of the operator H is infinite.

The critical value of v_0 can be explicitly calculated. Suppose that $2l < d$ and take into account the inequality (see [12])

$$\int_{\mathbb{R}^d} |x|^{-2l}|u(x)|^2 dx \leq C_l \int_{\mathbb{R}^d} |\xi|^{2l}|\hat{u}(\xi)|^2 d\xi, \tag{6}$$

where the best possible constant

$$C_l = 2^{-2l}\frac{\Gamma^2((d/2-l)/2)}{\Gamma^2((d/2+l)/2)}$$

(Γ is the gamma-function). It can be deduced from (6) that the negative spectrum of the operator $H = H_0 + V$ is finite if

$$v(x) \geq -C_l^{-1}|x|^{-2l},$$

for sufficiently large $|x|$. On the contrary, it is probably infinite (strictly speaking, we have not found this result in the literature) if for large $|x|$

$$v(x) \leq -v_0|x|^{-2l}, \quad v_0 > C_l^{-1}.$$

Of course, the Schrödinger or the perturbed polyharmonic operators can be reduced to the Friedrichs model. Let Φ be the Fourier transform. If $H_0 = (-\Delta)^l$, then the operator $\Phi H_0 \Phi^*$ acts (in the momentum representation) as multiplication by the function $|\xi|^{2l}$. The operator $\Phi V \Phi^*$ acts as convolution:

$$(\Phi V \Phi^* \hat{f})(\xi) = (2\pi)^{-d/2}\int_{\mathbb{R}^d} \hat{v}(\xi-\eta)\hat{f}(\eta)d\eta,$$

where $\hat{v} = \Phi v$ and $\hat{f} = \Phi f$. Passing to the spherical coordinates, considering the space $L_2(\mathbb{R}^d)$ as $L_2(\mathbb{R}_+;\mathfrak{N})$ with $\mathfrak{N} = L_2(\mathbb{S}^{d-1})$ and making the change of variables $\lambda = |\xi|^{2l}$, we reduce by a unitary transformation the operator H_0 to multiplication by λ and the operator V of multiplication by $v(x)$ to the operator (1) with kernel $g(\lambda,\mu) : \mathfrak{N} \to \mathfrak{N}$ defined by

$$(g(\lambda,\mu)a)(\omega) = (2\pi)^{-d/2}(2l)^{-1}(\lambda\mu)^{(d-2l)/(4l)}\int_{\mathbb{S}^{d-1}} \hat{v}(\lambda^{1/(2l)}\omega - \mu^{1/(2l)}\omega')a(\omega')d\omega'.$$

This kernel takes compact values and it is a Hölder continuous function of $\lambda, \mu > 0$. Moreover, it satisfies the boundary condition (4) if $2l < d$. On the contrary, it is singular at $\lambda = 0$ and $\mu = 0$ if $2l > d$. This explains the qualitative difference in the behaviour of the discrete spectrum of the perturbed polyharmonic operator from the point of view of the Friedrichs model.

Our aim here is to discuss a singular version of the Friedrichs model. Let $\mathbf{H}_0$ be multiplication by $|x|^{2l}$, $l > 0$, in the space $L_2(\mathbb{R}^d)$, let $\mathbf{V}$ be an integral operator defined by

$$(\mathbf{V}u)(x) = \int_{\mathbb{R}^d} v(\langle x, y\rangle)u(y)dy, \tag{7}$$

and $\mathbf{H} = \mathbf{H}_0 + \mathbf{V}$. We suppose that v is a real periodic function

$$v(t) = \sum_{n=-\infty}^{\infty} v_n e^{int}, \quad v_{-n} = \bar{v}_n, \quad v_0 = 0. \tag{8}$$

Under a simple condition on the behaviour of the Fourier coefficients v_n as $n \to \infty$, the operator $\mathbf{V}$ is bounded. The case of a constant function $v(t)$ corresponds [2] to the point interaction and is excluded by the condition $v_0 = 0$. The operator $\mathbf{V}$ can be viewed as a Fourier integral operator. For example, if

$$v(t) = \gamma \cos t, \quad \text{or} \quad v(t) = \gamma \sin t, \tag{9}$$

then (7) is a linear combination of the classical Fourier transform and of its adjoint. In this case the model we consider first appeared in [4], and the discrete spectrum of the corresponding operator $\mathbf{H}$ was studied in [10].

Of course, passing again to the spherical coordinates, considering the space $L_2(\mathbb{R}^d)$ as $L_2(\mathbb{R}_+;\mathfrak{N})$ with $\mathfrak{N} = L_2(\mathbb{S}^{d-1})$ and making the change of variables $\lambda = |x|^{2l}$, we reduce by a unitary transformation the operator $\mathbf{H}_0$ to multiplication by λ and the operator $\mathbf{V}$ to the operator (1) with kernel $g(\lambda,\mu) : \mathfrak{N} \to \mathfrak{N}$ defined by

$$(g(\lambda,\mu)a)(\omega) = (2l)^{-1}(\lambda\mu)^{(d-2l)/(4l)} \int_{\mathbb{S}^{d-1}} v((\lambda\mu)^{1/(2l)}\langle\omega,\omega'\rangle)a(\omega')d\omega'. \tag{10}$$

This kernel does not satisfy the assumptions of the Friedrichs model either for $\lambda = 0$ and $\mu = 0$ or for $\lambda \to \infty$ and $\mu \to \infty$ (or in both these cases). However, according to (10) it can be expected that the cases $2l < d$ and $2l \geq d$ are qualitatively different.

Due to oscillations of its kernel, the operator $\mathbf{V}$ is bounded in the space $L_2(\mathbb{R}^d)$, but it is not compact, even relatively with respect to $\mathbf{H}_0$. Nevertheless, as shown in [11], the essential spectra of the operators $\mathbf{H}$ and $\mathbf{H}_0$ coincide. This implies that the negative spectrum of the operator $\mathbf{H}$ consists of eigenvalues of finite multiplicity which may accumulate at the bottom of the essential spectrum (point zero) only. Our main goal is to study the discrete spectrum of the operator $\mathbf{H} = \mathbf{H}_0 + \mathbf{V}$. As we shall see below, it is very sensitive with respect to small variations of the parameter l and of the function v.

The model we consider is also rather interesting from the point of view of scattering theory. Recall that there are two completely different approaches, the smooth and the trace class, in scattering theory (this is discussed thoroughly in [9]). Normally, the smooth method works efficiently if the operator H_0 can be explicitly diagonalized; for example, the paper [5] relies on the smooth technique. On the contrary, if no information on eigenfunctions of H_0 is available, then only the trace-class method can be used. Nevertheless, although the operator $\mathbf{H}_0$ is diagonal only the trace class method seems to work [11] for the proof of the existence and completeness of the wave operators $W_\pm(\mathbf{H},\mathbf{H}_0)$ (see Theorem 2.2 below). The singular spectrum (or rather its absence) of the operator $\mathbf{H}$ has not been investigated.

For the study of the discrete spectrum, we proceed from the version of the Birman-Schwinger principle suggested in [10]. The corresponding operator $\mathbf{A}$ is not compact and its spectrum coincides with an interval $[-\mathbf{a}, \mathbf{a}]$. Since $\mathbf{V}$ is not sign-definite, in our case there are two "deficiency indices" r and r' corresponding, roughly speaking, to negative and positive singular parts of the form $(\mathbf{H}_0{}^{-1/2}\mathbf{V}\mathbf{H}_0^{-1/2}u, u)$. If $l - d/2$ is a nonnegative integer and $v^{(l-d/2)}(0) \neq 0$, then the spectrum of $\mathbf{A}$ covers the whole line. This implies that the negative spectrum of the operator $\mathbf{H}$ is infinite. If $l - d/2$ is not a nonnegative integer, then the negative spectrum of the operator $\mathbf{H}$ is infinite for $\mathbf{a} > 1$. Otherwise, it is finite and the total number of negative eigenvalues coincides with the negative "deficiency index" r. In our model it is determined by the set of p such that $v^{(p)}(0) < 0$.

From analytical point of view, we rely on separation of variables in spherical coordinates. Indeed, the space $L_2(\mathbb{R}^d)$ decomposes into the orthogonal sum of subspaces $\mathfrak{H}_m$, $m = 0, 1, 2, \ldots$ constructed in terms of the spherical functions of order m and invariant with respect to the operators $\mathbf{H}_0$ and $\mathbf{V}$. On the subspace $\mathfrak{H}_m$, the operator $\mathbf{H}$ reduces to operator $H^{(m)}$ acting in the space $L_2(\mathbb{R}_+)$. Operators $H^{(m)}$ have the form $H^{(m)} = H_0 + V_m$ where H_0 is again multiplication by r^{2l}, $l > 0$, and

$$(V_m f)(r) = \int_0^\infty v_m(rs) f(s) ds. \tag{11}$$

The functions v_m are expressed in terms of Bessel functions, and the corresponding operators A_m are diagonalized by the Mellin transform. We calculate the total number of negative eigenvalues of each operator H_m which implies the same result for the operator $\mathbf{H}$.

The paper is organized as follows. In Section 2 we give a precise definition of the operator $\mathbf{H}$ and recall the results of [11] on scattering theory for the pair $\mathbf{H}_0$, $\mathbf{H}$. The reduction of $\mathbf{H}$ to the orthogonal sum of the operators $H^{(m)}$ is described in Section 3. The modified form of the Birman-Schwinger principle is given in Section 4. In Section 5 we obtain a convenient representation for the form $(V_m H_0^{-1/2} u, H_0^{-1/2} u)$. All these results are used in Sections 6 and 7 to calculate the total number of negative eigenvalues of the operators $H^{(m)}$ and $\mathbf{H}$, respectively.

2.

It is convenient to rewrite (7) in operator notation. Let Φ,

$$(\Phi u)(x) = (2\pi)^{-d/2} \int_{\mathbb{R}^d} e^{-i\langle x, y\rangle} u(y) dy,$$

be the Fourier operator and let $\mathbf{T}_n$,

$$(\mathbf{T}_n u)(x) = n^{d/2} u(nx),$$

be the unitary operator of dilations in $L_2(\mathbb{R}^d)$. Then equality (7) can be rewritten as

$$\mathbf{V} = (2\pi)^{d/2} \sum_{n=1}^{\infty} n^{-d/2} \mathbf{T}_n \, (v_n \Phi^* + \bar{v}_n \Phi), \tag{12}$$

which we take for the precise definition of the operator $\mathbf{V}$. Under the assumption

$$\sum_{n=1}^{\infty} |v_n| n^{-d/2} < \infty \tag{13}$$

this series is convergent in norm so that the operator $\mathbf{V}$ is bounded. This implies of course that the operator $\mathbf{H}$ is self-adjoint on the domain $\mathcal{D}(\mathbf{H}_0)$ of the operator $\mathbf{H}_0$.

As a side remark, we note that the operator

$$\tilde{\mathbf{V}} = \Phi \mathbf{V} \Phi^* = (2\pi)^{d/2} \sum_{n=1}^{\infty} n^{-d/2} \mathbf{T}_{1/n} \, (v_n \Phi^* + \bar{v}_n \Phi)$$

has the structure similar to that of $\mathbf{V}$. In particular, if (8) contains only a finite number of terms, then $\tilde{\mathbf{V}}$ can be written as an integral operator (7) with kernel $\tilde{v}(\langle x, y \rangle)$ where $\tilde{v}(t)$ is again a periodic function of t (but of course its period is not necessarily 2π); for example, $\tilde{v}(t) = v(t)$ in the case (9). Thus, all our results can be reformulated in terms of the Hamiltonian $\tilde{\mathbf{H}} = \tilde{\mathbf{H}}_0 + \tilde{\mathbf{V}}$ where $\tilde{\mathbf{H}}_0 = (-\Delta)^l$.

The operator $\mathbf{V}$ is of course not compact and even not relatively compact. Nevertheless it is easy to deduce from representation (12) that the product

$$(\mathbf{H}_0 + cI)^{-1} \mathbf{V} (\mathbf{H}_0 + cI)^{-1}$$

is compact. By the Weyl theorem, this implies that the essential spectra of the operators $\mathbf{H}_0$ and $\mathbf{H}$ coincide. So we arrive at the following preliminary result.

Theorem 2.1. *Let condition* (13) *hold. Then the operator* $\mathbf{V}$ *is bounded and the operator* $\mathbf{H}$ *is self-adjoint on the domain* $\mathcal{D}(\mathbf{H}) = \mathcal{D}(\mathbf{H}_0)$. *Moreover,* $\sigma_{ess}(\mathbf{H}) = [0, \infty)$.

A more detailed analysis, based on the trace class method in scattering theory (see [11] for the proof), leads to the following result.

Theorem 2.2. *Let* $l > 1/2$ *and let the condition*

$$\sum_{n=1}^{\infty} |v_n| n^{-d/2+1+\varepsilon} < \infty, \quad \varepsilon > 0, \tag{14}$$

hold. Then the wave operators

$$W_{\pm}(\mathbf{H}, \mathbf{H}_0) = s - \lim_{t \to \pm\infty} \exp(i\mathbf{H}t) \exp(-i\mathbf{H}_0 t)$$

exist and are complete, that is their ranges coincide with the absolutely continuous subspace of the operator $\mathbf{H}$. *In particular, the absolutely continuous part of* $\mathbf{H}$ *is unitarily equivalent to the operator* $\mathbf{H}_0$.

We emphasize that both conditions (13) and (14) are getting less restrictive as d increases.

Note that in the case (9) the result of Theorem 2.2 was established in [4] under the assumption $2l > d$.

3.

Our study of the discrete spectrum of the operator $\mathbf{H}$ relies on separation of variables in the spherical coordinates (actually, this has already been used in [11] for the proof of Theorem 2.2). The operators $\mathbf{H}_0$ and $\mathbf{V}$ are rotation invariant. Let $\mathfrak{h}_m$ be the subspace of spherical functions $Y_m(\omega)$, $\omega \in \mathbb{S}^{d-1}$, of order m. Recall that

$$\dim \mathfrak{h}_m = (2m + d - 2)(m + d - 3)!((d-2)!m!)^{-1} =: \nu_m. \tag{15}$$

Let $\mathcal{K}$ be the L_2-space with weight r^{d-1} of functions defined on $\mathbb{R}_+$ and let $\mathfrak{H}_m = \mathcal{K} \otimes \mathfrak{h}_m$. To put it differently, $\mathfrak{H}_m \subset L_2(\mathbb{R}^d)$ is the subspace of functions u_m of the form

$$u_m(x) = |x|^{-\delta} f(|x|) Y_m(\hat{x}), \quad \hat{x} = x|x|^{-1}, \quad \delta = (d-1)/2, \tag{16}$$

where $f \in L_2(\mathbb{R}_+)$ and $Y_m \in \mathfrak{h}_m$. Then

$$L_2(\mathbb{R}^d) = \bigoplus_{m=0}^{\infty} \mathfrak{H}_m, \quad \mathfrak{H}_m = \mathcal{K} \otimes \mathfrak{h}_m,$$

and every subspace $\mathfrak{H}_m$ is invariant with respect to the Fourier operator Φ, which reduces (see, e.g., [8]) to the Fourier-Bessel transform on $\mathfrak{H}_m$. More precisely, set

$$(\Phi_m f)(r) = \int_0^\infty (rs)^{1/2} \mathcal{I}_{m+(d-2)/2}(rs) f(s) ds, \tag{17}$$

where $\mathcal{I}_{m+(d-2)/2}$ is the Bessel function. Then, for function (16),

$$(\Phi u_m)(x) = i^{-m} |x|^{-\delta} (\Phi_m f)(|x|) Y_m(\hat{x}). \tag{18}$$

The operator Φ_m is of course unitary on $L_2(\mathbb{R}_+)$ and $\Phi_m = \Phi_m^*$. Clearly,

$$(\mathbf{T}_n u_m)(x) = |x|^{-\delta} (T_n f)(|x|) Y_m(\hat{x}),$$

where T_n,

$$(T_n f)(r) = n^{1/2} f(nr), \tag{19}$$

is the unitary operator of dilations in $L_2(\mathbb{R}_+)$.

Let V_m be the operator (11) with

$$v_m(t) = (2\pi)^{d/2} \sum_{n=1}^{\infty} n^{-\delta} (v_n i^m + \bar{v}_n i^{-m}) (nt)^{1/2} \mathcal{I}_{m+(d-2)/2}(nt). \tag{20}$$

According to (17)

$$V_m = (2\pi)^{d/2} \sum_{n=1}^{\infty} n^{-d/2} (v_n i^m + \bar{v}_n i^{-m}) T_n \Phi_m. \tag{21}$$

This equality should be considered as the precise definition of the operator V_m. Let H_0 be the operator of multiplication by r^{2l}, and let $H^{(m)} = H_0 + V_m$ in the space $L_2(\mathbb{R}_+)$.

Obviously, $\mathfrak{H}_m$ is invariant with respect to the operator $\mathbf{H}_0$ which acts as H_0 on each of these subspaces. It follows from (16), (18) and (19) that every subspace $\mathfrak{H}_m$ is also invariant with respect to operator (12). It reduces on $\mathfrak{H}_m$ to the operator $\Omega V_m \Omega^* \otimes I_m$ where $\Omega : L_2(\mathbb{R}_+) \to \mathcal{K}$ is a unitary operator defined by $(\Omega g)(r) = r^{-\delta} g(r)$ and I_m is the identity operator on the space $\mathfrak{h}_m$. Then

$$\mathbf{H} = \bigoplus_{m=0}^{\infty} \Omega H^{(m)} \Omega^* \otimes I_m,$$

and hence the spectral analysis of the operator $\mathbf{H}$ reduces to that for the family of operators $H^{(m)}$.

We shall calculate the total number $N^{(m)}$ of negative eigenvalues (with multiplicity taken into account) of every operator $H^{(m)}$ and then the same quantity

$$\mathbf{N} = \sum_{m=0}^{\infty} \nu_m N^{(m)} \tag{22}$$

for the operator $\mathbf{H}$. Operators $H^{(m)} = H_0 + V_m$ where V_m is given by (11) with a sufficiently arbitrary function v_m were considered in Section 3 of [10]. However, since in our case v_m is defined by infinite sum (20), it is more convenient to calculate $N^{(m)}$ directly using the approach but not the results of [10]. Our calculation relies on the modified form of the Birman-Schwinger principle discussed in the next section.

4.

Let H_0 be an arbitrary self-adjoint positive operator with domain $\mathcal{D}(H_0)$ in a Hilbert space $\mathcal{H}$. We suppose for simplicity that the operator $V = V^*$ is bounded, so that the "full" Hamiltonian $H = H_0 + V$ is self-adjoint on the domain $\mathcal{D}(H) = \mathcal{D}(H_0)$. We suppose also that the operator

$$(H_0 + I)^{-1} V (H_0 + I)^{-1}$$

is compact and hence, by the Weyl theorem, the essential spectra of the operators H_0 and H coincide.

Our goal here is to calculate the total multiplicity

$$N = \dim E_H(-\infty, 0) \tag{23}$$

of the negative spectrum of the operator H. To that end we introduce some auxiliary objects. Let us consider the set $\mathcal{R} = \operatorname{Ran} H_0^{1/2}$ endowed with the norm

$$\|u\|_{\mathcal{R}}^2 = \|u\|^2 + \|H_0^{-1/2}u\|^2. \tag{24}$$

It is dense in $\mathcal{H}$ since $\operatorname{Ker} H_0^{1/2} = \{0\}$. Let us define the bounded quadratic form a on $\mathcal{R}$ by the relation

$$a[u,u] = (VH_0^{-1/2}u, H_0^{-1/2}u), \quad u \in \mathcal{R}. \tag{25}$$

In terms of the form (25), a simple version of the Birman-Schwinger principle can be formulated as follows.

Proposition 4.1. *Suppose that*

$$a[u,u] = (Au, u) \tag{26}$$

for some bounded self-adjoint operator A in the space $\mathcal{H}$ and all $u \in \mathcal{R}$. Let

$$M = \dim E_A(-\infty, -1) \tag{27}$$

be the total multiplicity of the spectrum of the operator A in the interval $(-\infty, -1)$. Then the numbers (23) and (27) are equal, that is $N = M$.

Corollary 4.2. *If the operator A is compact, then the negative spectrum of the operator H is finite.*

As a simple application of this result, we have

Proposition 4.3. *Suppose that estimate (3) holds for some $\alpha > 0$. Then the negative spectrum of the Friedrichs-Faddeev operator $H = H_0 + V$ is finite. Moreover, $H_\gamma = H_0 + \gamma V \geq 0$ if γ is small enough.*

Proof. – According to Proposition 4.1 it suffices to show that the operator $H_0^{-1/2}VH_0^{-1/2}$ is compact. To that end, we shall verify that

$$\int_0^\infty \int_0^\infty |g(\lambda,\mu)|^2 (\lambda\mu)^{-1} d\lambda d\mu < \infty. \tag{28}$$

It follows from (3), (4) that

$$|g(\lambda,\mu)| \leq C\lambda^\alpha (1+\mu)^{-\theta}$$

if $\lambda \in (0,1)$. A similar estimate holds for $\mu \in (0,1)$. This suffices to estimate integral (28) over the region where $\lambda \in (0,1)$ or $\mu \in (0,1)$. The convergence of this integral over the region where $\lambda \geq 1$ and $\mu \geq 1$ is a consequence of estimate (2). □

Proposition 4.1 is actually contained in the paper [3] because, under assumptions of Proposition 4.1, the form (Vf, f) is bounded in the space $\mathfrak{H}$ with the norm $\|f\|_{\mathfrak{H}} = \|H_0^{1/2}f\|$ and consequently is closable in this space. A more difficult case where the form (Vf, f) is not closable in $\mathfrak{H}$ was also investigated in [3]. Our aim here is to reconsider this situation and to formulate results in terms of the form (25). This is convenient in the case where the form (Vf, f) takes values of both signs. Actually, we suppose that equality (26) holds only up to some finite number of squares of (unbounded) functionals $\varphi_1, \dots, \varphi_n$, that is

$$a[u,u] = (Au,u) - \sum_{j=1}^{r} |\varphi_j(u)|^2 + \sum_{j=r+1}^{n} |\varphi_j(u)|^2. \tag{29}$$

Of course, one or both sums in (29) may be absent, that is we do not exclude the cases $r = 0$ and $r = n$. Note that $n - r = r'$ in notation of Section 1.

Let us give first sufficient conditions for the negative spectrum of the operator H to be infinite. In this case we do not assume that A is bounded and replace equality (29) by an estimate.

Theorem 4.4. *Let A be a self-adjoint operator with domain $\mathcal{D}(A)$. Suppose that a linear set $\mathcal{R}_0 \subset \mathcal{D}(A) \cap \mathcal{R}$ is dense in $\mathcal{D}(A)$ in the A-metrics. Assume that for all $u \in \mathcal{R}_0$ the form* (25) *satisfies the estimate*

$$a[u,u] \leq (Au,u) + \sum_{j=1}^{p} |\varphi_j(u)|^2,$$

where $\varphi_1, \dots, \varphi_p$ is a system of linear functionals defined on $\mathcal{R}_0$. Then the negative spectrum of the operator H is infinite provided $\dim E_A(-\infty, -1) = \infty$.

To calculate the number of negative eigenvalues of the operator H, we need to exclude the case where the sums in (29) are degenerate. To that end we introduce the following

Definition 4.5. *Let $\mathfrak{R} \subset \mathcal{H}$ be a linear set dense in $\mathcal{H}$ and let $\varphi_1, \dots, \varphi_n$ be linear functionals defined on $\mathfrak{R}$. We call $\varphi_1, \dots, \varphi_n$ strongly linearly independent if the inequality*

$$|\sum_{j=1}^{n} \alpha_j \varphi_j(u)| \leq C\|u\|, \quad \forall u \in \mathfrak{R},$$

(here C is some positive constant) implies that $\alpha_j = 0$ for all $j = 1, \dots, n$.

Thus, it is impossible to find a linear combination of strongly linear independent functionals which is a bounded functional on $\mathcal{H}$. Of course, the strong linear independence ensures the usual linear independence, and every functional from a strongly linear independent system is unbounded.

In Definition 4.5, $\mathfrak{R}$ is an arbitrary linear set but, if $\mathfrak{R}$ is a Banach space, we suppose that the functionals $\varphi_1, \dots, \varphi_n$ are bounded on this space (but not on $\mathcal{H}$, of course). In our study of the negative spectrum of the operator H, the role of $\mathfrak{R}$ is played by the space $\mathcal{R}$ endowed with the norm (24). Now we are in position to formulate a new form of the Birman-Schwinger principle (see [10] for the proof).

Theorem 4.6. *Let $\varphi_1, \dots, \varphi_n$ be strongly linearly independent functionals defined on $\mathcal{R}$ and let A be a bounded self-adjoint operator. Assume that equality* (29) *holds for all $u \in \mathcal{R}$. Then numbers* (23) *and* (27) *are related by the equality*

$$N = M + r.$$

Remark 4.7. *If the condition of the strong linear independence of $\varphi_1, \dots, \varphi_n$ is omitted, we can assert that $N \le M + r$.*

5.

Let us return to the operators H_0 and $H^{(m)} = H_0 + V_m$ in the space $\mathcal{H} = L_2(\mathbb{R}_+)$. The first step is to obtain representation (29) for the form (25). Our calculation relies on the Mellin transform $\mathbf{M}$. Let

$$(\mathbf{M}u)(\lambda) = (2\pi)^{-1/2} \int_0^\infty x^{-1/2-i\lambda} u(x)dx, \tag{30}$$

so that the operator $\mathbf{M} : L_2(\mathbb{R}_+) \to L_2(\mathbb{R})$ is unitary. The following assertion is an easy combination of the convolution formula and of the Parseval identity (see [10], for details).

Lemma 5.1. *Suppose that a function $b(t)$ is locally bounded on $(0, \infty)$ and the integral*

$$\int_0^\infty b(t)t^{-1/2-i\lambda}dt =: \beta(\lambda) \tag{31}$$

converges at $t = 0$ and $t = \infty$ uniformly in $\lambda \in \mathbb{R}$. Then for any function $u \in C_0^\infty(\mathbb{R}_+)$

$$\int_0^\infty \int_0^\infty b(xy)u(y)\overline{u(x)}dxdy = \int_{-\infty}^\infty \beta(\lambda)(\mathbf{M}u)(-\lambda)\overline{(\mathbf{M}u)(\lambda)}d\lambda. \tag{32}$$

Under assumptions of Lemma 5.1, the function $\beta(\lambda)$ is of course continuous and bounded.

Let us define a unitary mapping $U : L_2(\mathbb{R}) \to L_2(\mathbb{R}_+; \mathbb{C}^2)$ and a 2×2- matrix $\mathcal{B}(\lambda)$ by the equalities

$$(Uw)(\lambda) = \begin{pmatrix} w(\lambda) \\ w(-\lambda) \end{pmatrix}, \quad \mathcal{B}(\lambda) = \begin{pmatrix} 0 & \beta(\lambda) \\ \overline{\beta(\lambda)} & 0 \end{pmatrix}, \quad \lambda > 0. \tag{33}$$

If $\beta(-\lambda) = \overline{\beta(\lambda)}$, then

$$\int_{-\infty}^\infty \beta(\lambda)w(-\lambda)\overline{w(\lambda)}d\lambda = (\mathcal{B}Uw, Uw)_{L_2(\mathbb{R}_+;\mathbb{C}^2)}, \tag{34}$$

where $\mathcal{B}$ is the operator of multiplication by $\mathcal{B}(\lambda)$. Since eigenvalues of the matrix $\mathcal{B}(\lambda)$ equal $\pm|\beta(\lambda)|$, we have

Lemma 5.2. *Let $\beta(\lambda)$ be a continuous function of $\lambda \in (0, \infty)$ and*

$$\hat{\beta} = \max_{\lambda \in \mathbb{R}_+} |\beta(\lambda)|, \quad \check{\beta} = \min_{\lambda \in \mathbb{R}_+} |\beta(\lambda)|$$

(the case $\hat{\beta} = \infty$ is not excluded). Then the spectrum of the operator $\mathcal{B}$ consists of the union $[-\hat{\beta}, -\check{\beta}] \cup [\check{\beta}, \hat{\beta}]$.

To find the Mellin transform of the Bessel function, we use formula (19), section 7.7 of [1]:

$$\int_0^\infty \mathcal{I}_{m+(d-2)/2}(t)t^{-z}dt = 2^{-z}\Gamma((m+d/2-z)/2)\Gamma^{-1}((m+d/2+z)/2) \\ =: \mathfrak{B}_{m+d/2}(z), \tag{35}$$

valid for $\operatorname{Re} z \in (0, m+d/2)$. Recall also that

$$\mathcal{I}_{m+(d-2)/2}(t) = \sum_{k=0}^\infty (-1)^k \omega_{m+d/2,k} t^{m+(d-2)/2+2k} \tag{36}$$

where

$$\omega_{m+d/2,k} = 2^{-2k-m+1-d/2}(k!\Gamma(k+m+d/2))^{-1}.$$

Both sides of (35) are analytic functions of z in the half-plane $\operatorname{Re} z > 0$ except points $z = m+d/2+2p$, $p = 0,1,\dots$. Using (36) we can construct explicitly the analytic continuation of the left-hand side of (35) which leads to the identity

$$\int_0^\infty (\mathcal{I}_{m+(d-2)/2}(t) - \sum_{k=0}^p (-1)^k \omega_{m+d/2,k} t^{m+(d-2)/2+2k})t^{-z}dt \\ = 2^{-z}\Gamma((m+d/2-z)/2)\Gamma^{-1}((m+d/2+z)/2) \tag{37}$$

for $\operatorname{Re} z \in (m+d/2+2p, m+d/2+2p+2)$ and any $p = 0,1,\dots$.

Let us start calculation of the form $(V_m H_0^{-1/2}u, H_0^{-1/2}u)$.

Proposition 5.3. *Let*

$$l - d/2 - m \neq 0, 2, 4, \dots \tag{38}$$

and let $u \in C_0^\infty(\mathbb{R}_+)$. Then

$$\int_0^\infty \int_0^\infty (xy)^{1/2-l}\mathcal{I}_{m+(d-2)/2}(nxy)u(y)\overline{u(x)}dxdy \\ = n^{-1+l}\int_{-\infty}^\infty \mathfrak{B}_{m+d/2}(l+i\lambda)n^{i\lambda}(\mathbf{M}u)(-\lambda)\overline{(\mathbf{M}u)(\lambda)}d\lambda \\ + \sum_{2k<l-m-d/2} (-1)^k n^{-1+m+d/2+2k}\omega_{m+d/2,k}\left|\int_0^\infty x^{m+\delta-l+2k}u(x)dx\right|^2, \tag{39}$$

where the sum in (39) *is taken over all $k = 0,1,2,\dots$ satisfying $2k < l-m-d/2$ and should be dropped out if $l-m-d/2 < 0$.*

Proof. It suffices to check (39) for $n = 1$ only. Indeed, applying (39) for $n = 1$ to a function $u_n(r) = u(n^{-1/2}r)$ and making the change of variables $r \mapsto n^{1/2}r$, $s \mapsto n^{1/2}s$, we obtain (39) for an arbitrary $n > 0$.

Let first $l < m+d/2$. Remark that the function $b(t) = t^{1/2-l}\mathcal{I}_{m+(d-2)/2}(t)$ satisfies the assumptions of Lemma 5.1 and according to (35) the corresponding function (31) equals $\mathfrak{B}_{m+d/2}(l+i\lambda)$. Therefore (39) is a particular case of (32).

If $l > m + d/2$, we rewrite the left-hand side of (39) as

$$\int_0^\infty \int_0^\infty (xy)^{1/2-l} (\mathcal{I}_{m+(d-2)/2}(xy) - \sum_{2k<l-m-d/2} (-1)^k \omega_{m+d/2,k} (xy)^{m+(d-2)/2+2k}) u(y)\overline{u(x)} dx dy + \sum_{2k<l-m-d/2} (-1)^k \omega_{m+d/2,k} \left| \int_0^\infty x^{m+\delta-l+2k} u(x) dx \right|^2, \quad \delta = (d-1)/2.$$

Note that the function

$$t^{1/2-l}(\mathcal{I}_{m+(d-2)/2}(t) - \sum_{2k<l-m-d/2} (-1)^k \omega_{m+d/2,k} t^{m+(d-2)/2+2k})$$

satisfies the assumptions of Lemma 5.1. According to (37) the corresponding function (31) equals again $\mathfrak{B}_{m+d/2}(l + i\lambda)$. Therefore equality (32) gives (39) (for $n = 1$). □

Under the assumption

$$\sum_{n=1}^\infty |v_n| n^{l-d/2} < \infty, \tag{40}$$

we set

$$\mathcal{V}_{even}(\lambda) = 2(2\pi)^{d/2} \sum_{n=1}^\infty \operatorname{Re} v_n \, n^{l-d/2+i\lambda},$$
$$\mathcal{V}_{odd}(\lambda) = 2(2\pi)^{d/2} \sum_{n=1}^\infty \operatorname{Im} v_n \, n^{l-d/2+i\lambda}, \tag{41}$$

and

$$\left.\begin{aligned} \beta_m(\lambda) &= (-1)^{m/2} \mathcal{V}_{even}(\lambda) \mathfrak{B}_{m+d/2}(l + i\lambda) \quad \text{for} \quad m \quad \text{even}, \\ \beta_m(\lambda) &= (-1)^{(m+1)/2} \mathcal{V}_{odd}(\lambda) \mathfrak{B}_{m+d/2}(l + i\lambda) \quad \text{for} \quad m \quad \text{odd}. \end{aligned}\right\} \tag{42}$$

We emphasize that only the first factors $\mathcal{V}_{even}$ and $\mathcal{V}_{odd}$ in the definition of β_m depend on the function v or rather on its Fourier coefficients v_n. They are bounded continuous functions of λ. Of course, the functions β_m depend also on l and d, that is $\beta_m = \beta_m^{(l,d)}$, but we usually do not keep this dependence in notation.

Now we recall that the operator V_m was defined by equality (21). Let us multiply representations (39) by $(2\pi)^{d/2} n^{-d/2+1}(v_n i^m + \bar{v}_n i^{-m})$ and sum them over all $n = 1, 2, \ldots$. Taking also into account that under assumption (40) where $l \geq d/2$ the function $v \in C^q(\mathbb{R})$ for any $q \leq l - d/2$ and

$$v^{(q)}(0) = \sum_{n=1}^\infty (v_n i^q + \bar{v}_n i^{-q}) n^q, \tag{43}$$

we obtain

Proposition 5.4. *Let conditions* (38) *and* (40) *hold and let* $u \in C_0^\infty(\mathbb{R}_+)$. *Then*

$$(V_m H_0^{-1/2} u, H_0^{-1/2} u) = \int_{-\infty}^{\infty} \beta_m(\lambda)(\mathbf{M}u)(-\lambda)\overline{(\mathbf{M}u)(\lambda)}d\lambda$$
$$+ (2\pi)^{d/2} \sum_{2k<l-m-d/2} \omega_{m+d/2,k} v^{(m+2k)}(0)\Big|\int_0^\infty x^{m+\delta-l+2k} u(x)dx\Big|^2, \tag{44}$$

where the sum in (44) *is taken over all* $k = 0, 1, 2, \dots$ *satisfying* $2k < l - m - d/2$ *and should be dropped out if* $l - m - d/2 < 0$.

6.

By definition (24), the set $\mathcal{R} = \mathcal{R}^{(l)} = \operatorname{Ran} H_0^{1/2}$ consists now of functions u such that

$$\|u\|_{\mathcal{R}}^2 = \int_0^\infty (1 + x^{-2l})|u(x)|^2 dx < \infty.$$

Lemma 6.1. *The functionals*

$$\Phi_k(u) = \int_0^\infty x^{m+\delta-l+2k} u(x)dx, \quad 2k < l - m - d/2, \tag{45}$$

are bounded on $\mathcal{R}^{(l)}$ *and are strongly linearly independent.*

Proof. Clearly,

$$|\Phi_k(u)|^2 \le \int_0^\infty x^{2m+2\delta+4k}(1 + x^{2l})^{-1} dx \int_0^\infty (1 + x^{-2l})|u(x)|^2 dx$$

and the first integral in the right-hand side is convergent (at $x = \infty$) because $m + d/2 + 2k < l$. The functionals Φ_k are strongly linear independent because a function

$$\sum_{0 \le k < (l-m-d/2)/2} c_k x^{m+\delta-l+2k}$$

does not belong to the space $L_2(\mathbb{R}_+)$ unless all $c_k = 0$. □

Under assumption (38) $\mathfrak{B}_{m+d/2}(l + i\lambda)$ is a bounded continuous function of λ. Moreover, by the Stirling formula, $\mathfrak{B}_{m+d/2}(l + i\lambda) \to 0$ as $|\lambda| \to \infty$. According to (42) the function $\beta_m(\lambda)$ has the same properties, so that the spectrum of the corresponding operator $\mathcal{B}_m$ (see Lemma 5.2) consists of the interval $[-\mathbf{b}_m, \mathbf{b}_m]$ where

$$\mathbf{b}_m = \max_{\lambda \in \mathbb{R}} |\beta_m(\lambda)|. \tag{46}$$

Representation (44) extends by continuity to all $u \in \mathcal{R}^{(l)}$. Taking into account equalities (34), (45) and setting

$$A_m = (U\mathbf{M})^* \mathcal{B}_m U\mathbf{M}, \tag{47}$$

we write it in the form

$$(V_m H_0^{-1/2} u, H_0^{-1/2} u) \\ = (A_m u, u) + (2\pi)^{d/2} \sum_{2k<l-m-d/2} \omega_{m+d/2,k} v^{(m+2k)}(0) |\Phi_k(u)|^2. \tag{48}$$

This is exactly representation (29). The sum in (48) should be split up in two sums corresponding to the values $v^{(m+2k)}(0) < 0$ and $v^{(m+2k)}(0) > 0$. In particular, $r = r_m$ equals the number of k such that $v^{(m+2k)}(0) < 0$. Clearly, $r_m = 0$ if $l - m - d/2 \leq 0$ and

$$r_m = \#\{k = 0, \dots, [(l - m - d/2)/2] : v^{(m+2k)}(0) < 0\} \tag{49}$$

if $l - m - d/2 > 0$. By Theorem 4.6 the operators $H^{(m)}$ have the infinite discrete spectrum if $\mathbf{b}_m > 1$. In the case $\mathbf{b}_m \leq 1$ the total number of negative eigenvalues $N^{(m)}$ of the operator $H^{(m)}$ coincides with the number r_m. It is convenient to rewrite (49) in a slightly different way introducing the integer $n = [l - d/2]$ and the sets Ξ_{even} and Ξ_{odd} of all even and all odd p, respectively, such that $v^{(p)}(0) < 0$. Let us formulate the result.

Theorem 6.2. *Let conditions* (38) *and* (40) *hold. Then the negative spectrum of the operator* $H^{(m)}$ *is infinite if and only if* $\mathbf{b}_m > 1$. *In the case* $\mathbf{b}_m \leq 1$, *the total number of negative eigenvalues* $N^{(m)} = 0$ *if* $m > n$ *and is given by the formula*

$$\left.\begin{aligned} N^{(m)} &= \#\{\Xi_{odd} \cap [m, n]\} \quad \text{for} \quad m \quad \text{odd}, \\ N^{(m)} &= \#\{\Xi_{even} \cap [m, n]\} \quad \text{for} \quad m \quad \text{even} \end{aligned}\right\} \tag{50}$$

if $m \leq n$.

It remains to consider the exceptional case where $l - m - d/2 = 2p$ for some $p = 0, 1, \dots$. Then

$$\mathfrak{B}_{m+d/2}(l + i\lambda) = i(-1)^p \omega_{m+d/2,p} \lambda^{-1} + O(1) \tag{51}$$

as $|\lambda| \to 0$. We distinguish two cases $v^{(l-d/2)}(0) = 0$ and $v^{(l-d/2)}(0) \neq 0$.

According to (43) in the first case $\mathcal{V}_{even}(0) = 0$ for m even or $\mathcal{V}_{odd}(0) = 0$ for m odd. We suppose additionally that

$$\left.\begin{aligned} \sum_{n=1}^{\infty} |\operatorname{Re} v_n| n^{l-d/2} \ln n < \infty \quad \text{for} \quad l - d/2 \quad \text{even}, \\ \sum_{n=1}^{\infty} |\operatorname{Im} v_n| n^{l-d/2} \ln n < \infty \quad \text{for} \quad l - d/2 \quad \text{odd}. \end{aligned}\right\} \tag{52}$$

Then $\mathcal{V}_{even}(\lambda) = O(|\lambda|)$ for m even or $\mathcal{V}_{odd}(\lambda) = O(|\lambda|)$ for m odd and according to (42), (51) the function $\beta_m(\lambda)$ is bounded as $|\lambda| \to 0$. In this case representation (44) (or (48)) remains valid for $u \in \mathcal{R}^{(l)}$ and $l - m - d/2 = 2p$. Indeed, let us proceed from (44) for $u \in C_0^\infty(\mathbb{R}_+)$ and $l = m + d/2 + 2p - \varepsilon$, where $\varepsilon > 0$, and pass to the limit $\varepsilon \to 0$. The left-hand side is of course a continuous function of l.

The functions $\beta_m^{(l-\varepsilon,d)}$ converge as $\varepsilon \to 0$ to $\beta_m^{(l,d)}$ uniformly on any compact set $\mathcal{J}_\gamma = [-\gamma^{-1}, -\gamma] \cup [\gamma, \gamma^{-1}]$, $\gamma > 0$, and are bounded uniformly in ε and λ outside of $\mathcal{J}_\gamma$. It follows that

$$\lim_{\varepsilon \to 0} \int_{-\infty}^{\infty} \beta_m^{(l-\varepsilon,d)}(\lambda)(\mathbf{M}u)(-\lambda)\overline{(\mathbf{M}u)(\lambda)}d\lambda = \int_{-\infty}^{\infty} \beta_m^{(l,d)}(\lambda)(\mathbf{M}u)(-\lambda)\overline{(\mathbf{M}u)(\lambda)}d\lambda. \tag{53}$$

The corresponding operator $\mathcal{B}_m$ is again bounded. The passage to the limit $\varepsilon \to 0$ in the sum in the right-hand side of (48) is of course straightforward. By virtue of the condition $v^{(m+2p)}(0) = v^{(l-d/2)}(0) = 0$, this sum is actually taken over $k = 0, 1, \ldots, p-1$, so that all functionals Φ_k are again bounded on $\mathcal{R}^{(l)}$ and are strongly linear independent. Finally, (48) extends by continuity to all $\mathcal{R}^{(l)}$ and Theorem 4.6 again applies.

Theorem 6.3. *Let condition* (40) *be fulfilled. Suppose that* $l - m - d/2$ *equals a nonnegative even number,* $v^{(l-d/2)}(0) = 0$ *and condition* (52) *is satisfied. Then all conclusions of Theorem* 6.2 *hold.*

We emphasize that under the assumptions of Theorems 6.2 and 6.3 the number $N^{(m)}$ of total negative eigenvalues of $H^{(m)}$ depends on the function v and the integer part of the number $l - d/2$ only.

The case where $l - m - d/2 = 2p$ for some $p = 0, 1, \ldots$ but $v^{(l-d/2)}(0) \neq 0$ is qualitatively different. By virtue of (51) we have now that

$$\beta_m^{(l,d)}(\lambda) = i(2\pi)^{d/2} v^{(l-d/2)}(0)\omega_{m+d/2,p}\lambda^{-1} + O(1). \tag{54}$$

Let us check (44) (or (48)) on the set $\mathcal{R}_0$ of functions $u \in C_0^\infty(\mathbb{R}_+)$ such that

$$\int_0^\infty u(x)x^{-1/2}dx = 0. \tag{55}$$

We use again representation (44) for $l = m + d/2 + 2p - \varepsilon$, $\varepsilon > 0$, and pass to the limit $\varepsilon \to 0$. Compared to the previous case, there is no difference in passage to the limit in the left-hand side and in the sum in the right-hand side. Note however that the sum is now taken over $k = 0, 1, \ldots, p-1$ due to the condition (55) which implies that $\Phi_p(u) = 0$ on $\mathcal{R}_0$. To justify (53), we use that under assumption (55)

$$(\mathbf{M}u)(\lambda) = O(|\lambda|) \quad \text{as} \quad \lambda \to 0,$$

which compensates the singularity of function (51). Asymptotics (54) implies that the operators $\mathcal{B}_m$ defined by (33), as well as (47), are unbounded from below and hence

$$\dim E_{A_m}(-\infty, -1) = \infty.$$

Finally, we use the following elementary

Lemma 6.4. *The set* $\mathcal{R}_0$ *is dense in* $\mathcal{D}(A_m)$ *where* $m = l - d/2 - 2p$ *for some* $p = 0, 1, \ldots$.

Proof. – Recall that the function $\beta_m(\lambda)$ is bounded except the point $\lambda = 0$ where it satisfies (54). Therefore it follows from (33) and (47) that the inclusion $u \in \mathcal{D}(A_m)$ is equivalent to the bound

$$\int_{-\infty}^{\infty} (1 + \lambda^{-2})|w(\lambda)|^2 d\lambda < \infty \quad \text{for} \quad w = \mathbf{M}u. \tag{56}$$

Clearly, the Mellin transform (30) can be factorized as $\mathbf{M} = \Phi G$ where Φ is the Fourier transform in $L_2(\mathbb{R})$ and $(Gu)(t) = e^{t/2}u(e^t)$. In terms of $g(\lambda) = \lambda^{-1}w(\lambda)$, (56) is equivalent to the condition

$$\int_{-\infty}^{\infty} (|\tilde{g}(t)|^2 + |\tilde{g}'(t)|^2)dt < \infty, \quad \tilde{g} = \Phi^* g. \tag{57}$$

Of course, there exists a sequence $\tilde{g}_j \in C_0^\infty(\mathbb{R})$ such that $\tilde{g}_j$ converge to $\tilde{g}$ as $j \to \infty$ in the metrics (57) (that is in the Sobolev space $H^1(\mathbb{R})$). Set $\tilde{w}_j(t) = -i\tilde{g}_j'(t)$ so that $w_j(\lambda) = \lambda g_j(\lambda)$. Then w_j converge to w in the metrics (56). Moreover, $\tilde{w}_j \in C_0^\infty(\mathbb{R})$ and

$$\int_{-\infty}^{\infty} \tilde{w}_j(t)dt = -i\int_{-\infty}^{\infty} \tilde{g}_j'(t)dt = 0.$$

It follows that $u_j = G^*\tilde{w}_j \in C_0^\infty(\mathbb{R}_+)$, u_j satisfy (55) and $u_j \to u$ in $\mathcal{D}(A_m)$ as $j \to \infty$. □

Thus all assumptions of Theorem 4.4 are verified and we obtain the following result.

Theorem 6.5. *Let condition* (40) *hold. If $l - d/2 - m$ equals a nonnegative even number and $v^{(l-d/2)}(0) \neq 0$, then the negative spectrum of the operator $H^{(m)}$ is infinite.*

7.

To calculate the total number $\mathbf{N}$ of negative eigenvalues of the operator $\mathbf{H}$, we combine Theorems 6.2, 6.3 and 6.5 with equality (22).

Let us start with exceptional values of l. The following result is a direct consequence of Theorem 6.5.

Theorem 7.1. *Let condition* (40) *hold. If $l - d/2$ is a nonnegative integer and $v^{(l-d/2)}(0) \neq 0$, then the negative spectrum of the operator $\mathbf{H}$ is infinite.*

To consider other values of l, recall that the functions $\beta_m(\lambda)$ were defined by equalities (35), (41), (42). Since

$$\mathfrak{B}_{m+2+d/2}(z) = (m + d/2 - z)(m + d/2 + z)^{-1}\mathfrak{B}_{m+d/2}(z),$$

we have that

$$|\mathfrak{B}_{m+2+d/2}(z)| \le |\mathfrak{B}_{m+d/2}(z)|, \quad \operatorname{Re} z > 0,$$

and consequently $|\beta_{m+2}(\lambda)| \leq |\beta_m(\lambda)|$. It follows that numbers (46) are related, for any p, by the inequalities

$$\mathbf{b}_{2p} \leq \mathbf{b}_0, \quad \mathbf{b}_{2p+1} \leq \mathbf{b}_1.$$

Therefore, according to Theorems 6.2, 6.3, $N^{(2p)} < \infty$ if $N^{(0)} < \infty$ and $N^{(2p+1)} < \infty$ if $N^{(1)} < \infty$. Moreover, $N^{(m)} = 0$ if m is large enough $(m \geq l - d/2)$. It follows that $\mathbf{N} < \infty$ if and only if $N^{(0)} < \infty$ and $N^{(1)} < \infty$.

Theorem 7.2. *Let condition* (40) *hold and let one of the two following assumptions be fulfilled:*
1^0 *$l - d/2$ is not a nonnegative integer,*
2^0 *$l - d/2$ is a nonnegative integer, $v^{(l-d/2)}(0) = 0$ and condition* (52) *is satisfied. Then the negative spectrum of the operator $\mathbf{H}$ is finite if and only if $\mathbf{b}_0 \leq 1$ and $\mathbf{b}_1 \leq 1$.*

It remains to calculate the number $\mathbf{N}$ explicitly in the case where $\mathbf{b}_0 \leq 1$ and $\mathbf{b}_1 \leq 1$. Recall that the numbers ν_m are defined by formula (15).

Theorem 7.3. *Under the conditions of Theorem* 7.2 *assume additionally that $\mathbf{b}_0 \leq 1$ and $\mathbf{b}_1 \leq 1$. Let $n = [l - d/2]$ and let $N^{(m)}$ be determined by* (50). *Then the total number of negative eigenvalues of the operator $\mathbf{H}$ is zero if $l \leq d/2$, and it is given by formula*

$$\mathbf{N} = \sum_{m=0}^{n} \nu_m N^{(m)}$$

if $l > d/2$.

References

[1] H. Bateman and A. Erdélyi, *Higher transcendental functions*, vol. 2, New York; McGraw-Hill, 1953.

[2] F. A. Berezin and L. D. Faddeev, *A remark on Schrödinger's equation with a singular potential*, Soviet Math. Dokl., **2** (1961), 372–375.

[3] M. Sh. Birman, *On the spectrum of singular boundary value problems*, Amer. Math. Soc. Transl. Ser. 2, **53** (1966), 23–80.

[4] M. Sh. Birman and D. R. Yafaev, *On the trace-class method in potential scattering theory*, J. Soviet Math., **56**, no. 2 (1991), 2285–2299.

[5] L. D. Faddeev, *On the Friedrichs model in the theory of perturbations of the continuous spectrum*, Amer. Math. Soc. Transl. Ser. 2, **62** (1967).

[6] B. S. Pavlov and S. V. Petras, The singular spectrum of the weakly perturbed multiplication operator, Funct. Anal. Appl., **4** (1970), 136–142.

[7] M. Reed and B. Simon, Methods of modern mathematical physics , vol. 4, Academic Press, San Diego, CA, 1979.

[8] E. M. Stein, G. Weiss, *Introduction to Fourier analysis on Euclidean spaces*, Princeton Univ. Press, Princeton, N.J., 1971.

[9] D. R. Yafaev, Mathematical scattering theory, AMS, 1992.

[10] D. R. Yafaev, *The discrete spectrum in the singular Friedrichs model*, Amer. Math. Soc. Transl. Ser. 2, **189** (1999), 255–274.

[11] D. R. Yafaev, *A model in perturbation theory*, Operator Theory: Adv. and Appl., Birkhäuser Verlag, **108** (1999), 373–378.

[12] D. R. Yafaev, Sharp constants in the Hardy–Rellich inequalities, J. Funct. Anal., **168** (2000), 121–144.

2000 Mathematics Subject Classification: Primary: 35J10, 47A75, Secondary 81U20

Math. Depart.
Université de Rennes 1
Campus Beaulieu, 35042
Rennes
France

MIX
Papier aus verantwortungsvollen Quellen
Paper from responsible sources
FSC® C105338

If you have any concerns about our products,
you can contact us on
ProductSafety@springernature.com

In case Publisher is established outside the EU,
the EU authorized representative is:
Springer Nature Customer Service Center GmbH
Europaplatz 3, 69115 Heidelberg, Germany

Printed by Libri Plureos GmbH
in Hamburg, Germany